Developments in Petroleum Science, 11

drilling and drilling fluids

updated textbook edition

FURTHER TITLES IN THIS SERIES

1 A. GENE COLLINS
GEOCHEMISTRY OF OILFIELD WATERS

2 W.H. FERTL
ABNORMAL FORMATION PRESSURES

3 A.P. SZILAS
PRODUCTION AND TRANSPORT OF OIL AND GAS

4 C.E.B. CONYBEARE
GEOMORPHOLOGY OF OIL AND GAS FIELDS
IN SANDSTONE BODIES

5 T.F. YEN and G.V. CHILINGARIAN (Editors)
OIL SHALE

6 D.W. PEACEMAN
FUNDAMENTALS OF NUMERICAL RESERVOIR SIMULATION

7 G.V. CHILINGARIAN and T.F. YEN (Editors)
BITUMENS, ASPHALTS AND TAR SANDS

8 L.P. DAKE
FUNDAMENTALS OF RESERVOIR ENGINEERING

9 K. MAGARA
COMPACTION AND FLUID MIGRATION

10 M.T. SILVIA and E.A. ROBINSON
DECONVOLUTION OF GEOPHYSICAL TIME SERIES IN THE
EXPLORATION FOR OIL AND NATURAL GAS

Developments in Petroleum Science, 11

drilling and drilling fluids

updated textbook edition

by

G.V. CHILINGARIAN

Petroleum Engineering Department
University of Southern California, Los Angeles,
Calif., U.S.A.

and

P. VORABUTR

Texas Pacific Oil Company, Inc.,
Dallas, Texas, U.S.A.

with contributions from:
Daniel Acosta
James H. Allen
Frank R. Bagshaw
Baroid Petroleum Services
T.A. Bertness
Ryen Caenn
W.L. Clark
Eastman Whipstock Co.
Walter H. Fertl
B.S. Gupta
Dan Madick
Magcobar Division, Dresser Industries, Inc.
Samuel C. Murray
George S. Ormsby
Herman H. Rieke
Richard Seaver
Claude L. Thibault
N.K. Tschirley

ELSEVIER
AMSTERDAM — OXFORD — NEW YORK 1983

ELSEVIER SCIENCE PUBLISHERS B.V.
Molenwerf 1
P.O. Box 211, 1000 AE Amsterdam, The Netherlands

Distributors for the United States and Canada:

ELSEVIER SCIENCE PUBLISHING COMPANY INC.
52, Vanderbilt Avenue
New York, N.Y. 10017

First edition 1981
Updated textbook edition 1983
Second impression 1984

Library of Congress Cataloging in Publication Data

Chilingarian, George V., 1929-
Drilling and drilling fluids.

(Developments in petroleum science ; 11)
Includes bibliographies and indexes.
1. Oil well drilling. 2. Drilling muds. I. Vorabutr, P. (Paul), 1954- . II. Title. III. Series.
TN871.2.C52 1983 622'.338 83-1671
ISBN 0-444-42177-7 (pbk.)

ISBN 0-444-42177-7 (Vol. 11)
ISBN 0-444-41625-0 (Series)

Printed in The Netherlands

This book is dedicated to His Majesty King Bhumibol Adulyadej and Her Majesty Queen Sirikit of Thailand for their relentless support of all branches of engineering sciences in order to improve the well being of mankind. Their encouragement in writing this book is indeed greatly appreciated.

CONTRIBUTORS

D. ACOSTA	IMCO Services, P.O. Box 22605, Houston, TX 77027, U.S.A.
J.H. ALLEN	Sii Smith Tool, Division of Smith International, Inc., P.O. Box C-19511, Irvine, CA 92713, U.S.A.
F.R. BAGSHAW	Kelco Division, Merck and Co., Inc., 8355 Aero Drive, San Diego, CA 92123, U.S.A.
T.A. BERTNESS	14827 La Cuarta, Whittier, CA 90605, U.S.A.
N.L. Baroid Petroleum Services	Baroid Petroleum Services, N.L. Industries, Inc., P.O. Box 1675, Houston, TX 77001, U.S.A.
R. CAENN	14952 Geneva, Irvine, CA 92714, U.S.A.
G.V. CHILINGARIAN	Department of Petroleum Engineering, University of Southern California, University Park, Los Angeles, CA 90007, U.S.A.
W.L. CLARK	Hydril Company, 714 West Olympic Blvd, Los Angeles, CA 90015, U.S.A.
Eastman Whipstock, Inc.	P.O. Box 1675, Houston, TX 77001, U.S.A.
W.H. FERTL	Dresser Atlas Division, Dresser Industries, Inc., P.O. Box 1407, Houston, TX 77001, U.S.A.
B.S. GUPTA	Abadan Institute of Technology, Abadan, Iran
D. MADICK	Kuster Company, 2900 East 29th St., Long Beach, CA 90807, U.S.A.
Magcobar Division, Dresser Industries, Inc.	P.O. Box 6504, Houston, TX 77005, U.S.A.
S.C. MURRAY	Kelco Division, Merck and Co., Inc., 8355 Aero Drive, San Diego, CA 92123, U.S.A.
G.S. ORMSBY	Stonewall Associates, Inc., 1950 West Gray, Suite/2, Houston, TX 77019, U.S.A.
H.H. RIEKE	School of Mines, West Virginia University, Morgantown, W.Va. 26506, U.S.A.
R. SEAVER	Hydril Company, 714 West Olympic Blvd., Los Angeles, CA 90015, U.S.A.
C.L. THIBAULT	Sii Smith Tool, Division of Smith International, Inc., P.O. Box C-19511, Irvine, CA 92713-9511, U.S.A.
N.K. TSCHIRLEY	N.L. Baroid Petroleum Services, N.L. Industries, Inc., P.O. Box 1675, Houston, TX 77001, U.S.A.
P. VORABUTR	Texas Pacific Thailand, Inc., 1700 One Main Place, Dallas, TX 75250, U.S.A.

PREFACE

In recent years, with increasing depths of boreholes and higher formation temperatures, the attention of the drilling industry has been focussed on improving the technology of drilling and drilling fluids. The basic ingredients of water-base drilling fluids are water, clay, and polymers, but the precise physicochemical reactions occurring during drilling are not very well understood because they involve so many complicated factors. There is therefore, a vast, unknown area where practising engineers find it difficult to interpret adequately all the data obtained during the drilling operation. Emulsion, polymer, and other types of drilling fluids also require better understanding.

Drilling fluids have several significant functions. In actual drilling operation, a sustained effort plus a great deal of practical expertise are necessary to keep track of all the parameters and to drill under what may hopefully be considered as optimum conditions. The engineer attempts to identify the effects of many influencing factors, in the order of priority, and tries to understand and, if necessary, correct them within the predetermined order of their importance to operational process.

The specialized drilling companies have written their own manuals on drilling fluids and these vary from firm to firm. It is important to mention here that the first syllabus on the subject for instructional purposes was written by Professor George V. Chilingarian at the University of Southern California in 1949. The results and insights obtained thus far, however, have not yet been formulated into a book which can be used both as a textbook and a reference book.

The present book tries to present this broad body of material in a coherent, unified manner in order to embody the entire experience gained during the last half century. The perseverance of the authors in writing this book is praiseworthy, and it is hoped that petroleum engineering students and petroleum engineers and technicians, who are involved in the drilling operations, will greatly benefit from the contents of this monograph.

Dr. K. Kormi

CONTENTS

Chapter 15. CORROSION IN DRILLING OPERATIONS
T.A. Bertness and George V. Chilingarian

Chapter 16. DRILLING THROUGH OVERPRESSURED FORMATIONS
Walter H. Fertl and George V. Chilingarian

Chapter 1

HISTORY OF ROTARY DRILLING

JAMES H. ALLEN

INTRODUCTION

Drilling fluids were first used in the rotary drilling process some time between 1887 and 1901. Initially the primary purpose of drilling fluid (mud) was to remove cuttings continuously. As rotary drilling developed, more engineering attention was provided and what had commenced as a simple fluid evolved into a sophisticated, complex mixture of liquids, solids and chemicals. From laboratory and field tests it was found that the combination of two major factors, i.e., drilling mud and hydraulics, had a greater effect on rate of penetration than any of the other controllable variables in the rotary drilling process.

Interrelationships between drilling mud and jet bit hydraulics became obvious in the late 1930's and early 1940's. With a better understanding of jet bit hydraulics, bit hydraulic horsepower was related to rate of penetration, bit weight, and bit rotational speed. Bit weight and rotational speed, in turn, were related to the proper rock bit type for a particular formation hardness.

A brief history and description of rotary drilling, drilling fluid hydraulics and drilling fluids is deemed a proper introduction to this book entitled "Drilling and Drilling Fluids".

EARLY HISTORY

The first patents covering rotary drilling methods for boring holes in the earth were issued to Robert Beart in England in 1844 (see Brantley, 1971, p. 596). Some 2500 years prior to this time, wells had been drilled in China and later in Persia, Egypt and Europe by the percussion drilling method. The percussion drilling method consisted of a heavy metal bit that was suspended from the ground level to the hole bottom by a rope. The bit was repeatedly raised and dropped to crush rock at the hole bottom. Periodically the bit would be removed from the hole and a bailer would be lowered to remove the rock cuttings and water.

Basically, the rotary drilling method involved a rotating drill bit at the bottom of the hole or well to cut formation by a scraping or dragging action. Rotation of the drill bit was provided by an attached rotating drillpipe that extended down the hole from the well drilling machinery at the surface. The rotary drilling method normally required the continuous circulation of water through the drillpipe to cool the bit and to remove rock cuttings from the hole bottom up to the surface through the annulus between the drillpipe and the hole wall. The circulation system described in the Beart's patent involved the use of water which was injected down the annulus, with water and cuttings returning to the surface through the hollow drillrods. This method is now commonly called "reverse circulation".

The use of water to remove cuttings continuously from a hole bottom was reported by Fauvelle, a French engineer, in 1845. It is believed, however, that Fauvelle pumped water down the hollow drillrods or drillpipe connected to a percussion type bit.

In 1860, patents were issued to another French engineer, Leschot, for the use of diamond bits for drilling holes in hard rock for a tunnel construction project in the Swiss Alps. A rotary drilling system complete with a circulation system would have been necessary for diamond bits. A percussion drilling action would have shattered or broken diamonds and a good circulation of drilling fluid would have been required to dissipate the heat generated by the scraping action of diamonds on hard rock. Details of the Leschot circulation method are lacking.

Patents were issued to Catlin for a rolling cutter bit in the 1850's.

A United States patent was issued to Sweeney in 1866 for a small, hand-powered, rotary drill that was used to drill holes in stone quarries. This patent mentions the possible use of air as a circulating fluid to remove cuttings from the hole. The Sweeney patent depicted a drill bit with rolling cutters that would cut rock with a chipping—crushing action rather than a scraping action such as that required of diamond bits or blade-type bits. A blade-type or "fish-tail" bit similar to the type that was first used in rotary drilling from the mid-1850's to the 1930's is shown in Fig. 1-1.

In 1887, a United States patent was issued to Chapman covering the use of clay, bran, grain, cement, and similar materials to be used as drilling fluid additives to seal the wall of a borehole. The importance of drilling fluid to the rotary drilling process was recognized.

By 1890, there was a number of rotary drilling contractors drilling water wells in Texas and the Midcontinent area of the United States. In 1893, oil was discovered during rotary drilling of a water well near Corsicana, Texas. In the next ten years over 1000 wells were drilled near Corsicana by the rotary process. A rotary drilling rig manufacturing firm was opened in Corsicana and many of the earlier drilling tools were improved.

In 1901, the completion of the Lucas gusher at Spindletop near Gladys, Texas, received worldwide recognition when it flowed at the rate of 75,000—80,000 bbl of oil per day (Lucas, 1971). The well was drilled by the rotary drilling method and the drilling fluid that was circulated consisted of water, finely-ground cuttings from the hole, and sticky clays from surface deposits.

Fig. 1-1. Blade-type or "fish-tail" bit. (Courtesy of Sii Smith Tool, Division of Smith International, Inc.)

Captain Anthony F. Lucas, an Austrian engineer, had drilled an earlier well at Spindletop with cable (or percussion) tools and had abandoned the well due to formation caving. He employed the Hamill brothers who brought a rotary drilling rig from Corsicana to drill the second well. The previous experience of these men with the rotary drilling process and their understanding of using mud circulation for drilling soft formations subject to caving or collapse were the keys to success in the Lucas well completion. Drilling mud for this hole was obtained by driving cattle back and forth through a pit containing surface clay (Hamill, 1957). Mixing mud in a surface pit near a drilling well was a common practice in later years.

By 1906, a device became available for mixing surface clay or "gumbo" into the drilling fluid to make it "thick" and "heavy".

BEGINNING PERIOD: 1888—1928

The development of rotary drilling can be divided into four periods as follows: (A) 1888—1928 — beginning period, (B) 1928—1948 — development period, (C) 1948—1968 — scientific period, and (D) 1968—? — automation period.

The first rolling cutter rock bit with two cutters was introduced by Sharp and Hughes in 1908. This was a great improvement over the "fish-tail" drag bits previously used for rotary drilling harder formations. This bit development and the use of drilling muds to drill all types of formations caused the rotary drilling method to become popular and more widespread.

The rotary drilling method and tools have improved gradually after drilling at Spindletop. The first mud pumps developed exclusively for rotary drilling fluid circulation were introduced around 1910. They were steam-powered and probably pumped no more than 250 gal/min at 500 psi * discharge pressure. Steam-powered pumps suitable for 1000 psi discharge pressure were being manufactured by 1916. By 1928, 18-inch and 20-inch stroke length pumps for 1500 psi discharge pressure were available. The principal components of a rotary rig circulation system are shown in Figs. 1-2 and 1-3.

Drilling mud, an absolute necessity for the rotary drilling process, did not receive much engineering attention until the 1920's. In the early 1920's it was learned that iron oxide and barium sulfate (barite) could be used as a drilling fluid additive to increase density or weight above the density that could be obtained by mixing water and clay alone. Thus, drilling fluids were developed that had the ability to prevent encroachment or entry of high-pressure formation fluids into the borehole. The use of ground barite became the accepted method to control the density of drilling muds.

DEVELOPMENT PERIOD: 1928—1948

By 1930, the testing of muds for viscosity and density values in deep wells became common practice in California. The same mud tests on deep holes were used in the Texas and Louisiana Gulf Coast by 1935. The filtrate or water loss tests on drilling fluids were commenced about the same time.

In the early 1930's, drag bits with drilling fluid circulation nozzles or ports close to the cutting blade became popular when it was learned that a high-pressure stream of drilling fluid impinging on the hole bottom would sometimes more than double the rate of penetration. In 1933, three-cutter rock bits were introduced to the rotary drilling industry by Hughes Tool Company.

In 1935, Harth introduced the use of bentonite to suspend barite and this marked the commencement of the drilling mud industry.

Three-cutter rock bits equipped with internal streamlined passageways for directing drilling fluid circulation to the hole bottom through hydraulic jet nozzles were first tested in the early 1940's and introduced to the industry

* psi = pounds per square inch (lb/inch^2) = 7.031×10^{-2} kg/cm^2.

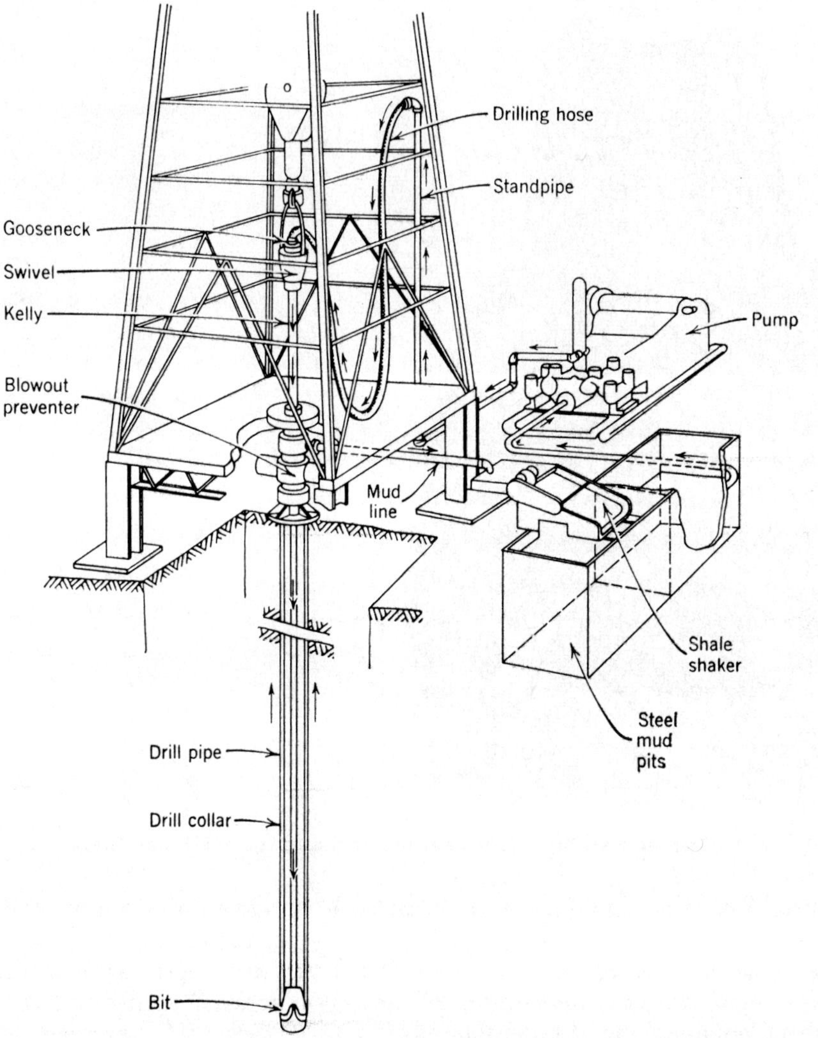

Fig. 1-3. Principal components of a rotary rig circulation system. (Courtesy of Sii Smith Tool, Division of Smith International, Inc.)

in the late 1940's and early 1950's. A jet bit is shown in Fig. 1-4. This development further increased the rate of penetration and initiated engineering attention to jet bit hydraulics and circulation systems. Drilling contractors found that holes could be drilled faster with high-powered pumps and that high-pressure mud circulation was needed to get hydraulic power to the bit.

Inasmuch as the successive insertions of casing reduce the hole diameter with increasing depth, the bit size which can be lowered through the smallest casing is limited. A special tool, which is called an "underreamer", was devel-

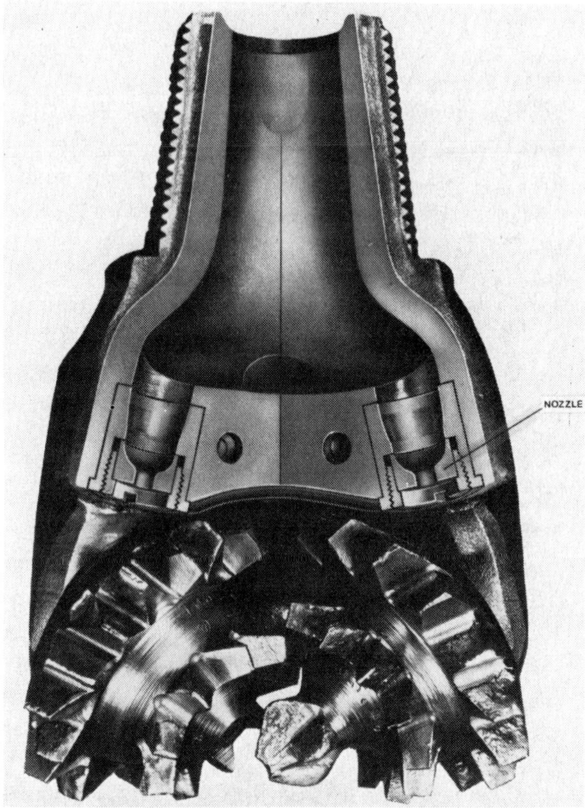

Fig. 1-4. Jet bit. (Courtesy of Sii Smith Tool, Division of Smith International, Inc.)

oped for use (Fig. 1-5) when a larger diameter hole was desired below the smallest casing, to provide more space for cementing, for example. Underreamer has expandable arms with cutters secured to them. Upon activation, the underreamer reams out the section of formation above the bit, resulting in a substantial enlargement of hole diameter.

ROTARY DRILLING CIRCULATION SYSTEM

During the end of the development period of rotary drilling, after higher pressure pumps and jet bits became available, considerable study and research were devoted to the circulation or hydraulic system. The basic components of the hydraulic system are shown in Fig. 1-6.

The mud pump is the heartbeat of the rotary drilling operation. There are usually two mud pumps per rig and both pumps are used for drilling surface holes larger than $12\frac{1}{4}$ inches in diameter. The function of the mud pump is to

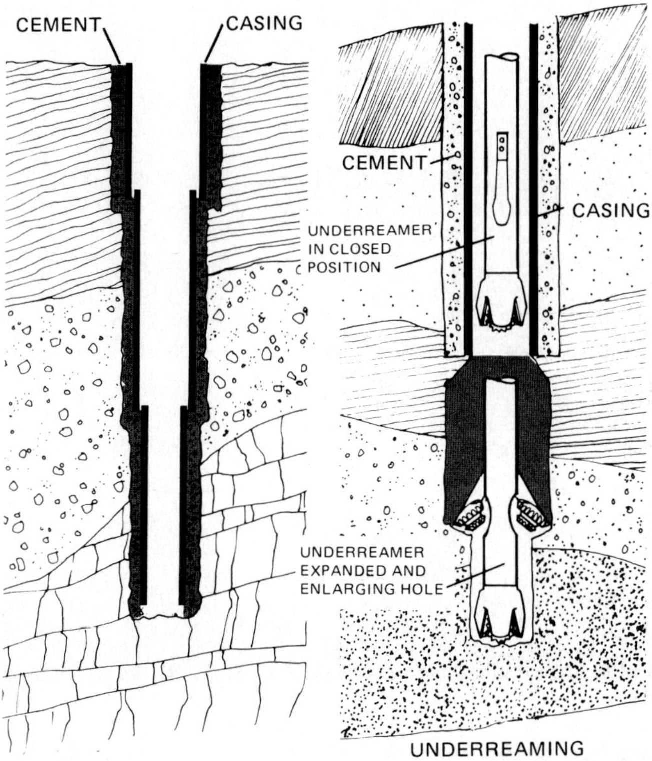

Fig. 1-5. Operation of underreamer. (Courtesy of Smith Tool, Division of Smith International, Inc.)

impart power (pressure) to the drilling fluid in order to move it from the mud tanks through the drillpipe to the bit, where hydraulic horsepower is expended for jetting, and then up the annulus back to the mud tanks. Mud pumps for rotary drilling in the 1970's have ratings up to 1750 input horsepower and higher. They are capable of pumping large volumes of fluid at pressures exceeding 3000 psi. The mud pumps are powered by gas or diesel engines or electric motors. Power transmission losses between the prime mover and power end of the pump are usually considered to be about 15%. The fluid end of a pump receives fluid from the suction line connected to the suction mud tank. Volumetric losses in the fluid end of the pump may vary from 0 to 15% depending on how the pump suction is arranged and the mechanical condition of the fluid end. Some pumps receive fluid from the discharge of a smaller centrifugal pump called a charge pump. Charge pumps usually increase the volumetric efficiency of the mud pump.

From the mud pump discharge, the drilling fluid is pumped under pressure through the surface piping, standpipe, rotary hose, swivel, and Kelly. These

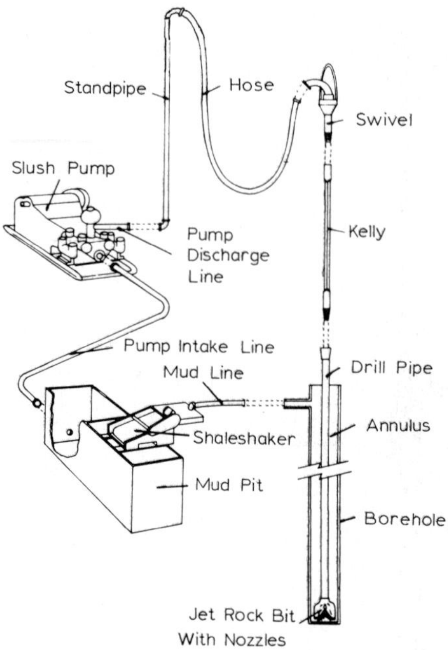

Fig. 1-6. Basic components of the hydraulic system. (Courtesy of Sii Smith Tool, Division of Smith International, Inc.)

components are referred to as the surface connections. A rotary hose and swivel arrangement is necessary to permit movement of the Kelly and drill-string down the hole while circulating fluid and rotating the Kelly.

From the Kelly the drilling fluid passes through the drillpipe and drill-collar bore to the bit. Rotational or torsional power to the bit is transmitted through the drillpipe and drillcollar. Drilling weight or force on the bit is provided by the drillcollars (Fig. 1-7). At the bit, pressure energy is converted to kinetic energy as fluid passes through the three abrasive-resistant (tungsten carbide) nozzles of small diameter. The pressure loss through the bit nozzles should range from 50 to 66% of the available surface pressure in a properly designed hydraulics program. The impingement of the drilling fluid on the hole bottom is directed just ahead of the oncoming teeth of the rotating rock bit. The high-velocity jetting action allows the drilling fluid to penetrate the fractures on the hole bottom and helps release rock chips generated by the bit teeth. Drilling fluid and rock chips rise in the annulus between the drillcollar and hole wall, and then in the annulus between the drillpipe and hole or casing wall. The ability to transport cuttings or rock chips up the hole annulus and to the settling tank depends partly on the annular velocity of the drilling fluid and partially on the mud properties which determine the carrying capacity characteristics. Rock chips fall or slip

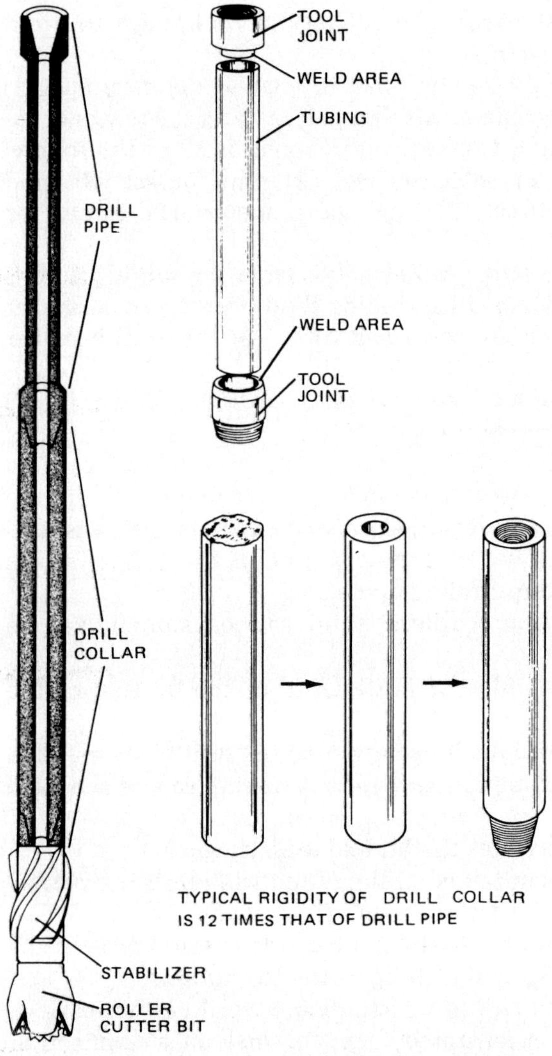

Fig. 1-7. Schematic diagram of drillpipe and drillcollar. (Courtesy of Sii Smith Tool, Division of Smith International, Inc.)

through fluid columns that are static, or not moving, and chips fall faster through a "thin" fluid than through a "thick", viscous fluid. In order to transport cuttings to the surface, the fluid velocity must be greater than the rock chip slip velocity. When the pump volumetric capacity is too low to provide sufficient fluid annular velocity to transport cuttings adequately, the viscosity of the drilling fluid can be increased to reduce the slip velocity of cuttings.

At the surface, the mud must release the cuttings or rock chips to avoid their circulation down the hole again.

The surface mud system should be equipped to enable: (1) cleaning; (2) cooling; (3) mixing; (4) adding chemicals, weight materials, etc.; and (5) removing air or gas from the mud. Consequently, some or all of the following equipment are required: (1) shale shaker; (2) mud tanks; (3) mud degasser; (4) desander; (5) desilter; (6) mud guns, mechanical mixers or stirrers; and (7) suction strainer.

The separation of some of the larger drilled solids from the drilling fluid is accomplished by passing the cutting-laden drilling fluid over the shale-shaker screen. Some of the solids, which are too small to be caught on the shaker screen, settle in the mud settling tank. Smaller solids, down to 75 microns in size, are removed by the desander hydrocyclones. Particles down to 30 microns in size are removed by the desilters.

From the mud settling tank, the relatively clean, cool, well-mixed, gas-free drilling fluid flows to the mud suction tank where it is recycled to the pump suction. The drilling fluid, originally regarded only as a vehicle for transporting cuttings to the surface, is now recognized as one of the major factors involved in the success of the rotary drilling operation.

The functions of a good rotary drilling fluid can be summarized as follows:

(1) Drilling fluid carries rock chips or cuttings from the bottom of the hole to the surface for disposal.

(2) Drilling fluid transmits hydraulic horsepower to the drill bit to assist in cleaning the hole bottom so that maximum penetration rate can be achieved at lower drilling costs.

(3) Drilling mud cools and lubricates the bit and drillstring.

(4) A properly designed and maintained drilling fluid will cover the permeable formation of the borehole with a thin, tough filter cake.

(5) The fluid column pressure (or hydrostatic head) exerted against the borehole wall helps prevent caving or sloughing of the formations.

(6) Drilling fluid suspends rock chips or cuttings and weight material (e.g., barite) when the circulation is interrupted, e.g., for making a connection (adding a joint of drillpipe).

(7) Sand and cuttings should be released readily by the drilling fluid at the surface.

(8) Part of the weight of the drillstring is supported through buoyancy effect while drilling.

(9) The hydrostatic pressure of the drilling fluid column should be of sufficient magnitude to prevent flow of formation fluids (oil, gas or water) into the borehole and blowouts.

(10) Drilling mud should permit the securing of all information necessary for evaluating the formations penetrated, using logging tools, for example.

SCIENTIFIC PERIOD: 1948–1968

One of the most significant developments in rotary drilling and the drilling industry has been the increase in drilling depth. Well depth records in the United States increased from 17,832 ft in 1947 to more than 31,000 ft in 1974.

Gray and Young (1973) reviewed 25 years of drilling technology advancement and accomplishment and concluded that even though the basic fundamentals of rotary drilling had not changed in that time, many basic improvements in the process and equipment had been introduced. Bit life or rotating time has increased twenty-fold through the introduction of improved design of jet bits, tungsten-carbide insert bits, sealed-bearing bits, and journal-bearing bits. Fundamental studies of rock breakage, rock chip transport, drilling fluids, and abnormal formation pressure prediction techniques, combined with improved equipment engineering, have been used to improve the effectiveness and economy of rotary drilling.

AUTOMATION PERIOD (SINCE 1968)

Great progress has been made in the development of rotary drilling rigs and drilling fluids since the 2000-ft depth capacity rigs and "muddy waters" were used by Lucas and the Hamill brothers at Spindletop in 1901. Proposals have been made and rigs have been designed for drilling to depths of around 50,000 ft with up to 9000 input horsepower.

The manufacture of materials necessary for formulating drilling fluids for well drilling is handled by a very large industry today. Mud companies maintain large warehouse stocks of bentonite, barite, polymers, chemicals, and many other additives near the principal oil drilling areas of the world.

In the late 1960's, many drilling engineers believed that the usage of computerized drilling control units would be more widespread than it actually is at the present time. The reasons for this development lag are complex, but contributing factors probably include large increases in drilling activity, shortages of experienced drilling engineers, and changes in drilling practices.

The highly successful introduction and widespread usage of soft-formation, tungsten-carbide insert, journal-bearing bits in the early 1970's resulted in many changes of bit weight and rotational speed determination practices. These operational practices were not compatible with the optimized bit weight and rotary speed recommendations provided by some of the earlier drilling prediction techniques.

Some of the early prediction techniques were developed for medium-hard and hard-formation milled tooth bits. These bit types have been replaced in virtually all drilling areas by tungsten-carbide insert bits. The earlier prediction techniques required solution by computers and were subject to many limiting assumptions, often ignoring the effects by hydraulic conditions and mud properties.

Outstanding drilling performance can be obtained from present-day journal-bearing insert bits if careful attention is given to the basic factors that affect the penetration rate. Contemporary methods for determining the parameters and variables that affect the rate of penetration should include the use of the latest prediction techniques and programs that result in minimum-cost drilling.

Drilling prediction techniques should be modified and updated and new calculation methods developed as the new drilling products and new technology are introduced.

REFERENCES

Brantley, J.E., 1971 (original 1892). *History of Oil Well Drilling.* Gulf Publishing Co., Houston, Texas, 1525 pp.

Gray, G.R. and Young, F.S., 1973. 25 years of drilling technology — A review of significant accomplishments. *J. Pet. Technol.*, 25 (12, Dec.): 1347—1354.

Hamill, C.G., 1957. *We Drilled Spindletop.* C.G. Hamill, Houston, Texas.

Lucas, A.F., 1971. The great oil-well near Beaumont, Texas. *J. Pet. Technol.*, 23 (11, Nov.): 1328—1333.

Smith Tool, 1977. *The Techniques, Tools, and Terminology of Drilling.* Division of Smith International, Inc., Publ., 13 pp.

Chapter 2

BASIC CONCEPTS OF DRILLING AND DRILLING-FLUID TECHNOLOGY

GEORGE V. CHILINGARIAN and PAUL VORABUTR

INTRODUCTION

Drilling-fluid technology became important after the success of the first rotary drilling well. Until 1863, drilling fluid was an unimportant item of cable-tool drilling. Stroud, who was probably the first "drilling-fluid engineer", started the modern history of drilling fluids in 1921 when he attempted to control drilling-fluid properties through additions of specially purchased additives (Rogers, 1953, p. 42).

Only three publications on drilling fluids appeared during the period 1916 to 1929. The number rose to 23 in 1930, thus signalling the awareness of drilling fluids (Tschirley, 1978, p. 11). In 1931, however, development of drilling-fluid technology gained real momentum.

The complexity of drilling fluids and the availability of numerous types of muds and additives at the present time indicates the progress that has been made in the field of drilling-fluid technology since the drilling of the famous Spindletop well. The latter was completed with drilling fluid prepared by driving cattle through a shallow water pit.

The high cost of drilling fluids and the time and attention devoted to them before and during drilling are indicative of their great importance.

In rotary drilling, the circuit of drilling fluid through a well is depicted in Fig. 2-1. The bit is pressed against the bottom of the hole and is rotated by the drillpipe extending all the way to the surface. The drilling fluid circulates down the inside of the drillpipe, out through the openings in the bit, and up through the annular space between the walls of the hole and the drillpipe.

The cuttings are lifted to the surface where they are separated from the drilling fluid. Drilling fluid passes through a screen, whereas the cuttings remain behind; it then flows through a mud ditch to a pit. Drilling fluid is pumped up from the pit by a mud-pump and is recirculated down the inside of the drillpipe to the bit.

The drillpipe with the attached bit is rotated from the surface by means of a rotary table, which has a square hole in the center. A square pipe (Kelly) at the top of the drillpipe passes through the hole in the rotary table and turns with it, along with the entire length of drillpipe and the bit. The Kelly can move down through the table as the bit digs deeper into the rocks. As soon as the top of the Kelly approaches the table, another length of pipe is added below the Kelly.

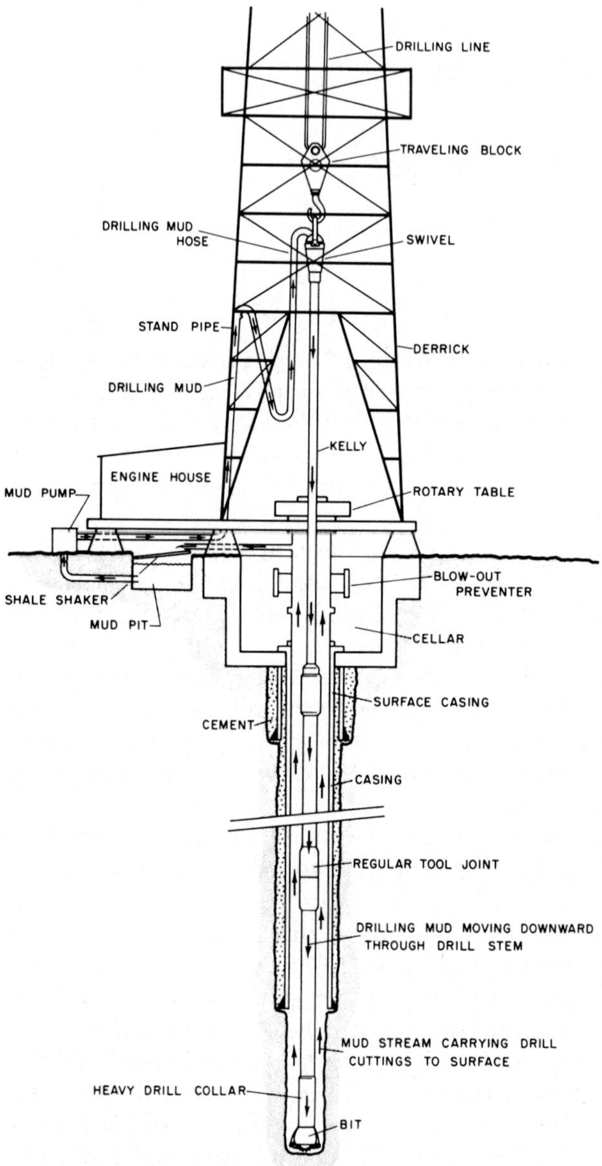

Fig. 2-1. Schematic diagram showing circuit of drilling fluid through a well in rotary drilling.

In some cases (e.g., to replace the worn-out bit), the entire length of drill-pipe must be removed from the hole. To achieve this, the pipe is gradually disconnected into convenient lengths and is racked in the derrick. Although the above information was presented previously in Chapter 1, it serves as an introduction to those interested in drilling fluids only.

FUNCTIONS OF DRILLING FLUIDS

Although at the present time numerous (over one hundred) brand names of drilling fluids are commercially available for all drilling purposes and conditions, the main function of all of these fluids is the successful completion of the wells.

Even though all drilling fluids have the same functions in rotary drilling, their properties may be altered as deemed necessary to facilitate a speedy, safe, and satisfactory completion of the well. These functions comprise:

(1) Removal of cuttings from the hole.
(2) Cleaning of the bottom of the hole.
(3) Controlling the subsurface pressure.
(4) Cooling and lubricating the bit, drillpipe, and drillcollar.
(5) Acting as medium for settling out of cuttings in the surface pit.
(6) Forming filtercake on walls of the borehole (wall-off permeable formations).
(7) Preventing caving-in of the formations.
(8) Avoiding damage to productivity of producing formations.
(9) Partial aid in supporting the weight of drillstring and casing.
(10) Reducing casing cost.
(11) Preventing corrosion fatigue of drillpipe.
(12) Allowing interpretation of electric logs.
(13) Transmission of the surface-available hydraulic horsepower to the bit.
(14) Suspension of cuttings and weight materials on stopping the circulation.

Obviously, all these functions must be performed without creating any hazard to the crew.

Removal of cuttings from hole

One of the most important and primary functions of drilling fluids is the removal of cuttings from the borehole. The fluid flowing through the exit nozzles of the bit exerts a jetting action (on the bottom of the hole), which cleans the face of the hole and carries the cuttings upward to the surface. Although gravity tends to pull these cuttings toward the bottom (slip velocity), if the volumetric rate of circulation is of sufficient magnitude and annular velocity (upward direction) is sufficiently high to exceed the slip velocity, cuttings can be removed.

In order for a cutting to reach the surface, the slip velocity must be lower than the average annular velocity. The annular velocity (which is a function of borehole size and condition, pump output, and drillpipe and drillcollar sizes) is commonly defined as:

$$\text{Annular velocity (ft/min)}, v_a = \frac{\text{pump output (bbl/min} \times 100)}{\text{annular volume (bbl/100 ft)}} \tag{2-1}$$

The above equation, however, does not take into consideration the well-bore conditions. The presence of thick washout sections can significantly reduce annular velocity.

As an approximate guide, the minimum annular velocities for hole sizes of 15, $12\frac{1}{4}$, $10\frac{5}{8}$, $8\frac{3}{4}$, $7\frac{7}{8}$, and 6 inches are 80, 90, 110, 120, 130, and 140 ft/min, respectively (Magcobar Services, 1972, p. 1, chapter 13, Hydraulics Section).

According to Tschirley (1978, p. 8, I(I)), the following equations may be used to estimate the slip velocities v_s in ft/sec, depending on the type of flow in the annulus:

$$\text{laminar flow:} \quad v_s = \frac{53.5(\gamma_s - \gamma_f)\, d_s^2 v_a}{6.65 YP(d_b - d_p) + v_a PV} \tag{2-2}$$

$$\text{turbulent flow:} \quad v_s = \frac{d_s(\gamma_s - \gamma_f)}{\gamma_f} \tag{2-3}$$

where γ_s = specific weight of cutting (lb/gal); γ_f = specific weight of drilling fluid (lb/gal); v_a = average annular fluid velocity (ft/sec); d_s = average cutting diameter (inches); YP = yield point (lb/100 sq ft); d_b = borehole diameter or bit size (inches); d_p = drillpipe O.D. (inches); and PV = plastic viscosity (cP).

According to Moore (1974, p. 234), the following equation for determining slip velocity may also be used:

$$v_s = \frac{175 d_s(\gamma_s - \gamma_f)^{0.667}}{(\gamma_f \mu_f)^{0.333}} \tag{2-4}$$

where d_s = particle diameter (inches); γ_s = specific weight of cuttings; γ_f = specific weight of fluid; and μ_f = fluid viscosity (cP). The Reynolds number, *Re*, may then be calculated by using the following equation:

$$Re = \frac{15.47 \gamma_f v_s d_s}{\mu_f} \tag{2-5}$$

For Reynolds numbers greater than 2000, which indicates a turbulent flow, slip velocity may be estimated by using the following equation:

$$v_s = 113.4 \left[\frac{d_s(\gamma_s - \gamma_f)}{C_D \gamma_f}\right]^{1/2} \tag{2-6}$$

where C_D = drag coefficient; 1.50 should be used in the absence of a more accurately determined value. γ_s and γ_f are in lb/gal.

The Pigott equation is also used to calculate slip velocity in a laminar flow:

$$v_s = \frac{2gd_s^2(\gamma_s - \gamma_f)}{92.6\mu_f} \tag{2-7}$$

where v_s = slip velocity (ft/sec); g = gravitational acceleration, which is equal to 32.2 ft/sec/sec; γ_s = specific weight of solids (lb/cu ft); γ_f = specific weight of drilling fluid (lb/cu ft); d_s = diameter of round particles or greatest dimension of flat particles (ft); and μ_f = viscosity [cP (0.0672)/100].

Thus, slip velocity in laminar flow can be decreased by (a) increasing the viscosity, (b) decreasing the particle size, and (c) increasing the mud weight. Bit geometry and fluid shear effect control the particle size.

In turbulent flow, the slip velocity can also be calculated by using the following equation (Magcobar, 1972, p. 5, Hydraulics Section):

$$v_s = \frac{3.5\sqrt{d_s(\gamma_s - \gamma_f)}}{\gamma_f} \tag{2-8}$$

The above equations represent some of the few that have been used in the Petroleum Industry. The choice depends entirely upon personal experience and familiarity with each equation.

Efficiency of removal of cuttings is a direct function of the carrying capacity of the drilling fluid, which depends on the following factors: (1) density of drilling fluid; (2) viscosity of drilling fluid; (3) type of flow (laminar or turbulent), i.e., velocity distribution; (4) torque; (5) size and shape of particles; (6) rotation of drillpipe; and (7) ratio of the specific gravity of solids to that of drilling fluid.

Specific weight of drilling fluids

Increase in drilling-fluid specific weight increases the buoyant forces acting on each particle in the direction opposite to the gravitational force. This, in turn, increases the lifting capacity of the drilling fluid, as illustrated in Fig. 2-2. The particle recovery is much higher in the case of heavier drilling fluid at the same circulation rate, other properties remaining constant.

Viscosity and gel strengths

An example of the effects of viscosity and gel strength on particle return are shown in Fig. 2-3. A mud having a low viscosity and low gel strength gives rise to a higher percent particle recovery at the same annular velocity and circulating time. According to Williams and Bruce (1951, p. 114), the higher carrying capacity of the low-viscosity—low-gel-strength mud is particularly noticeable for the medium-size particles, which closely resemble cuttings.

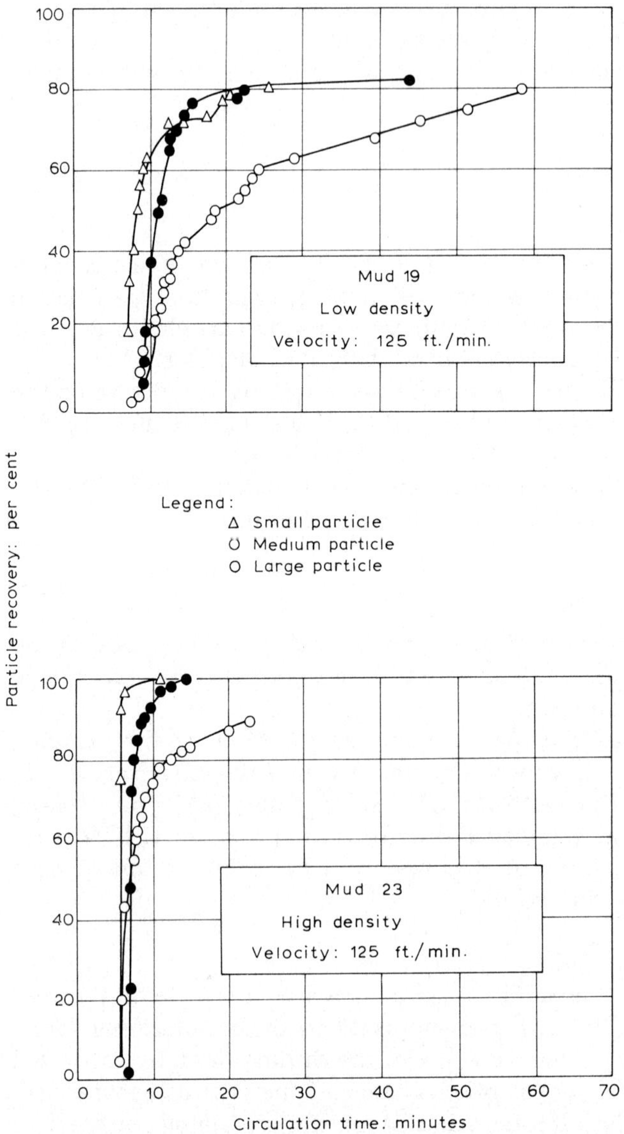

Fig. 2-2. Effect of mud weight on particle return, i.e., relationship between the particle recovery in percent and circulation time in minutes. Mud no. 19 — 9.6 lb/gal; mud no. 23 — 11.6 to 12.4 lb/gal. (After Williams and Bruce, 1951, p. 114, fig. 2; courtesy of Petroleum Transactions, AIME.).

Velocity distribution in the annulus

Higher carrying capacity may be achieved with turbulent flow than with viscous (laminar) flow for a drilling fluid having relatively low viscosity and

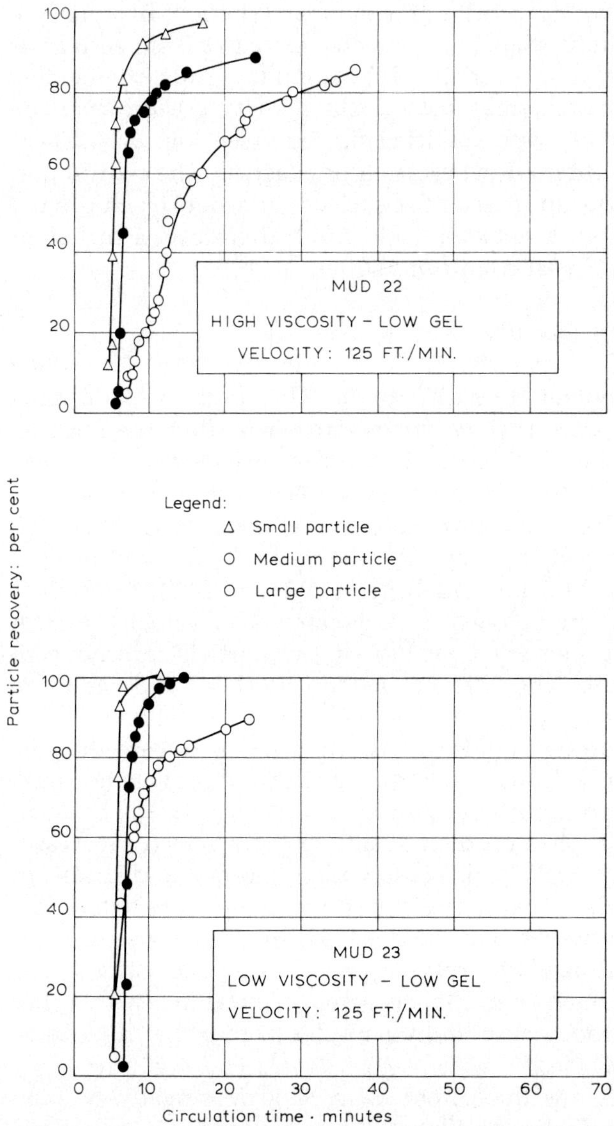

Fig. 2-3. Effect of mud treatment on particle return, i.e., relationship between particle recovery (%) and circulation time (min). Mud no. 22 — Stormer viscosity (600 rpm) = 30 cP; initial gel strength = 2- g. Mud no. 23 — Stormer viscosity (600 rpm) = 11 cP; initial gel strength = 0. (After Williams and Bruce, 1951, p. 116, fig. 4; courtesy of Petroleum Transactions, AIME).

gel strength compared to water in the case of non-rotating drillpipe. This is due mainly to the flatter velocity profile and the presence of "turbulence" in the drilling fluid, which tends to minimize the cutting slippage at or near the pipe and hole walls. The velocity profile of a laminar (viscous) flow, on the other hand, assumes a parabolic shape, i.e., velocity increases from zero at or near the wall to maximum at the center of the annulus. Thus, the cutting chips caught at or near the walls may have a slip velocity greater than the annular velocity at that point, with resulting slippage (see Fig. 2-4). These particles fall back through the fluid, whereas chips caught in the central portion of streamlines may be transported to the surface. In order to achieve the maximum carrying capacity in a relatively thin fluid, therefore, a turbulent flow may be more desirable if well conditions allow.

Torque effects on carrying capacity

Drillpipe rotation during drilling affects the carrying capacity of a drilling fluid having laminar or turbulent flow differently. The rotation of drillpipe induces a torque effect in viscous flow, which further hinders the removal of cuttings by a relatively thin drilling fluid. A torque effect arises from the tendency of a flattish particle to overturn when subjected to unequal forces, due to variation in fluid velocity as is the case in a viscous flow. Figure 2-5 depicts an isolated chip subjected to an unequal velocity distribution in a viscous flow. According to Williams and Bruce (1951, p. 117), force F_4 is greater than force F_2, because velocity v_4 is greater than velocity v_2. An induced torque around the center of gravity of the particle is produced, therefore, tending to overturn the particle. As the particle becomes more inclined, the force component F_H becomes effective and, combined with the turning moment (induced torque), tends to push the particle toward the area of even lower annular velocity. Thus, the largest area of the particle becomes parallel to the direction of transport. Inasmuch as a particle falling on-edge falls at a rate 1.5 times faster than on the flat side, the slip velocity increases greatly (Williams and Bruce, 1951, p. 116), with consequent decrease in effectiveness of upward particle transportation. Slippage is more severe in a wider annulus than in a narrower one, because of the higher viscous-flow velocity gradient. Combination of the force F_H and torque may cause cyclic overturn of the particles. Such a cyclic occurrence interferes with other upward-moving particles, which causes more particles to overturn, and, thus, reduces the transport effectiveness.

Inasmuch as in turbulent flow the difference in velocity is relatively small across the annulus (flatter velocity profile), the turning torque is minimized (Gatlin, 1960, p. 76). Theoretically, the carrying capacity of drilling fluid should be the highest in the case of plug flow.

Particle dimension

Rock bit designs determine the size and shape of cuttings generated. The physical dimensions of cuttings have been found to greatly affect the carrying capacity of a drilling fluid. Williams and Bruce (1951, p. 119) found that particles for which the thickness-to-diameter ratio is less than 0.3 are

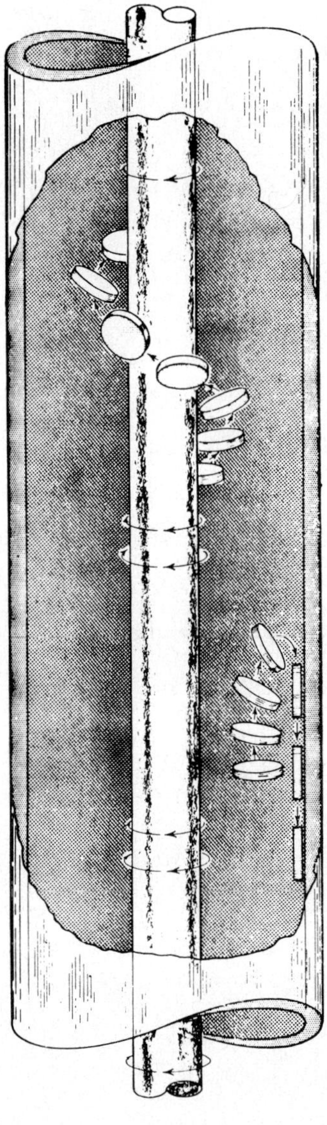

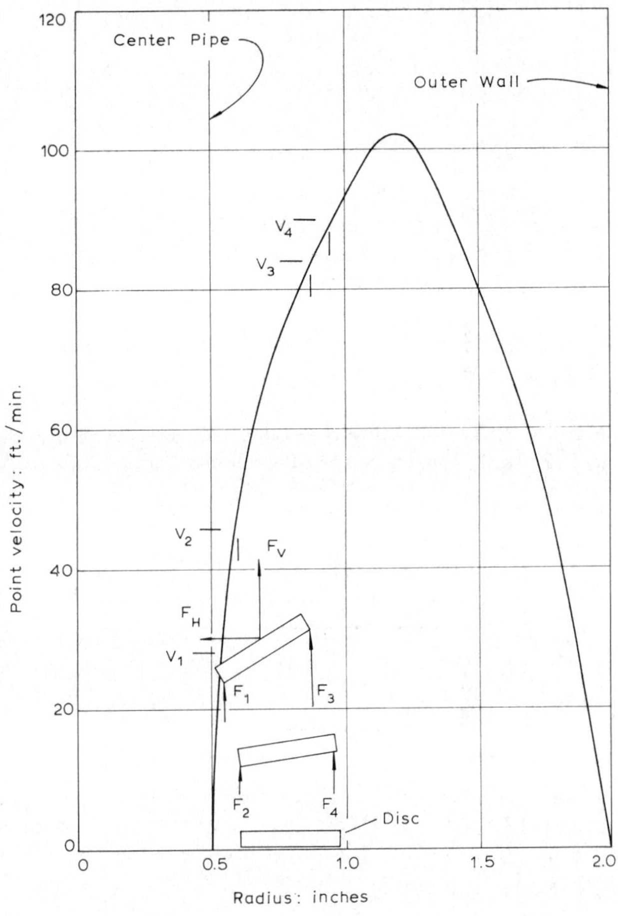

Fig. 2-4. Transportation of medium-size discs in a viscous (laminar) flow during pipe rotation. Thickness/diameter ratio of discs = 0.167. (After Williams and Bruce, 1951, p. 120, fig. 12; courtesy of Petroleum Transactions, AIME.)

Fig. 2-5. Torque effect on particles in a viscous flow. (After Williams and Bruce, 1951, p. 119, fig. 11; courtesy of the Petroleum Transactions, AIME.)

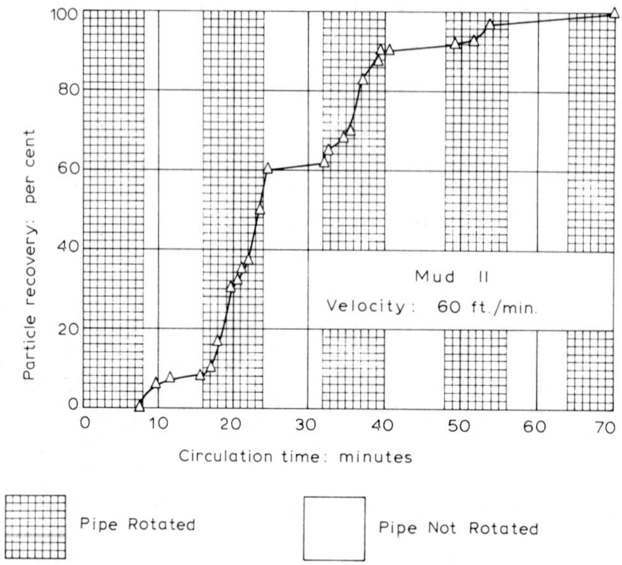

Fig. 2-6. Effect of drillpipe rotation on particle return. (After Williams and Bruce, 1951, p. 112, fig. 1; courtesy of the Petroleum Transactions of AIME.)

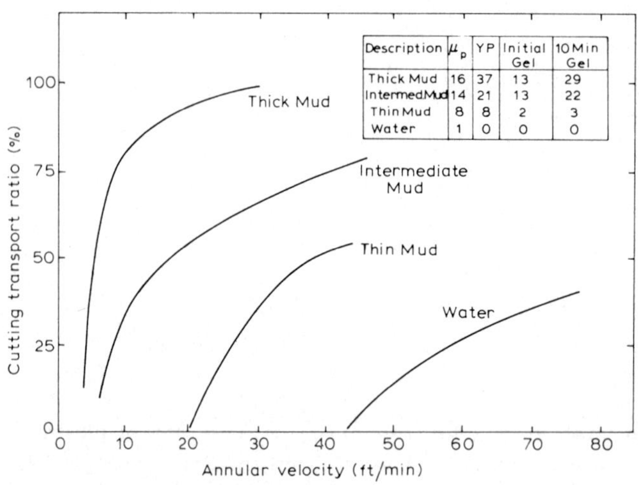

Fig. 2-7. Relationship between annular velocity and cutting transport ratio, showing improved carrying capacity of non-Newtonian drilling fluids at low annular velocities (laminar flow). (After Sifferman et al., 1974, p. 1299, fig. 4; courtesy of the Society of Petroleum Engineers of AIME.)

difficult to remove from the wellbore, because they tend to turn on-edge more readily.

Effects of drillpipe rotation

Drillpipe rotation increases the carrying capacity of drilling fluid, with the effect being more pronounced in the case of viscous flow than turbulent flow. This is due to the fact that cuttings tend to be thrown into a higher-velocity region of the annular velocity profile by centrifugal forces which are created by the rotating drillpipe (see Fig. 2-4; also see Fig. 2-6).

Bottomhole cleaning

Bottomhole cleaning is a function of the ability of drilling fluid to remove cuttings from the wellbore in order to prevent bit balling and floundering. Rock cuttings, which are continuously generated by the bit action, are removed by the drilling-fluid stream. Consequently, a drilling fluid having maximum carrying capacity at the existing conditions provides the most efficient bottomhole cleaning.

According to Williams and Bruce (1951) and Moore (1974, p. 229), on using water or other very thin fluids having a viscosity close to that of water, an annular fluid velocity of 100 to 125 ft/min is adequate for removing for-

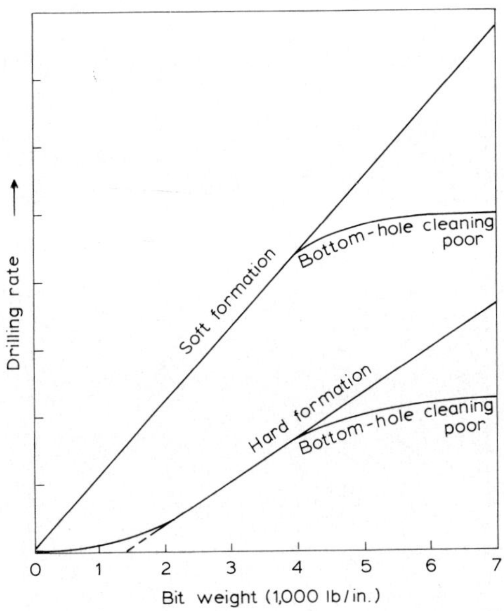

Fig. 2-8. Relationship between drilling rate and bit weight for soft and hard formations, and the effect of bottomhole cleaning efficiency. (After Moore, 1974, p. 9, fig. 2-1; courtesy of Petroleum Publishing Company.)

mation solids. Velocity of this magnitude would most likely give rise to turbulent flow in the annular space. Tests with shear-thinning fluids, however, have indicated that an annular velocity as low as 50 ft/min may be sufficient for adequate bottomhole cleaning, for reasons previously discussed (see Fig. 2-7). This is of great significance in drilling deep wells where it is often difficult or undesirable from the hole-erosion standpoint to maintain turbulent flow or use a thin fluid in the annulus.

The effect of degree of bottomhole cleaning on the rate of penetration is shown in Figs. 2-8 and 2-9. Reduced penetration rates in the case of inadequate bottomhole cleaning have been attributed to bit balling and floundering, when the cuttings are not swept away from bottom rapidly enough and are reworked by the bit teeth. Finely-ground material prevents bit teeth from attacking the newly-exposed bottomhole surface effectively, resulting in the waste of hydraulic power and a reduced penetration rate. A balled-up bit has been defined as a bit with a mass of sticky, finely-ground, and compacted material firmly attached to it (Kruse, 1976, p. 9), whereas floundering refers to regrinding of the cuttings by the bit.

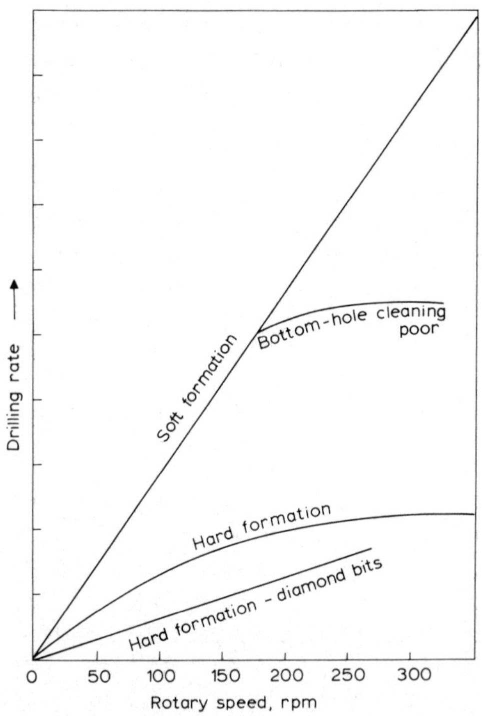

Fig. 2-9. Relationship between rotary speed and drilling rate, and the effect of inadequate bottomhole cleaning. (After Moore, 1974, p. 10, fig. 2-2; courtesy of Petroleum Publishing Company.)

Adequate bottomhole cleaning is necessary throughout any drilling operation and becomes extremely important with increasing depth.

Control of subsurface pressure

Drilling fluids must provide sufficient hydrostatic head to balance the formation pore pressure in order to restrict the influx of formation fluids into the wellbore. An "overbalanced drilling" condition exists when the pressure exerted by the drilling fluid column exceeds that of the formation pore pressure. On the other hand, the term "underbalanced drilling" is commonly used to describe the drilling conditions that exist when the pressure exerted by the drilling fluid column is lower than that of the formation pore pressure.

In order to be able to control the subsurface pressure by the drilling fluid, the following pressure concept must be understood. Overburden pressure originates from the combined weights of the formation solid matrix and the interstitial fluids (oil, gas, and/or water) in the rocks overlying the formation of interest (Fertl and Chilingarian, 1977, p. 347; and Chilingarian and Wolf, 1975). This concept may be summarized mathematically as follows:

$$p_o = D[(1 - \phi)\gamma_{rm} + \phi\gamma_{if}] \tag{2-9}$$

where p_o = overburden pressure (e.g., lb/sq ft); D = depth of the geologic column (e.g., ft); ϕ = porosity (fraction); γ_{rm} = specific weight of the rock matrix (e.g., lb/cu ft); and γ_{if} = specific weight of the interstitial fluids occupying the pore spaces (e.g., lb/cu ft).

Formation pore pressure, which is equal to the difference between the total overburden stress and grain-to-grain stress, is the pressure acting on the fluid in the pore spaces of the formation. Normally, formation pressure is equal to the hydrostatic head of water extending from the top of the water table to the subsurface formation. Any formation pressure greater than hydrostatic is classified as an abnormally high formation pressure ("surpressure"), whereas formation pressure lower than hydrostatic pressure is referred to as a subnormal pressure ("subpressure") (Fig. 2-10). The effective (grain-to-grain) stress is equal to:

$$p_e = D[(1 - \phi)\gamma_{rm} + \phi\gamma_{if} - \gamma_{if}] \tag{2-10}$$

The following factors must be considered in selecting the required specific weight of drilling fluid weight for successful completion of the well (L.P. Mitchell, 1972, personal communication (notes), p. 9): (1) fracture gradient; (2) pore pressure; (3) kick tolerance; (4) casing-shoe depth; (5) borehole stability; (6) surface-pressure control equipment; (7) annular circulating velocity; (8) pressure surges; (9) differential sticking of pipe; (10) magnitude and rate of filtration of drilling fluid; (11) filling of the hole; (12) gas cutting of the drilling fluid; (13) bit hydraulics; (14) drilling fluid cost; (15) drilling rate; (16) removal of drill solids; (17) formation porosity, permeability, and

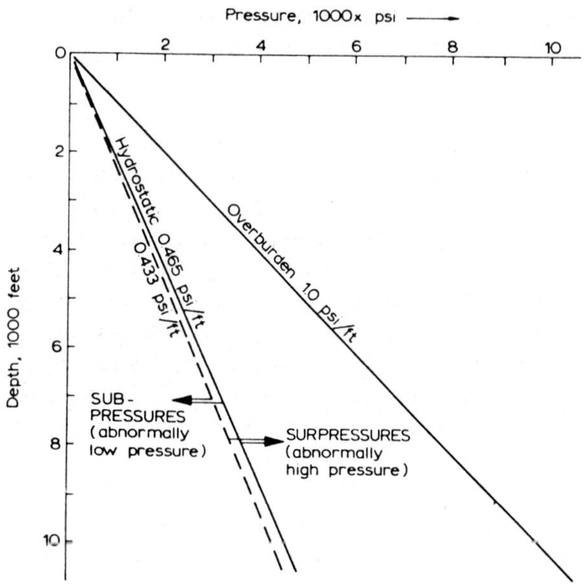

Fig. 2-10. Subsurface pressure concept. (After Fertl and Chilingarian, 1977, p. 348, fig. 1; courtesy of the Journal of Petroleum Technology, Society of Petroleum Engineers of AIME.)

type of fluids present; (18) safety margin over the pore pressure; (19) safety margin under the fracture gradient; (20) electric log interpretation; (21) geological setting; (22) reservoir engineering data (e.g., formation damage); and (23) cooperation with management.

How heavily each one of the above factors affects the final selection of drilling fluid specific weight, varies mainly with drilling conditions, experience in the area, and objectives.

The hydrostatic pressure that a column of a drilling fluid exerts at any depth in the hole can be calculated as follows:

Hydrostatic pressure (psi) = (0.052) × (depth, ft) × (drilling fluid weight, lb/gal) (2-11a)
= (0.0065) × (depth, ft) × (drilling fluid weight, lb/cu ft) (2-11b)
= (0.098) × (depth, ft) × (drilling fluid weight, g/cm^3) (2-11c)

The basic equation is $p = \gamma_f \times D$, where p is pressure in lb/sq ft, γ_f is specific weight of fluid in lb/cu ft, and D is depth in ft.

According to Moore (1974, p. 301), in the Gulf Coast of the United States, a drilling fluid specific weight of 9.0 lb/gal is required to balance most formation pore pressures. The required specific weight of drilling fluid, therefore, would be at least 9.2 lb/gal, whereas in the Rocky Mountains and

Midcontinent areas the required drilling fluid weight can be much lower than 9.0 lb/gal.

Abnormal fluid pressures *

Shales adjacent to permeable sands containing fluids at high pressures are much more porous than is normal for shales at comparable depths. This can be attributed to both stratigraphic and structural conditions. Expulsion of water from shale pores during compaction is sometimes restricted by overlying lithified (e.g., as a result of $CaCO_3$ deposition) or impermeable strata and by faults. Such conditions give rise to "overpressured" (or "abnormally high-pressure") reservoirs, whereas the associated shales are termed "undercompacted" or "overpressured". There is water movement from the overpressured shales into the sands as hydrocarbons are depleted owing to production. According to Hottman and Johnson (1965), fluid pressures within the pore space of shales can be determined by using data from both acoustic and resistivity logs.

During the past 20 years, problems related to the development of deep-drilling techniques and hydrocarbon production from reservoir rocks, mainly associated with thick sequences of argillaceous sediments in Tertiary basins, have become increasingly urgent in the petroleum industry. Successful drilling to depths greater than 20,000 ft and subsequent production depend to a great extent on knowledge of the basic mechanical properties and deformation characteristics of the encountered formations, especially shales, and the chemistry of their associated fluids. Changes in the physical and chemical properties of rocks and interstitial fluids with depth and the ability to understand and predict these changes are fundamental to successful petroleum exploration and development.

The relative salinities of interstitial waters in well-compacted and undercompacted shales and their associated sandstones, as envisioned by the writers, are presented in Fig. 2-11, if one assumes that fluids in the center of the capillaries (1) are more saline than fluids adjacent to the capillary walls and (2) are expelled first. In both cases (undercompacted and well-compacted shales), the salinities of interstitial fluids in shales should be fresher than those in associated sandstones if changes in all other variables are identical. Interstitial water in undercompacted shales, however, should be more saline than that in well-compacted shales, because in the former case a smaller portion of the more saline fluid present in the center of capillaries is squeezed into adjacent sandstones. Future field and laboratory investigations will show whether these relationships hold true or not.

In the worldwide search for oil and gas, fluids at pressures much higher

* This section is taken from the book entitled "Compaction of Argillaceous Sediments" by H.H. Rieke and G.V. Chilingarian, 1974, pp. 23—27.

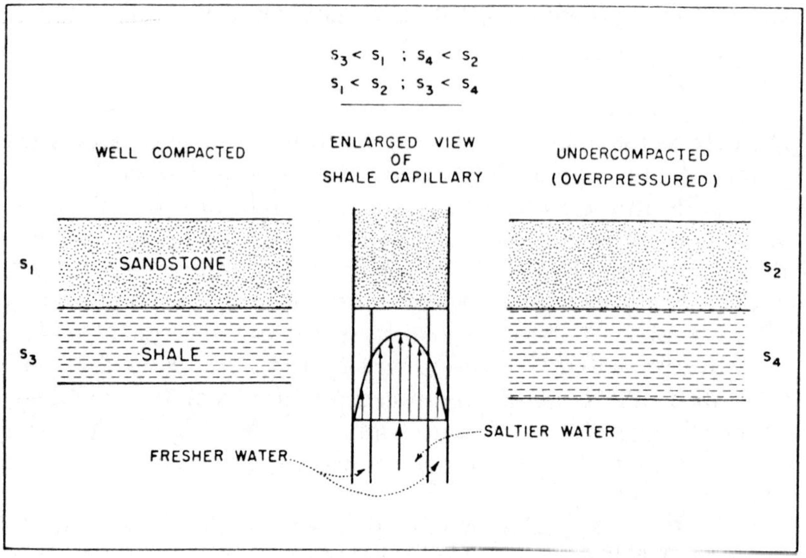

Fig. 2-11. Relative salinities of interstitial fluids in well-compacted and undercompacted (overpressured) shales, as envisioned by the writers. Arrows show direction of flow and velocity profile in an enlarged view of a capillary in a vertical direction. S = salinity. (After Rieke and Chilingarian, 1974, p. 25, fig. 9.) $S_2 < S_1$ in case of large influx of fresher water from associated undercompacted shales.

than normal hydrostatic pressures have been encountered in many countries (see Fig. 2-12). Abnormally high fluid pressures, which are often misnamed "geopressures", are defined by some geologists and petroleum engineers as any fluid pressure which exceeds the hydrostatic pressure of a column of water containing 80,000 ppm total solids (Fertl and Timko, 1970). Many deviations from this generalization, which holds true for the Gulf Coast depositional basin, readily occur in other areas. High hydrostatic gradients (change in pore pressure per unit depth), with respect to the normal brine gradient of 0.465 psi/ft or 0.106 bar/m, exist throughout the entire pre-Tertiary sedimentary sequence in many basins such as the Williston Basin of North America. Formation waters with approximately 356,000 ppm total solids occur in the latter basin, giving rise to an overall hydrostatic pressure gradient of 0.512 psi/ft (Finch, 1968). In order for fluid pressures to be classified as abnormal in the Williston Basin, they would have to yield a gradient larger than 0.512 psi/ft. A more comprehensive definition of abnormal pressure, therefore, is a fluid pressure that materially exceeds the weight of a column of interstitial fluid ($p_p = \gamma_f h$, where p_p is pressure in lb/sq ft, γ_f is specific weight of interstitial fluid in lb/cu ft, and h is height of column in ft).

It is not sufficient merely to recognize the presence of abnormal pressures.

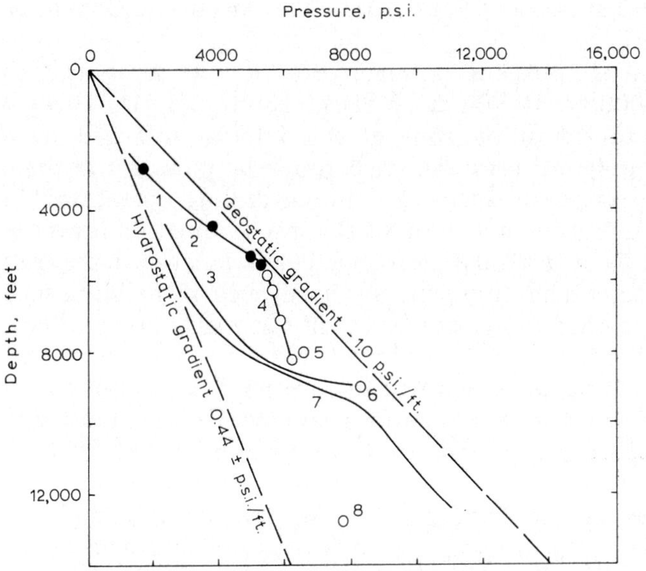

Fig. 2-12. Depth—pressure relationship in high-pressure areas. *1* = Khaur Field, Pakistan; *2* = Lost Hills, California; *3* = Ventura Avenue, California (pressures measured near crest); *4* = Chia-Surkh, Iraq (after Cooke, 1955); *5* = Iran; *6* = D-7 zone of Ventura Avenue Field, California; *7* = Gulf Coast fields, Texas—Louisiana; *8* = Church Buttes, Wyoming. Pressure gradient of 1 psi/ft corresponds to overburden specific gravity of 2.31. Hydrostatic gradient of water having specific weight of 62.4 lb/cu ft is equal to 0.433 psi/ft ($p = \gamma h = (62.4 \times 1)/144 = 0.433$ psi). (After Watts, 1948, fig. 2, p. 144. Courtesy of the SPE of AIME).

The magnitude of these fluid pressures must be known in order that drilling fluids can be selected to (a) prevent blow-out, (b) optimize penetration rate, (c) recognize hydrocarbon shows from productive formations, (d) lessen damaging fluid entry into potential pay sands, and (e) permit more reliable interpretation of formation log data (Griffin and Bazer, 1968).

Areas in the United States where abnormal fluid pressures have been reported in the literature include the Gulf Coast Basin (post-Cretaceous sediments), Anadarko Basin in Oklahoma (Pennsylvanian Morrowan sediments), Williston Basin in North Dakota (Devonian Sanish zone in the Antelope Oil Field), and in the Ventura area of California. Outside the United States, abnormal fluid pressures are known to exist in the Arctic Islands; Africa (Algeria, Morocco, Mozambique and Nigeria); Europe (Austria, the Carpathian region of eastern Europe, Rumania, France, Germany, The Netherlands, Ukraine, and in the Urals and Caucasian region of U.S.S.R.); Far East (Burma, China, India, Indonesia, Japan, Malaysia, and New Guinea); Middle

East (Iran, Iraq, and Pakistan); and South America (Argentina, Colombia, Trinidad, and Venezuela).

In the Gulf Coast area of Louisiana, normal hydrostatic pressures occur to depths ranging from 8000 to 12,000 ft (Wallace, 1965, p. 124). There is normally a short pressure-transition zone at this interval followed by a drastic increase in the wellbore pressure. High abnormal pressures in these oil or gas wells usually result in difficulties during drilling operations. In many of the oil fields, therefore, detection of abnormal formation pressures has become imperative. Geologists and petroleum engineers need to discover a way to detect and predict high fluid pressures in advance of the bit in such areas so that protective casing strings can be set in the wells during drilling. In many cases wells either have been abandoned before reaching the desired formation or have blown out, often burning the drilling rigs and injuring or killing the rig personnel. The numerous methods employed in the industry to detect and evaluate high-pressure zones are thoroughly discussed by Fertl and Timko (1970).

Thick Tertiary strata in the California basins have entirely different geologic characteristics from those along the Gulf Coast. In Californian basins, very thick and highly porous petroleum-saturated sandstones are interbedded with shaley sandstones having low permeability. In California, the problem of high pressures is not as paramount as along the Gulf Coast, but the problem of subsidence in connection with fluid withdrawal is of great concern to oil companies and local governmental agencies. In the opinion of Rieke and Chilingarian (1974, p. 27) all these problems are mainly related to compaction processes and can be resolved with increasing knowledge on this subject.

Cooling and lubricating the bit and drillstring

A considerable amount of heat is generated upon rotation of the bit and drillstring as a result of friction at the bit and at points where the drillstring comes in contact with the formation walls. The formation can absorb only a small portion of this heat due to its physical properties. The major portion of the heat is conducted away from the point of contact by the circulating drilling fluid up to the surface where it is dissipated.

The lubricating property of the drilling fluid is also of great importance because considerable saving in time and money can be achieved with sufficient lubrication of all moving parts of the drilling assembly. The life of equipment (e.g., the bit) can be prolonged with a resulting decrease in the number of trips required to change the bit, for example, due to premature wearing out of bearings or failure. This results in substantial saving of rig time, especially when making a trip (e.g., to change the bit) in deep and deviated holes. Hole problems such as excessive torque and drag and differential sticking of pipe can also be alleviated if the drilling fluid possesses sufficient lubricating properties.

The lubricating properties of drilling fluids depend on their basic composition (e.g., type of continuous phase) and additives which have been subsequently added. Thus, different drilling fluids possess different degrees of lubricity, which are discussed in detail in Chapter 13.

Formation of filter cakes on walls (wall-off) of permeable formations

Filter-cake formation (walling-off) on permeable rock sections exposed in the wellbore is directly related to the filtrate loss properties of the drilling fluid. Filtrate escapes into the formations exposed in the borehole, leaving behind associated solids (clays, etc.). Shale sections, which traditionally have been considered nonporous, have been found in many areas of the world to be highly permeable due to the existence of microfractures. Filtrate losses into microfissures along the structural cleavages in shales have led to severe borehole instability in many cases.

Prevention of caving-in of the wall

Opposite a permeable zone, a tough impermeable cake is formed on the face of a wellbore during the drilling of a well. This layer, commonly called mud cake, is the result of initial invasion of the liquid phase of the drilling fluid into the permeable zone and deposition of finely-divided solids (usually, plate-like clays) on the face of the formation. With increasing invasion, the thickness of mud cake also increases until formation of a tough impermeable cake which restricts further liquid invasion. This mud cake also helps strengthen the face of the wellbore, thereby preventing caving-in. This property of the mud cake can be improved by increasing the colloidal fraction of the drilling fluids and/or chemically treating them.

Obviously it is highly desirable for drilling fluid to form an impermeable mud cake quickly, thereby reducing the initial filtration loss and cake thickness (Gatlin, 1960, p. 77). Mud cake formation however, also has many disadvantages. With increasing mud cake thickness, it may become difficult to move the drillpipe in and out of the hole or to run casing. The presence of mud cake also reduces the effectiveness of sidewall coring. The quantity of the filtrate and its composition may be such that shale contacted by the filtrate may swell and then slough into the hole. The filtrate escaping into the pay zone may cause considerable damage to subsequent production, unless a proper drilling fluid is used in drilling the zone in question. In addition, gelled mud may remain in the liner slots, gravel pack, or between the liner and well walls. Cleaning operations may fail to remove (1) all the mud cake from the well walls or (2) clay particles that have been forced into the formation. Remaining mud cake may cause a poor cement job because of lack of a good bond between the formation and the cement.

The filtrate that has entered the formation may lower the relative permeability to oil by increasing the water saturation. In addition, precipitates may form as a result of reaction of the filtrate with the reservoir water or oil. Clays in the formation may be deflocculated (dispersed) by reaction with the filtrate and plug the pores. As a result of the introduction of an emulsion, globule resistance may develop in the pore channels. Water blocks may form as a result of introduction of water into the formation (see Appendix 2.I). The existence of water blocks, however, has been questioned by some petroleum engineers.

The cumulative volume of filtrate loss over a period of time may be calculated by using Darcy's law. One may consider the flow of fluid through the mud cake as the fine solids are being laid down on a circular filter paper in a filter press. In contrast to the usual case of flow through porous media, the thickness of the mud cake is increasing with time. Darcy's law may be used as a differential equation for this situation as follows (Rogers, 1963; Larsen, 1938):

$$\frac{\mathrm{d}V_w}{\mathrm{d}t} = \frac{kA\Delta p}{\mu_f L} \tag{2-12}$$

where V_w = cumulative volume of filtrate in time t, cm^3; $(\mathrm{d}V_w/\mathrm{d}t)$ = differential expression for the rate of flow of filtrate at time t in cm^3/sec; k = permeability of the mud cake (darcys); A = area of mud cake (cm^2); Δp = pressure drop across mud cake (atm); μ_f = viscosity of the filtrate (cP); and L = thickness of the cake (cm).

The area and pressure drop are considered constant in most filtrate-loss determinations. The permeability may be considered constant for a given filtrate-loss determination if the cake is incompressible and its character does not change with time. Although the cake thickness increases with time, it may be expressed in terms of other variables if the cake is incompressible. The volume of filtrate is proportional to the volume of the cake:

$$V_w = rAL \tag{2-13}$$

where r = constant of proportionality.

On substituting the value of L from eq. 2-13 into eq. 2-12 and separating the variables, one obtains:

$$V_w\,\mathrm{d}V_w = \frac{rkA^2\,\Delta p}{\mu_f}\,\mathrm{d}t \tag{2-14}$$

Integrating the left side of eq. 2-14 between the limits of 0 and V_w and the right side between the limits of 0 and t, and solving for V_w gives:

$$V_w = \frac{A(2rk\,\Delta pt)^{1/2}}{\mu_f^{1/2}} \tag{2-15}$$

Equation 2-15 shows that the cumulative volume of the filtrate is directly proportional to the square root of time, and this has been confirmed by experimental results. Changes in pressure, that alter the character of the cake, invalidate eq. 2-15.

According to Tschirley (1978, p. 16), another form of fundamental differential equation describing filtrate loss through a filtering medium is:

$$\frac{dV_w}{dt}=\frac{K^2\,\Delta p}{\mu_f V_s}=\frac{K^2\,\Delta pr}{\mu_f V_w} \tag{2-16}$$

On integration one obtains:

$$V_w=\frac{K\,\Delta p^{1/2}r^{1/2}t^{1/2}}{\mu_f^{1/2}} \tag{2-17}$$

where V_w = volume of the continuous phase; t = time; K = constant, characteristic of the drilling fluid; V_s = volume of solids; Δp = differential pressure across the permeable zone; r = ratio of V_w to V_s; and μ_f = viscosity of the filtrate or continuous phase.

The above equations provide a satisfactory means of relating the various parameters to the amount of fluid loss of any water- or oil-base drilling fluid. They may, however, indicate excessively high filtrate loss for emulsion systems (that is, oil-in-water and water-in-oil emulsion drilling fluids), although the existence of emulsion also provides an additional source of plugging (water or oil globules) of the pore spaces in the mud cake.

Typical standard API filtrate loss curves for high- and low-fluid-loss drilling fluids, illustrating the straight-line relationship between V_w and the square root of time ($t^{1/2}$) are presented in Fig. 2-13. The high-filtrate-loss curve shows slight deviation from a straight-line relationship at the initial stages of the test (≈4 min). This is due to the loss of fluid into the formation

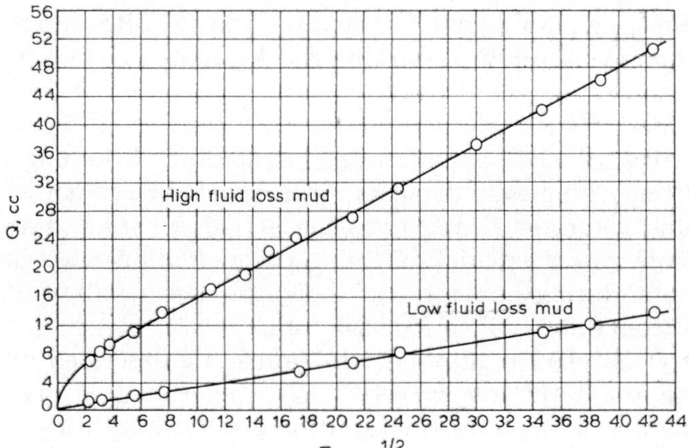

Fig. 2-13. Relationship between API fluid loss and square root of time. (Manual of Drilling Fluid Technology, Vol. II, Sec. 1, p. 16; courtesy of Baroid Petroleum Services.) $Q = K\sqrt{t}+e$; where Q = filtrate volume (cm^3); t = time (sec); K = constant, characteristic of the drilling fluid, cm^3/$\sqrt{\text{sec}}$ (slope of the line); and e = spurt loss (cm^3).

before a sufficient amount of solids is deposited to restrict rapid filtration. This initial high fluid loss which has been termed "spurt loss", explains the observed higher penetration rate using drilling fluids having high fluid loss.

Temperature and pressure are two uncontrollable factors which affect the static filtrate loss. Temperature affects the viscosity of the base fluid, i.e., the continuous phase, which decreases with increasing temperature and, thus, increases the filtrate loss. As a result of decreased viscosity, however, greater mobility is imparted to the solids in the fluid, with increased tendency to induce greater deposition of clay particles on the formation face, which slightly reduces the filtrate loss. Temperature affects viscosities of oil and water to varying degrees.

Inasmuch as water is highly incompressible at the pressure ranges to which most of the water-base drilling fluids are subjected, the effect of pressure may be insignificant. On the other hand, because oil is highly compressible, an oil-base drilling fluid would have a lower filtrate loss downhole than that indicated by the API static filtration test (Simpson, 1979, p. 645).

In deriving eq. 2-15, it has been assumed that the filter cake is incompressible, i.e., the cake permeability is not changing. This assumption, however, is invalid in actual conditions where varying degrees of compression occur in the filter cake. The filter cake composed entirely of bentonite clay will be more compressible than a cake containing a lower content of water-hydratable, plate-like bentonitic particles. The presence of silt and sand particles decreases the compressibility of filter cakes.

As pointed out by Ferguson and Klotz (p. 40, 1954), filtration from drilling fluid during drilling is of three kinds: (1) filtration beneath the bit, where the formation of filter cake is prevented by the bit action; (2) filtration at dynamic conditions, i.e., during mud circulation; and (3) filtration at static conditions in a quiescent mud. Ferguson and Klotz (p. 41, 1954) found that dynamic filtration rates are higher than static ones; only 10—30% of the total volume of mud filtrate enters the permeable formation during the static filtration, whereas 70—90% enters during circulation of the mud. Prokop (1952) showed that filter cake deposited in a quiescent cell shows little change in permeability during a long filtration period, whereas in an apparatus with a continuous mud circulation there is a gradual decrease in permeability. For example, during one run (3.7 days), the permeability of the filter cake decreased from $6.0 \cdot 10^{-4}$ to $2.0 \cdot 10^{-4}$ mD.

According to Horner et al. (1957), although the procedure described by the API for determining the fluid loss is very useful in evaluating mud properties, it is inadequate for comparing the filtration properties of various muds under dynamic condition, i.e., during mud circulation. They found no apparent relationship between dynamic filtration and API fluid loss. This is in close agreement with the findings of Outmans (1963), who concluded that the static filtration rate is not a reliable measure of the dynamic rate and vice versa.

Bezemer and Havenaar (1966) concluded that for most circulating drilling fluids, the equilibrium filtration rate is proportional to the rate of shear at the filter-cake surface, whereas the thickness of the equilibrium cake is inversely proportional to the rate of shear. The rate of shear at the cake surface thus appears to be a measure of the effect of mud flow on dynamic filtration. They also found that equilibrium cake thickness increases with increasing filtration temperature and is not significantly affected by the filtration pressure. If the cake permeability remains unaffected by temperature, equilibrium cake thickness is inversely proportional to mud filtrate viscosity (Bezemer and Havenaar, 1966, p. 298).

Outmans (1963) described filtration through compressible filter cakes by a nonlinear diffusion equation. The empirical data revealed that the nonlinearity is often mild and, therefore, can be neglected without great loss of accuracy. He also concluded that dynamic rate of filtration decreases with decreasing drilling-fluid viscosity and the rate of circulation. There is a high resistance of dynamically-deposited filter cakes against erosion. It is obvious that in any meaningful analysis of filtration, the compressibility of the filter cake must be considered as a variable. Tiller and Cooper (1962) developed equations relating average porosity of a cake and applied filtration pressure.

The equations for predicting the filter-cake build-up during dynamic—static filtration of drilling fluid were presented by Outmans (1963, p. 236). From analytical solutions of some simple cases, Outmans presented a qualitative manner in which the physical properties of a mud cake change with time.

Azari (1978) developed a numerical method for solving the equations which enables a quantitative estimation of filter-cake properties. This method consists of using the finite-difference approximation to the equations and solving them by the Crank-Nicholson method. The greatest difficulty in solving the equations is the unavailability of filter-cake thickness at every time step. Azari used an iterative procedure to arrive at the effective thickness present at each time step. Laboratory data relating cumulative filtrate volume to filter-cake thickness is used in the simulation approach. The numerical method allows the computation of k/h ratio (filter-cake permeability/thickness ratio). This ratio may be used for computation of the invasion radius of the mud filtrate as discussed by Ershaghi et al. (1978). As discussed by these authors for water sands, the computations of invasion radius and its comparison with the invasion radius obtained from Dual-Induction Laterolog may result in an estimation of the formation permeability profile with depth.

Sharma et al. (1979) conducted a systematic study on the effect of different mud additives on the filtration and wall-building properties of drilling fluids. The permeability of filter cakes was measured before and after addition of chemicals.

Generally, a straight-line relationship was found to exist between the

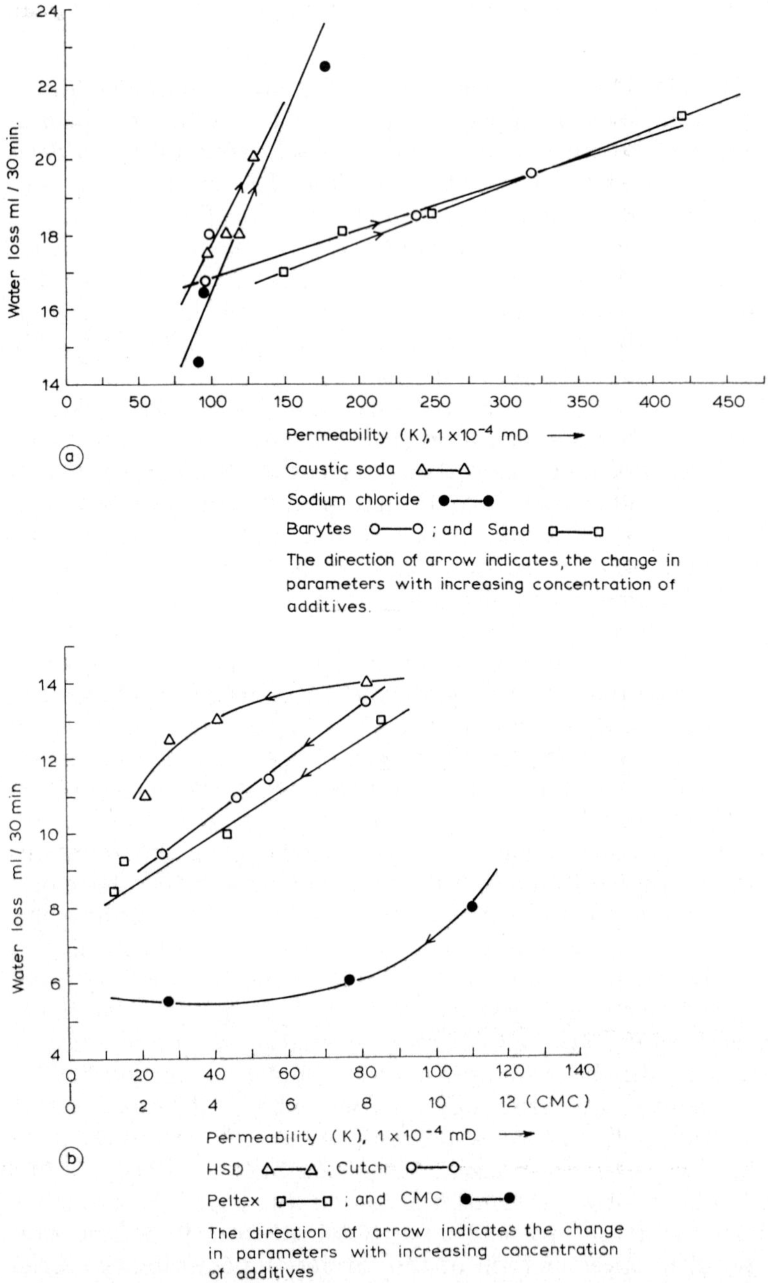

Fig. 2-14. Effect of mud additives on filter cake characteristics (relationship between filtrate loss and filter cake permeability). Original mud: 10% bentonite suspension. Range in caustic soda (NaOH) concentration: 0.02—0.3%; range in sodium chloride (NaCl) concentration: 4—20%; range in barite concentration: 10—25%; range in sand concentration: 2—10%; range in concentration of high-speed diesel (HSD): 5—15%; range in concentration of carboxy-methyl-cellulose (CMC): 0.5—2.0%; range in concentration of cutch/caustic-soda/potassium-dichromate (2 : 1 : 0.2): 0.4—1.5%; range in concentration of ferrochrome-lignosulfonate (Peltex)/caustic soda (5 : 1): 0.5—1.50%. (After Sharma et al., 1979; courtesy of Energy Sources.)

water loss and mud-cake permeability, except in the case of the presence of water-loss-reducing agents (Fig. 2-14). Different additives affect values of the slope and the intercept differently. Inert materials give rise to low values of slope as compared to electrolytes, whereas thinners impart intermediate values. Similar additives have comparable values of slope and intercept.

Although dynamic filtrate loss is of greater practical value, it defies accurate determination. The dynamic filtrate-loss test differs from the static one in that the mud cake is continually removed and redeposited while the drilling fluid is circulating. When the pressure differential at some point of interest along the length of the wellbore remains relatively constant, the dynamic filtrate loss is mainly a function of: (1) difference between rate of deposition and rate of erosion of mud cake; (2) flow pattern of the drilling fluid in the wellbore annulus; (3) annular fluid velocity; and (4) properties of the drilling fluid. Unlike static filtration, dynamic filtration also occurs at the bottom of the hole (Mitchell, personal communication (notes), 1972, p. 27). Inasmuch as at present there is no completely satisfactory way of measuring dynamic filtration, static filtration must be relied upon for evaluation and comparisons of various drilling fluids.

Suspension of cuttings and weight materials when circulation is stopped

One of the most important requirements of any good drilling fluid is its ability to suspend cuttings and weight materials when circulation is stopped for one reason or another. During drilling of a well, circulation may have to be stopped many times. In drilling a deep well, changing the bit often consumes many hours. If the solids are not kept in suspension during this time,

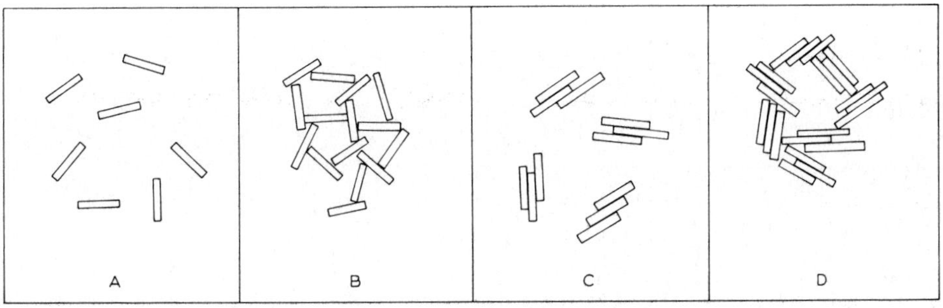

Fig. 2-15. Possible arrangements of clay platelets in thick clay-in-water suspensions. A. Deflocculated—dispersed. B. Deflocculated with edge-to-surface contacts. C. Flocculated—dispersed. D. Flocculated with edge-to-surface contacts. (After Rieke and Chilingarian, 1974, p. 10, fig. 4.)

their settling will result in recirculation and deposition of solids on the bit, which becomes stuck.

In order for the mud to have a suspension ability, it must (a) possess thixotropic (gelation) properties, when the fluid is not in motion, and (b) become fluid again upon movement (IMCO Services, 1977, p. 4). This thixotropic property is believed to be due to the formation of edge-to-surface contacts, i.e., owing to the attractive forces between the edges and surfaces of the clay particles suspended in the mud (see Fig. 2-15), because surfaces of plate-like clay particles are predominantly negatively charged, whereas their edges have positive charges. The thixotropic property can be altered by changing the content of the solids and by adding chemicals. In a heavier drilling mud, having more clay particles per unit volume, the attractive forces between the clay particles are stronger. Formation of edge-to-surface contacts of the clay particles can be visualized as formation of numerous truss structures, which build up one on top of another and support the weight of the cuttings and weight materials.

Partial support of drillstring and casing

The drillstring and casing in the borehole are buoyed up by the force equal to the weight of the displaced mud (IMCO Services, 1977, p. 4). This

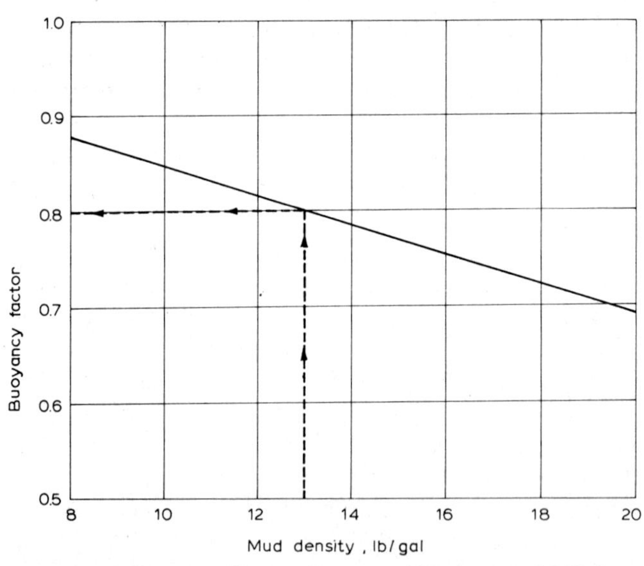

Fig. 2-16. Buoyancy factor diagram. (Courtesy of Milchem.) Example: If mud specific weight is 13 lb/gal and drillstring weight (in air) is 90,000 lb, then the drillstring weight in mud will be equal to 72,000 lb (0.8 × 90,000 = 72,000 lb, where buoyancy factor of 0.8 is obtained from graph).

results in considerable reduction in the loads which the surface equipment and structure must support. The buoyant force increases with increasing mud density and reduces the strain on the drillpipe and casing in deep wells. Buoyancy can be easily determined by using graphs such as Fig. 2-16.

Information obtained from the formation penetrated

A drilling fluid must not adversely affect the formation penetrated so that useful information leading to the location of producing zones can be obtained by using various tools. Information obtained from the analysis of drilling mud is almost instantaneous, with some delay due to the time required for circulation. For example, the presence of oil in a water-base mud (oil show) at the surface may indicate penetration of an oil zone. As wells are being drilled deeper and in more hazardous areas, pertinent information leading to the early detection of overpressured formations may be obtained from the mud analysis.

Often, drilling-fluid properties must be sacrificed in order to achieve a more accurate formation evaluation. For example, although salt will flocculate the water-base mud, causing an increase in viscosity, yield point, gel strength, and fluid loss, it may be added to control the resistivity for proper electrical log interpretation (Moore, 1974, p. 74).

Mud-gas shows are useful to indicate flow of gas into the wellbore if the formation permeability is extremely low (Fertl, 1976, p. 137). Depending on the geological environment and drilling practice being used, this mud-gas show may or may not be directly related to increased formation pressure (Fertl, 1976). Flowline mud checks for weight reduction may provide an additional indication for the possible presence of an overpressured formation. Monitoring the salinity of the water-base mud may also be a useful tool for detecting overpressured zones. Several investigators (see Rieke and Chilingarian, 1974) have shown that the salinity of formation water may decrease as an overpressured zone is approached. This is due to the fact that many overpressured sands are associated with thick undercompacted shale sequences which contain fresher water than that present in associated sands. The chloride content of the drilling mud may also be monitored; however, a reliable value is difficult to establish unless there is a close control over the mud properties in and out of the well. The writers of this chapter discovered that the chlorinity of shale cuttings, which can be correlated with sonic logs, for example, can be used for predicting overpressured formations.

One way of obtaining first-hand information about the formation penetrated is through coring and core analysis. Flushing of the core by drilling fluid may lead to incorrect fluid saturation data for the zone of interest. The water saturation of the core near the center, however, may experience little or no change at all due to flushing. Highly-permeable cores may be readily flushed with water and, therefore, may exhibit a higher than actual water saturation value. In such a case, disregarding the low cost and ease of using a water-base mud, one may have to change to oil-base mud before reaching the zones of interest in order to obtain more accurate fluid-saturation data.

Transmission of hydraulic horsepower to bit

A properly designed drilling-fluid hydraulic program is vital to the successful completion of any well. Drilling fluid is the medium which transmits the horsepower available at the surface to the bit where it cleans the face of the well. Experience has shown that good bottomhole cleaning increases the rate of penetration. A mud hydraulic program includes pressure-loss and mud-velocity calculations. For the maximum pressure drop across the bit, the total pressure losses must be as small as possible. Slide rules have been designed by various drilling-mud companies to facilitate the pressure-loss and velocity calculations. Computer programs have also been developed by drilling companies to perform and evaluate hydraulic programs. One such program has been included in Chapter 3.

CLASSIFICATION OF DRILLING FLUIDS

All drilling fluids may be divided into two main groups, i.e., weighted and unweighted, and further subdivided into several classes on the basis of specific weight, ranging from that of air to heavily-weighted drilling fluids (up to 150 lb/cu ft or 20 lb/gal) (Hutchison and Anderson, 1974, p. 84).

According to the classification of drilling fluids on the basis of specific weight (Fig. 2-17), any drilling fluid, the specific weight of which is 78 lb/cu

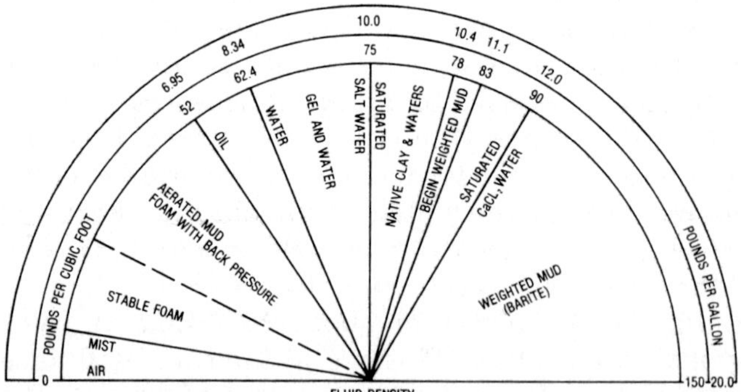

Fig. 2-17. Ranges in specific weights of various drilling fluid types. (After Hutchison and Anderson, 1974, p. 84, fig. 1, also see Tschirley, 1978, p. 1, fig. 1; courtesy of Baroid Petroleum Services.)

ft or less, may be considered as nonweighted, because it does not require the use of such weighting material as barite to achieve such weight, whereas any drilling fluid that is heavier than 83 lb/cu ft may be classed as weighted.

Nonweighted drilling fluids also include (1) practically solids-free drilling fluids, composed mainly of small concentration of gel and/or native clays and water, or polymers and water, and (2) water-in-oil or oil-in-water emulsions. Clear water has been long known to be an ideal drilling fluid which can achieve maximum drilling rates. Reasons for this behavior have been discussed previously.

Inasmuch as drilling with air or foam has been found to achieve maximum penetration rates, considerable interest has been generated in the field of these nonweighted fluids.

Many deep wells require the use of weighted drilling fluids to provide sufficient hydrostatic head in order to confine formation fluids to their native formations. Various weighting materials may be used to weight-up the drilling fluids, depending on the final weight desired and the type of fluid in use. For example, barite may be used to weight-up the water-base drilling fluid to as high as 20 lb/gal. On the other hand, oil-wettable weighting materials, such as limestone, dolomite, or pretreated (oil wettable) barite must be used to weight-up the oil-base drilling fluids. (See Appendix 2.I on "Fundamentals of surface and capillary forces".)

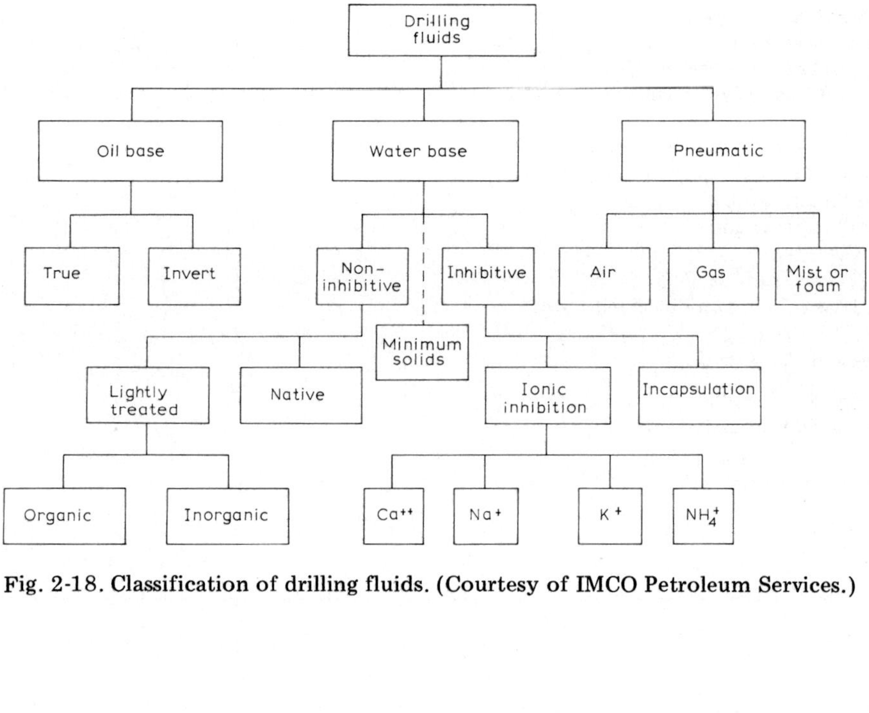

Fig. 2-18. Classification of drilling fluids. (Courtesy of IMCO Petroleum Services.)

Types of drilling fluids

Drilling fluids are expected to have physical and chemical properties that enable them to be used under greater varieties of well conditions, because attempts are being made to drill deeper wells in more hazardous areas. As a

TABLE 2-I

Classification of drilling fluids

(After API, "Principles of Drilling Fluid Control", 1972, p. 67; courtesy of American Petroleum Institute)

I. Water-base drilling muds

Fresh-water muds (little or no chemical treatment)
- 1. Inhibited muds
- 2. Spud muds
- 3. Natural muds

Chemically-treated muds (no calcium compounds added)
- 1. Phosphate muds
- 2. Organic-treated muds
  - a. Lignite
  - b. Quebracho and other extracts
  - c. Chrome-lignosulfonate

Calcium-treated muds
- 1. Lime
- 2. Calcium chloride
- 3. Gypsum

Salt-water muds
- 1. Sea-water muds
- 2. Saturated salt-water muds

Oil-emulsion muds (oil-in-water)

Special muds
- 1. Low-solids oil-emulsion muds
- 2. Low-clay-solids weighted muds
- 3. Surfactant muds
- 4. Low-solids muds
  - a. Clear water
  - b. Polymer muds
  - c. Low-solids, non-dispersed muds

II. Oil-base drilling muds

Oil-base muds
Invert emulsion muds (water-in-oil)

III. Gaseous drilling fluids

Air or natural gas
Aerated muds
Foams

result, many subtypes have been added to the generic classification of the drilling muds.

Drilling fluids are commonly classified, according to their base fluid, into three main groups: (1) water-base drilling fluids, (2) oil-base drilling fluids, and (3) gaseous drilling fluids.

A classification of the muds with the subtypes listed under each main class is presented in Table 2-I and Fig. 2-18. Each main group of drilling fluids and its subtypes are discussed in the following chapters.

The term "drilling fluid" generally refers to all fluids and includes air, gas, water, oil, and muds. On the other hand, "drilling mud" refers to a suspension of solids in water or oil, or of droplets of one of these liquids dispersed in the other (API, "Principles of Drilling Fluid Control", 1969, p. 1). The term drilling fluid is preferred by the authors of this book, although the term "mud" is used in the field almost exclusively.

PROPERTIES OF DRILLING FLUIDS

Three basic properties of drilling fluids that are most important for successful completion of a well are: (1) specific weight, (2) viscosity, and (3) filtrate loss. These properties are carefully monitored and controlled to maintain certain specified values at all time throughout the drilling operation.

Specific weight ("density") of drilling fluid

Specific weight, γ, is equal to density, ρ, times the gravitational acceleration, g, i.e., mass ρ is attracted toward the center of the earth with a force γ. For example, if the specific weight of water is equal to 62.4 lb/cu ft, its density (mass/unit volume) is equal to 62.4/32.2 = 1.94 slugs/cu ft (gravitational acceleration, g, is equal to 32.2 ft/sec/sec). In the drilling-fluids literature, "specific weight" is often erroneously referred to as density, giving rise to incorrect solutions of blade problems (turbodrill), for example.

Inasmuch as one of the main functions of drilling fluids is to confine all formation fluids (i.e., oil, gas, and water) to their native formations or beds, specific weight measurement is very important. The instrument used to measure specific weight ("density") of drilling fluids is discussed in Chapter 4. Weight materials may be added to increase the specific weight of the drilling fluid so that its hydrostatic head pressure becomes close to that of the formation pore pressure, whereas removal of solids, dilution with water, and/or addition of oil may be used to reduce the specific weight.

The control of drilling-fluid specific weight is critical, because an unnecessarily heavy drilling fluid can cause breakdown of formations, resulting in loss of circulation, or even a complete loss of a well, and reduction in drilling rate. Drilling-fluid specific weight usually must be controlled between narrow limits, often only ±0.1 to ±0.2 lb/gal with respect to the weight sufficient to confine the formation fluids (API, "Principles of Drilling Fluid Control", 1972, p. 68).

The specific weights of drilling fluids may range from that of air to up to 20 lb/gal for heavily-weighted, water-base drilling fluids. The ranges of drilling-fluid specific weights are graphically illustrated in Fig. 2-17.

Three types of specific weights (or "density") are commonly used in drilling-fluid calculations: (1) static; (2) equivalent circulating; and (3) annular. Static or surface specific weight is determined at the surface conditions with the mud balance. On the other hand, equivalent circulating specific weight refers to the weight of a circulating column of drilling fluid. Equivalent circulating specific weight at any depth in a hole is a function of annular pressure loss which, in turn, is related to factors such as circulation rate and hole condition.

Equivalent circulating specific weight at any depth may be calculated as follows:

$$\text{equivalent circulating specific weight} = \text{drilling-fluid specific weight} + \text{annular pressure loss} \times (1/\text{depth}) \tag{2-18}$$

When the weight of drilling fluid is expressed in lb/gal/ft of depth and pressure loss in psi, the following equation is used:

$$\text{equivalent circulating weight} = \text{drilling fluid specific weight} + \frac{\text{annular pressure drop}}{0.052 \times \text{depth}} \tag{2-19}$$

where 0.052 is the constant converting psi and ft to lb/gal.

Equivalent specific weight may be as much as 1 to 1.5 lb/gal greater than the static specific weight, depending on the amount of annular pressure drop. The additional "weight" obtained as a result of considering the annular pressure drop on drilling through fragile and soft formations is of considerable importance.

The annular specific weight, which refers to the total actual bottomhole pressure exerted on the formation being penetrated, seems to be the most important. The annular specific weight may be considerably higher than either the equivalent circulating or static ones, especially when the drilling rate is high and the well is deep. This is due to the incorporation of cuttings into the drilling fluid. More cuttings are incorporated into the drilling fluid at higher penetration rates. In addition, the deeper the hole, the more cuttings are present in the hole before being removed at the surface.

Annular specific weight may be defined as follows:

$$\text{annular specific weight} = \text{static specific weight} + \frac{\text{annular pressure drop}}{\text{depth}} + \text{additional pressure due to weight of cuttings} \times (1/\text{depth}) \tag{2-20}$$

Accurate determination of the annular pressure loss while drilling is not possible because the presence of drilled cuttings introduces an additional unknown amount of pressure drop. The procedure for calculating additional pressure drop due to the weight of rock cuttings is illustrated in Appendix 2.II.

As shown in Fig. 2-17, different types of drilling fluids possess a fixed range of usable specific weights. Such a range is a function of (1) specific weights of the base fluids and (2) the thixotropic properties and gel strengths of the final mixtures. Thixotropic property refers to the ability of a fluid to gel upon cessation of applied shear, which relates directly to the ability of the fluid to maintain the weighting materials and drilled cuttings in suspension when circulation stops. Water-base drilling fluids may be weighted to as high as 20 ppg (lb/gal) due to the high content of bentonitic particles in such systems. For more detailed discussion of thixotropic properties see Chapter 8.

Control of drilling-fluid specific weight may be achieved by several means depending on the type of drilling fluid and, thus, will be discussed in the later sections on individual drilling fluid types.

Viscosity

Viscosity is defined as the resistance of the fluid to flow or to being pumped, i.e., the ratio of shearing stress to rate of shearing strain. Drilling fluid (mud) engineers measure the viscosity with the Marsh funnel or rotational viscometers.

Viscosity is a very important property of drilling fluids because it affects the efficiency of their lifting capacity. Inasmuch as solids are continuously being picked up by the drilling fluid during a drilling operation, it is highly desirable to remove them at the surface before they are recirculated by the mud. During the early period, these solids were allowed to settle out by themselves at the surface pit; however, this has been found to be ineffective for drilling fluids having high weight or high viscosity. At the present time, mud screen equipment, desanding devices, centrifugal concentrators, and similar equipment are universally installed to remove the undesirable solids (see Chapter 12).

Viscosity is also involved in calculating the pressure drop in the annulus in the case of laminar flow by using the Bingham equation:

$$\Delta p = \frac{l\overline{YP}}{225d} + \frac{l\overline{PV}}{1500d^2} v_f \tag{2-21}$$

where Δp = pressure drop (psi); l = length or depth (ft); $\overline{YP}$ = yield point (lb/100 sq ft); $\overline{PV}$ = plastic viscosity (cP); d = hole diameter minus pipe diameter (inches); and v_f = velocity of drilling fluid (ft/sec). As shown by this equation, the largest portion of pressure losses is due to the yield point (stress) of the fluid. The viscosity of any drilling fluid is a function of 4 factors: (1)

viscosity of the base liquids; (2) size, shape, and number of solid particles per unit volume; (3) interparticle forces; and (4) degree of emulsification of oil in water or water in oil and stability of emulsions.

Temperature affects the viscosity of the base liquids, i.e., oil, water, or both. Inasmuch as the molecular spacings are small, molecular cohesion in liquids is very strong. With increasing temperature, this molecular cohesion decreases. This reduction results in a decrease in liquid viscosity (see Olson, 1973). Temperature affects the viscosity of oil more than that of water, which has a much lower viscosity than oil.

The magnitude of contact area between the solid particles and the liquid phase greatly affects the plastic viscosity. The total surface area to be wetted increases with (1) the reduction in particle size, (2) increase in the number of solid particles per unit volume, and (3) change in shape of the particles from spherical to flat. Plastic viscosity increases with increasing wetted surface area.

Attracting and repelling interparticle forces, which arise from the electrostatic fields that exist around the clay platelets, affect viscosity (see Chapter 8 for details).

RELATIONSHIP BETWEEN COMPOSITION AND PROPERTIES OF WATER-BASE DRILLING FLUIDS

The first drilling fluid used in rotary drilling was water. The term *mud* was probably applied to the mud which resulted from the actual drilling of the well.

Clay, which is the key component of a water-base drilling fluid, imparts many of the necessary properties, such as density, viscosity, plastering properties, and thixotropic characteristics. Clay is commonly the disintegration product of igneous rocks, with kaolinite mineral being the most abundant ($H_4Al_2Si_2O_9$). Clay minerals, which are used in fresh-water drilling fluids, are divided into three main groups: kaolinite, montmorillonite, and illite (see Chapter 5 for details).

Kaolinite group ($Al_2O_3 \cdot 2SiO_2 \cdot 2H_2O$)

Kaolinite clay has asymmetric structural units, bounded on one side by sheets with oxygen and on the other side by sheets with hydroxyl ions. The crystal lattice of kaolinite is, therefore, stable, and the structural units do not spread apart easily. Water cannot readily get in between the sheets to cause swelling and, as a result, kaolinite clays do not swell much. In addition, they have low base exchange capacity.

Montmorillonite group [$(OH)_4Al_4Si_8O_{20} \cdot nH_2O$]

The crystals of montmorillonite clay are composed of stacks of gibbsite sheets bounded between two silicon—oxygen—tetrahedron sheets. Silicon—

oxygen—tetrahedron consists of four oxygen atoms touching one enclosed silicon atom. The composition of the montmorillonite group, however, varies within wide limits. In tetrahedron layers, Si^{4+} can be replaced by R^{3+} (Al^{3+}, Fe^{3+}, and Cr^{3+}) up to 25%; and in montmorillonite minerals the SiO_2/R_2O_3 molecular ratio varies from 2 to 4. The structural units of montmorillonite are symmetrical and bounded by oxygen layers. These oxygen layers do not attract each other and the structural units are held together with water. The amount of water, however, varies and gives rise to the spreading apart of the lattice. As a result, the montmorillonite clay disintegrates in water solution into small particles (less than 0.1 μm in diameter), leading to good swelling property which enhances viscosity.

Illite group [$OH_4K_y(Al_4 \cdot Fe_4 \cdot Mg_4 \cdot Mg_6)(Si_{8-y} \cdot Al_y)O_{20}$, where y is around 1 to 1.5]

The crystal lattice of illite resembles that of montmorillonite; however, approximately 15% of the silica is replaced by aluminum. The resulting free valences are filled by the potassium ions which bind the structural sheets. The crystal structure of illite is, therefore, more stable than that of montmorillonite, resulting in lower swelling.

Base exchange

On the surfaces of the flakes of some clay minerals there are positions where exchangeable cations can be held. Different cations possess different hydration properties. For example, the degree of hydration of the sodium ion is much greater than that of calcium ion. Sodium-montmorillonite, therefore, would possess a tendency to build up a thicker water film than would calcium-montmorillonite. This is why sodium-bentonite swells markedly in water, whereas calcium bentonite does not swell as much.

A further influence of cations, which is vital to the understanding of some properties, is the disrupting influence of some cations. For example, the highly hydrating sodium ion tends to produce a spreading apart of the structural units and a consequent break-up of the clay particles into extremely minute flakes. On the other hand, potassium ion binds the structural sheets together, resulting in a lower swelling of the clay.

A cation of lesser attraction may also be released in exchange for one of greater attraction without altering the lattice structure of the clay building unit. This phenomenon is known as base exchange. Typical values of base exchange capacity for the various clay types discussed above have been listed in Table 2-II. The order of replacement of cations is as follows: $H > Ba > Sr > Ca > Mg > Rb > K > Na > Li$. For example, hydrogen ion will replace calcium ion, unless the latter is present at a higher concentration. The addition of calcium ions to the water—clay mixture (mud) results in the

TABLE 2-II

Base exchange capacity of main clay groups

(After Beeson et al., 1958)

Clay type	Degree of swelling and plasticity	Base exchange capacity, milliequivalents/100 grams of clay
Montmorillonite	High swelling High plasticity	60—100
Illite	Intermediate swelling Intermediate plasticity	20—40
Kaolinite	Low swelling Low plasticity	3—15

displacement of hydrogen ions from the surface of clay flakes and a corresponding decrease in the pH. For example, the addition of 2% $CaCl_2$ to a bentonite mud decreased the pH from an initial value of 8.8 to 7.4 (Beeson et al., 1958, p. 18). The sodium ions have a similar effect on the pH of bentonite muds. These results are in close accord with those of Rogers (1963, p. 442). (See Chapter 5 for details.)

Flocculation versus deflocculation

Flocculation is the formation of clusters of particles separable by a relatively weak mechanical force. It is produced by neutralization of the negative charges on the clay particles. On the other hand, *deflocculation* is the state of dispersion of a solid in a liquid in which each solid particle remains geometrically independent and unassociated with adjacent particles.

Clay particles can exist in the mud in four different states: (1) dispersed—deflocculated, (2) dispersed—flocculated, (3) aggregated—deflocculated, and (4) aggregated—flocculated (see Fig. 2-15).

In the dispersed—deflocculated state, the attractive forces between the clay particles are at a minimum, giving rise to an increase of surface area and resulting increase in plastic viscosity. This state of clay suspension results in the highest plastic viscosity, but low yield point and low gel strength. The filtrate loss is minimum among all four states of clay suspension.

As a result of flocculation in the dispersed—flocculated state, there is a reduction in surface area as a result of which the plastic viscosity of this state is lower than the first state. Clay platelets in this state also exhibit stronger attractive forces, leading to higher gel strength and higher yield value. Filtrate loss is high in this state because flocculated clay platelets form soft and thick filter cake which is quite permeable.

Loose stacking of clay platelets in the aggregated—deflocculated state further reduces the surface area of the clay platelets. This increases the interparticle attractive forces which results in even lower plastic viscosity and a higher yield value and higher gel strength than the two cases discussed above. The filtrate loss is also high for mud in this state (Baroid, 1975, p. 44).

The aggregated—flocculated state displays the minimum surface area of clay platelets. Consequently, it exhibits the lowest plastic viscosity and the highest yield value and highest gel strength of the four states. In all four cases, the weight of clay per unit volume of mud was assumed to be the same.

FACTORS AFFECTING RATE OF PENETRATION

Drilling-fluid properties, weight on bit, bit design, rotation speed, hydraulics, and formation characteristics are among several drilling parameters that affect the rate of penetration.

Based on laboratory and field studies, the following drilling-fluid properties affect the rate of penetration, to a varying degree: (1) specific weight, (2) viscosity, (3) filtrate loss, (4) solids content, and (5) oil content (see Fig. 2-19). Inasmuch as it is almost impossible to change one property of a

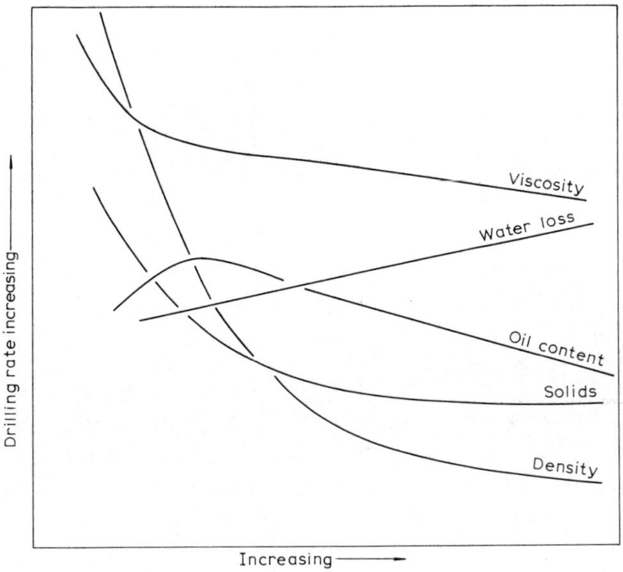

Fig. 2-19. Effects of various drilling-fluid properties on penetration rate. (After Moore, 1974, also see "Drilling Fluid Technology Manual", Vol. II, Sec. 1, fig. 1, p. 1; courtesy of Baroid Petroleum Services.)

drilling fluid without affecting the others, it is difficult to evaluate the true effect of an individual parameter on penetration rate.

Effect of specific weight on drilling rate

The fundamental function of the specific weight of a drilling fluid is to maintain sufficient counter pressure to confine formation fluids to their native formations. Excess of pressure overbalance, however, has been observed to reduce the penetration rate drastically.

Laboratory studies and field observations have indicated that penetration rate increases with decreasing difference between drilling-fluid column pressure (p_h) and formation pore pressure (p_p) (Fig. 2-20), i.e., it continues to increase as the pore pressure becomes greater than the drilling-fluid column pressure (Tschirley, 1978, Vol. II, Sec. I, p. 4).

Laboratory data collected with a miniature drilling bit at simulated reservoir conditions on various rocks, using various confining pressures, are illustrated in Figs. 2-21 and 2-22. Gatlin (1960, p. 116) attributed the decrease in drilling rate with increasing confining pressure to the increasing chip hold-down force associated with greater drilling-fluid specific weight or

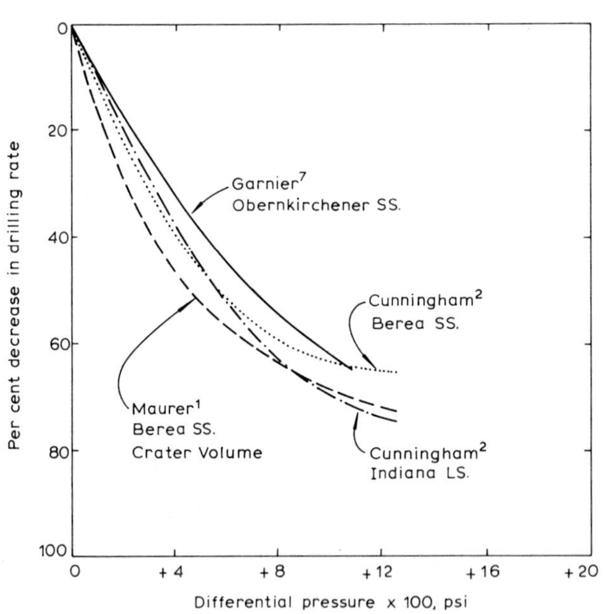

Fig. 2-20. Laboratory data on relationship between the penetration rate and differential pressure ($p_h - p_p$) for various rock types. (After Tschirley, 1979, p. 4, fig. 2; courtesy of Baroid Petroleum Services.)

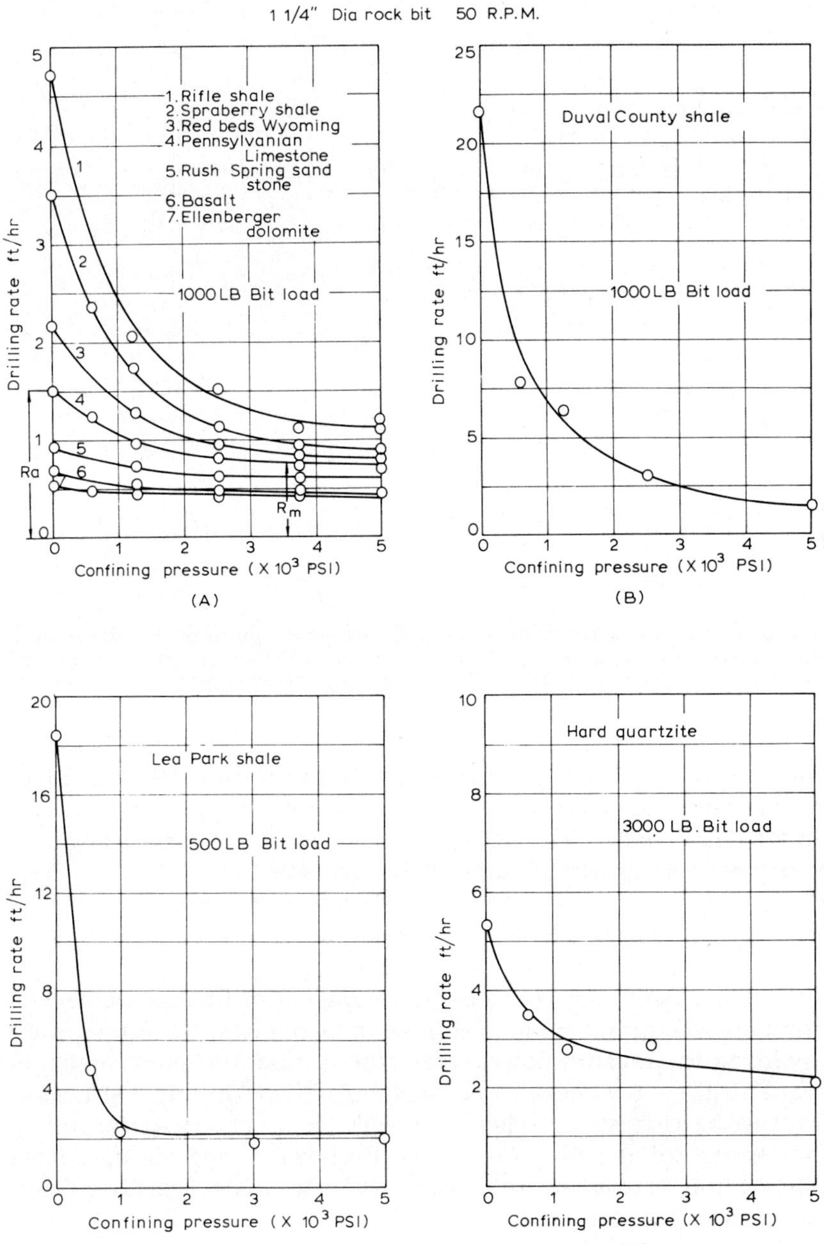

Fig. 2-21. Relationship between the drilling rate and specific weight of drilling fluid (confining pressure) for various rock types. (After Murray and Cunningham, 1955, p. 198; courtesy of the Petroleum Transactions of the AIME.) R_a = drilling rate at atmospheric condition; R_m = drilling rate at near horizontal portion of the curve.

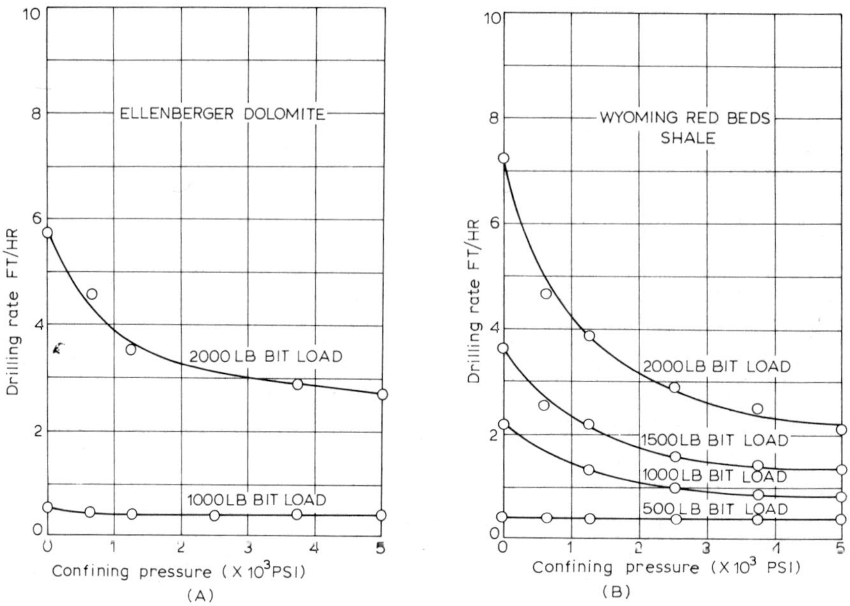

Fig. 2-22. Relationship between the drilling rate and specific weight of drilling fluid (confining pressure) for dolomite and shale at different bit loads. (After Murray and Cunningham, 1955, p. 199; courtesy of the Petroleum Transactions of the AIME.)

increased pressure differential ($p_h - p_p$). According to Gatlin (1960, p. 116), the penetration rate varies inversely with the rock compressive strength. Increasing differential pressure tends to prevent removal of the individual rock chip and to increase the rock compressive strength.

Inasmuch as not all rock types react similarly to increased differential pressure, the penetration rate at the same differential pressure may vary with rock type.

The above observations have led to the development of the underbalanced drilling technique, whereby a hole is drilled deliberately with drilling fluid having a hydrostatic pressure lower than the formation pore pressure. Underbalanced drilling has been used successfully in drilling hard, low-permeability, gas-bearing strata where the chance of encountering unexpected high-pressure zones is low. Air, foams, fresh water, and oils have been tried as drilling fluids in such areas. The use of air and foam as drilling fluids has been discussed in Chapter 13.

Effects of viscosity on penetration rate

The viscosity of the drilling fluid itself and that of its filtrate affect the drilling rate differently, as discussed in this and following sections.

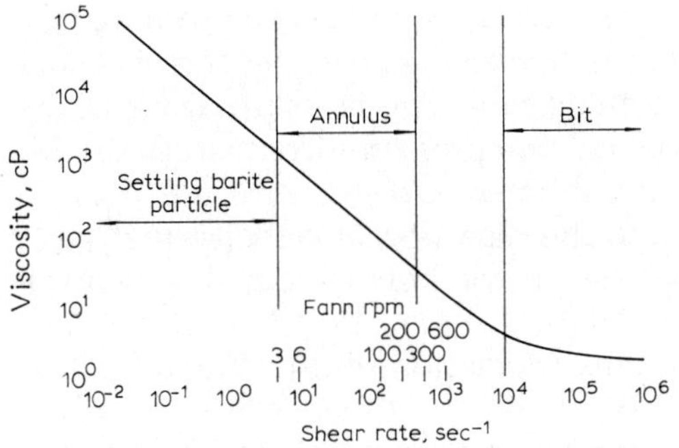

Fig. 2-23. Variation of optimum drilling-fluid viscosities at different areas of a circulating system. (After J.R. Eckel; courtesy of Oil and Gas Journal and Baroid Petroleum Services, 1979, Vol. 2, Sec. 1, p. 7, fig. 9.)

The viscosity of an ideal drilling fluid, from a viscosity standpoint, decreases with increasing shear (see Figs. 2-23 and 2-24). Such fluids include most of the water-base muds and are generally referred to as "shear-thinning fluids" (see Fig. 2-24). (See Chapter 7 for details.) Oil-base, emulsion, and inverted emulsion drilling fluids may not be considered as such, because in most cases their viscosity increases with increasing shear rate, with the exception of true oil-base drilling fluid which is practically devoid of water.

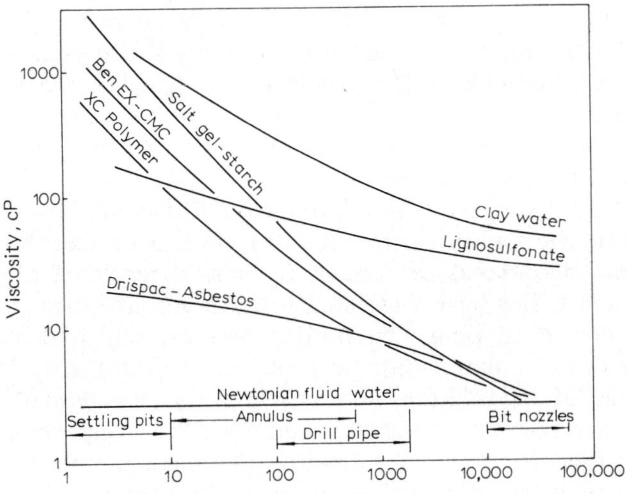

Fig. 2-24. Relationship between viscosity and shear rate at different areas of a circulating system for different drilling fluids. (After Kennedy and Meyer, 1968; courtesy of Oil and Gas Journal and Baroid Petroleum Services, 1975, Vol. 2, Sec. 1, p. 7, fig. 10.)

Unlike specific weight and filtrate loss, viscosity affects the drilling rate indirectly, because it affects the bottomhole cleaning, cuttings removal, and hydraulics which, in turn, affect the drilling rate. The pressure drops in the drillstring and in the annulus decrease and the pressure drop across the bit increases as the drilling fluid becomes thinner. Greater pressure drop is needed across the bit in order to create the high nozzle velocity that gives rise to greater impact force on the bottom of the hole to expedite removal of cuttings. As pointed out by Gatlin (1960, p. 127), the jet itself does not drill the hole. An example of pressure-drop calculations illustrating the effect of viscosity is presented in Appendix 2.II.

According to Gatlin (1960, p. 125), the bit teeth impart large shear to the drilling fluid, thin it, and displace it before rock is contacted. The energy expended for this purpose is viscosity dependent, e.g., very high-viscosity drilling fluids may provide an effective viscous cushion which softens and reduces the impact of the bit teeth with resulting slower penetration rate.

Effects of filtrate loss on rate of penetration

The effect of filtration rate on drilling rate has been extensively studied; however, the results obtained are still inconclusive. There are two schools of thought: (1) According to Moore (1974, p. 25) and others, drilling rates may or may not be affected by changes in filtration rate. The observed increase in drilling rates is due more to other factors, which are also affected when attempting to alter filtration rate. For example, lower viscosity is commonly associated with high filtrate loss and, therefore, the observed increase in penetration rate may be due partly or entirely to the low viscosity. (2) Gatlin (1960), Tschirley (1979), and others, however, attribute the increased drilling rate to the formation of a thick, fluffy, and soft cake, which is generally associated with high filtrate loss. Such a cake offers less resistance to the bit action in attacking the formation than does a tough, impermeable cake formed by a drilling fluid having low fluid loss.

High initial filtrate loss (spurt loss) tends to minimize the differential pressure which exists at (or near) the surface of the rock and, thus, decreases the rock compressive strength. Microcracks or fissures form as a result of bit action on the formation, and it has been suggested that a vacuum exists in these fissures for a short period of time. Overburden pressure and hydrostatic pressure of a drilling-fluid column tend to heal these fissures almost instantaneously. In the case of high filtrate loss, however, filtrate immediately fills these fissures. As a result of induced vacuum suction, very large forces are exerted that (1) tend to extend the fissures, (2) keep the fissures open, (3) reduce the pressure differential at the surface of the bottom of the borehole by equalizing pressure, and (4) facilitate removal of loosened cuttings.

The fact that the oil-base drilling fluids generally have very low fluid loss as compared to water-base drilling fluids and, yet, have a lower penetration rate, supports the above explanation.

Some oil-base drilling fluids, with relatively high fluid loss at high temperatures and pressures, have much higher rates of penetration than do lower-fluid-loss oil-base drilling fluids of similar properties.

With decreasing viscosity of the filtrate there is a more rapid fill-up of the cracks, which enhances the mechanisms discussed above. Viscosity of the filtrate is dependent on several factors: (1) temperature and (2) chemistry, i.e., content of salts, chemicals, polymers, etc.

Rate of penetration is also a function of wettability of rocks as affected by filtrates, due to the presence of surfactants (surface-active agents) (see Appendix 2.I). According to Gatlin (1960, p. 127), certain electrolytes and surfactants reduce rock hardness with a resultant 30—60% increase in drilling rate, when properly used. Further study, however, is required in this area. Increased rock wettability to the filtrate allows the filtrate to fill the microcracks created by the bit more rapidly and, thus, prevents their healing and increases the penetration rate.

Effect of solids content on drilling rate

Quantitative studies have been conducted to investigate the effects of solids content and the state of their dispersion on rates of penetration (see Figs. 2-25 and 2-26). It is a well-known fact that the drilling rate is the highest when using plain water and decreases with increasing concentration of solids incorporated into the fluid stream during drilling (Fig. 2-26). Initially, this reduction in drilling rate was attributed to the increase in the hydrostatic pressure of the drilling-fluid column due to the presence of solids. The fact that the observed reduction of drilling rate is much greater

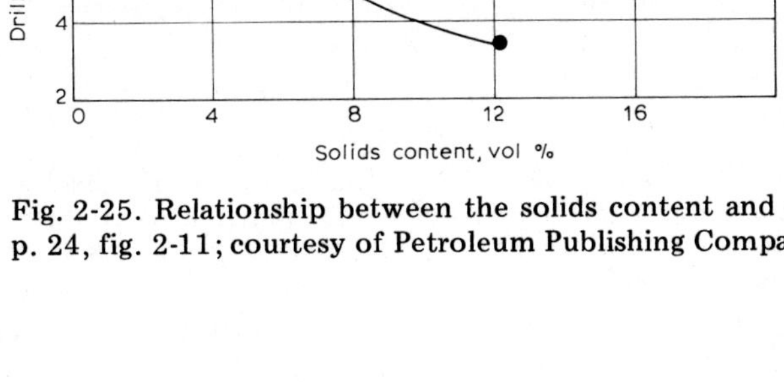

Fig. 2-25. Relationship between the solids content and drilling rate. (After Moore, 1974, p. 24, fig. 2-11; courtesy of Petroleum Publishing Company.)

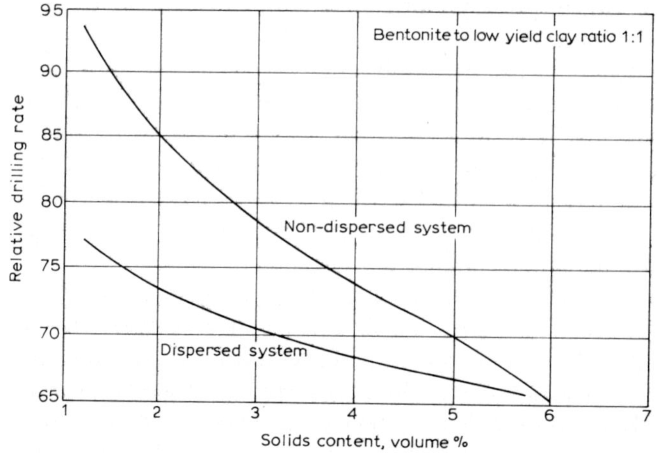

Fig. 2-26. Effect of solids content and their dispersion state on drilling rate. (After Moore, 1974, p. 24, fig. 2-12; courtesy of Petroleum Publishing Company.)

than that which could be attributed to the slight increase of hydrostatic pressure, suggests that solids interfere with the rate of penetration in more than one way. According to Moore (1976, p. 20), a much faster rate of drilling can be achieved on using 10 lb/gal saturated salt water than with a dispersed solid system in the same environment.

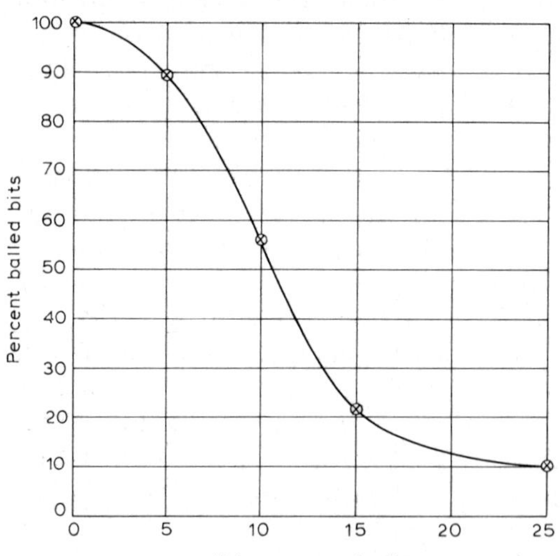

Fig. 2-27. Relationship between percentage of balled bits and oil content established in laboratory drilling of a Miocene shale. (After Cunningham and Goins, 1957, p. 119; courtesy of Petroleum Engineer.)

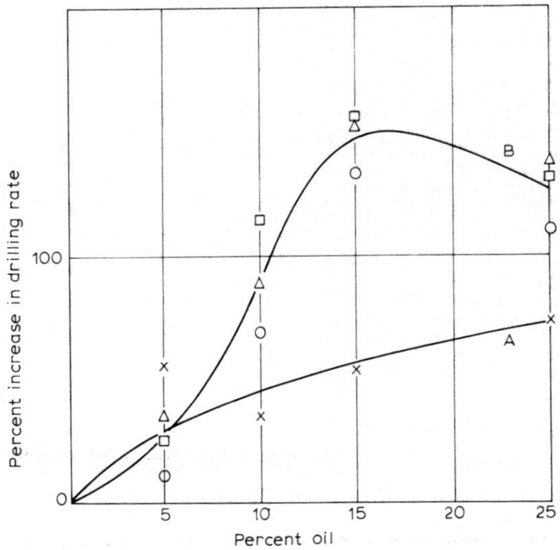

Fig. 2-28. Relationship between the percentage increase in drilling rate and oil content for the Vicksburg Shale (*A*) and Miocene shale (*B*). (After Cunningham and Goins, 1957; p. 119; courtesy of Petroleum Engineer.)

Gatlin (1960, p. 125) suggested that, perhaps, the solid particles cushion the contact between the bit-tooth and rock so that a clean, sharp impact is not achieved. Increase in the solids content also affects the hydraulics of the drilling-fluid system. A dispersed solids system affects the rheological properties and filtrate loss more than a non-dispersed system.

Effect of oil content on penetration rate

Cunningham and Goins (1957) and Eckel (1954) performed extensive studies on the effects of oil on the rate of penetration and percentage of balled bits. Correlations among oil content, percent of balled bits, and penetration rates for various rock types were not conclusive (see Figs. 2-27 and 2-28). At an oil content of about 15 to 25% by volume, the percent of balled bits is minimized, whereas drilling rate is decreased (Figs. 2-27 and 2-28). This may be due to the high shear mixing of the oil that forms emulsions which tend to increase viscosity, lower fluid loss, and increase pressure drop. As previously discussed, when carefully spotted (as one continuous phase), oil provides greater lubricity and gives rise to higher penetration rate.

Combined effect of drilling-fluid properties on drilling rate

It is important to relate the combined effect of various drilling-fluid properties on the drilling rate. Combined drilling-fluid properties may be

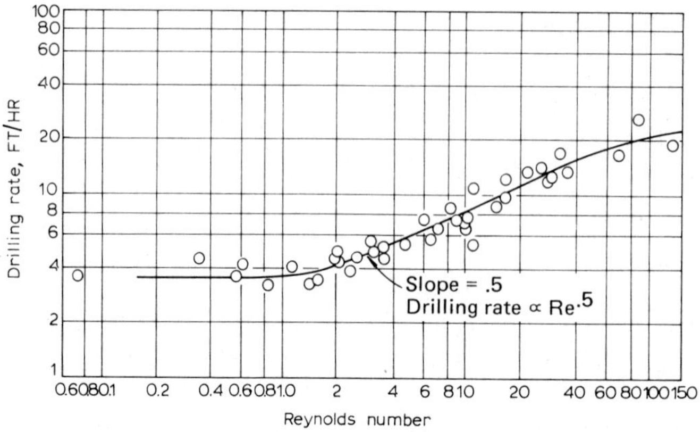

Fig. 2-29. Relationship between Reynolds number and drilling rate. (After Eckel, 1967, p. 544; courtesy of Journal of Petroleum Technology, Society of Petroleum Engineers of AIME.)

represented by a variable *Re* or Reynolds number, which may be expressed as follows (Milchem Services, 1979, p. 7):

$$Re = \frac{Kq\gamma}{d_n \mu_f} \tag{2-22}$$

where q = flow rate (gal/min); γ = specific weight (lb/gal); d_n = nozzle diameter (inches); μ_f = viscosity of drilling fluid moving through the nozzle (cP); and K = constant.

The Reynolds number, as presented by Eckel (1954), relates all principal drilling-fluid parameters which may have some effect on the drilling rate. Relationship between the drilling rate and Reynolds number is presented in Fig. 2-29. Drilling rate does not change with increasing Reynolds number until the latter reaches a value of about 2.0. Thereafter, it appears that the drilling rate increases with increasing Reynolds number. Consequently, the latter should be maintained at these moderate to high values in order to optimize the drilling rate attainable with the existing drilling-fluid parameters.

ADVERSE CONDITIONS AFFECTING DRILLING FLUIDS

Mud properties are adversely affected by many conditions encountered during a course of drilling a well. Contamination is a common term used to describe the effects of these adverse conditions, such as drilling through salt domes (NaCl), anhydrite, or gypsum. Fine solids picked up during drilling (drilled solids) increase the mud weight and viscosity.

Salt contamination of drilling fluids

Salt contamination is a result of addition of various salts to the drilling fluid during drilling of certain formations, completion, or workover. The commonly encountered soluble salts may be divided into two groups, i.e., monovalent and divalent.

The most common monovalent salts encountered are sodium chloride (NaCl) and, to a lesser extent, potassium chloride (KCl). The most common divalent salts that cause contamination are calcium sulfate ($CaSO_4$), calcium hydroxide (in cement) [$Ca(OH)_2$], calcium chloride ($CaCl_2$), magnesium sulfate ($MgSO_4$), and magnesium chloride ($MgCl_2$) (Baroid Petroleum Services, 1965, p. 44).

As a result of salt contamination, there is an increase in apparent viscosity, yield point, gel strengths, and filtrate loss. A salt-water flow can be detected by an increase in pit volume, increase in chloride ion content, strokes per minute of the mud pump, and decrease in pump pressure (Baroid Petroleum Services, 1965, p. 44).

Evaporite formations also supply sodium chloride to drilling fluids. These types of formations are found in several oil- and gas-producing areas, such as the "Red Beds" of west Texas. Salt is present in evaporite deposits, which are formed by the evaporation of water from ancient seas, or is dispersed in terrigenous (sands, silts, and clays) deposits. Water-base drilling fluids quickly dissolve salt from the formation, resulting in washed-out holes.

Salt domes constitute another major source of sodium chloride. Extreme *thickening* of the untreated water-base drilling fluid occurs rapidly after large influx of sodium chloride dissolved from the salt domes. Inasmuch as salt is softer than the surrounding rocks, the increase in penetration rate may indicate penetration of salt formations (beds or domes). In such cases, treatment should be initiated immediately.

Influx of saline formation waters during drilling of salt present in make-up water will have the same contaminating effects as those described above.

Divalent salts are introduced into the drilling-mud system during the process of drilling through different formations and/or through an unavoidable consequence of completing the well.

Calcium sulfate, which occurs in nature as gypsum ($CaSO_4 \cdot 2H_2O$) or anhydrite ($CaSO_4$), is found as thick sections, as stringers, in make-up water, embedded in sandstones, siltstones and shales, and, sometimes, as caprocks of salt domes (Baroid, 1965, p. 45).

Many phases of drilling and well-completion operations require the use of cement, which is mainly composed of calcium hydroxide. Calcium hydroxide is introduced into the drilling fluid as a result of such operations as squeeze cementing, drilling-out a cement plug, and running casing. The contamination effects of cement are similar to those resulting from contamination with a monovalent salt. Cement contamination of the untreated water-base drilling fluid consists of the following: (1) Ca^{2+} ions affect the ionic structure of the clay particles, and (2) introduction of additional hydroxyl ions raises pH of the drilling fluid (IMCO Services, 1974, p. 44).

The effects of contamination by salts on various properties of drilling fluids are described in detail in Chapter 8.

ADDITIONAL FACTORS INFLUENCING SELECTION OF DRILLING MUDS

Corrosion

Often, drilling fluids are used as packer fluids in the annular area of wells after completion. If not properly treated, this drilling fluid can cause costly corrosion problems late in the life of a well. Water-base and oil-emulsion drilling fluids (oil-in-water) may attack the metal surfaces in contact with it in the well if not properly treated. The type of corrosion involved is electrochemical or wet corrosion.

Numerous conditions encountered during drilling operations cause corrosion to the drilling equipment. Hydrogen sulfide (H_2S) and carbon dioxide (CO_2) are commonly encountered in drilling deep wells. Hydrogen sulfide, which reacts with iron at the pipe surface, forms iron sulfide (FeS), liberating two hydrogen atoms that permeate steel and lead to corrosion. Carbon dioxide, which forms carbonic acid (H_2CO_3) in water, has a similar reaction with steel, i.e., iron carbonate forms, releasing hydrogen atoms that attack steel (Magcobar Services, 1977, Corrosion, Sect. 19).

Oil-base drilling muds are very effective in reducing corrosion to the drilling equipment, because the continuous oil phase prevents the completion of the galvanic cell, which is necessary for corrosion to take place (Magcobar, 1977, Corrosion, Sect. 19). Certain chemicals, when added to the oil-base muds, make the steel surface oil wet and protect it against corrosion. (See Chapter 15 for details.)

Cost of drilling fluids

Because of different drilling conditions, drilling fluid cost varies widely. The most important factors which affect the cost of drilling fluids are as follows: (1) the type of formation to be drilled, (2) hole size, (3) total depth, and (4) logistics. Some drilling conditions which cause drilling-fluid contamination have been described in this chapter. Special drilling-fluid systems are available either for reducing the severity of these contaminations or for eliminating them completely. The cost of the drilling fluid is, however, increased as a result. For example, the use of oil-base drilling fluid in place of water-base mud completely eliminates the problems of cement and salt contaminations; however, it is very costly to change ("switch") the mud type.

The cost of drilling fluid increases with increasing size and depth of the

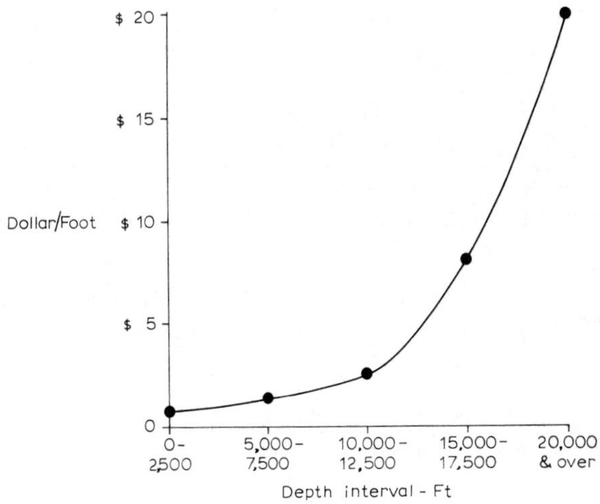

Fig. 2-30. Cost of drilling fluid by depth intervals,U.S.A., 1975 data (from Baroid and API — Joint Association Report, Feb. 1977).

well because of the higher volume of fluid required. Logistics is also responsible for high drilling-fluid cost, because materials which have to be transported a long distance will cost more than those that are available locally. Consequently, there are no typical drilling-fluid costs. For example, drilling-fluid costs alone for different wells in the west Texas area drilled to the same depths (15,000 ft) can vary from $ 20,000 to over $ 300,000 (Allen, 1977, p. 4).

From 1970 to 1977 (seven-year period) the cost of drilling fluids and additives have increased slightly over 100%. These costs have increased at a slightly higher rate than most other purchased items used in the drilling operations (Allen, 1977, p. 5). Figure 2-30 shows drilling-fluid cost (cost/ft) versus depth interval for wells drilled in the United States in 1975, whereas Fig. 2-31 shows cost of drilling and equipping wells by depth intervals. The cost/ft of drilling fluid itself varies from $ 0.50/ft in wells less than 2500 ft in depth to $ 20/ft in wells drilled to depths of 20,000 ft and deeper. Table 2-III shows drilling and completion costs for wells drilled in the United States and other countries. The cost of drilling fluid is as high as 8% of the total well cost (Allen, 1977, p. 7).

Drilling-fluid costs can be controlled in two ways:

(1) As much information as possible must be obtained about the formations to be penetrated so that early preventive actions can be taken to avoid contaminations and subsequent costly treatments of drilling fluids. An *ounce* of prevention is better than a *pound* of cure!

(2) Trained personnel must be used to supervise and direct any operations

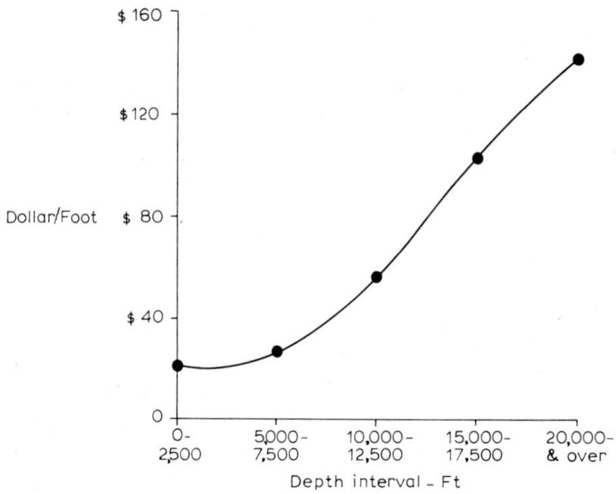

Fig. 2-31. Cost of drilling and equipping wells by depth intervals, U.S.A., 1975 data (from Baroid and API — Joint Association Report, Feb. 1977).

TABLE 2-III

Total drilling and completion costs for wells drilled in the U.S.A. and other countries in 1975

(Courtesy of Baroid Petroleum Services)

	U.S.A.		Other countries	
	1975 costs *	% of well cost	1975 costs *	% of total well cost
Payment to drilling contractors	2,366	36	1,750	34
Drilling mud (includes clay, barite, and chemicals)	460	7	430	8
Casing and tubing	1,120	17	3,023	58
Logging, perforating, cementing	788	12		
Drill bits, tools, miscellaneous	657	10		
All other	1,180	18		
Total	6,571	100	5,203	100

* in million $.

involving drilling-fluid treatments, such as treatment with chemical additives. For example, chemical thinner is usually added if the viscosity of a drilling fluid becomes higher than that desired, regardless of whether it reduces the viscosity or not. Trained mud engineers, on the other hand, can properly diagnose the situation and prescribe correct treatment. This will reduce unnecessary treatment and lower the drilling-fluid costs.

BOREHOLE PROBLEMS ASSOCIATED WITH DRILLING FLUIDS

The following three common borehole problems are related to drilling fluids: (1) sloughing and heaving shales, (2) differential sticking of pipe, and (3) lost circulation. Often, these hole problems can be avoided by (1) proper selection of drilling-fluid types, and (2) conditioning of these fluids in order to maintain good operating properties. Inasmuch as all three problems are equally important to a rapid and most economical completion of a well, the causes of each one must be well understood.

Sloughing shale

Wellbore instability due to the presence of sloughing and heaving shales is a function of (1) overburden pressure, (2) pore pressure, (3) degree of water adsorption by clays, and (4) local tectonic forces (Tschirley, 1979, p. 4).

Other factors which tend to promote hole instability and sloughing include: (1) high content of high-yield clays in shales; (2) presence of steeply-dipping shale beds; (3) existence of abnormally high-pressure formations with associated thick undercompacted shale sections; (4) presence of microfractures in shales, e.g., along cleavage planes; (5) age of the shales; (6) degree of compaction of formations; and (7) high water wettability of penetrated rocks.

Severe washouts, i.e., hole enlargement (see Fig. 2-32), bridging and filling during trips, and high solids content problems are usually associated with sloughing and heaving shales. Three types of shales are commonly recognized in oil-bearing provinces: (1) gumbo shales, (2) shales proper, and (3) indurated shales. Gumbo shales are considered to be soft and easily dispersed, forming highly plastic and sticky masses when wet, whereas shales are generally considered as being soft and fairly dispersible, but less susceptible to swelling and dispersion than gumbo shales. Indurated shales, which show slight to moderate dispersion tendency, are moderately hard (Milchem, 1979, p. 10.2).

Mechanisms of clay and shale hydration are discussed in Chapters 5 and 8. The chemistry of drilling fluid filtrates and adsorption ability of the clays composing the shales play the dominant roles. According to Tschirley (1979, p. 6) and others, prevention of water-wetting of shales can be most

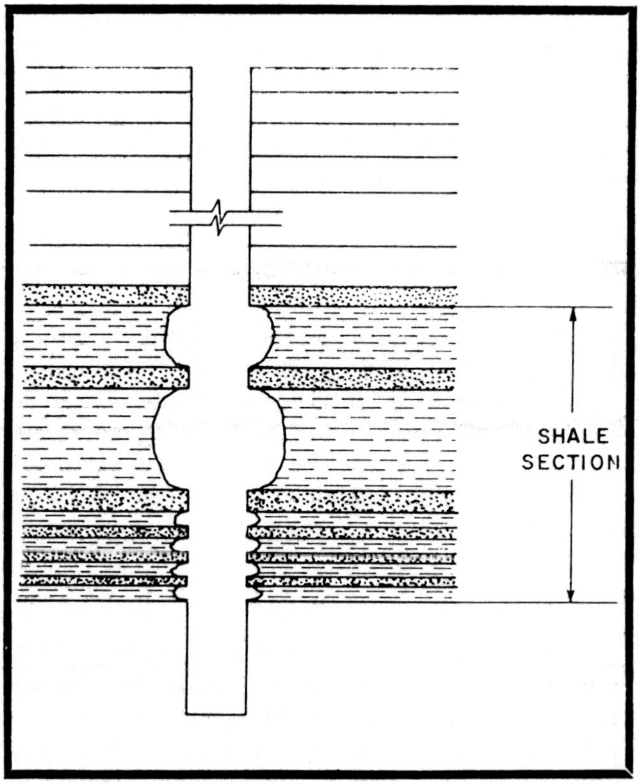

Fig. 2-32. Wellbore enlargement (e.g., due to washouts) caused by wellbore instability (e.g., due to filtrate invasion). (After Baroid, "Manual of Drilling Fluid Technology", II(II), p. 7, fig. 5; courtesy of Baroid Petroleum Services.)

helpful in combating sloughing shale. Some preventive measures include (1) chemical treatment of drilling-fluid filtrates with electrolytes and surfactants, (2) utilization of polymers, and (3) using oil-base drilling fluids which produce oil filtrate. Mechanisms of shale inhibition are discussed in detail in Chapter 8.

Numerous investigators have studied the effects of drilling fluids and their filtrates on the borehole stability under downhole conditions. Figure 2-33 illustrates one type of downhole simulator that has been developed. Equipment such as this yields valuable information and insight into control of borehole instability as a result of filtrate invasion. Such equipment also allows evaluation of the effects of exposure time of shale to drilling fluid and its filtrate under dynamic conditions.

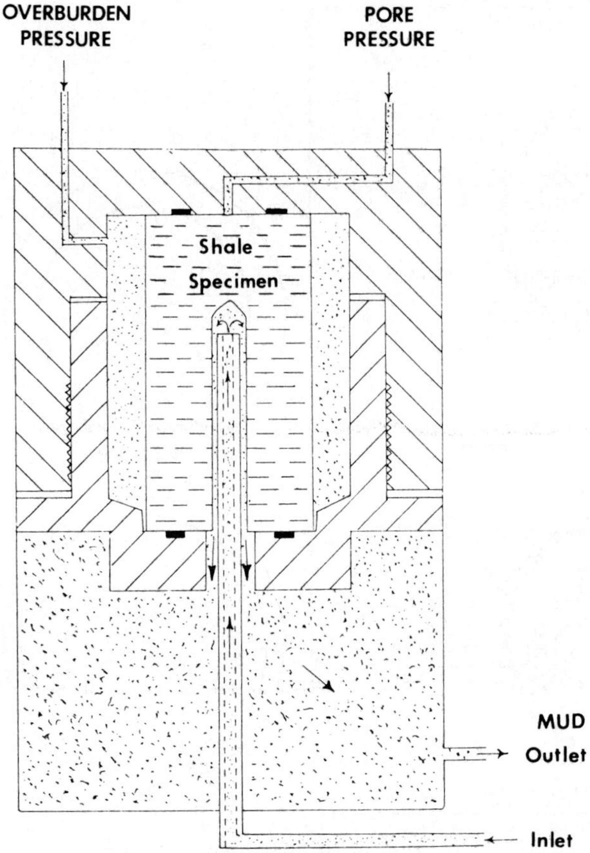

Fig. 2-33. An example of apparatus for testing (dynamic) the effects of various drilling fluids and their filtrates on shales at simulated downhole conditions. (After Simpson, 1971, p. 1295, fig. 2; courtesy of the Journal of Petroleum Technology and Baroid Petroleum Services.)

Differential pressure sticking

When drillstem (generally the drillcollar) becomes embeded and stuck in the filter cake, it is referred to as differential pipe sticking. Differential pressure sticking is related to the filter-cake thickness and friction between the steel surface and filter cake. A condition which initiates differential sticking is illustrated in Fig. 2-34. Differential pressure sticking usually occurs when pipe becomes motionless upon being embedded in the filter cake. As illustrated in Fig. 2-35, the contact surface area between the pipe and mud cake

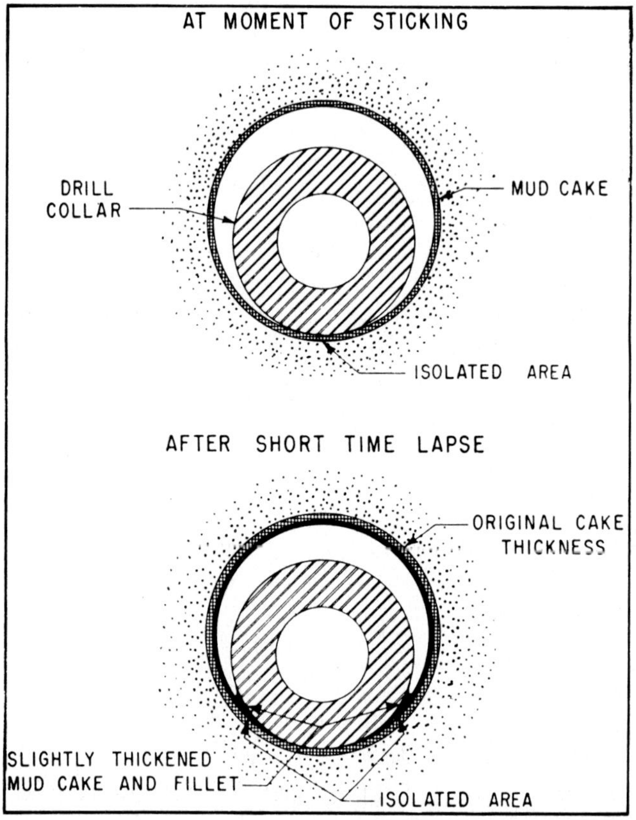

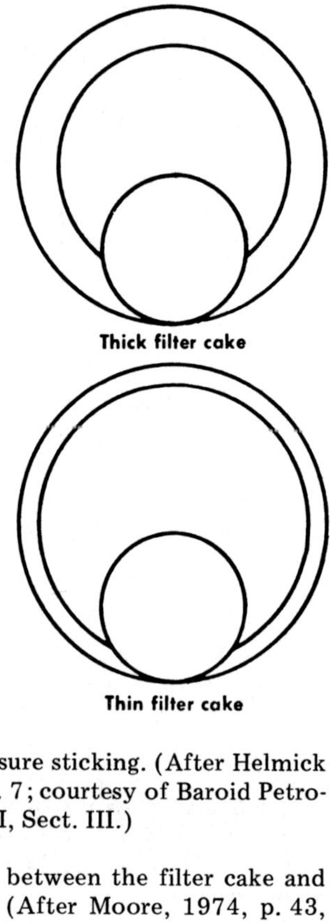

Fig. 2-34. Schematic diagram showing the mechanism of pressure sticking. (After Helmick and Longley, 1957, fig. 3; also see Tschirley, 1978, p. 6, fig. 7; courtesy of Baroid Petroleum Services, "Manual of Drilling Fluid Technology", Vol. II, Sect. III.)

Fig. 2-35. Schematic diagram showing greater contact area between the filter cake and pipe in the case of thick as compared to thin filter cake. (After Moore, 1974, p. 43, fig. 3-3; courtesy of Petroleum Publishing Company.)

increases with increasing thickness of the filter cake. Several formulae are used to calculate the sticking force between the pipe and the mud cake. It is always desirable to use a drilling fluid with low fluid loss (high-temperature, high-pressure), which will give rise to a thin filter cake. Filter cakes of certain drilling fluids have lower coefficients of friction (between the pipe and the filter cake). The presence of asphalt, graphite, detergents, and walnut hulls, for example, improves the lubrication characteristics.

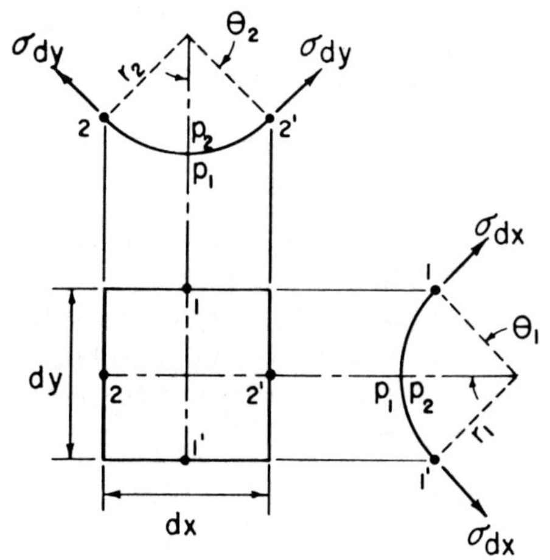

Fig. 2.I-3. Surface tension forces acting on a small element on the surface having double curvature. ($p_2 = p_1 + \gamma h$) (See Binder, 1962, and Vennard, 1961.)

Interfacial tension, σ, is caused by the molecular property (intermolecular cohesive forces) of liquids. It has the dimensions of force per unit length (lb/ft or dynes/cm), or energy per unit area (ergs/cm^2). On considering an element of a surface having double curvature (R_1 and R_2), the sum of the force components normal to the element is equal to zero (Fig. 2.I-3). The pressure difference, $p_2 - p_1$, is balanced by the interfacial tension forces:

$$(p_2 - p_1)\mathrm{d}y\mathrm{d}x = 2\sigma \mathrm{d}y \sin \theta_2 + 2\sigma \mathrm{d}x \sin \theta_1 \qquad \text{(I-1)}$$

If the contact angles θ_1 and θ_2 are small, the following simplifications may be made:

$$\sin \theta_1 = \frac{\mathrm{d}y}{2R_1} \qquad \text{(I-2)}$$

and:

$$\sin \theta_2 = \frac{\mathrm{d}x}{2R_2} \qquad \text{(I-3)}$$

Therefore, eq. I-1 becomes:

$$p_2 - p_1 = \sigma\left(\frac{1}{R_2} + \frac{1}{R_1}\right) \qquad \text{(I-4)}$$

For a capillary tube (Fig. 2.I-4):

$$R_1 = R_2 = R \qquad \text{(I-5)}$$

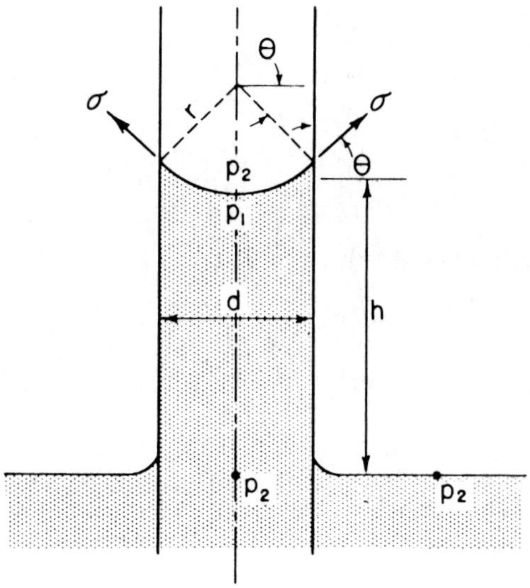

Fig. 2.I-4. Rise of water in a glass capillary tube. (See Binder, 1962, and Vennard, 1961.)

$$\cos \theta = \frac{d}{2R} \tag{I-6}$$

and:

$$p_2 = p_1 + \gamma h \tag{I-7}$$

where γ = specific weight of fluid, d = diameter of capillary tube, and h = height of capillary rise. Thus, eqs. I-4 through I-7 may be combined to yield the following expression for capillary rise, h:

$$h = \frac{4\sigma \cos \theta}{\gamma d} \tag{I-8}$$

Equation I-8 can also be derived on considering the equilibrium of vertical forces. The weight of fluid in the capillary tube, W, which is acting downward, is equal to:

$$W = \frac{\pi}{4} d^2 h \gamma \tag{I-9}$$

The vertical component of interfacial tension force acting upwards is equal to:

$$F_{\sigma y} = \pi d \sigma \cos \theta \tag{I-10}$$

Equating these two forces and solving for h gives rise to eq. I-8.

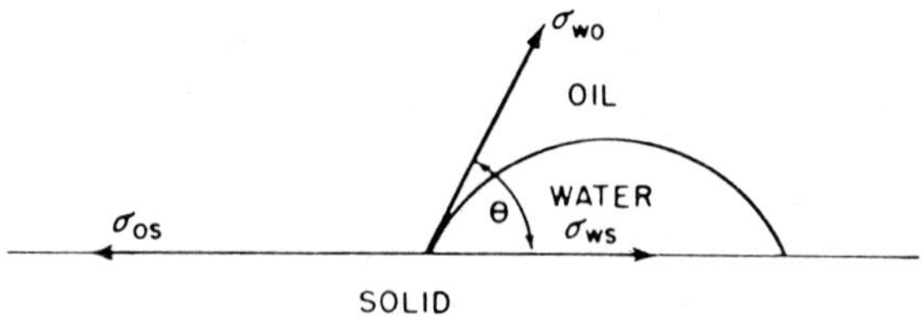

Fig. 2.I-5. Shape of water drop resulting from interfacial tension forces.

In reference to Fig. 2.I-5, the interfacial tensions can be expressed as

$$\sigma_{ws} + \sigma_{wo} \cos \theta = \sigma_{so} \tag{I-11}$$

where σ_{ws}, σ_{wo}, and σ_{so} = interfacial tensions at the phase boundaries water--solid, water—oil and solid—oil, respectively, or

$$\cos \theta = \frac{\sigma_{so} - \sigma_{ws}}{\sigma_{wo}} \tag{I-12}$$

As shown in Fig. 2.I-6A, when a solid is completely immersed in a water phase, $\theta = 0°$, $\cos \theta = 1$, and consequently

$$\sigma_{wo} = \sigma_{so} - \sigma_{ws} \tag{I-13}$$

When half of the solid is wet by water and the other half by oil (Fig. 2.I-6B), $\theta = 90°$, $\cos \theta = 1$ and thus

$$\sigma_{so} = \sigma_{ws} \tag{I-14}$$

On the other hand, if the solid is completely wetted by oil (Fig. 2.I-6C), $\theta = 180°$, $\cos \theta = -1$, and

$$\sigma_{so} = \sigma_{ws} - \sigma_{wo} \tag{I-15}$$

If $\theta < 90°$, the surfaces are called hydrophilic and when $\theta > 90°$ they are called hydrophobic. An interfacial-tension depressant lowers σ_{wo}, whereas a

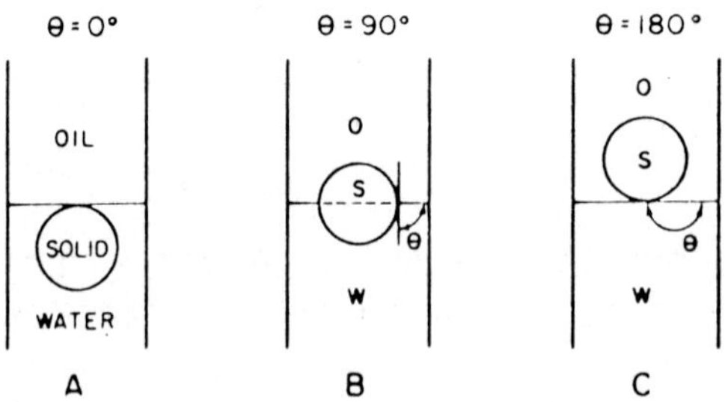

Fig. 2.I-6. Illustrations of 0°, 90°, and 180° contact angles.

wetting agent lowers θ or increases $\cos\theta$. A decrease in σ_{wo} does not necessarily mean an increase in $\cos\theta$, or vice versa, because of the changes in σ_{so} and σ_{ws}.

If a rock is completely water wet ($\theta = 0$), water will try to envelop all of the grains and force all of the oil out in the middle of the pore channel. Even though some oil may still be trapped in this case, the recovery would be high. On the other hand, if all of the solid surfaces were completely oil wet ($\theta = 180°$), oil would try to envelop all of the grains and force all of the water out into the center of the pore channel. In this extreme case, recovery would be very low by water drive. Many oil-wet reservoirs are known to exist.

In the usual case ($0° < \theta < 180°$), to improve waterflooding operations the contact angle θ should be changed from $>90°$ to $<90°$, through the use of surfactants. This would move the oil from the surface of the grains out into the center of the pore channels.

Contaminants or impurities may exist in either fluid phase and/or may be adsorbed on the solid surface. Even if present in minute quantities, they can and do change the contact angle from the value measured for pure systems. (See Marsden, 1968.)

Effect of contact angle and interfacial tension on movement of oil

For an ideal system composed of pure liquids, the advancing contact angle should equal the receding angle. Because of the presence of impurities within the liquids, however, the advancing contact angle is greater in most systems. The advancing contact angle is the angle formed at the phase boundary when oil is displaced by water. It can be measured as follows: The crystal plate is covered by oil and then the water drop is advanced on it. The contact angle is the limiting angle with time after equilibrium has been established (Fig. 2.I-7). The contact angle formed when water is displaced by oil is called the receding angle (Fig. 2.I-8). The contact angles during the movement of a water—oil interface in a cylindrical capillary having a hydrophilic surface are shown in Fig. 2.I-9.

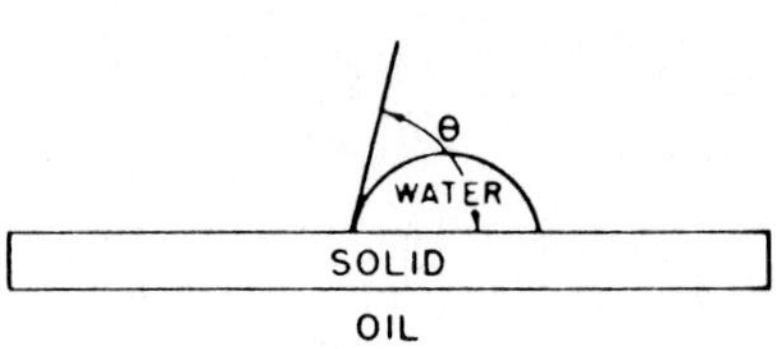

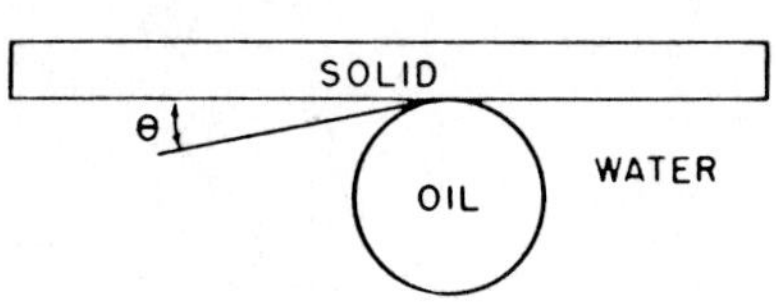

Fig. 2.I-7. Contact angle: plate first immersed in oil followed by the placement of water drop on top.

Fig. 2.I-8. Contact angle: plate first immersed in water followed by placing a drop of oil underneath.

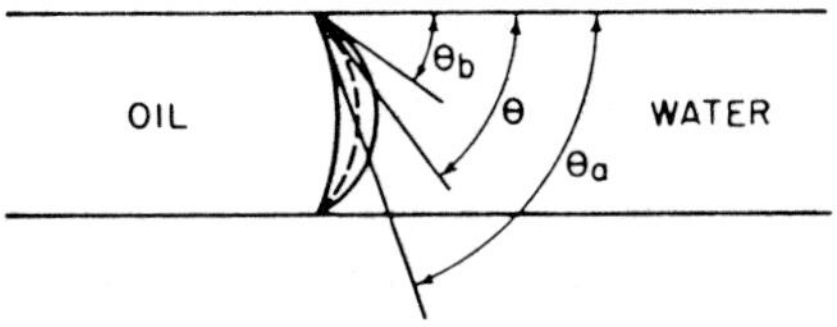

Fig. 2.I-9. Changes in contact angle as a result of movement of water—oil interface. θ = contact angle at static position; θ_a = contact angle when oil is displaced by water (advancing angle); and θ_b = contact angle when water is displaced by oil (receding angle).

Inasmuch as a reservoir is basically a complex system of interconnected capillaries of various sizes and shapes, an understanding of flow through capillaries is very important. In Fig. 2.I-10, a simple two-branch capillary system is presented. If a pressure drop is applied, the water will flow more readily through the large-diameter capillary than it will through the small-diameter one. Thus, a certain volume of oil may be trapped in the small capillary when the water reaches the upstream fork. Poiseuille's law states that:

$$q = \frac{\pi d^4 \Delta p_t}{128 \mu L} \tag{I-16}$$

and:

$$v = \frac{q}{A} = \frac{d^2 \Delta p_t}{32 \mu L} \tag{I-17}$$

where q = volumetric rate of flow (cm^3/sec); d = diameter of capillary (cm); Δp_t = total pressure drop (dynes/cm^2); A = cross-sectional area (cm^2); μ = viscosity (cP); L = flow path length (cm); v = velocity (cm/sec).

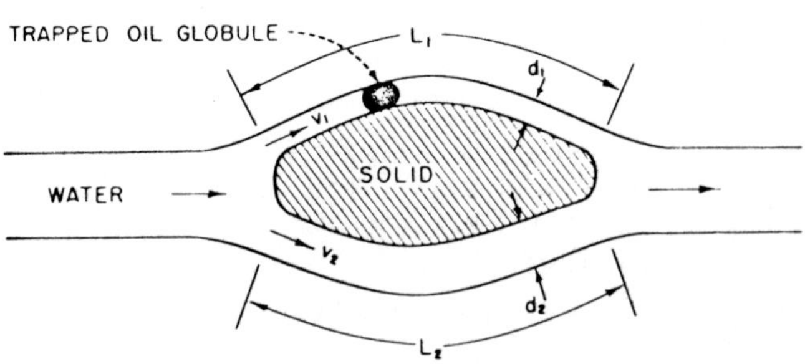

Fig. 2.I-10. Flow through a two-branch capillary and trapping of oil in a small-diameter capillary.

The capillary pressure, P_c, is equal to:

$$P_c = \frac{4\sigma \cos\theta}{d} \tag{I-18}$$

where: σ = interfacial tension between oil and water (dynes/cm); d = diameter of capillary (cm); θ = contact angle.

The total pressure drop, Δp_t, is equal to:

$$\Delta p_t = \Delta p_i + P_c \tag{I-19}$$

where Δp_i = applied pressure, dynes/cm^2. Solving for v in each capillary, by combining Eqs. I-17, I-18, and I-19 gives:

$$v_1 = \frac{d_1^2}{32\mu_1 L_1}\left(\Delta p_i + \frac{4\sigma\cos\theta}{d_1}\right) \tag{I-20}$$

and:

$$v_2 = \frac{d_2^2}{32\mu_2 L_2}\left(\Delta p_i + \frac{4\sigma\cos\theta}{d_2}\right) \tag{I-21}$$

Setting $L_1 = L_2$ and $\mu_1 = \mu_2$, and dividing Eq. I-20 by Eq. I-21 gives the following relationship:

$$\frac{v_1}{v_2} = \frac{d_1^2\,\Delta p_i + 4\sigma\cos\theta d_1}{d_2^2\,\Delta p_i + 4\sigma\cos\theta d_2} \tag{I-22}$$

Therefore, when $\Delta p_i >> P_c$:

$$\frac{v_1}{v_2} \approx \frac{d_1^2}{d_2^2} \tag{I-23}$$

and when $\Delta p_i << P_c$:

$$\frac{v_1}{v_2} \approx \frac{d_1}{d_2} \tag{I-24}$$

As shown in Fig. 2.I-11, the sum of forces acting on the trapped oil globule

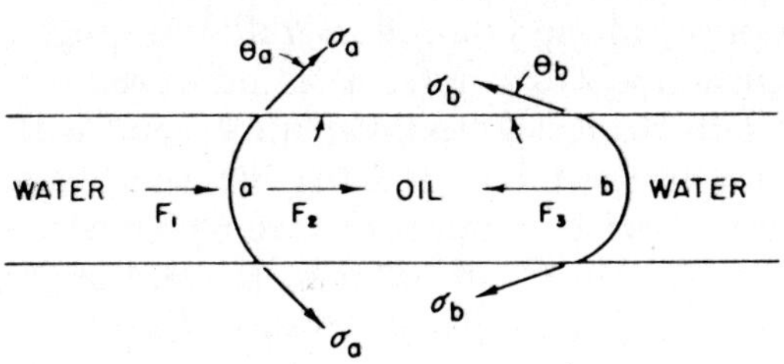

Fig. 2.I-11. Forces acting on a trapped oil globule in a capillary.

may be expressed as

$$\sum F = F_1 + F_2 - F_3 \tag{I-25}$$

where:

$$\sum F = \frac{\pi d^2}{4} \Delta p_t$$

$$F_1 = \frac{\pi d^2}{4} \Delta p_{\mathrm{i}}$$

$$F_2 = \pi d(\sigma_{\mathrm{a}} \cos \theta_{\mathrm{a}})$$

$$F_3 = \pi d(\sigma_{\mathrm{b}} \cos \theta_{\mathrm{b}})$$

thus:

$$\Delta p_t = \Delta p_{\mathrm{i}} + \frac{4\sigma_{\mathrm{a}} \cos \theta_{\mathrm{a}}}{d} - \frac{4\sigma_{\mathrm{b}} \cos \theta_{\mathrm{b}}}{d} \tag{I-26}$$

Because the receding angle is usually less than the advancing angle, the capillary pressure not only does not help but hinders the flow. The term $(4\sigma_{\mathrm{b}} \cos \theta_{\mathrm{b}}/d)$ is usually greater than $(4\sigma_{\mathrm{a}} \cos \theta_{\mathrm{a}}/d)$ because $\theta_{\mathrm{b}} < \theta_{\mathrm{a}}$. If a surfactant were added at the left to reduce σ_{a}, Δp_t would become less and the oil globule may eventually move to the left in Fig. 2.I-11 when Δp_t becomes negative. The quantity of trapped oil is dependent upon the value of $\sigma \cos \theta$ at each end of the globule as well as upon Δp_{i} (imposed pressure drop).

Inasmuch as the contact angle depends upon the interfacial tensions, which, in turn, may be influenced by surfactants, these chemicals may alter recovery by altering both the contact angle and interfacial tension. As the oil is displaced by water, which wets the rock surfaces, capillary pressure is a driving force. If, on the other hand, water does not wet the rock surface, capillary pressure is a retarding force which must be overcome.

The magnitude of capillary pressure in pores having a radius of around 15 microns is not large and, therefore, capillary pressure is not an important force during the movement of oil—water contact, providing there is no mixing. The movement of oil and water in a reservoir, however, results in the formation of water—oil and gas—water—oil mixtures. (See Muravyov et al., 1958.) The amount of gas coming out of solution during migration is greater with increasing amount of dissolved active substances, with increasing surface area of porous medium (i.e., with decreasing permeability), and with decreasing temperature. As the oil—water—gas mixtures move through pores, the gas bubbles and water droplets are deformed on passing through constrictions (Fig. 2.I-12). (See Muravyov et al., 1958.) In order to move, the gas bubble shown in Fig. 2.I-12 must overcome the capillary pressure equal to

$$\Delta p = p_1 - p_2 = \frac{2\sigma}{R_1} - \frac{2\sigma}{R_2} = 2\sigma\left(\frac{1}{R_1} - \frac{1}{R_2}\right) \tag{I-27}$$

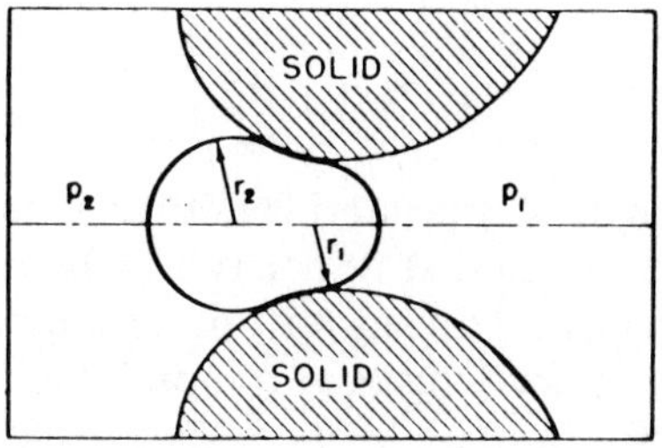

Fig. 2.I-12. Movement of gas bubble through a constriction. (After Muravyov et al., 1958.)

Although the Δp may be very small for a single bubble, the cumulative resistance of many bubbles may be large (Jamin effect). Additional resistance to flow is created by the polymolecular layers of oriented molecules of surface-active components in the oil, which are adsorbed on the rock surface and may be quite thick (10^{-3} to 10^{-4} cm). At a constant pressure differential, the rate of oil filtration through porous media diminishes with time and is more pronounced in the case of higher content of polar components in the oil.

Water block

The minimum pressure (P_{cwb}) required for oil to displace a globule of aqueous mud filtrate stuck in a pore throat between rock grains, providing the filtrate is the wetting phase, is equal to:

$$P_{cwb} = \sigma \left(\frac{1}{r_1} + \frac{1}{r_2} \right) \cos \theta \qquad \text{(I-28)}$$

where σ = interfacial tension between oil and water; r_1 and r_2 = radii of capillary in two perpendicular directions; and θ = contact angle. The above equation can be derived from eq. I-4:

$$p_2 - p_1 = \Delta p = \sigma \left(\frac{1}{R_1} + \frac{1}{R_2} \right) \qquad \text{(I-4)}$$

where Δp is the pressure difference across the oil—water interface, and R_1 and R_2 are curvatures of interface in two perpendicular directions, by substituting $r_1/\cos\theta$ for R_1 and $r_2/\cos\theta$ for R_2.

If σ = 25 dynes/cm, θ = 30°, r_1 = 0.0001 cm, and r_2 = 0.00001 cm, then P_{cwb} = 34.5 psi. By reducing interfacial tension between oil and water to 0.5 dynes/cm, using a surface-active agent (surfactant), for example, P_{cwb} will be lowered to 0.69 psi, and the water block can be easily broken.

APPENDIX 2.II — PRESSURE LOSSES

Magcobar Services

Whenever a fluid flows or is pumped through a pipe and whether it is water, gas, air, or drilling fluid, a pressure must be applied to the fluid which will decrease in the pipe along the direction of flow. The amount of pressure required to move the fluid has a direct relation to the length of the pipe, i.e., more pressure is required with increasing length of pipe. For a constant fluid velocity, pressure loss increases as pipe diameter decreases. The pressure loss increases as the velocity of the fluid increases, with constant-size pipe.

In a drilling-fluid circulating system, the pressure is created by a pump or compressor. The pressure loss will occur from pump discharge to flowline. Increased depth causes more circulating pressure to be required. For practical purposes, the pressure losses are calculated for (1) the surface equipment, which includes standpipe, rotary hose, swivel and kelly; (2) the flow inside the drillpipe; (3) the flow inside the drillcollars; (4) the flow through the bit nozzles; (5) the flow outside the drillcollars; and (6) the flow outside the drillpipe.

The total pressure loss in a mud system is affected by the properties of the mud, such as density, plastic viscosity and yield point. Turbulent flow generally exists in the surface equipment, inside the drillstring and through the bit nozzles. In the annulus the flow could be either turbulent or laminar. In addition, at different points in the annulus the fluid flow could be either turbulent or laminar. For example, the fluid flow could be turbulent around the drillcollars, but laminar around the drillpipe. Equations II-1 and II-2 can be used to determine whether the fluid flow is turbulent or laminar in the annulus:

$$\text{annular velocity (ft/min)} = \frac{\text{pump output (bbl/min)}}{\text{(hole capacity; bbl/ft)} - \text{(disp. and capacity of drillpipe, bbl/ft)}} \qquad \text{(II-1)}$$

$$\text{Reynolds number, } Re = \frac{928\gamma_f v d}{\mu} \qquad \text{(II-2)}$$

where ρ = specific weight of drilling fluid (lb/gal); v = velocity (ft/sec); d = flow diameter (inches); and μ = viscosity (cP).

The pressure-loss calculations are different for turbulent and laminar flows. In turbulent flow, as far as flow properties are concerned, plastic viscosity is the more important factor and yield point can be ignored. In laminar flow, on the other hand, yield point is the more important factor and the plastic viscosity effect is negligible. The pressure drop in the annulus for laminar flow can be calculated from the Bingham equation:

$$\Delta p = \frac{l\overline{YP}}{225d} + \frac{l\overline{PV}v}{1500d^2} \qquad \text{(II-3)}$$

where: Δp = pressure drop (psi); l = length or depth (ft); $\overline{YP}$ = yield point (lb/100 sq ft); $\overline{PV}$ = plastic viscosity (cp); d = hole diameter minus pipe diameter (inches); and v = velocity of fluid (ft/sec).

Inspection of the Bingham equation shows that the largest portion of the pressure losses is due to the yield stress of the mud. The term involving velocity is relatively small. Shear stresses in a mud, however, are related to the rate of shear.

The shear stresses predicted by the Bingham plastic equation are higher than the actual ones for the shear rates normal for annular flow. Thus, the Bingham pressure-loss equation yields higher values than actual values. The following method enables a more close approximation of the pressure losses in an annulus.

The pressure losses which occur in an annulus are due to the shear stresses at the wall of the hole and at the wall of the drillpipe. In order to calculate these pressure losses accurately, it is necessary to know the shear rate in the annulus and the shear stress of the mud at that specific shear rate. The shear stress values are plotted versus the various shear rates using the Fann V-G meter.

The shear rate in the annulus can be calculated by using the following equation:

$$RPM = \frac{1.4093\bar{v}}{d_h - d_p} \qquad \text{(II-4)}$$

where: $\bar{v}$ = average annular velocity (ft/min); d_h = diameter of hole (inches); and d_p = pipe diameter (inches).

The pressure loss gradient may be calculated using the following equation:

$$\frac{\Delta p}{l} = \frac{3.75\theta}{d_h - d_p} \qquad \text{(II-5)}$$

where $\Delta p/l$ = annular pressure loss (psi/1000 ft); θ = Fann dial reading at RPM equivalent to annular shear rate; d_h = diameter of hole (inches); and d_p = diameter of pipe (inches).

Inasmuch as oil muds and most treated bentonite muds follow a Power Law relationship at all but the lower shear-rate ranges, it is advisable to plot the Fann dial readings versus RPM on logarithmic paper, in order to obtain a straight line. The best way to draw the line through the data points is to draw one straight line through the higher RPM points and another one through the 3 and 6 RPM points and then connect the two lines (see Fig. II-1) as shown in the following example.

Example No. 1

Given

(1) Hole size = $8\frac{1}{2}$ inches; (2) drillpipe diameter = $4\frac{1}{2}$ inches; (3) annular

APPENDIX 2.I — FUNDAMENTALS OF SURFACE AND CAPILLARY FORCES

By George V. Chilingar, Gerald L. Langnes and John O. Robertson, Jr.

Introduction

Wettability may be defined as the ability of the liquid to wet, or spread over, a solid surface. Figure 2.I-1A shows a liquid wetting a solid. On the other hand, Fig. 2.I-1B shows the relationship between the liquid and solid when the liquid has little affinity for the solid. In Fig. 2.I-1C the liquid drop occupies an intermediate position. The fluid which wets the surface more strongly, occupies the smaller pores and minute interstices.

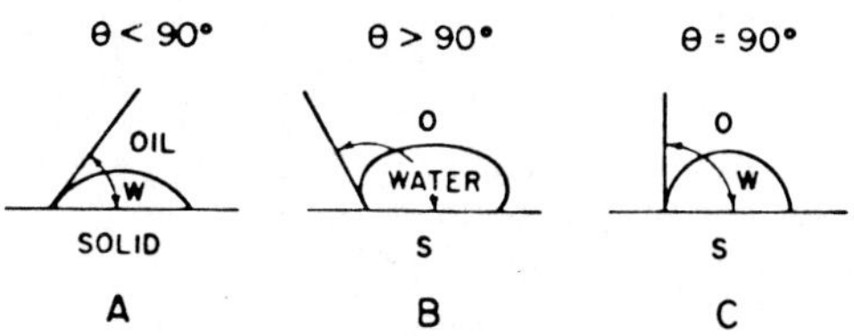

Fig. 2.I-1. Different degree of wetting of solid by a liquid.

Interfacial tension and contact angle

The angle which the interface makes with the solid is called the contact angle, θ. Usually, it is measured from the solid through the liquid phase (if the other phase is a gas) and through the water phase if oil and water are both present. In a capillary tube, shown in Fig. 2.I-2A, the angle between the side of the tube and the tangent to the curved interface (where it intersects the side of tube) is less than 90°. For a capillary depression, shown in Fig. 2.I-2B, the contact angle is greater than 90°. In the case of no rise or depression, the angle is 90° (Fig. 2.I-2C).

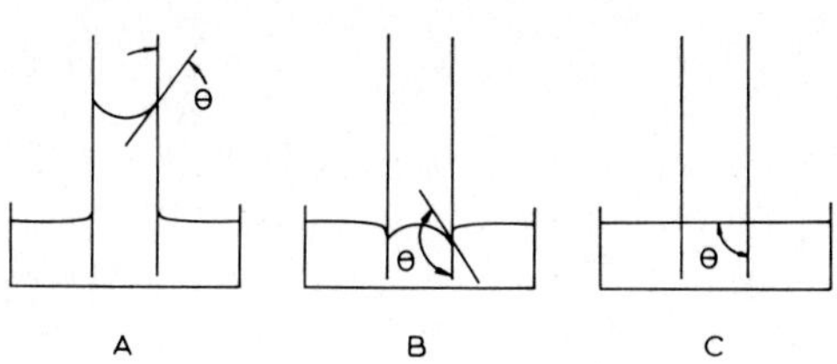

Fig. 2.I-2. Behavior of various fluids in glass capillary tubes. A. Water. B. Mercury. C. Tetrahydronaphthalene (when glass is perfectly clean and liquid is pure).

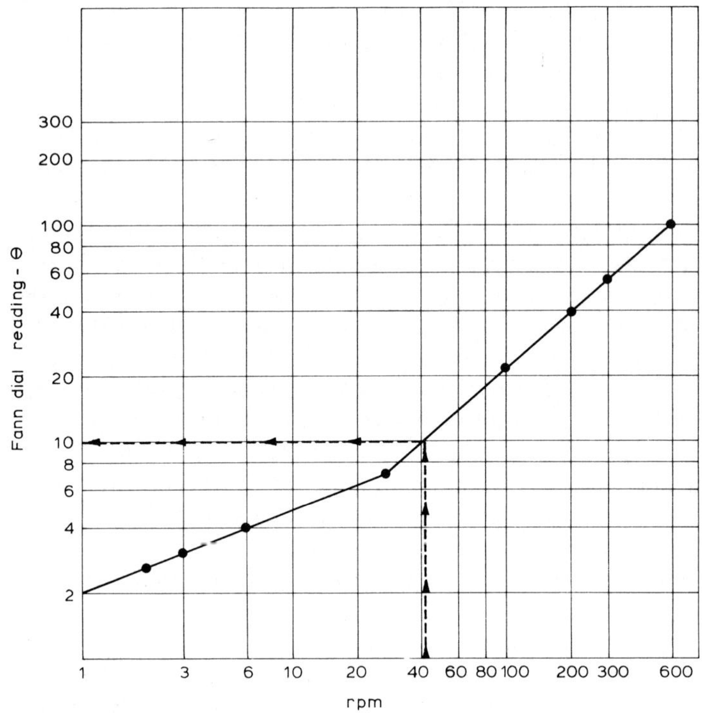

Fig. 2.II-1. Relationship between Fann dial reading (θ) and rotational speed in RPM for solving Example No. 1.

velocity = 120 ft/min; and (4) Fann viscometer data: $\theta_{600} = 100$, $\theta_{300} = 55$, $\theta_{200} = 39$, $\theta_{100} = 22$, $\theta_6 = 4$, and $\theta_3 = 3$.

Solution:

(1) Plot viscometer data on log-log paper.
(2) Draw a straight line through the 100, 200, 300, and 600 *RPM* points.
(3) Draw a straight line through the 3 and 6 *RPM* points.
(4) Connect these two lines (Fig. II-1).
(5) Using eq. II-4, calculate the *RPM* that is equivalent to the shear rate in the annulus:

$$RPM = \frac{(1.4093)(120)}{(8.5 - 4.5)} = 42$$

(6) Find the dial reading equivalent to 42 *RPM* from Fig. 2.II-1 (=10.5).
(7) Use eq. II-5 to calculate the pressure loss per 1000 ft:

$$\frac{\Delta p}{l} = \frac{3.75\theta}{d_h - d_p}$$

$$\frac{\Delta p}{l} = \frac{(3.75)(10.5)}{(8.5-4.5)}$$

$$\frac{\Delta p}{l} = 9.84 \text{ psi/1000 ft} .$$

(8) In order to obtain the total annular pressure loss, the pressure gradient calculated in the above step is multiplied by the length of annulus (of the dimensions used in calculations; in thousands of feet, e.g.). In the case of 10,000 ft of $8\frac{1}{2}$-inch hole, $4\frac{1}{2}$-inch drillpipe, and annular velocity of 120 ft/min, the total pressure loss would be:

$$\Delta p = \frac{(9.84)(10{,}000)}{(1000)} = 98.4 \text{ psi}$$

Obviously, sections of different geometry must be calculated separately.

Discussion

The advantages of the above procedure are twofold: (1) the simplicity of the calculation involved, and (2) a graphic illustration of how circulation rate affects the pressure losses after plotting the viscometer data.

The effect of treating the mud is readily seen from the plot (e.g., Fig. 2.II-1). If it is desired to reduce the shear stress and consequently the pressure requirements, the line must be shifted downward. This can be accomplished by removal or dilution of mechanical solids. Chemical treatment may be used to accomplish this if the mud is not completely deflocculated. The effect of temperature is to shift the line upward with decreasing temperatures and downward with increasing temperatures. The slope of the line should remain the same unless a temperature flocculation occurs. The Bingham equation would yield a value of 14.86 psi per 1000 ft in this example.

In addition to accuracy and simplicity, this method also gives an image of the fluid behavior of the mud. Thus, the effect of mud treatment may be easily observed.

In turbulent flow, yield point has no effect on pressure loss. Plastic viscosity and mud specific weight, however, are important in determining the friction factor of the fluid. Calculations for pressure losses of non-Newtonian fluids in turbulent flow can be made using the Fanning equation:

$$\Delta p = \frac{f l \gamma_m v^2}{25.6 d} \tag{II-6}$$

where: Δp = pressure drop (psi); f = friction factor (dimensionless); l = length or depth (ft); γ_m = specific weight of mud (lb/gal); v = velocity (ft/sec); d = diameter of pipe in inches, or difference in diameters of hole and drillpipe (inches).

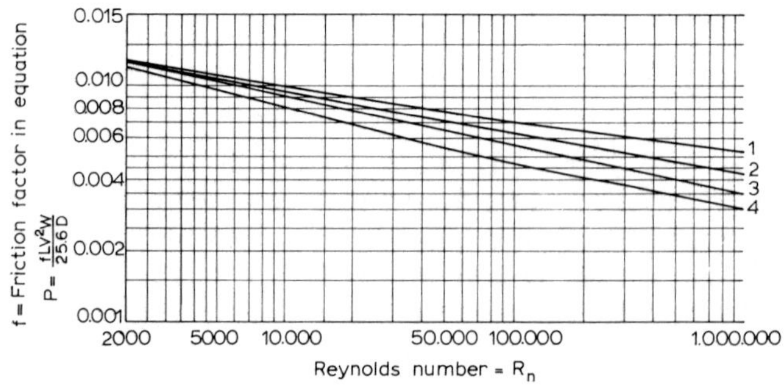

Fig. 2.II-2. Friction factor versus Reynolds number (Stanton type) diagram. *1* = lowest values for drawn brass or glass tubing (Walker, Lewis and McAdams); *2* = clean internal flush tubular goods (Walker, Lewis and McAdams); *3* = full-hole drillpipe annuli in cased holes (Pigott's data); *4* = annuli in uncased holes (Pigott's data).

All the data for the formula are known except velocity, v, and friction factor, f. The velocity can be calculated using the following equation:

$$v = \frac{q}{2.448(d_h^2 - d_p^2)} \tag{II-7}$$

where: q = flow rate (gpm); d_h = hole diameter (inches); and d_p = pipe diameter (inches).

The friction factor, f, is obtained by determining the Reynolds number, *Re*, first, and then using a diagram such as the one presented in Fig. 2.II-2. One of the formulas for determining the Reynolds number is as follows:

$$Re = \frac{2960 d v \gamma_m}{\overline{PV}} \tag{II-8}$$

where *Re* = Reynolds number; d = diameter (inches); v = velocity (ft/sec); γ_m = mud specific weight (lb/gal); and $\overline{PV}$ = plastic viscosity (cP).

Using the following data, the pressure loss inside the $4\frac{1}{2}$-inch drillpipe can be determined as follows: (1) determine the velocity in the pipe, e.g. eq. II-1; and (2) determine the Reynolds number (eq. II-8); (3) find the friction factor using the Reynolds number and Stanton's diagram, e.g., Fig. 2.II-2; and (4) calculate the pressure loss using eq. II-6.

Given

γ_m = 12 lb/gal; $\overline{PV}$ = 30 cP; q = 400 gpm; d = $4\frac{1}{2}$-inch O.D., 3.6-inch I.D.; and l = 1000 ft of drillpipe.

Solution

Using eq. II-7, velocity is equal to:

$$v = \frac{400 \text{ gpm}}{2.448(3.6)^2} = 12.6 \text{ ft/sec}$$

Reynolds number, Re, can then be calculated using eq. II-8:

$$Re = \frac{2960(3.6)(12.6)(12)}{30} = 53{,}600\ .$$

From Fig. 2.II-2, using line *3* for full-hole drillpipe, a friction factor is determined to be 0.0065. Pressure loss is then determined from eq. II-6:

$$\Delta p = \frac{0.0065(1000)(12.0)(12.6)^2}{25.6(3.6)} = 134 \text{ psi/1000 ft}\ .$$

Example No. 2

Given drilling conditions listed in the previous problem, calculate the equivalent annular circulating density. The following additional information is known:

Rate of penetration = 2 ft/min.
Pump output = 250 gal/min (corresponding to 120 ft/min annular velocity around the drillpipe).
Length of drillpipe = 10,000 ft.
Formation specific weight = 875 lb/bbl or 20.8 lb/gal.
Initial specific weight of drilling fluid = 10 lb/gal.
Buoyancy factor = 0.95.

It is assumed that drillcollar length is negligible compared to drillpipe length.

Calculation

$8\frac{1}{2}$-inch hole volume = 0.0702 bbl/ft.
Rate of cuttings generation = (2 ft/min) × (0.0702 bbl/ft) = 0.1404 bbl/min.
Rate of cuttings weight generation = 0.95(875 lb/bbl) × (0.1404 bbl/min) = 117 lb/min.

$$\text{Cuttings volume} = \frac{(117 \text{ lb/min})}{20.8 \text{ lb/gal}} = 5.61 \text{ gal/min}\ .$$

Annular pressure loss at 10,000 ft = 98.4 psi.

Equivalent circulating specific weight

$$= \frac{(10 \text{ gal/min})(0.052)(10{,}000 \text{ ft}) + 98.4}{(10{,}000 \text{ ft})(0.052)} = 10.2 \text{ lb/gal}\ .$$

Volumetric drilling-fluid circulating rate = 250 gal/min.
Weight rate of circulation = (10.2 lb/gal) (250 gal/min) = 2547 lb/min.
Equivalent annular circulating specific weight, therefore, is:

$$\frac{(2547 + 117) \text{ lb/min}}{(250 + 5.61) \text{ gal/min}} = 10.42 \text{ lb/gal}\ .$$

Thus, there is 0.42 lb/gal increase in specific weight as a result of annular pressure loss and generation of cuttings during drilling.

REFERENCES

Adams, J. and Algawi, M., 1970. Determination of mud cake permeability from filter press data. *Arab. Pet. Congr.*, 7th, Kuwait, 2: 16—22.

Allen, G.G., 1977. *Drilling Fluids.* Baroid Petroleum Services, 16 pp.

API Subcommittee, 1972. *Principles of Oil Well Drilling Fluids.* 12th ed., 230 pp.

Azari, M., 1978. *Numerical Solution of Dynamic-Static Filtration of Drilling Mud.* M.S. Thesis, Univ. of Southern California, 50 pp.

Baranov, V.S., 1955. *Clay Drilling Fluids Used in Complex Situations.* Gostoptekhizdat, Moscow, 212 pp.

Baroid Petroleum Services, 1975 (and other issues). *Baroid Mud Technology Handbook,* 160 pp.

Beeson, C.M., Chilingar, G.V. and Larsen, D.H., 1958. *Drilling Fluids Laboratory.* University of Southern California Publ., 73 pp.

Bezemer, C. and Havenaar, I., 1966. Filtration behavior of circulating drilling fluids. *Soc. Pet. Eng. J.*, 6(4, Dec.): 292—298.

Binder, R.C., 1962. *Fluid Mechanics,* 4th ed. Prentice-Hall, Englewood Cliffs, N.J., 453 pp.

Byck, H.T., 1940. The effect of formation permeability on the plastering behavior of drilling fluids. *API Drill. Prod. Prac.*, p. 40.

Chilingar, G.V., 1961. Notes on compaction. *J. Alberta Soc. Pet. Geol.*, 9: 158—161.

Chilingar, G.V. and Knight, L., 1960. Relationship between pressure and moisture content of kaolinite, illite, and montmorillonite clays. *Bull. Am. Assoc. Pet. Geol.*, 44: 101—106.

Chilingar, G.V., Rieke III, H.H. and Robertson Jr., J.O., 1963. Relationship between high overburden pressures and moisture content of halloysite and dickite clays. *Geol. Soc. Am. Bull.*, 74(8): 1041—1048.

Chilingar, G.V., Rieke III, H.H., Sawabini, S.T. and Ershaghi, I., 1969. Chemistry of interstitial solutions in shales versus that in associated sandstones. *Soc. Pet. Eng. Am. Inst. Min. Metall. Eng., 44th Annu. Fall Meet., Denver, Colo., SPE.* Pap., 2527: 8 pp.

Chilingarian, G.V. and Rieke III, H.H., 1968. Data on consolidation of fine-grained sediments. *J. Sediment. Petrol.*, 38(3): 811—816.

Chilingarian, G.V. and Wolf, K.H., 1975. *Compaction of Coarse-Grained Sediments,* I and II. Elsevier, Amsterdam, 552 pp. and 808 pp.

Cooke, P.W., 1955. Some aspects of high weight muds used in drilling abnormally high pressure formation. *Proc. World Pet. Congr., 4th, Rome,* Sec. II: 43—57.

Cunningham, R.A. and Goins, W.C., Jr., 1957. How mud properties affect drilling rates. *Pet. Eng.*, May: 119—126.

Eckel, J.R., 1954. Effects of mud properties on drilling rate. *API Drill. Prod. Prac.*, 2: 119—130.

Eckel, J.R., 1967. Microbit studies of effects of fluid properties and hydraulics on drilling rate. *J. Pet. Technol.*, 19 (April): 541—546; *Trans. AIME,* 240.

Ershaghi, I., Dougherty, E.L., Herzberg, D. and Ucok, H., 1978. Permeability determination in liquid dominated geothermal reservoirs using the dual induction laterolog. *SPWLA Trans.*, 19th Annu. Symp., June 13—16, El Paso, Texas, 20 pp.

Ferguson, C.K. and Klotz, J.A., 1954. Filtration from mud during drilling. *Trans. Soc. Pet. Eng. AIME,* 201: 29—42.

Fertl, W.F., 1976. *Abnormal Formation Pressures.* Elsevier, Amsterdam, 380 pp.

Fertl, W.H. and Chilingarian, G.V., 1977. Importance of abnormal formation pressure. *J. Pet. Technol.*, 29(4): 347—355.
Fertl, W.H. and Timko, D.J., 1970. Occurrence and significance of abnormal-pressure formations. *Oil Gas J.*, 68(1): 97—108.
Finch, W.C., 1968. Abnormal pressure in the Antelope Field, North Dakota. *Soc. Pet. Eng., 43rd Annu. Meet., Houston, Texas, SPE*, Pap. 2227: 4 pp.
Gatlin, C., 1960. *Petroleum Engineering — Drilling and Well Completions.* Prentice-Hall, Englewood Cliffs, N.J., 341 pp.
Griffin, D.G. and Bazer, D.A., 1968. A comparison of methods for calculating pore pressure and fracture gradients from shale density measurements using the computer. *Soc. Pet. Eng., 43rd Annu. Fall Meet., Houston, Texas, SPE*, Pap. 2166: 12 pp.
Helmick, W.E. and Longley, A.J., 1957. *Pressure-differential Sticking of Drill Pipe and How It Can Be Avoided or Relieved.* Meet. Pac. Coast Distr. Div. Prod., Los Angeles, Calif., May, pp. 55—61.
Horner, V., White, M.M., Cochran, C.D. and Deily, F.H., 1957. Microbit dynamic filtration studies. *Trans. Soc. Pet. Eng. AIME*, 210: 183—189.
Hottman, C.E. and Johnson, R.K., 1965. Estimation of formation pressures from log-derived shale properties. *J. Pet. Technol.*, 16(6): 717—722.
Hutchison, S.O. and Anderson, G.W., 1974. What to consider when selecting drilling fluids. *World Oil*, Oct.: 83—86.
IMCO Services, 1977. *Applied Mud Technology.* 6th ed., 210 pp.
Kennedy, W.A. and Meyer, R.L., 1968. Low-solids fluids reduce costs. *Oil Gas J.*, 66 (Jul. 15): 132—137.
Krueger, R.F., 1963. Evaluation of drilling-fluid filter-loss additives under dynamic conditions. *J. Pet. Technol.*, Jan.: 90—98.
Kruse, C., 1976. *A Dictionary of Petroleum Terms.* Univ. Texas Press, Austin, Texas, 145 pp.
Langnes, G.L., Robertson, J.O., Jr. and Chilingar, G.V., 1972. *Secondary Recovery and Carbonate Reservoirs.* Elsevier, New York, N.Y., 304 pp.
Larsen, D.H., 1938. Determining the filtrate characteristics of oil well drilling fluids. *Pet. Eng.*, Sept.: 42—51.
Magcobar Services, 1972. *Drilling Fluid Engineering Manual.* 200 pp.
Marsden, S.S., 1968. Wettability: The elusive key to waterflooding. *Pet. Eng.*, (April): 82—87.
Milchem Services, 1963. *Engineering Notebook.* Milwhite Mud Sales Company, Houston, Texas.
Milchem Services, 1979. *Drilling Fluid Reference Manual.* 260 pp.
Moore, L.P., 1974. *Drilling Practices Manual.* Petroleum Publ. Co., Tulsa, Okla., 448 pp.
Muravyov, I., Andriasov, R., Gimatudinov, Sh., Govorova, G. and Polozkov, V., 1958. *Development and Exploitation of Oil and Gas Fields.* Peace (Mir) Publishers, Moscow, 503 pp. (in English.)
Murray, A.S. and Cunningham, R.A., 1955. Effect of mud column pressure on drilling rates. *Trans. Soc. Pet. Eng. AIME*, 204: 196—204.
Olson, R.M., 1973. *Essentials of Engineering Fluid Mechanics*, 3rd ed. Intext Educational Publ., New York, N.Y., 636 pp.
Outmans, H.D., 1963. Mechanics of static and dynamic filtration in the borehole. *Soc. Pet. Eng. J.*, Sept.: 236—244.
Prokop, C.L., 1952. Radial filtration of drilling mud. *Trans. Soc. Pet. Eng. AIME*, 195: 5—10.
Rieke, H.H. and Chilingarian, G.V., 1974. *Compaction of Argillaceous Sediments.* Elsevier, Amsterdam, 424 pp.
Rogers, W.F., 1963. *Composition and Properties of Oil Well Drilling Fluids.* Gulf Publ. Co., Houston, Texas, 818 pp.

Sharma, S.M., Lall, R.C., Mathur, R.M. and Chilingarian, G.V., 1980. Effect of mud additives on the permeability of mud cakes. *Energy Sources*, 5(2): 171—181.

Shatsov, N.I., Federov, V.S., Kuliev, S.M., Ioannesyan, R.A., Shishenko, R.I., Glikman, L.S. and Balitskiy, P.V., 1961. *Drilling of Oil and Gas Boreholes.* Gostoptekhizdat, Moscow, 666 pp.

Sifferman, T.R., Myers, G.M., Haden, E.L. and Wahl, H.A., 1974. Drill-cutting transport in full-scale vertical annuli. *J. Pet. Technol.*, 26(11): 1295—1302.

Simpson, J.P., 1971. Drilling fluids — today and tomorrow. *J. Pet. Technol.*, 23(11): 1294—1298.

Simpson, J.P., 1979. A new approach to oil base muds for low cost drilling. *J. Pet. Technol.*, 31, May: 643—650.

Tiller, F.M. and Cooper, H., 1962. The role of porosity in filtration: Part V. Porosity variation in filter cakes. *AIChE J.*, 8(4): 445—449.

Tschirley, N.K., 1978. *Manual of Drilling Fluids Technology.* Baroid Petroleum Services Co., Houston, Texas, 200 pp.

Vadetskiy, Yu.V. and Shalimov, I.F., 1956. *Drilling of Oil and Gas Boreholes.* Gostoptekhizdat, Moscow, 418 pp.

Vennard, J.K., 1961. *Elementary Fluid Mechanics*, 4th ed. Wiley, New York, N.Y., 570 pp.

Wallace, W.E., 1965. Application of electric log measured pressure to drilling problems and a new simplified chart for wellsite pressure computation. *The Log Analyst*, 6: 4—10.

Watts, E.V., 1948. Some aspects of high pressures in the D-7 zone of the Ventura Avenue Field. *Trans. Am. Inst. Min. Metall. Eng., Pet. Div.*, 174: 191—200.

Williams, C.E., Jr. and Bruce, G.H., 1951. Carrying capacity of drilling fluids. *Trans. Soc. Pet. Eng. AIME*, 192 (3026): 111—120.

Williams, M., 1939. Radial filtration of drilling muds. *Trans. AIME*, 136: 55—68.

Williams, M. and Cannon, G.E., 1938. Filtration properties of drilling muds. *API Dril. Prod. Prac.*, also *Oil Weekly*, 90(2): 25.

Yatrov, S.N., 1963. *Natural Drilling Fluids.* Gostoptekhizdat, Moscow, 206 pp.

Zhukhovitskiy, S.Yu., 1955. *Clay Drilling Fluids.* Gostoptekhizdat, Moscow, 171 pp.

Chapter 3

FACTORS AFFECTING THE RATE OF PENETRATION *

JAMES H. ALLEN

INTRODUCTION

From 1948 to 1968, a great amount of study and research was conducted on the basic factors that affect the rate of penetration. Theoretical, experimental and field test studies have been made to maximize the drilling rate in order to minimize drilling costs and improve the operating efficiency. The basic factors affecting the penetration rate that have been studied and analyzed are: (1) bit selection, (2) bit weight or drilling force, (3) bit rotational speed, (4) bit hydraulics, (5) drilling fluid properties, and (6) formation or rock properties.

As new rotary drilling technology, materials, and equipment are developed, drilling industry engineers continue to review and reevaluate the above factors as long as possibilities exist for the improvement of drilling operations and the reduction of drilling costs.

BIT SELECTION

It has been stated that the best bit is that bit which, when utilized with the proper weight, rotational speed, hydraulics, and drilling fluid properties, will achieve the lowest cost per foot of hole drilled.

Bit selection has been complicated to some degree by the many changes in rock bit design since the early 1950's. Drilling personnel who are involved in bit selection should have a knowledge of the various basic design criteria that differentiate soft formation bit types from hard formation bit types. Such design factors as journal angle, cone offset, and cone profile variation all influence cutting action. The journal angle of a rock bit is defined as the angle formed by a line perpendicular to the axis of the journal and a line through the axis of the bit. Figure 3-1 illustrates a rock bit journal angle.

The ordinary (non-sealed bearing) rolling-cutter rock bit consists of a bit body and three cones or cutters. The cones are mounted on journals and are supported by ball and roller bearings, as shown in Fig. 3-2. The cones rotate freely when they are above the hole bottom. As soon as the bit touches the hole bottom, the cutter teeth engage the formation and rotation of the bit

* Edited by George V. Chilingarian.

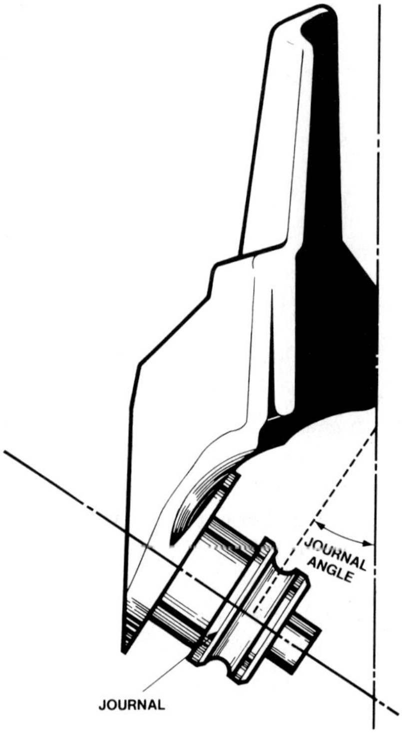

Fig. 3-1. Rock bit journal angle. (Courtesy of Sii Smith Tool, Division of Smith International, Inc.)

causes the cutters to rotate. The outer rows of cutter teeth have a true roll point greater than that of the inner rows of cutter teeth. This variation in cone profile (Fig. 3-3) produces a scraping action when the cutters rotate. Additional scraping action is caused by cutter offset. Offset is defined as the horizontal distance between a line through the axis of the bit and a vertical plane through the axis of the journal. The cutter journals are "offset" slightly forward of the bit center as shown in Fig. 3-4. Such design criteria as journal angle, cone offset, and cone profile variation have not changed since their first presentation by Bentson in 1956.

Soft formation bits with 33° journal angles, 3/8- to 1/4-inch cone offset, and maximum variation in cone profile drill rocks with a gouging—scraping action. Hard formation bits with 36° journal angles, little or no offset, and minimum variation in cone profile drill rocks with a chipping—crushing action.

The widespread usage of tungsten carbide insert, journal-bearing bits (Fig. 3-5) has reduced drilling costs since their introduction in the early 1970's. Improvements in metallurgy, thermal treatment techniques, bearing

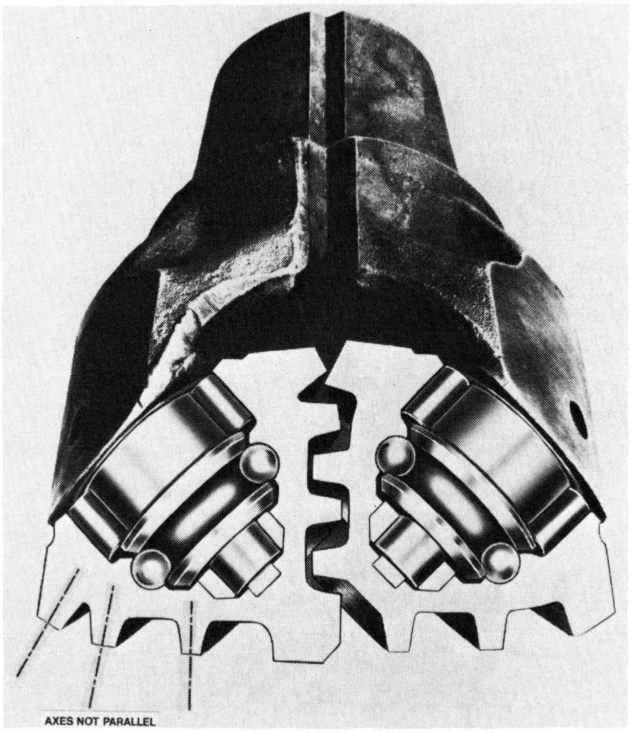

Fig. 3-2. Rolling-cutter rock bit. (Courtesy of Smith Tool, Division of Smith International, Inc.)

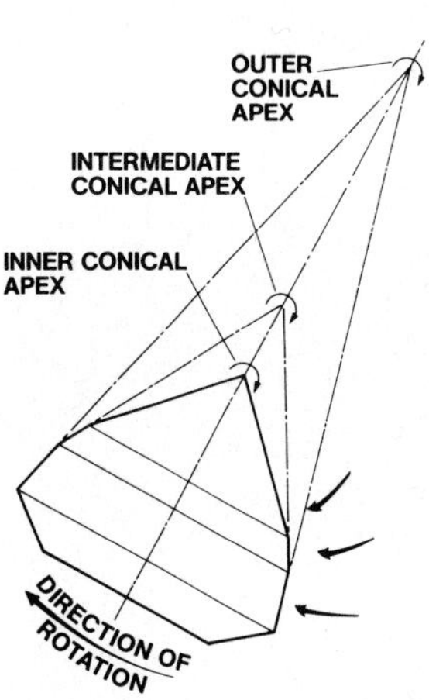

Fig. 3-3. True roll paths for cone profiles (variation in cone profile produces a scraping action upon rotation of cutter). (Courtesy of Smith Tool, Division of Smith International, Inc.)

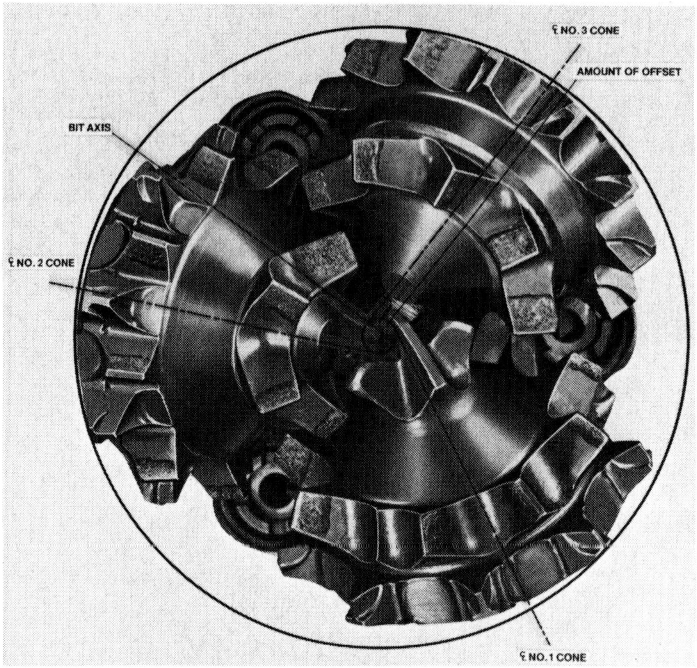

Fig. 3-4. "Offset" of cutter journals in rolling-cutter rock bit. (Courtesy of Smith Tool, Division of Smith International, Inc.)

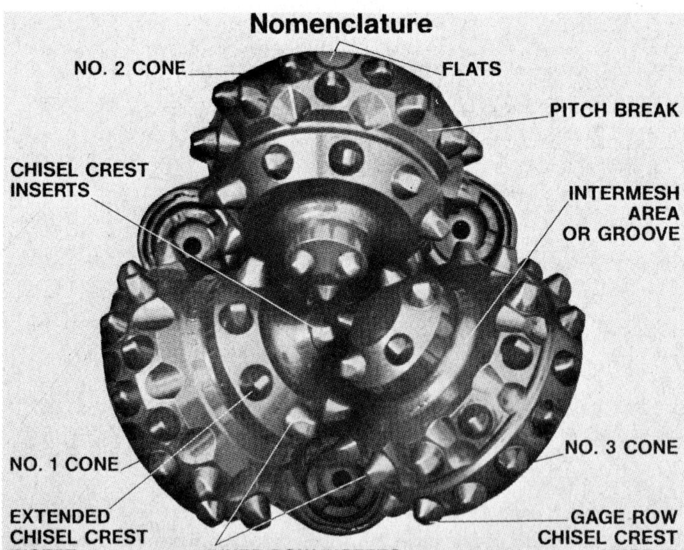

Fig. 3-5. Tungsten carbide insert, journal-bearing bit. (Courtesy of Smith Tool, Division of Smith International, Inc.)

seals, and tungsten carbide insert grades have made it possible to replace many milled tooth bit types. Many of the design features of milled tooth bits have been incorporated into the design of soft formation insert bits, so that a variety of insert bit cutting structures are now available for a wide range of formation types. Technological developments in journal bearing bits, such as bearing seals, pressure compensation and lubrication systems,

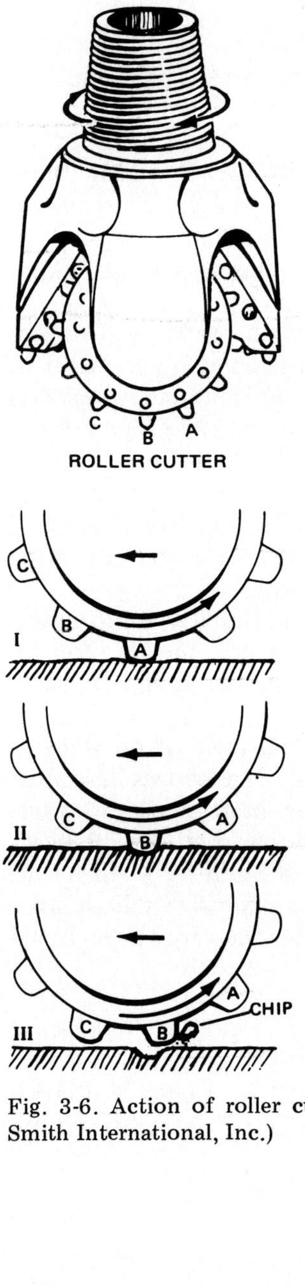

Fig. 3-6. Action of roller cutter on rock surface. (Courtesy of Smith Tool, Division of Smith International, Inc.)

and cutting structures, are described and depicted in the various bit manufacturers' catalogs.

The operation of roller cutter is presented in Fig. 3-6. The roller cutter tooth A, in a position I, is on the rock surface as rotation begins. As the tooth A rotates above the rock surface (position II), the tooth B rotates down into the rock and displaces a volume of rock by force. In position III, tooth B rotates upward and forces the particle of rock previously loosened to be dislodged. This particle becomes a "chip", which is carried to the surface by a drilling fluid.

At the present time there are at least eight separate methods which are employed around the world for bit selection. Quite often, combinations of these methods with literature aids are used. The following methods for bit selection are used at the present time: (1) dull bit evaluation, (2) offset well bit records, (3) offset well logs, (4) cost calculations, (5) seismic data, (6) computer programs, (7) mud properties, and (8) geologic time tables. Literature aids to bit selection include: (1) general product catalogs, (2) product feature bulletins, (3) product comparison charts, and (4) IADC bit classification code (Kastrop, 1972).

Dull bit evaluation is the oldest of the bit selection methods. Description of dulled milled tooth bit conditions were discussed by Bentson in 1956, by Bobo et al. in 1954, and by the various bit manufacturers. The IADC system of dull bit grading was first adopted in 1963 (Knowlton, 1961, 1964; Garner, 1971; and Jackson and Woods, 1973) for milled tooth bits and revised in 1971 for insert bits.

Use of offset bit records can be very valuable as a bit selection method if the dull bit condition is accurately reported. Offset bit records are the basis for 90% of the input data for drilling assistance programs.

Offset well logs are used not only for obtaining lithological information and types and amounts of various formation fluids present, but also for the evaluation of formation properties, such as hardness, porosity, and abrasiveness.

A method of using induction, sonic, and density logs from offset wells to select optimum bit types and predetermine their performance was described by Rios and Montiel (1974). Engineers from Pemex in Reynosa, Mexico, have concluded that information from sonic logs can be used for bit selection and to determine the bit life and expected penetration rate.

Cost calculations are necessary when comparing less expensive milled tooth bits with tungsten carbide insert bits. The basic cost per foot equation can be presented as follows:

$$C_1 = \frac{B_1 + R_1(T_1 + t)}{F_1} \tag{3-1}$$

where C_1 = cost per foot using control bit (dollars/ft); B_1 = cost of control bit (dollars); R_1 = rig cost or operating rate (dollars/hr); T_1 = rotating time for control bit (hr); t = round trip time (hr); and F_1 = footage drilled by control bit (ft).

A second cost calculation method to determine the bit performance required of a more expensive bit compared to a less expensive bit, based on equal cost per foot and rate of penetration, involves the following equation:

$$T_2 = \frac{B_2 + R_1(t)}{C_1(F_1/T_1) - R_1} \tag{3-2}$$

where T_2 = rotating time for proposed bit (hr) and B_2 = cost of proposed bit (dollars).

Seismic data from remote locations has been used for bit, hydraulic, and mud programs. The editor of Oilweek (1975) has described a method developed by Gulf Oil Canada Ltd., in which seismic data is correlated with lithology predictions and used for bit selection and formation drillability estimation.

Amoco Production Company of Tulsa, Oklahoma, has designed computer programs to provide the operator and contractor with the capability of analyzing conditions as drilling progresses, so that adjustments can be made in a basic well program to take into account unexpected formation changes or handle unforeseen drilling problems as they occur. These programs were also developed to provide the drilling engineer, toolpusher, or drilling foreman with the necessary tools to make logical day-to-day decisions as drilling progresses. They are based on large data banks which contain drilling and geological information and have the capability of generating recommendations based on a wide range of considerations.

About 25—30 individual programs are available through a timeshare network consisting of a computer terminal, telephone communication, and foreground and background computer systems. For some international locations, the earth satellite communications system is used in the network. The programs are divided into five categories: (1) drilling operations, (2) equipment evaluation, (3) economics, (4) deviation control, and (5) well control safety. Many of the programs calculate answers no differently than those that can be done by hand, and a simple question-and-answer format is used for the program language. The programs, designed to take into account changing conditions, can provide cost per foot answers using the following variables: hole size, mud type, pump pressure, bit type, pump volume, weight—rotational speed (rpm), nozzle size, depth, drilling fluid properties, and formation hardness.

At least three of the Amoco drilling assistance programs can be utilized for bit selection. These programs are the "BUTTON, "MUDHY" and "BITOP" programs; each one is briefly explained and demonstrated in the Appendix 3-I of this chapter.

Bits are often selected for a particular mud program, mud type, or mud property. In one southeastern area oil field in the United States, both water-base and oil-base muds are used in the same field. In one particular formation section, a hard formation insert bit is used with the water-base mud. If the oil-base mud is used in drilling this formation, penetration rate is lower and, consequently, a softer formation bit is used even though bit life is sacrificed to increase the rate of penetration.

In the absence of more reliable information, the geologic time table has

TABLE 3-I

Correlation of bit types with geologic time (Courtesy of Smith Tool, Division of Smith International, Inc.)

Era	Periods	Formations	Bit Types
	Quaternary		
	Present	Clay, sands, gravel	
	Pleistocene	Clays to sands and gravels	
	Tertiary		
Cenozoic	Pliocene	Hard clays, sticky clays, sand	Medium to long widely spaced teeth
	Miocene	Shales, sticky clays, soft sandstone, some limestone	
	Oligocene	Shales, sands, limestones, generally soft	
	Eocene	Soft to hard sands, shale and limestone	
		Siliceous rock	Medium to very short teeth
		Chert	
	Upper Cretaceous	Shale, sand; sandy, shale, limestone, chalk	Medium to long teeth widely spaced
	Lower Cretaceous	Calcareous shale, chalk, limestone	
	Jurassic	Shales	
Mesozoic	Triassic	Limestones	Medium to short teeth
		Hard sandy lime	Short teeth
		Iron pyrite	Very short, strong teeth, or, Tungsten-carbide compacts
		Cherts in hard limestones	
		Quartzitic sand	
	Permian	Red waxy shale	Long teeth, widely spaced
		Limestone	
	Pennsylvanian	Gypsum, sandstone	
	Mississippian	Salt	Short, strong teeth
	Devonian	Anhydrite, shale	Short teeth
Paleozoic	Silurian	Limestone	Close, short teeth
	Ordovician	Sandy limestone	
	Cambrian	Dolomite	
		Sandy dolomite	Very short, strong teeth, closely spaced
		Cherty dolomite	
		Novaculite	Tungsten carbide compacts
		Quartzite	

been used for bit selection in bit programs. A correlation of the bit types with the geologic age of formations is shown in Table 3-I.

Basic bit selection considerations are provided in most rock bit general catalogs. Articles that have been written on bit selection in recent years include those by Murphy (1969), Estes (1971, 1974), Jackson (1972), and Allen (1972, 1973, 1975a).

BIT WEIGHT AND ROTATIONAL SPEED

Inasmuch as bit weight and rotational speed considerations are interrelated, i.e., an increase in one usually necessitates a reduction in the other to achieve the minimum cost of drilling, they are usually discussed together.

The rate of penetration increases with increasing mechanical energy level on a bit (or increasing bit weight and rotational speed), if there is sufficient hydraulic energy (fluid volume and pressure loss) available for bottom-hole cleaning. Increases in bit weight and rotational speed, however, accelerate wear of bit cutting and bearing structures. In soft formations, a doubling of either bit weight or rotational speed will double drilling rate if sufficient bit hydraulic horsepower is available. In hard formations, the weight on the bit has to be sufficient to overcome the compressive strength of the rock before bit weight and rotational speed increases will increase the rate of penetration. After the rock compressive strength is exceeded, increasing bit weight by a factor of two doubles, or more than doubles, the penetration rate. The drilling rate is not linearly proportional to the rotary speed in hard formations because some finite time is required for a bit tooth to fracture the rock.

Most milled tooth drilling operations in the United States of America utilize a constant rotational speed with a variable or increasing bit weight for lowest drilling costs. Most shale formations are more responsive to rotational speed increase than they are to bit weight increases. Harder formations such as limestones are more responsive to the bit weight increase than they are to the rotational speed increases. For longest bit life in sandstone formations, rotational speed should be decreased and bit weight increased.

With the introduction of tungsten carbide, chisel crest insert bits in the late 1960's for drilling in softer formations, weight and rotational speed considerations were altered. Lower energy levels (less weight and lower rotational speed) were required to avoid insert chippage and breakage. Insert bits, however, have eliminated most of the tooth wear considerations that were necessary in optimized drilling prediction techniques for milled tooth bits. The lower energy levels required for insert bits dictated reevaluation of stabilization equipment, bottom hole assemblies, and straight-hole drilling practices.

Improvements in tungsten carbide insert shapes and grades have been made in the last few years. These improvements have eliminated a lot of the earlier insert chippage and breakage problems. Recent high rotary speed tests in the North Sea and Midcontinent in U.S.A. have proved that soft-formation bits with long insert extensions can achieve much higher rates of penetration in soft formations. Some of the test bits have been rotated at speeds as high as 160—180 rpm without insert breakage. In west Texas, higher weights are used with hard formation insert bits. At least two U.S.A. bit manufacturers have new facilities for manufacturing their own tungsten carbide inserts.

The exact number of contemporary methods for determining optimum bit weight and rotary speeds is difficult to ascertain at present. Many of the data acquisition and computation units, that are used basically for mud monitoring and detection of overpressured zones (see Chapter 16), also acquire data that may be used in bit weight—rpm calculations. Such units have been developed by the Analysts (Boone, 1976), Baker, Baroid (Young, 1969; Taylor, 1975), Magcobar, Halliburton, Martin-Decker, and other companies.

The most commonly used methods for bit weight and rotary speed determinations are as follows: (1) constant-energy drilling method, (2) Galle—Woods method, (3) computer programs, and (4) drill-off tests. Supplemental aids to bit weight and rotational speed determination methods are: (1) weight and rpm recommendations from bit suppliers in the form of bulletins, technical articles, bit catalogs, etc.; and (2) electric logs, such as the SP log, to provide information on formation type and changes in lithology.

The constant-energy drilling prediction technique developed by Fullerton and Fullerton (1965) and Fullterton and Mechem (1973) was first used for milled tooth bits in soft formations. It has also been used as a reliable technique since soft formation insert bits have been introduced. The method is predicated on the concept that a single expression of drilling energy level [the W (weight/inch of bit diameter, lb/inch) $\times$ R (bit rotational speed, rpm) product] can be related by graphical methods to the variables of rate of penetration, formation drillability, bit weight, rotary speed, well depth, mud pressure, and applied hydraulics. Within the normal bit weight—rotational speed operating ranges, any $W \times R$ value may be considered a constant, i.e., in a given interval, bit weight and speed may be varied considerably without changing the average rate of penetration as long as the product of the variables remains constant.

Drilling parameters chart (Fig. 3-7) provides the relationship between bit weight per inch of diameter (W, lb/inch), rotary speed (R, rpm), WR product, rate of penetration (ROP, ft/hr), and formation drillability factor (K_f). This chart is based on the following drilling rate equation:

$$ROP = K_f WR \tag{3-3}$$

A chart for the adjustment of formation drillability factor for differences

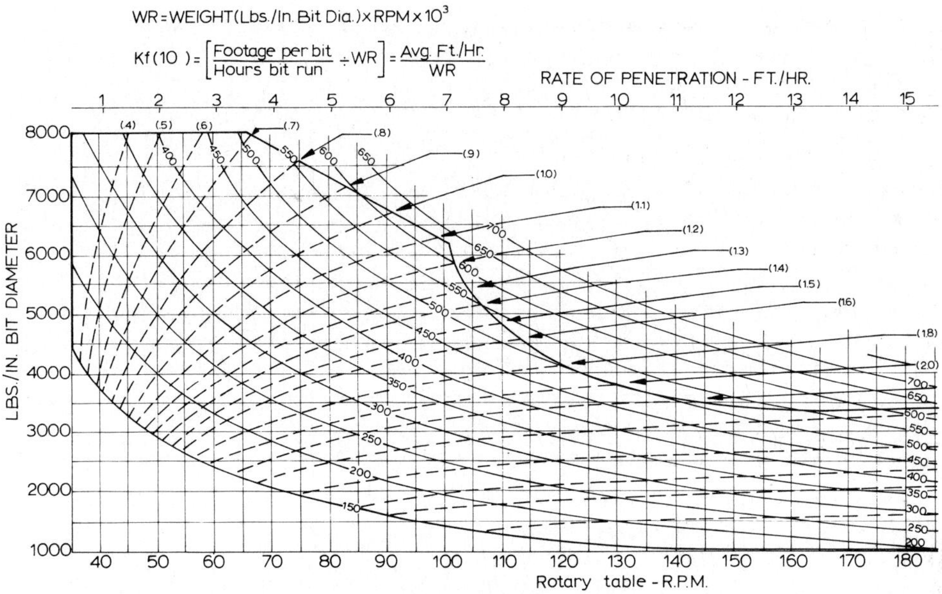

Fig. 3-7. Drilling parameters chart. (After Allen, 1977, fig. 5, p. 9; courtesy of Smith Tool, Division of Smith International, Inc.)

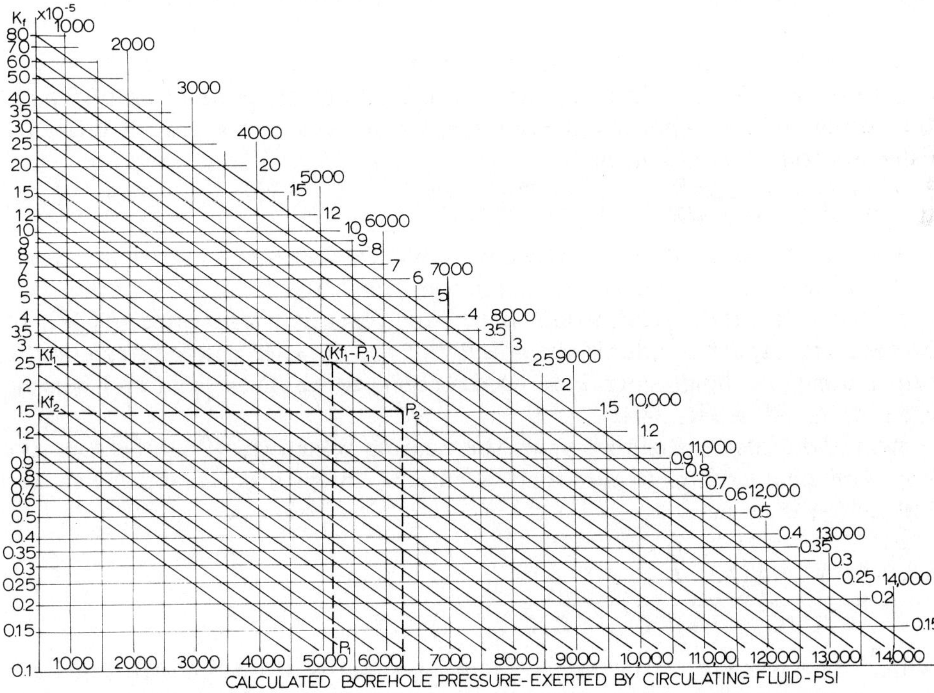

Fig. 3-8. Chart for adjustment of formation drillability factor for differences in mud specific weight or hole depth. (After Allen, 1977, fig. 6, p. 10; courtesy of Smith Tool, Division of Smith International, Inc.)

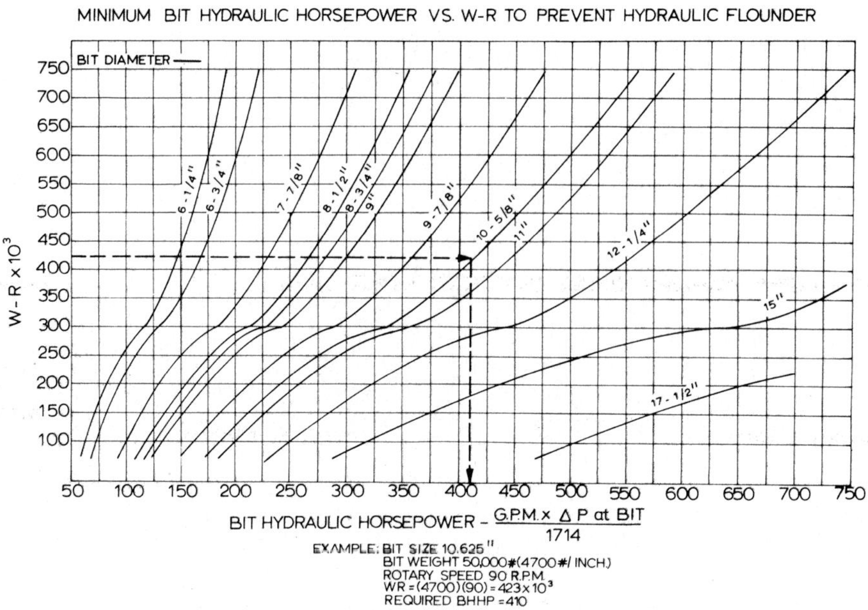

Fig. 3-9. Fullerton's relationship between hydraulic horsepower and *WR* to prevent hydraulic flounder. (After Allen, 1977, fig. 7, p. 10; courtesy of Smith Tool, Division of Smith International, Inc.)

in mud density or hole depth is shown in Fig. 3-8. The correction of formation drillability for changes in mud density or hydrostatic head is made by using the following equation:

$$\log_{10} K_{f_2} = 0.000208(p_1 - p_2) + \log_{10} K_{f_1} \tag{3-4}$$

where K_{f_1} = apparent formation drillability factor; K_{f_2} = corrected formation drillability factor; and p = bottomhole pressure (psi).

Fullerton's (1973) relationship between hydraulic horsepower and *WR* to prevent hydraulic flounder is presented in Fig. 3-9, whereas the interrelationship among bit weight per inch of diameter, rotational speed, and bearing life is depicted in Fig. 3-10.

A mathematical relationship for the bearing wear, which can be used for bits with bearing life greater than 50 hours (insert bits), is as follows (see Fig. 3-11):

$$C_B = \frac{L_B}{\left(\dfrac{100}{W/1000 \times R/1000}\right) - \dfrac{W}{1000}} \tag{3-5}$$

where C_B = bearing wear constant, and L_B = hours of bit life (hr). The bearing wear constant used in this relationship is premised on the assumption

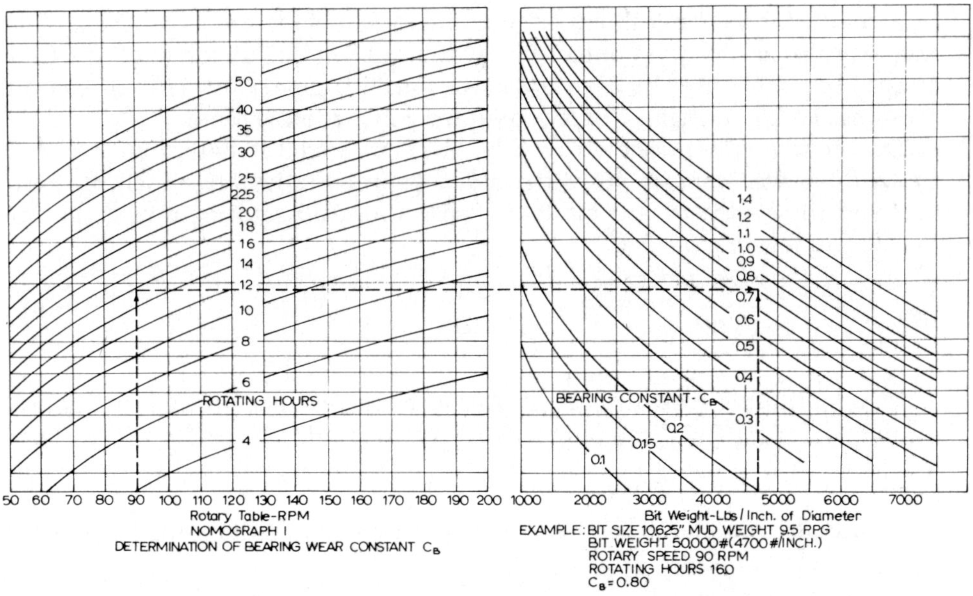

Fig. 3-10. Interrelationships among bit weight per inch of diameter, rotational speed, and bearing life. (After Allen, 1977, fig. 8, p. 11; courtesy of Smith Tool, Division of Smith International, Inc.)

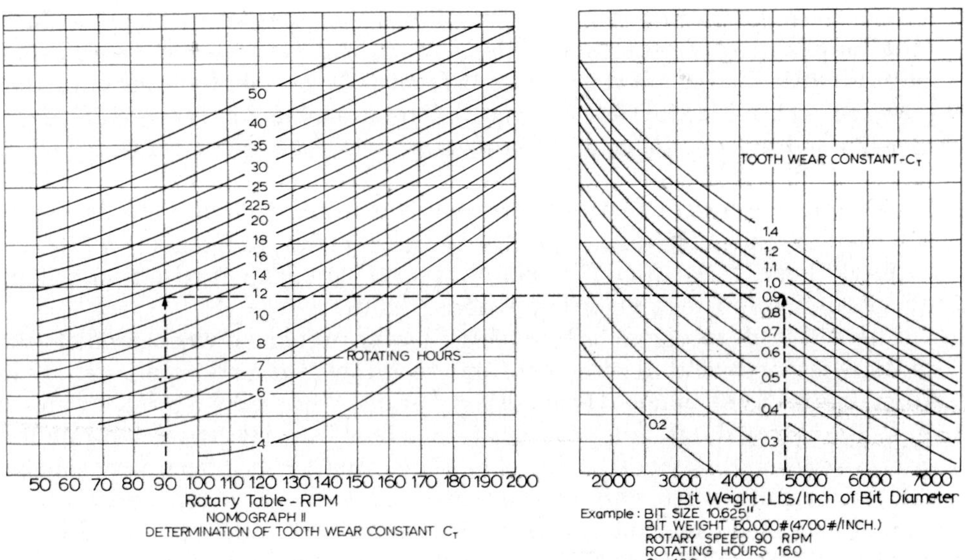

Fig. 3-11. Nomograph for determining milled tooth bit tooth wear or life with variations in bit weight per inch of diameter and rotary speed. (After Allen, 1977, fig. 9, p. 11; courtesy of Smith Tool, Division of Smith International, Inc.)

that the proposed bit will experience the same degree of bearing wear that was experienced by the control bit. If the dull bit evaluation grading of the control bit is T5—B7—I (T5 = 5/8's of tooth life gone; B7 = 7/8's of bearing life gone; I = bit in gage), the proposed bit will grade the same.

The nomograph presented in Fig. 3-11 is used for determining milled tooth bit tooth wear or life with variations in bit weight per inch of diameter and rotary speed. For bit tooth wear above 50 rotating hours, the following formula can be used:

$$C_T = \frac{L_T \times e^{(0.01R + 0.00032W)}}{189.2} \qquad (3\text{-}6)$$

where C_T = tooth wear constant, and L_T = hours of bit life (hr).

The constant-energy drilling system provides a good drilling prediction technique for soft and medium-soft formations that are responsive to both weight and rotational speed increases. A particularly good feature is the correlation between bit and hydraulic energies.

A drilling rate equation that has been used for medium-hard and hard formations is as follows:

$$ROP = K_f W^{1.2} R^{0.5} \qquad (3\text{-}7)$$

The Galle-Woods drilling prediction technique, discussed by Galle and Woods (1960), was developed primarily for medium-hard and hard formations using milled tooth bits and should not be used for optimizing the performance of soft formation insert bits. Insert bits used for soft and medium-soft formations are utilized with less bit weight and rotational speed than milled tooth bits to prevent insert breakage. This technique assumes that adequate bit hydraulic horsepower will be available for the predicted bit weights and speeds. In using the program, it is also assumed that bit selection and drilling fluid properties were correct in the control well.

Computer programs that are currently being used to determine optimum bit weight and rotational speed are the Amoco programs "WTRPM", "BITOP", and "DRILLOFF". The "WTRPM" program is based on drilling equations developed by Galle and Woods (1960).

The "BITOP" program is one of the newer programs which predicts the drilling response and recommends insert bit weight and rotational speed from input data obtained from drilling either with milled tooth or insert bits. Bit life and rate of penetration are based on statistical information collected from thousands of bit runs. This program has been very reliable for bit programs where the soft-formation insert bits have replaced the soft-formation milled tooth bits; an example is included in the Appendix 3-I.

The "DRILLOFF" program is a calculation method to determine the actual optimum bit weight. It is based on the original drill-off tests of Lubinski (1958), but modified to use 4000 lb drill-off weight multiples.

There are several types of drill-off tests and flounder point determin-

TABLE 3-II

Bit weight and rotational speed recommendations for journal-bearing insert bits (After Garner and Allen, 1976)

BIT TYPE—F2

Formation: Softest Shales, Clays, and Salts

Bit Size Range	High rpm— Low Wt. Range		Low rpm— High Wt. Range	
		lb.		lb.
7⅞	90	12,000	55	28,000
8½ to 8¾	90	13,000	55	31-32,000
9⅞	90	15,000	55	37,000
12¼	85	19,000	55	44,000

Formations: Unconsolidated Soft Shale with Sand Streaks

Bit Size Range	High rpm— Low Wt. Range		Low rpm— High Wt. Range	
		lb.		lb.
7⅞	60	20,000	40	37,000
8½ to 8¾	60	21-22,000	40	41-42,000
9⅞	60	25,000	40	47,000
12¼	60	31,000	40	55,000

BIT TYPE—F3

Formations: Medium Soft Shale

Bit Size Range	High rpm— Low Wt. Range		Low rpm— High Wt. Range	
		lb.		lb.
6¼ to 6½	70	15-16,000	55	20-21,000
7⅞	70	23,000	55	32,000
8⅜ to 8¾	70	27-28,000	55	34-36,000
9½ to 9⅞	70	30-32,000	55	39-40,000
11	70	33,000	55	44,000
12¼	70	34,000	55	47,000

Formations: Soft and Medium Soft Limestone and Dolomite

Bit Size Range	High rpm— Low Wt. Range		Low rpm— High Wt. Range	
		lb.		lb.
6¼ to 6½	65	20-21,000	45	27-28,000
7⅞	65	31,000	45	39,000
8⅜ to 8¾	65	34-35,000	45	42-44,000
9½ to 9⅞	65	38-40,000	45	48-49,000
11	65	43,000	45	50,000
12¼	65	47,000	45	53,000

Formations: Unconsolidated Medium Soft Shales, Limestones, and Sands

Bit Size Range	High rpm— Low Wt. Range		Low rpm— High Wt. Range	
		lb.		lb.
6¼ to 6½	55	20-21,000	40	25-26,000
7⅞	55	31,000	40	35,000
8⅜ to 8¾	55	34-35,000	40	39-40,000
9½ to 9⅞	55	38-40,000	40	44-45,000
11	55	43,000	40	50,000
12¼	55	47,000	40	54,000

ation tests that can be made at the rig. Flounder point tests can be made by maintaining constant hydraulic conditions and rotary speed, but increasing weight. Rate of penetration is recorded and plotted against weight on bit at several depths. Flounder point is defined as that point where either an increase in bit weight or bit speed, or both, will no longer cause a straight-line increase in the rate of penetration. Optimum bit weight and bit speed at a specific hydraulic condition is usually at, or slightly greater than, the energy level at the flounder point. Five-point (or spot) drilling rate tests used to determine the rotary speed exponent in the Galle—Woods drilling rate equation have been abandoned in favor of the Lubinski drill-off tests. Lithological changes during the five-point tests can provide erroneous data (Taylor, 1975).

Bit manufacturers have collected and developed a large amount of bit weight and speed data on journal bearing bits. The bit weight and speed recommendations developed by Garner and Allen (1976) are shown in Table 3-II.

BIT HYDRAULICS

The most significant development in improving drilling efficiency has been the acceptance and widespread use of jet bits. Bottomhole cleaning has been improved and this has enabled the efficient application of maximum bit weight for increased rate of penetration. Kendall and Goins (1960) made important contribution to the knowledge of (1) rotary drilling hydraulics by considering the proper distribution of surface fluid pressures through flow rate balance, and (2) nozzle area necessary to develop maximum jet velocity, bit hydraulic horsepower, and jet impact.

The two basic steps in designing a hydraulics program are:

(1) Determination of the required bottomhole cleaning or bit hydraulic horsepower to balance the energy level of bit weight and rotational speed.

(2) Maximizing the bit hydraulic horsepower in relation to the available surface horsepower.

Most hydraulic pressure drop calculations are computed with hydraulic calculators or slide rules. The following empirical equations have been developed to calculate hydraulic programs:

(a) Surface connections, drillpipe, and drillcollar bore losses:

$$\Delta p = \frac{0.0000765 Q^{1.82} \rho^{0.82} \overline{PV}^{0.18} L}{D_b^{4.82}} \quad \text{(turbulent flow)} \qquad (3\text{-}8)$$

(b) Annulus losses:

$$\Delta p = \frac{0.0000765 Q^{1.82} \rho^{0.82} \overline{PV}^{0.18} L}{(D_h - D_p)^3 (D_h + D_p)^{1.82}} \quad \text{(turbulent flow)} \qquad (3\text{-}9)$$

(c) Nozzle size determination:

$$\Delta p_{\mathrm{b}} = \frac{\rho Q^2}{12031 A_{\mathrm{n}}^2 C_{\mathrm{n}}^2}\,; \qquad A_{\mathrm{n}} = \frac{Q}{104.2}\,\frac{\rho}{\Delta p_{\mathrm{b}}} \tag{3-10}$$

(d) Bit hydraulic horsepower:

$$BHHP = \frac{\Delta p_{\mathrm{b}} Q}{1714} \tag{3-11}$$

where Q = flow rate (gpm); D_{b} = inside pipe diameter (inches); ρ = fluid specific weight (lb/gal); $\overline{PV}$ = plastic viscosity (cP); L = pipe length (ft); D_{h} = hole diameter (inches); D_{p} = outside pipe diameter (inches); Δp = pressure loss (psi); Δp_{b} = bit pressure loss (psi); A_{n} = total nozzle area (sq. inch.); and C_{n} = nozzle coefficient.

A hydraulic program calculation for an $8\frac{3}{4}$-inch bit in drilling to a depth of 10,000 ft with a 10-lb/gal mud is presented in Fig. 3-12. Hydraulic horsepower of 288 is available for bit cleaning at a penetration rate of 30 ft/hr. One should also note that the circulation rate used in the hydraulic calculation depicted in Fig. 3-13 is too high. This high pump volume has resulted in excessive pressure losses in the circulation system and the bit hydraulic horsepower is only 13% of the pump output horsepower. An optimum bit hydraulic horsepower ranges from 50 to 60% of the pump horsepower.

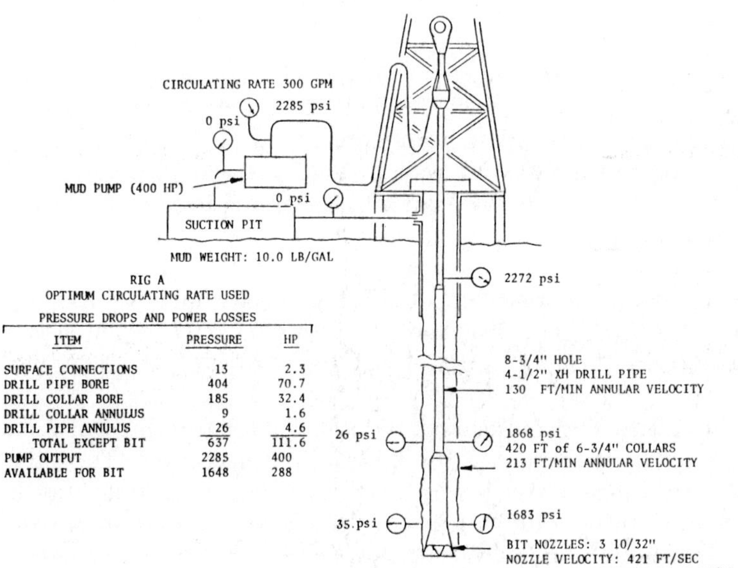

PRESSURE DROPS AND POWER LOSSES

ITEM	PRESSURE	HP
SURFACE CONNECTIONS	13	2.3
DRILL PIPE BORE	404	70.7
DRILL COLLAR BORE	185	32.4
DRILL COLLAR ANNULUS	9	1.6
DRILL PIPE ANNULUS	26	4.6
TOTAL EXCEPT BIT	637	111.6
PUMP OUTPUT	2285	400
AVAILABLE FOR BIT	1648	288

Fig. 3-12. A hydraulic program calculation for an $8\frac{3}{4}$-inch bit in drilling to a depth of 10,000 ft with a 10-lb/gal drilling fluid; circulating rate = 300 gpm at 2285 psi. (Courtesy of Sii Smith Tool, Division of Smith International, Inc.)

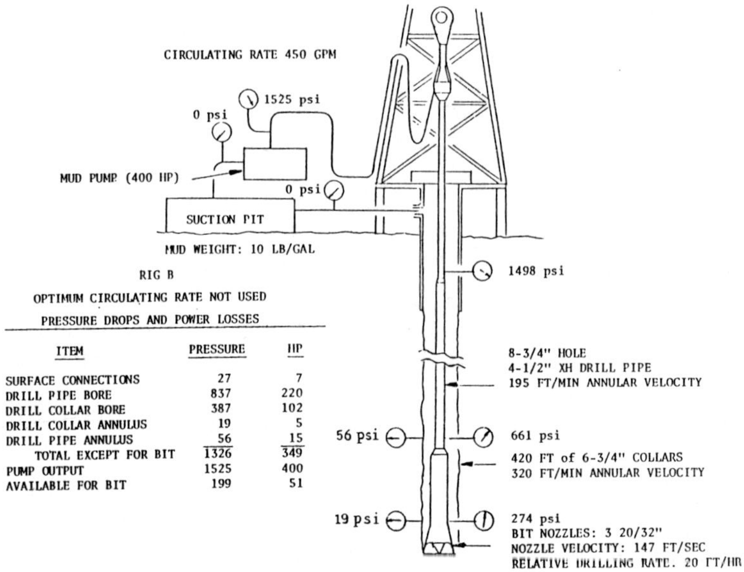

PRESSURE DROPS AND POWER LOSSES

ITEM	PRESSURE	HP
SURFACE CONNECTIONS	27	7
DRILL PIPE BORE	837	220
DRILL COLLAR BORE	387	102
DRILL COLLAR ANNULUS	19	5
DRILL PIPE ANNULUS	56	15
TOTAL EXCEPT FOR BIT	1326	349
PUMP OUTPUT	1525	400
AVAILABLE FOR BIT	199	51

Fig. 3-13. A hydraulic program calculation for a $8\frac{3}{4}$-inch bit in drilling to a depth of 10,000 ft with a 10-lb/gal drilling fluid; circulating rate = 450 gpm at 1525 psi. (Courtesy of Sii Smith Tool, Division of Smith International, Inc.)

Many optimum bit weight and rotary speed prediction methods assume that ample horsepower is always available. Fullerton and Mechem (1965) and Fullerton and Fullerton (1973) developed a correlation between minimum bit hydraulic horsepower and bit weight—rotary speed to prevent hydraulic flounder. This correlation provides means of predicting or calculating the rate of penetration based on the available bit hydraulic conditions (Fig. 3-9).

The "hydraulic drillability curves" shown in Fig. 3-14 were developed by Amoco Production Research and discussed by Lummus (1974). There is a direct correlation between drilling response and specific bit hydraulic horsepower.

Allen (1975b, Sii Smith Tool) discussed in detail the relationship between the bit weight—rotary speed and the bit hydraulic horsepower. Tungsten carbide insert bits are utilized with lower bit weights and rotational speeds than milled tooth bits. Insert bits with lower energy level requirements do not require maximum hydraulic conditions that can be provided with newer and larger mud pumps.

Randall (1975) discussed an approach to hydraulic optimization without the aid of available hydraulic computer programs. He also discussed the advantages of using only two jets to increase bottomhole cleaning.

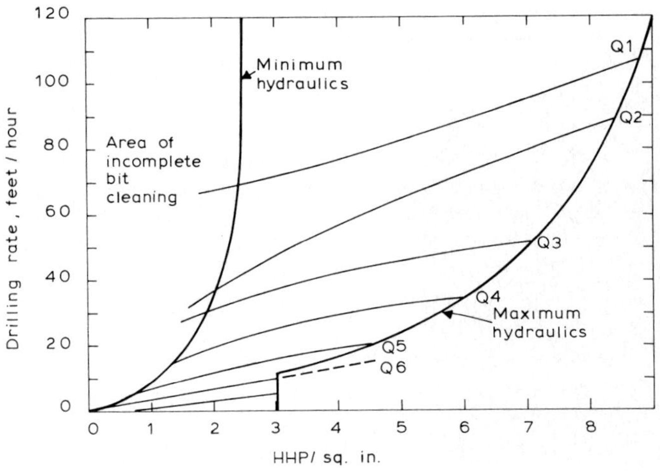

Fig. 3-14. Amoco "hydraulic drillability curves". (After Amoco Production Research.) First, a point is placed where control bit run is located. Then, using ship curve, the Q-line is fitted below this point. Finally, ship curve is moved up to the point. Thus, effect of a change in hydraulic conditions can be traced.

Crowley (1976) has written an article on how to update hydraulics with changing conditions during the drilling operations.

Various bit manufacturers and oil companies have developed and utilized extended jet tubes on bits to improve bottomhole cleaning. Figure 3-15 depicts a soft-formation bit with extended jet nozzles. A recently introduced two-cone insert bit with extended nozzles for soft-formation drilling with optimum bottomhole cleaning is shown in Fig. 3-16. These two bits were described and discussed by Garner and Millar (1976).

At present, hydraulic calculations are made by using the following tools: (1) bit hydraulic calculators (slide rules); (2) hydraulic tables or charts, and (3) computer programs. Bit hydraulic calculators are supplied by the Security, Reed, Smith, and IMCO companies, whereas hydraulic tables or charts are supplied by the Hughes and Security companies.

Most hydraulic calculators and tables enable calculation of the pressure losses through the use of the Fanning equation for turbulent flow. Inasmuch as turbulent flow may not occur in the drillpipe annulus in very deep wells, calculator errors could result in drillpipe annular losses.

Three excellent computer programs are provided by the Amoco Drilling Assistance Service:

(1) "HYD.ANAL" which computes actual hydraulic condition at the bit for the analysis of a particular run. Tapered drillstring feature is available.

(2) "HYDRAUL" which optimizes rig hydraulics, relating cross flow and hydraulic horsepower to penetration rate. Tapered drillstring feature is available.

Fig. 3-15. Soft-formation bit with extended jet nozzles. (Courtesy of Sii Smith Tool, Division of Smith International, Inc.)

(3) "MUDHY" which calculates the effects of mud properties and hydraulics on penetration rate and drilling cost.

Examples of the Amoco Drilling Assistance computer programs are presented in the Appendix 3.I of this chapter.

The types of hydraulic programs that are most commonly utilized are the maximum bit hydraulic horsepower method and the maximum jet impact method. These programs were discussed by Kendall and Goins (1960) and Randall (1975).

Often, rule-of-thumb values for annular velocity or flow rates are used as a starting point in commencing a hydraulics calculation. Figure 3-17, which is a plot of annular velocity versus hole diameter for various mud weights, was developed by Fullerton and Mechem (1965). Their plot is based on the relationship of mud density to (1) plastic viscosity, (2) yield point, and (3) hole size, which, in turn, are related to apparent viscosity. Other assumptions are that annular velocity can be determined from mud weight, hole size, chip diameter, and chip slip rate.

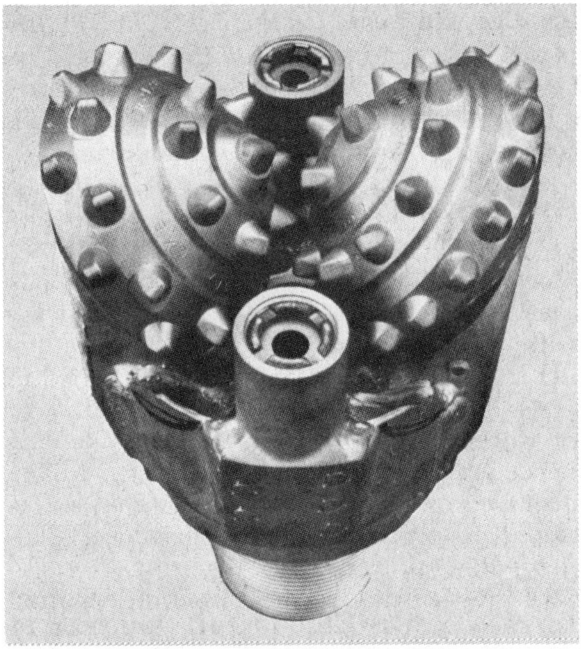

Fig. 3-16. Two-cone insert bit with extended nozzles for soft-formation drilling with optimum bottomhole cleaning. (Courtesy of Sii Smith Tool, Division of Smith International, Inc.)

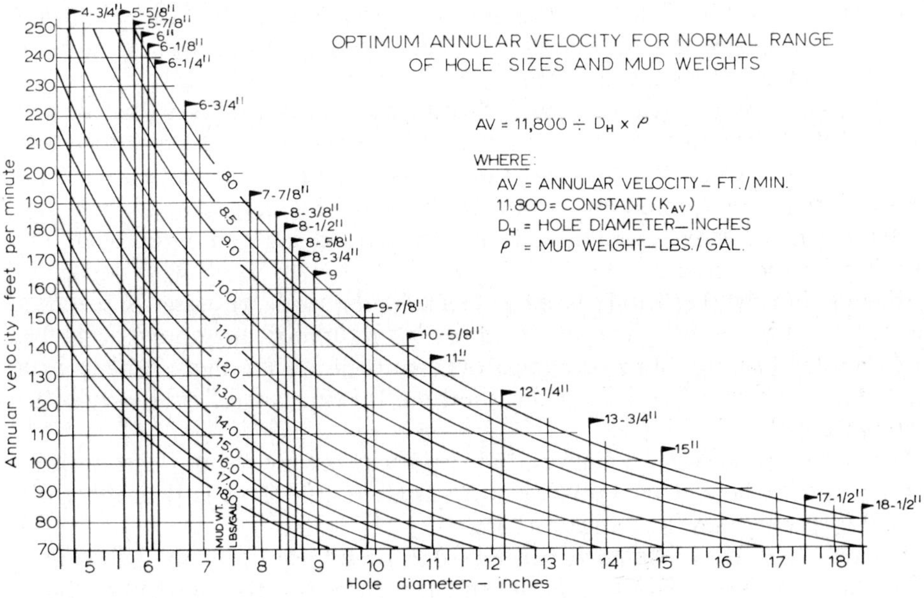

Fig. 3-17. Fullerton's graph relating annular velocity to hole diameter for various mud weights. (After Allen, 1977, fig. 10, p. 13; courtesy of Sii Smith Tool, Division of Smith International, Inc.)

Two-jet drilling, which was used advantageously in some areas in the early 1960's, has increased in several midcontinent areas of the U.S.A. in 1977—1978. The two-jet method (with the third jet plugged) is believed to provide better bottomhole cleaning in deep holes where high mud weight is used. The method is also used in the case of older or lower-pressure mud pumps.

DRILLING FLUID PROPERTIES

From 1958 to the present time, a large amount of research and study has been devoted to the determination of drilling fluid property effects on the rate of penetration (see Chapter 2). These studies are well documented in over 200 technical and trade journal articles. Moore (1974) stated that at various times papers have been written suggesting that almost any drilling fluid property has some influence on bit penetration rate. Many drilling fluid property studies have been based on laboratory tests; however, subsequent field tests (drilling operations) have failed to confirm laboratory findings.

As mentioned in Chapter 2, the most commonly recognized drilling fluid properties that affect penetration rate are: (1) density; (2) solids content; (3) viscosity; (4) water loss, especially spurt loss; and (5) oil content.

Unnecessarily high drilling fluid density causes a high differential pressure between the hydrostatic pressure (mud column) and the formation pore pressure. Rock chips that are generated by the bit teeth in contact with the hole bottom are "held down" and are not easily or quickly removed. Rapid chip removal below the bit allows each successive tooth impact to be more effective and improves drilling rate.

Solids content is related to fluid density and viscosity. Very fine or submicron bentonitic particles have more detrimental effect on penetration rate than larger particles. It is believed that these small particles plug the minute fractures that are formed by bit tooth impact. This plugging action delays pressure equalization over the chip thickness, delays chip removal, and decreases drilling rate.

Increases in drilling fluid viscosity cause higher pressure losses in the fluid circulation system, which result in lower bit hydraulic horsepower and less bottomhole cleaning. Some investigators also believe that high-viscosity drilling fluids will not clean chips from the hole bottom as well as low-viscosity drilling fluids.

Low-solids mud with a high initial fluid loss (or spurt loss) can improve rate of penetration. This spurt or surge loss of the fluid is believed to be more important than the filtration loss.

In some formations, penetration rate can be increased by the addition of oil to the water-base muds. Oil concentrations of 10—15% in drilling fluids help prevent bit balling by hydratable clays and shales in some formations.

The oil also lubricates the bottomhole assembly, reduces hole wall friction, and increases effective weight on the bit.

Many engineers believe that the combination of hydraulic conditions and drilling fluid properties has a greater effect on the rate of penetration than any of the other combinations of controllable variables. Hydraulic conditions can be calculated by several methods, whereas many of the drilling fluid effects are quite difficult to establish and very few programs are available. Drilling fluid properties constitute the most important factor that affects the penetration rate.

The present methods used to establish the effects of drilling fluid properties on the rate of penetration are: (1) computer programs, and (2) hand calculation methods based on curve fitting of the results of various laboratory tests.

The "MUDHY" program can be used to determine the effects of drilling fluid weight and contents of bentonitic solids, oil, barite, polymers, and chemical thinners in conjunction with consideration of hydraulic conditions. An example of application of "MUDHY" program is shown in the Appendix 3.I.

The individual effects that mud properties have on the rate of penetration are often meaningless, because most mud properties are interrelated. For example, if oil is added to a water-base mud with only a slight change in density, the effects of bentonitic solids, viscosity, and water loss on the rate of penetration might all change. The following individual drilling fluid property effects are based on curve fits of the results of laboratory tests:

(1) Plastic viscosity effects:

$$ROP_2 = ROP_1 \times 10^{0.003(\overline{PV}_1 - \overline{PV}_2)} \tag{3-12}$$

where ROP_1 = initial rate of penetration (ft/hr); ROP_2 = final rate of penetration (ft/hr); $\overline{PV}_1$ = initial plastic viscosity (cP); and $\overline{PV}_2$ = final plastic viscosity (cP).

(2) Bentonitic clay effects:

$$ROP_2 = ROP_1 \times e^{0.051(Vol\%_1 - Vol\%_2)} \tag{3-13}$$

where $Vol\%_1$ and $Vol\%_2$ are initial and final volume percents of bentonitic clay in muds, respectively.

(3) Total solids effect:

$$ROP_2 = ROP_1 \times 10^{0.0066(Vol\%_1 - Vol\%_2)} \tag{3-14}$$

(4) Water loss effects:

$$ROP_2 = ROP_1 \times \left(\frac{WL_2 + 35}{WL_1 + 35}\right) \tag{3-15}$$

where WL_1 and WL_2 are the initial and final water losses (cm^3/30 min, corrected), respectively.

(5) Oil content effects (Vol% oil < 30%):

$$ROP_2 = ROP_1 \left[\frac{\sin(10.6 Vol\%_2 - 48.3) + 10.33}{\sin(10.6 Vol\%_1 - 48.3) + 10.33}\right] \tag{3-16}$$

(6) Total drilling fluid effects (density, viscosity, solids content and pressure loss) for depths ranging from 8000 to 12,000 ft:

$$ROP_2 = ROP_1\, e^{0.382(\gamma_1 - \gamma_2)} \tag{3-17}$$

where γ_1 and γ_2 are the initial and final mud specific weights in lb/gal, respectively.

The effects of hole size on the rate of penetration can be estimated as follows:

(a) For bit diameters up to $17\frac{1}{2}$ inches:

$$ROP_2 = ROP_1(d_1/d_2) \tag{3-18}$$

where d_1 = smaller bit diameter (inches) and d_2 = larger bit diameter (inches).

(b) For bit diameters from $17\frac{1}{2}$–36 inches:

$$ROP_2 = ROP_1(d_1/d_2) \times 1.25 \tag{3-19}$$

A correction for the formation drillability factor when a flounder condition is reached is as follows:

$$K_{f_2} = K_{f_1} \sqrt{(WR)_1/(WR)_2} \tag{3-20}$$

where K_{f_2} = corrected formation drillability factor, K_{f_1} = calculated formation drillability factor, WR_1 = initial energy level, and WR_2 = energy level at flounder condition.

FORMATION PROPERTIES

The unalterable factors that affect the rate of penetration are formation characteristics or properties. Fundamental studies in rock mechanics continue to better define rock properties and failure mechanisms. Formation properties that have been investigated are as follows (Gatlin, 1960, p. 114): (1) compressive strength, (2) hardness and abrasiveness, (3) overburden pressure, (4) porosity, (5) permeability, (6) pore pressure, (7) stickiness, (8) elasticity, and (9) rock temperature.

In general, the penetration rate varies inversely with rock compressive strength. Unconfined compressive strength values for a particular rock, however, cannot be used as a true measure of rock drillability.

The hardness and abrasive characteristics of a rock affect the bit life. Excessive bit weights and rotational speeds cause tooth and insert breakage in hard formations, whereas in abrasive formations excessive rotational speed causes tooth, insert, and bearing seal wear.

Generally, the rock strength increases with depth, mainly because of increasing overburden pressures. The drilling rate is greater in a porous rock than in a dense rock. Porous zones of the same formation usually have lower compressive strengths than the less porous sections.

The degree of equalization of hydrostatic (drilling fluid column) and formation pressures increases with permeability of the rock. The drilling rate is greater in formations with high permeability because pressures across rock chip thickness can be equalized more quickly. Formations containing incompressible fluids have a quicker response to pressure equalization and will usually drill faster than a formation containing no interstitial fluids or having low pore pressure.

Some clays and shales form a sticky and plastic mixture when wetted with a water-base mud. This mixture becomes embedded between bit teeth and reduces tooth penetration and drilling rate. Many shale formations tend to become more plastic than brittle when subjected to high drilling fluid column pressures and high formation temperatures.

The following tests are performed on cores to determine the drillability characteristics: (1) laboratory tests, (2) microbit drilling tests, and (3) penetrator tests. The properties measured in the laboratory tests include the following: (1) relative hardness (from sonic measurements), (2) unconfined compressive strength, (3) porosity, (4) permeability, and (5) abrasiveness. In addition, the rate of penetration is determined by using large laboratory drilling rigs.

Microbit drilling tests are made with a $1\frac{1}{4}$-inch diameter rolling cutter bit usually rotated at 55 rpm using 200 lb of weight. Rate of penetration and tooth wear are recorded at short intervals and the results are compared with those from test curves prepared for formations of known drillability. Estimates of rate of penetration, bit life, bit weight—rotary speed, and drilling cost per foot can be provided along with recommended bit type. Drilling fluid can be air, water, oil, emulsion, or clay-in-water dispersion.

Penetrator or drillability tests are made by applying a load to a penetrator (tooth or insert) until a crater is formed in the rock sample. A "drillability index" is calculated by dividing the crater depth by the applied penetrator load. The penetrator index is compared to indices for rocks of known drillability. These tests enable determination of bit type, approximate bit life, recommended bit weights, and average penetration rates. Drilling costs can also be estimated. Penetrator tests probably provide better drillability estimates for hard rocks using air as a circulating fluid.

APPENDIX 3.I — BIT AND HYDRAULICS COMPUTER PROGRAM

Situation

A bit and hydraulics program is being prepared for a 14,000 foot well. To 12,500 ft, an $8\frac{1}{2}$-inch hole will be drilled. The drilling engineer is attempting to reduce drilling costs between 10,000 and 12,500 ft in the $8\frac{1}{2}$-inch hole. He has a bit record, an electric log and mud reports from an offset well.

From the bit record he has the following information:

Bit size	$8\frac{1}{2}$ inch.	Pump pressure	2200 psi
Bit type	134	Rig cost	$ 150/hr
Depth out	10,000 ft	Drill pipe	5 inch O.D., 19.5 lb/ft, XH
Footage	300 ft	Drill collars	22 JTS, $6\frac{1}{4}$ inch × $2\frac{1}{4}$ inch
Hours	18	Weight	40,000 lb
Nozzle sizes	3 × 12′s	RPM	80
Mud weight	10.2 ppg	Bit grading	T5, B7, I

From the mud reports he has determined the following:

(1) The mud weight can be reduced to 10.0 ppg.

(2) The methylene blue count can be lowered to 14 lb per bbl from 18 lb per bbl.

(3) The oil content in the offset well had been 6% by volume.

The BUTTON program is run on a computer terminal to determine if an insert bit can replace the good milled tooth bit run on the offset well. The BUTTON program indicates that an insert bit drilling at a penetration rate of 12.3 ft/hr for 85.6 hours would equal the mill tooth drilling cost.

The HYD.ANAL program is run to determine the flow rate and other pertinent hydraulic data.

The HYDRAUL program is run to optimize the hydraulics for a mud weight of 10.0 ppg and the program recommends three 11/32-inch jets rather than three 12/32-inch jets. Also the pump volume and pressure can be reduced.

The first MUDHY program is run to compare hydraulics in the offset well to hydraulics in the proposed well. By lowering mud weight, pump volume and jet nozzle size, the nozzle velocity and rate of penetration can be increased. Approximately $ 0.87 per foot can be saved. By lowering methylene blue count and oil content, $ 0.92 per foot can be saved. The total savings from improved hydraulics and fluid properties is $ 1.79/ft.

The WTRPM program is run to determine if improvements can be realized by changing weight and rpm. Drilling costs can be reduced from $ 15.98/ft to $ 15.39/ft by running the proposed bit at 75 rpm and 50,000 pounds of weight, but the available drilling weight is only 42,000 pounds (from HYDRAUL).

BITOP is run to determine what type of insert bit is necessary to replace the milled tooth bit. It is indicated that a 537 insert bit has two operating possibilities and that a 527 bit has three possibilities.

WELL PROGRAM FLOW CHART

(Insert Drilling Conditions)

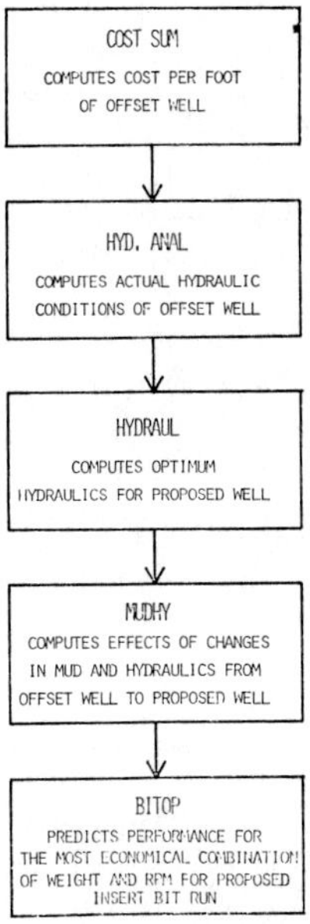

WELL PROGRAM FLOW CHART

(Milled Tooth Drilling Conditions)

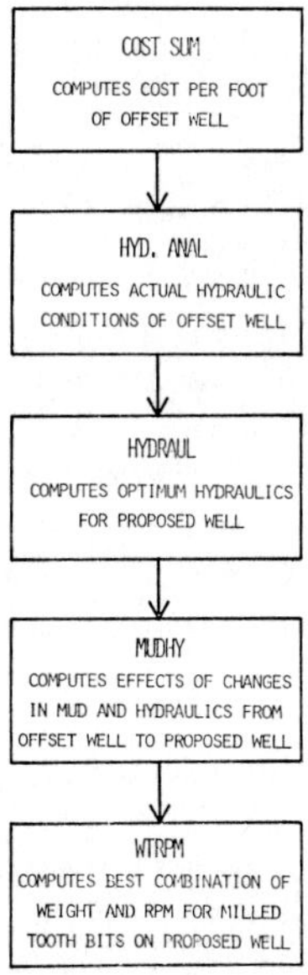

WELL PROGRAM FLOW CHART

(WHEN INSERT BITS CAN POTENTIALLY REPLACE MILLED TOOTH RUNS)

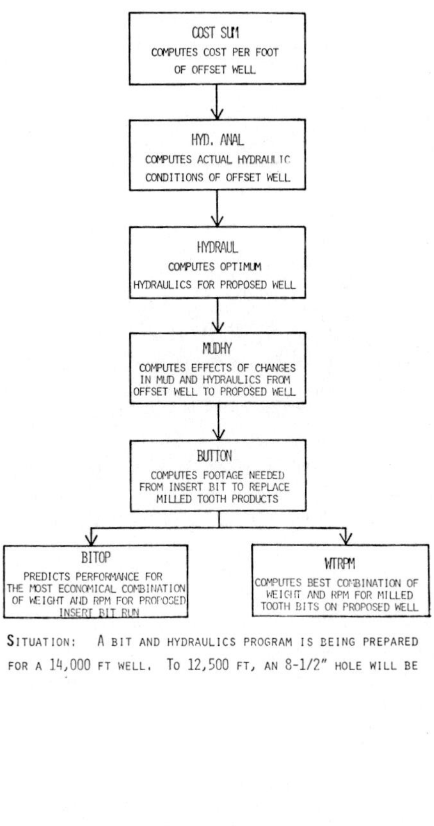

SITUATION: A BIT AND HYDRAULICS PROGRAM IS BEING PREPARED FOR A 14,000 FT WELL. TO 12,500 FT, AN 8-1/2" HOLE WILL BE DRILLED. THE DRILLING ENGINEER IS ATTEMPTING TO REDUCE DRILLING COSTS BETWEEN 10,000 AND 12,500 FT IN THE 8-1/2" HOLE. HE HAS A BIT RECORD, AN ELECTRIC LOG AND MUD REPORTS FROM AN OFFSET WELL.

FROM THE BIT RECORD, HE HAS THE FOLLOWING INFORMATION:

BIT SIZE	8-1/2 IN.
BIT TYPE	134
DEPTH OUT	10,000 FT
FOOTAGE	300 FT
HOURS	18
NOZZLE SIZES	THREE 12'S
MUD WEIGHT	10.2 PPG
PUMP PRESSURE	2200 PSI
RIG COST	$150/HOUR
DRILL PIPE	5" O.D., 19.5 LB/FT, XH
DRILL COLLARS	22, 6-1/4" X 2-1/4"
WEIGHT	40,000 LB
RPM	80
BIT GRADING	T5, B7, I

FROM THE MUD REPORTS HE HAS DETERMINED THE FOLLOWING:

1. THE MUD WEIGHT CAN BE REDUCED TO 10.0 PPG.
2. THE METHYLENE BLUE COUNT CAN BE LOWERED TO 14 LB/BARREL FROM 18 LB/BARREL.
3. THE OIL CONTENT IN THE OFFSET WELL HAD BEEN 6% BY VOLUME.

THE HYD.ANAL PROGRAM IS RUN TO DETERMINE THE FLOW RATE AND OTHER PERTINENT HYDRAULIC DATA.

```
OPTION? HYD.ANAL

760806

ANALYSIS OF HYDRAULICS (INPUT 'O' FOR UNKNOWNS)

BIT SIZE? 8.5

DRILL STRING DATA:  MINIMUM OR TOTAL? T

TAPERED STRING? N

DRILL PIPE OD, ID, TOOL JOINT OD, ID .. ?      5,4.28,6.25,3.75

NUMBER OF COLLARS TOP SET, COLLAR OD, ID .. ? 22,6.25,2.25

NUMBER OF BIG COLLARS, COLLAR OD, ID .. ?      0,0,0

MUD WEIGHT, PV, YV? 10.2,0,0

IS MUD DISPERSED? Y

HOLE SIZE  8.50 DP  5.0 X 4.28 TJ 6.25 X 3.75
 660.FT OF COLLARS 6.25 X 2.25 AND    0.FT 6.25 X 2.25

MUD WT.10.2 PV 14. YV 5.

DEPTH? 10000

JETS IN 32ND'S (EG.0,13,13)? 12,12,12

STANDPIPE PRESSURE, FLOW RATE? 2200,0

COLLAR WEIGHT AVAILABLE FOR DRILLING  41. THOUSAND POUNDS

REAL  ANNULUS  CHIP              JET   JET   HP/   %    PUMP
GPM     VEL.   RATE     JETS     VEL.   HP   SQIN  BIT  HP     ECD

361.   187.TT  170   12 12 12   350   238  4.2   51   545   10.5

CHANGE: HOLE, BHA, MUD, DEPTH, PSI, JETS, OR STOP? S
```

THE HYDRAUL PROGRAM IS RUN TO OPTIMIZE THE HYDRAULICS FOR A MUD WEIGHT OF 10.0 PPG.

```
OPTION? HYDRAUL

 760609

HYDRAULICS OPTIMIZATION (INPUT '0' FOR UNKNOWNS)

BIT SIZE, MAXIMUM PUMP PRESSURE, MAX FLOW RATE? 8.5,3000,0

DRILL STRING DATA: MIN/TOTAL? T

TAPERED STRING? N

DRILL PIPE OD, ID, TOOL JOINT, OD, ID .. ?      5,4.28,6.25,3.75

NUMBER OF COLLARS TOP SET, COLLAR OD, ID .. ?  22,6.25,2.25

NUMBER OF BIG COLLARS, COLLAR OD, ID .. ?      0,0,0

MUD WEIGHT, PV, YV? 10,0,0

IS MUD DISPERSED? Y

JETS, 2 OR 3, OR (0) WILL OPTIMIZE? 3

HOLE SIZE 8.50 DP 5.0 X 4.28 TJ 6.25 X 3.75
660.FT OF COLLARS 6.25 X 2.25 AND  0.FT 6.25 X 2.25

MUD WT. 10.0 PV 12. YV 4.

DEPTH, FT/HR  (EST)? 10000,16

 118.PSI TO START CIRC.(PIPE ROTATING)  10.3#/GAL.ECD AT 320.GPM

PRESSURE 2100.

COLLAR WEIGHT AVAILABLE FOR DRILLING  42. THOUSAND POUNDS

REAL  ANNULUS  CHIP    REC.     JET   HP/   %    PUMP
GPM     VEL.   RATE    JETS     VEL.  SQIN  BIT  MECH HP

300.  156.TT   130   10 11 11   367   4.0   62   432
310.  161.TT   136   10 11 11   379   4.1   62   446

320.  166TT    143   11 11 11   369   4.1   59   461  RECOMMENDED

330.  171.TT   149   11 11 12   358   4.0   56   475
340.  176.TT   155   11 12 12   348   3.9   53   490

RECOMMEND 2100.PSI WITH 11 11 11 JETS FOR 4.1 HHP/SQIN

CHANGE: HOLE, BHA, MUD, DEPTH, PUMP, JETS, OR STOP? S
```

THE PROGRAM RECOMMENDS THREE 11/32" JETS RATHER THAN THREE 12/32" JETS. IT ALSO INDICATES THAT THE PUMP VOLUME AND PRESSURE CAN BE REDUCED. THE PUMP PRESSURE AND FLOW RATE FROM HYDRAUL ARE USED IN THE INPUT DATA FOR MUDHY.

```
OPTION? MUDHY

770118      (INPUT '0' FOR UNKNOWNS)

EFFECT OF MUD AND HYDRAULICS ON DRILLING RATE

     CONTROL BIT RUN DATA:

SIZE, BIT COST, $/HR, TRIP TIME ? 8.5,595,150,10

DEPTH, FOOTAGE, HOURS ......... ? 10000,300,18

MUD WT, MBT, OIL, CL- ......... ? 10.2,0,0,0

DID MUD CONTAIN A POLYMER ..... ? N

WAS MUD DISPERSED ............. ? Y

DID MUD CONTAIN BARITE ........ ? N

3-JETS, PUMP PRESSURE ......... ? 12,12,12,2200

     PROPOSED MUD AND HYDRAULICS:

SIZE, BIT COST, $/HR, TRIP TIME ? 8.5,595,150,10

MUD WT, MBT, OIL, CL- ......... ? 10,0,0,0

WILL MUD CONTAIN A POLYMER .... ? N

WILL MUD BE DISPERSED ......... ? Y

WILL MUD CONTAIN BARITE ....... ? N

IS FORMATION GRADIENT NORMAL .. ? Y

3-JETS, PUMP PSI, MAX FLOW RATE ? 11,11,11,2100,320
```

	CONTROL BIT RUN	PROPOSED BIT RUN
DEPTH	10000	
FOOTAGE	300	317
HOURS	18.0	18.0
PENETRATION RATE	16.7	17.6
COST PER FOOT	15.98	15.11
JETS, 32ND INCH	12 12 12	11 11 11
HHP/SQIN AT BIT	3.6	3.6
JET VELOCITY	332	356
PCT. HHP AT BIT	46	54
ANNULAR VELOCITY	161	146
CALCULATED GPM	343	309
DIFFERENTIAL PRESSURE	838	734

THE FIRST MUDHY PROGRAM IS RUN TO COMPARE HYDRAULICS IN THE OFFSET WELL TO HYDRAULICS IN THE PROPOSED WELL. BY LOWERING MUD WEIGHT, PUMP VOLUME AND JET NOZZLE SIZE, THE NOZZLE VELOCITY AND RATE OF PENETRATION CAN BE INCREASED. APPROXIMATELY $0.87 PER FOOT CAN BE SAVED.

A SECOND MUDHY PROGRAM IS THEN RUN IN WHICH THE CHANGES IN METHYLENE BLUE COUNT AND OIL CONTENT ARE USED.

```
EFFECT OF MUD AND HYDRAULICS ON DRILLING RATE

     CONTROL BIT RUN DATA:

SIZE, BIT COST, $/HR, TRIP TIME ? 8.5,595,150,10

DEPTH, FOOTAGE, HOURS ......... ? 10000,300,18

MUD WT, MBT, OIL, CL- ......... ? 10.2,18,6,0

DID MUD CONTAIL A POLYMER ..... ? N

WAS MUD DISPERSED ............. ? Y

DID MUD CONTAIN BARITE ........ ? N

3-JETS, PUMP PRESSURE ......... ? 12,12,12,2200

     PROPOSED MUD AND HYDRAULICS:

SIZE, BIT COST, $/HR, TRIP TIME ? 8.5,595,150,10

MUD WT, MBT, OIL, CL- ......... ? 10,14,0,0

WILL MUD CONTAIL A POLYMER .... ? N

WILL MUD BE DISPERSED ......... ? Y

WILL MUD CONTAIN BARITE ....... ? N

IS FORMATION GRADIENT NORMAL .. ? Y

3-JETS, PUMP PSI, MAX FLOW RATE ? 11,11,11,2100,320
```

	CONTROL BIT RUN	PROPOSED BIT RUN
DEPTH	10000	
FOOTAGE	300	338
HOURS	18.0	18.0
PENETRATION RATE	16.7	18.8
COST PER FOOT	15.98	14.19
JETS, 32ND INCH	12 12 12	11 11 11
HHP/SQIN AT BIT	3.6	3.6
JET VELOCITY	332	356
PCT. HHP AT BIT	46	54
ANNULAR VELOCITY	161	146
CALCULATED GPM	343	309
DIFFERENTIAL PRESSURE	838	734

By lowering methylene blue count and oil content another $0.92 per foot can be saved. The total savings from improved hydraulics and fluid properties is $1.79/ft.

The WTRPM program is run to determine if improvements can be realized by changing weight and RPM.

OPTION? WTRPM

750306

BEST CONSTANT WEIGHT AND ROTARY SPEED PROGRAM

BIT COST, SIZE, RIG COST($/HR) ? 595,8.5,150

DEPTH, FOOTAGE, HOURS .... ? 10000,300,18

W.O.B., RPM, T, B .... ? 40,80,5,7

PROPOSED BIT COST, SIZE .... ? 595,8.5

FORMATION: SOFT, MEDIUM, OR HARD ? M

AVERAGE ROUND TRIP TIME = 7.3 HOURS
ENTER YOUR TRIP TIME IF DIFFERENT? 10

BCWR(1), BRGW(2), BWGR(3), NEW PROBLEM(4), CHANGE GRADINGS(5),
SET RPM-WT(6), STOP(0) ... WHICH? 1

CONTROL WELL: 15.98 $/FT AT 16.7 FT/HR

RPM	WEIGHT	$/FT	FOOTAGE	HOURS	FT/HR	T	B
133	51	12.51	264	8.1	32.8	8	8
127	50	13.10	260	8.8	29.6	7	8
120	48	13.84	257	9.8	26.4	6	8
113	47	14.79	253	11.0	23.0	6	8
75	50	15.39	257	12.5	20.7	5	7
67	49	17.02	222	11.2	19.8	4	5

WANT TO MODIFY FORMATION? N

The WTRPM program shows that drilling costs can be reduced from $15.98/ft to $15.39/ft by running the proposed bit at 75 RPM and 50,000 pounds of weight, but the available drilling weight is only 42,000 pounds (from HYDRAUL).

The BUTTON program is run on a computer terminal to determine if an insert bit can replace the good milled tooth bit run on the offset well.

OPTION? BUTTON

741212

FOOTAGE NEEDED FROM AN INSERT BIT TO REPLACE MILL TOOTH BITS

WHAT IS YOUR RIG COST ($/HR) .... ? 150

MILL TOOTH DATA:

DEPTH OUT, FEET, HOURS, BIT COST .. ? 10000,300,18,595

INSERT BIT DATA:

INSERT BIT COST .... ? 2490

AVG. TRIP TIME .... = 7.3

YOUR TRIP TIME IF DIFFERENT ? 10

EXPECTED PENETRATION RATE RANGE
(HIGH, LOW) OR ENTER 0,0 ...... ? 0,0

FOOTAGE	HOURS	FT/HR	$/FOOT
	MILL TOOTH		
300.0	18.0	16.7	15.98
	INSERT BIT		
512.4	28.0	18.3	
553.3	32.4	17.1	
609.2	38.3	15.9	
690.4	47.0	14.7	
818.9	60.7	13.5	
1053.3	85.6	12.3	
1615.5	145.5	11.1	
4796.7	484.5	9.9	
6000.0	612.7	9.8	

ANOTHER RUN, PLOT OR STOP ? S

The BUTTON program indicates that an insert bit drilling at a penetration rate of 12.3 ft/hr for 85.6 hours would equal the milled tooth drilling cost.

Since the program indicated that an insert bit might drill at a lower cost than the milled tooth offset bit, BITOP is run with two different insert bit types to see if a savings can be achieved.

OPTION? BITOP

760730

AMOCO PRODUCTION RESEARCH - DRILLING CHARACTERISTICS OF INSERT BITS

TYPE '0' FOR UNKNOWN BIT COSTS. DO NOT INPUT RUNS WITH OVER 100 RPM.
UNDER PROPOSED BIT RUN, 0,0 FOR WT,RPM WILL OPTIMIZE
EXPERIMENTAL IADC SERIES 5-1- IS NOW AVAILABLE

CONTROL BIT RUN DATA:
RIG COST($/HR), TRIP TIME ? 150,10
BIT SIZE, BIT COST ...... ? 8.5,595
BIT TYPE .... ? SDG
DEPTH, FOOTAGE, HOURS ... ? 10000,300,18
W.O.B., RPM, T, B .... ? 40,80,5,7

PROPOSED BIT RUN:
RIG COST ($/HR) .... ? 150
BIT SIZE, BIT COST ...... ? 8.5,2490
BIT TYPE .... ? F3
ADJUSTED DEPTH, TRIP TIME ? 10900,10.9
WT,RPM .... ? 0,0

CONTROL BIT RUN: SM SDG IADC 1-3-4 $ 595 10.0 HR. TRIP
PROPOSED BIT RUN: SM F3 IADC 5-3-7 $2490 10.9 HR. TRIP

	WT	RPM	FT	HR	FT/HR	$/FT
CONTROL:	40000	80	300.0	18.0	16.67	$ 15.98
PROPOSED:	40000	65	1239.2	91.1	13.60	$ 14.36 ***
	40000	50	1598.6	142.6	11.21	$ 15.96
	35000	65	877.8	79.7	11.01	$ 18.32 ***
	35000	45	1150.5	132.6	8.68	$ 20.87
	30000	65	883.5	93.6	9.44	$ 20.56 ***
	30000	40	1334.3	193.6	6.89	$ 24.86
	25000	70	628.5	82.5	7.62	$ 26.25 ***
	25000	35	647.3	142.8	4.53	$ 39.45
	20000	70	376.7	66.0	5.71	$ 37.23 ***
	20000	35	322.6	114.2	2.82	$ 65.89

***HOURS BASED ON MEDIAN BEARING LIFE. WT-RPM CONDITIONS ON THIS LINE MAY CAUSE BIT'S CUTTING STRUCTURE TO FAIL BEFORE BEARING IN NON-UNIFORM FORMATIONS. 'CHECK LOG'

CHANGE: CONTROL, RIG, BIT, DEPTH, WT, OR STOP? BIT

PROPOSED BIT RUN:
BIT SIZE, BIT COST .... ? 8.5,2490

BIT TYPE .... ? F2

ADJUSTED DEPTH, TRIP TIME ? 10900,10.9

WT, RPM .... ? 0,0

CONTROL BIT RUN: SM SDG IADC 1-3-4 $ 595 10.0 HR. TRIP
PROPOSED BIT RUN: SM F2 IADC 5-2-7 $2490 10.9 HR. TRIP

	WT	RPM	FT	HR	FT/HR	$/FT	
CONTROL:	40000	80	300.0	18.0	16.67	$ 15.98	***
PROPOSED:	40000	80	1275.6	91.1	14.00	$ 13.95	***
	40000	50	1660.4	142.6	11.64	$ 15.37	
	35000	60	1208.5	96.8	12.48	$ 15.43	***
	35000	45	1359.4	132.6	10.25	$ 17.67	
	30000	65	1047.2	93.6	11.19	$ 17.35	***
	30000	45	1525.7	173.6	8.79	$ 19.77	
	25000	65	920.2	104.6	8.80	$ 21.53	***
	25000	40	937.6	158.6	5.91	$ 29.77	
	20000	65	503.7	83.6	6.02	$ 33.10	***
	20000	35	426.5	114.2	3.73	$ 49.84	

***HOURS BASED ON MEDIAN BEARING LIFE. WT-RPM CONDITIONS ON THIS LINE MAY CAUSE BIT'S CUTTING STRUCTURE TO FAIL BEFORE BEARING IN NON-UNIFORM FORMATIONS. 'CHECK LOG'

CHANGE: CONTROL, RIG, BIT, DEPTH, WT, OR STOP? S

BITOP INDICATES THAT A 537 INSERT BIT HAS TWO OPERATING POSSIBILITIES AND THAT A 527 BIT HAS THREE POSSIBILITIES.

S U M M A R Y

THE BUTTON PROGRAM INDICATED THAT AN INSERT BIT COULD BE USED.

HYDRAUL INDICATED A LOWER FLOW RATE AND PUMP PRESSURE AND LOWERED THE MECHANICAL HORSEPOWER TO THE PUMP BY 84 HP WHICH WILL RESULT IN A FUEL COST SAVINGS.

MUDHY SHOWED A COST PER FOOT SAVINGS OF AS MUCH AS $1.79 PER FOOT JUST BY CHANGING THE FLUID PROPERTIES.

BITOP INDICATED FIVE BIT AND DRILLING PARAMETER COMBINATIONS WHICH COULD RESULT IN A SAVINGS.

BIT	WT/RPM	BIT SAVINGS	TOTAL SAVINGS
5-3-7	40/50	0.02	1.81
5-3-7	40/65	1.62	3.41
5-2-7	35/60	0.55	2.34
5-2-7	40/50	0.61	2.40
5-2-7	40/65	2.03	3.82

THESE FIGURES DO NOT REFLECT THE SAVINGS IN FUEL OR PUMP MAINTENANCE.

ACKNOWLEDGEMENT

The help extended by David B. Hilliard is greatly appreciated.

REFERENCES

Allen, J.H., 1972. How to obtain optimum performance with tungsten carbide insert bits. *Drilling-DCW*, 33(7, May): 40—41.

Allen, J.H., 1973. Tungsten carbide bit operating practices. *Europe and Oil*, (9, Sept.): 30—31.

Allen, J.H., 1975a. Eliminate trips to drill out casing cement. *Pet. Eng.*, 47(10, Sept.): 92—98.

Allen, J.H., 1975b. How to relate bit weight and rotary speed to bit hydraulic horsepower. *Drilling-DCW*, 36(8, May): 53—55.

Allen, J.H., 1977. *Contemporary Methods for Determining the Parameters that Affect the Rate of Penetration.* 3rd Adriatic Symp. on Oil Well Drilling, Porec, Yugoslavia, May 23—27, 20 pp.

Bentson, H.G., 1956. *Rock Bit Design, Selection and Evaluation.* Pap., Spring Meet., May, Los Angeles, Calif., Pacific Coast Distr., API Div. Prod., 36 pp.

Bobo, R.A., Hach, R.S. and Ormsby, G.S., 1954. Keys to successful competitive drilling. *World Oil*, 139(6): 172—177.

Bobo, R.A., Hach, R.S. and Ormsby, G.S., 1955. Keys to successful competitive drilling. *World Oil*, 140(1): 152—164; 140(4): 113—116.

Boone, D.E., 1976. Use of drilling data to cut costs. *Pet. Eng.*, 48(11, Sept.): 10—80.

Crowley, M., 1976. Procedure updates drilling hydraulics. *Oil Gas J.*, 74(7, Feb.): 59—63.

Estes, J.C., 1971. Selecting the proper rotary rock bit. *J. Pet. Technol.*, 23(11, Nov.): 1359—1367.

Estes, J.C., 1974. Guidelines for selecting rotary insert rock bit. *Pet. Eng.*, 46(10, Sept.): 30—34.

Fullerton, H.B. and Fullerton, H.B., IV, 1973. *Constant Energy Drilling System for Well Programming.* Sii Smith Tool, Irvine, Calif. (Unpubl.)
Fullerton, H.B. and Mechem, O.E., 1965. Computers invade the rig floor. *Oil Gas J.*, 63(41, Oct. 11): 125—140.
Galle, E.M. and Woods, H.B., 1960a. How to calculate bit weight and rotary speed for lowest-cost drilling. *Oil Gas J.*, 58(46, Nov.): 167—176.
Galle, E.M. and Woods, H.B., 1960b. *Variable Weight and Rotary Speed for Lowest Drilling Cost*, 20th Annu. Meet. AAODC, Sept. 25—27, New Orleans, La., Pap.
Galle, E.M. and Woods, H.B., 1970. *Best Constant Weight and Rotary Speed for Rock Bits.* Hughes Tool Co., Brochure No. 2M9/636.
Garner, C., 1971. *Is There a Better Way of Grading Insert Bits?* Spring AAODC Rotary Dril. Conf., Houston, Texas, Pap.
Garner, C.R. and Allen, J.H., 1976. How to achieve more economical bit runs. *Drilling-DCW*, 37(9, June): 42—43.
Garner, L.L. and Millar, D.J., 1976. *Rotary Drilling Bits 1976.* ASME Pap. 76-Pet-93, Mexico City, Mexico: 8 pp.
Gatlin, C., 1960. *Petroleum Engineering, Drilling and Well Completion.* Prentice-Hall Inc., Englewood Cliffs, N.J., 341 pp.
Gray, G.R. and Young, F.S., 1973. 25 years of drilling technology. A review of significant accomplishments. *J. Pet. Technol.*, 25(12, Dec.): 1347—1354.
Jackson, R.A., 1972. Cost per foot: key to rock bit selection. *Drilling-DCW*, 33(7, Apr.): 4 pp.
Jackson, R.A. and Wood, J., 1973. Running insert bits. *Oil Gas J.*, 71(42, Oct.): 119—122.
Kastrop, J.E., 1972. Proposed system simplifies rock bit coding. *Pet. Eng.*, 44(3, Mar.): 57—62.
Kendall, H.A. and Goins, W.C., Jr., 1960. Design and operation of jet-bit programs for maximum hydraulic horsepower, impact force or jet velocity. *Trans. AIME*, 219: 238—250.
Knowlton, J., 1961, 1964. Use of simple code on bit wear urged. *Dril. Contractor*, Oct. 1961, and Apr. 1964.
Lubinski, A., 1958. Proposal for further tests. *Pet. Eng.*, 30(1, Jan.): B-50—B-52.
Lummus, J.L., 1970. Drilling optimization. *J. Pet. Technol.*, (Nov.): 1379—1389.
Lummus, J.L., 1974. Drilling in the seventies, Part III, Bit selection. *Pet. Eng.*, 46(3, Mar.): 82—91.
Moore, P.L., 1974. *Drilling Practices Manual.* The Petroleum Publ. Co., Tulsa, Okla., 448 pp.
Murphy, D., 1969. Some field tips for cutting drilling costs. *Oil Gas J.*, 67(2, Mar. 24): 150—155.
Murphy, D., 1969. A practical engineering approach to running bits. *Oil Gas J.*, 67(5, Feb. 3): 88—92.
Oilweek (Editor), 1975. Seismic boosts drilling rate. *Oilweek (Calgary, Alberta)*, 25(50, Jan.): 11—14.
Randall, B.V., 1975. Optimum hydraulics in the oil patch. *Pet. Eng.*, 47(10, Mar.): 36—52.
Rios, F. and Montiel, D., 1974. Rock bits programmed from log data. *Oil Gas J.*, 72(3): 88—90.
Schlee, C., 1977. *News Bits.* Sii Smith Tool, Irvine, Calif., 2(2, Jan.)
Smith Tool, 1979. *The Technique, Tools and Terminology of Drilling.* Irvine, Calif., 92713: 11 pp.
Taylor, K.O., 1975. *Computer Control of Bit Weight and Rotary Speed.* ASME Annu. Winter Meet., Dec., Houston, Texas, Pap.
Young, F.S., Jr., 1969. Computerized drilling control. *J. Pet. Technol.*, 21(4, Apr.): 483—496; *Trans. AIME*, 246: 483—496.

Chapter 4

TESTING OF DRILLING FLUIDS *

NORMAN K. TSCHIRLEY

INTRODUCTION

Every drilling operation depends directly upon the properties of the drilling mud. Prepared as a guide, this chapter discusses field tests for evaluation of drilling fluid by measuring its certain properties and that of mud filtrate.

The apparatuses and procedures for field testing of drilling fluids are discussed here. The various committees of the American Petroleum Institute are to be commended for their untiring efforts to establish standard procedures. The standard test procedures recommended by the API have given the petroleum industry a basis for a more definite correlation of results among different fields. Many of the tests presented here are covered in the API code R.P. 10B, 13B, and 39.

In order to relate these tests more closely with drilling conditions, the significance of each test as a factor of proper mud control and, in turn, the importance of mud control to successful drilling, are pointed out briefly. It should be emphasized that the procedures given here are intended for field testing, and do not supplant the more detailed tests that properly find their places in the laboratory.

DENSITY

It is extremely important to know the density of the drilling fluid throughout most of the drilling operation. This is true whether drilling through (1) gas, oil, or salt water sands; (2) shale, where relatively high density may be required; or (3) low-pressure production zones where the use of low-density colloidal mud is advantageous. Frequent density tests aid in maintaining safe drilling conditions by disclosing any changes taking place in the unit weight of the mud.

Mud balance

The most practical instrument for determining the specific weight of drilling fluid is the mud balance, which is graduated in pounds per gallon,

* Edited by Paul Vorabutr and George V. Chilingarian.
Courtesy of Baroid Petroleum Services Division, NL Industries, Inc.

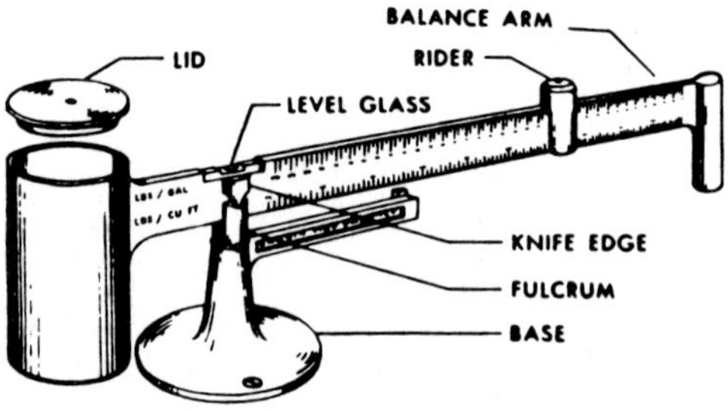

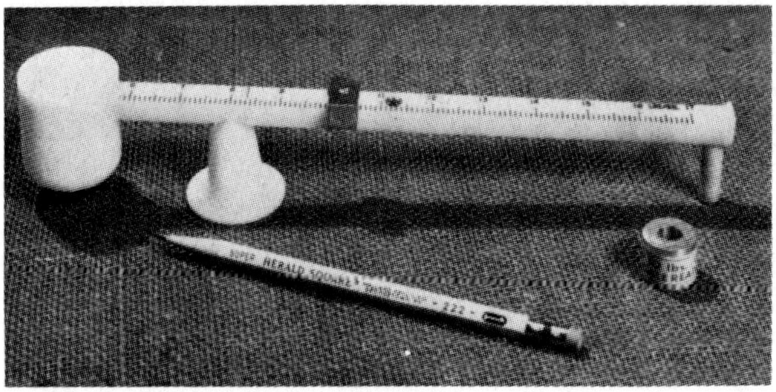

Fig. 4-1. Mud balance (top) and pocket mud balance (bottom). (Courtesy of Baroid Petroleum Services, NL Industries, Inc.).

pounds per cubic foot, pounds per square inch per 1000 feet of depth, and specific gravity. Pocket mud balance designed for use where portability and space savings are important factors, is also available.

The mud balance provides a simple method for the accurate determination of the specific weight of mud, and the temperature does not materially affect the accuracy of readings. The durable construction of the mud balance (it contains no easily breakable parts) makes it ideal for field use.

The mud balance, shown in Fig. 4-1, principally consists of a base and a graduated arm with cup, knife edge, rider, built-in spirit level, and counterweight. The constant volume cup is affixed to one end of the graduated arm, whereas counterweight is on the opposite end.

The range of scale readings of the 4-scale mud balance are as follows: (1) 310—1250 pounds per square inch per 1000 feet of depth (psi/1000 ft); (2) 6—24 pounds per gallon (lb/gal); (3) 45—180 pounds per cubic foot (lb/cu ft); and (4) 0.72—2.88 specific gravity.

The scale ranges of the 2-scale mud balance are as follows: (1) 8—18 (lb/gal), and (2) 58—136 (lb/cu ft).

A corrosion-resistant plastic carrying case is available which holds the balance intact and in working position. In addition, it acts as a protection against wind which may tend to disturb the natural oscillation of the sensitive balance during a test. The balance arm and cup provide a convenient dipper for taking a mud sample.

The mud balance must be cleaned and dried thoroughly after each use. The calibration of the mud balance can be easily checked by measuring the specific weight of clean water, which should be around 62.4 lb/cu ft. The fine balance of the instrument can be adjusted by adding or removing lead shot or by moving the set screw type counterweight under the cap screw at the end of the arm.

The pocket mud balance is a smaller and more portable version of the regular mud balance. When using the pocket mud balance, the procedures listed below should be followed for mud weights less than 17 lb/gal. Balance is indicated when the arm is horizontal. For mud weights over 17 lb/gal, the scale extended weight is needed. This weight, marked "add 10 lb/gal", is readily attached or removed by sliding it on or off the lead shot holder at the end of the balance. With this weight in place, 10 lb/gal are added to the indicated reading in order to give correct specific weight.

Because of the small cup volume, extra care should be taken when filling the cup to assure that no air bubbles remain trapped inside. Air bubbles occupy certain volume and, thus, decrease the specific weight of the material in the cup.

Procedure

(1) Remove lid and fill cup to the top with sample to be tested. If air bubbles have been trapped in the mud, tap cup briskly until they break out.

(2) Replace the lid and rotate until firmly seated, making sure some mud squeezes out the vent hole.

(3) Wipe mud from exterior of balance.

(4) Place balance on base of knife edge on fulcrum rest.

(5) Move rider until instrument is in balance, as determined by spirit level.

(6) Read mud weight and hydrostatic pressure or gradient at edge of rider nearest fulcrum.

Results

Report hydrostatic pressure gradient in pounds per square inch per thousand feet of depth (psi/1000 ft). Report specific weight of drilling mud in pounds per gallon (lb/gal), pounds per cubic foot (lb/cu ft), or as specific gravity (or g/cm^3).

VISCOSITY

Viscosity is equal to the ratio of shearing stress to the rate of shearing strain (see Fig. 4-2):

$$\mu = \frac{\text{shearing stress}}{\text{rate of shearing strain}} = \frac{F/A}{V/h} = \frac{\tau}{V/h} \tag{4-1}$$

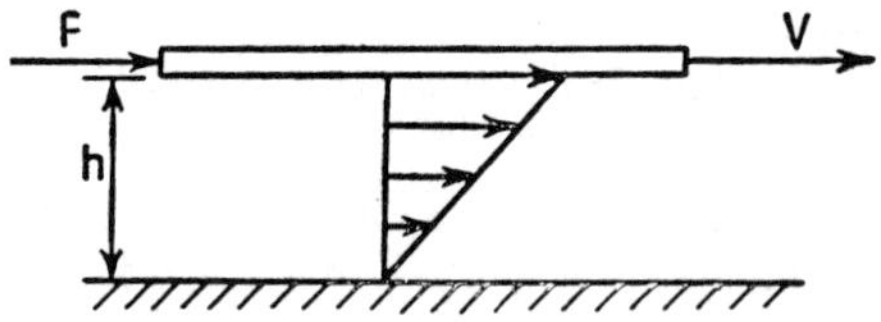

Fig. 4-2. Flow between parallel plates, the lower plate being stationary. (See Binder, 1962, pp. 77-85.) V = velocity of upper plate; a thin layer of fluid adhering to the lower plate is zero. F = force required to move the upper plate, having area A. h = distance between the plates. Velocity distribution is assumed to be linear here.

Thus, it is a measure of the internal resistance of a fluid to flow; the greater the resistance, the greater the viscosity of the drilling fluid (Chapter 7). For efficient drilling operations, viscosity of mud must be controlled and a standard means of measuring it must be provided. Hole size, hole condition, pumping rate, drilling rate, cutting size, cavings, presence or absence of the shale separators, mud weight, design of the pit system, and gel characteristics of the mud are factors influencing the specification of the viscosity of any given mud.

The following instruments may be used for determining mud viscosity: Marsh funnel, rheometer, and rotational viscometer.

Marsh funnel

The Marsh funnel viscometer is used for routine viscosity determinations on almost every drilling rig. The measurements obtained are influenced considerably by the rate and degree of gelation and by the density of mud which varies the hydrostatic head of the column of mud in the funnel. In addition, the hydrostatic head decreases with time as the mud flows out. Because of these variations, the viscosities obtained with the Marsh funnel cannot be correlated directly with those found using rotational viscometer and rheometer.

The Marsh funnel is 6 inches in diameter at the top and is 12 inches long. A 10-mesh screen fitted across one-half of the top removes foreign materials and cuttings from the mud to be tested. Figure 4-3 shows the Marsh funnel viscometer together with the graduated cup which is used to fill the funnel with drilling fluid.

Procedure

(1) Hold funnel in upright position with index finger over outlet.

(2) Pour the test sample freshly taken from the mud system, through the screen in top of funnel until mud level just reaches the underside of the screen.

(3) Immediately remove finger from the outlet tube and measure number

Fig. 4-3. Marsh funnel viscometer and graduated cup. (Courtesy of Baroid Petroleum Services, NL Industries, Inc.)

of seconds for a quart of drilling mud sample to run out. Report funnel viscosity in seconds (sec) API.

Like all other equipment, it must be cleaned and dried thoroughly after each use. Special care must be exercised not to bend or flatten the special brass orifice in the bottom of the funnel. Funnel is tested by filling it with clean fresh water and determining the time of efflux for a quart of this water; at 70° F it is equal to 26 ± 0.5 sec.

Rheometer

Values for the apparent and plastic viscosities, yield point, and gel or shear strength of the drilling fluids are determined using the rheometer. The rheometer is a rotational viscometer whereby viscosity at shearing rates of 300 and 600 rpm can be determined. This instrument is powered by the operator turning a crank which drives the spindle through a precision gear

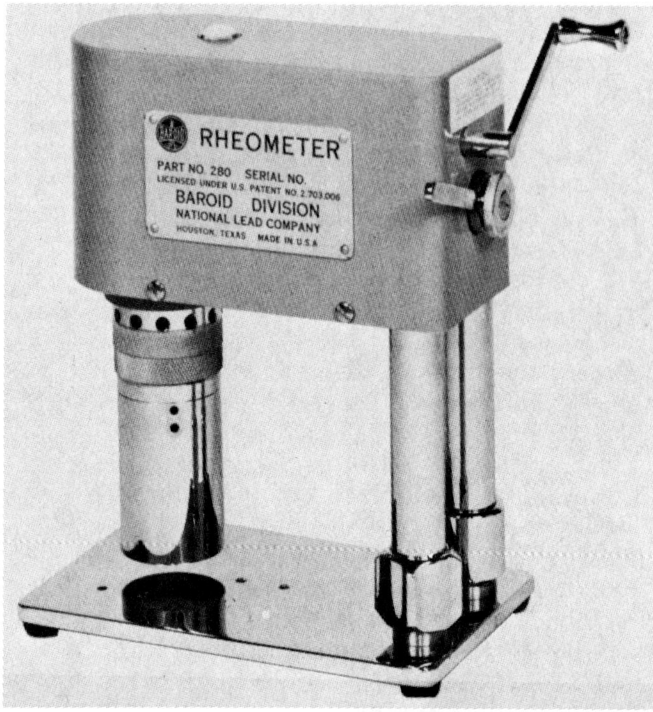

Fig. 4-4. Rheometer. (Courtesy of Baroid Petroleum Services, NL Industries, Inc.)

train. Thus, electric power is not required. Speeds of 300 and 600 rpm are selected by shifting the gear. Rotation for a gel strength is accomplished by knurled hand wheel. For portability, the support standards telescope to a total height of only $8\frac{1}{2}$ inches. The base is $4\frac{1}{2}$ inches wide by $8\frac{1}{2}$ inches long, and the instrument weighs $6\frac{1}{2}$ lb. The instrument shown in Fig. 4-4 has been manufactured by Baroid; other brands of this instrument are available, which operate on the same principle.

Operating procedure

In order to achieve rotation of 600 rpm, move the gear shift lever down until it catches in the detent and then turn the crank at the speed sufficient for slipping to be noticed. For 300 rpm, bring the gear shift lever all the way up and turn crank again with sufficient speed for slipping to be noticed. A high stirring speed is obtained by moving the shift lever all the way down past the detent and then rotating the crank.

In order to obtain the 300- and 600-rpm shearing stress readings for a mud sample the following procedure should be followed:

(1) Place a recently agitated sample in a suitable container and lower the

instrument head until the rotor sleeve is immersed exactly to the scribed line. To hold this position, tighten the lock screw on the left leg of the instrument. With the gear shift at high-speed level, rotate the crank for about 15 seconds, release the 600-rpm setting, and continue cranking.

(2) Wait for dial reading to come to a steady value. This is the high speed reading (600 rpm). Move the gear shift all the way up, crank and wait for the dial reading to come to a steady value. This is the slow-speed reading (300 rpm).

The apparent viscosity of true and plastic fluids may be determined with the rheometer by using a single-speed reading, i.e., by dividing the 600-rpm reading by 2:

$$\text{Apparent viscosity in centipoises} = \frac{600 \text{ reading}}{2} \tag{4-2}$$

For a true or Newtonian fluid, characterized by a straight-line flow curve (shear stress versus rate of shearing strain curve) that passes through the origin, the apparent viscosity is equal to the plastic viscosity. For plastic fluids, such as drilling fluids, the apparent viscosity is of limited value because the slope of the flow curve cannot be defined by measuring the shear rate at only one speed.

Values for plastic viscosity and yield point are obtained from the rheometer dial readings as follows:

$$PV = \text{plastic viscosity in centipoises} = (600\text{-rpm reading}) - (300\text{-rpm reading}) \tag{4-3}$$

$$YP = \text{yield point in lb per 100 sq ft} = (300\text{-rpm reading}) - (\text{plastic viscosity}) \tag{4-4}$$

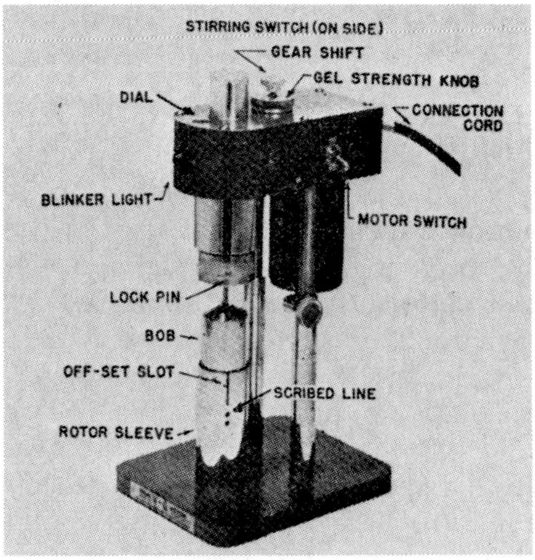

Fig. 4-5. Fann V-G meter.

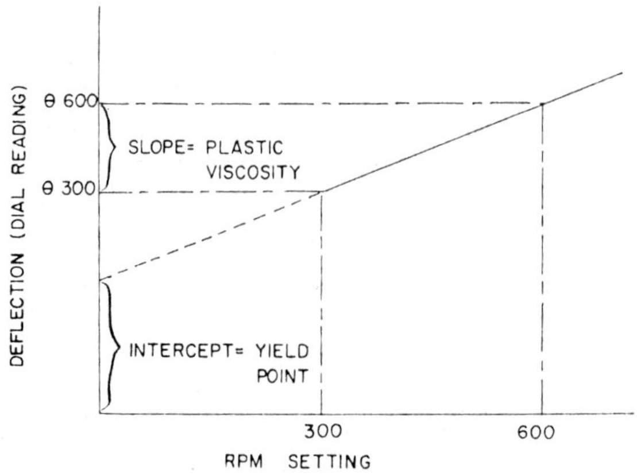

Fig. 4-6. Typical flow curve of drilling mud using a direct-indicating viscometer.

The use of hand-cranked rheometer has been mostly replaced in the past years by the rotary viscometer (Fig. 4-5); which differs from the rheometer essentially in that the mechanism is driven by an electric motor rather than a hand crank. Figure 4-6 shows a typical flow curve of mud using a direct-indicating viscometer.

Viscosity of muds (liquid) is a function of temperature more than pressure, and commonly it is necessary to measure viscosity at elevated bottom-hole temperature. This is done by using the viscometer cup heater, which is a thermostat-controlled unit for heating the mud sample directly on a viscometer.

GEL STRENGTH

The shear or gel strength of drilling fluids is a measure of the minimum shearing stress necessary to produce slip-wise movement of fluid. Two readings are generally taken: (1) immediately after agitation of the mud in the cup, and (2) after the mud in the cup has been quiescent for a period of ten minutes. The readings are referred to as the initial gel strength and the ten minute gel strength, respectively. Both gel strength readings are equal to zero for true fluids no matter how viscous, e.g., clarified honey, but the difference in the two readings may be appreciable for suspensions such as drilling fluids. This difference is considered to be the measure of thixotropy of a mud system. Hole size, type of formations, depth, temperature and pressure of formation fluids or gases, and the amount of weight material in the mud are factors that must be considered in determining necessary gel-strength properties of the drilling muds.

Procedure

The gel strength of a mud using the rheometer is determined as follows:

(1) Stir a mud sample at high speed for about 15 sec.

(2) Allow the desired rest time (10 sec or 10 min) and then turn the gel strength knob on the hub of the speed change lever clockwise slowly and steadily. The maximum deflection of the dial before the gel breaks is the gel strength in lb per 100 sq ft.

Stormer viscometer

Although the Stormer viscometer is not used at the present time, a considerable amount of valuable data was obtained and published in various technical journals and books by various investigators using this instrument. Consequently, it is imperative to acquaint mud engineers with the Stormer viscometer, which is illustrated in Fig. 4-7. The weight was adjusted so that the spindle revolved at a rate of shear of 600 rpm, as determined by the tachometer and stop watch. By means of a calibration chart, the weight in grams required to achieve this rate of shear was converted into centipoises. Sugar solutions of known concentrations were usually used for calibration.

Initial gel strength was a minimum weight in grams required to achieve a movement of approximately one-quarter revolution of the spindle after slowly releasing the brake. Drilling fluid was thoroughly agitated prior to the measurements. Ten-minute gel strength was a minimum weight in grams required to cause rotation after ten minutes of quiescence.

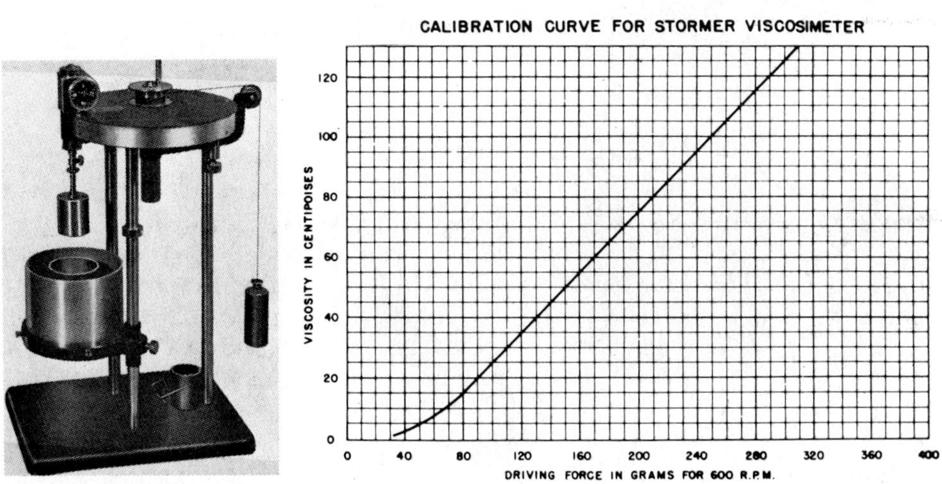

Fig. 4-7. Stormer viscometer and calibration curve. (Courtesy of American Petroleum Institute.)

It is critical for this equipment to be kept thoroughly clean and dry. The sleeve and the bob of the rheometer should be washed and dried after each operation. The instrument must also be kept in upright position when cleaning the bob and sleeve to prevent water running into the bearing.

Shearometer

The shearometer (Fig. 4-8) is an auxiliary instrument for use in determining gel strengths of drilling fluids. The readings are obtained directly from a calibrated scale and give gel strength in lb/100 sq ft of area. The readings, however, cannot be correlated with those obtained with the rotational viscometer or the rheometer. The shearometer is not recommended for use if gel strengths of drilling fluids are either very low or very high.

The shearometer consists of (1) a 3.5-inch long duraluminum tube, which has an internal diameter of 1.4 inches and weighs 5.0 g; (2) special shear scale graduated in lb/100 sq ft; and (3) sample cup which also serves to support the scale. Shearometer is sensitive to high-pH muds because of its

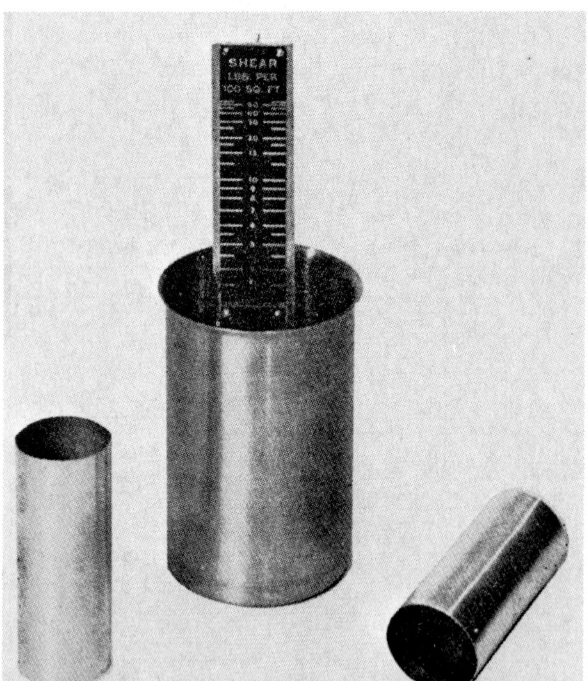

Fig. 4-8. Shearometer; cup, attached stem, calibrated in lb/100 sq ft, and two thin-walled aluminum cylinders. (Courtesy of Imco Services, Inc.)

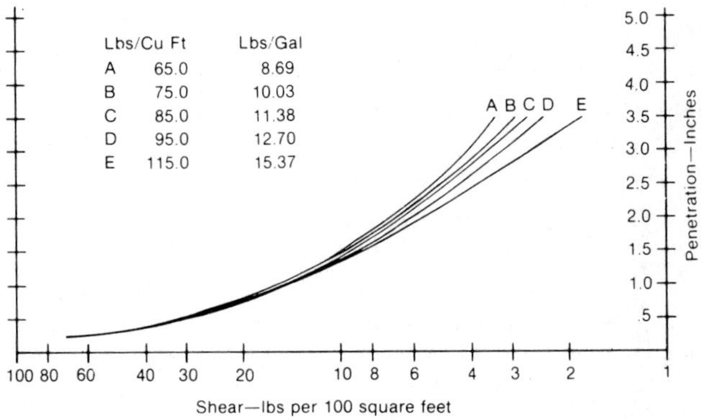

Fig. 4-9. Shearometer calibration curves (5 g). (Courtesy of Baroid Petroleum Services, NL Industries, Inc., 1978, Vol. 1, sec. 2, p. 8, fig. 10.)

duraluminum tube. In addition, the measured 10-min gel strength is lower than the true 10-min gel strength because duraluminum tube agitates the mud as it sinks in.

Procedure

(1) Wet the tube with water and wipe off excess water.

(2) Lower the shearometer tube over the scale support and place it on the surface of the freshly-agitated mud which has been poured into the container to bottom line across the scale. Allow the tube to sink vertically, guided by the fingers only if necessary.

(3) With a stopwatch measure the time from the instant the tube is released. After allowing the tube to sink for one minute, read the scale directly opposite the top of the tube and record the shear strength in lb/100 sq ft.

(4) Wait ten minutes and repeat the procedure outlined above to measure the ten-minute shear strength.

Sometimes drilling fluids may have such low shear strength that the tube sinks to the bottom in one minute or less. In this case, the shear strength is reported as zero with a superscript indicating the number of seconds of fall. Shearometer calibration curves for various mud weights are presented in Fig. 4-9.

FILTRATION PROPERTIES

The filtration properties of drilling fluids are indicative of (1) the ability of the solid components of muds to form a filter cake and (2) magnitude of cake permeability. The lower the permeability, the thinner the filter cake

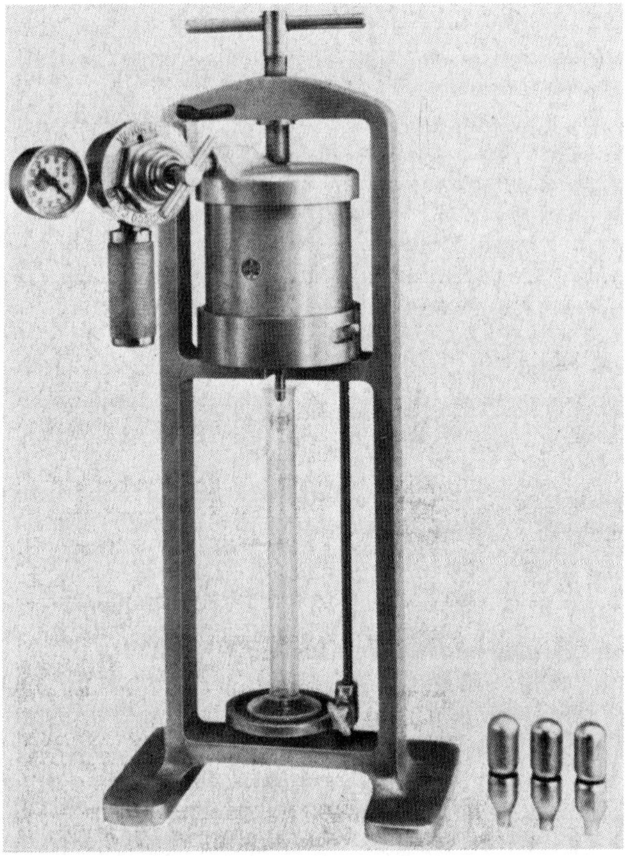

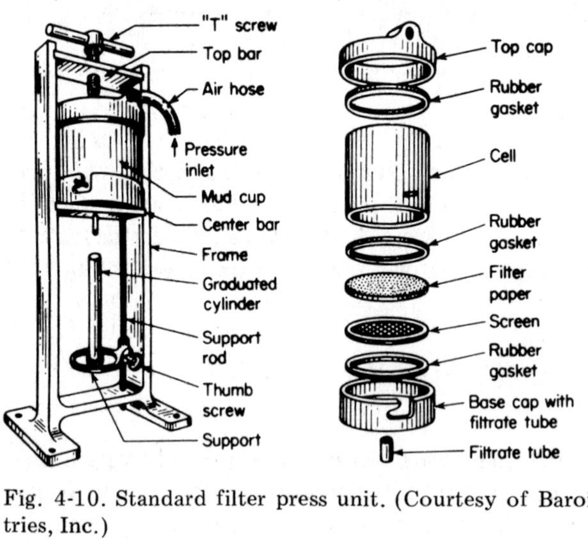

Fig. 4-10. Standard filter press unit. (Courtesy of Baroid Petroleum Services, NL Industries, Inc.)

and the lower the volume of filtrate from muds of comparable solids concentration. Filtration property is dependent upon the amount and physical state of colloidal material in the mud. When mud containing sufficient colloidal material is used, drilling difficulties are minimized. In contrast, a mud low in colloids content and high in inert solids content deposits a thick filter cake on the walls of the borehole. A thick filter cake restricts the passage of tools and allows an excessive amount of filtrate to move into the formation, creating a potential cause of caving. Lack of proper wall-building properties may result in further problems such as: (1) difficulty in running casing; (2) creating a swabbing effect, which may cause the formation to cave in or swab reservoir contents into the hole; and (3) difficulty in securing a water shutoff because of channeling of cement.

Filter presses are recognized as standard testing equipment for testing oil well cement additives, and evaluation of hydraulic fracturing fluids (see API R. P. 10B and R. P. 39).

Filter presses

The filtration properties of drilling fluid are determined by means of the standard filter press (Fig. 4-10). It consists of (1) mud reservoir mounted in a frame, (2) filtering medium, (3) means of catching and measuring the filtrate (e.g., graduated cylinder), and (4) pressure source. Cell, top cap, screen, rubber gaskets, and base cap make up the mud reservoir. The filtering medium is a sheet of specially hardened filter paper which fits on the screen in the base cap. A graduated glass cylinder is used to collect and measure the filtrate. Various sources of pressure may be used such as cylinders of compressed air or nitrogen, CO_2 cartridges, high-pressure air or water systems, or the dead weight hydraulic pressure assembly. Compressed oxygen must not be used as a pressure source, because of hazard of explosion. The standard filtration test requires 100 ± 5 psi pressure for 30 min.

The standard filter press unit which utilizes CO_2 cartridges is particularly useful where weight of equipment or difficulty in obtaining compressed air or nitrogen is a factor. The CO_2 cartridges must be kept out of direct sunlight and away from heat. Cartridge pressure at 80° F is about 900 psi when *properly* filled; it will reach a pressure of 2500 psi at 140° F. The cartridge cylinders are designed to withstand pressures up to 7000 psi and are tested at 140° F during manufacture. Some *overfilled* cartridges also will not present any problem as long as they are kept at fairly low temperatures. They should always be kept in the coolest place available to avoid danger of explosion.

Procedure

(1) Assemble the following dry parts of filter press in the following order: base cap, rubber gasket, screen, a sheet of filter paper, rubber gasket, and cell. Secure the cell to the base cap. (See Fig. 4-10.)

(2) Fill the cell with sample to be tested to within $\frac{1}{4}$ inch of the top. Filling the cell to this height with mud is necessary only for the purpose of conserving the compressed gas. (The volume of mud in the cell does not affect the water loss test results.) Set the unit in place in the filter press frame.

(3) Check the top cap to make sure the gasket is in place. Place the top cap on the cell and secure the unit in place with the T-screw.

(4) Place a dry graduated cylinder under the filtrate tube.

(5) Apply 100 psi pressure to the cell, and start timing of the test. At the end of 30 min, first close the pressure source valve and then open the safety-bleeder valve to release pressure from the entire system.

(6) Read the volume of filtrate collected in the graduated cylinder at the end of 2, 5, 10, 15, and 30 minutes.

Results

(1) Report the filtrate loss in ml as the API (30-min) filtrate loss of the mud.

(2) The filter cake thickness is determined after the filter cell has been disassembled. After the filter paper with the cake deposited on it is removed from the base cap, the excess mud is washed from the cake. The filter cake thickness is measured and reported in units of 1/32 of an inch. A sharp razor blade is necessary for cutting the filter cake prior to measuring its thickness.

(3) Properties of the filter cake, such as texture, hardness, flexibility, etc., may also be reported. The filtrate collected in the cylinder may also be tested for properties such as pH, total salinity, and contents of various ions. This is highly recommended by the writers.

It has been shown in Chapter 2 that the plot of the filtrate volume versus the square root of time is a straight line. It is, therefore, possible to run the test for a period of less than required 30 min. The volume of filtrate is noted at the end of 2, 5, 10, 15, and 30 min and values plotted on a special graph paper (see Figure 4-11A). The straight line, however, may not go through the origin, i.e., there may be a plus or minus value for the volume at zero time. For drilling fluids filtered through the proper filter paper, a plus value would indicate that excess amount of liquid was present in the filter paper or other parts of the equipment prior to the test, whereas a minus value would indicate that some filtrate had been lost in the filter paper or other parts of the equipment during the test. These plus or minus values, which are usually quite small, have been called the correction to the filtrate loss.

For drilling fluids with poor bridging qualities, relative to the filtering medium, the positive values may be quite large. This is especially true when filtrate losses are measured on sand beds or when mud with high sand content is being tested. These large positive water loss values have been termed mud spurts.

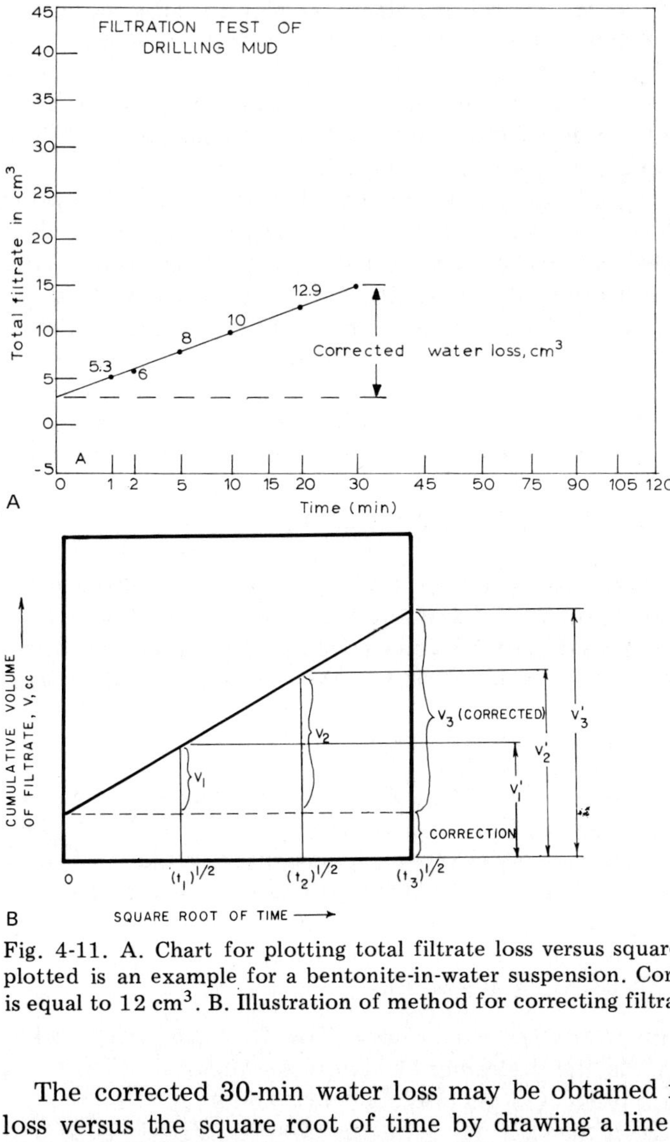

Fig. 4-11. A. Chart for plotting total filtrate loss versus square root of time. The curve plotted is an example for a bentonite-in-water suspension. Corrected, 30-min water loss is equal to 12 cm^3. B. Illustration of method for correcting filtrate loss.

The corrected 30-min water loss may be obtained from a plot of filtrate loss versus the square root of time by drawing a line that is parallel to the filtrate loss line previously plotted using the measured data and that passes through the origin. The volume of filtrate corresponding to the 30-min time on the latter line passing through the origin is, therefore, the corrected 30-min filtrate loss of the mud.

Another method of obtaining the corrected filtrate loss is illustrated in Fig. 4-11B. From the relations of the similar triangles, it follows that:

$$\frac{V'_3 - V'_2}{\sqrt{t_3} - \sqrt{t_2}} = \frac{V'_2 - V'_1}{\sqrt{t_2} - \sqrt{t_1}} = \frac{V_3}{\sqrt{t_3}} = \frac{V_2}{\sqrt{t_2}} = \frac{V_1}{\sqrt{t_1}} \quad (4\text{-}5)$$

where the volumes (V') with "prime" superscripts refer to uncorrected filtrate losses and the other values (V) refer to corrected filtrate losses.

When it is desired to test the filtration properties of muds containing extremely high sand content, the filter press may be tilted from its vertical position in such a manner as to prevent sand from settling on the filter paper. This will prevent obtaining erroneous results.

To meet the need for miniaturized testing equipment, a $\frac{1}{2}$-area filter press has been developed. Each unit can be used in the conventional manner or in the upside-down position to make measurements free of settling of the solids. Laboratory tests indicate that variation in filtration rate will be obtained due to solids settling. These units make possible a routine test to determine more nearly the true filtration and wall cake condition in the hole.

The $\frac{1}{2}$-area filter press has $\frac{1}{2}$ API filtration area and all filtrate volumes must be multiplied by 2 to correlate with API specifications. The $\frac{1}{2}$-area filter presses consist of a cell containing pressurizing inlet, pressure regulator, and gauge. A rubber boot is supplied to contain the mud and separate it from the pressurizing gas. The boot allows the filter cell to be operated in any position. A reversible frog mounting is provided on the cell to allow conventional filtration or upside-down filtration. A wall-mounting bracket is provided for supporting the filter press on wall or on a special stand. The operating procedure of the $\frac{1}{2}$-area filter press is similar to that of the conventional filter press unit.

Filtration of cement slurries

The low-pressure filter presses are also used for testing oilwell cements at room temperature. The filter medium is a 325-mesh stainless steel screen supported by a 60-mesh stainless steel screen. This is used in place of filter paper and 60-mesh screen which is standard for drilling fluid testing. In addition, a larger graduated cylinder may be required than is used in drilling fluid testing. The following operating procedures are in accordance with those recommended in the API RP 10B publication.

Preparation of filter press and cement slurry. The following preparation should be made prior to placing the cement slurry in the press:

(1) Assemble the filter press cylinder while all parts are dry.

(2) Prepare the slurry in accordance with Section 2, API RP 10B, and place it in the press as quickly as convenient handling will allow. The time between cessation of stirring and application of pressure must not exceed 2 min. Determine the slurry temperature in degrees Fahrenheit.

(3) When the test is used on field prepared slurries or on pressure- and/or temperature-treated slurries, state the method of preparation and handling.

Testing procedure of cement slurries. The testing procedure consists of the following steps:

(1) Fill the cylinder with at least two inches of cement slurry, cap the cylinder, and secure it in the frame. Place a dry graduated cylinder under the drain tube to collect the filtrate. Close the relief valve and apply 100 psi pressure within 5 sec.

(2) Time the test period from the instant of initial pressure application. Take filtrate readings at 15 sec, 30 sec, 1 min, 2 min, 5 min, and at 5-min intervals thereafter for 30 min. If dehydration occurs before the end of 30 min, record the time required for dehydration. At the end of 30 min, shut off the pressure supply and open the relief valve.

Results. Report the initial temperature of the slurry in degrees Fahrenheit. The volume of the filtrate should be recorded as follows:

(1) For 30-min test period, report the volume of filtrate as the fluid loss at 100 psi.

(2) For slurries which dehydrate in less than 30 min, or for tests shorter than 30 min, a hypothetical 30-min fluid loss value may be obtained by plotting the results on a log-log paper and extrapolating to the 30-min point or by using the following equation:

$$V_{30} = V_{\mathrm{t}} \times \frac{5.477}{\sqrt{t}} \qquad (4\text{-}6)$$

where V_{30} = the quantity of filtrate in 30 min, and V_{t} = the quantity of filtrate at the time of dehydration.

(3) Record the thickening time schedule employed.

All computed 30-min fluid loss values of cement should be so noted and should not be construed as the true fluid loss values.

High-pressure, high-temperature filter press

High-pressure, high-temperature filter press is especially designed for testing muds at elevated temperatures and pressures. It consists of a heating well with a thermostat, 250-ml filter cell, and pressure unit. The fluid loss value obtained with this apparatus more truly represents the actual fluid loss in the wellbore. Figure 4-12 shows the normal set-up of a high-pressure, high-temperature filter press. For tests which are to be run above 200° F, the back-pressure receiver must be used in place of the graduated cylinder to prevent evaporation of the filtrate. The 500-ml high-pressure, high-temperature filter press and a stainless steel aging cell are shown in Fig. 4-13.

In using the high-pressure, high-temperature filter press for cement testing, a special end cap and separate screen are used. The screen is a 325-mesh stainless steel screen supported by a 60-mesh stainless steel screen welded to a stainless steel ring that fits into the end cap. Testing procedure is similar to that used in room-temperature testing. Results are also noted and reported in the same manner.

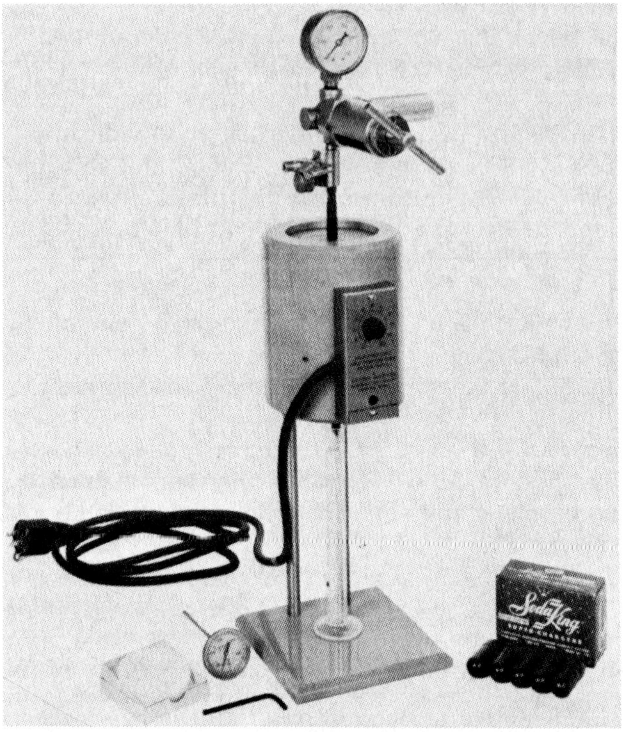

Fig. 4-12. Fully-assembled high-temperature, high-pressure filter press unit. (After Baroid Petroleum Services, 1978, Vol. 1, sect. 2, p. 14, fig. 18; courtesy of Baroid Petroleum Services, NL Industries, Inc.)

Some drilling muds, especially lime-treated muds, tend to thicken and, in some cases, to solidify when left under static conditions in a deep hot hole. This thickening impairs and sometimes prevents drilling and completion operations, such as logging, perforating, etc. High-temperature aging cells have been developed to aid in predicting the performance of drilling fluid under static, high-temperature conditions.

Research work on drilling fluids has indicated the desirability of applying pressure before heating the sample to elevated temperature in aging tests. Consequently, aging cells have been built which can be pressurized with nitrogen, carbon dioxide, or air to desired pressure in order to prevent boiling and evaporization of the mud before it reaches the test temperature.

Temperatures up to approximately 350°F and the resulting vapor pressures can be accomodated in aluminum-bronze aging cells which have been equipped with a valve for repressurizing from the same manifold used for the high-temperature, high-pressure filter press. Sufficient void space must be

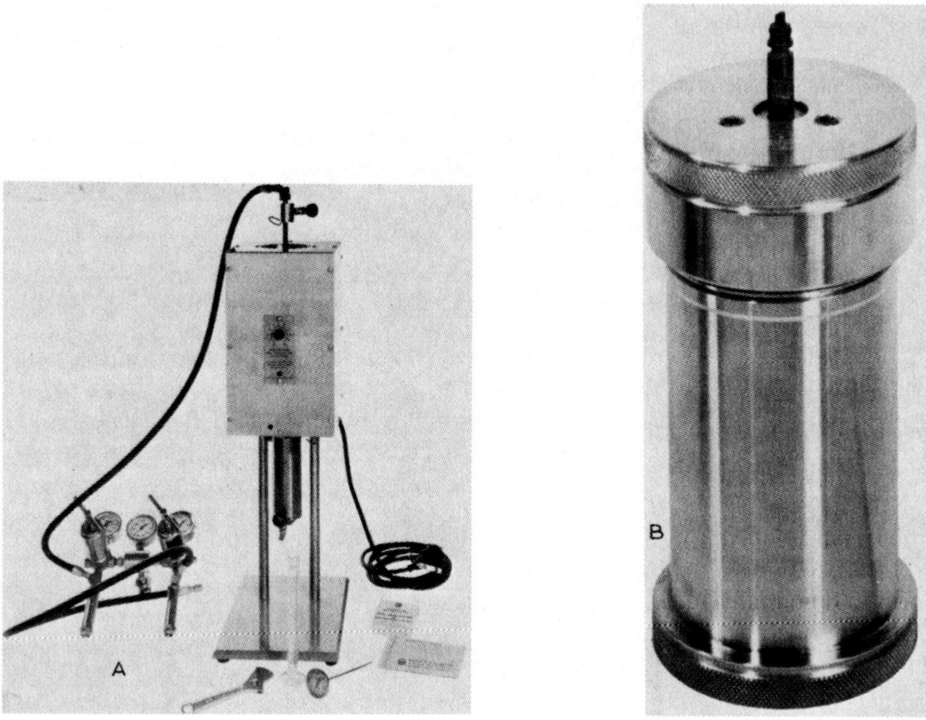

Fig. 4-13. A. A 500-ml high-temperature, high-pressure filter press with nitrogen manifold. It is used where higher fluid volumes, higher temperatures and greater volumes of filtrate are desired. B. A 500-ml stainless steel aging cell with valve for pressurized aging tests. (Courtesy of Baroid Petroleum Services, NL Industries, Inc.)

left in the cell to allow for the expansion of fluid. The chief limitation of the cell is the lack of sufficient volume for both the sample and the required void space.

For temperatures up to 600°F and pressures up to 1000 psi, a 500-ml stainless steel aging cell is available (Fig. 4-13). This cell utilizes a metal gasket seal and provides the extra safety factors required for this service. It can also be prepressurized from the filter press manifold, and can easily accomodate a 350-ml sample with adequate room for expansion. Shear tests can be run on the sample in the cell.

Procedure

The following procedures are applicable to this test:

(1) Pour the mud sample into a clean aging cell to within one-half inch of the lip. The cell holds approximately 200 ml of sample.

(2) Make sure the sealing edge of the cell is clean and put the inner cap in place. Then screw on the outer cap. With the Allen wrench tighten the small center screw in the middle of the screw cap.

(3) Place the cell in the portable aging oven and adjust to the desired aging temperature. A hot air oven may also be used provided a constant temperature may be maintained. Aging cells have a thermometer hole in the base for inserting a metal-stem thermometer.

(4) After the desired aging time, remove the cell from the oven and cool it to a temperature below 150° F.

(5) To open the cell, loosen the Allen screw, unscrew the outer cap, and remove the inner cap.

(6) Determine the condition of the aged mud and report as fluid, gelled, plastic, or hard. If a number is desired to demonstrate the condition of the aged mud, a shear measurement can be made as follows: A shear tube made from stainless steel is placed on the surface of the sample, and sufficient weight, if necessary, is placed on the tube to start its downward movement.

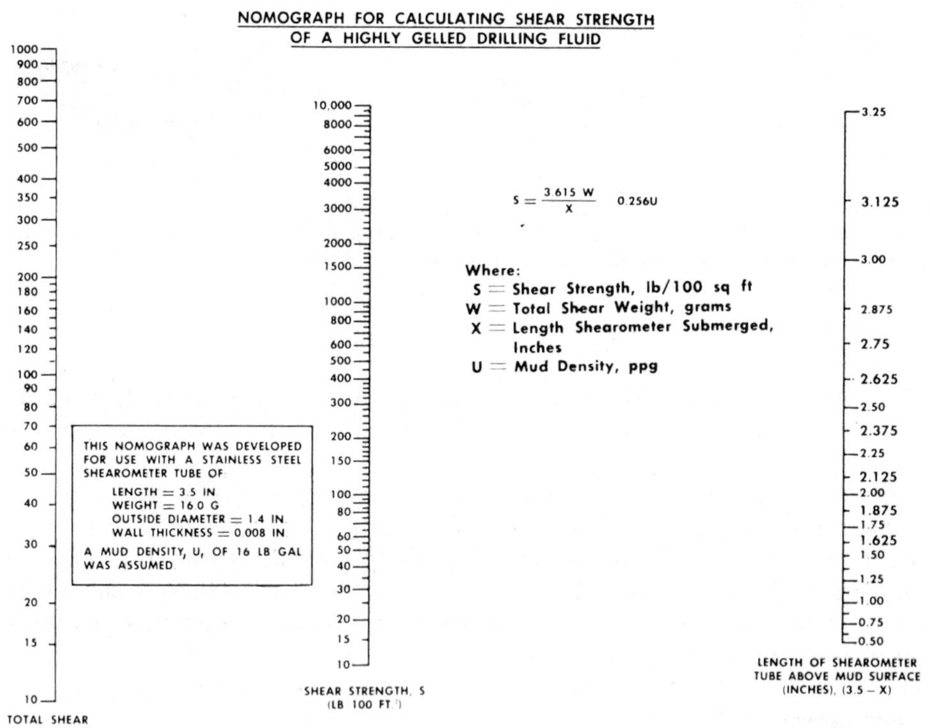

Fig. 4-14. Nomograph for calculating shear strength. (After Baroid Petroleum Services, 1978, p. 19, fig. 26; courtesy of Baroid Petroleum Services, NL Industries, Inc.)

TABLE 4-I

Mud volume and pressure for high-temperature aging tests.

(After Baroid Manual of Drilling Fluids Technology, 1978, vol. 1, sect. 2, p. 19, fig. 25; courtesy of Baroid Petroleum Services, NL Industries, Inc.)

Aging temperature (°F)	Water vapor pressure (psi)	Coefficient of expansion for water	Suggested applied pressure (psi)	Mud volume in cell (ml)	
				250-ml cell	500-ml cell
212	14.7	—	25	225	—
250	30	—	50	225	—
300	67	—	100	200	—
350	135	—	150	200	—
400	247	1.16	250	—	350
450	423	1.20	300	—	350
500	680	1.27	375	—	350
550	1100	1.36	500	—	300
600	1543	1.47	580	—	300

Unless too much weight has been placed on the tube, it will stop its downward motion where the shear strength of the gelled sample against the surface of the tube is sufficient to support the applied weight. The length of the tube exposed above the sample is measured. The shear strength in lb/100 sq ft is obtained from the graph shown in Fig. 4-14 by utilizing the force, in grams, applied to the shear tube, and the length of exposed tube after the tube reaches equilibrium.

To perform a pressurized test, caution must be exercised, with the following procedure being generally applicable to the above test:

(1) Determine the safe volume and safe initial pressure for the temperature at which the sample will be tested by using the data shown in Table 4-I. The cell must not be overfilled.

(2) Place safe volume of mud in cell, as determined from Table 4-I, assemble the cell, and pressurize it with nitrogen, carbon dioxide, or air to pressure shown under "applied pressure" column in Table 4-I to prevent vaporization. Normally, the manifold for the high-pressure filter press is used for pressurizing aging cells.

(3) Place cell in heating chamber and heat to desired temperature and for the desired time interval.

(4) Remove cell from heating chamber and air cool until sample temperature has been reduced to 300° F or less. The cell may then be cooled with air or water. Sample temperature must be less than 200° F before pressure is released and the cell opened. Do not handle the cell with bare hands because painful burns will result at temperatures in excess of 130° F.

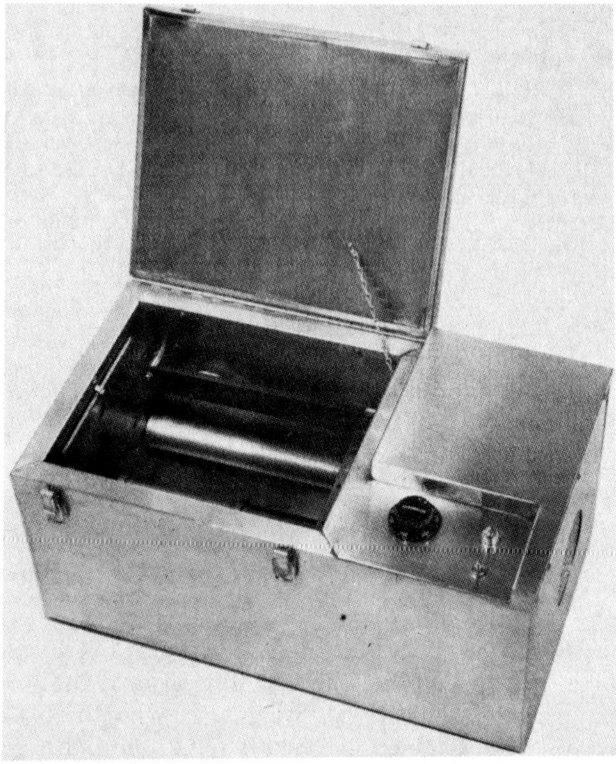

Fig. 4-15. Stainless steel portable roller oven. (After Baroid Services, 1978, vol. 1, sect. 2, p. 20, fig. 28; courtesy of Baroid Petroleum Services, NL Industries, Inc.)

(5) Examine the mud in open cell and report condition of the aged mud as fluid, gelled, plastic, or hard.

(6) Shear strength may be determined as described in step 7 under non-pressurized testing procedures.

In the investigation of performance properties of drilling muds and the evaluation of drilling mud additives, it is frequently desirable to simulate well conditions as closely as possible. The portable roller oven (Fig. 4-15) provides an excellent method of simulating the heating and agitation that the mud experiences in being circulated down the hole and back to the pit. When placed in the roller oven, a mud contained in a glass jar, the high-temperature aging cell, or the cell from the high-temperature, high-pressure filter press, is subjected to moderate heat and agitation on power-driven rollers. Such hot rolling is particularly of value in the investigation of mud in which a base exchange reaction occurs (such as lime-treated mud), in determining the stability of mud additives (such as thinners and organic colloids), and in determining the stability of muds (such as oil-base muds).

Most portable roller ovens are designed to provide combined heating and agitation of samples or either heating or agitation independently. The operation of the portable roller oven is very simple, involving plugging into the proper current source and setting thermostat dial to a desired temperature.

HYDROGEN-ION DETERMINATION

The degree of acidity or alkalinity of drilling mud is indicated by the hydrogen-ion concentration which is commonly expressed in term of pH, i.e., negative logarithm of H^+ concentration. A perfectly neutral solution has a pH of 7.0. Alkaline solutions have pH values ranging from just above 7 for slight alkalinity to 14 for the strongest alkalinity, whereas acid solutions range from just below 7 for slight acidity to less than 1 for strongest acidity.

The pH measurement is used as an aid in determining the need for chemical control of muds as well as indicating the presence of contaminants such as cement, gypsum, etc., which alter the hydrogen-ion concentration of the muds. The optimum pH for any drilling mud is dependent upon the type of mud being used.

Two satisfactory methods of measuring the pH of drilling mud are in use today: (1) a modified colorimetric method, such as the pHydrion dispenser of Baroid; and (2) an electrometric method using a glass-electrode instrument, such as the Beckman or analytical pH meter.

Figure 4-16 illustrates the pHydrion dispenser of Baroid, which provides a series of paper indicator strips that determine pH from 1.0 to 14.0. Changes in color or color intensity over the range of each indicator should be sufficient to allow the operator to read to within 0.5 pH unit. The following steps are followed in using the pHydrion dispenser:

(1) Remove about one inch strip of indicator paper from the pHydrion

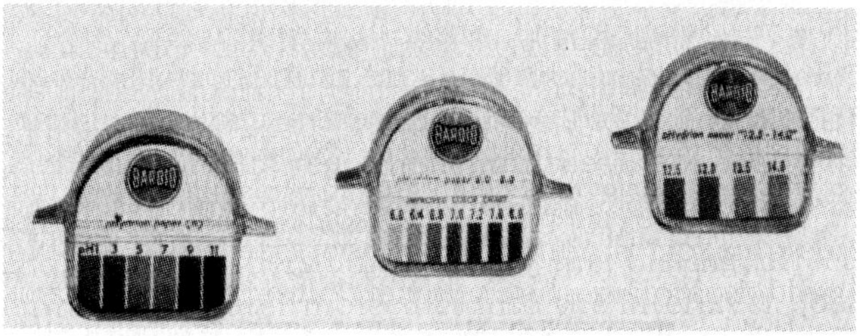

Fig. 4-16. pH-paper strip dispensers for different pH ranges. (After Baroid Petroleum Services, 1978, vol. 1, sect. 2, p. 22, fig. 31; courtesy of Baroid Petroleum Services, NL Industries, Inc.)

dispenser which is judged to be within the range required, and place it gently on the surface of the mud.

(2) Allow sufficient time to elapse for the paper strip to soak up filtrate and change color. This time laps may range from a few seconds to a couple of minutes.

(3) Match the color of the strip with the chart on the side of the dispenser from which the paper strip was taken and read the pH of the mud. If the color is off the scale and cannot be matched, repeat the test with a different indicator strip judged to be closer to the pH range required.

The pH measured by this method should be reported on a scale of 0 to 14 to the nearest 0.5 pH unit. It must be noted that the pHydrion dispenser does not give accurate pH measurement in muds having high salt concentrations. The electrometric pH meter is recommended for testing muds which contain Cl^- ion in excess of 10,000 ppm. The pHydrion dispenser should always be kept in a clean, dry place, preferably out of reach of direct sunlight.

The electrometric hydrogen ion determination offers a greater degree of accuracy than is possible with the colorimetric method. Glass-electrode instruments used for pH determinations should be of durable construction and combine simplicity of operation with the need for minimum of maintenance. Instruments satisfying these requirements are in widespread use. The analytical pH meter is one such instrument. The operation of all pH meters is based on the same principle, i.e., they involve the generation of an electrical potential in a glass and calomel electrode system by the solution under test. The electrical potential is amplified by means of a vacuum-tube system and the pH reading is indicated directly on a meter.

The procedure for determining pH of mud with these meters is as follows:

(1) Turn on the instrument and standardize according to the detailed directions supplied by the meter manufacturer. Buffer solutions having pH of 7, 9, and 12 are commonly used for standardizing pH meters.

(2) Wash the electrode tips under a stream of running water, or by stirring in clean water, and gently wipe dry with a piece of tissue paper.

(3) Insert electrodes into a 100-ml beaker containing a sample of the mud to be tested. Stir mud around electrodes by rotating the beaker.

(4) Measure the pH according to the detailed instructions attached to the instrument. Record the pH value after the meter reading becomes constant.

Meter reading is recorded in pH units on a scale from 0 to 14 to the nearest 0.1 pH unit. Inasmuch as the pH meter is a delicate instrument, the instrument must be thoroughly cleaned after each use by washing the electrodes under stream of running fresh water or by stirring in clean water and then gently wiping dry. Extreme care should be exercised in handling these electrodes. When not in use, the instrument should be stored in a cool, dry place and subjected to as little vibration and rough handling as possible.

The hydrogen-ion concentration of a drilling fluid filtrate may also be accomplished through titrating. Indicator solutions are available for determination of mud filtrate pH. The following procedure is generally followed:

(1) Collect mud filtrate using filter press method.

(2) Add 1 ml of mud filtrate to a clean titrating dish.

(3) Add 2 drops of phenolphthalein solution and observe the color. If no color develops, the pH is 8.2 or below. If a pink or red color appears, then the pH is 8.3 or higher.

(4) Add 1 ml of mud filtrate to a clean titrating dish. To this filtrate add 2 drops of thymolphthalein solution and observe color. If no color develops, the pH is below 10 and probably below 9.3, whereas if a light blue color appears, the pH is 9.3 to 10. A definite blue color indicates that the pH is above 10.

Filtrate alkalinity can be determined also in conjunction with this test: after observing any color change, the sample is titrated with standard sulfuric acid.

SAND CONTENT DETERMINATION

Periodic sand content determination of drilling mud is not only desirable, but sometimes imperative, because excessive sand may result in the deposition of a thick filter cake on the wall of the hole or may settle in the hole just above the tools when circulation is stopped, thus interfering with successful operation of drilling tools or setting of casing. A high sand content may also cause excessive abrasion of pump parts and pipe connections which results in costly repairs or replacements.

The initial sand content of drilling mud is usually nil; however, during drilling operations sand may be incorporated into the drilling fluid, thereby increasing its sand content. Different methods have been developed for determining the sand content: (1) elutriation method, (2) settling method, and (3) sieve analysis. Of these, sieve analysis is the most commonly used method because of its reliability and simplicity of equipment and operation. The volume of sand, including the void spaces between grains, is usually measured and expressed as percentage by volume of the mud. The sieve analysis method makes use of a 200-mesh sieve, funnel, and a glass measuring tube calibrated from 0 to 20% in order to read directly the percentage of sand content by volume (Fig. 4-17). The following procedure is followed in determining sand content:

(1) Pour mud into the sand content tube up to the mark labeled "Mud to Here" or similar marking. Then add water to the mark labeled "Water to

Fig. 4-17. Apparatus for sand content determination. (After Baroid Petroleum Services, 1978, vol. 1, sect. 2, p. 23, fig. 33; courtesy of Baroid Petroleum Services, NL Industries, Inc.)

Here" or other similar marking. Cover mouth of the tube with thumb and shake vigorously.

(2) Pour this mixture through the screen, being careful to wash everything out of the tube with clear water through the same screen. Wash sand retained on the screen with a stream of water to remove all mud and shale particles.

(3) Fit funnel down over top of the screen, invert slowly, turning tip of funnel into mouth of tube, and wash sand back into the tube with a fine spray of clear water on the back side of the screen. After allowing the sand to settle, observe the quantity of this sand in the calibrated tube recording it as the sand content of the mud.

The sand content of mud is reported in percent by volume. Obviously, the values obtained by this method are on the high side, because other coarse-grained materials are also measured as sand.

OIL, WATER, SOLIDS, AND CLAY CONTENT DETERMINATION

Knowledge of the liquid and solids contents of a drilling mud is essential for good control of mud properties. Such information will often explain poor performance of the mud and indicate if (1) the mud can best be conditioned by the addition of water, (2) treatment with chemical thinner is necessary, or (3) the removal of the contaminant is required. Similarly, a proper control of an oil emulsion mud depends upon a knowledge of its oil content.

For muds containing only water and solids, the quantity of each can be determined from the mud density or from the evaporation of a weighed sample of the mud. Oil and water contents can be obtained by distilling a measured volume of the mud and condensing and measuring the liquid (oil and water) fractions. The latter method can be used for oil emulsion muds.

The apparatus required for determining the oil, water, and solids content of the mud generally comes as a complete set known as the "Oil and Water Retort Kit". These kits, which are generally supplied with a metal portable carrying case, consist of (1) retort cell, (2) electrically-operated heater, (3) graduated cylinders, (4) a number of miscellaneous cleaning kits, (5) wetting agent, and (6) very fine steel wool.

The following steps are followed in calculating the contents of oil, water, and solids, all expressed on a volume percent basis:

(1) Fill mud chamber of retort cell with mud to the required level. Detailed instructions given by the manufacturer should be strictly followed.

(2) Retort is then heated to the desired temperature.

(3) Graduated cylinder is placed under the drain of the condenser to collect the liquids. It is generally recommended at this point that a drop of wetting agent be added to the graduated cylinder. This will improve the menisci for easier reading.

(4) Retorting should be continued until no more fluid is condensed and collected in the cylinder.

(5) Read the volume of oil and of water. Another drop of wetting agent added to the graduated cylinder at this time will greatly improve the menisci.

As a rule of thumb, for oil mud, the continuous phase of which is diesel oil, a retort time of 15—20 min may be required. For crude oil, a longer heating period may be required. Using this method, nearly 100% of refined oil may be recovered, whereas recovery for crude oil may be as low as 60%.

Cleanliness of equipment plays an important role in obtaining accurate results. The mud chamber and the lid must be free of dry mud to assure correct volume. The retort drain and condenser should be cleaned regularly with a pipe cleaner. The steel wool should be removed and replaced with fresh one before each test.

The following calculations are used in determining the solids content:

(1) Percent oil by volume = ml oil * × 10.

(2) Percent water by volume = ml water × 10.

(3) Percent solids by volume = 100 − [(ml oil * + ml water) × 10].

(4) Weight of oil in g = corrected ml oil × 0.8, where the specific gravity of oil is considered to be 0.8.

(5) Weight of water in g = ml water × 1.0, where the specific weight of water is considered to be 1.0 g/cm^3.

(6) Weight of mud in g = lb/gal (mud weight) × 1.2 **.

(7) Weight of solids in g = weight of mud in g − (weight of oil in g + weight of water in g).

(8) Volume solids in ml = 10 − (ml oil + ml water).

(9) Average specific gravity of solids $= \dfrac{\text{weight of solids, g}}{\text{volume of solids, cm}^3}$

(10) Solids percent by weight $= \dfrac{\text{weight of solids, g}}{\text{weight of mud, g}} \times 100.$

(11) High-gravity (sp. gr. = 4.3) solids percent by volume = (average sp. gr. of solids − 2.5) × 55.6.

(12) Low-gravity (sp. gr = 2.5) solids percent by volume = 100 − high-gravity solids percent by volume.

In the above calculations, the solid phase of the drilling fluid is considered to consist of two components: (1) high-specific gravity solids with a specific gravity of 4.3, and (2) low-specific gravity solids having specific gravity of 2.5. The total solids phase content, in volume %, is found by using the retort kit. It must be remembered that the results obtained from these equations are dependent on the accuracy of retort test results.

As a check on the above calculations, the average specific gravity of solids must fall between 2.5 and 4.3. If it is out of this range, an error has been made in tests or calculations.

For muds containing substantial quantities of salt, calculations of solids content will be in error unless a correction factor is applied, because any substantial amount of salt will greatly alter the mud weight without increasing the percent solids by volume. To obtain the content of salt by volume in mud, Fig. 4-18 may be used.

---

* When crude oil is used as a continuous phase in oil mud, a correction may need to be applied to determine the "ml of oil" in the formula. This correction is to account for the loss of crude oil (only 60% is recovered for oil mud using this method) during distillation. The correction factor is found by running sample of crude oil in the retort in order to determine the losses to be expected.

** 1.2 = Conversion of lb/gal to g/cm^3.

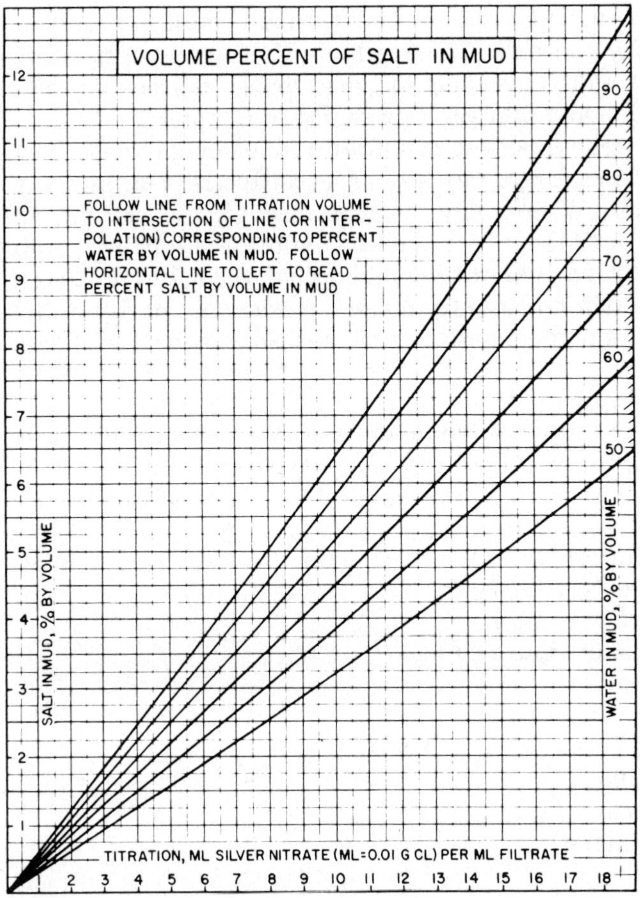

Fig. 4-18. Typical chart for finding the volume percent of salt in mud. (After Baroid Petroleum Services, vol. 1, sect. 2, 1978, p. 25, fig. 35; courtesy of Baroid Petroleum Services, NL Industries, Inc.)

ESTIMATION OF CATION EXCHANGE CAPACITY OF CLAYS

The methylene blue test is especially designed to determine the capacity of the clay to adsorb cations from solution. The method is particularly applicable to drilling fluids, but it can be effectively applied to aqueous dispersions of powdered clays, cores, drilled cuttings and foundry molding sands.

Like all cation exchange methods, the methylene blue test measures the total exchange capacity of the clay system, and is dependent on the type and content of clay minerals present. Only the reactive portions of solids, i.e., clays, are involved and such materials as finely-ground limestone, sand, or barite do not adsorb methylene blue. This method differs from most cation exchange methods in that a direct reading of the exchange capacity of clays is obtained. A detailed discussion of clays is presented in Chapter 5.

Inasmuch as the methylene blue test measures a fundamental clay property with ease and accuracy, it can be used as an analytical tool in many industrial clay applications. Where only a single clay type is present, an accurate estimate of the active clay content can be made. Moreover, even though unknown clay mixtures cannot be distinguished, the method offers a reasonable idea as to the predominant clay mineral present and purity of the clay sample in question. The methylene blue test can also provide useful data where determinations are made of the average specific gravity and volume percent of clay-containing solids.

In addition to its analytical aspects, the methylene blue test kit is useful as a control and research tool for relating exchange capacity to physical properties of clay-containing materials. A typical methylene blue testing kit consists of the following:

(1) Methylene blue solution (1 ml = 0.01 milliequivalents) containing 3.74 g USP-grade methylene blue per liter.
(2) Hydrogen peroxide, 3% solution.
(3) Dilute sulfuric acid, approximately 5 *N*.
(4) Distilled water.
(5) One 5-ml syringe or 1-ml pipette.
(6) One 10-ml pipette.
(7) Erlenmeyer flask, 250-ml.
(8) Medicine dropper, 1-ml.
(9) Graduated cylinder, 50-ml.
(10) Stirring rod.
(11) Hot plate or retort adapter.
(12) Filter paper.

The test procedure can be outlined as follows (see Fig. 4-19):

(1) Add 1 ml of mud sample, or suitable volume, to 10 ml of water in an Erlenmeyer flask. Add 15 ml of 3% hydrogen peroxide and 0.5 ml of sulfuric acid. Boil gently for 10 min and then dilute to 50 ml with distilled water.

(2) Add methylene blue solution to the flask. Use the 1-ml medicine dropper when approaching the endpoint. After each addition of 0.5 ml of dye, swirl the content of the flask for about 30 sec and, while the solids are still suspended, remove one drop of fluid with the stirring rod and place the drop on the filter paper. The endpoint of the titration is reached when the dye spreads as a greenish-blue ring or tint around the spot of the dyed solids.

(3) When the blue ring or tint is detected around the spot, shake the flask for an additional 2 min and place another drop on the filter paper. If the blue ring is again evident, the endpoint has been reached, whereas if the ring

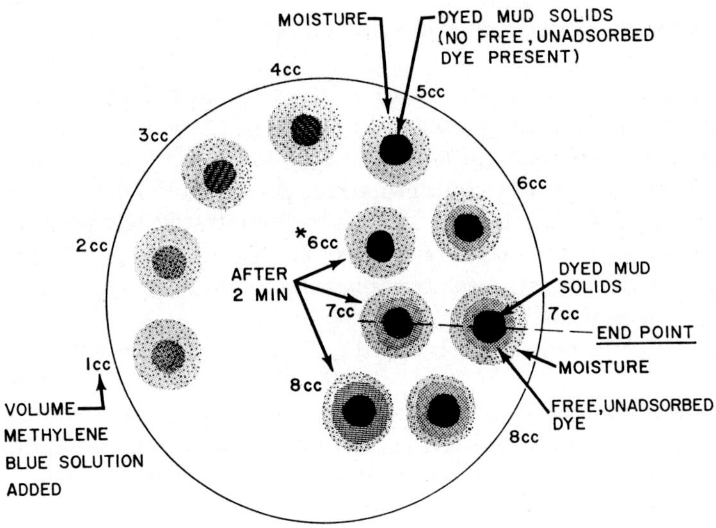

Fig. 4-19. Spot test for finding endpoint of methylene blue titration test for high-yield solids present in drilling fluids. (Courtesy of IMCO Services, 1974, fig. 9.1, p. 18) * = the free dye detected immediately after adding 6th cm^3 is adsorbed after two minutes, indicating that the endpoint has not been completely attained.

does not appear, continue the procedure until a drop taken after shaking for 2 min shows the blue tint.

(4) Report the cation exchange capacity of the mud as the methylene blue capacity, calculated as follows:

$$\text{Methylene blue capacity} = \frac{\text{vol. of methylene blue, ml}}{\text{vol. of mud, ml}} \tag{4-7}$$

Drilling fluids frequently contain substances other then bentonite clay that adsorb methylene blue. Treatment of the mud with hydrogen peroxide prior to titration is intended to remove the effect of organic materials such as CMC, polyacrylates, lignosulfonates, and lignites. If other adsorptive materials are not present in significant amounts, the bentonite content of the mud can be estimated as follows:

$$\text{Bentonite content in ppb mud} = 5 \times \text{methylene blue capacity} \tag{4-8}$$

Above procedure is also applicable to determine the cation exchange capacity of other clays, not necessarily used in drilling muds, such as shale clays incorporated into the mud during the process of drilling. A weighed quantity of that particular clay is dispersed into distilled water by a high-speed stirrer. The titration is carried out as above, except that hydrogen peroxide is not added. The cation exchange capacity of clays can be expressed as milliequivalents of methylene blue per 100 g of clay.

WATER AND MUD FILTRATE ANALYSIS FOR SALTS CONTENT

Frequently it is desirable or necessary to know the salt content of muds to explain certain aspects of their performance. Filtration, suspension, viscosity, and gel properties are adversely affected by salt, unless the mud is specifically designed to withstand salt contamination. Salt content determinations at regular intervals may be useful in identifying penetrated salt sections or influx of salt water into the mud system. In addition, changes in the salinity due to addition of formation waters may aid in identifying and predicting overpressured formations.

To determine the chloride content, which, in turn, represents the salt content of the drilling fluid, a sample of the mud filtrate is titrated with a standard silver nitrate solution, using potassium chromate as an indicator. The silver nitrate precipitates the chloride ion as silver chloride. When the chloride is completely precipitated, the addition of more silver nitrate produces a red color (due to silver chromate), which is taken as the endpoint. Results are reported in parts of chloride ion per million (ppm). This is a nominal value, as are all ppm values reported when a measured volume of sample is tested instead of a measured weight. The true ppm value can be obtained by dividing the titration value by the specific gravity of the sample. The difference is usually negligible, except for samples with high chloride concentrations. Reporting the nominal titration measurement (with no correction to density), however, conforms to API RP 13B — "Standard Field Procedure for Testing Drilling Fluids". All chlorides present in the filtrate, such as chlorides of magnesium and calcium, in addition to NaCl, are determined and reported as chlorides.

The following equipment and reagents are used for determining chloride content:

(1) Pipette, 1-ml.
(2) Pipette, 10-ml.
(3) Silver nitrate solution, 1 ml is equivalent to 0.01 or 0.001 g Cl^-.
(4) Distilled water.
(5) Potassium chromate solution.
(6) Polyethylene or porcelain titration dish and glass stirring rod.

The equipment and reagents are commonly assembled in a single unit for portability and ease of storage.

The procedure used to determine the chloride content using the above apparatus and reagents is as follows:

(1) Pipette 1.00 ml of sample into the titration dish and dilute to 40 or 50 ml with distilled water.

(2) Add four or five drops of potassium chromate indicator solution.

(3) Add standard silver nitrate solution from a pipette dropwise and slowly, while stirring continuously with a stirring rod, until the sample just

changes color from yellow to orange or brick red.

(4) The number of ml of standard silver nitrate used to obtain this endpoint is multiplied by 10,000 when using the 0.01-g silver nitrate solution or by 1000 when the 0.001-g silver nitrate solution is used. This gives results in parts per million (ppm) of chloride (Cl) ion. To express the results in terms of parts per million NaCl, multiply the ppm of Cl^- by 1.65.

In some cases the above procedure may have to be modified in order to obtain more reliable results. When the salt content is higher or lower than usual, the volume of the sample may be adjusted to use more practical volumes of standard silver nitrate solution, not to exceed 10.00 ml. If the volume of sample used is not equal to 1.00 ml, divide the result obtained in step No. 4 by the number of ml of sample used in order to obtain the correct concentration.

If the sample is alkaline prior to dilution in step No. 1, it should be made colorless to phenolphthalein by using $N/10$ sulfuric acid.

If the filtrate is so deeply colored by the thinner that the endpoint cannot be seen in the diluted sample, the following procedure may be used:

(a) Add 1 *N* nitric acid to a measured volume of the original filtrate until the color changes to tan.

(b) Add a small quantity of chloride-free, powdered, activated charcoal, stir thoroughly, and then filter.

(c) Wash with distilled water, collecting the washing with the first filtrate until the volume reaches 50 ml.

(d) Add an excess of powdered c.p. calcium carbonate so that some remains undissolved and proceed with step No. 2 of the procedure.

If sulfides are present, however, as indicated by a brownish black precipitate, take a new sample and add 1 ml of 1 *N* nitric acid with 2 ml of potassium chromate indicator. If it is also necessary to remove the thinner, it should be done at this point, i.e., starting with procedure (b) above. If removal of thinner is not necessary, add 1 g of powdered c.p. calcium carbonate, dilute to 50 ml, and proceed with step No. 3 of the procedure.

The procedure described above only pertains to the salt content determination of a water-base mud. For estimating the salt content of oil muds, a different procedure is required. It is recommended to report the following information for oil muds: (a) calcium chloride content in lb/bbl, (b) sodium chloride content in lb/bbl, (c) soluble sodium chloride content in lb/bbl, and (d) water-phase salinity in ppm.

Tests required to obtain the above-listed information are: (1) retort analysis in order to determine the water content, (2) test for calcium content, and (3) test for chloride content.

In order to determine the calcium chloride content, the following procedure is commonly followed:

(1) Transfer 10 ml of the mud to be tested into a stirring cup.

(2) Add 20 ml of diesel oil or solvent and mix thoroughly with the mud.

(3) Add 10 ml of suitable surfactant and 210 ml of distilled water to the solvent and mud mixture and stir for 15 min using a high-speed electric mixer (e.g., Hamilton Beach mixer).

(4) After stirring, pipette a 1.0-ml sample of the resulting emulsion into 50 ml of distilled water in a titrating dish.

(5) Add Versenate hardness buffer solution (10—15 drops) and Versenate hardness indicator solution (10—15 drops). Titrate with total hardness titrating solution (1 ml = 20 epm). (See Chilingar and Terry, 1954.)

(6) Calculate the pounds of calcium chloride per barrel of mud by using

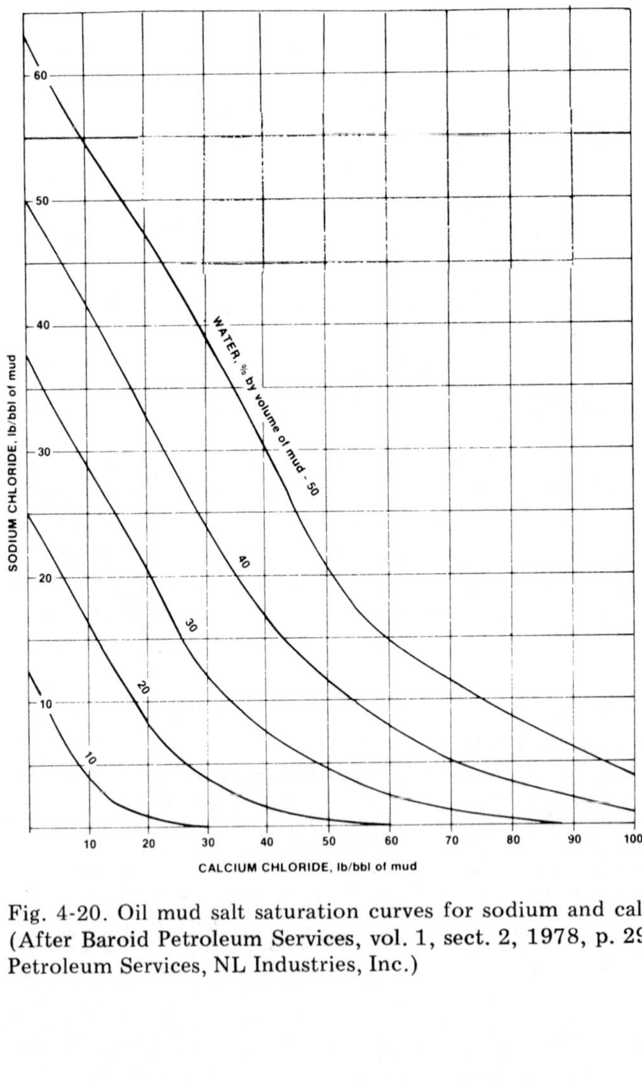

Fig. 4-20. Oil mud salt saturation curves for sodium and calcium chloride combinations. (After Baroid Petroleum Services, vol. 1, sect. 2, 1978, p. 29, fig. 39; Courtesy of Baroid Petroleum Services, NL Industries, Inc.)

the following formula:

$$\text{lb } CaCl_2\text{/bbl mud} = (9.6 \times \text{ml of total hardness titrating solution}) \quad (4\text{-}9)$$

The sodium chloride content may be determined by following the steps listed below:

(1) Transfer another 1.0-ml sample, from the solution prepared for calcium chloride content determination, to a clean titrating dish containing 50 ml of distilled water.

(2) Titrate the sample using the conventional procedure for determining the chloride content. Strength of silver nitrate solution: 1 ml neutralizes 0.001 g chloride ion.

(3) Calculate the pounds of sodium chloride per barrel of mud by using the following formula:

$$\text{lb NaCl/bbl mud} = (14.5 \times \text{ml silver nitrate solution}) - (1.06 \times \text{lb } CaCl_2\text{/bbl mud}) \quad (4\text{-}10)$$

The total soluble salts are determined as follows:

(1) Using Fig. 4-20, determine the maximum amount of soluble NaCl (ordinate) by entering the chart at the lb/bbl $CaCl_2$ scale (abscissa) and

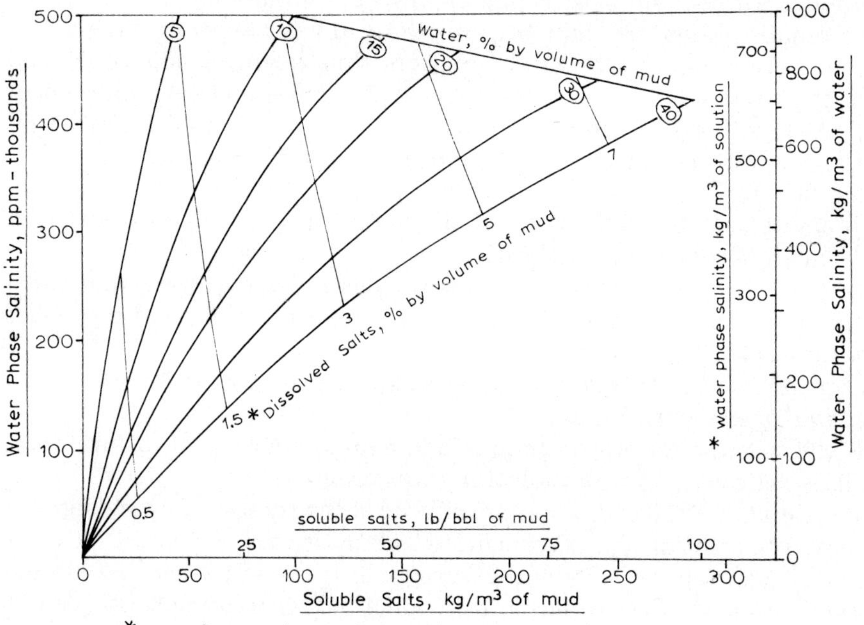

Fig. 4-21. General chart for estimating soluble salt, lb/bbl mud, and dissolved salts, % by volume of mud, knowing water phase salinity of mud and water content, % by volume of mud. (After Baroid Petroleum Services, vol. 1, sect. 2, 1978, p. 30, fig. 40; courtesy of Baroid Petroleum Services, NL Industries, Inc.)

moving up to the curve designating percent by volume of water, which was determined by retort analysis.

(2) If the NaCl content determined above is greater than the maximum soluble NaCl content determined from Fig. 4-20, use only the portion which is soluble for calculating the total soluble salts.

(3) Calculate the total concentration of soluble salts (lb/bbl) in the mud by using the following formula:

$$\text{Total soluble salts in lb/bbl mud} = \text{lb } CaCl_2\text{/bbl mud} + \text{lb soluble NaCl/bbl mud} \quad (4\text{-}11)$$

(4) Determine the water phase salinity from Fig. 4-21. Enter the lb of soluble salts/bbl mud scale (abscissa) and move up to the correct percent by volume of water curve. Then read the water phase salinity in ppm from the left hand scale (ordinate).

ALKALINITY OF OIL MUDS

Sometimes it may be desirable to determine the alkalinity of an oil mud. Back titration technique may be used to accomplish this goal, which consists of titrating 1.0 ml of mud to a phenolphthalein endpoint with acid. The titration requires about 30 min because the endpoint fades as the alkaline constituents are slowly leached into solution, thus requiring additional acid to reach the true endpoint. A procedure which utilizes an excess amount of acid to react with all of the alkaline materials in the oil mud, followed by back titration with caustic soda to obtain the amount of unreacted acid, is presented below:

(1) Measure a 50-ml portion of Exosol—isopropyl alcohol mixture (75 : 25) in an 8-oz glass sample bottle.

(2) Disperse 1.0 ml of oil mud into the solvent and add 100 ml of distilled water, 10 to 15 drops of phenolphthalein indicator solution, and 5.0 ml of $N/10$ sulfuric acid.

(3) Cap the bottle and shake vigorously for one minute, being careful to completely disperse the oil mud.

(4) If the solution is red, continue to add 5-ml increments of sulfuric acid, followed by agitation, until the solution turns clear.

(5) Titrate with $N/10$ sodium hydroxide until the solution first changes to a red color. The solution should be agitated while titrating.

(6) If less than 0.5 ml of sodium hydroxide is required to reach a red endpoint, then rerun the test using an additional 5.0-ml increment of sulfuric acid before titrating with sodium hydroxide.

(7) Calculate the alkalinity of the oil mud by using the following formula:

$$\text{Alkalinity} = \text{vol. of } N/10 \text{ sulfuric acid in ml} - \text{vol. of } N/10 \text{ sodium hydroxide in ml} \quad (4\text{-}12)$$

ALKALINITY DETERMINATION OF MUD FILTRATES

Alkalinity or acidity is measured by pH, as previously discussed, but the properties of a mud may fluctuate despite constant pH. This is due to the effect of the varying amounts and nature of the ions present. Analysis of the mud filtrate to determine the character of the alkalinity may assist in maintaining the desired physical properties of the drilling fluid.

The following equipment and materials, generally available as a complete kit, are necessary for alkalinity testing:

(1) Phenolphthalein indicator solution.
(2) Sulfuric acid, $N/50$ (0.02 N).
(3) Methyl orange indicator solution.
(4) Distilled water.
(5) Polyethylene or porcelain titration dish, 140-ml.
(6) Pipettes, 1 ml × 0.01 ml and 10 ml × 0.1 ml.
(7) Polyethylene or glass stirring rod.

The procedure described below is commonly followed in determining the alkalinity:

(1) Measure one or more ml of fresh filtrate into a titration dish, without diluting with water.

(2) Add 2 or 3 drops of phenolphthalein indicator solution.

(3) Add 0.02 normal ($N/50$) sulfuric acid from a pipette, stirring continuously until the color changes from red to that of original sample. If the sample is so colored with thinner that this endpoint is masked, the endpoint is taken at pH of 8.3 using a pH meter. The volume (ml) of 0.02 normal sulfuric acid divided by the volume of sample taken (ml) is called the P alkalinity of the filtrate or P_f.

(4) To the sample that has been titrated to the P endpoint add 2 or 3 drops of methyl orange indicator solution. Methyl purple may be used instead of methyl orange on dark filtrates.

(5) Now continue to add 0.02 normal sulfuric acid, while stirring continuously, until the sample changes color from yellow to a salmon pink. If the sample is so colored with thinner that this endpoint is masked, the endpoint is then taken at pH of 8.3 using the pH meter. Record as M_f the total volume of acid per ml of sample used to reach the methyl orange endpoint (including that for the P endpoint).

(6) Report P_f alkalinity as ml of 0.02 normal acid per ml of filtrate required to reach phenolphthalein endpoint.

(7) Report the M_f alkalinity as the total ml of 0.02 normal acid required per ml filtrate to reach methyl orange endpoint.

LIME CONTENT ESTIMATION

An estimation of lime content can be based on an alkalinity titration of the filtrate and of the whole mud. The titration of the mud must be made rapidly to permit titration of calcium hydroxide and sodium hydroxide without interference from calcium carbonate.

The same testing kit listed for "Alkalinity Determination" may be used for this purpose with the addition of a 1-ml or 2-ml hypodermic syringe. The testing procedure differs, however:

(1) Measure 1 ml of mud into a casserole with syringe and dilute to approximately 50 ml with distilled water. Stir well to disperse mud.

(2) Add 2 or 3 drops of phenolphthalein indicator solution.

(3) Add 0.02 normal ($N/50$) sulfuric acid (or 0.1 normal acid if the lime content is known to be above 10 lb/bbl) rapidly from a pipette, stirring continuously, until the sample first turns from red to the color of the mud. The volume of 0.02 N acid (ml) divided by the volume of sample (ml) taken is called the P of the mud (P_m). Results must be multiplied by 5 if 0.1 N acid is used.

(4) Determine the P alkalinity of the filtrate (P_f) following the procedure for "Alkalinity Determination".

(5) The lime content is calculated as follows:

$$\text{lime in lb/bbl} = 0.26 \times (P_m - \text{water fraction of mud} \times P_f) \qquad (4\text{-}13)$$

(6) The water fraction of the mud (% water by volume divided by 100) can be determined following the procedure for "Oil and Water Determination". If the water content is not known, the estimated lime content may be obtained, using the nomograph, on the basis of the average amount of the water volume for a given mud weight.

(7) Report the estimated lime content in lb/bbl of mud. (Also see Fig. 4-22.)

A different procedure must be followed when determining lime content or alkalinity of oil mud, i.e., an aliquot of oil mud is dispersed into a solvent–water solution and then titrated with acid.

The following equipment and materials, in addition to those necessary for water-base mud alkalinity and lime tests, are required:

(1) Graduated cylinder, 50-ml.

(2) Exosol–isopropyl alcohol mixture (75 : 25).

(3) $N/10$ sulfuric acid.

The procedure can be outlined as follows:

(1) Measure 20 ml of Exosol–isopropyl alcohol mixture into a 250-ml glass beaker.

(2) Disperse 1 ml of oil mud into the solvent and add 75 ml of distilled water.

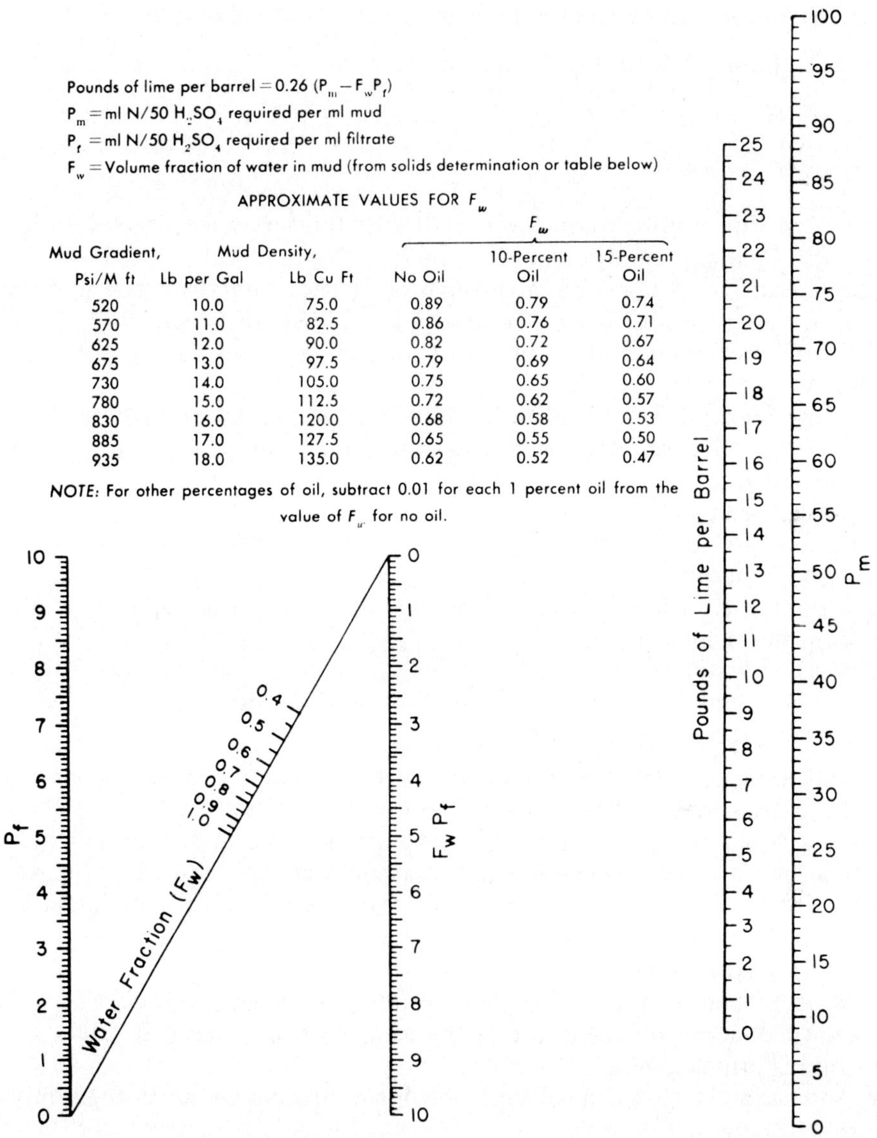

Pounds of lime per barrel = 0.26 ($P_m - F_w P_f$)

P_m = ml N/50 H_2SO_4 required per ml mud

P_f = ml N/50 H_2SO_4 required per ml filtrate

F_w = Volume fraction of water in mud (from solids determination or table below)

APPROXIMATE VALUES FOR F_w

Mud Gradient, Psi/M ft	Mud Density, Lb per Gal	Mud Density, Lb Cu Ft	F_w No Oil	F_w 10-Percent Oil	F_w 15-Percent Oil
520	10.0	75.0	0.89	0.79	0.74
570	11.0	82.5	0.86	0.76	0.71
625	12.0	90.0	0.82	0.72	0.67
675	13.0	97.5	0.79	0.69	0.64
730	14.0	105.0	0.75	0.65	0.60
780	15.0	112.5	0.72	0.62	0.57
830	16.0	120.0	0.68	0.58	0.53
885	17.0	127.5	0.65	0.55	0.50
935	18.0	135.0	0.62	0.52	0.47

NOTE: For other percentages of oil, subtract 0.01 for each 1 percent oil from the value of F_w for no oil.

Fig. 4-22. Nomograph for estimating lime content of lime-treated muds. (Courtesy of Magcobar Services, 1972, fig. 15, p. 14, chapter 1.)

(3) Add phenolphthalein indicator solution (8—10 drops).

(4) Titrate slowly with $N/10$ sulfuric acid until red color disappears permanently from both the solvent and the water phase, while stirring vigorously with stirring rod.

(5) Oil mud alkalinity may be expressed as ml of required $N/10$ sulfuric

acid or as lime content calculated by using the following formula:

$$\text{lb/bbl (ppb) lime} = 1.5 \times \text{ml } N/10 \text{ sulfuric acid} \tag{4-14}$$

CALCIUM ION CONTENT

Proper control and maintenance of a drilling fluid may require determination of the quantity of calcium ion present. Contamination with calcium ions may result from (1) drilling through cement or anhydrite and gypsum formations, (2) addition of hard make-up water, (3) treatment with chemicals containing calcium, and (4) influx of formation waters having high calcium content.

Versenate method (see Chilingar and Terry, 1954) may be used to determine accurately the calcium and magnesium ion contents in a sample of water or mud filtrate. The following apparatus and reagents are used in determining total hardness by the versenate method:

(1) Hardness buffer solution.

(2) Hardness indicator solution.

(3) Hardness titrating solution, 1 ml is equivalent to 20 epm or 2 epm (Ca + Mg) using 1 ml sample.

(4) Distilled water.

(5) Polyethylene or porcelain titration dish and stirring rod.

(6) Graduated pipette, 10.00-ml.

The following procedure is applicable for the versenate test for hardness:

(1) To approximately 50 ml of distilled water in a titration dish add about 2 ml of hardness buffer solution and 5—10 drops of hardness indicator solution. If a red color develops, indicating hardness in the distilled water, add hardness titrating solution dropwise until the water first turns to blue. Do not include this volume of titrating solution in calculating hardness of the sample in step No. 4 below.

(2) Measure one or more ml of sample into the casserole. A pink to wine red color (depending on the color of the sample) will develop if calcium or magnesium is present.

(3) Add hardness titrating solution, stirring continuously, until the sample first turns to blue.

(4) The total hardness is calculated as follows:

$$\text{ml of titrating solution} \times 20 = \text{epm of Ca + Mg (if hardness titration solution of 1 ml = 20 epm is used)} \tag{4-15a}$$

and

$$\text{ml of titrating solution} \times 2 = \text{epm of Ca + Mg (if hardness solution of 1 ml = 2 epm is used)} \tag{4-15b}$$

(5) Larger or smaller sample can be used to obtain a titration between 1 and 10 ml, in which case the answer obtained in step 4 should be divided by the ml of sample used. Report the concentration of total hardness in equivalent parts per million (epm).

It must be noted that although the final result is reported as calcium content (API RP 13B), both calcium and magnesium ions are included. Test results are also sometimes erratic at low temperatures, probably due to the change in solubility of the ingredients.

To overcome the above problem, versenate test for calcium ion content may be run by following the procedure listed below:

(1) To approximately 50 ml of distilled water in a titrating dish, add a pinch of calcium versenate indicator powder (murexide-sodium chloride). Then add 1 ml of calcium buffer solution. If a pink color develops, indicating presence of calcium ion in the distilled water, add hardness titrating solution until appearance of a purple color. This volume of titrating solution should not be included in calculating calcium content of the sample in step No. 4.

(2) Measure 1 ml of sample into the titrating dish. A pink color will appear if calcium is present.

(3) Add hardness titrating solution until the water turns purple.

(4) The calcium is then determined as follows:

ml of titrating solution × 20 = epm Ca, if hardness titration solution of 1 ml = 20 epm is used (4-16a)

and

ml of titration solution × 2 = epm Ca, if hardness titration solution of 1 ml = 2 epm is used (4-16b)

(5) One can vary the size of the sample with resulting variation in the volume of titrating solution (1—10 ml). In such a case, the answer obtained in step No. 4 should be divided by the volume (ml) of sample used.

(6) To determine the Mg content, subtract the obtained Ca content from the total hardness (Ca + Mg), which can be determined by versenate test. Report the concentration of magnesium and calcium ions separately in equivalent parts per million (epm).

For complete understanding of tests for total hardness and Ca^{2+} and Mg^{2+} contents, the reader should consult Chilingar and Terry (1954) and Chapter 1 in both Magcobar (1972) and IMCO Services (1979) manuals.

SULFIDE ION DETERMINATION

Sulfide ion content of drilling fluid can be determined by the color chart test. This method involves comparing the color of the test paper to a standard sulfide color chart, as described in the following procedure:

(1) Place 2.5 ml of filtrate or mud in test bottle and add 22.5 ml of fresh water. Soluble sulfides can be determined by using mud filtrate, whereas total sulfides can be determined using mud.

(2) Insert test paper inside the cap of the test bottle.

(3) Place a film of defoamer above the water-line in the test bottle to eliminate excessive foaming.

(4) Add one ml of sulfuric acid (1 : 6 strength).

(5) Add one *fresh* Alka-Seltzer tablet and immediately screw the cap containing the test paper onto the bottle.

(6) After the tablet has dissolved, remove the test paper and compare with the Hach color chart. Multiply result by 10.

ANILINE POINT DETERMINATION

The aniline point of oil in oil-base drilling fluids is the guide to whether damage can be expected to occur to rubber parts of a drilling rig circulating system.

Oils with a high aromatics content are more detrimental to rubber products than those having low aromatics content. The relative aromatics content of an oil is indicated by its aniline point. Inasmuch as oils having low aromatics content have high aniline points and vice versa, the oils with high aniline points are the most desirable for use in drilling fluids. As discussed in Chapter 11, the aniline point of the oil-base muds should be 140° F or above to minimize damage to the rubber parts.

Aniline point testing kit generally consists of the following equipment and materials:

(1) Test tube.
(2) Stopper with hole.
(3) Thermometer, 0—220° F range.
(4) Aniline solution.
(5) Heating source.

The aniline point of an oil is defined as the lowest temperature at which equal volumes of fresh aniline and oil are completely miscible (see Chilingar, 1956). This lends itself to a simple testing procedure:

(1) Clean and dry the apparatus (Fig. 4-23). Measure equal portions of the aniline solution and the oil to be tested, commonly 4 ml of each, into the test tube.

(2) Place stopper in test tube and insert the thermometer, making sure its bulb does not touch the bottom or sides of the tube.

(3) Heat the tube slowly while stirring the mixture. The stirring can be done by moving the thermometer up and down or using a stirrer. Continue

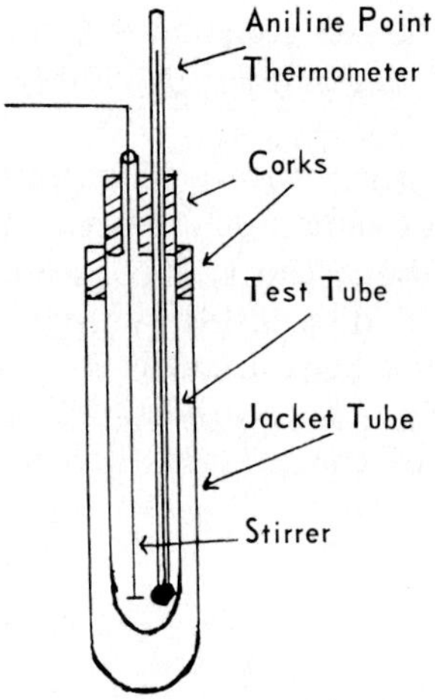

Fig. 4-23. Schematic diagram of an apparatus for determining the aniline point. (Courtesy of IMCO Services, 1976, fig. 10, p. 22.)

heating and stirring until complete miscibility occurs, i.e., when the mixture becomes clear.

(4) Remove the tube from heat source and continue stirring until aniline—oil mixture becomes cloudy. Record thermometer temperature in °F at cloud point (appearance of cloudiness), which is the aniline point.

If the aniline and oil are completely miscible at room temperature, the mixture will have to be cooled below this temperature to obtain the aniline point. Inasmuch as aniline is an aromatic, this indicates that oil has a high content of aromatics.

Great care should be exercised during testing, because aniline solution is extremely poisonous and hazardous. Contact with skin and clothing, and breathing the vapor must be avoided. In case of contact, immediately remove all contaminated clothing, including shoes, and flush skin with water for at least 20 minutes. If taken internally and the patient is still conscious, induce vomitting by giving a tablespoon of salt with warm water and repeat until the vomitting fluid is clear.

RESISTIVITY DETERMINATION

Control of resistivity of a mud and mud filtrate while drilling may be desirable and, at times, necessary to permit better evaluation of formation characteristics from electric logs. The determination of resistivity is essentially the measurement of the resistance to flow of electrical current through a sample of known configuration. Measured resistance is converted to resistivity in ohm-meters by use of a cell constant. The cell constant is fixed by the configuration of the sample in the cell and is determined by calibration with standard solutions of known resistivity.

Any type of cell and instrument which is sufficiently accurate to permit determination of resistivity to within 5% of the correct value may be used. If the instrument measures the sample resistance in ohms, the cell constant must be known. The resistivity in ohm-meters is obtained by multiplying the resistance in ohms by the cell constant in m^2/m. If the instrument is a type of direct-reading resistivity meter, the cell constant has been adjusted to a particular value or accounted for in the circuitry of the meter. Such an

Fig. 4-24. A typical resistivity meter used to measure the resistivity of filtrate, mud, and mud cake. (After Baroid Petroleum Services, 1978, vol. 1, sect. 2, p. 38, fig. 48; courtesy of Baroid Petroleum Services, NL Industries, Inc.)

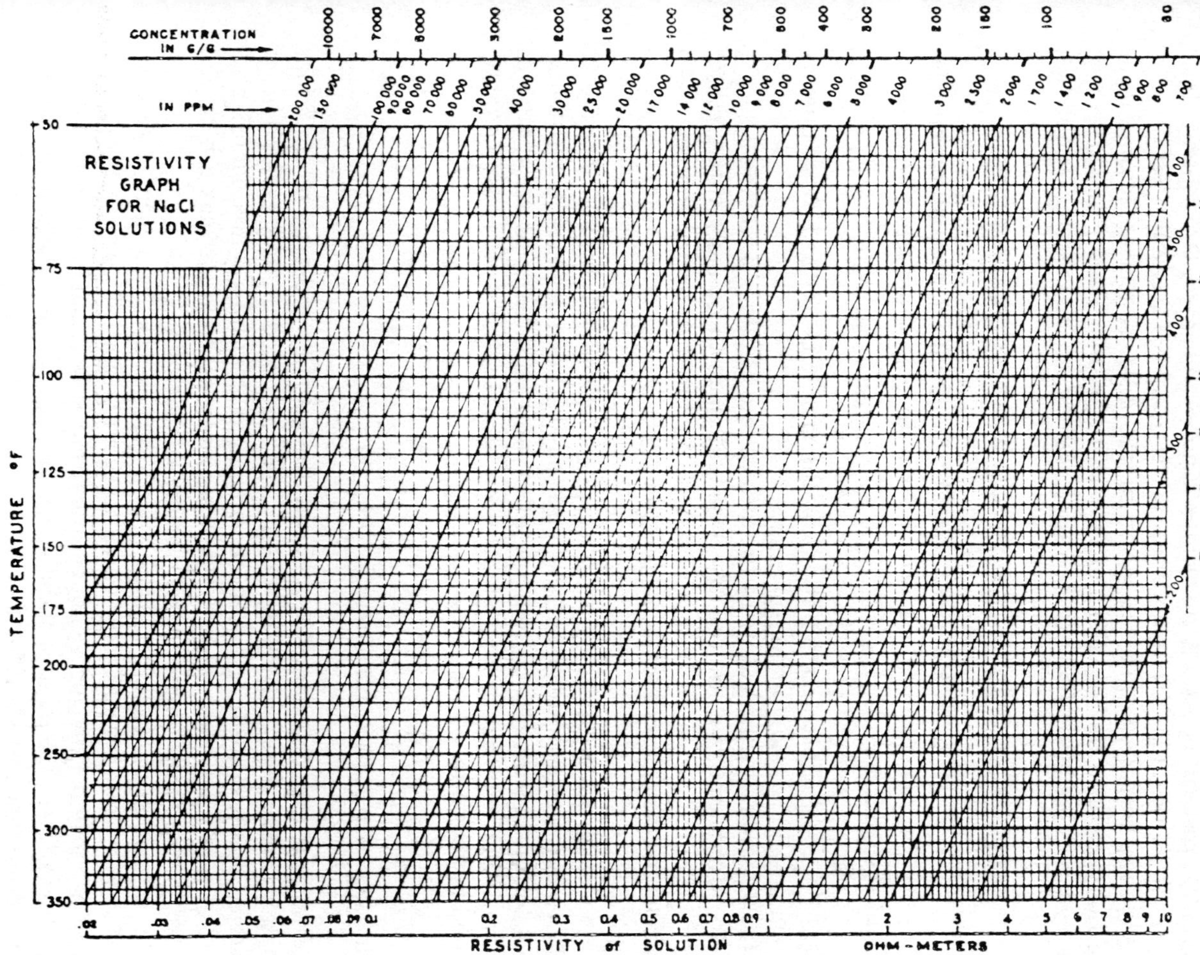

Fig. 4-25. Resistivity graph for NaCl solutions. (After Baroid Petroleum Services, 1978, vol. 1, sect. 2, p. 39, fig. 49; courtesy of Baroid Petroleum Services, NL Industries, Inc.)

instrument measures the sample resistance and converts it to the resistivity so that the reading is taken directly in ohm-meters.

The resistivity meter is a transistorized meter for measuring the resistivity of fluids, slurries, or semi-solids having resistivities of 0.01 to 10 ohm-meters. Measurements can be made quickly and accurately. Conductivity of the medium being measured is equal to the reciprocal of the resistivity measurement. Insofar as testing the drilling fluid is concerned, two separate tests are generally run: (1) filtrate and/or mud test, and (2) mud cake test.

A typical resistivity meter is shown in Fig. 4-24. A special feature of this resistivity meter is its fluid cell and the suction bubble. During measurement, the cell is filled with fluid to be tested. Filling of the cell is accomplished with the use of the suction bubble.

The resistivities of filtrate and mud are measured by using the following procedure:

(1) Fill the cell with fluid; one must be careful to remove all air bubbles by filling and exhausting fluid two or three times to thoroughly wet the surface of the cell.

(2) Connect cell to the resistivity meter.

(3) Follow the instruction put out by the manufacturer on how to operate the meter and to take the measurements.

(4) Record the temperature of the cell at time of taking a reading.

(5) Always remove the cell after use and clean thoroughly with fresh or distilled water. Be careful not to scratch the inner surface of the cell when using a pipe cleaner. Solvents must not be used.

Mud cake obtained from the filter press is used for testing. Be sure to remove any excess water from the mud cake before measuring its resistivity. The slot on top of the cell must be filled with mud cake. Interrelationships between NaCl concentration, temperature, and resistivity of solution are presented in Fig. 4-25.

EMULSION STABILITY DETERMINATION

Emulsion testers are used to indicate the stability and type of emulsions whether water-in-oil or oil-in-water. They are used in the evaluation of inverted emulsion drilling fluids, cements, and fracturing fluids. Time stability and resistance to electrolyte contamination of these systems can be predicted from a measurement of relative emulsion stability.

In the emulsion tester, voltage is imposed on the fluid using two electrodes and then gradually increased until initiation of current flow. Inasmuch as current flow (or a change in current flow) is an indication of emulsion breakdown, the relative stability is indicated by the voltage at the breakdown point (see IMCO, 1974).

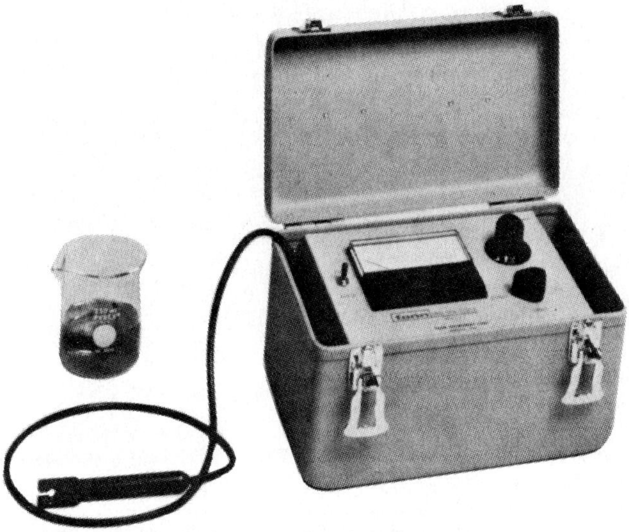

Fig. 4-26. **Emulsion tester with extended probe.** (After Baroid Petroleum Services, 1978, vol. 1, sect. 2, p. 39, fig. 50; courtesy of Baroid Petroleum Services, NL Industries, Inc.)

Emulsion testers available on the market have many different designs. For detailed testing procedure, therefore, the reader must consult the operating manual of each tester. Figure 4-26 illustrates a typical emulsion tester with a probe and a meter box. In performing the test, care should be exercised to make sure that the probe is completely submerged in the sample and that the sample is well stirred.

ACKNOWLEDGEMENT

The help extended by Dr. J.P. Rupert is greatly appreciated.

REFERENCES

American Petroleum Institute, 1969. *Standard Field Procedure for Testing Drilling Fluids.* API Div. Proc., Dallas, Texas, RP 13B.

Baroid Petroleum Services, 1978. *Manual of Drilling Fluids Technology. Mud Testing — Tools and Techniques.* Vol. I, Sec. II, 43 pp.

Binder, R.C., 1962. *Fluid Mechanics,* 4th ed. Prentice-Hall Inc., Englewood Cliffs, N.J., 453 pp.

Chilingar, G.V., 1956. *Technology of Testing Petroleum Products.* USAF Publ., Petrol. and Chem. Lab., Wright-Patterson AFB, Ohio, 76 pp.

Chilingar, G.V. and Terry, R.D., 1954. Simplified techniques of determining calcium and magnesium content of carbonate rocks. *Pet. Eng.,* 26 (12): 368—370.

IMCO Sevices, 1974. *Applied Mud Technology,* 5th ed. Responsive Printing Co., Houston, Texas, 226 pp.

Magcobar, 1972. *Drilling Fluid Engineering Manual.* Dresser Industries, Inc., Houston, Texas.

Chapter 5

CLAYS

HERMAN H. RIEKE, GEORGE V. CHILINGARIAN and PAUL VORABUTR

INTRODUCTION

What are clay minerals?

As pointed out by Weaver and Pollard (1973) in their classical book entitled "The Chemistry of Clay Minerals", clay minerals which occur in all types of sediments and sedimentary rocks, constitute the most abundant class of minerals in these rocks, comprising 40% of all the minerals present. Clay minerals mostly belong to the group of silicates having layer structures, but differ from most layer silicates by having hydrous nature.

Over 50% of the clay minerals in the earth's crust are illites. The order of relative abundance of other clay minerals is as follows: (1) montmorillonite and mixed-layer illite--montmorillonite; (2) chlorite and mixed-layer chlorite--montmorillonite; (3) kaolinite and septechlorite; and (4) attapulgite, palygorskite, and sepiolite.

Although classically the term "clay" has included the connotation of small particle size (<2 μ), it is not a correct basis for definition of a mineral family, because most layer silicates which are usually microscopic in size also occur as submicroscopic particles. Most chlorites and vermiculites frequently occur as microscopic crystals and are not seperated from minerals such as kaolinite and montmorillonite in the classification of the hydrous layer silicates (Warshaw and Roy, 1961, p. 1456).

Clay minerals, i.e., hydrous aluminum silicates, have been thoroughly studied because of their importance in various fields, especially in ceramics, agriculture, and drilling fluids. Colloid chemistry of clays was actively investigated during this century, being supplemented in the past 40 years by an intensive study of crystal chemistry of clay minerals. Earlier research yielded a considerable amount of practical information on the properties of clay minerals; however, it was hampered by the extremely fine-grained nature of clays and the fact that petrographic microscope was of limited use in the recognition and identification of the phases. Pauling (1930) used X-ray diffraction methods to establish a basis for classification and identification of clay minerals, and the X-ray diffractometer was instrumental in the rapid expansion of the systematic study of clay minerals.

In the present chapter, the writers attempt to familiarize mud engineers with the physical and chemical properties of clay minerals used in drilling fluids.

History of utilization of clays in drilling fluids

The rotary method of drilling is the most widely used system in the petroleum industry for drilling to very great depths, and drilling fluid is present in the hole while it is being drilled. The cuttings produced by the action of the bit are removed from the hole continuously by the circulating drilling fluid. Drilling may be interrupted, however, because of shut-downs occasioned by adding new lengths of drillpipe, the necessity of changing bits, setting casing, and occasional drilling accidents.

Although initially water was used as the carrying fluid, the advantages of using suspensions of clay-in-water became apparent as clay minerals were inadvertently incorporated in the fluid as a result of drilling through argillaceous strata. Subsequently, it became the usual practice to add surface clays to water in order to prepare a mud for circulation. Inasmuch as the primary function of the circulating mud is to remove cuttings, a thick slurry facilitated this action as compared to a liquid of low viscosity and devoid of shear strength, e.g., water. In addition, inasmuch as formation fluids are usually under a pressure roughly corresponding to or slightly exceeding that of a column of water of equal depth, the specific gravity of 1.1 to 1.2 of ordinary clay-in-water suspensions served to keep the formation fluids confined to their respective formations during drilling operations. Clay-in-water suspensions also impeded the loss of fluid into permeable, low-pressure formations. Clay forms a filter cake on the face of the rock and, thus, retains the drilling fluid in the borehole. Consequently, selection of a clay, which will impart the desired properties to the drilling fluid and form the most impermeable filter cake, is of utmost importance.

STRUCTURE OF CLAY MINERALS

Numerous investigators studied the structure of clay minerals. Pauling (1930), who is considered the pioneer in this field, investigated the structure of micas and related layer minerals. As pointed out by Grim in 1953, most of the clay minerals have two structural units that are the building blocks of their atomic lattices. The first unit is built of silica tetrahedrons, whereas the second unit consists of two sheets of closely-packed oxygens or hydroxyl group in which Al, Mg, or Fe atoms are embedded in octahedral coordination. A brief survey presented in this chapter is intended to acquaint the petroleum engineer with the crystalline structure of clay minerals and their chemistry.

Crystalline structure of clays

As pointed out by Van Olphen (1963, p. 59), the principal building elements of the clay minerals are (1) two-dimensional arrays of silicon—oxygen tetrahedra and (2) two-dimensional arrays of aluminum—oxygen—hydroxyl or magnesium—oxygen—hydroxyl octahedra. These sheets of tetrahedra and of octahedra are superimposed in different fashions in most clay minerals.

The silicon atoms are coordinated with four oxygen atoms in the silicon—oxygen sheets. As shown in Fig. 5-1a, the oxygen atoms are located on the

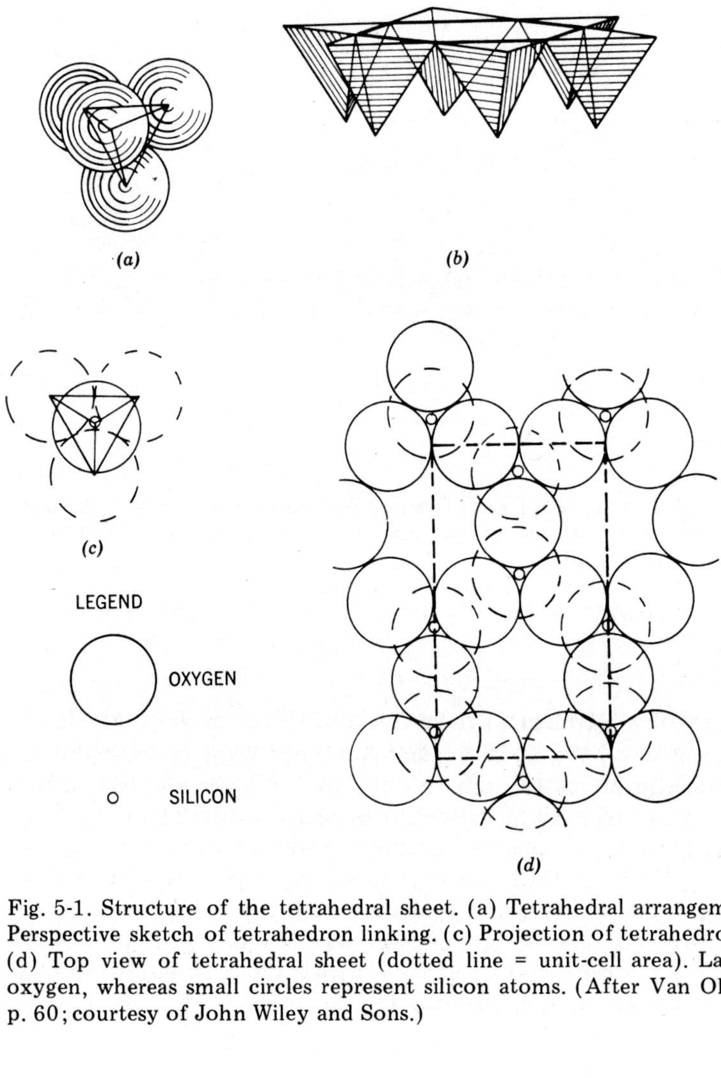

Fig. 5-1. Structure of the tetrahedral sheet. (a) Tetrahedral arrangement of Si and O. (b) Perspective sketch of tetrahedron linking. (c) Projection of tetrahedron on plane of sheet. (d) Top view of tetrahedral sheet (dotted line = unit-cell area). Large circles represent oxygen, whereas small circles represent silicon atoms. (After Van Olphen, 1963, fig. 15, p. 60; courtesy of John Wiley and Sons.)

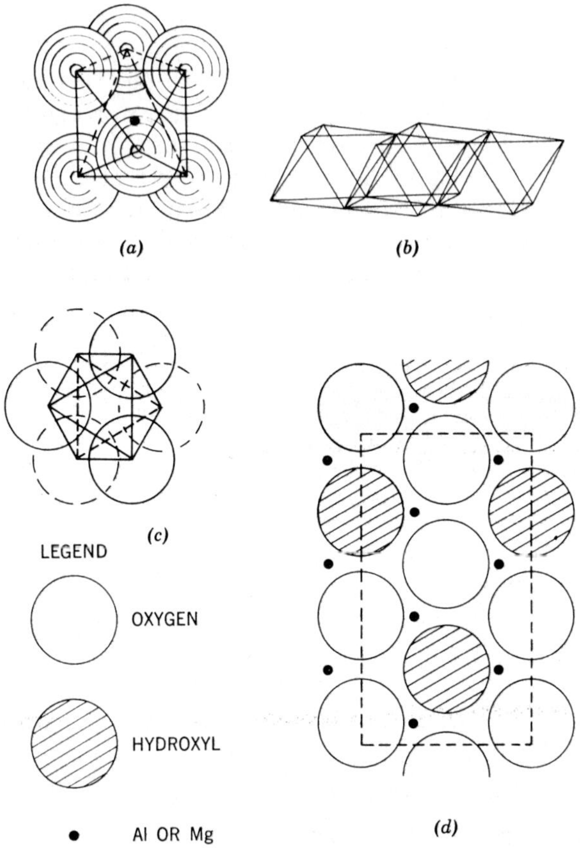

Fig. 5-2. Structure of the octahedral sheet. (a) Octahedral arrangement of Al or Mg with O or OH. (b) Perspective sketch of octahedron linking. (c) Projection of octahedron on plane of sheet. (d) Top view of octahedral sheet (dotted line = unit-cell area). Large circles represent oxygen; shaded circles = hydroxyl; and small black circles = aluminum or magnesium. (After Van Olphen, 1963, fig. 16, p. 61; courtesy of John Wiley and Sons.)

four corners of a regular tetrahedron, whereas the silicon atom is located in the center. Thus, three of the four oxygen atoms of each tetrahedron are shared by three neighboring tetrahedra in one sheet. Figure 5-1b shows that the fourth oxygen atom of each tetrahedron is pointed downward. In Figs. 5-1c and 5-1d, projections of this arrangement demonstrate the hexagonal symmetry of such a sheet. Rings of six oxygen atoms appear in these sheets, which are called tetrahedral sheets or the silica sheets.

The Al or Mg atoms in the Al—O—OH or Mg—O—OH sheets are coordinated with six oxygen atoms or OH groups, which are located around the Al or Mg atom, with their centers forming the six corners of a regular octahedron (Fig. 5-2) (Van Olphen, 1963, p. 61). A sheet, such as the one shown in perspective in Fig. 5-2b, results from the sharing of oxygen atoms by neighboring octahedrons. The Al or Mg atoms are located between two parallel planes of (1) oxygen atoms and (2) hydroxyl groups. Oxygen atoms and hydroxyl groups form a hexagonal close packing, as shown by the projection of the sheet in Fig. 5-2c. The name assigned to this sheet is octahedral, or alumina or magnesia sheet, which are also called gibbsite layer or brucite layer, respectively (Van Olphen, 1963, p. 62).

Because of the analogous symmetry and the almost identical dimensions in the tetrahedral and the octahedral sheets, they can share oxygen atoms. Octahedral sheet shares the fourth oxygen atom protruding from the tetrahedral sheet. The sharing of atoms between one silica and one alumina sheet occurs in the two-layer minerals. One alumina or magnesia sheet shares oxygen atoms with two silica sheets, one on each side, in the three-layer minerals. The combination of one or two tetrahedral sheets and an octahedral sheet is called a unit layer. These layers are stacked parallel to each other in most clay minerals.

As pointed out by Van Olphen (1963, p. 62), a certain unit of structure repeats itself in a lateral direction within each unit layer; this unit structure can be called the unit cell. Crystallographically-defined unit cell, however, extends from a certain plane in one unit layer to the corresponding plane in another parallel unit layer of the crystal. Simplified schematic crystal structures of kaolinite, illite, and montmorillonite are presented in Fig. 5-3.

CLASSIFICATION SCHEMES OF CLAYS

Several attempts have been made to classify the hydrous aluminum silicates, some of which are presented here. As an introduction, an elegant classification of Degens (1965, p. 16) is presented first (Table 5-I). Brindley (1955) presented one of the first utilitarian classifications of clay minerals, which can serve both the practical needs of industry and mineralogists (Table 5-II). Successive subdivisions are based alternately on structural features and on chemical compositions, which is of utmost importance in clay-mineral identification. Obviously, it is necessary for the methods employed to reflect both structure and composition.

Classification schemes of clay minerals have been based on atomic structure, chemical composition, and properties, such as swelling and morphology. As pointed out by Warshaw and Roy (1961, p. 148), the earliest classifications of clays were compositional, whereas later ones have tended to overemphasize the structural details while neglecting gross variations in chemistry. It is imperative to keep a proper balance between structure and composition, which are the two basic properties that must be incorporated in a classification scheme.

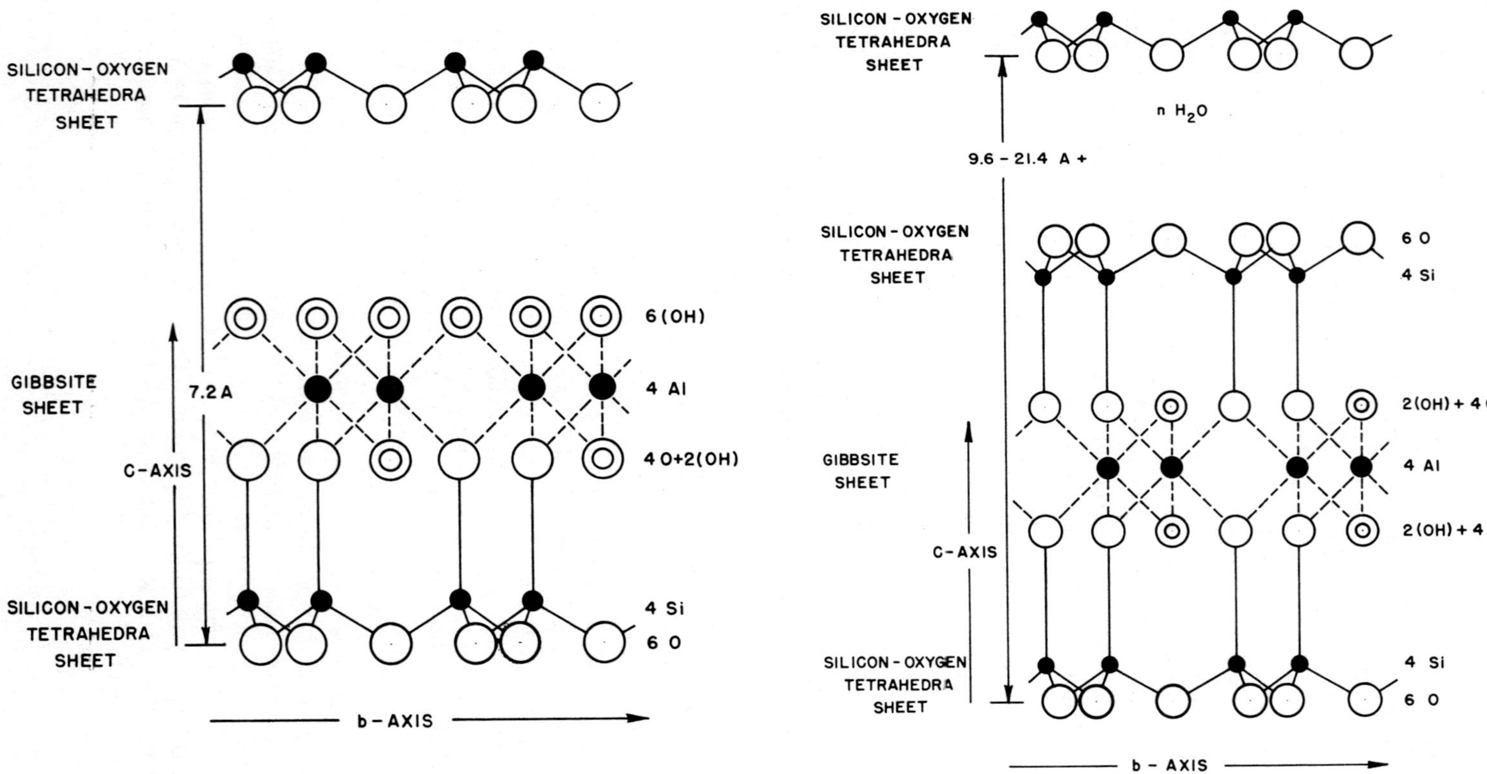

a KAOLINITE $(OH)_8\ Al_4\ Si_4\ O_{10}$

b MONTMORILLONITE $(OH)_4\ Al_4\ Si_8\ O_{20} \cdot n\ H_2O$

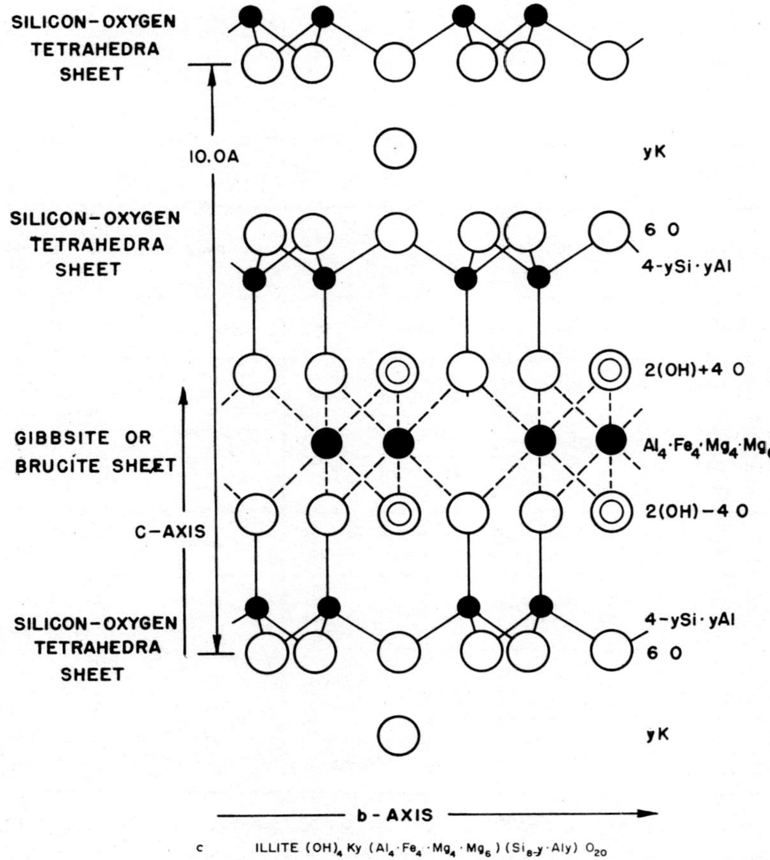

Fig. 5-3. Simplified schematic crystal structures of (a) kaolinite (after Gruner, 1932, and Grim, 1942); (b) montmorillonite (after Hofmann et al., 1933); and (c) illite (after Grim et al., 1937). (After Hughes, 1950, figs, 1, 2, and 3.)

TABLE 5-I

Classification of principal clay minerals in sediments (excluding impure minerals and amorphous phases) (After Degens, 1965, table 1, p. 16; courtesy of Prentice Hall, Inc.)

LAYERS	POPULATION OF OCTAHEDRAL SHEET	EXPANSION	GROUP	SPECIES	CRYSTALLOCHEMICAL FORMULA	SPACE GROUP	STRUCTURE (SCHEMATIC)
Two-Sheet (1:1)	Dioctahedral	Non-Swelling	Kaolinite	Kaolinite Dickite Nacrite	$Al_4(OH)_8\,[Si_4O_{10}]$	*Cc* or $P\bar{1}$ *Cc* *Cc* or *R3c*	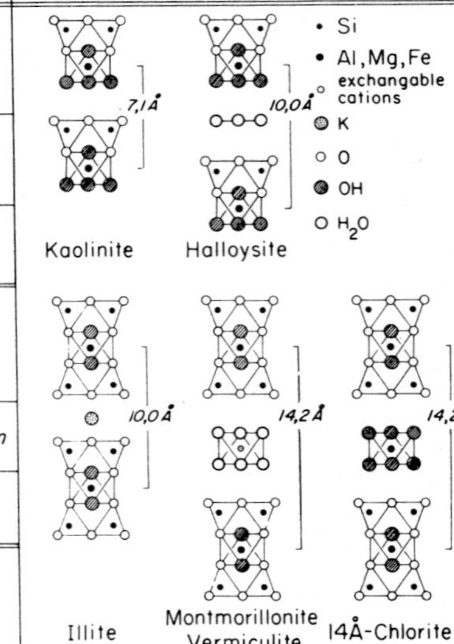
		Non-Swelling and Swelling	Halloysite	Halloysite Metahalloysite	$Al_4(OH)_8\,[Si_4O_{10}]\cdot(H_2O)_4$ $Al_4(OH)_8\,[Si_4O_{10}]$	*Cm*	
	Trioctahedral	Non-Swelling	7Å-Chlorite (Septechlorite)	Berthierine ("Kaolin-Chamosite")	$(Fe^{\cdot\cdot}, Fe^{\cdot\cdot\cdot}, Al, Mg)_6\,(OH)_8[(Al,Si)_4O_{10}]$	*Cm*	
Three-Sheet (2:1)	Dioctahedral	Swelling	Montmorillonite* (Smectite)	Montmorillonite Beidellite Nontronite	$\{(Al_{2-x}Mg_x)(OH)_2[Si_4O_{10}]\}^{-x}\,Na_x\cdot nH_2O$ $\{Al_2(OH)_2[(Al,Si)_4O_{10}]\}^{-x}\,Na_x\cdot nH_2O$ $\{(Fe^{\cdot\cdot\cdot}_{2-x}Mg_x)(OH)_2[Si_4O_{10}]\}^{-x}\,Na_x\cdot nH_2O$	Monoclinic	
		Non-Swelling	Illite* (Hydromica)	Illite-Varieties	$(K,H_3O)\,Al_2(H_2O,OH)_2[Al\,Si_3O_{10}]$	*C2/m* or *Cm*	
	Trioctahedral	Swelling	Vermiculite**	Vermiculite	$(Mg,Fe)_3(OH)_2[Al\,Si_3C_{10}]\,Mg\cdot(H_2O)_4$	*Cc*	
Three-Sheet+ One-Sheet (2:2)	Trioctahedral	Non-Swelling+	14Å-Chlorite** (Normal Chlorite)	Chlorite-Varieties	$(Al,Mg,Fe)_3(OH)_2[(Al,Si)_4O_{10}]\,Mg_3(OH)_6$	Monoclinic	

* also trioctahedral varieties } less common
** also dioctahedral varieties } less common
\+ swelling chlorites are rare and intermediate forms between vermiculite and chlorite

TABLE 5-II

A scheme of classification of clays and related silicate minerals
(After Brindley, 1955, table 1, p. 34; courtesy of Clays and Clay Technol. Bull., Dept. Nat. Resources, Division of Mines, Calif.)

CHEMICAL CATEGORIES	SILICATES					
STRUCTURAL GROUPS	Layer silicates					Chain silicates
SUB-GROUPS	Kaolin type	Mica type	Chlorite type	Mixed-layer type	Chloritoid	
CHEMICAL SPECIES	Kaolin minerals Serpentine minerals Chamosite Amesite Greenalite Cronstedtite etc.	Talc Pyrophyllite Muscovite Phlogopite Biotite Glauconite Illite(s) Montmorillonoids Vermiculite etc.	Penninite Clinochlore Prochlorite Daphnite etc.	Anauxite Bravaisite etc.		Polygorskite (attapulgite) Sepiolite
STRUCTURAL VARIETIES	Halloysite Kaolinite Dickite Nacrite Chrysatile(s) Antigorite and hydrated forms	Polymorphic varieties and hydrated forms	Polymorphic varieties			

The work of Pauling (1930) on the general structural features of clay minerals resulted in the basic advance in the crystal chemistry of the layer silicates. Several books and articles on crystal chemistry and mineralogy describe detailed features serving as a basis of classification (e.g., Brindley, 1951; Brindley and MacEwan, 1953; Grim, 1953; MacKenzie, 1957). The various classifications proposed up to 1959 were reviewed by MacKenzie (1959). Warshaw and Roy (1961) proposed a classification scheme which divided the layer silicates into three groups, i.e., two-layer (1 : 1), three-layer (2 : 1) (expanded 2 : 1 minerals), and three-layer (2 : 1) (micas).

Two-layer (1 : 1) silicates

As pointed out by Warshaw and Roy (1961, p. 1462), the names of the minerals within the two-layer (1 : 1) silicates group (Table 5-III) are well established. Although designation by polytype (variations in stacking, either

TABLE 5-III

Two-layer (1 : 1) silicates
(After Warshaw and Roy, 1961, table 2, p. 1466; courtesy of the Geol. Soc. of America)

Dioctahedral—KAOLINITE MINERALS

Composition and name	Polytype	
$Al_4Si_4O_{10}(OH)_8$		
Nacrite[1]	6-layer monoclinic	(kaolinite-6M)
Dickite[2]	2-layer monoclinic	(kaolinite-2M)
Kaolinite[3]	1-layer triclinic	(kaolinite-1T)
Kaolinite$_d$[4]	1-layer (monoclinic?) *b* axis disordered	(kaolinite$_d$)
Halloysite[5]	1-layer *a* and *b* disordered	(kaolin$_d$)
$Al_4Si_4O_{10}(OH)_8 \cdot 4H_2O$		
Endellite[5]	1-layer + H_2O (d_{001} = ca. 10 A)	

Trioctahedral—SEPTECHLORITES[13]

	Composition (idealized)	Name	Polytype
SERPENTINES	$Mg_6Si_4O_{10}(OH)_3$	Chrysotile[6,7]	3 2-layer types
		antigorite[6]	1-layer monoclinic (with a large a_0)
		lizardite[6]	1-layer orthohexagonal
	$Mg_{5\cdot5}Al_{0.5}Si_{3.5}Al_{0.5}O_{10}(OH)_8$		1-layer orthohexagonal
	$Mg_{5.0}Al_{1.0}Si_{3.0}Al_{1.0}O_{10}(OH)_8$	Clinochlore[7]	1- and 6-layer
	$Mg_{4.0}Al_{2.0}Si_{2.0}Al_{2.0}O_{10}(OH)_8$	Amesite[7,8]	6-layer orthohexagonal
	$Fe_6Si_4O_{10}(OH)_8$	Greenalite[9]	
	$Fe''_{4.0}Fe'''_{2.0}Si_{2.0}Fe_{2.0}0_{10}(OH)_8$	Cronstedtite[10]	
	$Fe''_{5.0}Al\ Si_3\ Al\ O_{10}(OH)_8$	chamosite[11]	
	$(Ni,Mg)_6Si_4O_{10}(OH)_8$ (variable)	garnierite[12] (Ni-serpentine)	
	$Mn_5\ Al\ Si_3\ Al\ O_{10}(OH)_8$	bementite (and grovesite)	

Note: Family names are in capital letters; names of minerals which have close to idealized end-member compositions are written with initial capitals; others vary in composition between certain limiting end members.

[1] Hendricks (1939a)

[2] Newnham and Brindley (1956)

[3] Brindley and Robinson (1946a). Name "kaolinite T" proposed by Holdridge and Vaughan *in* Mackenzie (1957)

[4] Called "fireclay" in structural study by Brindley and Robinson (1946b). Name "kaolinite pM" proposed by Holdridge and Vaughan in Mackenzie (1957); "ψM-kaolinite" proposed by Mackenzie" (1959). The subscript "*d*" used in this classification signifies a disordered mineral. This subscript has been used by one of the authors in previous publications (*e.g.*, Hill and Roy, 1958).

[5] The relation between structure and tubular morphology has been discussed by Bates, Hildebrand, and Swineford (1950). The nomenclature above agrees with the propcsals of Faust (1955).

[6] Whittaker and Zussman (1956)

[7] Gillery (1959a)

[8] Brindley, Oughton, and Youell (1951)

[9] Gruner (1936)

[10] Hendricks (1939b)

[11] Brindley and Youell (1953); the name "berthierine" is preferred by the Clay Minerals Group Subcommittee on Nomenclature (Brown, 1955).

[12] Noll and Kircher (1952)

[13] Prefix "septe" may be added to names of septechlorites, *e.g.* septeclinochlore, only if it should be desirable to differentiate between these minerals and the normal chlorites.

structural or compositional) need to be used, differences in stacking may constitute the principal distinction among minerals of this group. A logical system of nomenclature depending on polytypes, however, was given by Warshaw and Roy (1961) for comparison with other groups of clay minerals. A similar system has been proposed by MacKenzie (1959). The two-layer silicates are divided into the kaolinite (dioctahedral) and septechlorite (trioctahedral) families (Warshaw and Roy, 1961, p. 1462).

The dioctahedral minerals are pure hydrous aluminum silicates, and incorporation of minor amounts of other constituents found in samples of these minerals, especially kaolinite, in their structure is not likely.

The trioctahedral 1 : 1 minerals are called septechlorites, as proposed by Nelson and Roy (1954). The serpentines are magnesium-septechlorites containing only minor amounts of aluminum and negligible amounts of iron. They may also contain nickel (Warshaw and Roy, 1961, p. 1463). Many of the septechlorites are macroscopically and microscopically similar to chlorites, and "septe" refers to the 7-Å basal spacing in the X-ray diffraction pattern.

There is a minor (trioctahedral versus dioctahedral) difference in structure between kaolinites and septechlorites; however, they differ markedly in composition. Kaolinites exhibit no ionic substitutions, whereas extensive solid solution occurs among the septechlorites. According to Warshaw and Roy (1961, p. 1463), in addition to the simple ion-for-ion replacement, the solid solution is of the specific type, i.e., simultaneous entry of Al^{3+} for Mg^{2+} in the octahedral and Al^{3+} for Si^{4+} in the tetrahedral layer. Magnesium, manganese, nickel, ferrous and ferric iron, chromium, and aluminum are found in the octahedral positions, whereas aluminum and ferric iron may substitute for as much as half of the silicon content (Warshaw and Roy, 1961, p. 1463).

Three-layer (2 : 1) silicates (expanded 2 : 1 minerals)

Three-layer (2 : 1) silicates (expanded 2 : 1 minerals) are intermediate in structure between the pyrophyllite—talc minerals and the micas, i.e., well-known concept of Hofmann-Endell-Wilm-Marshall-Hendricks structure of montmorillonite. Included is the hypothesis that these minerals have a two-dimensional lattice with no order in the stacking of layers (Warshaw and Roy, 1961, p. 1465). Unlike the other 2 : 1 layer silicates, the expanded minerals include loosely-bound cations and layers of water between the silicate sheets. Polar organic molecules can partially or completely replace the water. Inasmuch as the amount and thickness of water layer and organic compounds in the interlayer position can vary, the spacing at the first basal X-ray diffraction maximum of these clay minerals is also variable (Warshaw and Roy, 1961, p. 1467). Much larger amounts of water than that which could occupy interlayer space in an oriented manner are present in many clay minerals of this group, especially in bentonites. These clay—water com-

TABLE 5-IV

Three-layer (2 : 1) silicates — expanded 2 : 1 minerals
(After Warshaw and Roy, 1961, table 3, p. 1468; courtesy of the Geol. Soc. of America)

	Dioctahedral		Seat of charge		Trioctahedral	
Group	Name	Formula of 2:1 Layer*		Group	Name	Formula of 2:1 Layer*
MONTMORILLONITE GROUP	Montmorillonite[1,3]	$Al_{2-x}Mg_xSi_4O_{10}(OH)_2$	octahedral only (essentially)	MONTMORILLONITE GROUP	Hectorite[3]	$Mg_{3-x}Li_xSi_4O_{10}(OH)_2$
			x > y, y near 0		Stevensite[1,3,6]	$Mg_{3-x}\square_xSi_{4-y}\square_yO_{10}(OH)_2$
			octahedral dominant x > y		SAPONITE[1,3]	$Mg_{3-x}Al_xSi_{4-y}Al_yO_{10}(OH)_2$
					pimelite[2,3]	$Ni_{3-x}Al_xSi_{4-y}Al_yO_{10}(OH)_2$
	(BEIDELLITE)				sauconite[2,3]	$Zn_{3-x}Al_xSi_{4-y}Al_yO_{10}(OH)_2$
	nontronite[4]	$(Fe,Al)_2Si_{4-y}Al_yO_{10}(OH)_2$	tetrahedral dominant y > x		medmontite	$Cu_{3-x}Al_xSi_{4-y}Al_yO_{10}(OH)_2$
	volchonskoite[2]	$(Al,Cr)_2Si_{4-y}Al_yO_{10}(OH)_2$				
	(synthetic) beidellite[1,3]	$Al_2Si_{4-y}Al_yO_{10}(OH)_2$	tetrahedral only y < 0.5		(synthetic)[1,3] saponite	$Mg_3Si_{4-y}Al_yO_{10}(OH)_2$
	dioctahedral vermiculite[5]	$(Al,Fe)_{2-x}Mg_xSi_{4-y}Al_yO_{10}(OH)_2$	tetrahedral dominant y near 1		VERMICULITE	$Mg_{3-x}Fe_xSi_{4-y}Al_yO_{10}(OH)_2$

* Idealized; interlayer cations omitted because variable

Note: Family names are in capital letters; names of minerals that have close to idealized end-member compositions are written with initial capitals; others vary in composition between certain limiting end members.

[1] The expandable 2:1 silicates in the system MgO-Al_2O_3-SiO_2-H_2O (Roy and Roy, 1955) have been synthesized many times. An example of the systematic studies on variation of properties with composition is that of Koizumi and Roy (1959). The natural minerals of the montmorillonite group have been described in detail by Ross and Hendricks (1945)

[2] Synthesis reported by Roy and Sand (1956)

[3] Synthesis reported by Mumpton and Roy (1956)

[4] Synthesis reported by Ewell and Insley (1935)

[5] Properties of natural material described by, *e.g.*, Brown (1953), Hathaway (1955)

[6] Detailed mineralogical study of stevensite made by Faust, Hathaway, and Millot (1959)

For the references in these notes see original publication, Warshaw and Roy (1961).

plexes exhibit many remarkable properties, which have been studied by many colloid chemists, electrochemists, and mud engineers. (See Warshaw and Roy, 1961).

The expanded (or expandable) 2 : 1 layer silicates are presented in Table 5-IV. The interlayer cations have been omitted from the formulae by Warshaw and Roy (1961), because they may be changed by a simple cation exchange. Several families are included within this important group of clay minerals. Smectite (montmorillonite), hectorite, and stevensite show less variations in composition than do the other minerals in this group, because of only minor amounts of substitution in their tetrahedral layer (Warshaw and Roy, 1961, p. 1467). The most common of these three minerals is smectite. Beidellite, saponite, and vermiculite are three families which have both octahedral and tetrahedral substitution.

As pointed out by Warshaw and Roy (1961, p. 1467), it is generally agreed that the minerals in the upper portion of Table 5-IV should be grouped together. Ross (1960) has included all of them, except beidellite, in his revised classification of this group. Inasmuch as vermiculites have larger amounts of tetrahedral substitution and can be distinguished from other minerals in Table 5-IV by heat treatment and/or by treatment with ethylene glycol, it is convenient to assign a subgroup name to them. Although montmorillonite is used as a group name by many American clay mineralogists, smectite is considered to be the correct name. Both names, however, are used interchangeably in this chapter.

All the three-layer 2 : 1 silicates that can be expanded with ethylene glycol to about 17 Å are included in the montmorillonite group. According to Warshaw and Roy (1961, p. 1467), the vermiculites should not be separated from the montmorillonite-type minerals, because the properties are continuously variable from saponites to vermiculites. Vermiculites have been classified as a separate subgroup of the expandable 2 : 1 layer silicates, because of (1) their high tetrahedral charge, (2) their usually coarse particle size, and (3) their failure to expand to 17 Å with ethylene glycol at room temperature, particularly when magnesium is the interlayer cation (Walker, 1958).

A natural expanded 2 : 1 mineral which is considered to be an aluminum-rich variety with predominantly tetrahedral charge was originally named beidellite. Inasmuch as the type mineral has been found to be impure, however, the term became unacceptable to Ross (1960) and others (Warshaw and Roy, 1961, p. 1467). Greene-Kelly (1955), however, cited evidence for the existence of pure natural beidellite. In addition, an aluminum-rich montmorillonite, can be easily synthesized (Roy and Osborn, 1954). Thus, the name "beidellite" as originally defined, can be applied to this material.

Swelling chlorite, which was first described by Stephen and MacEwan (1950), may also be included in the 2 : 1 group of clay minerals with variable basal spacings, including vermiculites. Stephen and MacEwan (1951) compared this "swelling chlorite" and a chlorite—montmorillonite mixed-layer mineral (later named "corrensite") with the chlorite-like material synthesized by Caillère and Hénin (1949). The term "swelling chlorite" is thus applied to all these chloritic minerals (Warshaw and Roy, 1961, p. 1470). Caillère et al. (1954) also included some iron-rich chlorites that can be transformed to "montmorillonite" by acid treatment, and used the terms "swelling chlorite" and "pseudo-chlorite" interchangeably for all these chlorite-like minerals. According to Warshaw and Roy (1961, p. 1470), the latter name is a better choice because some of these minerals inhibit swelling.

Except vermiculite, all the expanded 2 : 1 silicate minerals can be synthesized under hydrothermal conditions (see Table 5-IV for references to the hydrothermal syntheses). In contrast to the situation which exists in the case of pyrophyllite and kaolinite, substitutions are possible in the dioctahedral minerals inasmuch as the resulting layer charges can be balanced by the interlayer cations. Karšulin and Stubičan (1954), Ames and Sand (1958), Koizumi and Roy (1959), and Gillery (1959) made systematic studies of the variation in properties of expanded 2 : 1 layer minerals with types and amounts of substitution. Koizumi and Roy (1959) showed that the amounts of substitution in "montmorillonites" are variable and that corresponding variations in cation exchange capacity are obtained.

According to Mumpton and Roy (1956), experimental evidence indicates that the solid solution between synthetic dioctahedral and trioctahedral expanded 2 : 1 silicates is incomplete.

Three-layer (2 : 1) silicates (micas)

A structural unit of micas consists of two tetrahedral layers with a layer of octahedrally coordinated cations between them, together with a layer of large cations (interlayer) bonding this 2 : 1 structural unit to the next. According to Warshaw and Roy (1961, p. 1470), there is one interlayer cation per formula unit (two per unit cell of a single-layer polytype). Substitutions of ions of smaller charge for some of those in the octahedral layer and/or tetrahedral layer balance the interlayer positive charge. Inasmuch as the interlayer cations fill all possible positions and are not hydrated, the bonding between sheets is much stronger in the micas than in the other clay minerals (Warshaw and Roy, 1961, p. 1470).

As in the case of other layer silicates, micas are grouped according to whether they are dioctahedral or trioctahedral, and their classification is based primarily on composition. The separation of micas into heptaphyllites and octaphyllites by Winchell (1925, pp. 427—428) is the same type of grouping, i.e., heptaphyllites are dioctahedral and octaphyllites are trioctahedral ("hepta" and "octa" refer to the total number of cations in the mica "formula").

Micas, which show large variations in composition, can be categorized according to the number of silicon ions in the tetrahedral positions, i.e., disilicic, monosilicic, trisilicic, and tetrasilicic. Many more varieties of mica have been named, but they have the same types of cationic substitution as shown by Warshaw and Roy (1961). The sericite, however, has not been included, because it merely implies the fine-grained nature of a mica (Warshaw and Roy, 1961, p. 1471).

BASIC TYPES OF CLAY MINERALS

As pointed out by Ries (1927), the literature contains a description of a large number of hydrous aluminum silicates. Naming of separate species, however, has been based sometimes on a chemical analysis of a limited amount of material.

Howe (1914) gave the names of 61 hydrous aluminum silicates, which had been described in the literature up to 1914; however, the validity of many of the species described may be questioned. Wherry (1925) has added to the list presented by Howe.

Smectite (montmorillonites)

In 1953, Grim stated that minerals in the montmorillonite group occur as extremely small particles which are hard to study. Structural concepts were deduced from powder data and by other means that reduced the uncertainty of the structure of smectites. The most accepted structure for the smectite minerals is that suggested by Hofmann et al. in 1933. Weaver and Pollard (1973, p. 55) reported that most of these montmorillonites were formed by the alteration of volcanic material and have a relatively small amount of Al substituting for Si in the tetrahedral sheet. Many chemical analysis of smectites are presented in the literature.

Grim and Kulbicki (1961) described two types of montmorillonite, which differ primarily in the amount of octahedral Mg. In the Cheto-type montmorillonite, approximately one-fourth of the Al is replaced by Mg (0.55 octahedral positions on the average). It is suggested that the Mg^{2+} ions are regularly distributed in a hexagonal arrangement (Weaver and Pollard, 1973, p. 60). The Wyoming-type montmorillonites have considerably lower Mg content (0.19 octahedral positions on the average). As pointed out by Weaver and Pollard (1973, p. 61), there is no significant difference in the tetrahedral makeup of the two types. Most of the measurements made by Grim and Kulbicki (1961) indicate that these two types of montmorillonite are significantly different; however, many of the samples studied were intermediate in composition and were believed to be mixtures of the two end member types. Size fractionation of one sample indicated that the coarser fraction exhibited the characteristics of the Cheto-type montmorillonite (when heated), whereas the finer fraction had the characteristics of the Wyoming-type montmorillonite.

Ross (1946), as well as some other investigators, have concluded that Mg is essential for the formation of montmorillonite. This cation increases the size and charge of the octahedral sheet and tends to decrease layer strain. Its major contribution, however, may be in enabling Al to take six-fold coordination under basic conditions. As pointed out by Dekimpe et al. (1961), at low temperatures Al tends to take six-fold coordination in an acid environment and four-fold coordination in a basic environment. On the other hand, Mg takes only the six-fold coordination. It is likely that Al tends to resist six-fold coordination in the basic environments in which montmorillonite and illite could form, and presence of Mg may be necessary to nucleate the octahedral coordination.

According to Keller (1964), Ca^{2+} is the dominant exchangeable cation in montmorillonites which are in equilibrium with river water. In seawater, Mg^{2+} tends to replace Ca^{2+} and Na^{+} cations, thus becoming the dominant exchangeable cation (Carrol and Starkey, 1960). Inasmuch as in many ancient clays Na^{+} is the most abundant exchangeable cation, this abundance of Mg^{2+} appears to be inconsistent with the above data (Weaver and Pollard, 1973, p. 72). Hanshaw's (1964) experiments showed that the order of cation exchange selectivity is dependent upon whether a clay is dispersed or compacted. He found that compacted montmorillonite showed the following order of preference for cations: $K^{+} > Na^{+} > H^{+} > Ca^{2+}$. Consequently, in dispersed marine montmorillonites, maybe Mg is the predominant exchange cation, whereas in compacted muds as a result of burial, Na^{+} replaces a portion of the Mg^{2+}. This has been confirmed by Weaver and Beck (1971). It is important to mention here that Na-base montmorillonites swell more than Ca-base ones.

Ross and Hendricks (1945) laid the foundation for the understanding of the montmorillonite group of clay minerals. No basic modifications have been made since that time, even though refinements have been made and some new minerals added to this group. Montmorillonite, beidellite, and volchonskoite were considered as the dioctahedral members, whereas saponite and hectorite were considered as the trioctahedral members by Ross and Hendricks (1945). Ross (1946) later added sauconite, the zinc member, and discussed the iron-rich griffthite.

In 1945, Ross and Hendricks redefined beidellite as the aluminum-rich end member of the dioctahedral montmorillonites. Many of their samples, however, were later found to be impure and for a long period of time the concept of a high-aluminum montmorillonite was in considerable disrepute (Weaver and Pollard, 1973, p. 63). In 1962, Weir and Greene-Kelly made a careful analysis of purified material from the Black Jack Mine from Beidell, Colorado, and definitely established that it is monomineralic and an ideal Al-rich end member (Table 5—V). They stated that "beidellites and mont-

TABLE 5-V

Beidellite analyses and structural formulas
(After Weaver and Pollard, 1973, table XXVIII, p. 60; courtesy of Elsevier Sci. Publ. Co.)

	1	2	3	4		1	2	3	4
					Octahedral				
SiO_2	49.01	55.80	59.30	44.02	Al	1.34	1.85	1.98	1.80
Al_2O_3	20.5	28.60	36.11	28.90	Fe^{3+}	0.39	0.02	0.02	0.29
Fe_2O_3	6.85	0.41	0.50	5.15	Fe^{2+}	–	–	–	0.02
FeO	–	–	–	0.33	Mg	0.28	0.20	0.01	0.05
MgO	2.08	2.03	0.10	0.50	Σ	2.01	2.07	2.01	2.16
CaO	0.17	2.23	0.02	0.68					
Na_2O	0.58	2.09	3.98	0.17	Tetrahedral				
K_2O	0.95	0.48	0.11	0.50	Al	0.27	0.35	0.52	0.74
TiO_2	0.17	0.26	–	0.53	Si	3.73	3.65	3.48	3.26
H_2O^+	12.1	9.70							
H_2O^-	10.5	–		18.64	Interlayer				
					Ca/2		0.31		
	103.0	99.60	100.12*	99.42	K		0.04		
					Na		0.01		
					C.E.C.				
					(mequiv./100 g)	115		130**	
					Layer charge	0.52	0.34	0.50	0.33

1. Sawhney and Jackson (1958): Houston soil, less than 0.08 microns.
2. Heystek (1962): Castle Mountain, Ivanpah, Calif., U.S.A., hydrothermally altered.
3. Weir and Green-Kelley (1962): Black Jack Mine, Beidell, Colo., U.S.A., Gouge clay.
4. Ross and Hendricks (1945): Ancon, Canal Zone, laminated organic clay.

*Loss on ignition; 6.3% (ignited weight).

**Mequiv./100 g ignited weight).

For the references in the notes see Weaver and Pollard, 1973.

morillonites should be divided at the composition at which the lattice charges from octahedral and tetrahedral substitution equal one another". Beidellite layer charge does not appear to be any larger than that for montmorillonites. (See Weaver and Pollard, 1973, p. 63.)

According to Weaver and Pollard (1973, p. 72), many montmorillonites, particularly the beidellite-type, have units with a sufficiently high layer charge to fix K^+ in the interlayer position and cause the layers to contract to 10 Å. Using the normal methods of determining exchange capacity, this K is not usually exchangeable. On exposure to KOH or seawater, Wyoming montmorillonite loses approximately 15% of its exchange capacity and a similar percentage of layers contract to 10 Å. As pointed out by Weaver (1958), beidellite-type clays may fix several percent of K^+ in the interlayer positions and have most of their layers contracted to 10 Å. The above discussion may shed light upon the reason for popularity of potassium-base drilling fluids.

McAtee (1958) discussed the heterogeneity of smectites. He concluded that the ion-exchange capacity of montmorillonites, which ranges from 70 to 130 mequiv./100 g, is one of its most important properties. In most cases, the use for which a given montmorillonite is employed depends on the particular exchangeable ions which are present. In montmorillonites, about 80% of the cation exchange is attributed to lattice substitutions, with this portion of the exchangeable cations being held by the basal plane surfaces. The remaining 20% of exchangeable cations are probably adsorbed on the edges

TABLE 5-VI

Structural formulas for smectites from Cretaceous bentonites of Black Hills Region of Wyoming and South Dakota
(After Weaver and Pollard, 1973, table XXXV, p. 73)

Sample		Formula
1	bluish gray	$(Al^{3+}_{1.60}Fe^{3+}_{0.06}Fe^{2+}_{0.18}Mg^{2+}_{0.22})(Si^{4+}_{3.91}Al^{3+}_{0.09})O_{10}(OH)_2$
	olive green	$(Al^{3+}_{1.61}Fe^{3+}_{0.13}Fe^{2+}_{0.06}Mg^{2+}_{0.20})(Si^{4+}_{3.92}Al^{3+}_{0.08})O_{10}(OH)_2$
2	bluish gray	$(Al^{2+}_{1.62}Fe^{3+}_{0.00}Fe^{2+}_{0.13}Mg^{2+}_{0.20})(Si^{4+}_{3.92}Al^{3+}_{0.08})O_{10}(OH)_2$
	olive green	$(Al^{3+}_{1.62}Fe^{3+}_{0.13}Fe^{2+}_{0.05}Mg^{2+}_{0.19})(Si^{4+}_{3.90}Al^{3+}_{0.10})O_{10}(OH)_2$
3		$(Al_{1.64}Fe^{3+}_{0.15}Fe^{2+}_{0.02}Mg_{0.19})(Si_{3.90}Al_{0.10})O_{10}(OH)_2$
4		$(Al_{1.54}Fe_{0.16}Mg_{0.33})(Si_{3.91}Al_{0.09})O_{10}(OH)_2$
5		$(Al_{1.55}Fe_{0.15}Mg_{0.33})(Si_{3.92}Al_{0.08})O_{10}(OH)_2$
6		$(Al_{1.63}Fe_{0.17}Mg_{0.25})(Si_{3.93}Al_{0.07})O_{10}(OH)_2$
7		$(Al_{1.61}Fe_{0.18}Mg_{0.23})(Si_{3.87}Al_{0.13})O_{10}(OH)_2$
8		$(Al_{1.55}Fe^{3+}_{0.19}Fe^{2+}_{0.02}Mg_{0.26})(Si_{3.88}Al_{0.12})O_{10}(OH)_2$
9		$(Al_{1.56}Fe^{3+}_{0.03}Mg_{0.19})(Si_{3.85}Al_{0.15})O_{10}(OH)_2$

1,2. Knechtel and Patterson (1962): Clay spur bentonite bed, Crook Co., Wyoming; analyst M.D. Foster.
3. Grim and Kulbicki (1961): Crook Co., Wyoming.
4. Kerr et al. (1950): Clay spur, Wyoming; 5. Upton, Wyoming; 6. Bell Fourche, S.D. A portion of the octahedral Mg undoubtedly belongs in the interlayer position.
7. Foster (1954): Belle Fourche, South Dakota.
8. Ross and Hendricks (1945): Upton, Wyoming.
9. Mackenzie (1963): Wyoming bentonite.

For the references in the notes see Weaver and Pollard (1973).

of sheets as a result of broken bonds (see Hendricks, 1942). The most common interlayer cations found in nature are Na^+, Ca^{2+}, Mg^{2+}, and H^+

Naturally-occurring montmorillonites contain as exchangeable cations various amounts of sodium, calcium, hydrogen, and magnesium, and lesser amounts of other cations. Some of the montmorillonites in natural bentonites carry mainly sodium ion, whereas others carry mainly calcium and/or magnesium ions. The former bentonites exhibit considerably greater degree of swelling.

Table 5-VI contains a number of structural formulas for smectites obtained from the Cretaceous bentonite beds of the Black Hills region of Wyoming, Montana, and South Dakota (see Weaver and Pollard, 1973, p. 75). The first four formulas were obtained from analyses of two partially weathered blocks from one mine. According to Knechtel and Patterson (1962, p. 74), the major difference is due to the oxidation of the iron as a result of weathering; however, the analyses are quite similar. Failure of some of the analysts to distinguish between octahedral and exchangeable Mg^{2+} cations accounts for much of the small variation that does exist. Uniformity in composition is probably indicative that this is the most stable composition for montmorillonite. Variability perhaps would have been greater if noncommercial samples were included together with commercial samples tested.

Illite

As observed by Warshaw (1959) and other investigators, illite (a clay mineral substance structurally related to the micas) is the dominant constituent of many fine-grained sediments. Most of the oil-producing sandstones and sands contain variable amounts of interstitial illite clay of detrital or authigenic origin. Griffiths (1952), MacKenzie (1955), and others described possible formation of illite by degradation or aggradation processes. Low-temperature reactions suggested by various investigators are based on considerable amount of geologic evidence and on a relatively few experimental studies on (1) "potassium fixation" by expanding-type clay minerals and (2) cation exchange.

A very valuable research work was done by Warshaw (1959) in order to determine how natural illites can be synthesized from amorphous gels and to delineate the upper temperature stability limits of such phases. This type of research work is of value in understanding the aggradation process suggested for the formation of illite.

In 1937, Grim et al. proposed to use "illite" as a "general term for the clay mineral constituents of argillaceous sediments belonging to the mica group". According to them, with further accumulation of data, specific names would be given to some of the clay micas and "illite" would remain the group name. This did not happen, however, and the term "illite" became restricted almost entirely to minerals resembling practically in every detail

TABLE 5-VII

Percent MgO in illites formed in different environments
(After Weaver and Pollard, 1973, table IV, p. 10)

No.	Environment	%MgO	No.	Environment	%MgO
6	weathered feldspar	1.09	2	Atchafalaya Bay mud	3.33
11	weathered granite	1.37	3	Mississippi delta, marine mud	3.32
22	veneer in shear planes	0.44	4	Gulf of Mexico, marine mud	2.90
27	post-depositional alteration		13	slate from marine shale	3.56
	of feldspar in sandstone	1.49	14	slate from marine shale	3.05
			23	clay from marl	4.48

the Illinois illites. Because of necessity for a different group name, the term "hydrous mica" was proposed by Brown (1951) to include all the clay micas. Clay micas with interstratified layers of montmorillonite were also included. Minerals the 10-Å reflections of which were not significantly affected by mild thermal or chemical treatments were termed illite.

Weaver and Pollard (1973) made an intensive study on the structure of illite, collecting numerous samples from soils, marine muds, limestones, sandstones, and shales. According to them (Weaver and Pollard, 1973, p. 7), it is difficult to show any relation between composition and environmental factors because of (1) the small number of analyses, (2) the limited amount of geological and mineralogical information, and (3) the fact that most illites in sediments are at least second-cycle. There is a decided tendency for MgO to be more abundant in "marine" illites, however, than in those formed under nonmarine conditions (Table 5-VII). According to Weaver and Pollard (1973, p. 7), the MgO content of the "marine" illites is similar to that of montmorillonites and may indicate that these illites formed from montmorillonites, probably during burial; the low-MgO illites are primarily formed from the alteration of feldspar. Consequently, Mg content may be used to give information on the origin of illites (Weaver and Pollard, 1973, p. 7). Illites having low MgO content all have more than 1.70 octahedral positions filled with Al (Table 5-VII), which is within the range of low-Al muscovites.

The structural formulas of 29 different illites are presented in Table 5-VIII. The structural formula which is based on an average of 7 Fithian illites was also included by Weaver and Pollard (1973, p. 7). Tetrahedral Al, which is perhaps the most significant compositional factor, has a larger range of values for the 7 Fithian illites than for all 29 different illites.

Weaver and Pollard (1973, p. 12) showed that the average K_2O content of the first 24 illites is 7.02%. Weaver (1965) has reported that the average K_2O content of 16 Paleozoic and 2 Precambrian, predominantly 2M-illites, is

8.75%. Although it was necessary to correct composition of the latter samples for non-illite impurities, the latter average K_2O value is probably more likely characteristic of 10-Å 2M-illites than that obtained from the analyses in the literature (Weaver and Pollard, 1973, p. 13).

Samples No. 6 and 27 (Table 5-VIII) apparently formed from the weathering of feldspar, whereas samples No. 28 and 29 are of hydrothermal origin. Samples No. 6, 27, and 28 have a relatively low content of divalent cations in the octahedral sheet. On the basis of total charge and K_2O content, these polytypes are similar to the 2M-polytype (Weaver and Pollard, 1973, p. 13). Limited data, however, suggest that the octahedral sheet of the 2M-illites (high-temperature) has more R^{2+}, largely Mg^{2+}, than the other polytypic forms. Apparently, R^{2+} ions are not only not necessary for the formation of illite, but their presence may even inhibit its development. The presence of R^{2+} ions in montmorillonite may account for the relatively high temperature ($\sim$200°C) necessary to convert it to illite. Weaver and Pollard (1973, p. 13) offered two possible explanations for this: (1) the thermal energy may be needed to allow the R^{2+} ions to migrate and form chloritic layers interspersed among muscovite-type layers; or (2) more energy is required to force the silica tetrahedrons to rotate enough to lock-in the K^+ ions, because the large R^{2+} ions increase the size of the octahedral sheet, making it more similar in size to the tetrahedral sheet.

Although several other varieties of "illite" have been described, they appear to be relatively rare; analyses of two varieties are shown in Table 5-IX. Andreatta (1949) described an illite, which he called illidromica, of hydrothermal origin. This material, which is deficient in alkalis and has a high water content, may be a mixed-layer illite—montmorillonite (Weaver and Pollard, 1973, p. 18).

Bannister (1943) described a sodium-illite (brammallite) which contains 5.22% Na_2O and 2.58% K_2O. Analysis 3 in Table 5-IX was taken from Meyer (1935) and, according to Weaver and Pollard (1973, p. 18), presumably represents a sodium-illite. The former sample was obtained from a coal seam and the latter from a nodule in a conglomerate. Although sodium-illite is not common, sodium-rich layers probably are interlayered with potassium-rich layers in some illites (Weaver and Pollard, 1973, p. 18). Walker (1950) and Weiss et al. (1956) described trioctahedral illites. Walker's analysis, which he considered only as a rough approximation, is given in Table 5-IX. Although the "clay biotite" occurs in a Scottish soil and is believed to be of authigenic origin, it weathers so easily to vermiculite that unweathered material is difficult to find. The existence of clay-sized biotite in large amounts is unlikely because of its instability; however, trioctahedral biotite-like layers may occur interlayered with dioctahedral illite layers (Weaver and Pollard, 1973, p. 18). This type of interlayering has been reported by Bassett (1959).

TABLE 5-VIII

Illite structural formulas
(After Weaver and Pollard, 1973, table V, pp. 10, 11)

	1	2	3	4	5	6	7	8	9	10	11	12	13
Octahedral													
Al	1.51	1.53	1.65	1.65	1.53	1.71	1.39	1.51	1.40	1.56	1.75	1.81	1.46
Fe^3	0.23	0.23	0.12	0.12	0.22	0.16	0.09	0.03	0.39	0.15	0.12	0.04	0.30
Fe^2	0.10	0.07	0.00	0.00	0.04	0.00	0.00	0.03	0.00	0.00	0.03	0.10	0.00
Mg	0.29	0.27	0.34	0.34	0.29	0.11	0.37	0.43	0.22	0.30	0.14	0.17	0.37
	2.13	2.10	2.11	2.11	2.08	1.98	1.85	2.00	2.01	2.01	2.04	2.12	2.13
Tetrahedral													
Al	0.55	0.54	0.66	0.66	0.50	0.53	0.38	0.57	0.62	0.54	0.60	0.78	0.68
Si	3.45	3.46	3.34	3.34	3.50	3.47	3.62	3.43	3.38	3.46	3.40	3.22	3.32
Interlayer													
Ca	0.02	0.02	0.03	0.03	0.02	0.04	0.18	0.09	0.16	0.06	0.00	0.01	0.00
Na	0.02	0.02	0.02	0.02	0.02	0.01	0.03	0.02	0.02	0.02	0.02	0.14	0.07
K	0.52	0.53	0.62	0.62	0.55	0.65	0.98	0.89	0.64	0.71	0.66	0.54	0.60
Layer charge													
Octahedral	0.00	0.04	0.01	0.01	0.09	0.17	0.82	0.46	0.19	0.27	0.05	0.09(+)	0.02(+)
Tetrahedral	0.55	0.54	0.66	0.66	0.50	0.53	0.38	0.57	0.62	0.54	0.60	0.78	0.68
Total	0.55	0.58	0.67	0.67	0.59	0.70	1.20	1.03	0.81	0.81	0.65	0.69	0.66

1. Grim et al. (1937): Pennsylvania underclay, near Fithian, Ill., U.S.A.; analyst O.W. Rees.
2. Whitehouse and McCarter (1958): soil Point Chevrecil, La., U.S.A.
3. Whitehouse and McCarter (1958): Recent marine mud, Atchafalaya Bay, La, U.S.A.
4. Whitehouse and McCarter (1958): Recent marine mud, Mississippi delta region, La., U.S.A.
5. Whitehouse and McCarter (1958): Recent marine mud, Gulf of Mexico.
6. Weaver (1953): weathered from feldspar in Ordovician graywacke, State College, Pa., U.S.A.; analyst G. Kunze
7. Weaver (unpublished): Cambrian marine shale, central Texas, U.S.A.
8. Weaver (unpublished): Precambrian shale bed in Belt Limestones, Glacier National Park, Mont., U.S.A.; analyst E.G. Oslund.
9. Weaver (unpublished): Silurian, red shale in limestone, Mont., U.S.A.
10. Weaver (unpublished): Devonian, Bloomsburg Red Beds, Central Penn., U.S.A.: analyst E.G. Oslund.
11. Mackenzie et al. (1949): decomposed granite, Aberdeenshire, Scotland; analyst R.C. Mackenzie.
12. Nagelschmidt and Hicks (1943): Coal Measures shales of South Wales; analyst W.N. Adams.
13. Bates (1947): Paleozoic slate, northeastern Penn., U.S.A.; analyst R.J. Grace.
14. Bates (1947): Paleozoic slate, northeastern Penn., U.S.A. analyst R.J. Grace.
15. Mackenzie (1957b): from Upper Old Red Sandstone, Roxburgshire, Scotland; analyst J.B. Craig.
16. Grim et al. (1937): Pennsylvanian shale, near Petersburg, Ill., U.S.A.; analyst O.W. Rees.
17. Grim et al. (1937): Cretaceous shale, near Thebes, Ill., U.S.A.; analyst O.W. Rees.
18. Grim et al. (1937): slightly weathered till, Ford Col, Ill., U.S.A.; analyst O.W. Rees.
19. Grim et al. (1937): Ordovician, Mawuokita shale, Gilead, Ill., U.S.A.; analyst O.W. Rees.
20. Mankin and Dodd (1963): Silurian, Blaylock shale (alternating thin beds of shale and fine sandstone), Beavers Bend State Park, Okla., U.S.A.,; analyst J.A. Schleicher.

14	15	16	17	18	19	20	21	22	23	24	25	26	27	28	29
1.38	1.52	1.34	1.55	1.19	1.40	1.41	1.43	1.92	1.29	1.86	1.51	1.61	1.76	1.87	1.50
0.38	0.16	0.64	0.20	0.58	0.26	0.35	0.26	0.06	0.32	0.00	0.04	0.08	0.11		0.12
0.09	0.05	0.00	0.00	0.00	0.09	0.11	0.04	0.02	0.07	0.00	0.07	0.11	0.00		
0.32	0.34	0.26	0.27	0.39	0.40	0.19	0.30	0.04	0.46	0.19	0.36	0.20	0.15	0.10	0.43
2.17	2.07	2.24	2.02	2.16	2.15	2.06	2.03	2.04	2.14	2.05	1.98	2.00	2.02	1.97	2.05
0.64	0.49	0.89	0.49	0.62	0.60	0.53	0.46	0.74	0.46	0.69	0.46	0.43	0.82	0.55	0.51
3.36	3.51	3.11	3.51	3.38	3.40	3.47	3.54	3.21	3.54	3.31	3.54	3.57	3.18	3.45	3.49
0.00	0.10	0.02	0.09	0.03	0.05	0.02	0.01	0.04	0.00	0.08	0.10	0.07	0.05		0.03
0.02	0.03	0.01	0.04	0.00	0.01	0.05	0.04	0.03	0.04	0.07	0.03	0.03		0.02	0.04
0.53	0.57	0.43	0.56	0.53	0.60	0.59	0.68	0.66	0.53	0.57	0.70	0.58	0.80	0.72	0.69
0.10(+)0.18		0.46(+)0.21		0.09(+)0.04		0.12	0.25	0.06(+)0.11		0.04	0.49	0.31	0.09	0.19	0.28
0.64	0.49	0.89	0.49	0.62	0.60	0.53	0.64	0.74	0.46	0.69	0.46	0.43	0.82	0.55	0.51
0.54	0.67	0.43	0.70	0.53	0.64	0.65	0.89	0.68	0.57	0.73	0.95	0.74	0.91	0.74	0.79

21. U.S.G.S., Chinle Formation, U.S.A.; analyst W.W. Brannock.
22. Brammall et al. (1937): veneer on shear planes, Ogofau, Wales; analyst H. Bennett.
23. Maegdefrau and Hofmann (1937): in marl, probably contains interlayers of montmorillonite, Goeschwitz, Germany.
24. Maegdefrau and Hofman (1937): in kaolin deposits, Saraspatak, Hungary.
25. Gaudette (1965): in laminated pockets in Silurian dolomites, Marblehead, Wisc., U.S.A.; analyst J. Witters.
26. Gaudette et al. (1966): pockets in Ordovician limestone near Geneseo, Ill., U.S.A.; analyst J. Witters.
27. Triplehorn (1967): post-depositional in Cambro-Ordovician sandstone, Algeria.
28. Tien (1969): in hydrothermal lead–zinc ore, Silverton, Colo., U.S.A.
29. Levinson (1955): in hydrothermal kaolin deposit, St. Austell, England.

For references in the notes see Weaver and Pollard (1973).

Attapulgite and palygorskite

Attapulgite clay has a fibrous texture and a chain structure, which is also true for palygorskite. Attapulgite and palygorskite appear to be two distinct fibrous species. A 2 : 1 layer structure with five octahedral positions (four filled) was proposed by Bradley (1940). Four Si tetrahedra occur on either side of the octahedral sheet with their apices directed towards the octahedral sheet (see Fig. 5-4). Inasmuch as these structural units alternate in a checker-

TABLE 5-IX

Compositions of atypical illites
(After Weaver and Pollard, 1973, table XI, p. 18)

	1	2	3		1	2	3
SiO_2	53.12	40.87	47.30	Octahedral			
Al_2O_3	27.36	20.45	36.31	Al	1.67	0.89	1.84
Fe_2O_3	2.64	12.81	2.17	Fe^{3+}	0.13	0.72	0.01
FeO	1.06	0.00	–	Fe^{2+}	0.06		
MgO	2.62	6.86	–	Mg	0.26	0.72	
CaO	0.53	0.89	–	Σ	2.12	2.38	1.85
Na_2O	0.49	0.70	5.27				
K_2O	3.51	3.25	2.70	Tetrahedral			
TiO_2	0.00	2.13	–	Al	0.47	0.93	0.94
H_2O^+	8.60	11.84	5.80	Si	3.53	3.07	3.06
H_2O^-	0.00	0.00	–				
	99.93	99.80	99.55	Interlayer			
				Ca/2	0.08	0.14	
				Na	0.06	0.10	0.66
				K	0.30	0.31	0.23

1. Andreatta (1949): Illidromica, Capalbio, Italy.
2. Walker (1949): Weathered biotite, Scotland.
3. Meyer (1935): Na illite, Missouri.

For references in the notes see Weaver and Pollard (1973).

board pattern, a series of channels are left between them. Weaver and Pollard (1973) suggested that such channels contain water molecules.

Table 5-X contains a list of seven chemical analyses of attapulgite from Attapulgus, Georgia. The results are reasonably similar when compared with the seven analyses of the Fithian illite; the Al_2O_3/MgO ratio is relatively constant, varying from 0.64 to 1.08 (Weaver and Pollard, 1973, p. 124).

Three forms of water appear in the structure of palygorskite as proposed by Bradley in 1940 [$Mg_5Si_8O_{20}(OH)_2(OH_2)_4 \cdot 4\,H_2O$]: (1) zeolitic water, (2) bound water (at the edges of the octahedral sheet), and (3) structural hydroxyls. Using thermogravimetric curves, Caillère and Hénin (1961) attempted to measure the amount of these three types of water for several attapulgites (Table 5-XI). The sum of hydroxyls and bound water differs considerably from that calculated on the basis of the ideal structure of palygorskite (column 5), which suggests that there are more structural hydroxyls than proposed for the ideal structure.

Exchange values of 5—30 mequiv./100 g seem to be most characteristic of the relatively pure attapulgite samples. Cation exchange capacity values for attapulgite range from 10 to 99 mequiv./100 g; the higher values either

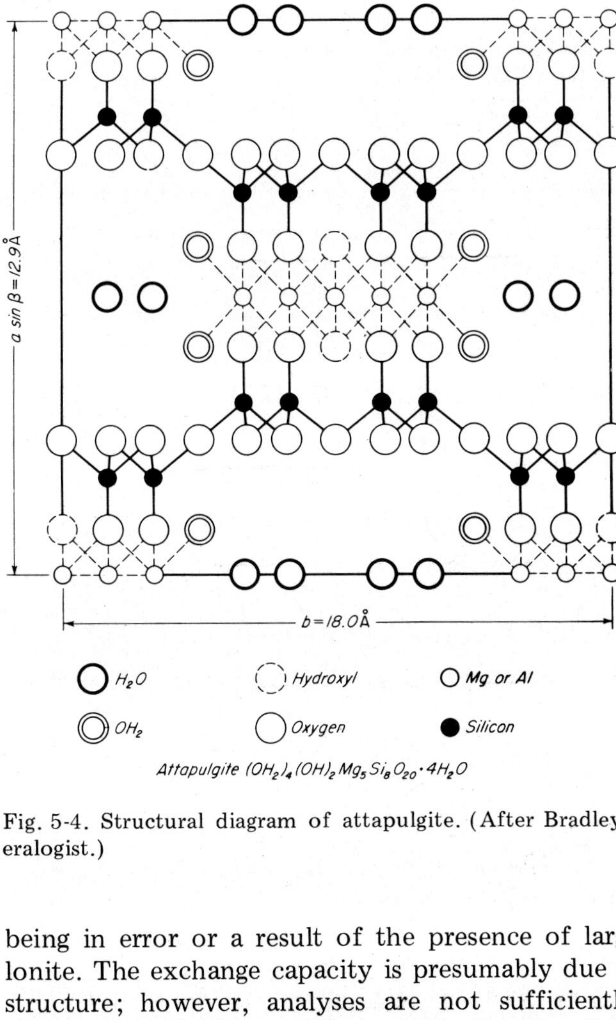

Fig. 5-4. Structural diagram of attapulgite. (After Bradley, 1940; courtesy of Am. Mineralogist.)

being in error or a result of the presence of large amounts of montmorillonite. The exchange capacity is presumably due to deficits of charge in the structure; however, analyses are not sufficiently accurate to prove this (Weaver and Pollard, 1973, p. 125). Usually, CaO, Na_2O, and K_2O are all reported in analyses of attapulgite and at least some of Ca^{2+}, Na^+, and K^+, along with some of the Mg^{2+}, are probably present as exchangeable cations. In most samples, however, these cations are more abundant than can be accounted for by the cation exchange values. Calcium cation is usually most abundant, amounts of CaO reported ranging from 1 to 2%. Examination of hundreds of samples of attapulgite from Georgia, U.S.A., indicate that illite and calcite are present in most samples (Weaver and Pollard, 1973, p. 125). The presence of illite and calcite would account for a large portion of the K_2O and some of the CaO content. After acid-leaching an attapulgite from Ipswich, Loughnan (1959) found that CaO content was lowered from 5.1 to 0.9%. In addition, CO_2 was not present in the acid-treated sample. Inasmuch

TABLE 5-X

Chemical analyses of attapulgite clay from Attapulgus, Georgia, U.S.A. (After Weaver and Pollard, 1973, table LIV, p. 123)

	1	2	3	4	5	6	7
SiO_2	55.04	53.67	55.03	53.64	54.04	53.96	56.00
Al_2O_3	8.96	8.63	10.24	8.76	9.83	8.56	7.98
Fe_2O_3	3.43	3.04	3.53	3.36	3.52	3.10	2.51
FeO				0.23	0.19	0.19	
MgO	9.52	11.06	10.49	9.05	9.07	10.07	12.40
CaO	1.55	1.55		2.02	1.69	2.01	
Na_2O	0.05	0.42	0.47	0.83	0.08	0.03	
K_2O	0.50	0.35		0.75	0.57	0.39	
TiO_2	0.48	0.49		0.60	0.32	0.24	
H_2O^+	19.60	19.49	10.13	10.89	10.93	11.51	
H_2O^-			9.73	9.12	10.00	9.79	
P_2O_3	0.80	1.30		0.79			
Total	100.46	100.42	99.62	100.04	100.24	99.85	
C.E.C. (mequiv./100 g)	10.60	10.8	21		(99.0)	35.0	

1. Minerals and Chemicals Corp. of America: typical mine-run sample containing montmorillonite.
2. Minerals and Chemicals Corp. of America: 1.0–0.1μ fraction contains 5% montmorillonite, increase in Na_2O and P_2O_5 due to dispersant.
3. Bradley (1940): 0.1–0.05μ fraction.
4. De Lapparent (1935).
5, 6. Kerr et al. (1950): 1–3% impurity.
7. Abdul-Latif and Weaver (1969): purified sample from Minerals and Chemicals Corp. of America.

For references in the notes see Weaver and Pollard (1973).

TABLE 5-XI

Observed water (zeolitic, bound, hydroxyl, and total) losses in percent compared with those for ideal palygorskite

(After Caillère and Hénin, 1961; courtesy of London Mineral. Soc., London)

	1	2	3	4	5
Zeolitic water, below 200°C	10.8	10.7	8.0	9.8	8.60
Bound water, 250—400°C	3.6	2.7	4.0	4.0	8.60
Hydroxyl water, above 400°C	4.2	6.2	6.2	6.5	2.15
Total water	18.6	19.6	18.2	20.3	19.35

1. Attapulgus, Ga., U.S.A.
2. Taodeni, Algeria.
3. Tafraout, Morocco.
4. Meyssonial en Mercoeur, Upper Loire, France.
5. Ideal palygorskite, $(Si_8)(Mg_5)O_{20}(OH)_2(OH_2)_4 4H_2O$.

as acid treatment would have removed exchangeable Ca^{2+} cations, probably some of the Ca^{2+} is an integral part of the structure.

Invariably, attapulgite is associated with montmorillonite. In some cases, it is believed by some investigators (1) to weather to montmorillonite (Barshad et al., 1956) and (2) to be transformed to montmorillonite by the addition of magnesium (Loughnan, 1959). Although occurrence of neither one of these two transformations is certain, attapulgite does alter to montmorillonite under hydrothermal conditions (Mumpton and Roy, 1958).

Sepiolite

According to Weaver and Pollard (1973, p. 127), sepiolite is a magnesium-rich, lath-shaped clay mineral having a structure similar to that of attapulgite (see Fig. 5-5). Nagy and Bradley (1955) assigned the following structural formula to sepiolite: $[(Si_{12})(Mg_9)\,O_{30}(OH)_6(OH_2)_4\ 6\ H_2O]$. The Mg fills 90–100% of occupied octahedral positions; the MgO content of sepiolites ranges from 21 to 25%.

Martin-Vivaldi and Cano-Ruiz (1955) found a mean value of 53.9% SiO_2 based on 50 analyses (standard error = ±1.09). The moles of MgO/100 g range from about 0.4 to 0.7 (modal value ≈0.6), whereas moles of (Al_2O_3 + Fe_2O + FeO + Mn) range from 0 to 0.15. In contrast to palygorskite, where Al (a smaller cation) commonly fills half of the octahedral positions, in sepi-

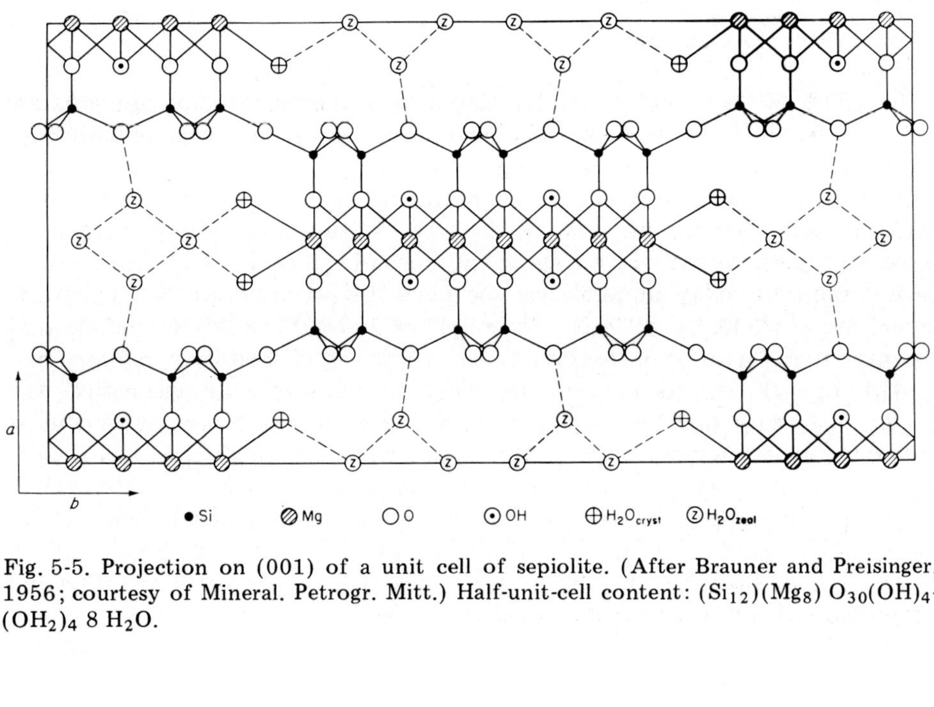

Fig. 5-5. Projection on (001) of a unit cell of sepiolite. (After Brauner and Preisinger, 1956; courtesy of Mineral. Petrogr. Mitt.) Half-unit-cell content: $(Si_{12})(Mg_8)\,O_{30}(OH)_4$-$(OH_2)_4\ 8\ H_2O$.

TABLE 5-XII

Observed water losses in % compared with those calculated from ideal formulas (After Caillère and Hénin, 1961; courtesy of London Mineral. Soc., London)

	1	2	3	4	5
Zeolitic water, <250°C	11.43	13.6	11.7	8.1	11.1
Bound water, 250—620°C	5.7	5.6	5.2	5.4	5.5
Hydroxyls, 620—1000°C	2.4	2.4	2.6	4.1	2.7
Total	19.53	71.6	19.5	17.6	19.3

1. Ampandrandava, Madagascar.
2. Eski Chehir, Asia Minor.
3. Salinelles, France.
4. Ideal sepiolite (Nagy and Bradley, 1955).
5. Ideal sepiolite (Brauner and Preisinger, 1956).

For references in the notes see Caillère and Hénin (1961).

olites Al is relatively uncommon. Most of the cations in the octahedral positions are of the larger variety (Weaver and Pollard, 1973, p. 127). Sepiolite-type minerals exhibit wide range of compositions (accommodated mostly in the octahedral sheet), even though the Mg-rich variety is the most common. The reported exchange capacity ranges from 20 to 45 mequiv./100 g, and appears to be due to the replacement of silicon by trivalent ions. Observed water losses compared with those expected from ideal formulas are presented in Table 5-XII.

Mixed-layer clay minerals

In 1973, Weaver and Pollard stated that mixed-layer clay minerals are those in which individual crystals are composed of unit cells or basic unit layers of two or more types. The great majority of clay minerals are probably composed of interstratified layers having different composition. These differences, however, are not detected by routine methods of analysis in most instances. As stated by Weaver and Pollard (1973, p. 107), from a practical standpoint, clay minerals can be classified as interstratified when the layers are sufficiently different in character and sufficiently abundant that the presence of two or more layer types can be established by X-ray analysis.

Although the layers may be regularly or randomly interstratified, the latter type is by far the more common and gives rise to the second most abundant clay mineral species, after illite, which in most cases is a mixed-layer clay. More regularly interstratified clay minerals have a definite periodicity and some are given specific mineral names (Weaver and Pollard, 1973, p. 107). The randomly interstratified clay minerals, many of which exhibit some degree of regular interlayering, are described in terms of the type and proportion of the two or more types of layers.

TABLE 5-XIII

Chemical analyses and structural formulas of rectorite-like clays (After Weaver and Pollard, 1973, table XLV, p. 108)

	1	2	3	4	5
SiO_2	53.7	54.11	55.2	49.93	54.15
Al_2O_3	38.2	40.38	38.7	25.97	36.12
Fe_2O_3	0.77	0.15	0.59	2.01	2.19
FeO	0.65	–	–	0.30	–
MgO	0.42	0.78	–	2.77	0.65
CaO	1.19	0.52	0.33	1.11	1.50
Na_2O	2.68	3.87	4.40	0.09	–
K_2O	1.31	0.29	0.11	3.46	5.54
TiO_2	0.36	0.01	–	0.42	0.06
H_2O^+	–	–	–	7.45	–
H_2O^-	–	–	–	6.49	–
Total	99.46[a,b]	100.24[a,c]	99.33[a]	100.00	100.47[a,d]

	1	2	3	4	5
Octahedral					
Al	1.93	2.01	2.01	1.64	1.88
Fe^{3+}	0.04	0.01	0.03	0.11	0.10
Fe^{2+}	0.03	–	–	0.02	–
Mg	0.04	0.07	–	0.29	0.02
Σ	2.04	2.09	2.04	2.06	2.00
Tetrahedral					
Al	0.77	0.80	0.71	0.50	0.70
Si	3.23	3.20	3.29	3.50	3.30
Interlayer					
Ca and Mg	0.16	0.07	0.04	0.17	0.12
Na	0.31	0.44	0.51	0.01	–
K	0.10	0.02	0.01	0.31	0.45
Charge					
octahedral	+0.05	+0.20	+0.12	–0.13	–0.02
Total	–0.72	–0.60	–0.59	–0.63	–0.72

1. Hénin et al. (1954): allevardite, Allevard, France.
2. Kodama (1966): Fort Sandeman district, Baluchistan, Pakistan.
3. Brown and Weir (1963): Na-saturated sample from Garland Co., Ark., U.S.A.
4. Rateyev et al. (1969): K-rectorite, weathered clay, Samarskaya Luka, U.S.S.R.
5. Cole (1966): rectorite-like clay formed from weathered feldspar, Surges Bay, Tasmania.

[a]ignited weight basis; [b]includes 0.18% P_2O_5; [c]includes 0.13% SrO; [d]includes 0.26% P_2O_5.

For references in the notes see Weaver and Pollard (1973).

Illite—montmorillonite

Weaver and Pollard (1973) described illite—montmorillonite as the most common type of mixed-layer clay, which is composed of expanded, water-bearing layers and contracted, non-water-bearing layers, i.e., illite—montmorillonite, chlorite—montmorillonite, and chlorite—vermiculite. Most of these clays form by the partial leaching of K^+ or $Mg(OH)_2$ from between illite or chlorite layers and by the incomplete absorption of K^+ or $Mg(OH)_2$ on montmorillonite- or vermiculite-like layers. Illite—montmorillonite clays most commonly form during weathering or after burial, but are frequently of hydrothermal origin (Weaver and Pollard, 1973, p. 107).

Regular mixed-layer illite—montmorillonites (Table 5-XIII), which are composed of alternating mica-like and montmorillonite-like layers, have been described by several investigators. Brown and Weir (1963) suggested, however, that clays which had previously been described under other specific names, such as allevardite, were forms of the mineral rectorite. Rectorite is composed of high-charge (mica-like) layers with fixed interlayer cations alternating in a regular manner with low-charge (montmorillonite-like) layers that have exchangeable cations capable of hydration (Weaver and Pollard, 1973, p. 107). According to Kodama (1966), the expandable layers are both beidellitic and montmorillonitic.

The structural formulas for the Na-rectorites, which are presumably an average of at least two or more types of layers, indicate that the negative charge originates in the tetrahedral sheet and that the octahedral sheets have a slight positive charge. According to Weaver and Pollard (1973, p. 109), the octahedral positions are almost entirely filled with Al, which suggests that these minerals are of hydrothermal origin or have formed from the leaching of paragonitic material. The homogeneity of the octahedral sheets may favor the development of regularly-alternating layers (Weaver and Pollard, 1973, p. 109).

Random mixed-layer illite—montmorillonite is by far the most abundant of the mixed-layer clays. As pointed out by Weaver and Pollard (1973, p. 110), illite and montmorillonite layers occur intergrown in all proportions. Table 5-XIV contains chemical data for a series of mixed-layer illite—montmorillonites with content of expanded layers ranging from less than 10 to 60%. It is difficult to accurately determine the number of illitic layers when expandable layers comprise more than 60% and less than 10% of clay. In addition, Hower (1967) has shown that many of these clays are inter-stratified partially regularly and partially randomly, which also complicates the interpretation of the X-ray data.

According to Weaver (1953), Ormsby and Sand (1954), Bystrom (1956) and Hower and McWatt (1966), the cation exchange capacity is closely related to the proportion of expanded layers. If all the exchange capacity is assigned to the montmorillonite-like layers, values range from 96 to 144 mequiv./100 g, which are within the range obtained for pure montmoril-

TABLE 5-XIV

Structural formulas of some mixed-layer illite—montmorillonite clays (From Weaver and Pollard, 1973, table XLVII, p. 110)

	1	2	3	4	5	6	7	8	9	10
Octahedral										
Al	1.66	1.45	1.47	1.64	1.57	1.49	1.62	1.27	1.96	1.50
Fe^{3+}	0.03	0.16	0.20	0.02	0.09	0.33	0.07	0.38	0.01	0.10
Fe^{2+}	0.05	0.07	0.10	0.00	0.03	0.04	0.02	0.07	0.00	0.04
Mg	0.30	0.34	0.21	0.36	0.30	0.14	0.30	0.27	0.15	0.38
Σ	2.04	2.02	2.00	2.02	1.99	2.00	2.01	1.99	2.12	2.02
Tetrahedral										
Al	0.57	0.31	0.40	0.45	0.32	0.42	0.42	0.20	0.66	0.17
Si	3.43	3.69	3.60	3.55	3.68	3.58	3.55	3.80	3.34	3.83
Interlayer										
Ca and Mg	0.02	0.05	0.07	0.10	0.09	0.13	0.20	0.17	0.02	0.15
Na	0.01	0.01	0.01	0.02	0.00	–	0.02	–	0.13	0.01
K	0.77	0.62	0.65	0.64	0.43	0.48	0.50	0.39	0.29	0.22
Charge										
octahedral	0.23	0.35	0.33	0.30	0.36	0.18	0.29	0.37	+0.21	0.36
tetrahedral	0.57	0.31	0.40	0.45	0.32	0.42	0.42	0.20	0.66	0.17
total	0.80	0.66	0.73	0.75	0.68	0.62	0.71	0.57	0.45	0.53
C.E.C. (mequiv./100 g)	12	20	24	27	46	37	–	50	48	77
Percent expanded layers	<10	10	15	20	30	33	40	45	50	60

For legend, see Table XLVI.

1. Hower and Mowatt (1966): Silurian Interlake well core, Williston Basin, Mont., U.S.A.
2. Hower and Mowatt (1966): Cambrian, Eau Claire, Illinois Basin, U.S.A.
3. Hower and Mowatt (1966): Upper Cambrian, Gros Ventre, Wind River Canyon, Wyo., U.S.A.
4. Weaver (1953): Ordovician, Salona, Pa., U.S.A.; analyst J.A. Maxwell.
5. Byström (1956): Ordovician, Sweden.
6. Hower and Mowatt (1966): Upper Cretaceous, Two Medicine, Bowman's Corners, Mont., U.S.A.
7. Bradley (1945): bravasite.
8. Hower and Mowatt (1966): Upper Cretaceous Colorado, Bozeman Pass, Mont., U.S.A.
9. Heystek (1955): hydrothermal alteration of illite, South Africa; analyst D. Sampson.
10. Byström (1956): Ordovician, Sweden.

For references in the notes see Weaver and Pollard (1973).

lonites (Weaver and Pollard, 1973, p. 113).

Mixed-layer clays form as a result of (1) alteration of pre-existing micas and illites, (2) hydrothermal action, (3) alteration of volcanic glass, and (4) diagenetic alteration of montmorillonite. Upon being carried to the sea, these clays may absorb K^+ and revert partially or entirely to illite. The expanded layers can persist until K^+ becomes available if they remain in a continental environment which may be devoid of K^+. When formed in this manner, expanded layers have a higher layer charge than those formed in a marine environment (Weaver and Pollard, 1973, p. 114).

Chlorite—montmorillonite

Weaver and Pollard (1973, p. 114) stated that the second most abundant type of mixed-layer clay is the one composed of the intergrowth of chlorite and 2 : 1 water-bearing layers, which may be montmorillonite-like (two layers of ethylene glycol) or vermiculite-like (one layer of ethylene glycol). Although chlorite and expanded layers occur in all proportions, only samples in which the ratio is approximately 1 : 1 have been analyzed chemically. The name corrensite or swelling chlorite has been applied to some of such clays, which are regularly interstratified (Weaver and Pollard, 1973, p. 114). According to recommendation by MacEwan et al. (1961), the name corrensite should be restricted to the type mineral described by Lippman (1954). According to Weaver and Pollard (1973, p. 112), the expandable layers in this mineral appear to contain a sufficient amount of brucite material to prevent the layers from contracting to 10 Å when heated. Bradley and Weaver (1956), Earley et al. (1956), and others described mixed-layer clays which readily contract to 10 Å. Inasmuch as most of these clays, for which chemical analyses are available, adsorb two layers of ethylene glycol, they probably should be called chlorite—montmorillonites rather than chlorite—vermiculites. Most of the clays of this type are trioctahedral. Sudo and Kodama (1957), however, described a sample in which the octahedral sheet of the 2 : 1 layer is dioctahedral and the hydroxide sheet is trioctahedral. Shimoda (1969) described a regular mixed-layer, di/trioctahedral chlorite—montmorillonite and assigned the specific name of tosudite to it.

As pointed out by Weaver and Pollard (1973, p. 116), there appears to be no difference between the tetrahedral and octahedral sheets of the chlorite layer and the montmorillonite layer. Magnesium is the dominant (60—90%) octahedrally-coordinated cation in the trioctahedral clays, whereas in the dioctahedral clays Al^{3+} is in the dominant octahedral cation. Chemical analyses of some mixed-layer chlorite—montmorillonites are presented in Table 5-XV.

Chlorite and montmorillonite layers are intergrown in a variety of proportions. The 1 : 1 regular intergrowth, however, appears to be the most well developed from a crystalline viewpoint and probably is in equilibrium with the Mg-rich environments in which it forms (Grim et al., 1961; Peterson,

TABLE 5-XV

Chemical analyses of some mixed-layer chlorite—montmorillonite clays (After Weaver and Pollard, 1973, table L, p. 115)

	Trioctahedral					Dioctahedral	
	1	2	3	4	5	6	7
SiO_2	41.2	43.1	37.2	36.12	33.44	39.94	42.14
Al_2O_3	12.1	16.6	15.5	13.67	14.33	33.17	37.38
Fe_2O_3	1.74	6.32	6.7	2.14	1.66	1.34	0.30
FeO	0.39	–	–	6.85	3.68	0.18	
MgO	22.0	17.65	18.9	25.16	22.44	6.44	0.08
CaO	1.4	0.94	1.0	1.70	2.83	1.30	1.65
Na_2O	0.07	0.52	0.2	0.37	0.16	0.52	0.15
K_2O	0.22	2.72	1.4	0.10	tr	0.24	1.40
TiO_2	0.04	0.73	0.4	0.23	–	0.74	–
H_2O^-	6.80	–	–	4.81	5.50	4.39	6.16
H_2O^+	15.4	7.40	18.4	9.25	8.16	11.64	11.22
Total	101.36	100.48[a]	99.7	100.40	92.36[b]	100.67[c]	100.48
C.E.C. (mequiv./100 g)	40			24	47		

1. Bradley and Weaver (1956): Upper Mississippian limestone, Juniper Canyon, Colo., U.S.A., contains approximately 20% quartz.
2. Earley et al. (1956): Permian, Yates siltstone, Ward Co., Texas, U.S.A., contains dolomite, mica and organic material; analyst B.B. Osthaus.
3. Peterson (1961): Upper Mississippian dolomite, Sparta, Tenn., U.S.A., contains illite and amorphous Fe_2O_3; analyst J. Ito.
4. Alietti (1957): hydrothermal with serpentine, Taro Valley, northern Appennines, contains serpentine; analyst A. Alietti.
5. Mongiorgi and Morandi (1970): secondary hydrothermal alteration of Al-saponite; breccia associated with diabase, near Rossena, Italy.
6. Sudo and Kodama (1957): hydrothermal, altered halo around pyrite ore body in volcanic rocks, contains pyrite.
7. Shimoda (1969): alteration mineral of Tertiary tuff and tuffaceous sediments around gold-silver-quartz veins.

[a]includes 0.03% MnO, 0.26% P_2O_5, 0.81% CO_2, 3.40% C; [b]includes 0.69% S, 0.08% P_2O_5; [c]includes 0.16% MnO; analysis corrected for 10% datolite.

For references in the notes see Weaver and Pollard (1973).

1961). Grim et al. (1961) gave an explanation for the cause of this regularity.

Other mixed-layer clay minerals

According to Weaver and Pollard (1973, p. 118), although there is a variety of other types of mixed-layer clays, identifications and chemical analyses are not numerous. A regular mixed-layer talc—saponite mineral from Italy was described by Alietti (1956). Another talc—saponite from Italy was described by Veniale and Van der Marel (1968). These investigators suggested that this regular mixed-layer talc—saponite clay mineral formed as a result of hydrothermal serpentinization of the parent ultramafic rock, after which the clay accumulated as a residual mineral inside the surface weathering products (Weaver and Pollard, 1973, p. 118). Guenot (1970) reported that interstratified talc—saponite mineral is of sedimentary origin. An iron-bearing chlorite (classified as daphnite) from Cornwall, England, was reported by Brindley and Gillery (1954) to have a mixed-layer serpentine—chlorite structure. Table 5-XVI gives chemical data for these mixed-layer clays.

METHODS OF IDENTIFYING CLAYS

Although the structure of clay minerals is identified by several methods, the X-ray analysis is considered as the most reliable one. Chemical tests, however, are used as supplementary methods to achieve accuracy. Differential thermal (Grim and Rowland, 1944), the electron probe, and infrared methods are also used extensively to identify structures of clay minerals.

X-ray diffraction method

As a pioneering genius, Brindley (1951) recognized the usefulness of X-ray diffraction analysis in identifying clay minerals and in formulation of various classification schemes (e.g., see Table 5-II). In these schemes, minerals are subdivided successively into chemical categories, structural groups and subgroups, chemical species, and structural varieties. Inasmuch as X-ray diffraction method is concerned primarily with the structural aspects of clay minerals, this method enables the recognition of structural groups and varieties. The recognition of chemical species generally requires knowledge of the manner in which the details of X-ray diffraction patterns depend on chemical constitution (atoms differ in size and in scattering power). Structurally similar minerals, but having different chemical composition, may show characteristic differences in the lattice parameters and/or in the intensities of their X-ray diffraction effects (see Brindley, 1955).

Quality of X-ray diffractograms depends on (1) the presence of well-

TABLE 5-XVI

Chemical analyses and structural formulas of some regular mixed-layer clays (After Weaver and Pollard, 1973, table LI, p. 117)

	1	2	3		1	2**	3
SiO_2	54.31	49.42	22.0	Octahedral			
Al_2O_3	1.90	4.97	27.6	Al	–	–	2.12
Fe_2O_3	–	4.81	4.7	Fe^{3+}	–	0.20	0.38
FeO	2.95	1.54	30.2	Fe^{2+}	0.17	0.09	2.74
MgO	27.44	28.56	4.7	Mg	2.86	2.80	0.76
CaO	1.01	0.60	–		3.03	3.09	6.00
Na_2O	0.32	0.38	–	Tetrahedral			
K_2O	0.86	0.04	–				
TiO_2	–	0.21	–	Al	0.16	0.45	1.41
H_2O^+	9.48	8.35	10.6	Si	3.80	3.48	2.39
H_2O^-	2.59	–	0.10	Fe^{3+}	–	0.07	–
Total	100.89	98.91*	99.9	Interlayer			
				Mg	–	0.006	–
				Ca/2	0.07	0.02	–
				K	0.07	–	–
				Na	0.04	0.01	–

1. Allieti (1956): mixed-layer talc-saponite, Monte Chiaro, Italy.
2. Veniale and Van der Marel (1968): mixed-layer talc-saponite, Ferriere, Nure Valley, Italy.
3. Brindley and Gillery (1954); Brindley (1961): mixed-layer serpentine-chlorite, Cornwall, England; analyst R.F. Youell.

*analysis before correction for estimated 10% chrysotile; also contained 0.03% P_2O_5 and trace MnO.
**after correction for 10% chrysotile.

For references in the notes see Weaver and Pollard (1973).

focused lines with good resolution, (2) ability to record long spacings up to about 24 Å and higher, and (3) absence of background scattering and white radiation anomalies.

The subject of X-ray identification of clay minerals has been reviewed up to 1950 in a classical symposium edited by Brindley (1951). In addition to the discussion of principles of identification, lists of X-ray powder data are presented in this publication. The American Society of Testing Materials published a card index of X-ray diffraction data. As an illustration, X-ray diffractogram of Na- and Ca-montmorillonites are presented in Figs. 5-6 and 5-7.

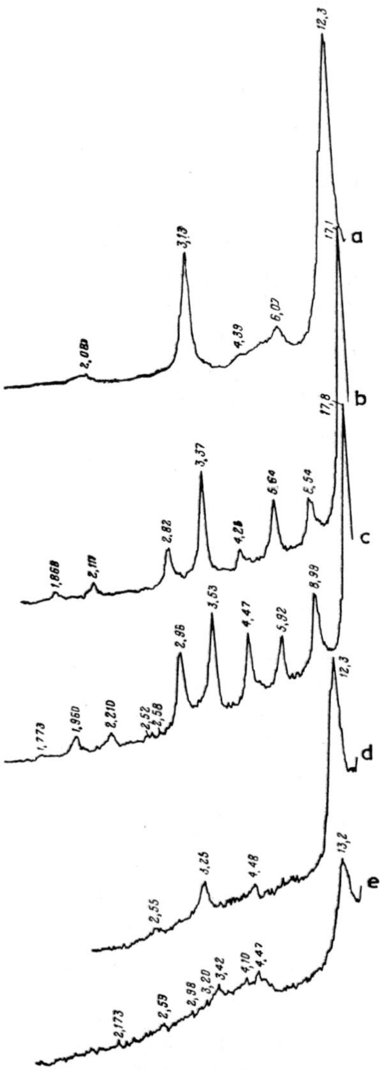

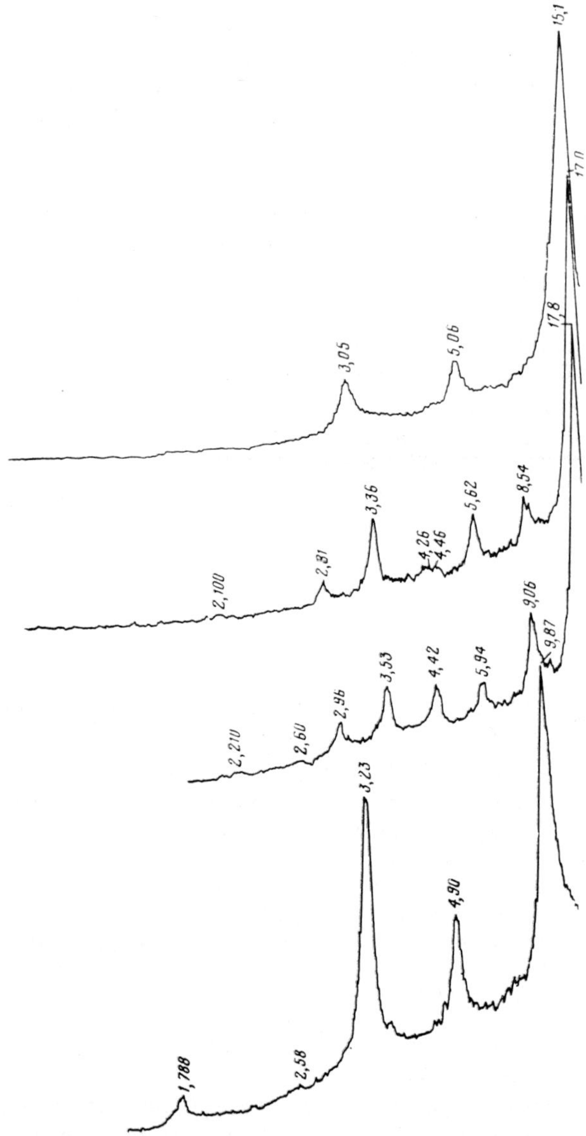

Fig. 5-6. X-ray diffractograms of Na-montmorillonite from Askan, Georgian S.S.R., U.S.S.R. Fraction <0.001 mm in size, Eocene in age. From top to bottom: *a* = original sample; *b* = original sample saturated with ethylene-glycol; *c* = original sample saturated with glycerine; *d* = original sample treated with 1N solution of KOH; *e* = sample *d* saturated with glycerine. (After Sarkisyan and Kotel'nikov, 1971, p. 26, fig. 13.)

Fig. 5-7. X-ray diffractograms of Ca-montmorillonite from Askan, Georgian S.S.R., U.S.S.R. Fraction <0.001 mm in size, Eocene in age. From top to bottom: *a* = original sample; *b* = original sample saturated with ethylene-glycol; *c* = original sample saturated with glycerine; and *d* = original sample heated at 600°C for two hours.

CHEMICAL ANALYSIS OF CLAYS

As pointed out by Kelley (1955), chemical analysis is a valuable supplement to other methods used for distinguishing between the broad groups of clay minerals.

Three-layer clays, such as the montmorillonites and hydrous micas, can be readily distinguished from the two-layer clay minerals, typified by kaolinite, by the $SiO_2 : Al_2O_3$ ratio as determined by the chemical analysis (Kelley, 1955, p. 92). The $SiO_2 : Al_2O_3$ ratio, however, cannot be relied upon as a definite criterion for determining what specific mineral is present in a given sample. According to Kelley (1955, p. 92), this is true even when the analyzed specimen is completely devoid of impurities. For example, the same ratio is characteristic of kaolinite, dickite, nacrite, and halloysite. Owing to the fact that the montmorillonites form an isomorphous series, montmorillonites from different sources have a rather wide variation in this ratio. Consequently, the $SiO_2 : Al_2O_3$ ratio, SiO_2 : sesquioxide ratio, or the $SiO : (Al_2O_3 + Fe_2O_3 + MgO)$ ratios do not constitute a dependable criterion for identifying clay-mineral species (Kelley, 1955, p. 92).

Isomorphism in common clays

According to Kelley (1955, p. 92), chemical analysis of the montmorillonite group of clay minerals may be used to determine the essential nature of isomorphism. Consequently, the chemical analysis is indicative of the origin and location of the charge on the lattice, and could show the isomorphous character of the members of montmorillonite group of minerals. The specific nature of isomorphism can be calculated from the results of chemical analysis. The specific pattern of isomorphism in montmorillonites from different sources is widely variable and may involve: (1) substitution of Al for Si in tetrahedral positions in the lattice, (2) substitution of Fe for Al in octahedral coordination, (3) substitution of Mg for Al in octahedral positions, and (4) substitution of various cations by other cations in the interlayer positions (Kelley, 1955, p. 92). Basically, the variations in the first three substitutions impart the greatest value to a chemical analysis.

Impurities and their effect on chemical analyses

Another shortcoming in the interpretation of chemical analysis of montmorillonite is the fact that samples as analyzed may not have been free from impurities (Kelley, 1955, p. 92). Evidence on this point was presented by Osthaus (1955, p. 95) who showed that reasonably pure montmorillonite contains only minor amounts of non-exchangeable Ca^{2+}, Na^+, and probably little, if any, of K^+. According to Osthaus (1955), if the sample contains significant amounts of non-exchangeable Ca^{2+}, Na^+, or K^+, the sample probably contains impurities.

According to Kelley (1955, p. 93), if the sample of montmorillonite is merely hand selected, even with the aid of the petrographic microscope, significant amounts of various impurities may be overlooked. Inasmuch as the greater part of the impurity is likely to be some form of SiO_2, the effect of SiO_2 as impurity on the calculations must be considered.

The SiO_2 impurity has the following effects on the calculated results: (1) increase in the number of calculated Si^{4+} ions in a unit quantity of the sample; (2) decrease in the calculated total octahedral cations; and (3) decrease in the total charge per unit cell (Kelley, 1955, p. 93). Inasmuch as the Si^{4+} ion has four charges, whereas practically all other constituent ions have three or fewer charges, if the sample is contaminated with SiO_2, an undue portion of the total cation charge will be borne by Si. Consequently, the charge of all other cations is reduced (Kelley, 1955, p. 93). In the presence of tetrahedral Al, the effect of SiO_2 is to increase the calculated number of Si ions. Inasmuch as this decreases the number of Al^{3+} ions required to fill the remaining tetrahedral positions in the lattice, the calculated charge in these layers is decreased. The extent to which some trivalent ion, such as Al^{3+}, is present in tetrahedral coordination, determines the unbalanced charge in the tetrahedra. Consequently, as far as the tetrahedral layers of the lattice are concerned, the calculated isomorphism is diminished by SiO_2 as impurity (Kelley, 1955, p. 93). According to Kelley (1955, p. 93), inasmuch as SiO_2 impurity increases the number of calculated Si^{4+} ions per unit of sample, its presence has the effect of reducing the calculated numbers of all other cations. The fact that Si^{4+} has a higher valency than the other cations further magnifies this effect. Thus the total number of calculated Al^{3+}, Fe^{3+}, and Mg^{2+} cations available for octahedral coordination is reduced in the presence of SiO_2 impurity, which increases the calculated charge in this layer of the structure (Kelley, 1955, p. 94).

As pointed out by Kelley (1955, p. 94), the presence of Al^{3+} and Fe^{3+} impurities has an effect opposite to that of SiO_2, because they increase the number of trivalent cations required to supplement Si^{4+} in tetrahedral coordination. This increases the calculated tetrahedral charge. Inasmuch as Al^{3+} and Fe^{3+} impurities increase the total octahedral cations, the octahedral charge is reduced.

The SiO_2, Al_2O_3, and Fe_2O_3 impurities, however, have one effect in common, i.e., all of them reduce the total calculated charge on the lattice (dilution effect) (Kelley, 1955, p. 94).

Differential thermal analysis

According to Rowland (1955, p. 151), differential thermal analysis (DTA) began to be used soon after the development of the thermocouple. Differential thermal analysis (DTA) has found an important place among techniques which enable identification of clays. Differential thermal curves record trans-

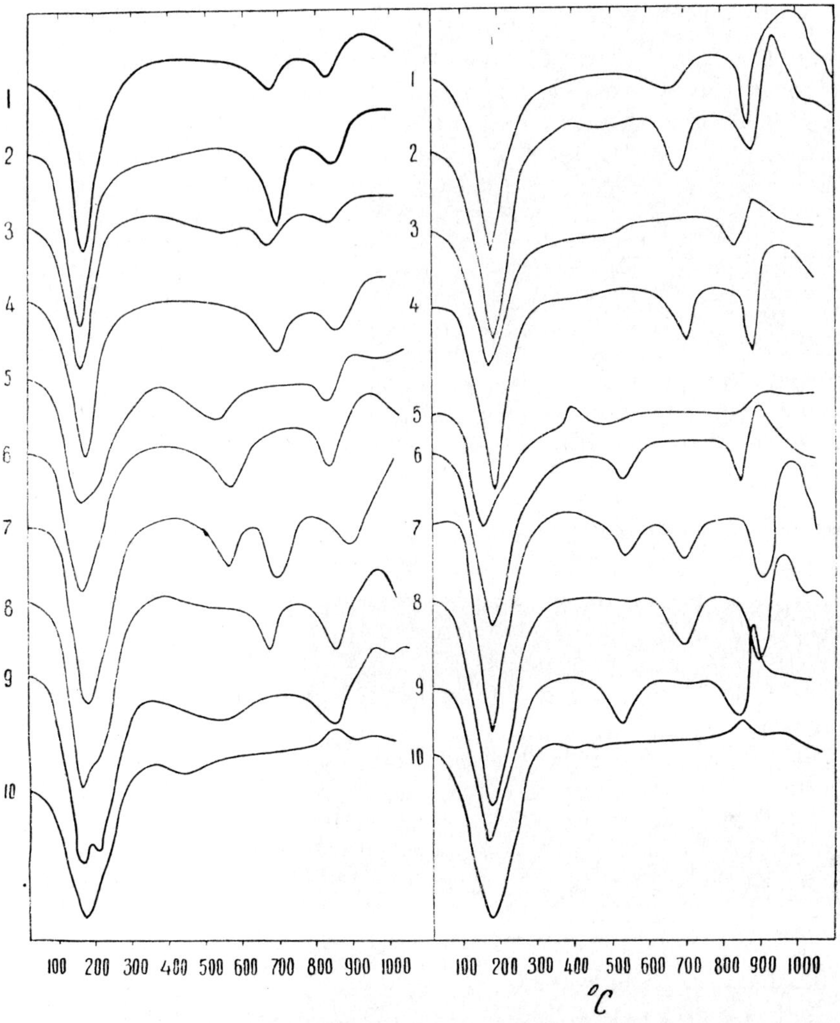

Fig. 5-8. DTA curves of montmorillonite clays from different localities in U.S.S.R. Left-hand side = natural clays; right-hand side = clays treated according to Grim's technique. *1* = Ascangel (a colloidal variety of askanite clay found in Georgian S.S.R.); *2* = Azkamar, white; *3* = Azkamar, green; *4* = Mangyshlak; *5* = Kyngrak; *6* = Tagan, grey; *7* = Tagan, pink; *8* = Pyzhevsk; *9* = gumbrin; and *10* = nontronite. (After Eyrish et al., 1970, p. 52, fig. 1.) In samples *1*—*6*, containing appreciable amount of exchangeable Na^+ cation, exothermic reactions are weak.

formations in which heat is taken up (endothermal reaction) or given off (exothermal reaction). Reactions include (1) dehydration of clays, (2) decarbonation of carbonates, (3) reversible change from α- to β-quartz, (4) burning of materials, and (5) recombination of elements into more stable forms

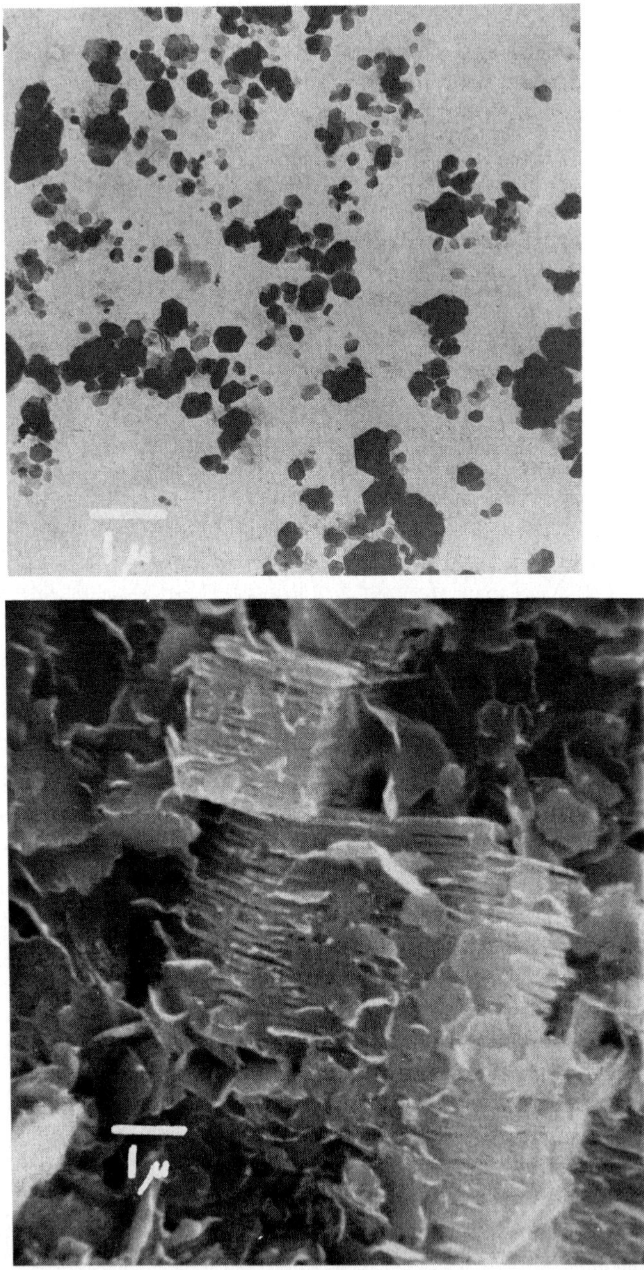

Fig. 5-9. Electron micrographs of kaolinite, Macon, Georgia (API No. 3). Top diagram was made by dispersion technique in 1949 in the laboratory of Phillips Petroleum Company showing only a few pseudohexagonal crystal outlines. Lower photograph is a scanning electron micrograph of a different aliquot of the same sample (same magnification); it shows bonding relationships between kaolinite books and crystals: edge-to-face and edge-to-edge relationships in three dimensions. (After Borst and Keller, 1969, p. 875, figs. 3 and 4; courtesy of the International Clay Conference.)

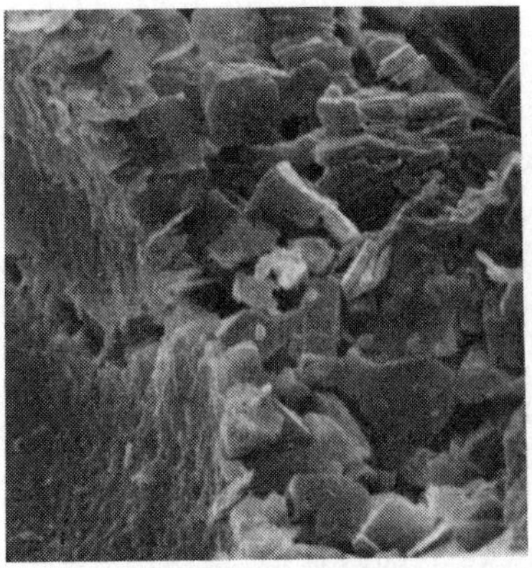

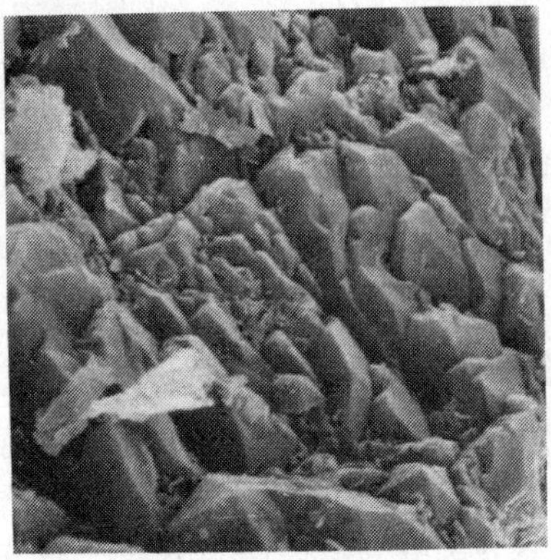

Fig. 5-10. Electron photomicrograph of secondary allotigenic kaolinite cement in Franskiy Formation, Volga-Ural region. Magnification = ×1500. Note the random arrangement of different particles and packets, resulting in a considerable number of intercommunicating micropores. (Courtesy of Professor S.G. Sarkisyan.)

Fig. 5-11. Electron photomicrograph of clayey cement (authigenic kaolinite) in quartz sandstones of Vorob'evskiy horizon of Zhivetskiy Formation, Volgograd, U.S.S.R. Mishinskaya area, borehole 37, depth = 3061—3062 m, porosity = 10.75%, permeability = 0.18 mD. Magnification: ×3000. (After Sarkisyan and Kotel'nikov, 1971, p. 150, fig. 52.)

(Rowland, 1955, p. 151). If the DTA technique is used in conjunction with X-ray diffraction analysis, microscopy, and chemical analysis, very complex identifications can be made. Excellent treatment of differential thermal analysis appears in Chapter 9 of Grim's (1968) classical book entitled "Clay Mineralogy". As an illustration, DTA curves for montmorillonite clays are presented in Fig. 5-8.

Electron microscopy

One of the pioneers, Bates (1955) conducted several electron microscope studies and showed that most of the clay minerals have morphological characteristics which can be used in their identification. Significant structural information can be obtained through detailed studies at high magnification.

If electron micrographs of two clays, which have the same X-ray pattern and the same differential thermal curve, exhibit different morphological traits, fundamental differences in depositional environment, nature of growth, and/or subsequent alteration (diagenetic and/or epigenetic) may exist. Details of morphology aid in distinguishing one clay mineral from another e.g., see Figs. 5-9—5-16.

1μ

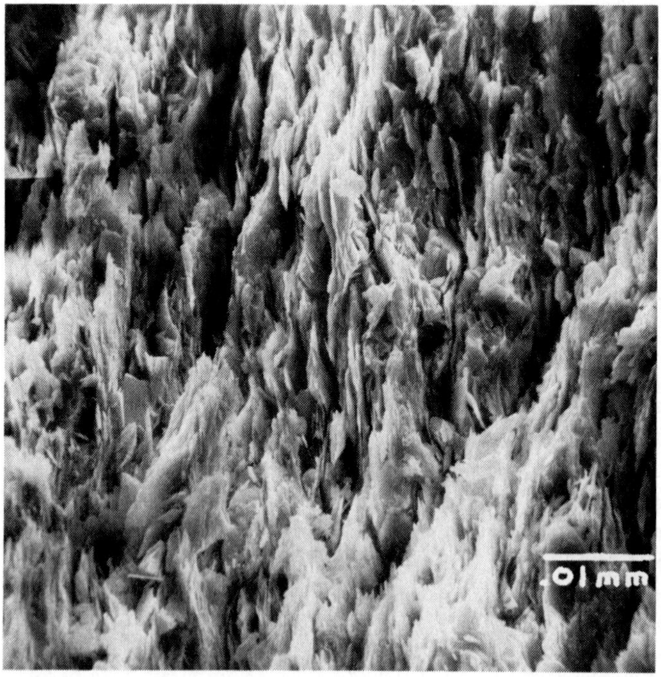
.01mm

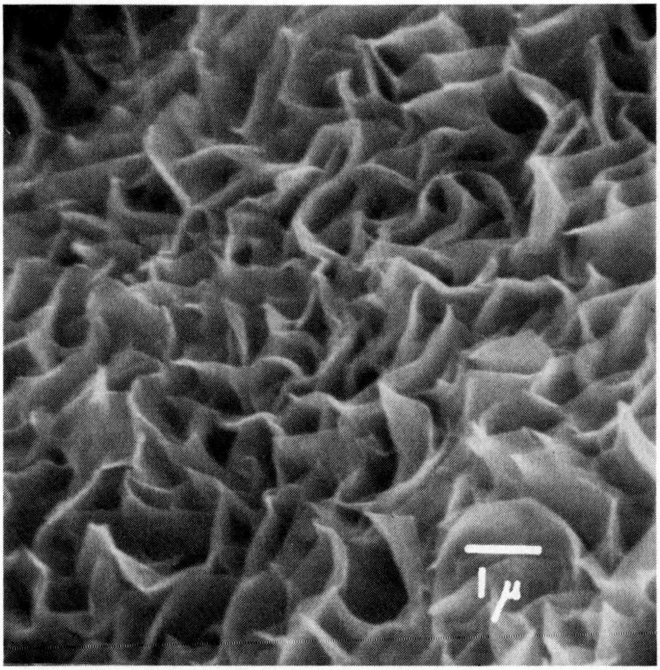

Fig. 5-13. Micrograph of API montmorillonite 21 from Burns, Mississippi, which shows a strong development of cusp-like crystals larger than those of the Chambers montmorillonite (Fig. 5-15). (After Borst and Keller, 1969, p. 895, fig. 33; courtesy of the International Clay Conference.)

Infrared analysis

Although, according to Nahin (1955a, p. 112), in some instances it is sufficient to use only one of the many methods available to identify clay minerals in a pure form or mixed with other minerals, usually the complex composition of the mineral sample necessitates the use of two or more supplementary analytical techniques for establishing the nature of the minerals and their relative amounts. As pointed out by Nahin (1955a, p. 112), the petrographic microscope, chemical analysis, dye adsorption, determination of

---

Fig. 5-12. Scanning electron micrograph of well-crystallized illite from Beavers Bend State Park, Oklahoma. Top photograph represents a view normal to sample laminae, showing that illite consists of irregularly-shaped plates and packets of plates. Lower photograph, at a lower magnification, represents a view parallel to sample laminae, showing subparallel alignment of packets which results in sample lamination. (After Borst and Keller, 1969, p. 883, figs. 14 and 15; courtesy of the International Clay Conference.)

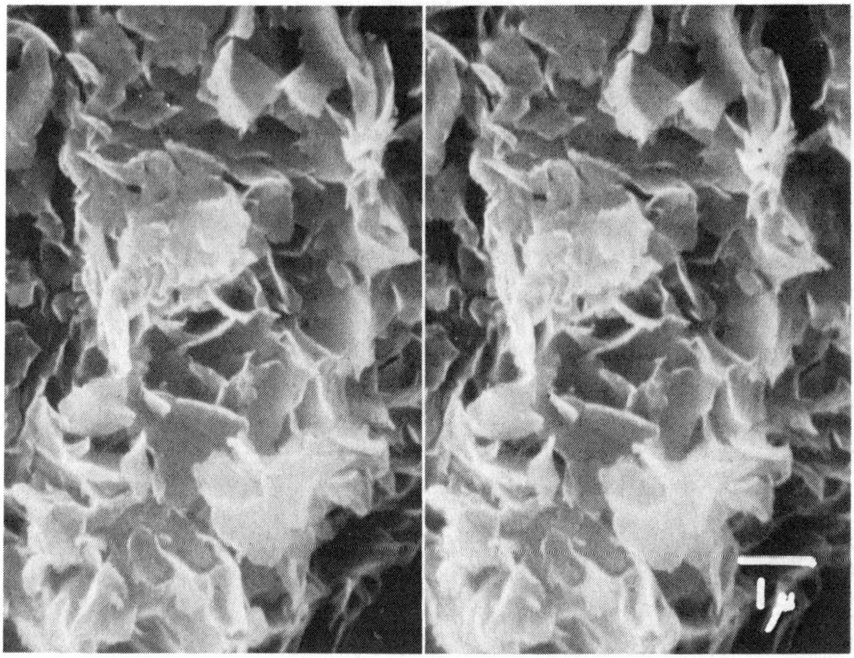

Fig. 5-14. Scanning electron micrograph of API sodium-montmorillonite 26 from Clay Spur, Wyoming. Sample consists of both crumpled and planar clay flakes, some of which show edge-to-face bonding whereas others appear to be aggregated into packets. Surfaces are relatively smooth. (After Borst and Keller, 1969, p. 894, fig. 31; courtesy of the International Clay Conference.)

cation-exchange capacity, differential thermal analysis, X-ray diffraction method, and the electron microscope provide a variety of methods which, when employed in conjunction with each other, are adequate for identification of most complex mineral mixtures. Another method of analysis has been described by Hunt (1950), Keller and Pickett (1949, 1950), Nahin et al. (1951) and others, which is based on the absorption of infrared radiation by minerals.

The infrared spectrum comprises the portion of the electromagnetic spectrum lying between the long-wavelength edge of the visible and the short end of the radio-microwave spectrum. The infrared region can be roughly divided into two sections: the near-infrared subregion ranging from 0.75 to 25 μm and the far-infrared subregion ranging from 25 to 1000 μm (Nahin, 1955a, p. 112). As pointed out by Nahin (1955a, p. 112), the near-infrared spectra arise from the resonance absorption of monochromatic radiations of wavelengths corresponding exactly to those vibrational frequencies of specific atoms (or groups of atoms) involved in a change of molecule dipole

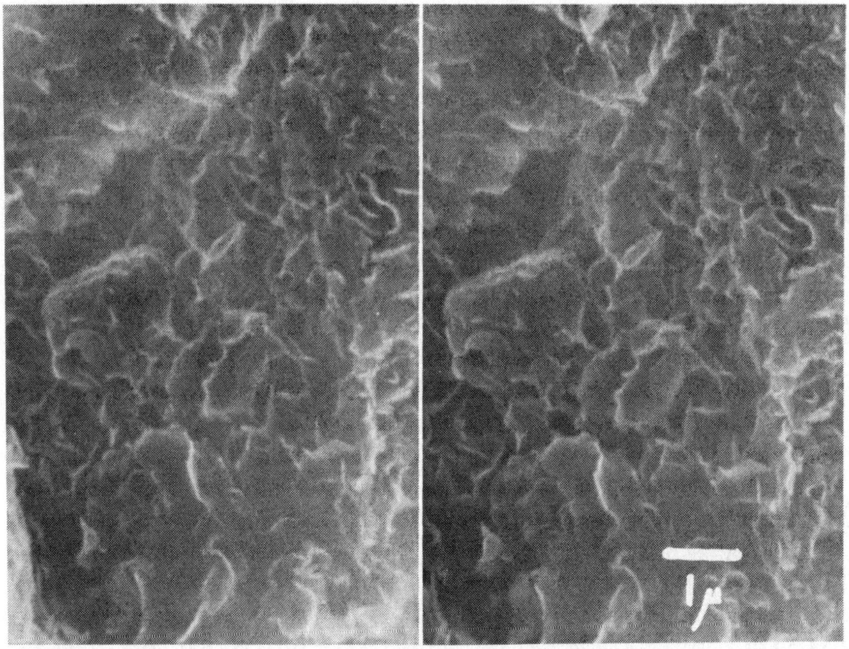

Fig. 5-15. Scanning electron micrograph of calcium-montmorillonite from Chambers, Arizona (API No. 23). Surface irregularities are curly or cusp-like. (After Borst and Keller, 1969, p. 894, fig. 32; courtesy of the International Clay Conference.)

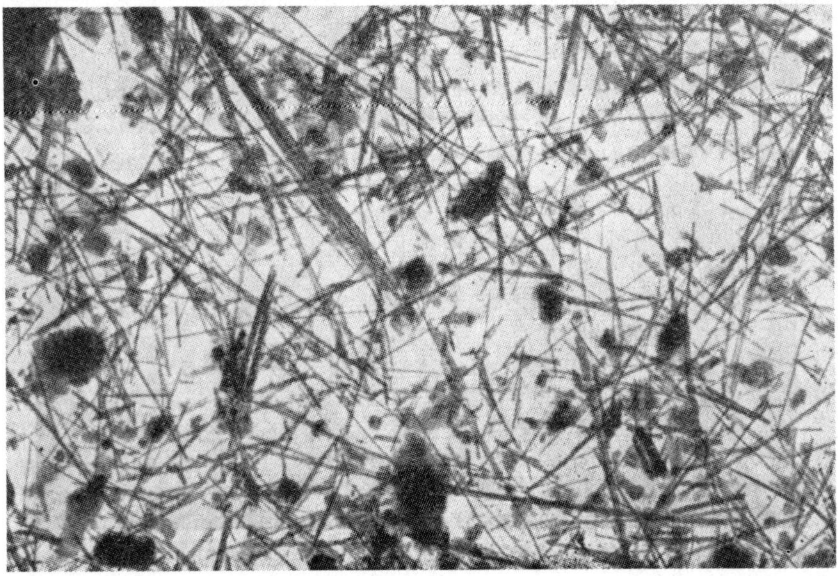

Fig. 5-16. Electron photomicrograph of palygorskite, Paleogene, Kazakhstan, U.S.S.R. Fraction <0.001 mm; ×12,500. (After Sarkisyan and Kotel'nikov, 1971, p. 84.)

moment. Absorption of radiation in the far-infrared subregion reflects the presence of complex molecular rotational or massive lattice vibrational energy levels. The infrared spectrum region which is of greatest interest in the case of most types of molecules, including those of minerals, lies in the fundamental vibrational region ranging from 2 to 16 μm (Nahin, 1955a, p. 112). Detailed discussions of the origin of spectra are presented in the textbooks on spectroscopy (e.g., Barnes et al., 1944; Harrison et al., 1948; and Mellon, 1950). In some simple instruments, the radiation from an infrared glower (single beam) is focused on the sample, the transmitted portion of the beam is collimated and monochromatized by a prism and mirror optics and, then, measured by a thermal detector.

In the petroleum industry, the utilization of infrared spectra of minerals proved to be an important supplement to the data obtained by using X-ray diffraction method, electron microscopy, and other methods for identifying minerals.

PROPERTIES OF CLAY MINERALS

Two characteristic physical properties of clay minerals are size and shape of particles. The physicochemical properties of clays which are of great interest to the petroleum engineers and petroleum geologists are: (1) base exchange capacity, (2) adsorption and retention of water, and (3) deflocculation and flocculation. The montmorillonites exhibit properties (1) and (2) to a marked degree, the illites to a much lesser degree, and the kaolinites to a relatively unimportant degree.

Size and shape of clay particles

Determination of the variety of sizes and shapes of clay particles was made possible by the use of electron micrographs. Numerous electron micrographs of various clay minerals are available in the literature, and electron-microscopic techniques as applied to clay mineralogy have been discussed by many investigators. Obviously, the electron micrographs of dry clay particles may not reflect the size and shape of the particles in the suspended condition (Van Olphen, 1963, p. 76).

As pointed out by Van Olphen (1963, p. 76), when the particles of the suspension are of greater than amicronic size ($\approx 1 \cdot 10^{-7}$ cm), their number can be counted directly in the ultramicroscope. If the weight concentration of the suspension and the number of particles in a given volume are known, the average weight of a single particle may be computed (Van Olphen, 1963, p. 76).

Factors determining particle size and shape

According to Van Olphen (1963, p. 78), the size of the clay particles is probably limited by the creation of tensions in the lattice. The breakdown of the particles could be a result of tensions created in the altered lattice, if clays are formed from micas in the solid phase. In the case of crystallization from solution by nucleation and growth process, the growth of crystallites not only will be limited by such tensions, but also the course of the nucleation process will affect the final dimensions of the particles (Van Olphen, 1963, p. 78).

As pointed out by Van Olphen (1963, p. 78), studies on possible causes of lattice tension show that the lattice imperfections due to isomorphous atom substitutions may introduce weak spots in the lattice. Misfit between the oxygen-sharing superimposed octahedral and tetrahedral sheets is another cause of tensions. According to Van Olphen (1963, p. 78), recent X-ray studies of some layer silicates have shown that the silicate sheet is able to adapt its dimensions to those of the octahedral sheet by a slight rotation of the tetrahedrons around an axis perpendicular to the sheet. The holes in the hexagonal oxygen rings of the ideal lattice become smaller and the total area of the tetrahedral sheet is reduced as a result of this rotation (Radoslovich and Norrish, 1962).

The mechanical strength of a clay particle depends on (1) whether or not water can penetrate between the units layers and (2) the average number of unit layers stacked in the particle (Van Olphen, 1963, p. 79).

Physicochemical properties of clays

Base exchange, swelling, and flocculation—deflocculation, which are the three characteristic physicochemical properties of clays that are of particular interest to the petroleum engineers and petroleum geologists, are discussed below.

Base exchange

Hughes (1950, p. 5) stated that all the structural building units of clays try to maintain a balance and stability between the cationic and anionic charges. Attachment of certain metallic cations such as sodium, calcium, iron, magnesium, and hydrogen on clay building units satisfy any deficiency in cationic charges. As pointed out by Hughes (1950, p. 5), the degree of attraction between a clay building unit and different cations varies widely, and a cation having a lesser attraction may be released in exchange for one of greater attraction. This does not alter the lattice structure of the clay building unit.

Base exchange is defined as cation-for-cation exchange between the clay building unit and a water solution, which contains the cation of greater attraction. Base exchange capacity, which is reported in terms of milliequivalents per 100 g of clay, is the relative ability of clays to carry exchangeable cations. Approximate ranges in the base exchange capacity of different clay

TABLE 5-XVII

Base-exchange capacity of various clays

Clay group	Milliequivalents per 100 g of clay
Montmorillonites	70—130 *
Illites	20—40
Kaolinites	3—15
Attapulgites	5—99
Sepiolites	20—45

* According to Weaver and Pollard (1973, p. 69).

mineral groups are presented in Table 5-XVII.

Beeson et al. (1958) have shown that the addition of calcium ions to the clay-in-water dispersions results in the displacement of H^+ from the surface of clay particles, with a corresponding decrease in pH. For example, addition of 2% by weight of $CaCl_2$ to a 15-cP bentonite-in-water dispersion decreased the pH from an initial value of 8.8 to 7.4. Sodium ions have a similar effect on the pH of bentonite muds.

The following series shows the relative ease of replacement of one cation by another (Grim, 1942):

$$Li^+ < Na^+ < K^+ < Rb^+ < Cs^+ < Mg^{2+} < Sr^{2+} < Ba^{2+} < H^+$$

For example, magnesium cation in general will replace sodium cation which, in turn, will replace lithium. The firmness with which cations are held in the clay structure increases with the valence of the cation, with the exception of potassium cation in illites. As pointed out by Grim (1942), the order of replaceability of cations varies with the kind of clay, the size of the clay particle, the size and type of the cation, and the experimental conditions. These factors, however, are more effective in illites than in montmorillonites (Hughes, 1950, p. 6). In addition, one should keep in mind that if there is a great predominance of one particular ion in solution, the base-exchange order does not hold true ("mass effect") for that ion, i.e., it will simply "force its way in". A good example of this is conversion of Ca-base montmorillonites to Na-base montmorillonites during secondary recovery operations by using injection water containing high Na^+ content, i.e., if Na/Ca ratio of injection water exceeds Na/Ca ratio of formation water. In the opinion of the writers, there is no single replaceability series and a large amount of research work still remains to be done on the subject (see Grim, 1968, p. 212).

According to Weaver and Pollard (1973, p. 69), the cation exchange capacity of montmorillonites ranges from 70 to 130 mequiv./100 g. Most of the exchange capacity (80%) is due to substitution within the structure, and

TABLE 5-XVIII

Base-exchange capacity of untreated montmorillonite clays

(After Osthaus, 1955, table 1, p. 96; courtesy of Department of Natural Resources, Division of Mines, San Francisco, California)

Location	Cations (mequiv./100 g)					Anions (mequiv./100 g)			Base-exchange capacity cations—anions	Base-exchange capacity NH_4^+ dist.	pH *
	Na^+	K^+	Ca^{2+}	Mg^{2+}	H^+	SO_4^{2-}	Cl^-	CO_3^{2-}			
Merritt, British Columbia	23.88	1.49	32.50	10.00	—	—	7.80	3.50	56.57	57.5	7.5
Santa Rita, New Mexico	2.26	1.91	62.9	29.0	—	—	—	—	96.07	97.6	7.9
Belle Fourche, South Dakota	84.80	1.70	7.86	8.50	—	18.30	5.70	—	78.86	79.8	8.4
Little Rock, Arkansas	1.29	1.27	2.14	22.50	48.70	—	—	—	75.90	78.8	2.9
Polkville, Mississippi	2.00	1.00	95.30	21.00	—	—	1.60	—	117.70	113.7	8.1
Amory, Mississippi	1.09	2.76	70.00	28.00	—	1.70	—	29.00	71.15	71.4	6.1
Otay, California	40.70	1.27	20.00	70.40	—	1.50	8.40	—	122.47	122.5	6.9
Plymouth, Utah	0.45	1.36	99.00	39.00	—	—	—	33.00	106.80	111.8	8.3
Clay Spur, Wyoming	82.60	1.74	12.45	8.80	—	13.00	3.60	10.50	78.49	75.5	9.5
Chambers, Arizona	5.50	1.20	100.00	20.00	—	1.00	4.40	2.00	119.30	117.8	8.5

* pH determined on 1 to 10 suspension in distilled water.

a lesser amount (20%) to the charges at the edges of the sheets (Hendricks, 1942). The most common interlayer cations found in nature are Na^+, Ca^{2+}, Mg^{2+} and H^+. The types of exchangeable cations from a variety of montmorillonites is presented in Table 5-XVIII.

As determined by Hofmann et al. (1956), the *anion* exchange capacities (milliequivalents/100 g of clay) of kaolinites, montmorillonites, beidellite, nontronite, saponite, and vermiculite are 6.6—20.2, 23—31, 21, 12—20, 21, and 4, respectively.

N.A. Maksimovich and N.I. Maksimova (1956, in: Chilingar, 1957c, p. 417) developed a rapid field method for determining mineralogic type of clays, based on the maximum amounts of OH^- anions the various clay minerals can adsorb. The pH of the clay suspension filtrates at maximum adsorption point was around 8.1—8.2.

Swelling of clays

Some clays do not swell upon hydration, e.g., kaolinite clay exhibits little or no swelling on hydration. Sodium-montmorillonite, on the other hand, swells in water to many times its dry volume. Calcium- and magnesium-montmorillonites and illites have intermediate swelling characteristics. Swelling properties of different clays are a function of (1) structure, (2) chemical composition, and (3) the amounts and types of exchangeable cations.

Barshad (1955) discussed the subject of the relationship between adsorption of water (and swelling) and properties of the clay—water systems. In studying swelling of clays as a result of hydration, one can recognize two types of swelling: (1) First type of swelling is due to the expansion of the crystal lattice itself (interlamellar or interlayer expansion), which can be measured by X-ray analysis. For example, Na-montmorillonite exhibits this type of swelling. (2) The second type of swelling is due to the adsorption of water on the surfaces of the clay particles. In the case of clay platelet aggregates, one should also consider interstitial (interplatelet) water.

Kelley et al. (1936) stated that adsorption and retention of water is associated with clays in three ways: (1) crystal water, (2) broken-bond water, and (3) planar water.

(1) *Crystal water*, which is an integral part of the clay-building unit, is so firmly bound that high temperatures are required for its release: kaolinites require 550°—600°C; illites require 500°—650°C; and montmorillonites require 600°—700°C.

(2) *Broken-bond water* is adsorbed on the corners and edges of clay-building units in order to satisfy the charges of unsaturated ions in those positions, which result when a clay particle is parted or broken. In illites, most of the adsorbed water is believed by some investigators to be of this type. The capacity of a clay to adsorb broken-bond water increases with decreasing particle-size. Temperatures required to drive off broken-bond water are usually

around 100°—200° C (Hughes, 1950, p. 5).

(3) *Planar (interlamellar) water* is adsorbed on the surfaces of the structural building units. Whereas montmorillonites are superior for drilling muds due to the adsorption of very large amounts of planar water, the same property causes clays and shales of the montmorillonite-type to heave and cave while drilling. This is probably due to the ease with which water may pry apart the individual crystal plates of montmorillonites and become firmly attached as planar water. Planar water of montmorillonites requires temperatures between 100° and 200° C to be driven off. In illites, potassium cations form bonds between individual crystal plates having sufficient strength to resist the prying apart of the crystal surfaces. This makes potassium-base muds very popular. Illites, even when finely divided, adsorb much less planar water than montmorillonites, whereas kaolinites adsorb very small amounts of planar water (Hughes, 1950, p. 5).

In the case of attapulgite, palygorskite, and sepiolite, the water occurs within the tubular openings between the elongate structural units (Grim, 1968, p. 234). Nature of adsorbed water in clays has been discussed in detail by Grim (1968, chapter 8).

As pointed out by Grim (1968, p. 246), there is an overwhelming evidence that water molecules adsorbed directly on clay mineral surfaces and for some distance outward have organized arrangement. This quasi-crystalline water has a structure similar to that of ice, with properties different from those of liquid water. Some experimental data on compaction of water-saturated clays (e.g., Chilingar and Knight, 1960; Chilingar et al., 1963; Rieke and Chilingarian, 1974) also suggest (1) difference between free, liquid water and adsorbed water, and (2) that transition from free to adsorbed water appears to be gradual. There is disagreement among various investigators, however, as to the distance adsorbed water extends from the clay-mineral surface, ranging from a very few molecular layers to over 30. The results obtained by Low and Anderson (1958), for example, indicated that the specific volume of water adsorbed by montmorillonite is different from normal water out to distances in excess of 60 Å. Grim and Cuthbert (1945), on the other hand, concluded that in the case of Ca-montmorillonite the thickness of nonliquid water is 4 molecular layers ($\approx$10 Å), whereas for Na-montmorillonite, the thickness is 3 molecular layers. They also suggested that for Ca-montmorillonite the transition to liquid water is abrupt, whereas in the case of Na-montmorillonite there is a gradual transition to liquid water beyond the 7.5-Å thickness. With increasing amount of Na^+ as the adsorbed cation, the thickness of oriented water required to develop plasticity is reduced. The thickness of adsorbed water appears to increase from kaolinite to illite to smectite.

Swelling properties of clays are of great practical importance because of their widespread occurrence in reservoir rocks. Inasmuch as clays have relatively large surface areas, they can significantly affect the flow of hydrocarbons. The degree of hydration of clays in the vicinity of the borehole affects the efficiency of (1) primary production, which greatly depends on the magnitude of permeability to oil and the maximum drainage area immediately surrounding the borehole; and (2) secondary recovery, which largely depends on the degree of injectivity of various fluids into the injection wells. Consequently, it is of utmost importance to understand the mechanisms of swelling of clays and its degree as various fluids are introduced into the wells during drilling, completion, production, and secondary recovery operations.

Swelling of clays is affected by exposure to high temperatures as a result of certain reactions, mainly dehydration and recrystallization. These reactions were studied by various investigators by using X-ray diffraction, and other techniques. In the experience of the writers, usually exposure of clays to heat results in lower swelling characteristics. Rapid transformations can take place within a short range of temperature. Chilingar et al. (1970) also discovered that application of direct electric current may result in driving out (electroosmosis) of interlayer and adsorbed water and cause collapse of clay structure. They suggested this method as a stimulation technique to suppress swelling of clays.

Nahin (1955b) presented some preliminary results on the swelling of Wyoming bentonite under constraint in several media. Fractionated and powdered clay was mixed with an equal weight of non-swelling microporous porcelain, which is used for the purpose of distributing the swelling agent rapidly and uniformly. This mixture was then compacted in a cell at high pressure and permitted to imbibe fluid through a porous porcelain plate under simultaneously applied almost equal pressures ranging from 0 to 10,000 psig on both sides of the fluid—bentonite interface. Unidirectionally constrained volume change of the sample was measured by displacement of mercury in a steel buret. Degree of swelling was considered by Nahin (1955b) to be the ratio of swollen clay volume to the volume of dried (at 105°C) clay.

The swelling ratio, SR, was defined by Nahin (1955b) as follows:

$$SR = \frac{V_g}{V_c} = \frac{V_c + P + \Delta V_0}{V_c} = 1 + \frac{\rho_c}{M_c}(P + \Delta V_0) \tag{5-1}$$

where: V_g = volume of clay gel at the end of test; V_c = volume of clay solid phase at 105°C; P = void volume of clay—porcelain mixture which is available to clay gel; ΔV_0 = volume change of clay—porcelain mixture upon swelling as observed on buret; ρ_c = solid phase specific weight of clay determined by helium displacement; and M_c = weight of 105°C-dried clay in sample.

In all Nahin's (1955b) experiments, the initial rate of swelling was very great (70—90% of swelling occurring within a few minutes), which led him to conclude that the rate of swelling is dependent mainly on the rate of accessibility of clay surfaces. According to him, in the pressure range of 0 to

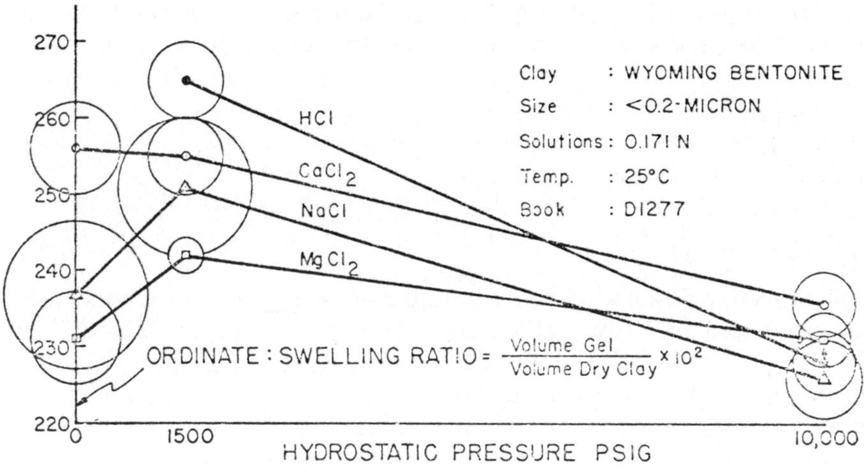

Fig. 5-17. Relationship between swelling and hydrostatic pressure. (After Nahin, 1955b, fig. 4, p. 181; courtesy of Natl. Acad. Sci. — Natl. Research Council.)

10,000 psig, the colloidal magnesium-sodium-montmorillonite swells from about 93 to 231% (volume ratio of 1.93 to 3.31) depending on the composition of the swelling agent.

In Fig. 5-17, the effect of hydrostatic pressure on swelling of Wyoming bentonite in 0.171 *N* solutions of hydrochloric acid and three common salts is presented. The 0.171 *N* concentration was chosen because it corresponds approximately to the ionic strength of a 10,000 ppm NaCl brine, which is characteristic of certain California, U.S.A., reservoir formation waters.

Einstein's equation and clay hydration

Einstein's equation for the viscosity of a dilute dispersion of spherical particles is:

$$\mu_s = \mu_0(1 + 2.5R) \tag{5-2}$$

where μ_s is the viscosity of the dispersed system in centipoises, μ_0 is the viscosity of the dispersion medium in centipoises, and R is the ratio of the volume of the dispersed phase to the volume of the sol *. Although this equation does not apply to viscous clay + water mixtures, it is useful when considering low clay contents (2—5% of clay by volume). Extensions to Einstein's equation that are presented in Kruyt (1949) take into consideration higher concentrations of both spherical and non-spherical particles (see also Robertson et al., 1965).

* A colloidal suspension of a solid in a liquid.

Kunitz (as quoted in: Abramson, 1934, p. 129) showed that the following expression holds very well for R up to 0.50 for sugar solutions and sulfur suspensions:

$$\mu_s = \mu_0 \frac{1 + 0.5R}{(1 - R)^2} \tag{5-3}$$

or:

$$\mu_s = \mu_0(1 + 4.5R + 12R^2 + ...) \tag{5-4}$$

Several other extensions of the Einstein's equation to higher concentrations are presented in Kruyt (1952, pp. 350—354) both for spherical and non-spherical particles. It is not known which one of these equations is the most accurate. Vand (as quoted in: Kruyt, 1952, p. 351), for example, gave a seemingly reasonable equation for spherical particles on considering both hydrodynamic and mechanical interaction between particles:

$$\frac{\mu_s - \mu_0}{\mu_0} = 2.5R + 7.349R^2 \tag{5-5}$$

The knowledge on the dilute and concentrated suspensions of elongated particles is even scarcer, and the pertinent equations are complex (Kruyt, 1952; pp. 346, 353).

Inasmuch as charges on particles may influence the viscosity by making the apparent volume of the particles greater as a result of introducing an additional resistance owing to electrokinetic effects, Von Smoluchowski (as quoted in: Kruyt, 1949, p. 205) gave an extension of Einstein's formula:

$$\frac{\mu_s - \mu_0}{\mu_0} = KR\left[1 + \frac{1}{\lambda\mu_0 r^2}\left(\frac{D\zeta}{2\pi}\right)^2\right]$$

where λ = specific conductivity, r = particle radius, D = dielectric constant, K = constant, and ζ = electrokinetic potential (difference in potential between the layers of charges affixed to the particles and the layer of balancing charges in the liquid bathing the particles). Robertson et al. (1965) estimated that the increase in viscosity owing to charges on clay particles is probably of the order of 10%.

On determining the viscosity of clay dispersions of known concentration, one can calculate the approximate ratio of the volume of dispersed phase to the volume of the sol on using eq. 5-2. Assuming that the viscosity of the dispersion medium, i.e., water, is 1 cP, Einstein's equation may be written as:

$$R = (\mu_s - 1)/2.5 \tag{5-6}$$

Thus, any discrepancy between the actual and calculated R enables one to estimate the increase in volume of the dispersed phase, namely, amount of clay swelling. For example, according to the results obtained by Chilingar and Knight (1960), a sol containing 1.2% by weight of gum tragacanth (natural organic colloid) had a viscosity of 37 cP, whereas according to the Einstein's formula the viscosity should be only 1.03 cP. This discrepancy indicated that the tragacanth gum has taken up large amounts of water, thus increasing the volume of the dispersed phase. The viscosity of the sol usually increases with time, indicating further swelling and expansion of the particles. In accordance with eq. 5-6 in the foregoing case of 37 cP viscosity of the sol, it is necessary for the total volume of the particles of the dispersed phase to be about 1200 times that of the dry tragacanth gum. It is not known with certainty whether the particles of tragacanth gum take up the water like a sponge or whether they become surrounded by thick layers of water molecules.

Similarly, various mud samples, containing a certain weight percent of clay, were prepared. The laboratory procedure was described in detail by Chilingar et al. (1963) and Robertson et al. (1965). The swelling index, SI_v of clays, which is indicative of the degree of hydration, can be mathematically expressed as:

$$SI_v = \frac{R_c \text{ calculated from eq. 5-6}}{R_a \text{ actual (dry basis)}} = \frac{\text{volume of hydrated clay}}{\text{volume of dry clay}} \tag{5-7}$$

A mud sample containing 25% by weight of kaolinite clay had a viscosity of 3 cP, showing that the volume of the dispersed phase (hydrated clay) is about 7.3 times that of the dry clay. For a mud sample containing 2.7% by weight of bentonite, the viscosity was 2.9 cP. This indicated that the volume of hydrated bentonite is about 28.2 times that of the volume of dry clay. Table 5-XIX shows the ratio of the volume of hydrated clay to dry clay to vary from 1.78 in the case of dickite to 3.96 in the case of halloysite. In the foregoing calculations a specific gravity of 2.5 g/cm^3 was assumed for all clays. Obviously, the calculated values are not absolute and would vary from clay to clay.

TABLE 5-XIX

Measured and calculated (from Einstein's equation) viscosities of clay + water mixtures *

(After Chilingar et al., 1963)

Volume of water (cm^3)	Weight of clay (g)	Measured viscosity (cP)	Calculated viscosity (cP) using R_{dry}	Vol. % of clay		Type of clay	Swelling index SI_v
				actual (dry)	calculated (hydrated)		
95.0	5.4676	1.1	1.0563	2.25	4.0	dickite	1.78
90.0	9.1775	1.2	1.098	3.92	8.0	dickite	2.04
95.0	4.8973	1.2	1.0505	2.02	8.0	halloysite	3.96
90.0	10.0349	1.21	1.11	4.26	8.4	halloysite	1.97

* In these calculations, a specific weight of 2.5 g/cm^3 was assumed for clays.

In the opinion of Ralph E. Grim (personal communication, 1960) the water adsorbed immediately on the surfaces of the clay minerals is composed of oriented water molecules. The distance that such "oriented" water extends from the clay mineral surfaces depends on the nature of clay mineral, the nature of the adsorbed cations, and, perhaps, on the presence of organic matter. The water present in pores at considerable distances from the surface is not oriented, and is termed "liquid" water. In some cases there is a gradual transition from the non-oriented water on proceeding toward the clay mineral surface; whereas in the other cases there is a sharp transition in the character of the water. For example, E. Grim (personal communication, 1960) believes that there is a gradual transition in the case of sodium-montmorillonite and a more abrupt transition for calcium-montmorillonite.

Van Olphen (1963a, b) has shown that the last few mono-molecular layers of oriented water adsorbed on clay surfaces are not removed upon deep burial. According to Van Olphen (1963b), even on heating sodium Llano vermiculite clay to 50°C, it takes a pressure of about 65,000 psi to remove the last water layer. The resistance of clay to compaction is equal to the swelling pressure, which, in turn, is equal to the net repulsive pressure between the unit layers of clays. Figure 5-18, which is based on work-of-removal values reported by Van Olphen (1963b), was prepared by Powers (1967) to show the approximate pressures required for the removal of successive monomolecular layers of water from a calcium-montmorillonite. The work-of-removal for water layers located between unit layers of montmorillonite was determined from desorption isotherms. Approximate depths of burial corresponding to the desorption pressures also are given in Fig. 5-18 by Powers (1967).

Robertson et al. (1979) described a simple apparatus for measuring the degree of swelling of clays. They also found that at a temperature range of 75—160°F, the rate of hydration appreciably increases, without altering the ultimate total volume of sorbed interlamellar water to any appreciable degree. These authors described the swelling index *SI* as the weight of interlamellar sorbed water divided by the weight of dry clay (dried at 100°C). The *SI* value for Wyoming bentonite, for example, was found to be equal to about 10, using distilled water, and about 3.5, using $CaCl_2$, KCl, and NaCl solutions (1.59, 1.43, and 2.5% by dry weight, respectively).

Flocculation and deflocculation

Deflocculation is defined as the state of a dispersion in which solid particles in a liquid remain geometrically independent and unassociated with adjacent particles (Fischer and Gans, 1946). In good drilling fluids, clay is in a state of deflocculation (well dispersed).

Flocculation is defined as the state of dispersion in which there is a formation of clusters of particles separable by relatively weak mechanical forces. This may be caused by a change in chemical composition of the dispersion

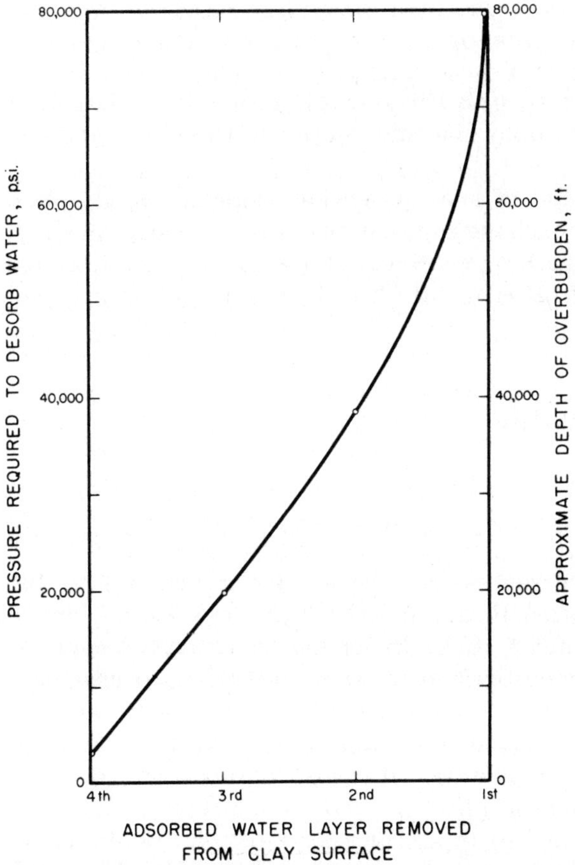

Fig. 5-18. Pressure (or approximate equivalent of depth of burial) necessary to remove each of the four layers of water from interlayer positions on montmorillonite clay. (After Powers, 1967, fig. 2, p. 1243; courtesy of American Association of Petroleum Geologists.)

phase as a result of neutralization of negative charges on clay particles. For example, a good drilling fluid may become flocculated as a result of contamination by formation salt water or drilling through salt domes. Permeability of filter cake of a well-dispersed drilling mud is much lower than the filter cake of a flocculated mud.

According to Hughes (1950, p. 6), flocculation and deflocculation of clays are reversible under certain conditions, depending upon base exchange capacity and the kind and quantity of exchangeable cations in the various waters with which they may come into contact. Such conditions, however, rarely exist in drilling fluids. As pointed out by Van Olphen (1963, p. 17), the agglomeration of particles is an irreversible process and the particles cannot disengage spontaneously faster (or as fast) than they associate. Whereas macromolecular sols, which are true solutions, are reversible systems, hydrophobic sols are irreversible systems. According to Van Olphen (1963, p. 17), colloidal systems are either *irreversible* (including hydrophobic colloids) or *reversible* (including the macromolecular and the association colloids, i.e., hydrophilic colloids).

Flocculation or deflocculation of clay suspensions depends chiefly upon the kind and concentration of exchangeable cations, base exchange capacity, and the pH. Generally, pH values below 8 and above 11 promote flocculation, whereas the existence of pH range of 8 to 11 tends to promote dispersion (Hughes, 1950, p. 6).

MODIFICATION OF CLAY PROPERTIES

One of the several reasons to modify the properties of clay minerals is to produce a drilling fluid which would result in safe and speedy drilling, with a subsequent maximum productive capacity after completion of a well.

Modification of clays in drilling fluids in large measure depends on diagenetic and epigenetic (or catagenetic) processes which clays had previously undergone in sediments and later in rocks. In general, there are two types of modification: (1) controlled modification in which alteration of clays and drilling fluids is done in the laboratory under specific conditions, and (2) diagenetic and epigenetic modifications in nature. According to Chilingar et al. (1967, p. 313), diagenesis includes all modifications that sediments undergo between deposition and lithification under conditions of temperature and pressure that are normal to the surface or outer part of the crust. Epigenesis (or catagenesis) includes those changes that take place after lithification but prior to metamorphism (see Bissell and Chilingar, 1967, p. 157).

Chemical changes in clay minerals that occur during sedimentation or those that are imparted by the ionic or other constituents of seawater along the transport route to the site of deposition (or at the site of deposition) may control later possible diagenetic and epigenetic alterations of the source material.

CLAYS USED IN DRILLING FLUIDS

According to Larsen (1946), the usual composition by volume of many clay—water drilling fluids is as follows: water = 65—98%; clay = 2—30%; and substances arising from drilling operations, such as rock cuttings and gas = 0—10%. There are, however, many drilling fluids of quite different compositions, which are used under special conditions. In addition, as a result of shortage of high-yield clays, at the present time polymers are used extensively instead of clays or mixed with them.

Grim (1958) pointed out that a vast amount of valuable research on colloid chemistry of clays has been done by many investigators studying the properties of drilling fluids and the fundamental factors which control them. These efforts can be appreciated by reading excellent papers and extensive bibliographies of Larsen (1946 and 1955) and Van Olphen (1950), and a classical book of Rogers (1963).

The suitability of clays for use in drilling fluids may be determined by (1) the yield, i.e., the number of barrels of mud of a given viscosity (e.g., 15 cP) obtained from a ton of clay in fresh water; (2) suspension capacity in salt water; (3) plastic viscosity; (4) apparent viscosity; (5) yield strength; (6) thixotropic properties, i.e., the difference in gel strength determined immediately after agitation and after quiescence (usually 10 min); (7) the wall-building properties as measured by the water loss through a filter paper when a 15-cP mud is subjected to a pressure of 100 psi for 30 min and at higher temperatures and pressures which exist at depth; and (8) the thickness of filter cake produced in the standard API water-loss test (static) or dynamic tests. Although other tests, such as the influence of electrolytes on viscosity, are frequently used, the foregoing tests are usually sufficient to evaluate clays for use in drilling fluids. Figures 5-19, 5-20, and 5-21 and Table 5-XX give the characteristics of a variety of clays as determined by drilling-mud tests.

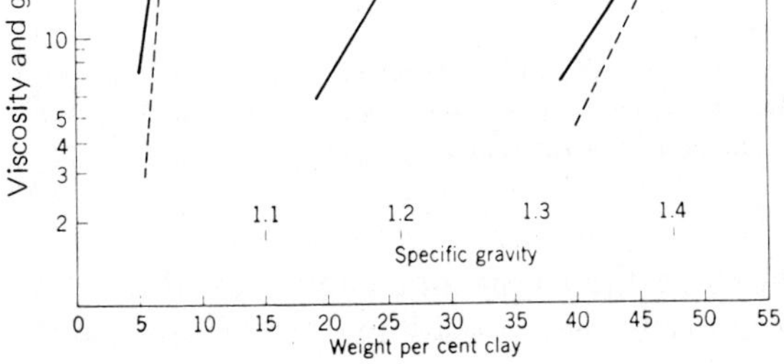

Fig. 5-19. Viscosity and gel strength of some clay minerals determined with Stormer viscosimeter (600 rpm, see Chapter 4). (After Larsen, 1955; also see Grim, 1962, fig. 1, p. 234.) As designated in American Petroleum Institute, *a* = sodium-montmorillonite clay; *b* = calcium-montmorillonite and illite clays; *c* = impure clay with montmorillonite and attapulgite. Broken line = shear strength after 10 min (g); dashed line = shear strength immediately after agitation (g); and solid line = Stormer viscosity at 600 rpm (cP).

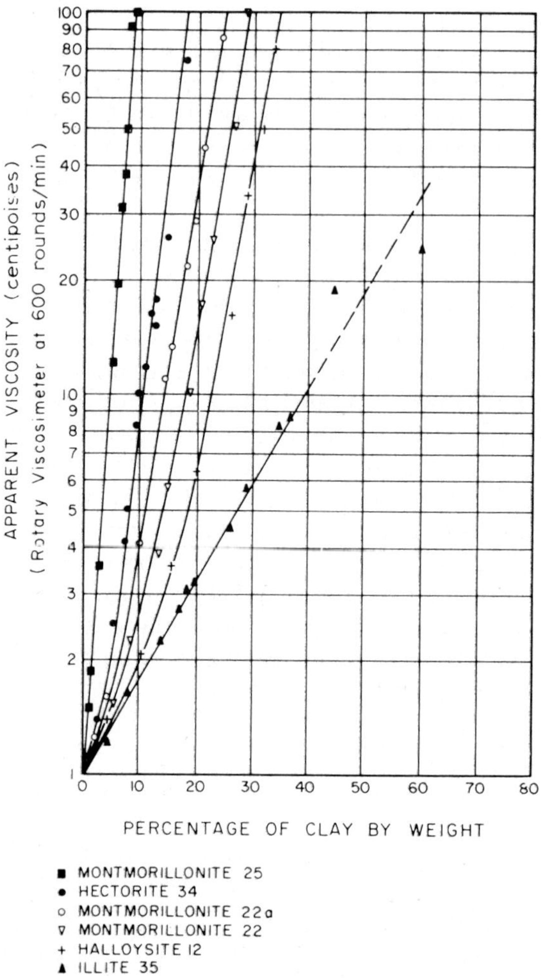

Fig. 5-20. Relationship between apparent viscosity and weight percent of various clays in clay—water systems. All measurements were made with a Fann VG meter. (After Robertson et al., 1965, p. 183, fig. 2; courtesy of Sedimentology.)

Clays and solids which contain more than very small amounts of nonclay minerals, particularly of silt and sand sizes, are not suitable for drilling fluids, because these materials have abrasive action on pumps and other drilling equipment and impart undesirable properties to muds. Local clays and soils having low nonclay mineral content, however, are frequently used to start drilling a well. Drilling of a well may also be started with water alone if considerable section of argillaceous material (e.g., shales) having suitable properties for drilling fluids, is drilled through initially. The surface clays (often dry

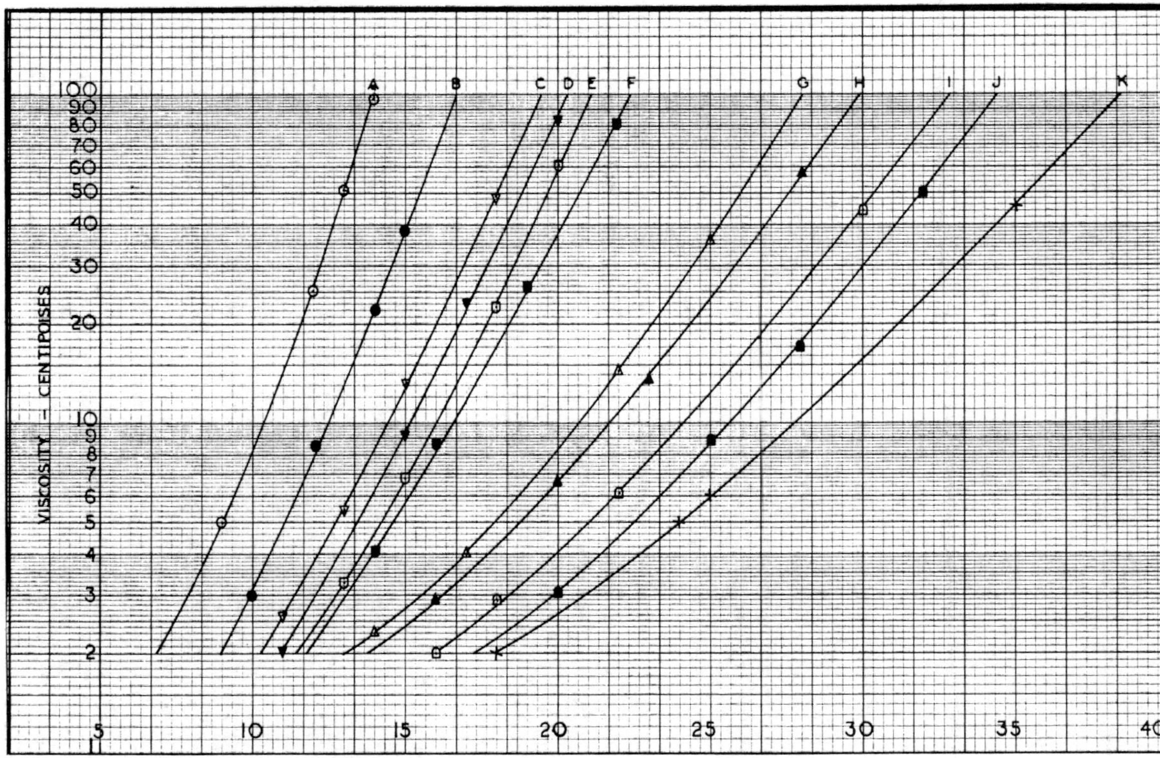

Fig. 5-21. Relationship between apparent viscosity and weight percent of various clays in clay—water suspensions. *A* = montmorillonite No. 22a (yellow), Amory, Mississippi; *B* = montmorillonite No. 24 (bentonite), Otay, California; *C* = montmorillonite No. 21 (bentonite), Chisholm Mine, Polkville, Mississippi; *D* = kaolinite No. 7, Dixie Rubber Pit, Bath, South Carolina; *E* = 50% montmorillonite No. 21 and 50% kaolinite No. 9; *F* = 33.3% montmorillonite No. 21, 33.3% kaolinite No. 9, plus 33.4% illite No. 35; *G* = 50% illite No. 35 plus 50% kaolinite No. 9; *H* = kaolinite No. 2, Birch Pit, Macon, Georgia; *I* = kaolinite No. 9, Canada del Camino, Mesa Alta, New Mexico; *J* = illite No. 36, Morris, Illinois; and *K* = illite No. 35, Fithian, Illinois.
(After Coughlin and Chilingar, 1962, p. 98.)

TABLE 5-XX

Yield and water loss data for various types of clay minerals

(After Larsen, 1955; courtesy of California Division of Mines)

	Yield, bbl of 15-cP mud/ton clay (fresh water)	Solids in filter cake (wt %)	API water loss of 15-cP mud	Filter cake permeability, (mD)	pH
Hectorite, California	160	6.5	7	0.85	8.6
Na-montmorillonite, Wyoming	125	10	11	1.8	8.2
Ca-montmorillonite, California	71	16	15	2.1	8.7
Ca-montmorillonite, Texas	18	50	11	1.5	7.5
Illite, Illinois	13	67	57	38	7.4
Kaolinite, Georgia	14	70	190	285	7.0
Attapulgite, Georgia	105	23	105	63	7.1
Halloysite, Colorado	17	60	35	15	7.7

* Stormer viscosity (600 rpm).

lake clays) frequently require concentrations of 20—40% by weight in order to give adequate mud viscosity (see Fig. 5-21). The resulting muds usually have low gel strength and high water loss. Drilling with such clay-water muds, however, cannot be continued after encountering formations containing saline water, soft formations and/or reaching great depths. In these cases, where special mud properties are required, special types of clays must be added, e.g., bentonites. Obviously, local and surface clays containing smectite (montmorillonite) in the case of fresh-water muds and attapulgite in the case of salt-water muds are most desirable. The latter type of clay swells in salt water.

REFERENCES

Abramson, H.A., 1934. *Electrokinetic Phenomena and Their Application to Biology and Medicine.* Chemical Catalog Co., New York, N.Y., 331 pp.

Ahrens, P.L., 1950. Differential thermal analysis, a conventional method. *Inter. Congr. Soil Sci., 4th, Trans.*, 4: 26—27.

Alietti, A., 1956. Il mineral a strati misti saponite-talco di Monte Chiaro. *Accad. Naz. Lincei, Rend, Cl. Sci. Fis. Math. Nat., Ser. 8*, 21 (3/4): 201—207.

Ames, L.L. and Sand, L.B., 1958. Factors affecting the maximum hydrothermal stability in montmorillonites. *Am. Mineral.*, 43: 641—648.

Anderson, T.F., 1951. Techniques for the preservation of three-dimensional structures in preparing specimens for electron microscopy. *Trans. N.Y. Acad. Sci.*, 13: 130—134.

Andreatta, C., 1949. Studio di un intersante giacimento di riempimento di argille montmorillonite idrotermali. *Mem. Accad. Sci. Inst. Bologna, Cl. Sci. Fiz.*, Ser. 10,5: M.A. 11—547.

Anonymous, 1957. *Standard Field Procedure for Testing Drilling Fluids,* 4th ed., Am. Pet. Inst., Rep. 29.

Anonymous, 1962. *Standard Procedure for Testing Drilling Fluids.* Am. Pet. Inst., Rep. 138: 15 pp.

Bannister, F.A., 1943. Brammalite (sodium-illite), a new mineral from Llandebie, South Wales. *Mineral. Mag.*, 26: 304—307.

Barnes, R.B., Gore, R.C., Liddel, U. and Williams, V.Z., 1944. *Infrared Spectroscopy, Industrial Applications and Bibliography.* Reinhold, New York, N.Y., p. 11.

Barshad, I., 1955. Adsorptive and swelling properties of clay-water system. *Clays Clay Technol. Bull.*, 169: 70—77.

Barshad, I., Halevy, E., Gold, H.A. and Hagin, J., 1956. Clay minerals in some limestone soils from Israel. *Soil Sci.*, 81: 423—434.

Basset, W.A., 1959. Origin of the vermiculite deposit at Libby, Montana. *Am. Mineral.*, 44: 282—299.

Bates, T.F., 1955. Electron microscopy as a method of identifying clays. *Clays Clay Technol. Bull.*, 169: 130—150.

Beeson, C.M., Chilingar, G.V. and Larsen, D.H., 1952, 1958. *Drilling Fluids Laboratory.* Univ. of South. Calif. Publ., Los Angeles, Calif., 73 pp.

Berkelhamer, L.H., 1945. An apparatus for differential thermal analysis. *U.S. Bur. Mines, Tech. Pap.*, G64: 38—55; *U.S. Bur. Mines Rep. Invest.*, 3762 (1944): 11 pp.

Bertness, T.A., 1953. *Observations of Water Damage to Oil Productivity.* Am. Pet. Inst. Prod. Pract., Spring Meet., Pac. Coast Distr., Los Angeles, Calif., June 2—3, Pap. 801-29J, pp. 287—295.

Bissell, H.J. and Chilingar, G.V., 1967. Classification of sedimentary carbonate rocks. In: G.V. Chilingar, H.J. Bissell and R.W. Fairbridge (Editors), *Carbonate Rocks. Origin, Occurrence and Classification. Dev. Sedimentol.*, 9A. Elsevier, Amsterdam, pp. 87—168.

Borst, R.L. and Keller, W.D., 1969. Scanning electron micrographs of API reference clay minerals and other selected samples. In: L. Heller (Editor), *Proc. Int. Clay Conf.*, Tokyo, 1: 871—901.

Bradley, W.F., 1940. The structural scheme of attapulgite. *Am. Mineral.*, 25: 405—410.

Bradley, W.F. and Weaver, C.E., 1956. A regularly interstratified chlorite-vermiculite clay mineral. *Am. Mineral.*, 41: 497—504.

Brauner, K. and Preisinger, A., 1956. Structure of sepiolite. *Mineral. Petrogr. Mitt.*, 6: 120—140.

Brindley, G.W. (Editor), 1951. *X-Ray Identification and Crystal Structures of Clay Minerals.* London Mineral. Soc., Clay Miner. Group, 345 pp.

Brindley, G.W., 1955. Identification of clay minerals by X-ray diffraction analysis. *Clays Clay Technol. Bull.*, 169: 119—129.

Brindley, G.W. and Gillery, F.H., 1954. A mixed-layer kaolin-chlorite structure. *Clays Clay Miner.*, 2: 349—353.

Brindley, G.W. and MacEwan, D.M.C., 1953. *Structural Aspects of the Mineralogy of Clays, Ceramics, A Symposium.* British Ceramics Soc., Stoke-on-Trent.

Brown, G., 1951. Nomenclature of the mica clay minerals. II. In: G.W. Brindley (Editor), *X-ray Identification and Crystal Structures of Clay Minerals.* London Mineral. Soc., Clay Miner. Group, Chap. 5, part II: 155—172.

Brown, G. and Weir, A.H., 1963. The identity of rectorite and allevardite. *Proc. Int. Clay Conf.*, 1: 27—35.

Browning, W.C., 1964. *The Hydroxyl Factor in Shale Control.* Annu. Meet. SPE, Dallas, Texas, March 23—24, SPE Preprint 795, 18 pp.
Browning, W.C. and Perricone, A.C., 1963. *Clay Chemistry and Drilling Fluids.* 1st Conf. on Drilling and Rock Mechanics, cosponsored by Pet. Branch of AIME, Pap. SPE 540: 11 pp.
Burst, J.F., 1969. Diagenesis of Gulf Coast clayey sediments and its possible relation to petroleum migration. *Bull. Am. Assoc. Pet. Geol.*, 53 (1): 73—93.
Bystrom, A.M., 1956. Mineralogy of the Ordovician bentonite beds at Kinnekulle, Sweden. *Sver. Geol. Unders., Arsb.*, 48: 1—62.
Caillère, S. and Hénin, S., 1949. Experimental formation of chlorite from montmorillonite. *Mineral. Mag.*, 28: 612—620.
Caillère, S. and Hénin, S., 1961. Palygorskite. In: G. Brown (Editor), *The X-Ray Identification and Crystal Structures of Clay Minerals.* London Mineral. Soc., London, pp. 325—342.
Caillère, S., Hénin, S. and Esquevin, J., 1954. Transformation experimentale de chlorite en montmorillonite. *Clay Miner. Bull.*, 2: 166—170.
Cayeux, L., 1916. Introduction à Etude Pétrographique des Roches Sédimentaires. *Mém. Carte Géol. Fr.*, pp. 229—233.
Carroll, D. and Starkey, H.C., 1960. Effect of sea water on clay minerals. *Clays Clay Miner.*, 7: 327—341.
Chesser, B.C. and Perricone, A.C., 1973. *A Physicochemical Approach to the Prevention of Balling of Gumbo Shales.* 48th Annu. Fall Meet., Pet. Branch, AIME, Preprint SPE 4515.
Chilingar, G.V., 1954. The mineralogy of the montmorillonite group. *Compass*, 31 (4): 288—295.
Chilingar, G.V., 1957a. Evaluation of clays for drilling fluids. *Compass*, 34 (3): 174—185.
Chilingar, G.V., 1957b. Addendum to evaluation of clays for drilling fluids. *Compass*, 35 (1): 75.
Chilingar, G.V., 1957c. Rapid method of determining mineralogic type of clays. *Soil Sci.*, 83 (5): 417.
Chilingar, G.V. and Knight, L., 1960. Relationship between overburden pressure and moisture content of silicic acid and gum ghatti. *Int. Geol. Congr. Proc., Norden, 21st Sess.*, Part 18: 384—388.
Chilingar, G.V., Rieke, H.H. and Robertson, J.O., Jr., 1963. Degree of hydration of clays. *Sedimentology*, 2 (4): 341—342.
Chilingar, G.V., Bissell, H.J. and Wolf, K.H., 1967. Diagenesis of carbonate rocks. In: G. Larsen and G.V. Chilingar (Editors), *Diagenesis in Sediments. Dev. Sedimentol.*, 8. Elsevier, Amsterdam, pp. 179—322.
Chilingar, G.V., El-Nassir, A. and Stevens, R.G., 1970. Effect of direct electrical current on permeability of sandstone cores. *J. Pet. Technol.*, 22 (Jul.): 830—836.
Christensen, R.W. and Kim. J.S., 1969. Rheological model studies in clay. *Clays Clay Miner.*, 17 (2): 83—93.
Coleman, N.T. and Harward, M.E., 1953. The heats of neutralization of acid clays and cation exchange resins. *J. Am. Chem. Soc.*, 75: 6045—6046.
Correns, C.W., 1949. *Einführung in die Mineralogie.* Springer, Berlin, 414 pp.
Coughlin, H. and Chilingar, G.V., 1962. Easy and rapid method of determining mineralogic type of clays. *Compass*, 39 (2): 97—98.
Degens, E.T., 1965. *Geochemistry of Sediments (A Brief Survey).* Prentice-Hall, Inc., Englewood Cliffs, N.J., 342 pp.
De Kimpe, C.R., Gastuche, M.C. and Brindley, G.W., 1961. Ionic coordination in alumino-silicic gels in relation to clay mineral formation. *Am. Mineral.*, 46: 1370—1381.
Earley, J.W., Brindley, G.W., McVeagh, W.V. and Van den Heuvel, R.C., 1956. A regularly interstratified montmorillonite-chlorite. *Am. Mineral.*, 41: 258—267.

Eitel, W., 1954. *The Physical Chemistry of the Silicates.* Univ. Chicago Press, Chicago, Ill., 1592 pp.
Eyrish, M.V., Dvorechenskaya, A.A. and Pshenichnaya, N.F., 1970. Influence of composition of sorbed cations on formation of high-temperature phases in montmorillonite clays. In: P.T. Tazhibaeva et al. (Editors), *Investigation and Utilization of Clays and Clay Minerals.* Nauka, Kazakh. SSR, Alma-Ata, U.S.S.R., pp. 51—61.
Fischer, E.K. and Gans, D.M., 1946. Dispersions of finely-divided solids in liquid media. In: J. Alexander (Editor), *Colloid Chemistry.* Reinhold, New York, N.Y., 6: 286—287.
Garrels, R.M. and Christ, C.L., 1956. Application of cation-exchange reactions to the beidellite of the Putnam silt loam soil. *Am. J. Sci.*, 254: 372—379.
Gatlin, C., 1957. Formation damage. *Pet. Eng.*, Nov.: B-102—B-120.
Geuze, E.C.W.A. and Tan, T.K., 1953. The mechanical behavior of clays. *2nd Int. Congr. Rheol., Proc.*, Oxford, p. 247.
Gillery, F.H., 1959. X-ray study of synthetic Mg-Al serpentines and chlorites. *Am. Mineral.*, 44: 143—152.
Greene-Kelly, R., 1955. Dehydration of the montmorillonite minerals. *Mineral. Mag.*, 30: 604—615.
Griffiths, J.C., 1952. Reaction relation in the finer-grained rocks. *Clay Miner. Bull.*, 1: 251—257.
Grim, R.E., 1942. Modern concepts of clay minerals. *J. Geol.*, 50: 225—275.
Grim, R.E., 1953. *Clay Mineralogy.* McGraw-Hill, New York, N.Y., 384 pp.
Grim, R.E., 1958. *Applied Clay Mineralogy.* McGraw-Hill, New York, N.Y., pp. 278—306.
Grim, R.E., 1961. Montmorillonite, high-temperature reactions and classification. *Am. Mineral.*, 46: 1329—1369.
Grim, R.E., 1962. *Applied Clay Mineralogy*, McGraw-Hill, New York, N.Y.
Grim, R.E., 1968. *Clay Mineralogy.* McGraw-Hill, New York, N.Y., 596 pp.
Grim, R.E. and Cuthbert, F.L., 1945. The bonding action of clays. I. Clays in green molding sands. *Ill. State Geol. Surv., Rep. Invest.*, 102.
Grim, R.E. and Güven, N., 1978. *Bentonites. Geology, Mineralogy, Properties and Uses. Dev. Sedimentol.*, 24. Elsevier, Amsterdam, 256 pp.
Grim, R.E. and Johns, W.D., 1954. Clay mineral investigation of sediments in the northern Gulf of Mexico. *Proc. Natl. Conf. Clays Clay Miner. 2nd, — Natl. Acad. Sci. Natl. Resour. Counc., Publ.*, 327: 81—103.
Grim, R.E. and Kulbicki, G., 1961. Montmorillonite high-temperature reactions and classification. *Am. Mineral.*, 46: 1329—1369.
Grim, R.E. and Rowland, R., 1944. Differential thermal analysis of clays and shales, control and prospecting method. *Am. Ceram. Soc. J.*, 27: 65—76.
Grim, R.E. and Rowland, R.A., 1942. Differential thermal analysis of clay minerals and other hydrous materials. *Am. Mineral.*, 27: 746—761, 801—818.
Grim, R.E., Bray, R.H. and Bradley, W.F., 1937. The mica in argillaceous sediments. *Am. Mineral.*, 22 (7): 813—829.
Grim, R.E., Dietz, R.S. and Bradley, W.F., 1949. Clay mineral composition of some sediments from the Pacific Ocean off the California coast and the Gulf of California. *Bull. Geol. Soc. Am.*, 60: 1785—1808.
Grim, R.E., Droste, J.B. and Bradley, W.F., 1961. A mixed-layer clay mineral associated with an evaporite. *Clays Clay Miner., Proc. Natl. Conf. Clays Clay Miner.*, 8 (1959): 228—236.
Gruner, J.W., 1932. The crystal structure of kaolinite. *Z. Kristallogr.*, Abt. A, 83: 75—88.
Gruver, R.M., 1948. Precision method of thermal analysis. *Am. Ceram. Soc. J.*, 31: 323—328.
Guenot, B., 1970. Etude d'un minéral argileux du type interstratifique talc-saponite trouvé dans le Précambrien du Cong Kinshasa. *Bull. Groupe Fr. Argiles*, 22: 97—104.

Hanshaw, B.B., 1964. Cation-exchange constants for clays from electrochemical measurements. *Clays Clay Miner., Proc. Natl. Conf. Clays Clay Miner.*, 12: 397—422.

Harrison, G.R., Lord, R.C. and Loofbourow, J.R., 1948. *Practical Spectroscopy*. Prentice-Hall, Englewood Cliffs, N.J., 605 pp.

Harvey, H.W., 1957. *The Chemistry and Fertility of Sea Waters*, 2nd ed. Cambridge University Press, Cambridge, 234 pp.

Hendricks, S.B., 1939. Minerals present in soil colloids. I. Description and methods for identification. *Soil Sci.*, 48: 257—271.

Hendricks, S.B., 1942. Lattice structure of clay minerals and some properties of clays. *J. Geol.*, 50: 267—290.

Hendricks, S.B., 1946. A portable differential thermal analysis unit for bauxite exploration. *Econ. Geol.*, 41: 64—76.

Hendricks, S.B. and Ross, C.S., 1941. Chemical composition and genesis of glauconite and celadonite. *Am. Mineral.*, 26: 683—708.

Herold, P. and Planje, T.J., 1948. Modified differential thermal analysis apparatus. *J. Am. Ceram. Soc.*, 31: 20—22.

Hofmann, U., Endell, K. and Wilm, D., 1933. Crystalline structure of montmorillonite. *Z. Kristallogr.*, 86: 340—348.

Hofmann, U., Endell, K. and Wilm, D., 1934. Röntgenographische und Kolloidchemische Untersuchungen über Ton. *Angew. Chem.*, 47: 539—547.

Hofmann, U., Weiss, A., Koch, G., Mehler, F. and Scholz, A., 1956. Intracrystalline swelling, cation exchange, and anion exchange of minerals of the montmorillonite group and of kaolinite. *Natl. Acad. Sci., — Natl. Res. Counc. Publ.*, 456: 273—287.

Howe, J.A., 1914. Handbook of kaolin china clay and china stone. *Mus. Prac. Geol., London.*

Hower, J., 1967. Order of mixed layering in illite/montmorillonite. *Clays Clay Mineral., Proc. Natl. Conf. Clays Clay Miner.*, 15: 63—74.

Hower, J. and McWatt, T.C., 1966. The mineralogy of illites and mixed-layer illite/montmorillonite. *Am. Mineral.*, 51: 825—854.

Hughes, R.V., 1950. *The Application of Modern Clay Concepts to Oil Field Development and Exploitation*. Soc. Pet. Eng., Los Angeles, Calif., May 11—12, Preprint, SPE 801-26A.

Hunt, J.M., 1950. Infrared spectra of clay minerals. *Am. Pet. Inst. Proj. 49, Prelim. Rep.*, New York, Columbia Univ., 8: 105—121.

Jarusov, S.S., 1937. On the mobility of exchangeable cations in the soil. *Soil Sci.*, 43: 285—303.

Karšulin, M. and Stubičan, V., 1954. Über die Struktur und die Eigenschaften synthetischer Montmorillonite. *Monatsh. Chem.*, Ser. B., 85: 343—358.

Kaufman, A.J., Jr. and Drilling, E.D., 1950. Differential thermal curves of certain hydrous and anhydrous minerals, with a description of the apparatus used. *Econ. Geol.*, 45: 222—244.

Keller, W.D., 1964. Processes of origin and alteration of clay minerals. In: C.I. Rich and G.W. Kunze (Editors), *Soil Clay Mineralogy*. Univ. of North Carolina Press, Chapel Hill, N.C.

Keller, W.D. and Pickett, E.E., 1949. Absorption of infrared radiation by powdered silicate minerals. *Am. Mineral.*, 34: 855—864.

Keller, W.D. and Pickett, E.E., 1950. The absorption of infrared radiation by clay minerals. *Am. J. Sci.*, 248: 264—273.

Kelley, W.P., 1955. Interpretation of chemical analyses of clays. *Clays Clay Technol. Bull.*, 169: 92—94.

Kelley, W.P. and Liebig, G.F., Jr., 1934. Base exchange in relation to composition of clay with special reference to effect of sea water. *Bull. Am. Assoc. Pet. Geol.*, 18: 358—367.

Kelley, W.P., Jenny, H. and Brown, S.M., 1936. Hydration of minerals and soil colloids in relation to crystal structure. *Soil Sci.*, 41: 259—274.

Kennedy, J.L. (Editor), 1972. Inhibitor aids gumbo-shale drilling. *Oil Gas J.* (Oct.): 79—81.

Kistler, S.S., 1932. Coherent expanded aerogels. *J. Phys. Chem.*, 36: 52—64.

Klubova, T.T., 1973. *Clay Minerals and Their Role in Genesis, Migration, and Accumulation of Petroleum.* Nedra, Moscow, 255 pp.

Kodama, H., 1966. The nature of component layers of rectorite. *Am. Mineral.*, 51: 1035—1055.

Koizumi, M. and Roy, R., 1959. Synthetic montmorillonite with variable exchange capacity. *Am. Mineral.*, 44: 788—805.

Knechtel, M.M. and Patterson, S.H., 1962. Bentonite deposits of the Northern Black Hills District, Wyoming, Montana and South Dakota. *U.S. Geol. Surv. Bull.*, 1082-M: 893—1030.

Kruyt, H.R. (Editor), 1949. *Reversible Systems. Colloid Sci.*, II. Elsevier, Amsterdam, 733 pp.

Kruyt, H.R., 1952. *Irreversible Systems. Colloid. Sci.*, I. Elsevier, Amsterdam, 368 pp.

Larsen, D.H., 1946. Colloid features of drilling fluids. *Colloid Chem.*, 6: 509—530.

Larsen, D.H., 1955. Use of clays in drilling fluids. *Calif. Div. Mines Bull.*, 169: 269—281.

Le Châtelier, H., 1887. The constitution of clay. *Z. Phys. Chem.*, 1: 396—402.

Le Châtelier, H., 1887. *Compt. Rend.*, CIV, p. 1443; *Polytech. J.*, CCLXV: 94.

Le Châtelier, H., 1887. The action of heat on clays. *Bull. Soc. Fr. Minéral.*, 10: 204—211.

Lippmann, F., 1954. Uber einen Keuperton von Zaiserweiker bei Maulbronn. *Heidelb. Beitr. Mineral. Petrogr.*, 4: 130—134.

Loughnan, F.C., 1959. Further remarks on the occurrence of palygorskite at Red Banks Plains, Queensland. *Proc. R. Soc. Queensl.*, 71: 43—50.

Low, P.F., 1955. The role of aluminium in the titration of bentonite. *Soil Sci. Soc. Am., Proc.*, 19: 135—139.

Low, P.F. and Anderson, D.M., 1958. The partial specific volume of water in bentonite suspensions. *Soil Sci. Soc. Am., Proc.*, 22: 22—24.

MacEwan, D.M.C., Ruiz, A.A. and Brown, G., 1961. Interstratified clay minerals. In: G. Brown (Editor), *The X-ray Identification and Crystal Structures of Clay Minerals.* London Mineral. Soc., London, pp. 393—445.

MacKenzie, R.C., 1955. Potassium in clay minerals. *Potassium Symp.*, Int. Potash Inst., Berne.

MacKenzie, R.C. (Editor), 1957. *The Differential Thermal Investigation of Clays.* London Mineral. Soc., London, 456 pp.

MacKenzie, R.C., 1959. The classification and nomenclature of clay minerals. *Clay Miner. Bull.*, 4: 52—66.

Marshall, C.E., 1954. Multifunctional ionization as illustrated by the clay minerals. *Proc. Natl. Conf. Clays Clay Miner., 2nd — Natl. Acad. Sci. Natl. Res. Counc., Publ.*, 327: 364—385.

Martin-Vivaldi, J.L. and Cano-Ruiz, J., 1955. Contribution to the study of sepiolite, 2. Some considerations regarding the mineralogical formula. *Clays Clay Miner.*, 2: 254—263.

McAtee, J. Jr., 1958. Heterogeneity in montmorillonite. *Proc. Natl. Conf. Clays Clay Miner., 5th — Natl. Acad. Sci. Natl. Res. Counc. Publ.*, 566: 279—288.

Mellon, M.G. (Editor), 1950. *Analytical Absorption Spectroscopy.* New York, Wiley, New York, N.Y., 440 pp.

Meyer, D.B., 1935. A sericite of unusual composition. *Am. Mineral.*, 20: 384—388.

Mitchell, J.K., Campanella, R.G. and Singh, A., 1968. Soil creep as a rate process. *J. Soil Mech. Found. Div., Am. Soc. Civ. Eng.*, No. 94 SMI, Proc. Pap. 5751: 231—253.

Mumpton, F.A. and Roy, R., 1956. The influence of ionic substitution on the hydrothermal stability of montmorillonites. *Proc. Natl. Conf. Clays Clay Miner., 4th — Natl. Acad. Sci. Natl. Res. Counc., Publ.*, 456: 337—339.

Mumpton, F.A. and Roy, R., 1958. New data on sepiolite and attapulgite. *Proc. Natl. Conf. Clays Clay Miner., 5th — Natl. Acad. Sci. Natl. Res. Counc., Publ.*, 566: 136—143.

Murayama, S. and Shibata, T., 1958. On the rheological characteristics of clay, Part 1. *Disaster Prev. Res. Inst., Kyoto Univ., Bull.*, 26.

Nagy, V. and Bradley, W.F., 1955. The structural scheme of sepiolite. *Am. Mineral.*, 40: 885—892.

Nahin, P.G., 1955a. Infrared analysis of clays and related minerals. *Clays Clay Technol. Bull.*, 169: 112—118.

Nahin, P.G., 1955b. Swelling of clay under pressure. *Proc. Natl. Conf. Clays Clay Miner., 3rd — Natl. Acad. Sci. Natl. Res. Counc.*, 395: 174—185.

Nahin, P.G., Merrill, W.C., Grenall, A. and Crog, R.S., 1951. Mineralogical studies of California oil-bearing formations, I. Identification of clays. *Trans. Am. Inst. Min. Metall. Pet. Eng.*, 192: 151—158.

Nash, V.E. and Marshall, C.E., 1956. The surface reactions of silicate minerals, Part II, Reactions of feldspar surfaces with salt solutions. *Univ. Mo. Coll. Agric. Res. Bull.*, 614: 36 pp.

Nelson, B.W. and Roy, R., 1954. New data on the composition and identification of chlorites. In: A. Swineford and N. Plummer (Editors), *Clays and Clay Minerals. Natl. Acad. Sci., Natl. Res. Counc. Publ.*, 327: 335—348.

Norton, F.H., 1939. Critical study of the differential thermal method for identification of clay minerals. *Am. Ceram. Soc. J.*, 22: 54—63.

Nowak, T.J. and Krueger, R.F., 1951. The effect of mud filtrate and mud particles upon the permeabilities of cores. *API Dril. Prod. Prac.*, p. 164.

Oakes, D.T., 1958. Filtration theory for oil-well drilling fluids. *Proc. Natl. Conf. Clays Clay Miner., 5th — Natl. Acad. Sci. Natl. Res. Counc., Publ.*, 566: 46—60.

Orcel, J., 1935. Use of differential thermal analysis for determination of the constituents of clays, laterites and bauxites. *Congr. Int. Mines Métall. Géol. Appl., 7 Sess., Paris, Oct. 1935*; also *Geologie*, 1: 359—373.

Ormsby, W.C. and Sand, L.B., 1954. An analytical tool for mixed-layer aggregates. *Clays Clay Miner.*, 2: 254—263.

Osthaus, B.B., 1955. Interpretation of chemical analyses of montmorillonites. *Proc. 1st Natl. Conf. Clays Clay Technol., Div. Mines, San Francisco, Calif. — Clays Clay Technol. Bull.*, 169: 95—100.

Ovcharenko, F.D., 1963. *Colloid Chemistry of Palygorskite.* Acad. Sci. Ukranian SSR, Kiev, 120 pp.

Ovcharenko, F.D., Kruglitskiy, N.N., Nichiporenko, S.P. and Agabal'yants, E.G., 1966. *Palygorskite in Drilling.* Technika, Kiev, 159 pp.

Patterson, S.H. and Hosterman, J.W., 1962. Geology and refractory clay deposits of the Haldeman and Wrigley Quadrangles, Kentucky. *U.S. Geol. Surv. Bull.*, 1122-F: 113 pp.

Pauling, L., 1930. The structure of micas and related minerals. *Proc. Natl. Acad. Sci. U.S.*, 16: 123—129.

Perricone, A.C. and Browning, W.C., 1962. *Formation Clay Chemistry.* 37th Annu. Fall Meet., Los Angeles, Calif., SPE, Pap. 433: 7 pp.

Peterson, N.M.A., 1961. Expandable chlorite clay minerals from Upper Mississippian carbonate rocks of the Cumberland Plateau in Tennessee. *Am. Mineral.*, 46: 1245—1269.

Powers, M.C., 1954. Clay diagenesis in the Chesapeake Bay area. *Proc. Natl. Conf. Clays Clay Miner., 2nd — Natl. Acad. Sci. Natl. Res. Counc., Publ.*, 327: 68—80.

Powers, M.C., 1957. Adjustment of land-derived clays to the marine environment. *J. Sediment. Petrol.*, 27: 355—372.

Powers, M.C., 1967. Fluid release mechanisms in compacting marine rocks and their importance in oil exploration. *Bull. Am. Assoc. Pet. Geol.*, 51 (7): 1240—1253.

Radoslovich, E.W. and Norrish, K., 1962. The cell dimensions and symmetry of layer lattice silicates. I. Some structural considerations. *Am. Mineral.*, 47: 599—616.

Rieke, H.H., 1976. *Drilling Fluids Laboratory Experiments.* Vandalia Publishers, Morgantown, W.Va., pp. 207—216.

Rieke, H.H. and Chilingarian, G.V., 1974. *Compaction of Argillaceous Sediments.* Elsevier, Amsterdam, 424 pp.

Ries, H., 1927. *Clays: Their Occurrence, Properties, and Uses*, 3rd ed. Wiley, New York, N.Y., 613 pp.

Robertson, J.O., Jr., Rieke, H.H. and Chilingar, G.V., 1965. Viscosity measurements of aqueous clay suspensions as a tool for determining mineralogic types of clays. *Sedimentology*, 4: 181—187.

Robertson, J.O., Jr., Chilingarian, G.V. and Stone, R.O., 1979. "Swellmeter": an apparatus for measuring the degree of swelling of clays. *Energy Sources*, 4 (3): 299—311.

Rogers, W.F., 1963. *Composition and Properties of Oil Well Drilling Fluids*, 3rd ed. Gulf Publ. Co., Houston, Texas, 818 pp.

Ross, C.S., 1946. Sauconite — a clay mineral of the montmorillonite group. *Am. Mineral.*, 31: 411—424.

Ross, C.S., 1960. Review of the relationships in the montmorillonite group of clay minerals, pp. 225—229. In: A. Swineford (Editor), *Clays and Clay Minerals, Proc. 7th Natl. Conf. Clays Clay Miner.* Pergamon, London, 369 pp.

Ross, C.S. and Hendricks, S.B., 1945. Minerals of the montmorillonite group. *U.S. Geol. Surv., Prof. Pap.*, 205-B: 79 pp.

Rowland, R.A., 1955. Differential thermal analysis of clays and carbonates. *Clays Clay Technol. Bull.*, 169: 151—163.

Rowland, R.A. and Lewis, D.R., 1951. Furnace atmosphere control in differential thermal analysis. *Am. Mineral.*, 36 (1, 2): 80—91.

Roy, R. and Osborn, E.F., 1954. The system Al_2O_2-SiO_3-H_2O. *Am. Mineral.*, 39: 853—885.

Sarkisyan, S.G. and Kotel'nikov, D.D., 1971. *Clay Minerals and Problems of Petroleum Geology.* Nedra, Moscow, 183 pp.

Saunders, H.L. and Giedroyc, V., 1950. Differential thermal analysis in controlled atmosphere. *Trans. Br. Ceram. Soc.*, 49: 365—374.

Serdyuchenko, D.P., 1948. Chemical composition and classification of micas. *Dokl. Akad. Nauk SSSR*, 59 (3): 545—548.

Serdyuchenko, D.P., 1953. Concerning some questions about mineralogy of silicates. *Izv. Akad. Nauk SSSR, Geol. Ser.*, 1953 (3): 135—146.

Serdyuchenko, D.P. and Shashkina, V.P., 1958. Sepiolite, palygorskite, and attapulgite. *Miner. Sb. L'vovsk. Geol. Obshch.*, 12: 396—404.

Schofield, R.K. and Taylor, A.W., 1955. Measurements of the activities of bases in soils. *J. Soil Sci.*, 6: 137—147.

Searle, A.A. and Grimshaw, R.W., 1959. *The Chemistry and Physics of Clays.* Ernest Benn Ltd., p. 425.

Shimoda, S., 1969. New data for tosudite. *Clays Clay Miner.*, 17: 179—184.

Stephen, I. and MacEwan, D.M.C., 1950. Swelling chlorite. *Geotechnique*, 2: 82—83.

Stephen, I. and MacEwan, D.M.C., 1951. Some chloritic clay minerals of unusual types. *Clay Miner. Bull.*, 1: 157—162.

Stone, R.L., 1952. Differential thermal analysis of kaolin group minerals under controlled partial pressure of H_2O. *Am. Ceram. Soc. J.*, 35 (4): 90—99.

Stremme, H., 1908. *Z. Prakt. Geol.*, 16: 124.

Sudo, T. and Kodama, H., 1957. An aluminian mixed-layer mineral of montmorillonite-chlorite. *Z. Kristallogr.*, 109: 380—387.

Taylor, R. and Smalling, D., 1973. *A New and Practical Application of Annular Hydraulics.* 48th Annu. Meet. SPE, Las Vegas, NV, Sept.—Oct., SPE Preprint 4518, 8 pp.
Tazhibaeva, P.T. (Editor), 1970. *Investigation and Utilization of Clays and Clay Minerals. Proc. Symp.* Nauka, Alma-Ata, Kazakh, SSR, 307 pp.
Tchillingarian, G., 1952a. Study of the dispersing agents. *J. Sediment. Petrol.*, 22 (4): 229—233.
Tchillingarian, G., 1952b. Possible utilization of electrophoretic phenomenon for separation of fine sediments into grades. *J. Sediment. Petrol.*, 22 (1): 29—32.
Tchillingarian, G. and Beeson, C.M., 1950. *Iranian Gums as Treating Agents for Drilling Fluids.* Pet. Branch AIME Meet., Los Angeles, Oct., No. 78-G: 8 pp.
Tchillingarian, G. and Beeson, C.M., 1952. Natural organic colloids of Iranian origin as mud treating agents. *Pet. Eng.*, 24 (4): 345—352.
Twenhofel, W.H., 1950. *Principles of Sedimentation,* 2nd ed. McGraw-Hill, New York, N.Y., 673 pp.
Van Olphen, H., 1950. Pumpability, rheological properties, and viscometry of drilling fluids. *J. Inst. Petrol.*, 36: 223—234.
Van Olphen, H., 1963a. *An Introduction to Colloid Chemistry.* Interscience, New York—London, 149 pp.
Van Olphen, H., 1963b. Compaction of clay sediments in the range of molecular particle distances. In: W.F. Bradley (Editor), *Clays and Clay Minerals. Proc. Natl. Conf. Clays Clay Miner.*, 11: 178—187.
Veniale, F. and Van der Marel, H.W., 1968. A regular talc-saponite mixed-layer mineral from Ferriere, Hure Valley (Piacenza Province, Italy). *Contrib. Mineral. Petrol.*, 17: 237—254.
Walker, G.F., 1950. Trioctahedral minerals in the soil clays of northeast Scotland. *Mineral. Mag.*, 29: 72—84.
Walker, G.F., 1958. Reactions of expanding lattice clay minerals with glycerol and ethylene glycol. *Clay Miner. Bull.*, 3: 302—313.
Warshaw, C.M., 1959. Experimental studies of illite. *Proc. Natl. Conf. Clays Clay Miner., 7th — Natl. Acad. Sci. Natl. Res. Counc., Publ.*, 7 (1958): 303—316.
Warshaw, C.M. and Roy, R., 1961. Classification and scheme for the identification of layer silicates. *Geol. Soc. Am. Bull.*, 72: 1455—1492.
Weaver, C.E., 1953. A lath-shaped non-expanded dioctahedral 2 : 1 clay mineral. *Am. Mineral.*, 38: 279—289.
Weaver, C.E., 1958. The effects and geologic significance of potassium "fixation" by expandable clay minerals derived from muscovite, biotite, chlorite, and volcanic material. *Am. Mineral.*, 43: 839—861.
Weaver, C.E., 1965. Potassium content of illite. *Science*, 147: 603—605.
Weaver, C.E. and Beck, K.C., 1971. Clay-water diagenesis during burial: how mud becomes gneiss. *Geol. Soc. Am., Spec. Pap.*, 134: 96 pp.
Weaver, C.E. and Pollard, L.D., 1973. *The Chemistry of Clay Minerals.* Elsevier, 213 pp.
Weir, A.H. and Green-Kelly, R., 1962. Beidellite. *Am. Mineral.*, 47: 137—146.
Weiss, A., Scholz, A. and Hofmann, U., 1956. Zur Kenntniss von Trio Ktaedrischen Illite. *Z. Naturforsch.*, 11: 429—430.
Wherry, E.T., 1915. *Proc. U.S. Natl. Mus.*, 49: 463.
Wherry, E.T., 1925. A tabulation of the aluminum silicate minerals. *Am. Mineral.*, 10 (6): 140—145.
Winchell, A.N., 1925. Studies in the mica group. Part II. *Am. J. Sci.*, 209: 415—430.

Chapter 6

DRILLING FLUID ADDITIVES OTHER THAN CLAYS

B.S. GUPTA, GEORGE V. CHILINGARIAN and PAUL VORABUTR

INTRODUCTION

In order to achieve the optimum performance of any mud during drilling operations, the physical as well as the chemical properties of the mud must be carefully controlled. Viscosity, gel strengths (initial and 10-min), and fluid loss are of particular importance because they are related to the removal of cuttings from the borehole to the surface, holding rock cuttings and weighting material in suspension during the period of no circulation, and releasing cuttings at the surface. The importance of the presence of thin filter cake along the sides of the borehole is discussed in Chapter 2 and elsewhere in this book.

The clay-surface chemistry and that of various drilling fluids constitute quite a complicated subject researched by many investigators. Although the study and control of all the parameters simultaneously is indeed very difficult, it is possible. The gel-forming components in drilling fluids, which are either suspensions of colloids (in water or oil) and/or emulsions, receive the utmost attention. Of equal importance is the particle size of the basic components, e.g., clays, and of the additives. Contamination of muds results from drilling through formations containing salt (NaCl), gypsum ($CaSO_4 \cdot 2H_2O$), anhydrite ($CaSO_4$), limestones, dolomites, and other various carbonates and sulfates. Thus, it becomes necessary to use various chemicals in order to restore the desired drilling fluid properties. Higher temperatures and formation pressures are encountered at greater depths. The temperature and pressure gradients, however, vary from place to place. Whereas normally the pressure effect is not critical, the temperature effect is extremely significant due to the thermal instability of most drilling fluids. For example, a drilling fluid, which might exhibit excellent properties at the surface temperature, may become completely useless at a depth of 20,000 ft due to the higher formation temperature, causing severe complications. The control and conditioning of different types of drilling fluids are treated in different chapters in this book. Inasmuch as the water-base drilling fluids are used most extensively, more attention is usually paid to their conditioning and characteristics than to the other types of drilling fluids (see Chapter 8). The main components of water-base drilling fluids are water and suspended clays. The latter are added for the purpose of imparting required properties, whereas some clays are incorporated into the drilling fluid as dispersed drilled solids as a result of drilling through shales, etc. Usually, the commercial clays that are used for drilling fluids are bentonite (montmorillonite), attapulgite, and sepiolite, whereas the clays that enter the mud as drilled solids include montmorillonite, mixed-layer clays, illite, and kaolinite. Bentonite (1) improves the suspension of weighting materials, (2) increases the carrying capacity of drilling fluid due to the increased viscosity and gel strength, and (3) improves the filtration and filter cake properties. Attapulgite and sepiolite, on the other hand, do not improve filtration and filter cake properties and, therefore, drilling fluids containing these clays require additives for the reduction of water loss.

WEIGHTING MATERIALS

During drilling operations, all formation fluids (i.e., oil, gas, and water) must be confined to their original locations. In order to achieve this, the hydrostatic pressure of the drilling fluid column should be slightly higher than the formation pressure (see Chapter 16). The following problems are created if the mud is unnecessarily heavy: (1) lost circulation and (2) swabbing and surge pressures during pulling or lowering of drillpipes and casing from or into the borehole. Consequently, it is accepted practice to keep a check on the specific weight of the drilling fluid in the pits and the flowline at regular intervals of time in order to:

(a) Prevent inflow of fluids from the formations being drilled.

(b) Maintain constant bottomhole pressure, which must be slightly greater than the formation pressure.

(c) Detect any inflow of fluids from the formation, which is manifested by a sudden drop in mud weight. Gain in pit level and/or a change in the salinity may also indicate an inflow of formation fluids.

(d) Prevent excessive drilling fluid weight. Density of a drilling fluid may be increased by adding soluble salts and/or suspended solids.

Soluble salts

Sodium chloride, calcium chloride, potassium chloride, sodium carbonate, and sodium nitrate yield higher density solutions upon dissolving in water. Solutions of these salts are used as workover fluids or packer fluids because they do not contain solids that might settle and make the workover job difficult to perform. A saturated solution of sodium chloride weighs about 10 lb/gal at room temperature, whereas calcium chloride and sodium nitrate solutions can weigh 11.5—12 lb/gal. These solutions, however, must be treated with inhibitors to prevent corrosion.

Suspended solids

Suspended-solids weighting materials include: (1) ground carbonates [ground limestone, mainly $CaCO_3$, or dolomite, mainly $CaMg(CO_3)_2$]; (2) ground barite (barium sulfate); and (3) ground galena (lead sulfide).

Carbonates

Limestone is an inert material which, if composed of pure calcite ($CaCO_3$), has a specific gravity of 2.71, whereas dolomite [$CaMg(CO_3)_2$] has a specific gravity of 2.87. In nature, however, carbonate rocks are composed of various proportions of calcite and dolomite (see Chilingar et al., 1967). The maximum density obtained by using limestone, therefore, is about 12.0 lb/gal in water-base drilling fluids and about 11.5 lb/gal in oil muds. Carbonates are usually used in nondrilling operations (oil-mud packer fluids and casing packs) rather than drilling operations. In such applications, carbonates result in reduced costs and easier suspension in drilling fluids as compared to barium sulfate (barite) and other weighting materials. It is important to mention here that some carbonates are more oil-wet than water-wet.

Barite

Inasmuch as clays and drilled solids give slurries of only limited density but high viscosity, in order to attain densities above 11.5 lb/gal, yet maintaining low viscosities and keeping content of clay at a minimum, barite ($BaSO_4$) is used as a weighting material. The specific gravity of barite varies from 4.2 to 4.3, i.e., about 4.2 times heavier than water and about 1.6 times heavier than drilled solids and clays. Inasmuch as barite is insoluble in water, it does not react with clays or drilled solids. Drilling fluids weighing as high as 22 lb/gal have been prepared with barite.

Galena

Galena (lead sulfide) is used alone or together with barite in order to attain drilling fluid weights in excess of 22 lb/gal. The specific gravity of commercially available galena is of the order of 6.8.

Methods of calculating amounts of weighting materials

When two substances having different specific weights are mixed, the specific weight of the mixture is a function of the quantities and specific weights of the components of the mixture:

$$V_i\gamma_i + V_w\gamma_w = (V_i + V_w)\,\gamma_f \tag{6-1}$$

where V_i = volume of mud before weighting; V_w = volume of weighting material added; γ_i = initial specific weight of mud; γ_w = specific weight of weighting material; and γ_f = final mud specific weight. The unknown quantity is V_w, i.e., the amount of weighting material that is required to increase the specific weight of mud from γ_i to γ_f. Simplifying eq. 6-1 gives the following relation:

$$V_w = \frac{V_i(\gamma_f - \gamma_i)}{(\gamma_w - \gamma_f)} \tag{6-2}$$

or:

$$V_w/V_i = \frac{\gamma_f - \gamma_i}{\gamma_w - \gamma_f} \tag{6-3}$$

Inasmuch as barite, for example, is packed in 100-lb sacks for commercial use, it would be easier to express results of calculations in units of 100 lb of barite required for 100 bbl mud. If V_i is equal to 100 bbl, then eq. 6-3 becomes:

$$\text{Sacks of barite/100 bbl of mud} = \frac{42\gamma_b(\gamma_f - \gamma_i)}{(\gamma_b - \gamma_f)} \tag{6-4}$$

where γ_b = specific weight of barite.

Sample problem

How many sacks of barite (sp. gr. = 4.3) are required to increase the density of 500 bbl of mud from 13 to 14 lb/gal? (Specific weight of water in lb/gal = 8.34.)

$$\text{Sacks of barite/100 bbl of mud} = \frac{42(4.3 \times 8.34)(14-13)}{(4.3 \times 8.34 - 14)} = 69.1$$

Sacks of barite/500 bbl = 69.1 × 5 = 345.5 or 346 sacks

DISPERSANTS (THINNING AGENTS)

The rheological properties of Bingham-plastic fluids can be described by two main parameters, i.e., plastic viscosity and yield point. In addition, most of the drilling fluids are thixotropic, i.e., they develop gel strength upon quiescence (when drilling is stopped). Plastic viscosity is a function of (1) friction between the solid particles present in the drilling fluid, (2) amount of charges on these particles, and (3) viscosities of the dispersed phase and the dispersion medium. High viscosity gives rise to reduced hydraulic efficiency, increased pressure surges, and accelerated swabbing action. On the other hand, if viscosity is kept too low, failure to clean the borehole properly may result, with consequent settling of cuttings.

Apparent viscosity is a function of the viscosity of base liquid, size and number of the dispersed particles, and the forces between the particles at a single condition of flow (e.g., only one particular shear strain — rpm). Two such viscosities are (1) the Marsh funnel viscosity and (2) Stormer viscosity determined at 600 rpm. The yield point is a function of forces between the solid particles in the drilling fluid and is the imaginary intercept on the shear stress axis (see Fig. 4-5). The initial gel strength of a drilling fluid is a measure of the minimum shearing stress necessary to initiate strain (rotation, or movement of drilling fluid). The ten-min gel strength is a measure of mud thixotropic property. During the initial stage of development of drilling fluids, the compounds which could reduce the viscosity and gel strength were known as *thinners*, whereas at the present time they are called *dispersants*.

Numerous dispersing (thinning) agents have been proposed by various investigators; however, only those which are effective, economical, widely accepted, or of historical value are mentioned in this chapter. According to Garrison et al. (1940), the function of thinner is to reduce that portion of the viscosity which is due to the development of the structure, without substantially affecting portion of the viscosity owing to the hydration of clay minerals. Results obtained by Garrison et al. (1940) are based on viscometric studies of water-base drilling fluids containing clay treated with dehydrated phosphates and plant tannins. According to Loomis et al. (1941), the effect of viscosity reducers is to prevent the formation of high viscosity and gel strength by preventing formation of a "brush-heap" structural configuration of clay particles. These conclusions are in accordance with the conclusions reached by Garrison et al. (1940). Loomis et al. (1941) studied the effect of various quantities of sodium tannate and dehydrated phosphates on the viscosity of water-base clay drilling fluids in California.

Fancher and Whiting (1943) studied the effect of phosphate compounds on a mud from an oilfield in Texas. They concluded that sodium-acid-pyrophosphate gives rise to the most permanent viscosity reduction, whereas the compounds with increasing number of sodium ions exhibited the least permanent viscosity reduction effects.

Melrose and Lilienthal (1952) investigated the effect of quebracho (thinner) on a mud prepared by mixing 30% by weight of bentonite and kaolinite in the laboratory to attain high plastic viscosity and high yield point. They established that with increasing quantities of quebracho the yield value decreases markedly. The plastic viscosity, however, increased instead of decreasing. With the initial addition of a thinner, the Stormer viscosity (600-rpm) was first reduced and later increased with further additions. Possibly, the yield value is reduced because of the adsorption of thinner on the charge-carrying (mainly positive) broken edges of the clays, which prevents the formation of a brush-heap structure. In addition, the formation of the latter may be prevented by the greater repulsion of clay particles as they are "plated" with extensive negative anions. The following reasons may account for the increase in plastic viscosity: (1) increased charge of the ionic atmosphere resulting from the ionized thinner, and (2) increase in the number of

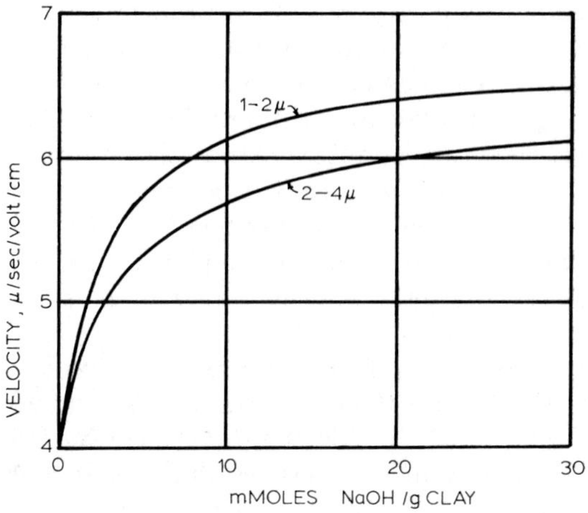

Fig. 6-1. Relationship between velocity of kaolinite clay particles toward a positive electrode and the amount of NaOH added to the aqueous phase. (After Tchillingarian, 1952a, fig. 1, p. 230; courtesy of the Journal of Sedimentary Petrology.)

particles because of the dispersion effect of the thinner.

Dispersion (or thinning) involves (1) the breaking-up of flocculated lumps of clays and (2) the creation of additional clay surfaces as a result of separation of aggregated or stacked sheets of clay. The drilled solids have some base exchange capacity to various degrees; however, majority of interstitial clays in sands and clays in shales are Ca-base rather than Na-base. The polyphosphates, tannins, lignins, and lignosulfonates are used in fresh-water muds as dispersants. The latter three are used together with sodium hydroxide in order to be effective, i.e., to achieve greater dissociation of these weak acids in order to form extensive negative molecules necessary for dispersion purposes.

Inasmuch as in the past, NaOH was used as a dispersing agent in very small quantities, a brief discussion of its dispersion capability (and mechanism) is presented here. The dispersing action of NaOH in very small amounts is probably due to the adsorption of the OH^- ions on the clay particles. As shown in Fig. 6-1, the cataphoretic velocity (velocity towards the positive electrode) of the kaolinite clay particles increases with increasing concentration of NaOH. This effect is decreased at greater concentrations of NaOH, probably because unsaturated bonds of the kaolinite clay particles are eventually satisfied on addition of OH^- anions. Eventually, flocculation of the system results with further addition of NaOH, because of the excess influx of Na^+ ions which neutralize the negative charges on clay particles.

Inorganic complex phosphates

Molecularly dehydrated polyphosphates, which can be used in fresh-water systems at low temperatures (<130°F) and which are effective only in small concentrations, include the following:

(1) Sodium-acid-pyrophosphate (SAPP): $Na_2H_2P_2O_7$ (pH = 4.8)
(2) Sodium-hexametaphosphate (Calgon): $(NaPO_3)_5$ (pH = 6.8)
(3) Sodium-tetraphosphate (BARAFOS): $Na_6P_4O_{13}$ (pH = 7.5)
(4) Tetra-sodium-pyrophosphate (TSPP): $Na_4P_2O_7$ (pH = 10.0)

"Calgon" and "BARAFOS" are the commonly-adopted commercial trade names. Although these chemicals are not filtration control agents, their dispersive action on clay aggregates results in lower water loss, especially in the presence of moderate water "hardness". It should be stressed here that, generally, improved dispersion reduces filtration losses. When the temperature exceeds 130°F, the polyphosphates revert into simpler phosphates in the presence of water and, thus, lose their thinning ability. In addition, upon contamination with NaCl in excess of 5000 ppm, the polyphosphates do not function efficiently as thinners because of a strong flocculating effect of the Na^+ cation of salt. Contamination with calcium and magnesium ions also limits the use of polyphosphates as thinners.

A pH of 8—9.5 is generally considered suitable for proper function of polyphosphate muds. If the pH of the system exceeds 10, however, as a result of addition of alkaline substances, the phosphate thinners do not perform efficiently.

Upon dissociation, molecularly-dehydrated phosphates yield spatially extensive anions, which plate the particles and give rise to deflocculation due to the repulsion of negatively-charged clay particles. Some phosphates also have the ability to sequester Ca^{2+} ions, which enhances the deflocculating power of a compound.

According to Tchillingarian (1952a), who made an extensive study of dispersants, the amount of polyphosphate required for maximum reduction of viscosity varies from 0.025 to 0.06 g/100 cm^3 of mud. The degree of reduction of viscosity increases with the increasing number of phosphate radicals per molecule, with sodium hexametaphosphate producing the most rapid reduction in viscosity.

It is very important to remember that on adding polyphosphates, which are electrolytes, the eventual flocculating effect due to the neutralization of negative charges on clay particles by Na^+ ions, for example, is only delayed.

Great care should be exercised, therefore, in controlling the amount of polyphosphate necessary to achieve optimum results. Adverse results are obtained if a certain limit is exceeded (e.g., around 2 g/100 cm^3 of mud in the case of sodium-acid-pyrophosphate, dispersing action of which is close to that of hexametaphosphate). Based on the experience of the writers, the effective range of concentrations of polyphosphates is 0.1 to 2 g/100 cm^3 of water-base clay drilling fluids.

Organic thinners

Organic thinners have wider applications than the inorganic thinners, because these thinners can be used in deep wells where high formation temperatures are encountered. Organic thinners include the following groups: lignins, tannins and lignosulfonates.

Lignin and tannin are two related substances which are characterized by aromatic (phenolic) structures. It should be stressed here that the names lignin and tannin are collective terms, which do not refer to clearly defined substances.

Lignins

Naturally-occurring lignin compounds are classed as humic acids which are obtained as degradation products of decaying vegetation. In general, lignin is

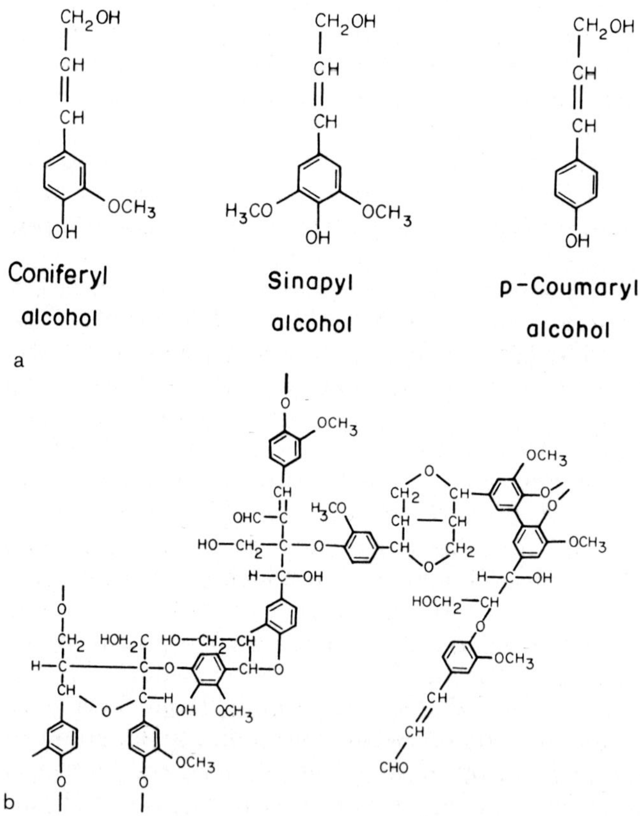

Fig. 6-2. Part of a lignin molecule (b) and aromatic plant alcohols (a), which are important building blocks of lignin. (After Metzner, 1973; courtesy of Enkel Publ. Co., Stuttgart.)

a high-molecular-weight polyphenol, which is made up of units constructed from phenyl-propane derivatives. It occurs as a three dimensional network located between the cellulose micelles of supporting tissues of plants (see Tissot and Welte, 1978, p. 44), and is very abundant in higher plants. Lignins are synthesized by dehydration and condensation of aromatic alcohols, such as coumaryl, sinapyl, and coniferyl (Fig. 6-2).

Lignins are purified from lignite which can be strip-mined and is quite abundant. Dried and ground lignin constitutes dark brown-to-black free-flowing powder, and lignin solutions have a neutral pH. They are comparatively cheap and are effective in controlling viscosity, yield point, and gel strength of water-base muds. Lignins also reduce fluid loss. Because of thermal stability they can be used in deep wells where higher temperatures are encountered.

Carbonox (trade name of a Baroid product) is an organic acid which has been used along with caustic soda (NaOH) in wells having bottomhole temperatures above 450°F. The alkali is added in order to maintain the pH of the mud above 8, which results in more soluble sodium lignite. Weak organic acids are not effective as thinners at low pH values. In the opinion of the writers, this is due to the fact that at low pH values, weak acids do not dissociate to a degree necessary to form sufficient amount of extensive negative anions to plate clay particles and cause their greater repulsion and consequent dispersion. In addition, in the presence of large amounts of NaCl, Carbonox is rendered ineffective due to the strong flocculating effect of the salt. Presence of divalent cations, such as calcium and magnesium, also renders Carbonox inefficient. Commonly, Carbonox cannot be used at high salt concentrations, because it may act as a colloidal material, increasing rather than decreasing viscosity. One form of lignin was prepared by Baroid by mixing one part of alkali and six parts of lignite and prereacted under special conditions. It performs better than Carbonox in hard water and higher salt concentrations.

Tannins

Tannins are intermediate in composition (and behavior) between lignin and cellulose. Although their chemical composition varies considerably, they are all complex derivatives of gallic and ellagic acids (Fig. 6-3), and possess carboxyl and phenolic groups on their aromatic structures. The name *tannin* is derived from ability of tannins to tan hides.

Tannin extracts from plants, concentrated mainly in the bark (up to 17%) and in leaves (up to 6.5%), can be used as viscosity-reducing agents. They can be obtained from the quebracho tree, which grows in abundance in Argentina, South America. The manufacturing process involves: (1) extraction of the soluble material from the trunks and limbs of the quebracho tree, and (2) evaporation which leaves the tannin residue that is ground to a desired size. In contrast to tannic acids, the tannins are composed of mixtures of five

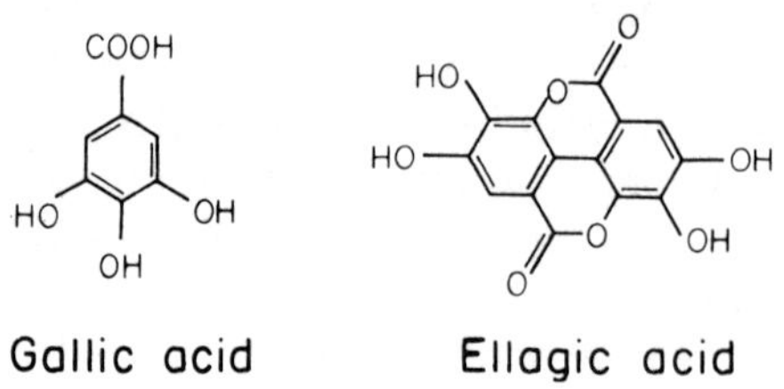

Fig. 6-3. Structures of gallic and ellagic acids, which are the principal building blocks of tannins. (After Francis, 1954; courtesy of Edward Arnold Publ. Co., London.)

molecules of digallic acid and one molecule of glucose: $5(C_{14}H_9O_9) \cdot C_6H_7O$. Rogers (1963, p. 259) presented the transformation of tannin upon hydrolysis as follows:

$$5(C_{14}H_9O_9) \cdot C_6H_7O + 5\ H_2O \rightarrow 5\ C_{14}H_{10}O_9 + C_6H_{12}O_6$$

(digallic acid)

$$C_{14}H_{10}O_9 + H_2O \rightarrow 2\ C_7H_6O_5$$

(digallic acid) (gallic acid)

After final hydrolysis, the acid anhydride/gallic acid weight ratio is 33/35.

Although only 16.4 lb of NaOH are needed per 100-lb sack of quebracho for proper effectiveness, in the actual field use the NaOH/quebracho ratio ranges from 1 : 1 to 1 : 2. At low pH values (<9), quebracho does not have effective thinning capacity because tannic acids are weak and do not dissociate for proper plating of clay particles with negative anions.

Lignosulfonates

As previously mentioned (see Chapter 2), lignosulfonates have been the most widely used members of the lignin family of organic polyelectrolytes in formulating the chemistry of inhibited muds. The effectiveness of these compounds to inhibit clay swelling depends on their ability to adsorb on the clay structural surfaces, and strong adsorption would imply strong inhibitive ability. In order to achieve this, lignosulfonates have been treated with various compounds. These modified lignosulfonates may be of sodium-treated, calcium-treated, and chromium-treated types.

Modification of lignosulfonates by reactions with compounds involving chromium and mechanism for increasing their adsorption capacity, has been discussed by Browning and Perricone (1962) and others. Figure 6-4 depicts the structural formula of one form of chrome-modified lignosulfonate where L represents the lignosulfonate molecules, whereas Fig. 6-5 shows the adsorption of the chrome-modified lignosulfonate on a silicate surface of a clay structure, involving the hydrogen bonding.

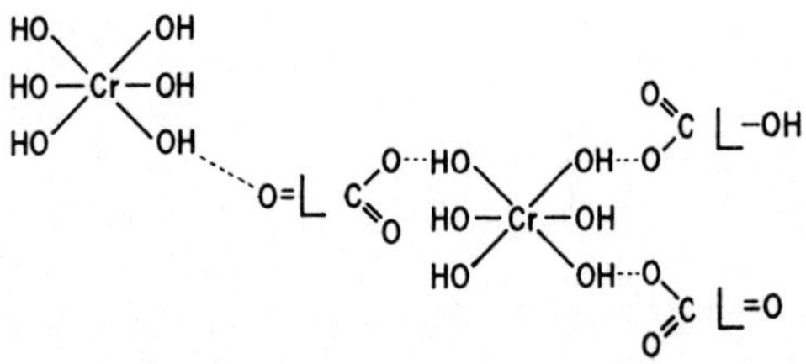

Fig. 6-4. Structural formula of chrome-lignosulfonate molecule where L represents the basic lignosulfonate molecule. (After Browning and Perricone, 1962, p. 10; courtesy of the Society of Petroleum Engineers of AIME.)

Several different modified lignosulfonates are available, and their effectiveness as inhibiting agents and drilling fluid conditioners varies greatly depending on the type of modification. Environmental consideration, particularly in offshore areas, preclude the use of certain modified lignosulfonates. One should also keep in mind that although some lignosulfonates are excellent drilling fluid conditioners, during cementing job they hinder formation of good cement—rock bond. Consequently, this problem should be thoroughly evaluated prior to initiating drilling fluid conditioning with a particular lignosulfonate. Fig. 6-6 illustrates effectiveness of various chrome-treated lignosulfonates as deflocculants in non-contaminated clay-in-water suspension. *DOIL* refers to dichromate-oxidized iron-lignosulfonate; *CSCL* to lignosulfonate treated with chrome sulfate complex; $CaLSO_3$ to calcium-treated lignosulfonate; and $CrLSO_3$ to chromium-treated lignosulfonate.

Factors affecting effectiveness of lignosulfonates. Figures 6-7, 6-8, and 6-9 show the effects of temperature on the adsorption ability and, consequently,

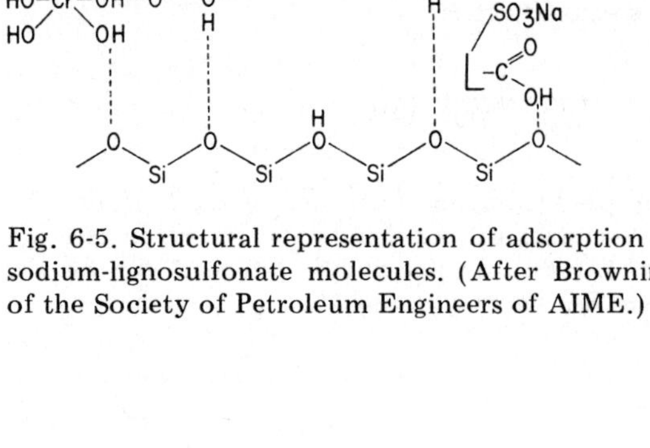

Fig. 6-5. Structural representation of adsorption on the silica surface of chrome-modified sodium-lignosulfonate molecules. (After Browning and Perricone, 1962, p. 10; courtesy of the Society of Petroleum Engineers of AIME.)

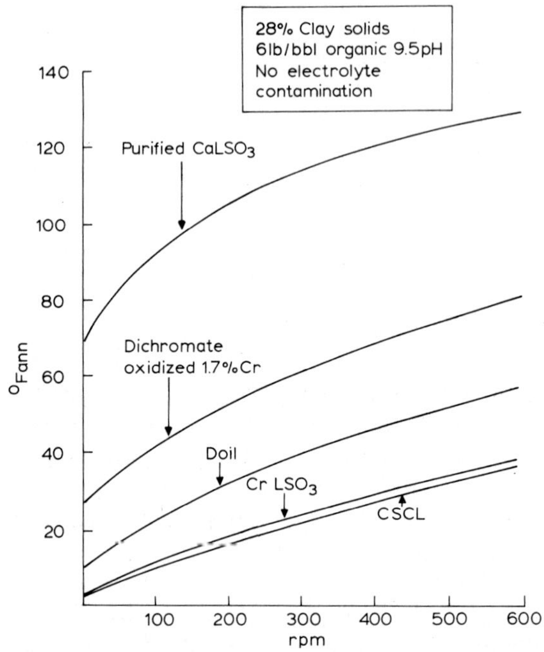

Fig. 6-6. Relationship between rotational speed (rpm) and Fann viscometer deflection in degrees for various chrome-treated lignosulfonates, establishing their effectiveness as deflocculants in a non-contaminated clay-in-water suspension. (After Browning and Perricone, 1962, fig. 16, p. 21; courtesy of the Society of Petroleum Engineers of AIME.)

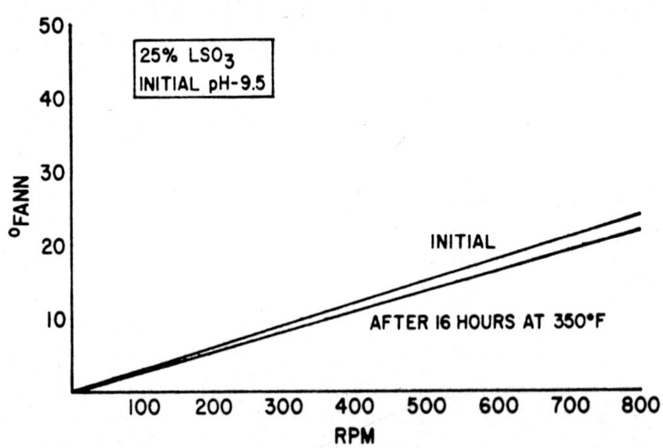

Fig. 6-7. Relationship between rotational speed (rpm) and Fann viscometer deflection for a lignosulfonate solution treated with chromium sulfate complex at room temperature and after treatment at 350°F for 12 hr. (After Browning and Perricone, 1962, fig. 20, p. 14; courtesy of the Society of Petroleum Engineers of AIME.)

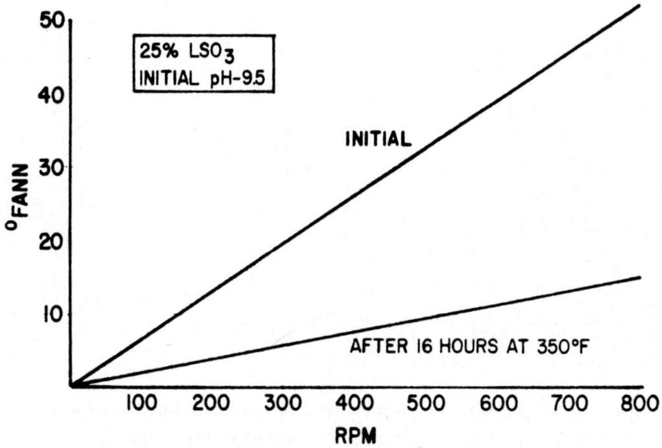

Fig. 6-8. Relationship between rotational speed (rpm) and Fann viscometer dial deflection in degrees for dichromate-oxidized iron-lignosulfonate solution at different temperatures. (After Browning and Perricone, 1962, fig. 19, p. 12; courtesy of the Society of Petroleum Engineers of AIME.)

inhibiting ability of various modified lignosulfonate compounds. Apparently, the lignosulfonate solution treated with chromium sulfate complex is barely affected by temperature even after aging at 350° F for 16 hours (Fig. 6-7), whereas a marked effect on flow properties occurs in the case of dichromate-oxidized iron-lignosulfonate (Fig. 6-8).

The effect of pH on the effectiveness of various modified lignosulfonates at different temperatures is presented in Fig. 6-9. The effectiveness of these

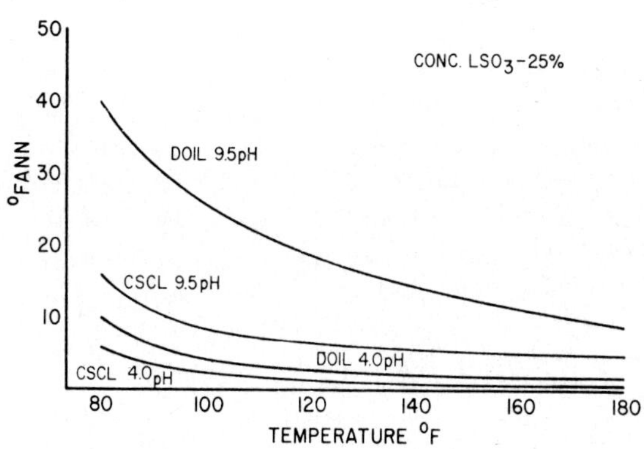

Fig. 6-9. Relationship between temperature and Fann viscometer dial deflection in degrees for different chrome-treated lignosulfonate solutions. (After Browning and Perricone, 1962, fig. 22, p. 26; courtesy of the Society of Petroleum Engineers of AIME.)

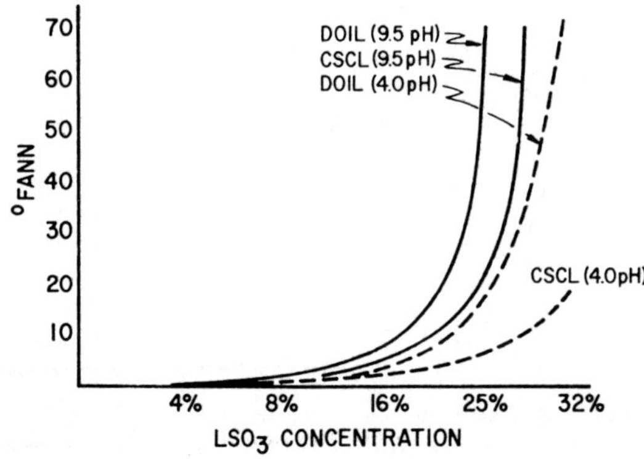

Fig. 6-10. Relationship between chrome-modified lignosulfonate concentration and Fann viscometer dial deflection in degrees at various pH values, adjusted with NaOH. (After Browning and Perricone, 1962, fig. 12, p. 12; courtesy of the Society of Petroleum Engineers of AIME.)

modified lignosulfonates decreases with increasing pH. The effectiveness of dichromate-oxidized iron-lignosulfonates is affected more by the increase in pH than the lignosulfonate treated with chromium sulfonate complex. In general, the effect of pH, however, is minimized with increasing temperature.

The effect of common contaminants on the effectiveness of various modified lignosulfonates are presented in Figs. 6-10, 6-11, and 6-12.

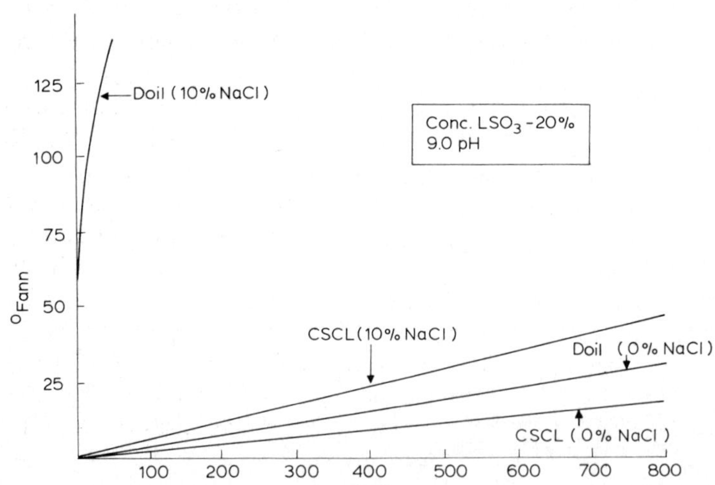

Fig. 6-11. Effect of sodium chloride on rheology of 20% chrome-treated lignosulfonate solutions. (After Browning and Perricone, 1962, fig. 15, p. 13; courtesy of the Society of Petroleum Engineers of AIME.)

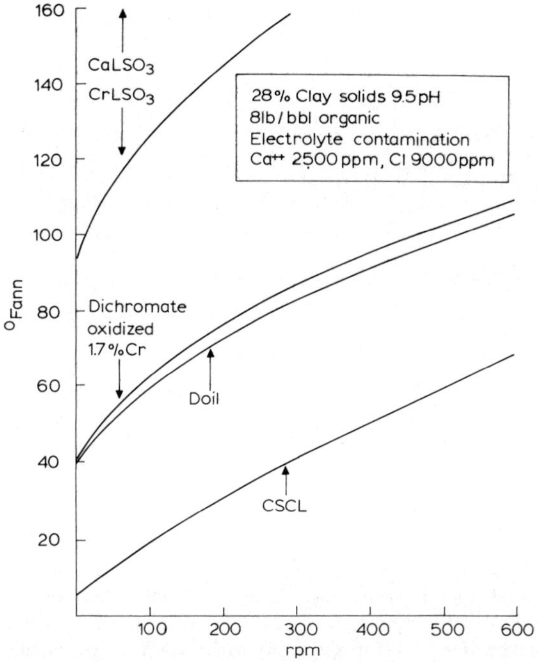

Fig. 6-12. Effect of various chrome-treated lignosulfonates on the rheology of contaminated clay suspensions. (After Browning and Perricone, 1962, fig. 17, p. 13; courtesy of the Society of Petroleum Engineers of AIME.)

The effects of NaOH on the viscosity of the chrome-modified lignosulfonate solutions are presented in Fig. 6-10. The viscosity of lignosulfonate treated with chromium sulfate complex at a pH of 4.0 is substantially lower than at a pH of 9.5 when the concentration of lignosulfonate is above 16%. Similar effect is true for dichromate-oxidized iron-treated lignosulfonate. According to Browning and Perricone (1962, p. 11), in the case of chromium sulfate complex, the increase in viscosity is due to hydrous oxide bonding, whereas chrome-linked lignosulfonate units are mainly responsible in the case of dichromate-oxidized iron-treated lignosulfonates.

The ability of various chrome-treated lignosulfonate solutions to deflocculate clay particles in the presence of sodium chloride are presented in Fig. 6-11. Dichromate-oxidized iron-treated lignosulfonate becomes completely ineffective at NaCl concentration of 10%, whereas chromium-sulfate-complex-modified lignosulfonate shows relatively high tolerance to NaCl contamination. This is attributed to the presence of viscous multi-layers of chromium-sulfate-complex-modified lignosulfonate on clay surfaces even at high NaCl concentrations. In the case of dichromate-oxidized iron-lignosulfonate, however, these layers are present in fresh water, but are absent at

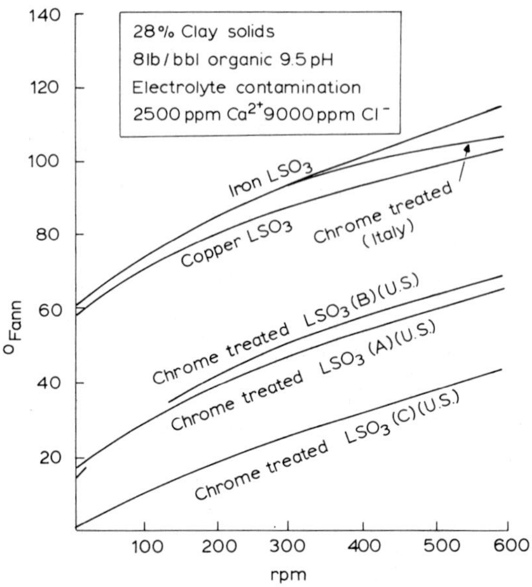

Fig. 6-13. Relationship between rotational speed (rpm) and Fann viscometer deflection in degrees for contaminated calcium-bentonite suspension deflocculated with various chrome-treated lignosulfonates manufactured by different companies. (After Browning and Perricone, 1962, fig. 4, p. 13; courtesy of the Society of Petroleum Engineers of AIME.)

higher salt concentrations due to salting out of lignosulfonate molecules (Browning and Perricone, 1962, p. 11).

Chromium-sulfate-complex lignosulfonate is the most efficient deflocculant and inhibiting agent in the presence of calcium ion contamination (Fig. 6-12). At a Ca^{2+} cation concentration of 2500 ppm and Cl^- anion concentration of 9000 ppm, unmodified calcium and chromium lignosulfonates are ineffective as deflocculants and inhibiting agents.

It has been shown previously that aging at high temperature has very little effect on the deflocculating and inhibiting ability of lignosulfonates. Aging at high temperature in the presence of common electrolytes, however, show a reduction in the efficiency of chromium-sulfate-complex lignosulfonate and dichromate-oxidized iron-lignosulfonate.

Manufacture of chrome-modified lignosulfonates from normal lignosulfonate molecules involves double decomposition and complex reactions. Great care must be exercised, therefore, when selecting chrome-lignosulfonates offered by various companies for utilization as drilling fluid conditioner, because their effectiveness may vary greatly depending upon the completeness of the processes used in obtaining the final products (see Fig. 6-13). As shown in Fig. 6-13, chrome-treated lignosulfonate supplied by company (C) is the most efficient.

Fluid loss control by lignosulfonates. Lignosulfonates reduce fluid loss as a result of formation of viscous layers on the clay-particle surfaces. Inasmuch as the lignosulfonate molecules show strong lateral attraction among themselves due to the hydrogen bonding, they have a rapid tendency to form impermeable viscous multi-layers when clay particles are laid down on the formation face. In addition, these layers are deformable and adjust themselves to the size and shapes of openings between the clay particles, thereby reducing the amount of filtrate that must be lost to the formation before sufficient number of clay particles are deposited in order to form impermeable filter cake. As a result, thin and firm filter cakes having good texture are generally associated with this type of chemical treatment.

Special applications of lignosulfonates. Other applications of lignosulfonates which are a direct consequence of the ability to adsorb on the clay surfaces are: (1) use as an anticorrosion agent, and (2) stabilization of oil-in-water emulsions.

Some water-base drilling fluids may promote corrosion of the drillpipe and other surface tubular goods (see Chapter 15). A properly-prepared chrome-modified lignosulfonate has been found to have strong oxygen scavenging property (Chesser, 1968, p. 3). It has been established that oxygen is a major cause of drillpipe corrosion, fatigue, and eventual failure. There is a definite correlation between the oxygen content of drilling fluid, with resulting corrosion and pitting tendency of the drillpipe, and the amount and type of lignosulfonate present (Bradley, 1970, p. 50).

Figure 6-14 shows a dissolved oxygen concentration in water having various concentrations of chrome-modified sodium-lignosulfonate. At a lignosulfonate concentration of 2 g/250 cm^3, the oxygen concentration is reduced by as much as 8 ppm, with further reduction occurring on increasing lignosulfonate concentration. Figure 8-20 shows a reduction in corrosion rate as a result of converting oxygen-rich water-base drilling fluid system to an oxygen scavenging lignosulfonate and lignite system.

Under static conditions, lignosulfonates form a thin protective film on the steel surface, which has tendency to protect the steel against attack by sulfides, and prevent stress cracking (Chesser, 1968, p. 7).

The corrosion inhibition of lignosulfonates and modified lignosulfonates has been improved by addition of chromate. In general, aged untreated lignosulfonate solutions tend to be less effective as corrosion inhibitors than chromate-treated lignosulfonate solutions (Skelly and Dieball, 1969, p. 3). Table 6-I shows the reduction in corrosion rate as a result of adding chromate to a 4% chrome-lignosulfonate solution. Even though some anomalies have been observed, the general trend indicates an improvement in corrosion rate reduction with addition of chromate. Table 6-II illustrates the reduction in the corrosion rate on treating chrome-lignosulfonate with chromate for 5% by weight bentonite slurries.

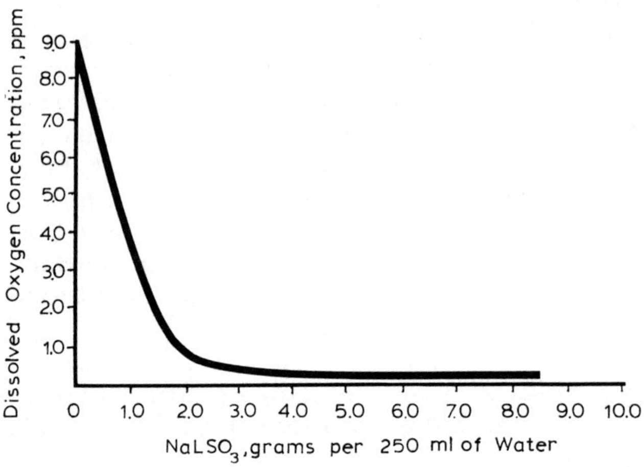

Fig. 6-14. Relationship between dissolved oxygen content and concentration of chrome-modified sodium-lignosulfonate. (After Chesser, 1968, fig. 1, p. 14; courtesy of Milchem Inc.)

TABLE 6-I

Effect of chromate concentration on corrosion rates of a 4% chrome-treated lignosulfonate solutions

(After Skelly and Dieball, 1969, table 5, p. 5; courtesy of the Society of Petroleum Engineers of AIME)

Chromate ($Na_2CrO_4 \cdot 10\ H_2O$) concentration (ppm)	Aging		Corrosion rate (mils/yr)
	time (hrs)	temperature (°F)	
0	0	75	5.5
0	0	75	6.7
0	0	75	7.4
2850	0	75	3.0
2850	0	75	1.9
2850	0	75	3.8
0	0	75	6.7
0	672	75	8.0
2850	0	75	1.9
2850	672	75	7
2850	0	75	3.9
2850	0	75	2.2
2850	0	75	3.4
2850	3	350	6
2850	3	350	8.3
2850	3	350	6.9

TABLE 6-II

Effect of chromate concentration on corrosion rates of 5% bentonite slurries containing 4% chrome-treated lignosulfonate

(After Skelly and Dieball, 1969, table 7, p. 6; courtesy of the Society of Petroleum Engineers of AIME)

Chromate ($Na_2CrO_4 \cdot 10\ H_2O$) concentration (ppm)	Duration of aging (days) temperature = 75°F	Corrosion rate (mils/yr)
0	0	5.4
0	0	3.5
500	0	1.8
500	0	2.2
2850	0	1.2
2850	0	0.8
0	10	5.4
0	10	5.5
500	10	7.8
500	10	7.8
2850	10	0.7
2850	10	1.0

The protective films formed in the presence of chrome-modified sodium-lignosulfonate are principally composed of Fe_3O_4 (Chesser, 1968, p. 5). Formation of this film may be due to the low content of oxygen and reduced potential of the water-base drilling fluid in the presence of chrome-modified sodium-lignosulfonates.

At high bottomhole temperatures and high bacterial activity, severe corrosion may result in the presence of lignosulfonates. As far as corrosion of tubular goods is concerned, carbon dioxide and hydrogen sulfide are considered to be the most potent offenders. Lignosulfonates themselves can generate CO_2 and H_2S in situ if the drilling fluid is left in the annulus for long periods of time. This is due to (1) high bacterial activity, (2) high-temperature degradation of lignosulfonates, and (3) electrochemical reactions (Simpson and Barbee, 1970, p. 42).

Lignin molecules are susceptible to degradation by several types of bacteria. The degradation products include organic acids and carbon dioxide (Simpson and Barbee, 1970, p. 42). Thermal degradation of an organic compound occurs at a temperature at which the compound starts to break down generating degradation products. In the case of lignosulfonates, the degradation is not only a function of temperature but also of the pH of the solution.

One of the common byproducts of lignosulfonate degradation is H_2S (Chesser, 1968, p. 4). Simpson and Barbee (1970) and others have found that thermal degradation of caustic lignosulfonate drilling fluid may cause release of H_2S within several hours at temperatures in excess of 400° F. Thermal degradation of lignosulfonates also produces CO_2. Increase in the pH of the solution tends to lower the thermal degradation temperature. According to Simpson and Barbee (1970, p. 45), lignosulfonate-treated fluids with an initial pH of 10 break down generating H_2S and CO_2 at a temperature as low as 330° F.

Sulfur-containing organic lignosulfonates may be reduced to sulfide even at room temperature and in the absence of bacterial activity if there is a high-potential corrosion cell which may form on tubing or casing (Simpson and Barbee, 1970, p. 45). Sulfide formation in the presence of lignosulfonates has been found to be a highly localized surface phenomenon. Based on laboratory tests, sulfide does not form in a lignosulfonate-treated drilling fluid at a corrosion cell potential of 0.12 Volt. At a potential of 0.40 Volt and higher, however, there was evolution of H_2 and the cathode was coated with sulfide in 16 hours, leading to severe sulfide corrosive environment (Simpson and Barbee, 1970, p. 45).

Stabilization of oil-in-water emulsions by lignosulfonates. Lignosulfonate molecules stabilize oil-in-water emulsion as a result of adsorption at the oil—water interface. Strong lateral attraction forces between lignosulfonate molecules give rise to a strong multilayer film around each oil droplet, preventing coalescence of droplets.

Quantitative determination of lignosulfonate content in drilling fluids. Inasmuch as the effectiveness of lignosulfonate as a drilling fluid conditioner (e.g., as a deflocculant) is a function of concentration of undegraded lignosulfonate molecules, quantitative determination of their effective concentration becomes imperative. Knowledge of the effective concentration of lignosulfonates at any particular time allows most economical and effective treatment in order to achieve optimum control of flow properties and greatest degree of shale inhibition.

According to Carney et al. (1971, p. 1), an optical density difference property based on the ultra-violet adsorption measurements of the lignosulfonate may be related to the effective concentration of lignosulfonate in an aqueous solution. This method is valid even at extremely low concentrations, in the presence of thermally degraded modified lignosulfonates, and at various pH of the solution.

A family of optical density versus temperature curves for different lignosulfonate types and having different solution concentrations may be prepared, as shown in Fig. 6-15. The following procedure can be used to determine the active lignosulfonate concentration from optical density of the lignosulfonate-treated drilling fluid samples (Carney et al., 1971, p. 2):

(1) Pipette 1 ml of drilling fluid from a well-stirred suspension.

(2) Dilute with deionized water to 200 ml, add 5 ml 1*N* ammonium ace-

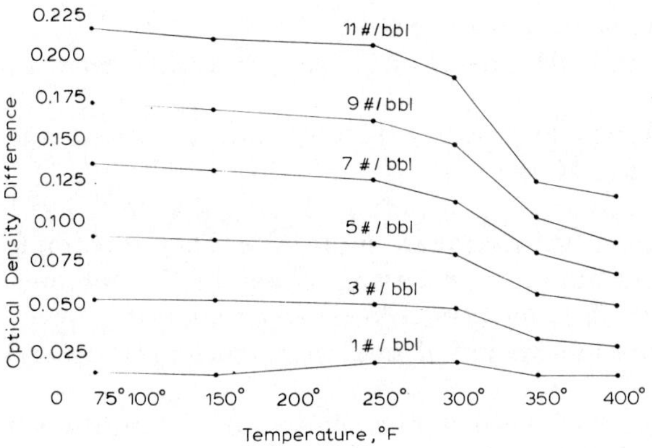

Fig. 6-15. Aging (16 hr in hot-roll oven) effect at various temperatures expressed as optical density differences of lignosulfonate solutions having different concentrations. (After Carney et al., 1971, fig. 4, p. 8; courtesy of the Society of Petroleum Engineers of AIME.)

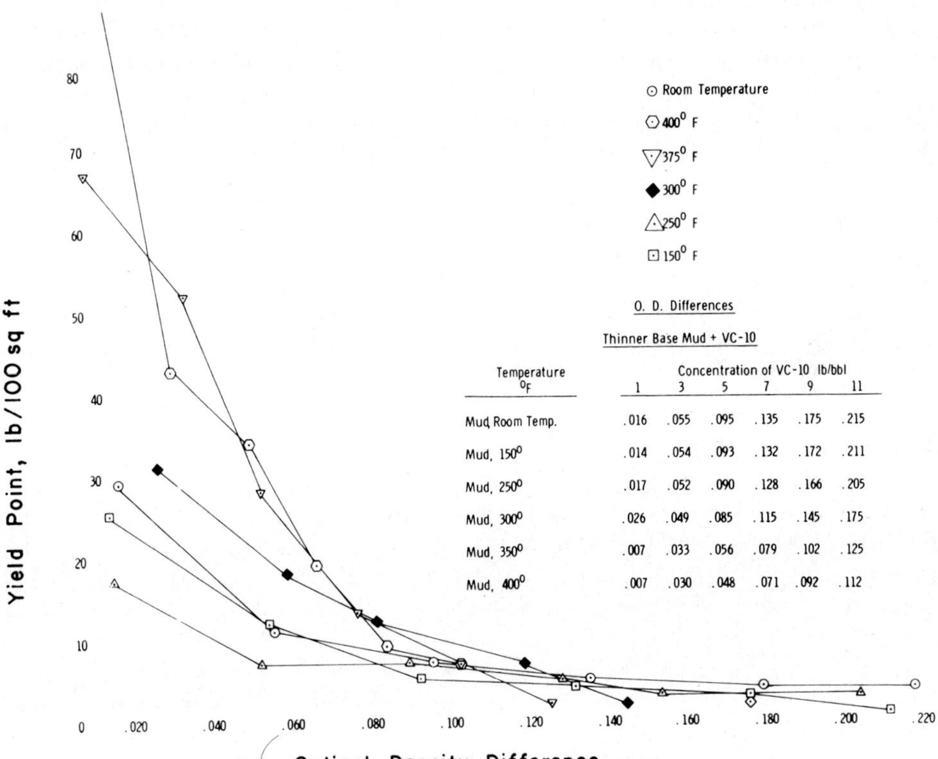

Fig. 6-16. Relationship between yield point and optical density differences for various lignosulfonate concentrations at different temperatures. (After Carney et al., 1971, fig. 5, p. 5; courtesy of the Society of Petroleum Engineers of AIME.)

tate, and then dilute with deionized water to 250 ml.

(3) Centrifuge aliquotes of the well-mixed suspension at 13,000 rpm for 10—15 min.

(4) Place a portion of a clear solution in a 10-mm silica cell. Scan and record the spectrum from 320 to 240 nm.

(5) From chart, measure the difference in optical density due to the 281 nm peak, i.e., that of undegraded lignosulfonate molecules. Then determine the active lignosulfonate concentration by referring to one of the commercially available standard curves. Care must be exercised in choosing appropriate base curves based on the known initial treatment concentration and bottomhole temperature.

Inasmuch as temperature causes thermal degradation of lignosulfonate molecules, additional amount of lignosulfonate must be continuously added. The amount required may be determined from such predetermined curves as shown in Fig. 6-16. For example, on considering a temperature of 300° F and desired yield point of 10 lb/100 sq ft, from Fig. 6-16 an optimum optical density value should be approximately 0.11. This value corresponds to a modified lignosulfonate treating concentration of approximately 7 lb/bbl.

As previously shown, the effects of modified lignosulfonates on drilling

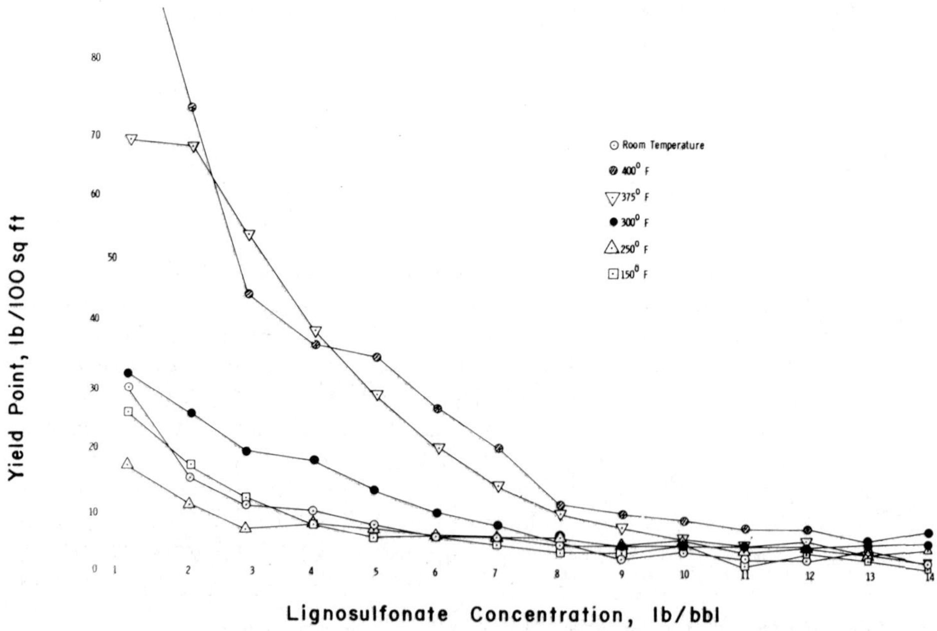

Fig. 6-17. Relationship between yield point and lignosulfonate concentration at different temperatures. (After Carney et al., 1971, fig. 11, p. 14; courtesy of the Society of Petroleum Engineers of AIME.)

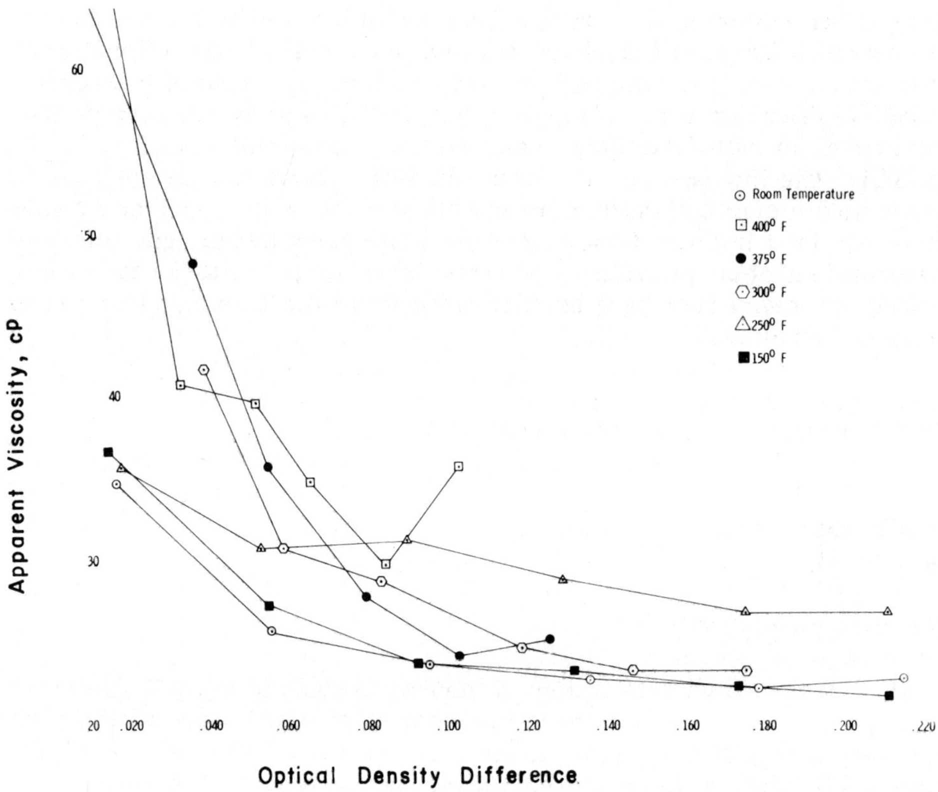

Fig. 6-18. Relationship between apparent viscosity and optical density difference at various temperatures. (After Carney et al., 1971, fig. 12, p. 14; courtesy of the Society of Petroleum Engineers of AIME.)

fluids may vary greatly from one manufacturer to another. Standard curves of optical density versus lignosulfonate concentration at various temperatures, therefore, should be prepared for a particular brand being used.

Other useful curves may also be preplotted for various lignosulfonate-treated drilling fluid systems at various temperatures. Two such curves are presented in Figs. 6-17 and 6-18. Figure 6-17 depicts yield point of the system as a function of lignosulfonate concentration at various temperatures. The effectiveness of lignosulfonates in reducing apparent viscosity in terms of optical density differences is presented in Fig. 6-18. The lignosulfonates tested showed erratic flow properties control at temperatures above 375° F, even at high concentration. The latter temperature thus appears to represent the upper limit at which lignosulfonates are effective.

Environmental impact of modified metal complex lignosulfonates. According to Hollingworth and Lockhent (1975, p. 115), the projected consumption of lignosulfonate type drilling fluid conditioners in U.S.A. would be in excess of 50,000 tons annually. Such a large amount of lignosulfonates will obviously present some disposal problem. Fortunately, lignosulfonate chemicals have been shown to belong to a "low-toxicity" group of chemical additives used in oilwell drilling fluids (Hollingworth and Lockhent, 1975, p. 121). The low toxicity of chrome-modified lignosulfonate has made it acceptable for use both at onshore and offshore locations. Some local regulations, however, prohibit discarding of the solutions containing lignosulfonates overboard in offshore drilling operations, even though little or no adverse effects on marine lives have been observed in most offshore platform areas (Ray and Shinn, 1975, p. 53).

FILTRATE LOSS REDUCTION ADDITIVES

Besides clays, other additives can be used to lower the fluid loss such as: (1) starch, (2) gums, (3) polyanionic cellulose, (4) sodium polyacrylonitrile, and (5) oil.

Starch (pregelatinized)

Most of the starch used in the oil industry is made from corn. The starch grains, which are separated from corn, are processed in such a manner that they can swell and gelatinize rapidly in fresh or salt water (either hot or cold) at various pH values. Upon rupturing outside amylopectin shell on using heat, the amylose inside can absorb water and swell, forming sponge-like bags. As a result, fluid loss can be reduced because of: (1) loss of free water from the system, and (2) plugging of openings in the filter cake by these sponge-like bags. In addition to lowering the fluid loss, starch also stabilizes the mud. As visualized by the writers, starch covers clay particles with hydrophilic colloids which protect clays against contamination.

In order to prevent fermentation one must use: (1) saturated salt solutions (260,000 ppm NaCl or higher), (2) pH of 11.5 or higher, and (3) preservatives such as paraformaldehyde (concentrations of around 0.5 lb/bbl).

Gums

Extensive research on possible use of natural gums of Iranian and Indian origin in drilling fluids has been conducted by Tchillingarian and Beeson (1950, 1952). These gums included gum tragacanth, gum shiraz, and gum ghatti. It was found that the addition of gums improved markedly the filtration characteristics of all muds tested, including the high-pH red lime-base mud. Subsequent to that research, natural water-dispersible gums have been used successfully in drilling operations in Iran.

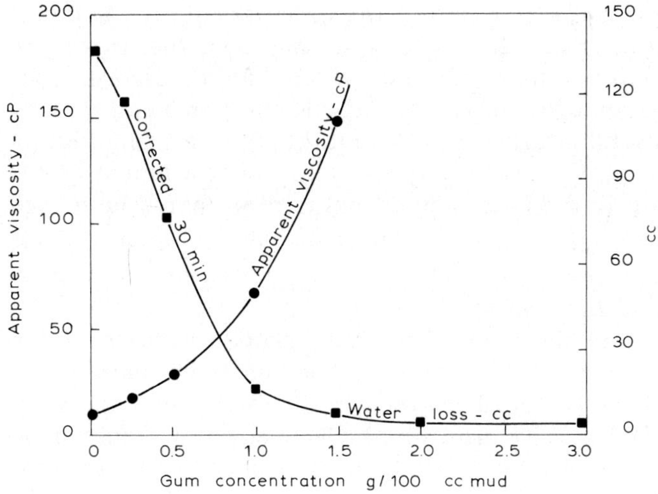

Fig. 6-19. Effect of gum guar HV on saturated salt-water mud. Aged for two days at room temperature; pH ~8.4. Viscosity was measured with Fann V-G Meter. (After Chilingar and Croushorn, 1964, fig. 4, p. 20; courtesy of Canadian Petro Engineering.)

In 1964, Chilingar and Croushorn studied the behavior of gum guar in drilling fluids (e.g., see Figs. 6-19 and 6-20).

Apparently gums play a double role in reducing the water loss, acting both as plugging agents and as protective colloids. The molecular size of gums

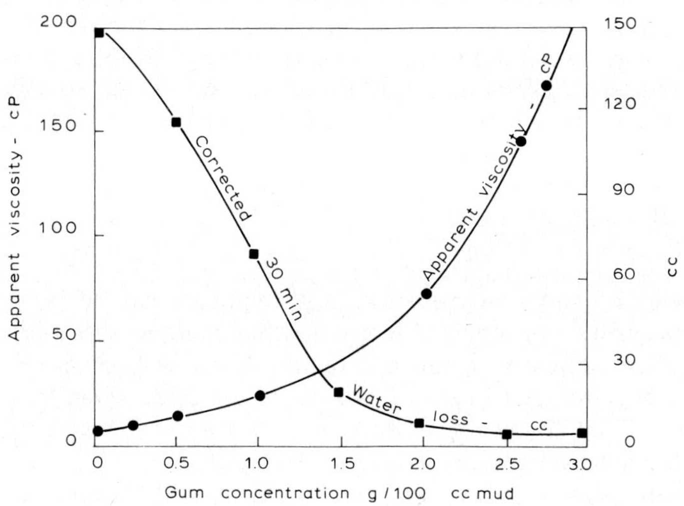

Fig. 6-20. Effect of gum guar HV on saturated salt-water mud. Aged for two days at 200°F; pH ~8.4. Viscosity was measured with Fann V-G Meter. (After Chilingar and Croushorn, 1964, fig. 5, p. 20; courtesy of Canadian Petro Engineering.)

seems to be an important factor in reducing the water loss of drilling fluids. The permeability of the average filter cake, resulting from the packing of clay particles, is of the order of one-thousandth of a millidarcy. Hence, if the gum molecules are large enough, they can plug off the pores and thereby reduce the water loss (Tchillingarian and Beeson, 1950, p. 2). In addition, the stabilizing action of protective colloid is due to the attachment of the colloid to the particles giving rise to a hydrophilic dispersion where each particle has a clay nucleus.

Source and chemistry of gum guar

Gum guar is derived from the seed of the guar plant, *Cyamopsis tetragonolobus*, family Leguminosa. It has grown for centuries in Pakistan and India, where it is one of the principal crops (A.M. Goldstein and E.N. Alter, p. 322, in Whistler, 1959). It is also grown in the Kennedy and Vernon areas of Texas. It is important to note here that there is no practical limit to the potential amount of gum guar which could be made available should the demand arise (Goldstein and Alter, in Whistler, 1959, p. 325).

The guar plant is a pod-bearing, nitrogen-fixing legume. The leguminous seed contains a high percentage of water-soluble materials, which are also soluble in alkali solutions. The gum (non-ionic, neutral polysaccharide) is a galactomannan. The molecules are apparently D-mannose and D-galactose copolymers having a high molecular weight. D-mannopyranose forms the principal chain in the polymer's molecules with β-(1 $\rightarrow$ 4) linkages, whereas D-galactopyranose forms regular side chains with an α-(1 $\rightarrow$ 6) linkage (Montecatini, 1960, p. 1). For every two D-mannopyranose units in the principal chain there is a D-galactopyranose unit as a side chain. The proposed structure of a guaran is given by Goldstein and Alter (in Whistler, 1959, p. 331), and its molecular weight is around 220,000. This molecular structure is apparently responsible for the remarkable water-solubility of gum guar both in low and high pH mediums.

Sodium-carboxy-methyl-cellulose

Long-chain molecules of sodium-carboxy-methyl-cellulose can be polymerized into different grades and lengths: high-viscosity, medium-viscosity, and regular-viscosity. This additive is more effective in water-base muds. Complete degradation of sodium-carboxy-methyl-cellulose occurs when temperatures are in excess of 250°F, and its effectiveness as a water loss control agent is reduced at salt concentrations exceeding 50,000 ppm.

The reduction in water loss may be attributed to (1) plating (filming) of clay particles by long-chain molecules, (2) formation of ball-like plugs as a result of curling-up of long-chain molecules, and (3) wedging and plugging action of long-chain molecules.

Polyanionic cellulose

Polyanionic cellulose is a long-chain polymer of high molecular weight. Small amounts of this additive can both impart viscosity to the drilling fluid and reduce water loss of either fresh- or salt-water muds. It is used successfully in low-solids muds, usually in conjunction with fiberized asbestos or attapulgite clay.

Sodium-polyacrylonitrile

Sodium-polyacrylonitrile is a high-molecular weight polymer, sometimes called sodium polyacrylate with a Magcobar Co. trade name of "Cypan". Long-chain molecules can be adsorbed on the individual edges of the clay particle and, thus, link it to the other particles (Fig. 6-21a). One should keep in mind, however, that if this polymer is added to a flocculated system (Fig. 6-21b), this additive will fail to control the fluid loss and it will be very difficult to disperse the system because the flocks are held together.

This polymer is added rapidly to the system in the pits in concentrations of at least 0.5 lb/bbl. This concentration should be maintained in the surface system at all times to reduce the viscosity hump that may occur in the case of high-solids muds.

This additive does not reduce fluid loss of drilling fluids which contain an excess of calcium salts. Consequently, calcium ion should be precipitated out by addition of chemicals prior to adding sodium-polyacrylonitrile. This additive is operative at high temperatures.

Detailed discussion on polymer drilling fluids appears in Chapter 11.

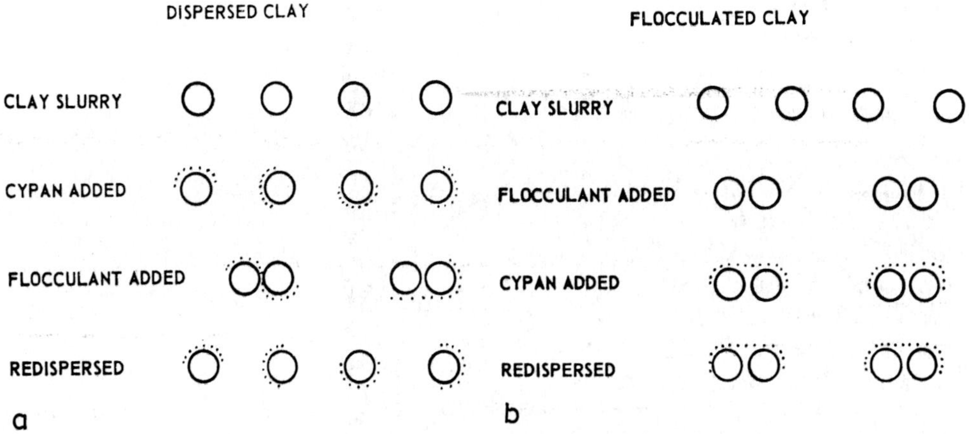

Fig. 6-21. Schematic diagram of behavior of sodium-polyacrylonitrile in dispersed and flocculated clay—water systems. (After Magcobar Services, 1978, figs. 9 and 10, p. 7.)

VARIOUS DRILLING FLUID PRODUCTS (Courtesy of IMCO Services)

Comparable drilling fluid products by tradenames are presented below.

COMPARABLE MUD PRODUCTS BY TRADENAMES

Description or Principal Component	IMCO SERVICES	Baroid	Magcobar	Milchem	Primary Application
WEIGHTING AGENTS AND VISCOSIFIERS					
Barite	IMCO BAR	Baroid	Magcobar	Mil-Bar	For increasing mud weight up to 20 ppg.
Calcium Carbonate	IMCO WATE		Lo-Wate	W.O. 30 W.O. 50	For increasing weight of oil muds up to 10.8 ppg.
Bentonite	IMCO GEL	Aquagel	Magcogel	Milgel	Viscosity and filtration control in water-base muds.
Sub-Bentonite	IMCO KLAY	Baroco	High Yield Blended Clay	Green Band Clay	For use when larger particle size is desired for viscosity and filtration control.
Attapulgite	IMCO BRINEGEL	Zeogel	Salt Gel	Salt Water Gel	Viscosifier in saltwater muds.
Beneficiated Bentonite	IMCO HYB	Quick-Gel	Kwik-Thik	Super-Col	Quick viscosity in fresh-water, upper-hole muds with minimum chemical treatment.
Asbestos Fibers	IMCO SHURLIFT	Flosal	Visquick	Flosal	Viscosifier for fresh-water or saltwater muds.
Bacterially Produced Polymer	IMCO XC	XC Polymer	Duovis	XC Polymer	Viscosifier and fluid-loss control additive for low-solids muds.
Sepiolite	IMCO DUROGEL				Viscosifier in all water-base muds, especially high-temperature drilling fluids.
DISPERSANTS					
Sodium Tetraphosphate	IMCO PHOS	Barofos	Magco-Phos	Oil Fos	Thinner for low pH fresh-water muds where temperatures do not exceed 180°.
Sodium Acid Pyrophosphate	SAPP	SAPP	SAPP	SAPP	For treating cement contamination.
Quebracho Compound	IMCO QBT	Tannex	M-C Quebracho	Tanco	Thinner for fresh-water and lime muds.
Modified Tannin	DESCO	Desco	Desco	Desco	Thinner for fresh-water and saltwater muds alkalized for pH control.
Processed Lignite	IMCO LIG	Carbonox	Tann A Thin	Ligco	Dispersant, emulsifier and supplementary additive for fluid-loss control.
Causticized Lignite	IMCO THIN	CC-16	Caustilig	Ligcon	1-6 ratio caustic-lignite dispersant, emulsifier and supplementary fluid-loss additive.
Modified Lignosulfonate	IMCO VC-10	Q-Broxin	Spersene	Uni-Cal	Dispersant and fluid-loss control additive for water-base muds.
Blended Lignosulfonate Compound	IMCO RD-111				Blended multi-purpose dispersant, fluid-loss agent and inhibitor for IMCO RD-111 mud systems.
Chrome-Free Lignosulfonate	IMCO RD-2000			X-KB Thin	Dispersant and fluid-loss control additive for water base muds.
FLUID-LOSS REDUCERS					
Organic Polymer	IMCO PERMALOID	DEXTRID			Controls fluid loss in water-base systems.
Pregelatinized Starch	IMCO LOID	Impermex	My-Lo-Jel	Milstarch	Controls fluid loss in saturated salt water, lime and SCR muds.

COMPARABLE MUD PRODUCTS BY TRADENAMES

Description or Prinicpal Component	IMCO SERVICES	Baroid	Magcobar	Milchem	Primary Application
Sodium Carboxy-methyl Cellulose	IMCO CMC (Regular)	Cellex (Regular)	Magco CMC (Regular)	Milchem CMC (Med-Vis)	For fluid-loss control and barite suspension in water-base muds.
Sodium Carboxy-methyl Cellulose	IMCO CMC (Hi-Vis)	Cellex (Hi-Vis)	Magco CMC (Hi-Vis)	Milchem CMC (Hi-Vis)	For fluid-loss control and viscosity building in low-solids muds.
Polyanionic Cellulosic Polymer	DRISPAC	Drispac	Drispac	Drispac	Fluid-loss control additive and viscosifier in salt muds.
Polyanionic Cellulosic Polymer	DRISPAC SUPERLO	Drispac Superlo	Drispac Superlo	Drispac Superlo	Primary fluid-loss additive, secondary viscosifier in water-base muds.
*Sodium Polyacrylate	IMCO SP-101				Fluid-loss control in calcium-free low solids and nondispersed muds.

LUBRICANTS, DETERGENTS, EMULSIFIERS, SURFACTANTS

Description or Prinicpal Component	IMCO SERVICES	Baroid	Magcobar	Milchem	Primary Application
Extreme Pressure Lubricant	IMCO EP LUBE	EP Mud Lube	Bit Lube	Lubri-Film	Used in water-base muds to impart extreme pressure lubricity.
Processed Hydrocarbons	SOLTEX	Soltex	Soltex	Soltex	Used in water-base muds to lower downhole fluid loss and minimize heaving shale.
Water Dispersible Asphalts	IMCO HOLECOAT		STABIL-HOLE	ITI-WD	Lubricant and fluid-loss reducer for water-base muds that contain no diesel or crude oil.
Oil Dispersible Asphalts	IMCO MUD OIL	Baroid Asphalt	Pave-A-Hole	Carbo-Seal	Lubricant and fluid-loss reducer for water-base fluids that contain diesel or crude oil.
Oil Soluble Surfactants	IMCO FREEPIPE	Skot-Free	Pipe Lax	Petrocote	Nonweighted fluid for spotting to free differentially stuck pipe.
Detergent	IMCO MD	Con Det	D-D	Milchem MD	Used in water-base muds to aid in dropping sand. Emulsifies oil, reduces torque and minimizes bit-balling.
Blend of Anionic Surfactants	IMCO SWS	Trimulso	Salinex	Atlosol & Atlosol S	Emulsifier for saltwater and fresh-water muds.
An Organic Entity Neutralized with Amines	IMCO LUBRIKLEEN	Torq Trim	DOS-3	Mil-Plate 2	Supplies the lubricating properties of oils without environmental pollution.
Blend of Fatty Acids, Sulfonates, and Asphaltic Materials	IMCO SPOT	SF 100			Invert emulsion that may be weighted to desired density for spotting to free differentially stuck pipe.

DEFOAMERS, FLOCCULANTS, BACTERICIDES

Description or Prinicpal Component	IMCO SERVICES	Baroid	Magcobar	Milchem	Primary Application
Aluminum Stearate	Aluminum Stearate	Aluminum Stearate	Aluminum Stearate	Aluminum Stearate	Defoamer for lignosulfonate muds.
Liquid Surface-Active Agent	IMCO DEFOAM L				Defoamer for all water-base muds.
Surface-Active Dispersible Liquid Defoamer	IMCO FOAMBAN	W200 W300		LD-7	All-purpose defoamer.
Flocculating Agent	IMCO FLOC	Barafloc	Floxit	Separan	Used to drop drilled solids where clear water is desirable for a drilling fluid.
Blended Solutions	IMCO CIDE				Bactericide used to prevent fermentation.

*Cypan and WL-100 are sold by American Cyanamid and Rotary Engineering, respectively.

COMPARABLE MUD PRODUCTS BY TRADENAMES

Description or Principal Component	IMCO SERVICES	Baroid	Magcobar	Milchem	Primary Application
LOST CIRCULATION MATERIALS					
Fibrous Material	IMCO FYBER	Fibertex	Mud Fiber	Mil-Fiber	Filler as well as matting material.
Nut Shells:					
Fine	IMCO PLUG	Wall-Nut	Nut-Plug	Mil-Plug	Most often used to prevent lost circulation.
Medium	IMCO PLUG	Wall-Nut	Nut-Plug	Mil-Plug	Used in conjunction with fibers or flakes to regain lost circulation.
Coarse	IMCO PLUG	Wall-Nut	Nut-Plug	Mil-Plug	Used where large crevices or fractures are encountered.
Ground Mica:					
Fine	IMCO MYCA	Micatex	Magco-Mica	Milmica	Used for prevention of lost circulation.
Coarse	IMCO MYCA	Micatex	Magco-Mica	Milmica	Forms a good mat at face of wellbore.
Cellophane	IMCO FLAKES	Jel Flake	Cell-O-Seal	Milflake	Used to regain lost circulation.
Combination of granules, flakes, and fibrous materials of various sizes in one sack.	KWIKSEAL	Kwik-Seal	Kwik-Seal	Kwik-Seal	Used where large crevices or fractures are encountered.
CORROSION INHIBITORS					
Zinc Compound	IMCO SULF-X II				For use as a hydrogen sulfide scavenger in water-base and oil-base muds.
Liquid Corrosion Inhibitor	IMCO CRACK CHEK				Prevent stress cracking of drill strings in an H_2S environment.
A Catalyzed Ammoninium Bisulfite	IMCO XO_2	Coat 777	A-202		For use as an oxygen scavenger.
Filming Amine	IMCO X-CORR				Corrosion inhibitor.
Filming Amine	IMCO PERMAFILM				Corrosion inhibitor.
Organic Polymer	IMCO SCALECHEK				Scale inhibitor.
SPECIALTY PRODUCTS					
Bentonite Extender	IMCO GELEX	Benex	Benex	Benex	Increases yield of bentonite to form very low-solids drilling fluid.
Inhibiting Agent	IMCO IE PAC				Imparts high-temperature fluid-loss control, temperature stability and increased inhibition.
Synergistic Polymer Blend	IMCO POLY Rx		Resinex		Rheological stabilization and filtration control.
Biodegradable Surfactant	IMCO FOAMANT				Foaming agent in air or mist drilling.
High-Temperature Polymer	IMCO DRILTHERM				High-temperature fluid-loss control.
Multipurpose Polymer	IMCO POLYSAFE				Polymer for fluid-loss control.
COMMERCIAL CHEMICALS					
Sodium Chromate	Sodium Chromate	Sodium Chromate	Sodium Chromate	Sodium Chromate	Used in water-base muds to prevent high-temperature gelation.
Sodium Hydroxide	Caustic Soda	Caustic Soda	Caustic Soda	Caustic Soda	For pH control in water-base muds.
Sodium Carbonate	Soda Ash	Soda Ash	Soda Ash	Soda Ash	For treating out calcium sulfate in low pH muds.
Sodium Bicarbonate	Sodium Bicarbonate	Sodium Bicarbonate	Sodium Bicarbonate	Sodium Bicarbonate	For treating out calcium sulfate or cement in high pH muds.
Barium Carbonate	Barium Carbonate	Anhydrox	Barium Carbonate	Barium Carbonate	For treating out calcium sulfate (PH should be above 10 for best results).

COMPARABLE MUD PRODUCTS BY TRADENAMES

Description or Principal Component	IMCO SERVICES	Baroid	Magcobar	Milchem	Primary Application
Calcium Sulfate	Gypsum	Gypsum	Gypsum	Gypsum	Source of calcium for formulating gyp muds.
Calcium Hydroxide	Lime	Lime	Lime	Lime	Source of calcium for formulating lime muds.
Sodium Chloride	Salt	Salt	Salt	Salt	For saturated salt muds and resistivity control.
Chrome Alum (chromic chloride)	Chrome Alum	Chrome Alum	Chrome Alum	Chrome Alum	For use in cross-linking XC Polymer systems.
OIL-MUD ADDITIVES					
Primary Emulsifier	IMCO KENOL-S (L) and IMCO KEN-X Conc. 1 (L)	Invermul	Vertoil	Carbo-Tec (D) and Carbo-Tec (L)	Primary additives to form stable water-in-oil emulsion.
Viscosifier and Gelling Agent	IMCO KEN GEL and IMCO KEN-X Conc. 2	Gel-Tone and # Petro-Tone	VG-69	Carbo-Gel	Provides viscosity, weight suspension, and filtration control.
High-Temperature Stabilizer	IMCO KEN-X Conc. 3	Duratone HT	DV-22 and DV-33		Improves emulsion under high-temperature conditions.
Stabilizes Borehole Conditions	IMCO VR				Stabilizes running shale, improves emulsion, weight suspension, and fluid loss under high-temperature conditions.
Dispersant	IMCO KEN THIN	OMC and Driltreat	Verthin and SE-11		Dispersant for reducing rheological properties.
Calcium Oxide	IMCO KENOX				Calcium source for saponification.
Fatty Acid Emulsifier	IMCO KEN SUPREME Conc. A		Oilfaze		Primary emulsifier and stabilizer for oil-base drilling fluids.
Emulsion Stabilizer	IMCO KEN SUPREME Conc. B		DG-55		Imparts gels, contributes to viscosity for weight suspension, and provides filtration control.
Specially Modified Saponified Fatty Acid Chemicals	IMCO KEN PAK	SF-100			Gelling agent for formulating high-gelation casing packs.
Powdered Wetting Agent	IMCO KEN CAL				Dispersing agent in KEN-X systems with a $CaCl_2$ water internal phase.

REFERENCES

Beeson, C.M., Chilingar, G.V. and Larsen, D.H., 1958. *Drilling Fluids Laboratory*. Pet. Eng. Dep. Publ., University of Southern California, 73 pp.

Bissell, H.J. and Chilingar, G.V., 1967. Classification of sedimentary carbonate rocks. In: G.V. Chilingar, H.J. Bissell and R.W. Fairbridge (Editors), *Carbonate Rocks. Origin, Occurrence and Classification. Dev. Sedimentol.*, 9A. Elsevier, Amsterdam, pp. 87—168.

Bradley, B.W., 1970. Oxygen: cause of drill-pipe erosion. *Pet. Eng.*, 42 (12): 50—57.

Browning, W.C. and Perricone, A.C., 1962. *Lignosulfonate Drilling Mud Conditioning Agents.* Fall Meet., Soc. Pet. Eng. AIME, Oct. 2—10, SPE 432: 20 pp.

Carney, L.L., Skelly, W.G. and Gullickson, R., 1971. *Quantitative Determination of Lignosulfonates in Drilling Fluids by Ultraviolet Adsorption Analysis.* Fall Meet, Soc. Pet. Eng. AIME, Oct. 3—6, SPE 3531: 10 pp.

Chesser, B.G., 1968. *Corrosivity of Chrome Treated Sodium Lignosulfonate Packer Fluid System.* Spring Meet., Southwestern Distr. API, March 20—22: 15 pp.

Chilingar, G.V., 1957. Evaluation of clays for drilling fluids. *Compass,* 34 (3): 174—85.

Chilingar, G.V. and Croushorn, A.L., 1964. Gum guar proves useful as mud additive. *Can. Pet. Eng.*, 5 (11): 19—21.

Chillingar, G.V., Bissell, H.J. and Fairbridge, R.W., 1967. *Carbonate Rocks. Physical and Chemical Aspects. Dev. Sedimentol.*, 9B. Elsevier, Amsterdam, 413 pp.

Fancher, G. and Whiting R.L., 1943. Response of a Gulf Coast drilling mud to chemicals, temperature and heat treatment. *Trans. AIME*, 151: 233—247.

Francis, W., 1954. *Coal, its Formation and Composition.* Edward Arnold, London.

Garrison, A.D. and Ten Brink, K.C., 1940. A study of some phases of chemical control in clay suspensions. *Trans. AIME,* 136: 175—194.

Garrison, A.D., Ten Brink, K.C. and Elkin, P.B., 1940. Dispersion of clays and shales by fluid motion. *Trans. AIME*, 136: 195—205.

Hollingworth, W.D. and Lockhent, D.H., 1975. *Environmental Aspects of Chemical Use in Well Drilling Operations.* EPA Publ. No. 56051-75-004: 200 pp.

Loomis, A.G., Ford, T.E. and Fidiam, J.F., 1941. Colloidal chemistry of clay drilling fluids. *Trans. AIME,* 142: 86—97.

Magcobar Services, 1972. *Drilling Fluid Engineering Manual.* Dresser Industries, Inc., Houston, Texas.

Melrose, J.C. and Lilienthal, W.B., 1952. Meter aids exact mud control. *World Oil,* July 1: 136.

Metzner, H., 1973. *Biochemie der Pflanzen.* Enke, Stuttgart, 150 pp.

Montecatini, I'Instituto di Ricerche "G. Donegani", 1960. *Report on the Use of Dealca HV for Drilling Fluids.* Sezione Geochimica, Novara, Rep. 60066: 14 pp.

Ray, R.F. and Shinn, L.D., 1975. *Environmental Aspects of Chemical Use in Well Drilling Operations.* EPA Publ. No. 56051-75-004, May, Houston, Texas, 200 pp.

Rogers, W.F., 1962. *Composition and Properties of Oil Well Drilling Fluids,* 2nd ed. Gulf Publ. Co., Houston, Texas, 818 pp.

Simpson, J.P. and Barbee, R.D., 1970. How corrosive are water base completion muds. *Pet. Eng.*, 42 (1, Dec.): 42—48.

Skelly, W.G. and Dieball, D.E., 1969. *Behavior of Chromate in Drilling Fluids Containing Chromate.* Fall Meet., Soc. Pet. Eng. AIME, Sept. 28—Oct. 1, SPE 2539: 6 pp.

Tchillingarian, G., 1952a. Study of the dispersing agents. *J. Sediment. Petrol.*, 22 (4): 229—233.

Tchillingarian, G., 1952b. Possible utilization of electrophoretic phenomenon for separation of fine sediments into grades. *J. Sediment. Petrol.*, 22 (1): 29—32.

Tchillingarian, G. and Beeson, C.M., 1950. *Iranian Gums as Treating Agents for Drilling Fluids.* Pet. Branch of AIME Meet., Los Angeles, Calif., Oct. 12—13, No. 78-G: 8 pp.

Tchillingarian (also Chilingar), G. and Beeson, C.M., 1952. Natural organic colloids of Iranian origin as mud treating agents. *Pet. Eng.*, 24 (4): B45—B52.

Tissot, B.P. and Welte, D.H., 1978. *Petroleum Formation and Occurrence, A New Approach to Oil and Gas Exploration.* Springer, Berlin—Heidelberg—New York, 538 pp.

Whistler, R.L. (Editor), 1959. *Industrial Gums, Polysaccharides and Their Derivatives.* Academic Press, New York—London, 766 pp.

Chapter 7

RHEOLOGY

RYEN CAENN*, FRANK R. BAGSHAW, SAMUEL C. MURRAY and
GOERGE V. CHILINGARIAN

INTRODUCTION

Rheology is derived from the Greek words *rheo*, meaning flow, and *logi*, meaning science. It can be defined as the science of the deformation and flow of solids, liquids, and gases. Rheology, in essence, deals with the stress—strain—time relationships of any matter. The action of a force (i.e., a stress, e.g., pump output) is discussed in this chapter. The chapter is further limited to a discussion of single-phase flow of non-pneumatic drilling fluids (i.e., liquid and not air or gas drilling fluids) and is focussed on the practical applications of the "state of the art" of today's drilling fluids.
of the "state of the art" of today's drilling fluids.

Rheological data can be evaluated and used to control fluid flow without requiring a thorough knowledge of rheological theory. Unfortunately, the rheological data available are often misinterpreted or under-utilized during data evaluation. Familiarity with the basic theory of rheological properties, especially as it applies to drilling fluids, makes the available data more useful on the drilling rig. Ideally, familiarity with rheological properties will stimulate the gathering of more data, which could ensure more efficient drilling operations.

IMPORTANCE OF RHEOLOGY IN DRILLING OPERATIONS

The major applications of rheological properties for evaluating drilling fluid behavior are in solving problems of (1) hole cleaning and hole erosion, (2) suspension of cuttings, (3) hydraulic calculations, and (4) mud treatment. Although each one of these applications is of equal importance, drilling requirements vary with time so that one problem may take precedence over another at a particular time. For example, in drilling a borehole with unstable shale, hole cleaning and erosion problems are of prime importance; whereas, in areas of lost circulation, pressure drop calculations to obtain the equivalent circulating density (ECD) are more important.

There are certain aspects of rheology that need to be understood to make optimum use of collected data. Relationship between shear stress and shear

---

*Formerly: Robert D. Carico.

rate and its effects on properties of drilling fluids is an important aspect discussed in the following section.

RELATIONSHIP BETWEEN SHEAR STRESS AND SHEAR RATE

The concepts involving shear stress and shear rate and their measurement enables the mathematical description of the flow of drilling fluids. The amount of force applied to a fluid determines the shear rate, which in oil-field terms is determined by the flow rate of the fluid through a particular geometrical configuration. Resistance of the fluid to the applied rate of shear or force is called the shear stress, which in oil-field terms is analogous to the pump pressure. Physically, the shear rate relates to particles or planes of the fluid moving or sliding past one another, as illustrated in Fig. 7-1 (see Eirich, 1956, p. 21), which depicts a fluid flowing in a pipe. In this case, the fluid velocity is 1.0 cm/sec at a distance of 1.0 cm from the wall of the pipe. Inasmuch as, theoretically, at the pipe wall the velocity will be zero, the fluid velocities will vary between 0 and 1.0 cm/sec. The ratio of velocity to distance from the pipe wall (velocity gradient) is called the shear rate. If velocity is expressed in cm/sec and distance in cm, then the units of shear rate is in 1/sec (sec^{-1}) (see Fig. 7-1). The magnitude of the shear rate depends on the geometrical configuration of the pipe (or annulus, i.e., area between two concentric pipes), the overall velocity of the fluid, and the viscous properties of the fluid. The shear rate, $\dot{\gamma}$, is defined as follows:

$$\dot{\gamma} = dv/dr \tag{7-1}$$

where v is the velocity and r is the distance from the pipe wall. The shear

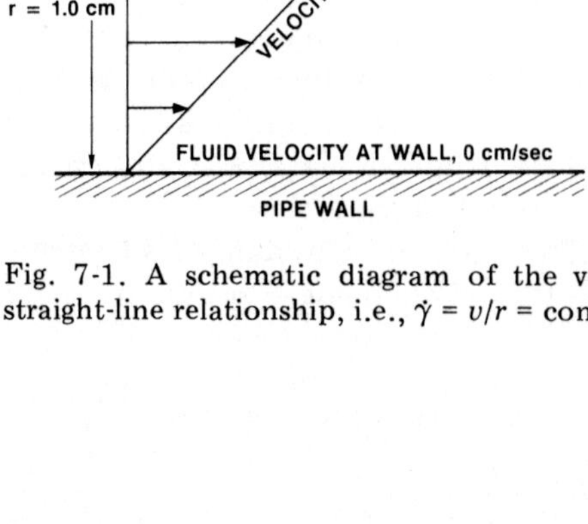

Fig. 7-1. A schematic diagram of the velocity gradient in a flowing fluid assuming a straight-line relationship, i.e., $\dot{\gamma} = v/r$ = constant, which in this case is equal to 1.0 sec^{-1}.

stress, τ, is defined as follows:

$$\tau = F/A \quad (7\text{-}2)$$

where F is the applied force and A is the surface area exposed to that force.

Shear rate and shear stress are the two basic measurements that are widely utilized in the petroleum industry and their relationship defines the type of fluid flow.

MEASUREMENT OF SHEAR-STRESS—SHEAR-RATE RELATIONSHIP

The most common instrument for measuring relationship between the shear stress and shear rate in drilling fluids is a cup-and-bob viscometer, which was first described by Melrose and Lilienthal (1950) (see also Chapter 4). A representation of a typical, concentric-cylinder viscometer used in the oil field is presented in Fig. 7-2 (API Bull. 13D, tentative). Drilling-fluid sample is first placed between the rotor and bob. As the rotor is turned at a constant speed, the fluid's resistance to flow imparts a torque on the bob that deflects the dial proportionally to the viscous properties of the fluid. The magnitude of dial deflections is a function of (1) the strength of the bob spring, and (2) the viscous properties of the fluid being tested. More versatile

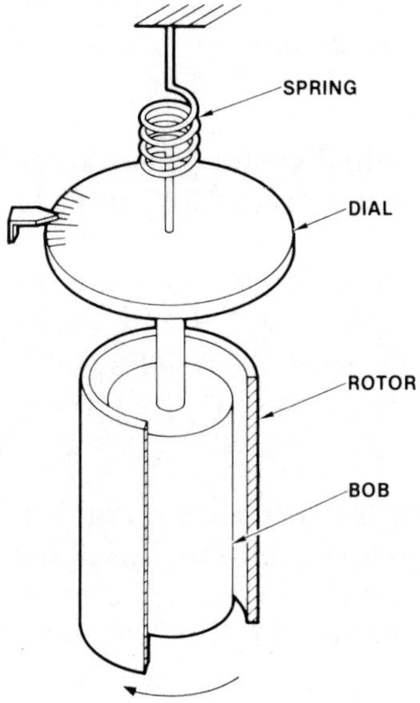

Fig. 7-2. Schematic diagram of a concentric cylindrical viscometer (after API Bulletin 13D).

viscometers have springs that can be changed to accomodate thicker (e.g., cement slurries) or thinner (e.g., polymer muds) fluids. Sophisticated instruments are connected to a transducer which records electrical impulses and displays them digitally or on a pin recorder.

The specific geometrical configuration of rotor and bob determines the shear rates obtainable with these viscometers. In addition, the shear rate is dependent, to some extent, upon the viscous properties of the fluid. For the standard rotor bob combination (i.e., 0.1170 cm gap), the conversion from rpm to shear rate (in sec^{-1}) is given by the following formula (Fann Instrument Corp., personal communication):

$$\dot{\gamma}(sec^{-1}) = rpm \times 1.7034 \qquad (7\text{-}3)$$

The shear stress, τ, from this concentric-cylinder viscometer is taken from the dial reading, R, which is in degrees of a circle. The formulas for converting R to τ are as follows (Fann Instrument Corp., personal communication):

$$\tau = (lb/100\ ft^2) = R \times 1.067 \qquad (7\text{-}4)$$

$$\tau = (dynes/cm^2) = R \times 5.1 \qquad (7\text{-}5)$$

Field viscometers

The most common viscometer in the field is the two-speed model. Fann Instrument Corp., Houston, Texas, manufactures two models: the 34A motor-driven model and the HC34A hand-cranked model. Baroid Petroleum Services Division, NL Industries, Inc., Houston, Texas, manufactures a model 280 two-speed viscometer (rheometer) which is hand cranked (see Chapter 4).

These viscometers measure the shear stress of the fluid at two fixed speeds of 600 and 300 rpm, which impart to the mud shear rates of 1022 and 511 sec^{-1}, respectively (calculated from eq. 7-3). For field use, shear stress is usually read directly (dial reading) in lb/100 ft^2, without the necessity of correcting by use of eq. 7-4. The correction factor (see eqs. 7-4 and 7-5) is normally ignored because it is considered so small as to be insignificant.

Multi-speed viscometers

Measuring the shear stress at only two shear rates has certain limitations, because it does not describe the rheological properties accurately over the shear rate ranges encountered in a mud-circulating system. For this reason, multi-speed viscometers have been developed to measure the fluids at other rates.

A common multi-speed viscometer is the six-speed Fann model 35A. The gear system allows the operator to use six different speeds (3, 6, 100, 200, 300, and 600 rpm) for shear stress measurement. This viscometer utilizes the same rotor, bob, dial, and spring arrangement as the previously discussed two-speed viscometers. The speeds of the model 35A viscometer and their corresponding shear rates in sec^{-1} (calculated from eq. 7-3) are as follows: 600 rpm (1022.0 sec^{-1}), 300 rpm (511.0 sec^{-1}), 200 rpm (340.7 sec^{-1}), 100 rpm (170.3 sec^{-1}), 6 rpm (10.22 sec^{-1}), and 3 rpm (5.11 sec^{-1}).

NL Baroid developed the model 286 viscometer, with a continuously variable speed from 2 to 600 rpm. It also has a switch for selection of the same speeds as the Fann model 35A viscometer. Both the Fann and the N.L. Baroid viscometers are rugged enough for field use. Although rpm readout of Baroid's variable-speed instrument is digital, it incorporates the same spring, dial, rotor, and bob arrangement as the previously described viscometers.

In order to obtain accurate correlations, because the dial readings obtained for most fluids can vary markedly with temperature changes (Annis, 1967; Bartlett, 1967), it is important to take the readings at the same temperature.

NEWTONIAN FLUIDS

One of the first attempts to describe fluid flow was made by Isaac S. Newton in his classical book entitled, "Philosophiae Naturalis Principia Mathematica", in which he stated: "The resistance which arises from the lack of slipperiness originating in a fluid — other things being equal — is proportional to the velocity by which the parts of the fluid are being separated from each other" (Van Wazer et al., 1963, frontispiece). In oil-field terms, viscosity is the resistance of fluid to flow, i.e., the shear stress, τ, of a fluid is proportional to the shear rate, $\dot{\gamma}$:

$$\tau = (\text{constant})\dot{\gamma} \tag{7-6}$$

Newton defined the constant in eq. 7-6 as the coefficient of viscosity, μ (Van Wazer et al., 1963, p. 13):

$$\tau = \mu\dot{\gamma} \tag{7-7}$$

This equation describes a Newtonian fluid, e.g., water, salt waters, glycerine, ink, silicon oils, honey, etc. The most important concept involved is that a Newtonian fluid has a constant viscosity at all shear rates and is calculated by using the following equation:

$$\mu = \tau/\dot{\gamma} \tag{7-8}$$

This is a straight-line relationship for Newtonian fluids as shown in Fig. 7-3 (straight line passes through the origin). The slope of this straight line is the coefficient of viscosity, μ. Typical plots of a thick and a thin Newtonian fluid are presented in Fig. 7-3. Inasmuch as the units for shear stress are dynes/cm^2 or lb/100 ft^2 and the units for the shear rate are in sec^{-1}, viscos-

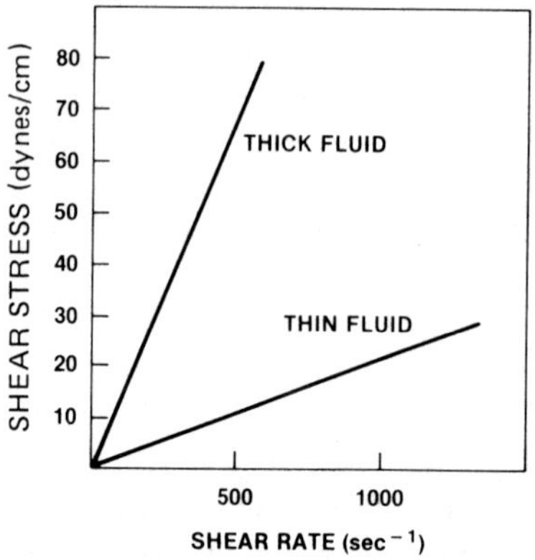

Fig. 7-3. Examples of relationships between shear stress and shear rate for Newtonian fluids.

ity is expressed (see eq. 7-8) in dynes-sec/cm^2 or lb-sec/100 ft^2. One poise (P) is equal to one dyne-sec/cm^2. The more common unit of viscosity, however, is the centipoise (cP), which is equal to 1/100 of a poise. The following equation is used to convert from lb/100 ft^2 to dynes/cm^2:

$$5.1 \times \text{lb}/100\ \text{ft}^2 = \text{dynes/cm}^2 \qquad (7\text{-}9)$$

Example:

If a 300-rpm Fann viscometer dial reading is 32, then the shear stress obtained by using eq. 7-5 is equal to:

$$\tau = 32 \times 5.1 = 163.2\ \text{dynes/cm}^2$$

Using eq. 7-3, shear rate at 300 rpm is equal to:

$$\dot{\gamma} = 300 \times 1.7034 = 511\ \text{sec}^{-1}$$

The viscosity of this fluid (see eq. 7-8) is thus equal to:

$$\mu = \frac{163.2}{511} = 0.32\ \text{P}$$

or:

$$\mu = 0.32\ \text{P} \times 100 = 32\ \text{cP}$$

This fluid would have viscosity of 32 cP at all other shear rates if it was Newtonian. For example, if at 600 rpm the dial reading of this fluid was equal to 64, the shear stress then is equal to:

$$\tau = 64 \times 5.1 = 326\ \text{dynes/cm}^2$$

The shear rate from eq. 7-3 is equal to:

$\dot{\gamma} = 600 \times 1.7034 = 1022\ \text{sec}^{-1}$

Thus, the viscosity is:

$$\mu = \frac{327}{1022} = 0.32\ \text{P or } 32\ \text{cP}$$

NON-NEWTONIAN FLUIDS

The shear-rate/shear-stress ratio of a non-Newtonian fluid is not constant, which is true for most drilling fluids. In Fig. 7-4, lines A and B represent low- and high-viscosity Newtonian fluids, whereas curves C, D, and E represent non-Newtonian drilling fluids. The two most popular mathematical models for describing non-Newtonian drilling fluids are called the Bingham-plastic model and the Power Law model.

Bingham-plastic model

The Bingham-plastic model is being successfully used to describe drilling fluids (line C of Fig. 7-4). As shown in Fig. 7-4, a Bingham-plastic fluid is similar to a Newtonian fluid in that both exhibit straight-line relationship between shear stress and shear strain. The Bingham-plastic fluid, however, has an intercept on the shear-stress axis, which is called the yield point ($\overline{YP}$). The slope of shear stress versus shear strain line is the plastic viscosity ($\overline{PV}$).

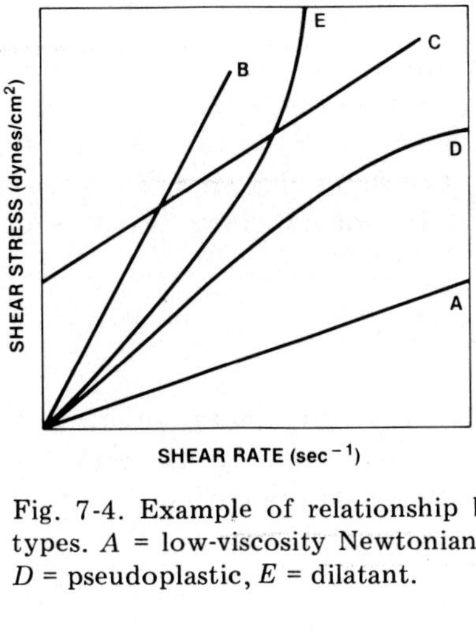

Fig. 7-4. Example of relationship between shear stress and shear rate for various fluid types. A = low-viscosity Newtonian, B = high-viscosity Newtonian, C = Bingham-plastic, D = pseudoplastic, E = dilatant.

TABLE 7-I

Shear rate ranges encountered in a drilling fluid at various locations

Location	$\dot{\gamma}$ (sec^{-1})	Equivalent rpm
Bit	10,000—1,000,000	none
Solid removal equipment and inside drillpipe and collars	170—10,000	100 to none
Annulus	5—170	3—100
Pits	<1—3	none to 3

Thus, mathematically, a Bingham-plastic fluid can be expressed by the following equation:

$$\tau = \overline{PV}\dot{\gamma} + \overline{YP} \tag{7-10}$$

Bingham-plastic fluid is additionally defined as a fluid the yield point and plastic viscosity of which are the intercept and slope, respectively, of a line drawn through the shear-stress measurements at 600 and 300 rpm on the standard viscometer. The concepts of measuring plastic viscosity and yield point of drilling fluids and description of the rotational viscometer was presented by Melrose and Lilienthal in 1951 and later by Savins and Roper in 1954. The main advantage of using the Bingham-plastic model is the simplicity of calculating *PV* and *YP*. The plastic viscosity is obtained by subtracting the 300-rpm dial reading from the 600-rpm dial reading, whereas the yield point is obtained by subtracting the *PV* from the 300-rpm dial reading:

$$\overline{PV} = R_{600} - R_{300} \tag{7-11}$$

$$\overline{YP} = R_{300} - PV \tag{7-12}$$

The major shortcoming of using the Bingham-plastic model is that it only describes the fluid flow over a very narrow shear-rate range of 300 to 600 rpm, i.e., 511 to 1022 sec^{-1}. The shear-rate ranges encountered in a typical drilling situation are presented in Table 7-I. Consequently, the Bingham-plastic model may not accurately describe fluid rheological characteristics in all drilling situations.

Power Law model

The Power Law model more accurately describes flow characteristics of drilling fluid over the shear-rate ranges normally encountered in the annulus of a wellbore. Lines *D* and *E* in Fig. 7-4 show two examples of shear-stress versus shear-rate plots that follow the Power Law model, i.e.:

$$\tau = K\dot{\gamma}^n \tag{7-13}$$

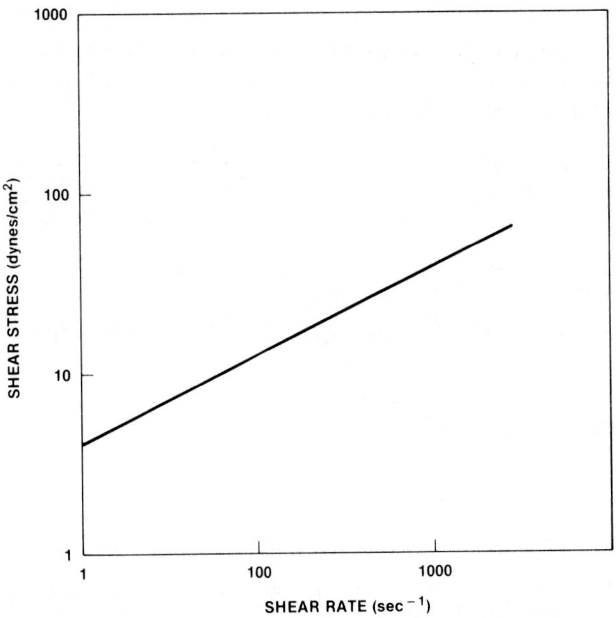

Fig. 7-5. Example of a log-log plot of shear rate versus shear stress for a pseudoplastic fluid.

For the special case of $n = 1$, eq. 7-13 becomes equal to eq. 7-7, i.e., fluid is Newtonian, with K being equal to μ. If n is less than one, represented by line D, the fluid is called pseudoplastic. Line E represents a case when n is greater than one (dilatant fluid). Inasmuch as dilatant fluids rarely occur in drilling operations, only pseudoplastic fluids ($n < 1$) are discussed here.

Description of n and K

Equation 7-13 can be presented in the following form:

$$\log \tau = n \log \dot{\gamma} + \log K \tag{7-14}$$

A logarithmic plot of τ versus $\dot{\gamma}$, which is a straight line for a pseudoplastic fluid is shown in Fig. 7-5. The slope of the line is n, whereas K is the intercept of the line at $\dot{\gamma} = 1.0\ \text{sec}^{-1}$.

The term n can be considered as a measure of the degree of deviation of a fluid from Newtonian behavior. As n decreases, the fluid becomes more shear-thinning or pseudoplastic. The n may also be calculated by the following equation where R_{2w} and R_{1w} are the dial readings at 2w and 1w rpm, respectively:

$$n = \frac{\log_{10} R_{2w}/R_{1w}}{0.301} \tag{7-15}$$

The term K is called the consistency index and is indicative of the pumpability or overall thickness of the fluid, and is equal to the viscosity of the fluid at $\dot{\gamma} = 1.0 \text{ sec}^{-1}$. The K value may also be calculated by using the following equation if the R_{1w} reading is taken at 300 rpm:

$$K = \frac{5.1 R_{1w}}{511^n} \tag{7-16}$$

Example:

If dial readings at 600 and 300 Fann rpm are equal to 38 and 28, respectively, then using eqs. 7-15 and 7-16:

$$n = \frac{\log_{10} 38/28}{0.301} = 0.44 \qquad \text{and} \qquad K = \frac{(5.1)(28)}{(511)^{0.44}} = 9.18$$

In actual practice, a plot involving many different shear rates is usually a curved line. Consequently, a drilling fluid can have an infinite number of values for n and K, which are relevant only when associated with a specific shear rate. Two straight lines, one for low shear rate and one for high, adequately describe most drilling fluid conditions.

A slide-rule calculator that provides a simple and quick means for obtaining n and K values is available *. It can be used over the full range of shear rates from the available viscometers. This calculator can be used to determine n and K values for any two shear-stress measurements, providing the higher reading is taken at a shear rate twice the lower (see Fig. 7-6).

Sample calculations

Dial readings of a typical drilling fluid measured with Fann model 35 VG meter are 38, 28, 23, 17, 5.5, and 4.5 at 600, 300, 200, 100, 6, and 3 rpm. $\overline{PV}$ and $\overline{YP}$ are calculated as follows:

$$\overline{PV} = R_{600} - R_{300} = 38 - 28 = 10 \text{ cP}$$

$$\overline{YP} = R_{300} - PV = 28 - 10 = 18 \text{ lb/100 ft}^2$$

It should be remembered that $\overline{PV}$ and $\overline{YP}$ can only be calculated from 600- and 300-rpm readings (straight-line portion of the curve), whereas the Power Law parameters can be calculated using any two dial readings and eq. 7.13. In the latter case, it is necessary to change rpm to sec^{-1} ($\dot{\gamma}$) and dial readings to either lb/100 ft^2 or dynes/cm^2 (τ). These conversions are presented in Table 7-II. For 600- and 300-rpm readings, the calculations are as follows (see

---

*The calculator can be obtained from Kelco Rotary, P.O. Box 45286, Tulsa, OK 74145. This calculator has been revised so that n and K values can be determined for any two shear-stress measurements, regardless of the shear rates at which they were taken.

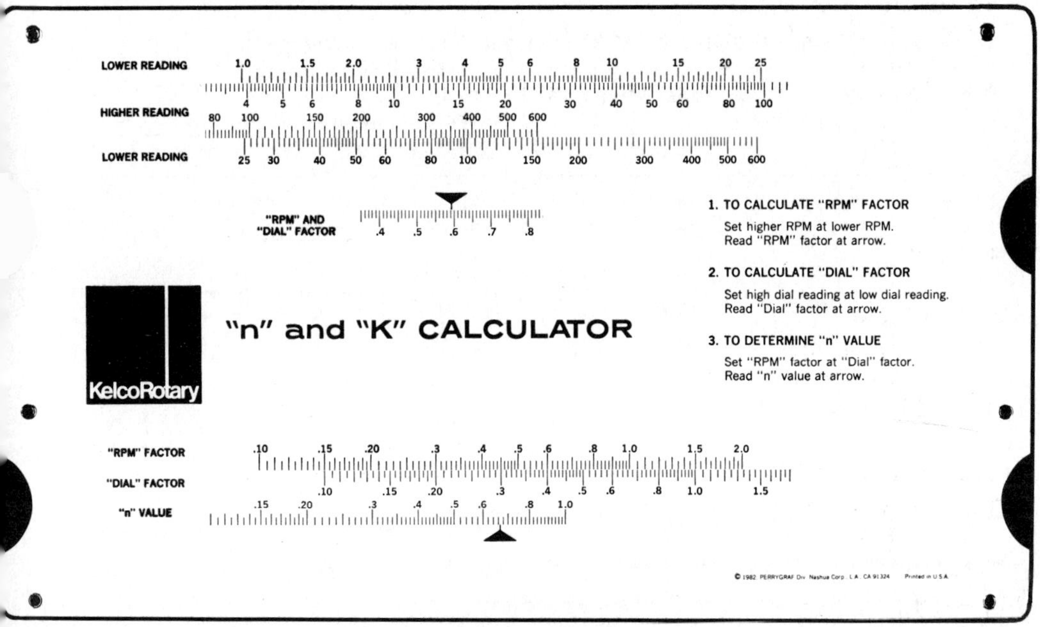

FRONT SIDE

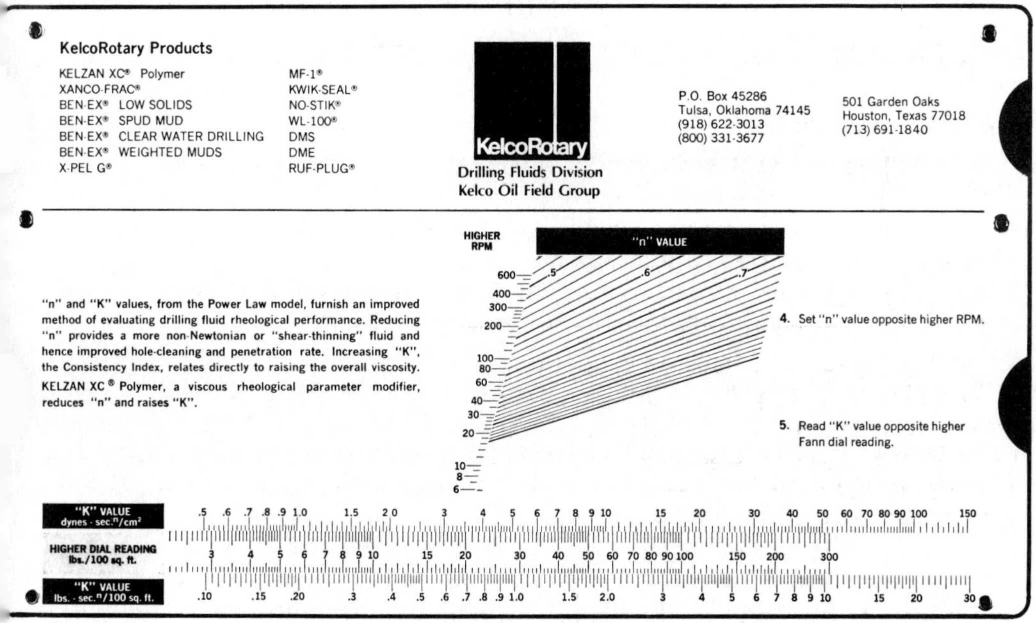

BACK SIDE

Fig. 7-6. New slide-rule calculator for n and "K" values.

TABLE 7-II

Conversion of Fann VG Meter model 35 shear rate and shear stress readings

rpm	$\dot{\gamma}$ (sec^{-1})	Dial reading	τ (lb/100 ft^2)	τ (dynes/cm^2)
600	1022	38	40.7	194
300	511	28	30.0	143
200	340	23	24.6	117
100	170	17	18.2	87
6	10.2	5.5	5.9	28
3	5.1	4.5	4.8	25

eq. 7-14):

$$\log (194 \text{ dynes/cm}^2) = n \log (1022 \text{ sec}^{-1}) + \log K$$

$$\log (143 \text{ dynes/cm}^2) = n \log (511 \text{ sec}^{-1}) + \log K$$

Solving these two equations simultaneously gives:

$n = 0.43$ and $K = 9.8$ dynes-sec/cm^2

Repeating the calculation for 200 and 100 rpm and 6 and 3 rpm gives:

$n(200/100) = 0.43$ and $K(200/100) = 9.1$ dynes-sec/cm^2

$n(6/3) = 0.29$ and $K(6/3) = 57$ dynes-sec/cm^2

FLOW PROPERTY ADJUSTMENT BY ADDITIVES

In general, it is difficult to determine the effect of specific mud additives outside of the context of the total mud system. The following general guidelines, however, can be used to alter the flow properties of mud systems with respect to the Bingham-plastic and Power Law models.

Bingham-plastic model

Changes in plastic viscosity can best be accomplished by adjustments to the base liquid of the mud, i.e., water or oil, or the solids content of the mud. Increasing the liquid content of the mud (dilution) decreases the plastic viscosity, assuming that the liquid being added has the same viscosity as the base liquid in the mud, whereas increasing the solids content normally results in an increased plastic viscosity. Similarly, increasing the viscosity of the base liquid increases the plastic viscosity. The yield point decreases as a result of dilution (increasing liquid content) or by addition of dispersants

TABLE 7-III

General effects of mud additive on rheological properties of clay-water muds

Additive	Properties *				
	$\overline{PV}$	$\overline{YP}$	n	K	Funnel viscosity
Bentonite	+	+	—	+	+
Drilled solids	+	0	0	0	+
Xanthan gum	0	+	—	+	0
Polyacrylates	+	+	0	+s	+
Cellulose fiber	+	+	0	+s	+
Thinners and dispersants	—	—	0	—	—
Salts	+	+s	0	+s	+

* + = increase, — = decrease, +s = increase slightly, and 0 = no effect.

and deflocculants, whereas it increases by addition of viscosifiers and flocculating agents.

Power Law model

Inasmuch as the K value is the viscosity of the fluid at $\dot{\gamma} = 1.0\ \text{sec}^{-1}$, any material that affects low shear-rate viscosities (or gel strengths) will affect K. The most effective materials for raising K values are bentonite and xanthan gum. Flocculation or beneficiation of bentonite also raises K, whereas deflocculation or dispersion lowers K. Dilution is the most effective method of lowering K. It is important to note that some materials, such as cellulose polymers, while increasing $\overline{YP}$, have little effect on increasing the K.

Most common mud additives lower the n of a water-base mud to some degree (water has an n value of 1.0). The n values below 0.5, however, usually can be reached only by adding bentonite or xanthan gum. Dilution and addition of solids usually cause an increase in the n value.

The general guidelines on the effects of most common mud additives on the flow properties of muds are presented in Table 7-III. Inasmuch as the chemistry of muds is very complex, these guidelines are obviously not absolute.

TIME-DEPENDENT VISCOSITY

Neither of the aforementioned models take into account the effect of time. Generally, most drilling fluids contain clays that exhibit thixotropic properties. Thixotropic fluids form gels upon quiescence (static condition) and regain their fluidity under dynamic conditions.

This formation and breakdown of gel structure is time dependent. Most

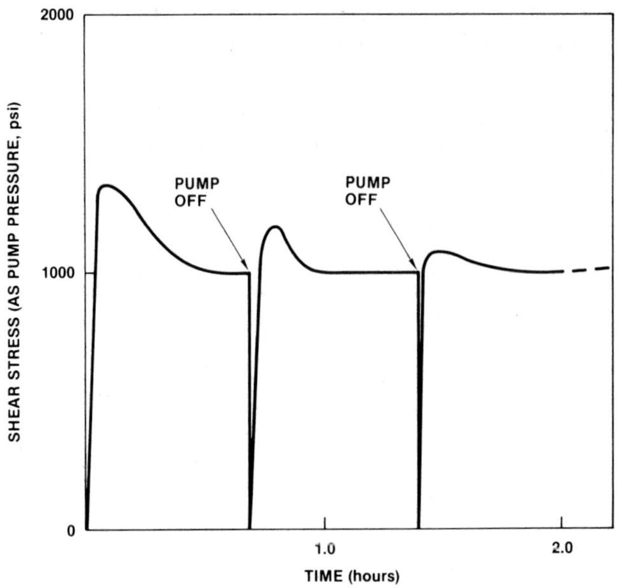

Fig. 7-7. Typical variation of shear stress values with time.

water-base muds exhibit this property because of electrically charged clay particles (edges are positively charged, whereas surfaces are charged negatively) that tend to link together (edge-to-surface contacts) at low shear rates. The strength of the gel structure formed is a function of the amount and type of solids, time, temperature, and chemical constituents of the mud.

Gel strengths are normally classified as progressive (strong) or fragile (weak). The less desirable progressive gel strength increases with time, whereas the fragile gel, even though it may be initially high, only increases slightly with time (see Fig. 7-7). Inasmuch as the latter gel is easily broken, the pump pressure required to start circulation is less. Gel strengths of drilling fluids are measured by a viscometer at approximately 3 rpm at two intervals of time (right after agitation and after ten minutes; see Chapter 4) and are reported in lb/100 ft^2.

EFFECTIVE VISCOSITY

There are many different geometrical configurations in the mud-circulating system (see Fig. 7-8, for example). A mud being pumped at a constant volume will experience different flow rates at these different shear rates. As previously described (Table 7-I), if the mud is non-Newtonian, the shear-rate/shear-stress ratio is not constant. In different zones, therefore, the mud has different viscosities, which are best termed effective viscosities, μ_e. The effective viscosity is defined as the viscosity exhibited by the fluid at the existing shear rate.

A log-log graph paper for plotting viscosity of the mud versus the shear rate is presented in Fig. 7-9. Vertical lines show the rpm equivalents of shear rates in sec^{-1} for the viscometers now in use in the drilling fluid industry. This graph is used by some oil-field engineers as a worksheet in the field. The bottom right corner contains spaces for the regular mud check data, whereas the upper right corner has spaces for recording and determining the viscosity at each viscometer rpm reading.

At 600 and 300 rpm only a very small portion of the shear-rate range of a drilling mud in a circulating system is being measured. Using only these two measurements can result in substantial calculation errors when attempting to extrapolate the data for other shear rates, as shown previously.

A viscosity versus shear rate plot for a drilling mud is shown in Fig. 7-10. As mentioned previously, drilling fluids are usually pseudoplastic, i.e., shear-

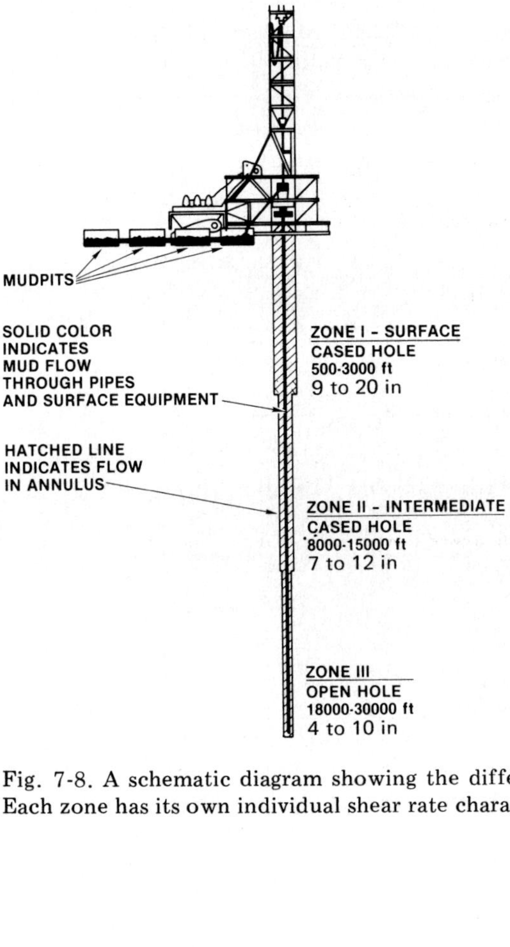

Fig. 7-8. A schematic diagram showing the different zones of a mud circulating system. Each zone has its own individual shear rate characteristic.

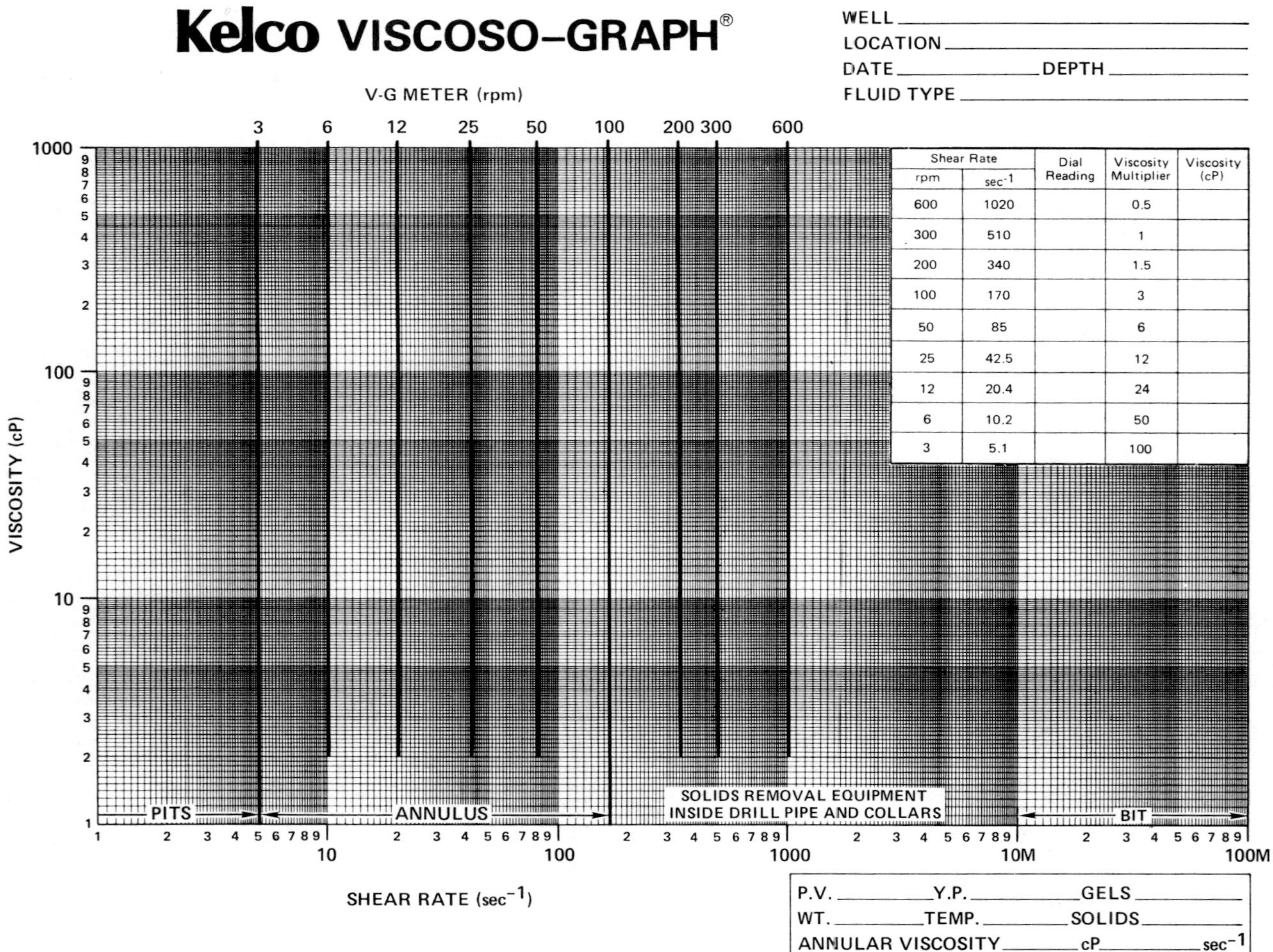

Shear Rate		Dial Reading	Viscosity Multiplier	Viscosity (cP)
rpm	sec^{-1}			
600	1020		0.5	
300	510		1	
200	340		1.5	
100	170		3	
50	85		6	
25	42.5		12	
12	20.4		24	
6	10.2		50	
3	5.1		100	

Fig. 7-9. Example of a worksheet used in the field for recording viscometric data and plotting viscosities vs. shear rate.

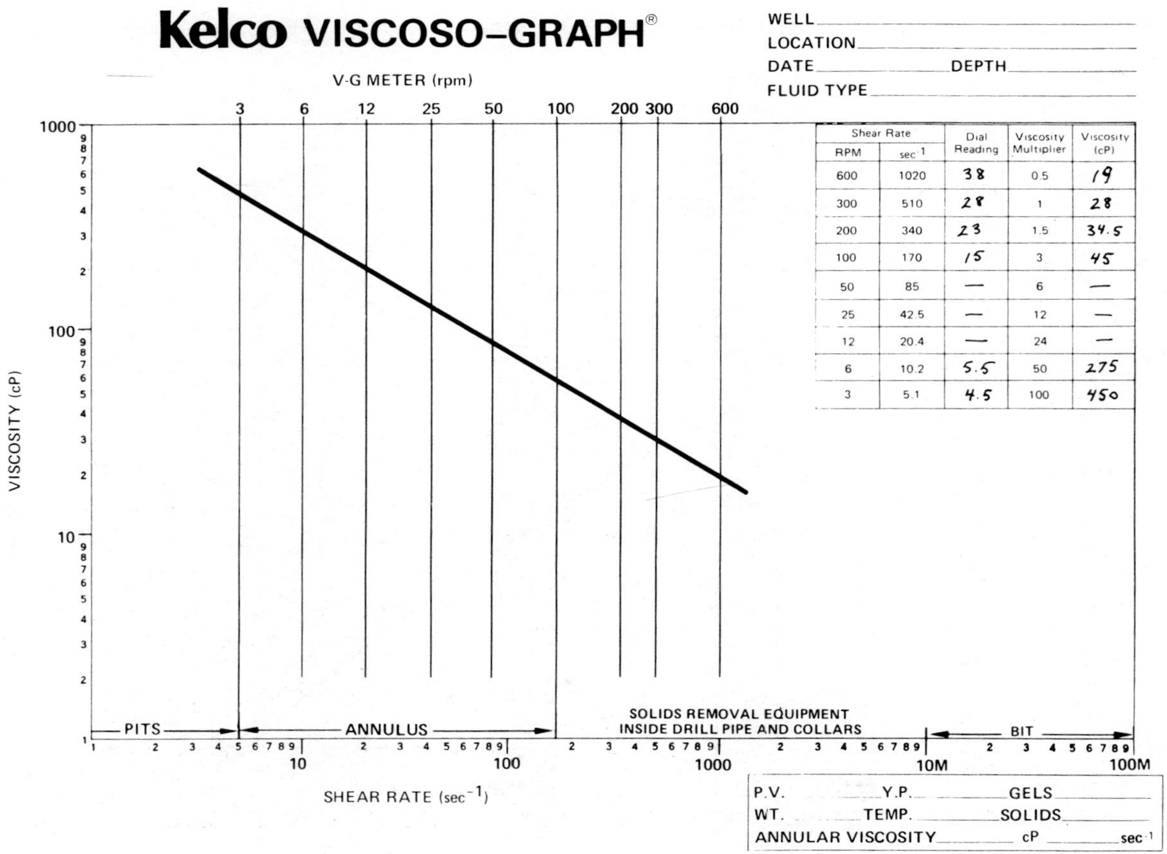

Shear Rate RPM	Shear Rate sec^{-1}	Dial Reading	Viscosity Multiplier	Viscosity (cP)
600	1020	38	0.5	19
300	510	28	1	28
200	340	23	1.5	34.5
100	170	15	3	45
50	85	—	6	—
25	42.5	—	12	—
12	20.4	—	24	—
6	10.2	5.5	50	275
3	5.1	4.5	100	450

Figs. 7-10. Example of actual plot of viscometric data on viscosograph showing a pseudoplastic fluid with a constant n.

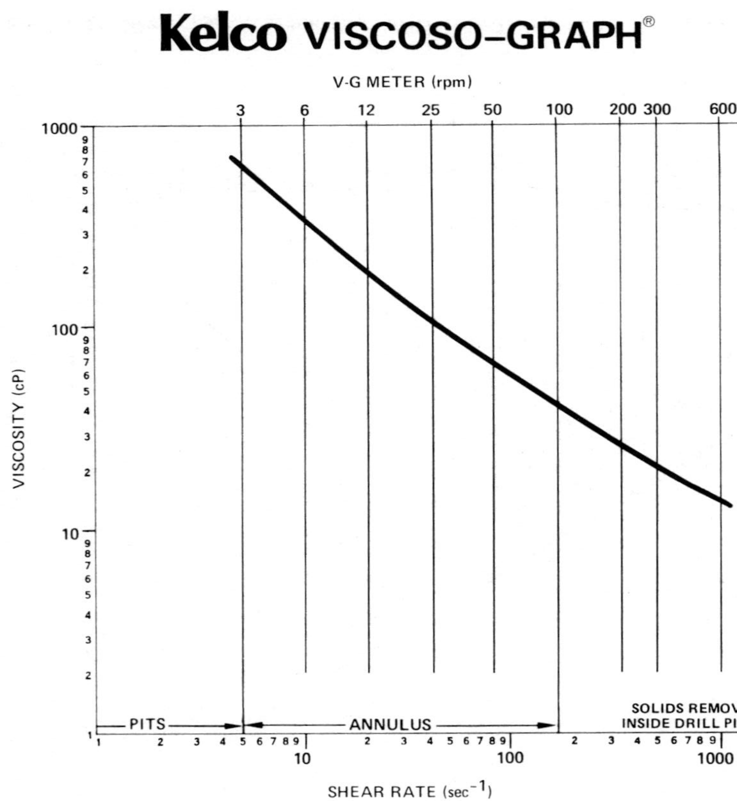

WELL
LOCATION
DATE DEPTH
FLUID TYPE

Shear Rate RPM	Shear Rate sec^{-1}	Dial Reading	Viscosity Multiplier	Viscosity (cP)
600	1020	27.0	0.5	13.5
300	510	20.5	1	20.5
200	340	17.8	1.5	26.7
100	170	14.2	3	42.6
50	85	—	6	—
25	42.5	—	12	—
12	20.4	—	24	—
6	10.2	7.2	50	360
3	5.1	6.6	100	660

P.V. Y.P. GELS
WT. TEMP. SOLIDS
ANNULAR VISCOSITY cP sec^{-1}

Fig. 7-11. Example of actual plot of viscometric data on viscosograph showing a pseudoplastic fluid with a variable n.

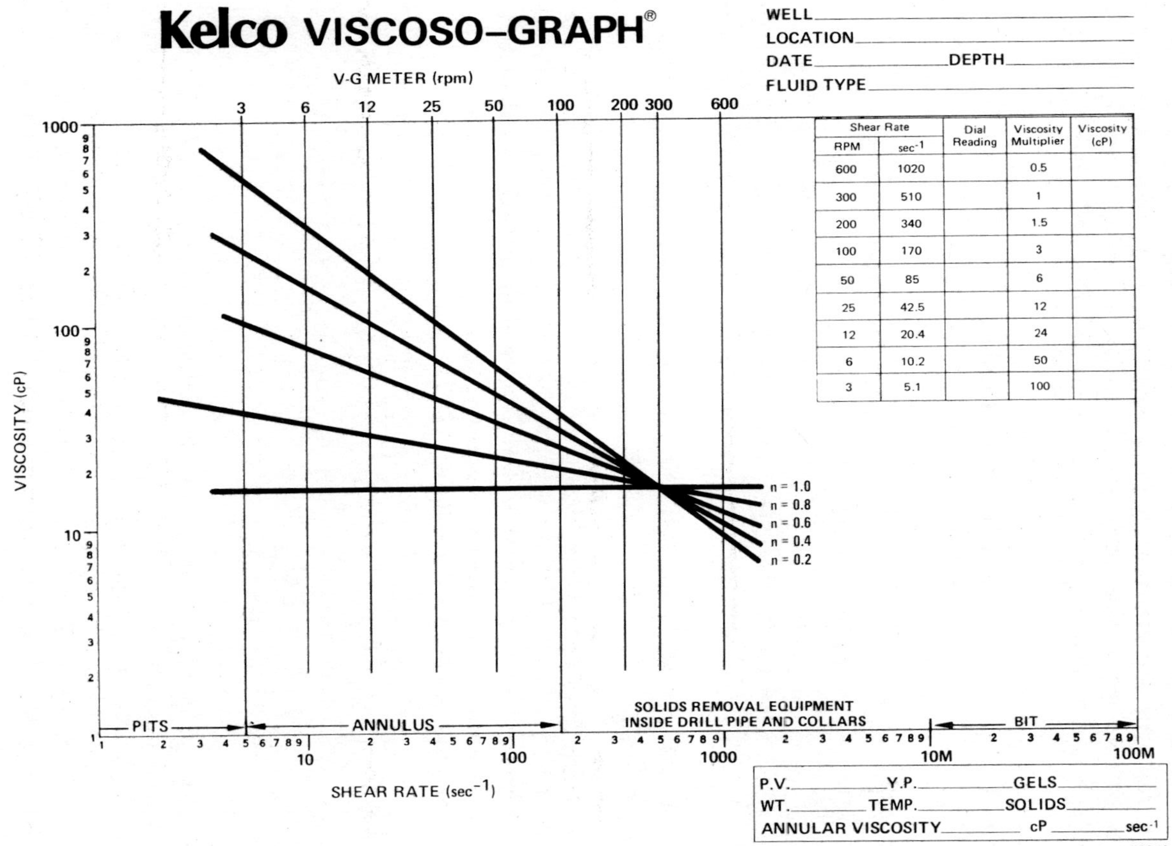

Shear Rate		Dial Reading	Viscosity Multiplier	Viscosity (cP)
RPM	sec^{-1}			
600	1020		0.5	
300	510		1	
200	340		1.5	
100	170		3	
50	85		6	
25	42.5		12	
12	20.4		24	
6	10.2		50	
3	5.1		100	

Fig. 7-12. Relationships between viscosity and shear rate for drilling fluids having different *n* values.

thinning fluids. As shown in Figs. 7-10 and 7-11 effective viscosity of the mud decreases with increasing shear rate. The degree of shear-thinning of the mud is determined by the value of n. Relationship between viscosity and shear rate for various drilling fluids having different n values is presented in Fig. 7-12. By converting the dial readings using the previously developed equations gives the viscosities, which are listed in the upper, right-hand corner of Figs. 7-10 and 7-11.

Inasmuch as most of the variables of a multi-speed viscometer are constant, except for the dial reading, a series of conversion factors can be calculated for converting dial readings to viscosity, assuming the standard rotor and bob. For the most commonly used rotational speeds these conversion factors are as follows:

Rotational speed (rpm)	Conversion factor (multiplier): dial reading to viscosity
600	0.5
300	1.0
200	1.5
100	3.0
6	50.0
3	100.0

These conversion factors are also listed in the upper right corner of Figs. 7-9 through 7-12.

Determination of effective viscosity

The shear rate at a particular rotational speed must first be calculated in order to determine the effective viscosity. The following formulas can be used to calculate shear rates (Skelland, 1967):

$$\text{Drillpipe: } \dot{\gamma} = \left(\frac{3n+1}{4n}\right)\frac{8v}{D} \tag{7-17}$$

$$\text{Annulus: } \dot{\gamma} = \left(\frac{2n+1}{3n}\right)\frac{12v}{D_2 - D_1} \tag{7-18}$$

where velocity, v, is in cm/sec and the diameters D, D_1, and D_2 are in cm; D = internal diameter of drillpipe, D_1 = outside diameter of drillpipe, and D_2 = inside diameter of casing. In order to determine the effective viscosity, the viscosograph must be entered at the calculated shear rate and read from the appropriate scale. For example, using the mud properties, which were plotted in Fig. 7-11, and if the mud were flowing at 18.30 cm/sec in an annulus with D_2 = 22 cm and D_1 = 11 cm, then using eq. 7-18 the shear rate, $\dot{\gamma}$, would be equal to:

$$\dot{\gamma} = \left(\frac{(2)(0.41)+1}{(3)(0.41)}\right) \times \left(\frac{(12)(18.3)}{(22-11)}\right) = 29.5 \text{ sec}^{-1}$$

Using Fig. 7-11, at a shear rate of 29.5 sec^{-1}, the μ_e is approximately equal to 150 cP. This viscosity can then be used in pressure drop calculations and in comparing the relative carrying capacity of this mud with another mud in that particular section of the hole. The n value used in the above calculation should be determined at a shear rate as close as possible to the actual one.

SUSPENSION OF CUTTINGS

The settling rate of a particle falling through a quiescent fluid is dependent on (a) size and shape of the particle, (b) density of the particle, (c) shear-rate viscosity of the fluid, and (d) density of the fluid. Only one of these parameters can be easily controlled by engineers, i.e., viscosity of the fluid. The factors pertaining to the particle size and shape are beyond control at the rig site. Although under control, the density of the fluid is usually determined by other considerations, mainly balancing formation pressure. The viscosity of the fluid, however, usually changes throughout the drilling operation.

Tables 7-IV and 7-V show statistical correlation of various rheological parameters with settling rates of 30-mesh and 10- to 20-mesh sand particles. For the 30-mesh sand, the K values (Power Law model) correlated well with settling rates, whereas for the 10- to 20-mesh particles there was a good correlation for 3-rpm dial readings. In order to accurately describe the suspending ability of a fluid, therefore, the viscosity of fluid at or close to the shear rate experienced by the falling particle must be used. In most cases, this viscosity is determined at low shear rates. In the two cases presented in Tables 7-IV and 7-V, the 30-mesh sand was probably experiencing a shear rate of

TABLE 7-IV

Settling rate correlation with rheological parameters for 10- to 20-mesh sand

Rheological parameter	Curve-fit correlation
3-rpm dial reading	0.9273
K (200—100 rpm)	0.7823
n (200—100 rpm)	0.7394
Yield point (lb/100 ft^2)	0.6470
Viscosity at 600 rpm (cP)	0.3632

TABLE 7-V

Settling rate correlation with rheological parameters for 30-mesh sand

Rheological parameter	Coefficient of correlation
K (200—100)	0.8825
3-rpm dial reading	0.8347
K (600—300 rpm)	0.8050
Yield point (lb/100 ft^2)	0.7669
n (200—100 rpm)	0.5353
n (600—300 rpm)	0.4644
Viscosity at 600 rpm (cP)	0.1638

about 1.0 sec^{-1} during its fall (K is the viscosity of the fluid at $\dot{\gamma} = 1.0$ sec^{-1}), whereas the shear rate experienced by the 10- to 20-mesh sand was about 5.0 sec^{-1} (3-rpm dial reading is shear stress at $\dot{\gamma} = 5.1$ sec^{-1}).

FUNNEL VISCOSITY

The Marsh funnel has been used (and misused) since the late 1920's when H.N. Marsh, a field engineer of the General Petroleum Corporation, developed the instrument (Am. Petrol. Inst., 1961, p. 385) (also see Chapter 4). The following description of the Marsh funnel is presented in the API RP 13B publication (1980, p. 5) as follows: "The Marsh funnel is dimensioned so that, by following standard procedures, the outflow time of one quart (946 cm^3) of fresh water at a temperature of 70 ± 5° F (21 ± 3° C) is 26 ± 0.5 seconds". Fluids thicker than water will, therefore, require longer efflux time than 26 sec. Although the Marsh funnel is a useful tool, the results obtained should be used with great caution. The usefulness of the Marsh funnel can be exemplified as follows: for example, instead of saying that mud A looks thicker than mud B, one could state that mud A is 20 sec thicker than mud B. It should be remembered that the measurements of viscosity by the Marsh funnel are only qualitative (thick or thin) and do not reveal the cause of the variations in viscosity nor dictate the treatment of a mud. It should be kept in mind that the head of drilling fluid, h, in a Marsh funnel decreases with time (velocity of outflow $v = \sqrt{2gh}$) and gel strength of mud increases with time during the test. In addition, the funnel viscosity measurements cannot be manipulated mathematically like the data obtained from viscometers. In general, the funnel should be used only to identify trends in viscosity variations over various time periods during the drilling operation.

SUMMARY AND CONCLUSIONS

The rheological properties of drilling fluids play a very important role in determining the success of overall drilling operations. In order to obtain a

LIST OF SYMBOLS

A	area
D	diameter (pipe or hole)
F	force
K	Power Law constant, consistency index.
n	Power Law coefficient, measure of deviation from Newtonian behavior.
P	poise
cP	centipoise
$\overline{PV}$	plastic viscosity (Bingham-plastic)
R	dial reading
R_w	dial reading at a particular rpm (w)
r	distance from pipe wall
v	velocity
$\overline{YP}$	yield point (Bingham-plastic)
$\dot{\gamma}$	shear rate
τ	shear stress
μ	viscosity
μ_e	effective viscosity

properly functioning drilling fluid, rheological properties should be continuously measured and the following facts should be considered:

(1) Viscosity is not constant in most drilling fluids, which are non-Newtonian, and varies with shear-stress/shear-rate ratio in different parts of circulating systems.

(2) The 600- and 300-rpm viscometer dial readings are not necessarily sufficient to calculate the actual viscosity of a drilling fluid.

(3) Temperature control during viscosity measurements is of utmost importance.

(4) Use of the effective viscosity of the fluid gives the most accurate results.

REFERENCES

American Petroleum Institute, 1961. *History of Petroleum Engineering*, pp. 271—453.

American Petroleum Institute, 1978. *The Rheology of Oil-Well Drilling Fluids.* API Bull. 13D (tentative), 26 pp.

Annis, M.R., 1967. High-temperature flow properties of water-base drilling fluids. *J. Pet. Technol.*, (Aug.): 1074—1080.

Bartlett, L.E., 1967. *Effect of Temperature on the Flow Properties of Drilling Fluids.* 42nd Annu. Fall Meet. Soc. Petr. Eng. of AIME, SPE Pap. 1861: 12 pp.

Eirich, F.R., 1956. *Rheology, Theory and Applications.* Academic Press, New York, N.Y., pp. 18—27.

Melrose, J.C. and Lilienthal, W.B., 1951. Plastic flow properties of drilling fluids — measurement and application. *Trans. AIME*, 192: 159—165.

Odeh, A.S., 1977. *Flow of non-Newtonian Power Law Fluids through Porous Media.* 52nd Annu. Fall Tech. Conf., Soc. Pet. Eng. AIME, Denver, Colo., Oct. 9—12, SPE 7150: 9 pp.

Savins, J.G. and Roper, W.F., 1954. A direct indicating viscometer for drilling fluids. *API Drill. Prod. Prac.*, 1954: 7—22.
Skelland, A.H.P., 1967. *Non-Newtonian Flow and Heat Transfer.* Wiley, New York, N.Y., pp. 36—38.
Van Wazer, J.R., Lyons, J.W., Kim, K. and Colwell, R.E., 1963. *Viscosity and Flow Measurement.* Interscience, New York, N.Y., 406 pp.

Chapter 8

WATER-BASE MUDS

GEORGE V. CHILINGARIAN and PAUL VORABUTR

INTRODUCTION

Drilling fluids in which the continuous phase is water, may be termed water-base muds. Water-base muds basically consist of four basic phases: (1) water, (2) active colloidal solids, (3) inert solids, and (4) chemicals (McCray and Cole, 1973).

Proportions of various phases and interactions among them lead to the diverse and varied properties of water-base muds. Consequently, composition is the key factor in controlling the properties of water-base muds.

Many water-base mud types have been formulated and tried (see Chapter 2). The authors believe, however, that any attempt to discuss each individual type would be infutile because numerous variations of each type are available depending on the area of operation, the prevailing trends, and types of formations to be penetrated. The authors firmly believe that a solid understanding of the mechanisms that lead to the success or failure and wide acceptance of various mud types would allow the readers to better evaluate and select a drilling fluid type that is most suitable for their particular needs.

WATER PHASE

Water, which is the continuous phase of any water-base mud, may be fresh, hard, or salty. Final mud cost obviously depends on the availability of the water selected for use.

The primary function of the liquid or continuous phase is to provide the initial viscosity which can be modified to obtain any desirable rheological properties. At standard conditions (i.e., 14.7 psi and 60°F), the viscosity of fresh water is equal to 1.1 centipoises.

The viscosity of water is a function of temperature (indirect), pressure (direct function), and the concentration of dissolved salts. As the temperature increases, the volume expands resulting in lower molecular friction and, consequently, a lower resistance to flow (Olson, 1973, p. 28). The effect of temperature on viscosity of fresh water is shown in Fig. 8-1a. As pointed out by Rieke (1978, p. 80), temperature changes in the range of 50°—500°F do not have any significant effects on water viscosity. As shown in Fig. 8-1b,

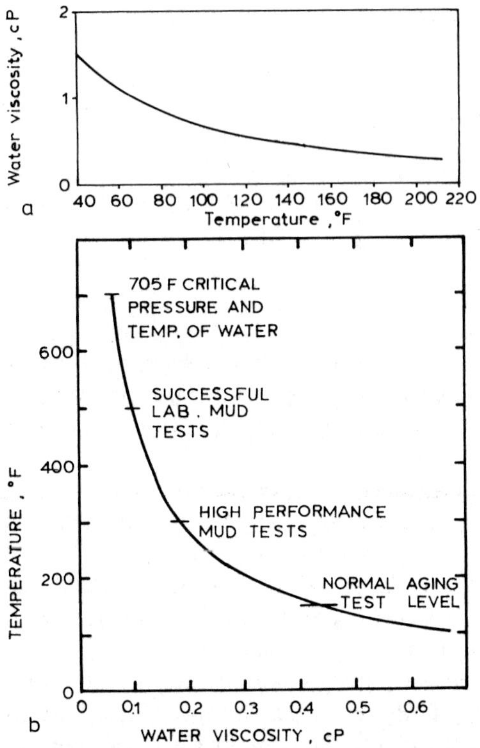

Fig. 8-1. Relationship between temperature and viscosity of distilled water. a. After Gatlin, 1960, fig. 6-8, p. 74; courtesy of Prentice-Hall, Inc. b. After Rieke, 1978, fig. 2, p. 81.

the approximate viscosity range corresponding to the above temperature change is from about 1 cP to 0.1 cP.

Inasmuch as water is highly incompressible at pressures to which water-base muds are usually subjected, increased resistance to flow due to total volume reduction is negligible. The overall effects of temperature and pressure on the continuous phase are so small that they are normally neglected. Temperature, however, greatly affects the properties of the actual drilling fluids.

The viscosity of brine water increases with the salt concentration and at saturation may be as high as 1.7 times that of fresh water at the same temperature (API, 1974, p. 83).

The second function of the continuous phase (water) is to suspend the reactive colloidal solids, such as bentonites, inert solids, such as barites, and other drilled solids. Water also acts as a medium for transferring the surface-available hydraulic horsepower to the bit on bottom of the hole. This third function is termed "jetting action".

Water (continuous phase) also serves as a massive heat sink for absorbing heat generated in the borehole during drilling operations due to rubbing friction of drill bit, drillpipe, drillcollars, and stabilizers against the walls of the hole.

Water is also a solution medium for all conditioning chemicals which are added to the drilling fluid. Obviously, properties of the water, e.g., pH and salinity, affect the effectiveness of the added chemicals.

Many functions of drilling muds in effect are the functions of the water. As previously mentioned, different types of water may be used as the continuous phase in water-base muds. The selection of the type used is controlled by the following factors:

(1) *Availability.* Availability of water depends on location, e.g., fresh water which may be abundant in one area may not be available in others. In offshore drilling, for example, seawater has often been substituted for fresh water in drilling fluids (salt- or seawater muds). Otherwise, costly generating equipment may have to be installed if fresh water must be used. Transportation of fresh water is also costly.

(2) *Types of geologic formations.* Inasmuch as some of the drilled formations are sensitive to fresh water, continued use of fresh water may result in formation damage and severe hole enlargement problems. Fresh-water filtrate causes in-situ clay particles in the productive zones to swell and, possibly, to migrate. This leads to plugging of the pore throats, causing significant permanent reduction in permeability.

(3) *Types of chemicals.* Solubility and consequent effectiveness of chemicals in the continuous phase in large measure determine the efficiency of mud conditioning. The salinity and pH of the continuous phase greatly affect the solubility of common mud-conditioning chemicals.

(4) *Types of data-collecting tools or techniques.* Many commonly used logging tools rely on the continuous phase of the mud as their operating medium, e.g., the SP and the electric logs. The accuracy of the results obtained with these tools is a function of salinity of the continuous phase and the temperature. Inasmuch as the SP and the resistivity tools provide accurate readings only up to certain salinities (e.g., see Schlumberger manuals), a careful selection of the continuous phase is important.

The above-mentioned criteria must be carefully weighed against one another in order to arrive at a final compromise. Economics (i.e., cost of fresh water) plays a greater role in water selection than in the mud type (e.g., seawater versus fresh-water muds) selection. From the operational point of view, particularly in remote offshore operating areas, the cost of transporting fresh water to certain areas may be prohibitive.

REACTIVE SOLIDS PHASE

In water-base muds, clay is added to increase density, viscosity, gel strength, and yield point, and to decrease fluid loss. Clays have been previously discussed in Chapter 5. Mud engineers usually divide the clays used in fresh-water drilling fluids into three main types: montmorillonites, kaolinites, and illites. Montmorillonite (smectite) is the most commonly used clay because of its superior ability to swell uniformly in fresh water upon shear application, resulting in a more homogeneous clay—water mixture. In drilling fluid literature manuals, montmorillonite (smectite) is commonly referred to as bentonite. Interchangeable usage of these terms is acceptable because bentonite is presently defined as the clay type, the major mineral type of which is montmorillonite (smectite).

Attapulgite and palygorskite clays, which can swell in salt water, are used in salt-water muds. Clays present in water-base muds have two different sources: (1) Clays are added purposely to impart rheological properties and to control water loss; they are collectively referred to as commercial clays. (2) Natural clays, which are encountered when drilling through various argillaceous formations, can also be incorporated into the drilling mud.

As pointed out by Rogers (1963, p. 344), bentonite is a highly reactive colloid (sodium-montmorillonite or calcium-montmorillonite) the physical and chemical properties of which greatly affect the preparation and handling of water-base muds. Wyoming bentonite, the predominant mineral type of which is Na-montmorillonite, swells more than montmorillonites having lower Na-montmorillonite and higher Ca-montmorillonite contents. Rogers (1963) attributed this difference in swelling to the fact that if replaceable cation is tightly adsorbed, as are calcium and hydrogen, there is a smaller tendency for the adjacent clay structural units to push each other away. The hydrated Ca-montmorillonite platelets are still held together by weak van der Waals' forces, whereas upon hydration Na-montmorillonite platelets are spread apart to a considerable distance.

Different forces acting on the clay particles in aqueous suspension are shown in Fig. 8-2. Forces F_1 and F_3 result from repulsion of the like charges

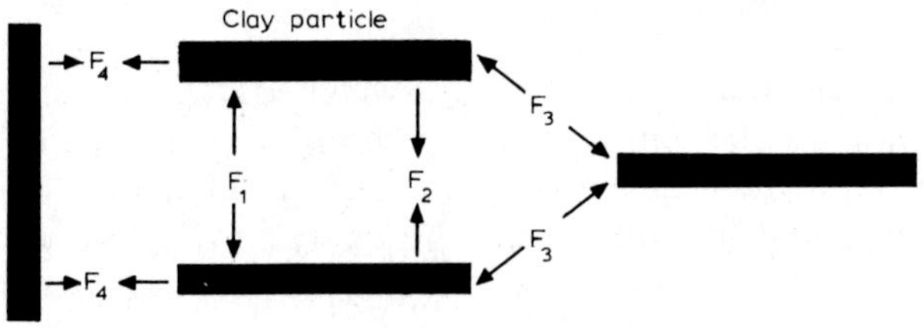

Fig. 8-2. Different forces acting on the clay particles in aqueous suspensions. (After Baroid, 1975, fig. 2, p. 21; courtesy of Baroid Petroleum Services.)

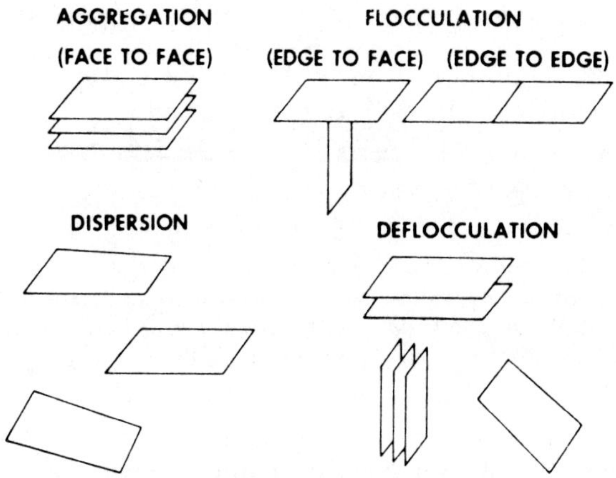

Fig. 8-3. Possible equilibrium states of montmorillonite clay particles in water. (After Magcobar, 1977, Chapter 3, fig. 3, p. 2.)

(— or +) on the clay planar surfaces or edges, whereas forces F_2 and F_4 are due to the attraction of different charges of the platelets. Distribution of these forces, which determine different properties of water-base muds, is controlled by the types of electrolytes present and pH of the aqueous phase. Distribution of these forces and resulting mud properties are controlled through addition of various drilling fluid additives.

Wyoming bentonite, which swells markedly in fresh water, is mined exclusively from deposits in Wyoming and South Dakota, U.S.A. (Baroid, 1975, p. 19). Wyoming bentonite platelets are extremely small in size and the sodium ions are held loosely to them. In the aqueous phase, as Na^+ cations leave the clay structures, negative charges are left behind causing repulsion of nearby platelets (dispersed state).

The ability of bentonite to hydrate and become dispersed is reduced in the presence of electrolytes in water. According to Baroid (1975, p. 43), when bentonite is added to fresh water, four equilibrium states between bentonite and water may exist (Fig. 8-3). In the presence of various contaminants several different combinations of these states can prevail: (1) dispersed with edge-to-surface contacts, (2) completely dispersed, (3) aggregated with edge-to-surface contacts, and (4) dispersed-aggregated. More than one of these equilibrium states may be assumed by a clay-water system at the same time, depending on chemical environment.

Schematic diagrams of clay particles in adherence and nonadherence states in water and in aluminum complex solution, respectively, are presented in Fig. 8-4, whereas a schematic representation of attraction between two clay particles and proton distribution, showing "rigid" water layer, is presented in Fig. 8-5.

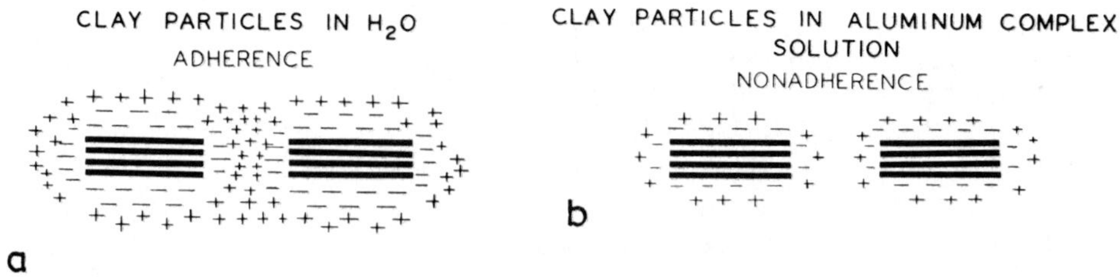

Fig. 8-4. Schematic diagram of clay particles in adherence and nonadherence state in water and in aluminum complex solution, respectively. (After Chesser and Perricone, 1973, p. 11, fig. 5; courtesy of the Society of Petroleum Engineers of AIME.)

Inasmuch as bentonites do not hydrate sufficiently in water having high contents of electrolytes, other clays must be used to impart rheological properties. (Electrolytes dissociate into charged positive and negative ions in the aqueous phase.)

Attapulgite and palygorskite clays are used instead of bentonite to develop rheological properties in waters containing large amounts of electrolytes. These clays differ from bentonite in the shape of their particles, which are small, cylindrical and needle-like rather than being platelet-like. Figure 8-6 illustrates dispersion of attapulgite particles in salt water. The viscosity-building properties of attapulgite depend entirely on the entanglement of these needle-shaped particles (Baroid, 1975, p. 22). Deposition of these particles on porous formation face, however, does not result in impermeable filter cake to prevent water movement into the formation (mud filtrate).

Sepiolite clays and asbestos fibers also have been tested and slurries have been prepared to withstand temperatures as high as 500° F (Carney and Myer, 1976).

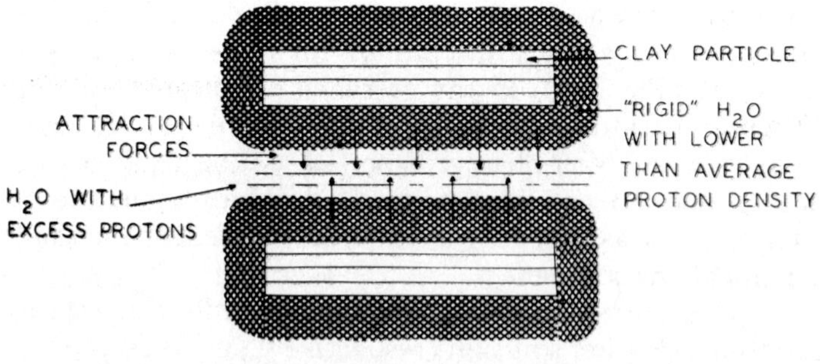

Fig. 8-5. Schematic representation of attraction between two clay particles and proton distribution, showing "rigid" water layers. (After Chesser and Perricone, 1973, p. 11, fig. 2; courtesy of the Society of Petroleum Engineers of AIME.)

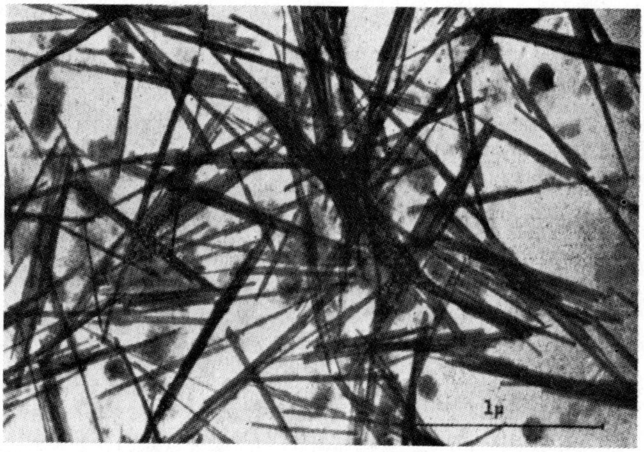

Fig. 8-6. Manner of entanglement of attapulgite particles which imparts viscosity and gel strength to the base fluid. (Courtesy of Baroid Petroleum Services, 1968, fig. 4, p. 52.)

Inasmuch as many clays may be used to impart rheological properties to muds, the yield of clays has been used for the purpose of classifying them and determining their quality. The yield of clay has been defined as the number of barrels of a slurry having certain viscosity (generally 15 cP is considered as standard) that can be prepared from one ton of clay. As shown in Fig. 8-7, the 15-cP point has been chosen as a standard reference because it is located at the critical part of the curve for all clay types. Up to 15 cP, addition of clay results in low viscosity increase. Above 15 cP, however, even slight addition of clay causes an exponential increase in viscosity. Attapulgite and palygorskite impart as much viscosity to salt water as Na-montmorillonite does to fresh water. It is important to note that attapulgite and palygorskite impart same rheological properties to both fresh water and water containing electrolytes (Baroid, 1975, p. 23).

Commonly, clays are grouped into (1) high-yield clays (Na-montmorillonite, attapulgite, palygorskite, asbestos, etc.); (2) medium-yield clays or sub-bentonitic clays (Ca-montmorillonite, etc.); and (3) low-yield or native clays (e.g., dry lake clays). On the basis of commonly accepted standard, a high-yield clay yields 45 bbl or higher of 15-cP mud per ton of clay, whereas a medium-yield clay yields only 20—40 bbl of such mud per ton of clay. Clays yielding less than 20 bbl/ton of clay are considered low-yield (see Fig. 8-7).

The following formula, which can be easily derived, can be used to calculate the yield of the mud:

$$\text{Yield (bbl mud/ton clay)} = \frac{2000}{Wt_f \times \gamma_m \times 5.62} \tag{8-1}$$

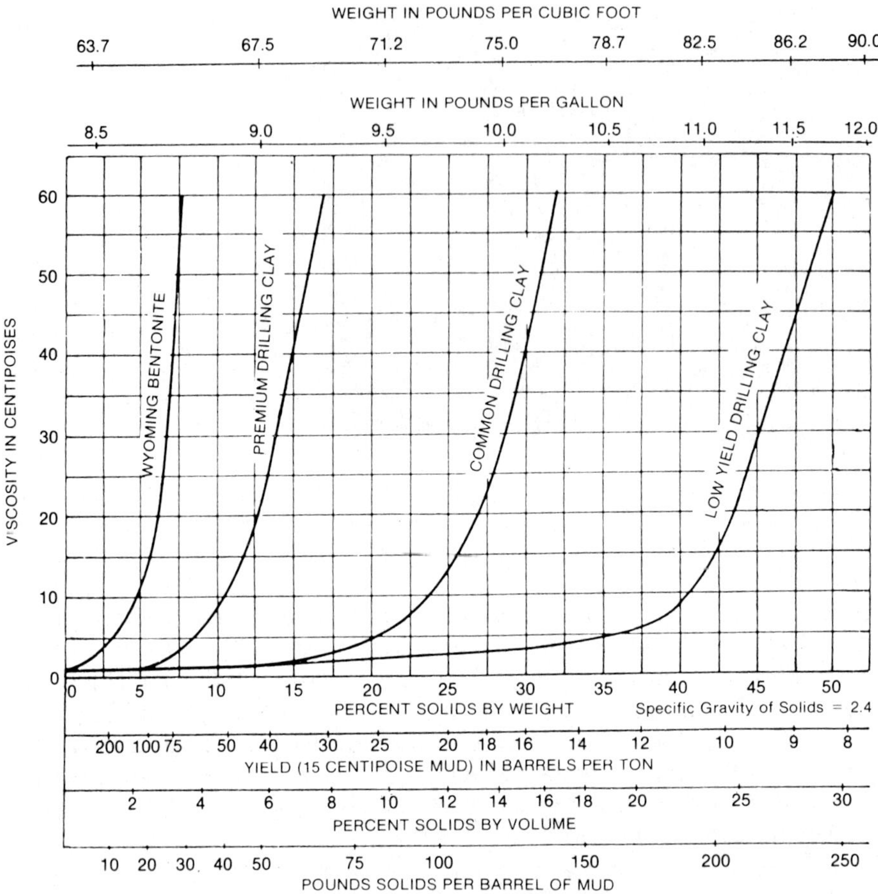

Fig. 8-7. Viscosity, API yield of clay (bbl of 15-cP mud/ton clay), and specific weight of mud in relation to percentage of various types of clay in fresh water. (Courtesy of IMCO Services, 1978, Chapter 5, fig. 2, p. 4.)

where Wt_f = weight fraction of clay in the mud, and γ_m = specific weight of mud in lb/cu ft. 2,000 = conversion factor from lb to ton and 5.62 = conversion factor from cu ft to bbl.

Natural clays may be introduced into the mud system through the bit action while penetrating argillaceous (clayey) formations, such as shales. These argillaceous formations may contain water-sensitive clays. Upon their introduction into the mud stream, the reactive solids content is increased resulting in higher apparent and plastic viscosities, gel strength, yield point, and mud weight. Natural clays are considered as contaminants and are treated as such in many instances. Spud mud and natural mud, however, make use of the natural clays found in shallow argillaceous formations for providing rheological properties.

Effect of electrolytes on yield of clays

The yields of sodium- and calcium-bentonites are usually severely lowered in the presence of salts in excess of 10,000 ppm. When the make-up water contains large amounts of electrolytes (⩾10,000 ppm) attapulgite clay must be used to impart the desired rheological properties. Attapulgite clays, unlike smectites, develop rheological properties equally well in fresh and salt waters. According to Rogers (1963, p. 203), one ton of attapulgite clay yields approximately 90 bbl of 15-cP mud. According to Magcobar (1977, p. 5), attapulgite clay develops rheological properties in aqueous phase as a result of two processes:

(1) Development of "brush heap" structures, i.e., entanglement of needle-shaped particles, as a result of strong shearing action during mixing of clays in the aqueous phase.

(2) As a result of entanglement of particles, which brings particles closer together, strong attractive forces develop between particles because of the presence of broken bond charges on the edges of needle-shaped particles. This stabilizes the brush heap structures.

Although attapulgite clay develops equally high yields per ton of clay in fresh water and in salt water, its usage is limited owing to: (1) greater cost than smectites, (2) lack of filtrate loss control owing to its particle shape, (3) difficulty in maintaining rheological characteristics once they are developed, and (4) development of a new method of obtaining viscosity in salty water using bentonite.

The restricted use of attapulgite clay is mainly due to the lack of filtrate loss control (see Fig. 8-6) and development of a new method of achieving bentonite hydration in salt water. Bentonite swells sufficiently in salty water if it is allowed to hydrate properly in fresh water prior to addition to salt water (Magcobar, 1977, p. 5).

Rheological properties development in water-base muds

States of hydration and dispersion of bentonite or other clay particles in the aqueous phase determine the flow properties of water-base muds. The most important rheological properties are viscosity (apparent, plastic, and effective), gel strength, and yield point. (Also see Chapter 8.)

Viscosity development in water-base muds

The viscosity of fresh-water muds is determined mainly by (1) viscosity of base liquid; (2) mineralogy, size, shape, and number of particles in the aqueous suspension medium; and (3) interparticle forces.

The effects of viscosity of base liquid has been discussed previously. The effect of the size, shape, and number of clay particles and interparticle forces on viscosity is closely linked to (1) the state and degree of clay particles hydration, (2) degree of dispersion through mechanical agitation (shear), (3) the base exchange capacity of the clay minerals, and (4) pH of the aqueous phase.

Plastic viscosity is primarily determined by the size, shape, and number of clay particles and their concentration, whereas the yield point is mainly controlled by the interparticle forces.

Mechanisms causing high viscosity increase

Particle linkage is an important process that must be considered in studying viscosity of water-base drilling fluids. This process refers to association of bentonite particles in face-to-face, edge-to-edge, and edge-to-face fashion (Magcobar, 1977, p. 2), as shown in Fig. 8-3. The term aggregation is often used to describe the face-to-face orientation of the clay particles. It is caused by the predominance of the attractive van der Waals' forces with the repulsive forces being at a minimum. Flocculation is referred by many investigators to

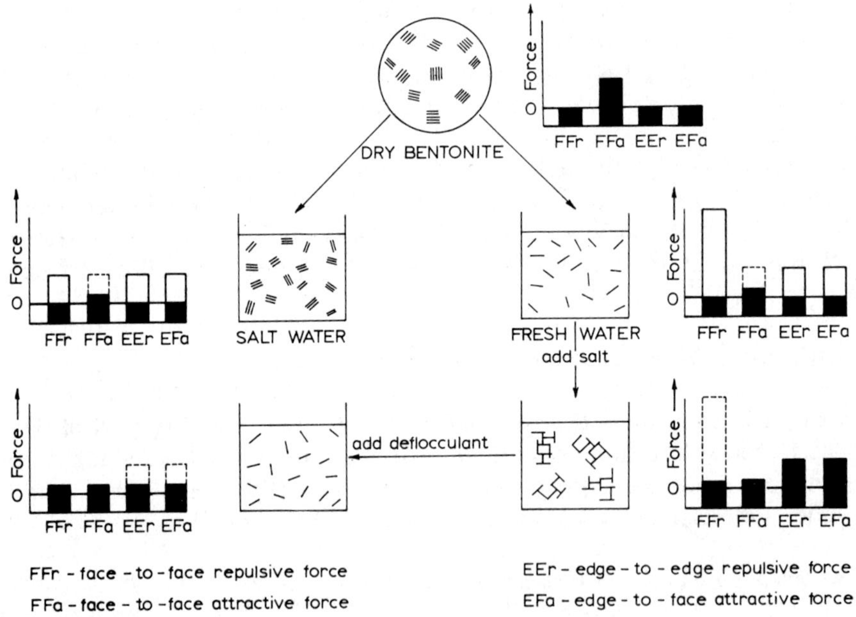

Fig. 8-8. Relative magnitude of forces F_1, F_2, F_3, and F_4 acting on clay particles (see Fig. 8-2). (After Baroid, 1975, fig. 3, p. 21.)

an edge-to-edge and edge-to-surface orientation of the clay platelets; there is a predominance of edge-to-edge repulsive forces and edge-to-surface attractive forces between clay particles. Dispersion of clay particles occurs when all four existing forces cancel the effect of one another. Dispersion is often referred to as the reverse aggregation (Magcobar, 1977, p. 2). In this state, the magnitudes of all four forces, which are not necessarily zero, are approximately equal. The state of deflocculation is opposite to that of flocculation, i.e., the clay platelets are separated as a result of neutralization of the edge-to-edge repulsive forces and the edge-to-surface attractive forces (Baroid, 1975, p. 43). Figure 8-8 depicts the relative magnitudes of the four forces acting in each case.

It is important to mention here that many petroleum engineers define deflocculation as the state of dispersion of clay platelets because of their repulsion, resulting from the predominance of negative charges on clay particles. On the other hand, flocculation is simply defined as the state of aggregation of clay particles as a result of neutralization of negative charges on clay platelets by excessive amounts of cations, e.g., during drilling operations through salt domes (NaCl).

Effects of NaOH on prehydrated bentonite mixture

Many investigators have studied the effects of pH on the viscosity of a prehydrated bentonite suspensions. The drilling fluid viscosity increases with the increasing pH value of the aqueous phase (above pH value of 8), as shown in Fig. 8-9. The minimum values of viscosity occur at a pH value of about 8. The portion of the curve below a pH of 7.0 is usually not considered because water-base muds are always maintained at pH above 7.0 in order to (1) inhibit corrosion of drillpipe and other metal parts, and (2) maintain safety of personnel. The pH of water-base muds is controlled by addition of caustic soda (NaOH), which dissociates into sodium (Na^+) and hydroxyl (OH^-) ions. Thus, the effects of pH on water-base muds must take in consideration not only the effect of hydroxyl ions but also that of sodium ions. In effect, most investigators were studying the effect of NaOH on properties of muds rather than that of pH alone.

The hydroxyl ions probably adsorb on the bentonite clay platelets (see Tchillingarian, 1952) increasing the total negative charges on the clay sheets. The consequent increase in negative charges causes the clay platelets to repulse each other more and to become better dispersed. Thus, NaOH in very small concentrations may actually act as a dispersant and decrease viscosity of a mud.

Sodium ions of NaOH, which are introduced into the mud stream, cause the viscosity to increase, possibly as a result of base exchange. Sodium-base clays swell more than Ca-base clays, for example. Excessive addition of NaOH, however, may result in complete separation of clay and water phases

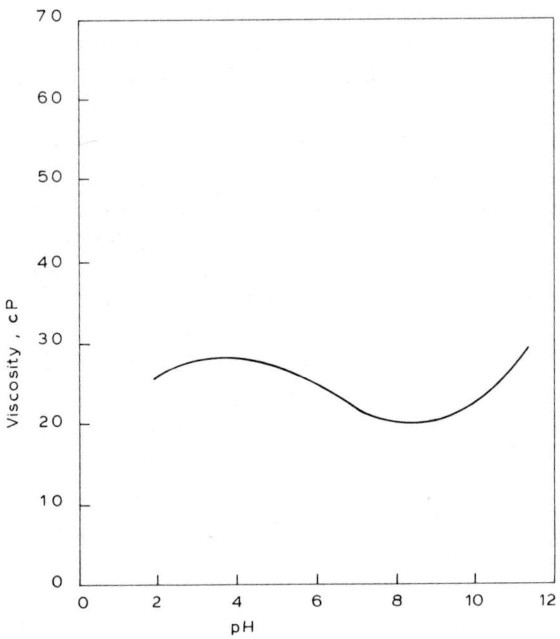

Fig. 8-9. Relationship between pH and viscosity of a prehydrated bentonite slurry. (After Magcobar, 1968, fig. 6-2, p. 78; courtesy of Magcobar Petroleum Services.)

because of neutralization of negative charges of clay platelets by the Na^+ cation.

The same ion replacement mechanism may also be applicable to natural clays incorporated into the mud stream in the course of drilling (see Rogers, 1963, p. 351). Influx of Na^+ ions on addition of NaOH may convert the low-yield clay picked up during drilling to high-yield-clay.

INERT SOLIDS PHASE

Inert solids in drilling fluids may be classified as either weighting materials or drilled solid contaminants.

Inasmuch as the weight of a clay plus water in water-base muds may not be sufficient to control formation pressure, weighting materials must be added to attain the desired mud weight. Weighting materials are chemically inert solids, the specific gravity of which is high enough to develop necessary hydrostatic head. Various weighting agents and their specific gravities are presented in Chapter 6. Barite (barium sulphate, $BaSO_4$; sp. gr. = 4.0—4.6) has enjoyed widespread usage because of its high specific gravity and water wettability. Weighting materials used in water-base mud must be water-wet for proper suspension in the continuous phase. Thin water films form on the surfaces of the water-wet barite particles, thereby increasing the lubricity of the mud. Barite is available in various particle sizes.

Addition of weighting materials increases total mud volume, which is a function of the specific gravity of a particular material. Considerable volume increase may result from the addition of weighting materials. In order to provide ample mud handling facilities, a volume increase calculation must be performed prior to the commencement of weighting-up operation. A sample calculation is presented in Appendix A at the end of the book.

Inert solids, which may contribute to the total solids content of the mud, may greatly affect the final mud properties.

CHEMICAL PHASE

Although water-base muds traditionally may be considered as consisting of three previously-discussed components, at the present time complex formulation of water-base muds have made chemicals their integral part. These chemicals may be either (1) organic or (2) inorganic. Each group may be subdivided into subgroups based on their specific functions: (1) thinners (organic and inorganic), (2) dispersants, and (3) deflocculants.

Nearly all chemicals utilized in water-base muds affect the properties of the mud in more than one way, and usually have both primary and secondary functions. Appendix I in Chapter 6 lists primary and secondary functions of most commercially available mud conditioning chemicals.

TYPES OF CONTAMINATIONS OF UNTREATED FRESH-WATER MUDS AND TREATMENTS

Inasmuch as while drilling in new areas the knowledge of the subterranean formations is always limited, preventive steps must be taken to insure proper and continuous conditioning of the drilling fluid. The petroleum engineer must have frequent consultations and discussions with the petroleum geologist to obtain "while drilling" data and geologist's forecast of the formation types to be penetrated. Frequent testing of the drilling fluid properties is also imperative. Testing at regular, specified intervals will indicate any changes occurring in the muds. The testing interval may be determined by the rate of penetration, for example. Mud testing provides necessary information on the type and degree of contamination and the required treatment program.

A common type of contamination due to human error is cement contamination. For example, when drilling out cement plugs, cement must be sufficiently dry, because wet or "green" cement introduces more $Ca(OH)_2$ into the mud stream than does dry cement.

The profound statement "an ounce of prevention is better than a pound of cure" suggests the best method of treating contamination, i.e., contamination must not be allowed to occur. If, however, contamination does occur, it is necessary to restore the original properties of drilling fluid through treatment.

The preventive treatments center around conditioning the muds to be immune to certain types of contaminants, which are expected to be encountered during the drilling. Utilization of the more inert saline and polyvalent muds is an example of this approach.

The second main type of treatment is the "on the spot" treatment of the contamination after its occurrence. Common contaminants which are encountered during drilling include: (1) drilled solids (high content), (2) salt (NaCl), (3) gypsum and anhydrite, and (4) cement. Hardness contamination is an important type of contamination. Contaminating effects of high temperatures and gases may also be quite pronounced.

Contaminating effects of high solids content

High contents of solids give rise to increased viscosity, gel strength, yield point, mud weight, water loss, and mud cake thickness. Inasmuch as water-base muds contain both reactive and inert solids, their contaminating effects are considered separately.

Extremely high viscosity, gel strength, and yield point result from high content of reactive solids. As shown in Fig. 8-7, the viscosity of water-base muds, the reactive solids content of which consists of attapulgite, Na-montmorillonite or Ca-montmorillonite clays, falls within reasonable operating range up to a mud weight of 8.6 lb/gal. Up to a mud weight of 12 lb/gal, reasonably good viscosity, yield point, and gel strength are still maintained by mud on addition of inert solids. Magnitudes of changes of properties in previously untreated water-base muds depend on the yield of the incorporated clays. Inasmuch as Na-montmorillonite clays swell markedly more than Ca-montmorillonite clays, higher concentrations of the latter can be tolerated.

Contamination with solids may also be a result of an attempt to weight up the mud by adding too much of the clay. As a rule of thumb, a viscosity of 15 cP is considered sufficient to keep barite or other weighting materials in suspension. The clay content of the mud should be kept below that necessary to attain this viscosity. The effects of drill-solids concentration (mainly reactive solids) on the plastic viscosity of a 3% by volume bentonite slurry are presented in Fig. 8—10. For example, a 15-lb/gal mud, having 27% total solids content, may contain 2% drilled solids, 3% bentonite, and 22% barite, and have a plastic viscosity of approximately 90 cP at room temperature. As

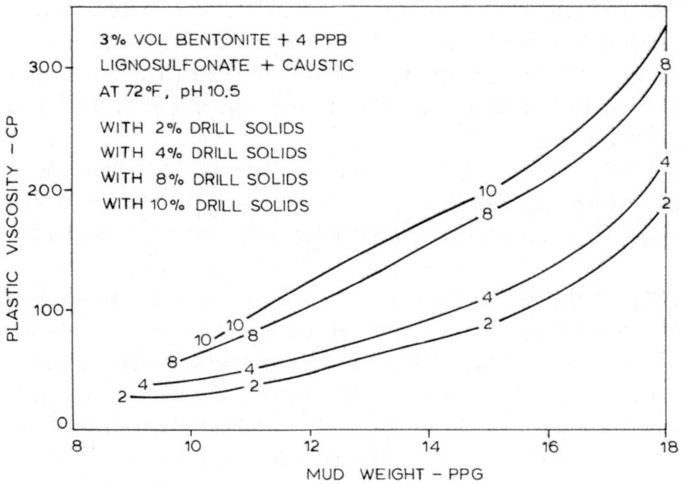

Fig. 8-10. Relationship between the plastic viscosity and drilled solids content for a 3% by volume bentonite-in-water suspension. (Courtesy of Milchem Petroleum Services.)

the drilled solids concentration is increased to 10%, with the concurrent reduction of barite content to maintain the same mud weight, the plastic viscosity increases to 200 cP. Thus, the plastic viscosity more than doubles as the total solids content increases from 27% to 31%.

Some reactive solids are incorporated into the mud system in the form of finely-ground drilled argillaceous formation solids.

The viscosity increase due to the introduction of non-reactive solids into the mud stream is lower than that in the case of reactive solids, because the inert solids do not hydrate and become dispersed in water. They increase the internal friction of the fluid, however, which increases the plastic viscosity to some extent depending on concentration.

Water loss is affected by particle size, shape, size distribution, and concentration of dispersed clays. Incorporation of finely-ground drilled solids creates non-uniform particle-size distribution, which results in the formation of fluffier, thicker, and more permeable filter cake, with concurrent high water loss.

Treatment of high solids contamination

In order to prescribe a proper treatment for contamination due to incorporation of drilled solids into the drilling fluid, it is necessary to know the percentage of total solids in mud and the concentrations of inert and reactive solids (i.e., high- and low-specific-gravity solids concentrations, respectively).

Several methods of treatment are available for restoration of the original properties of contaminated muds. Some methods may be more practical than others in a given case and the choice is dictated by the previous experience in the area, availability of treating agents and equipment, and the degree of conditioning desired.

According to Rogers (1963, p. 43), chemical treatment to correct the effects of high solids content contamination is effective only where mud weight is lower than 10 lb/gal and only minor property changes are to be corrected.

More practical treating methods involve physical removal of the excess solids from the mud. Proper design of the mechanical solids removal system for effective removal of all undesirable solids in order to restore the original properties of mud necessitates knowledge of the total solids content and types of solids (see Chapter 12). Other methods of treating contamination include: (1) dilution with water, (2) addition of chemical flocculants to settle out the solids at surface, and (3) conversion to mud types that do not allow hydration and dispersion of drilled solids (see Magcobar, 1977, p. 1).

The treatment may be accomplished by discarding portion of the existing contaminated drilling fluid, adding fresh water, and then rebuilding the desired drilling fluid properties by adding such additive as barite to regain the initial specific weight, for example. Dilution with fresh water is often not practical because of lack of water availability. This method is also unattractive in offshore areas where the excess mud cannot be discarded into the seawater and has to be moved away, increasing the overall operating cost. In order to reduce the amount of water required for dilution, when mud specific weight is low and contamination is not severe, thinning agents may also be added with water. Reduction in the amount of weighting material needed to restore the original drilling fluid specific weight may offset the cost of chemical thinning agents.

Flocculating agents or flocculants can be used to flocculate the reactive solids (i.e., to form "flocs") in order to increase their settling rate. Large surface settling pits are required to increase the retention time. This method may be used successfully in hard-rock country (Magcobar, 1977, p. 1). In areas having thick argillaceous formations, however, this method may not be effective because of excessively large amounts of flocculants and long retention time required to settle the clays. Settled solids must be regularly removed from the bottom of the pits in order to insure success of this treating method. As discussed in detail in Chapter 12, properly designed and sized solids-removal equipment constitutes the most efficient method of removing excess solids.

Contamination with monovalent salts

Contamination of drilling fluids with monovalent salts is commonly called "salt contamination". Monovalent salts dissociate in water to singly-charged

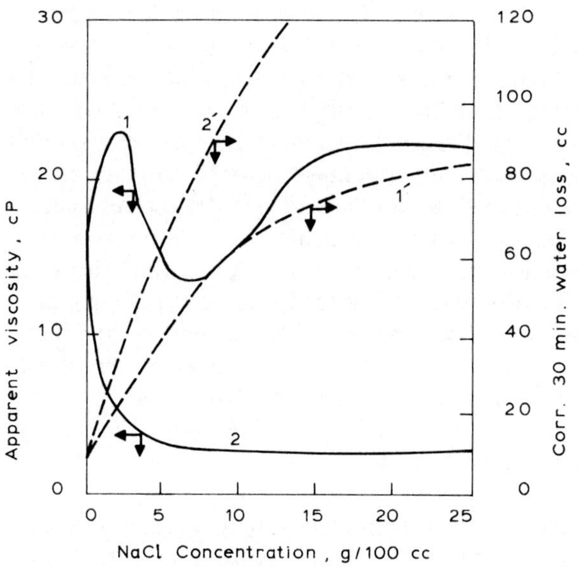

Fig. 8-11. Effects of NaCl concentration, on the viscosity and water loss of suspensions of prehydrated and dry bentonite (5 wt.%). (Also see IMCO, 1976, p. 42, fig. 25; and Rogers, 1963, fig. 7-4, p. 356.) *1* and *1'* = apparent viscosity and water loss of suspensions of bentonite prehydrated in distilled water, respectively; *2* and *2'* = apparent viscosity and water loss of suspensions of previously dry bentonite, respectively.

cations and anions such as Na^+, K^+, and Cl^-. Sodium chloride (NaCl) and, to a much lesser extent, potassium chloride (KCl) are the most common monovalent salts which are encountered during drilling (Baroid, 1975, p. 44).

Salts are encountered in drilling through salt domes, salt beds, salt stringers, and formations containing salty water. Some salt stringers may range in thickness from a few inches to a few feet, and they may not be noticed until sufficient amount of salt accumulates in the mud system to cause adverse changes in properties.

Concentration of salts in formation waters may be many-fold that of seawater (3.5% or 35,000 ppm). Bentonite clay swells in water the salt content of which is less than 1% (<10,000 ppm). Introduction of salt in excess of 1% into the mud stream may result in large increases in apparent viscosity, gel strength, yield point, and water loss, and in a loss of thixotropic property.

An example of the effects of sodium chloride on prehydrated bentonite suspensions is presented in Fig. 8-11. Relationship between the concentration of NaCl in solutions to which dry bentonite was added later and the properties (water loss and viscosity) of resulting suspensions is also presented in Fig. 8-11. In both cases, the water loss increases with increasing concentration of NaCl in water probably due to the neutralization by Na^+ cations of negative charges on the clay platelets and their subsequent aggregation (referred to as flocculation by many investigators), which results in the formation of a permeable filter cake by "flocs". The viscosity of prehydrated bentonite—water suspension first increases probably as a result of base exchange (Na^+ for any Ca^{2+} cation present), because Na-base clays swell more than Ca-base ones. Then the viscosity decreases due to the neutralization of negative charges on clay platelets by additional Na^+ cations and aggregation of clay platelets. Subsequently, with increasing Na^+ cation concentration the viscosity increases, possibly as a result of edge-to-surface flocculation of aggregated flocs. Finally, decrease in viscosity at very high concentrations of Na^+ cations may be due to separation of liquid and solid phases due to complete neutralization of negative charges on surfaces of clay platelets by Na^+ cations and positive charges on the edges by Cl^- anions.

Treatments of contamination by monovalent salts

Inasmuch as NaCl and KCl cannot be precipitated out of the mud, other methods must be utilized to treat the effects of contamination. These include: (1) dilution with fresh water, (2) addition of chemical thinners, and (3) conversion to salt-water muds.

The first method is not often used because of the following reasons: (1) disposal of some portion of the mud creates problems, especially in offshore area; and (2) additional costs of clay, weighting materials, water loss control agents, and other chemicals. In addition, according to Zwicker (1979, p. 46), shortages of barite as primary weighting material in the future may severely restrict the utilization of this method.

A more commonly used treatment involves addition of chemical thinners to salt-contaminated water-base mud. It has the following advantages: (1) There are many different chemical thinners available. (2) Chemical treatments, which require small concentrations of chemicals, are efficient. (3) Ease of handling, because most chemical thinners are available either in dry-powdered or platelet form in 100-lb sacks. Consequently, their storage, care, and use are facilitated, particularly in offshore operations.

The commonly available chemical thinners, which are used in treating salt contamination, are listed in Chapter 6. The concentrations of chemical thinners required for treatment vary depending on the (1) availability, (2) amount of salt concentration in the mud, and (3) degree to which mud properties must be restored. Phosphate thinners (see Chapter 6) may reduce the apparent viscosity, yield point, and gel strength at low salt concentrations (up to 2%), and may maintain reasonable mud properties. Tannin compounds may also be used as thinners at low salt concentrations. Thinning effect of these chemicals is probably due mainly to the deflocculating effect of the highly charged phosphate and tannin anions, which plate the clay particles and impart strong repulsive forces to them. Limiting temperatures

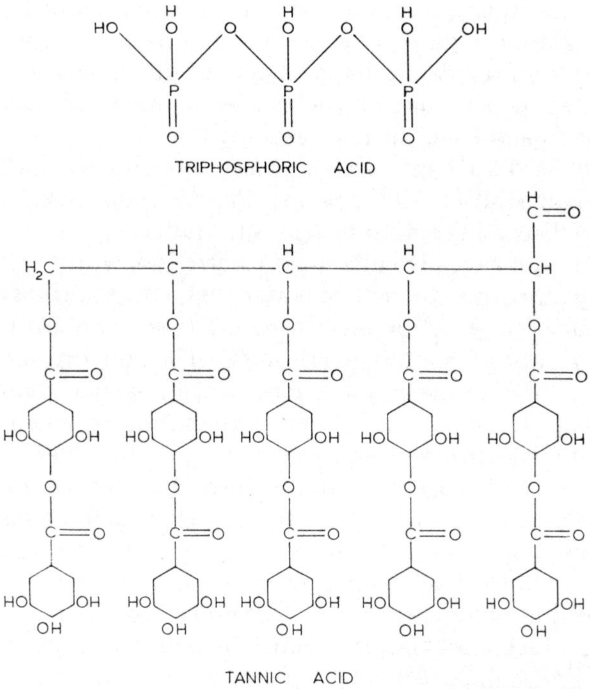

Fig. 8-12. Probable structures of triphosphoric and tannic acids. (After Beeson et al., 1958.)

of phosphates and tannins are about 175° and 275° F, respectively.

Phosphate thinners can be effective in a low-pH environment, whereas tannin compounds, such as quebracho which is a weak acid, require a high-pH environment (about >9). The latter must be provided to enable OH^- ions to strip the H^+ ions from the quebracho molecules and form H_2O. This creates highly charged, extensive anions which can plate the clay particles. The structure of the quebracho is presented in Fig. 8-12.

Inasmuch as NaCl contamination reduces the pH of the mud due to displacement of the hydrogen ions from the clays through mass action of sodium ions, phosphate treatment may aggravate the problem by further reduction of pH. Although reduction in pH to acidic values may control viscosity and gel strength (see Fig. 8-9), muds having low pH value may be highly detrimental to rubber parts of the circulating system, aggravate acid corrosion of the drillpipe and tubular goods, and be harmful to the drilling personnel. In the case of excessive pH reduction, NaOH (caustic soda) may be used to increase the pH of the mud.

Phosphates or quebracho can be used together with lignite when the NaCl concentration is high. At high temperatures, a 50 : 50 mixture of quebracho and lignite (with NaOH to maintain high pH) has been found to be an effective treatment. This may be due to the dispersing property of quebracho and the large size of lignite particles, which position themselves between the clay platelets dispersed by quebracho and keep them spread apart.

Lignosulfonates have been used to treat salt contamination up to NaCl concentration of 10% by weight (IMCO, 1977, p. 44). Caustic soda (NaOH) may be added to augment the dispersing action of lignosulfonates.

As mentioned before, there is an excessive increase in water loss with NaCl contamination. Consequently, in order to reduce water loss, the water-loss control agents must be added in conjunction with thinners (see Chapter 6). Pregelatinized starch, sodium-carboxy-methyl-cellulose (CMC), and organic polymers can be used in water-base muds as water-loss control agents. Caution must be exercised in using starch to prevent its fermentation by maintaining high pH (alkaline range), adding preservatives, or using bactericides.

Inasmuch as more than one treating agent is usually needed to restore the mud properties to acceptable values, laboratory testing of the contaminated mud before and after treatment is of utmost importance to insure proper, prompt, and most economical control of drilling fluid. Pilot tests must be performed prior to treatment and after each successive addition of different treating agents. The order in which the chemicals must be added also must be established.

The amount of solids in the mud also must be maintained at an optimum value to achieve best results on treatment with various chemicals.

Inasmuch as chemical treatment of NaCl contamination at concentrations above 10% is extremely difficult and uneconomical, conversion to salt-water-base mud may be necessary (IMCO, 1977, p. 44). "An ounce of prevention is better than a pound of cure", however, and contamination must be avoided prior to its occurrence, e.g., salt-water drilling fluid must be used in drilling through salt formations or those containing NaCl.

In converting fresh-water-base drilling fluid to salt-water drilling fluid through addition of salt, it is critical to know the initial and the desired final salt concentrations. The salt concentration is usually estimated by silver nitrate titration of the mud filtrate (see Chapter 4).

On converting fresh-water drilling fluid to salt-water drilling fluid, sometimes it is advisable to save as much of the original water-base drilling fluid as possible to provide viscosity to the final drilling fluid. Addition of salt may cause excessive increase in viscosity. According to Magcobar (1977), the usual practice is: (1) to discard about $\frac{1}{3}$ to $\frac{1}{2}$ of the fresh-water mud, (2) to add salt, and (3) at the same time to add continuously fresh water to maintain pumpable mud. Addition of some clay, which swells in salt water, and water-loss control agents is commonly necessary to improve viscosity and reduce water loss, respectively.

Effects of divalent cations

Divalent cations increase the apparent viscosity, gel strength, yield point, and water loss, and reduce plastic viscosity of prehydrated bentonite mixtures. The magnitude of changes depends on (1) type and concentration of divalent cation, (2) type of clay in the mixture, and (3) condition of clay prior to addition of divalent cation.

In general, all divalent cations affect the properties of the bentonite drilling fluid similarly. Inasmuch as calcium is the most common cation encountered during drilling, its effects on properties of drilling fluids are of greatest interest to petroleum engineers.

Figure 8-13 shows the effect of calcium ion on the apparent viscosity of prehydrated bentonite. The most probable mechanism most responsible for the initial increase in viscosity with increasing concentration of Ca^{2+} cation is base exchange. Upon introduction of calcium cation, the base exchange capacity of which is higher than that of sodium cation, a high-swelling Na-montmorillonite is easily converted to a lower-swelling Ca-montmorillonite. According to Rogers (1963, p. 362), this conversion leads to new charge distribution among the clay platelets, which assume one or more of the equilib-

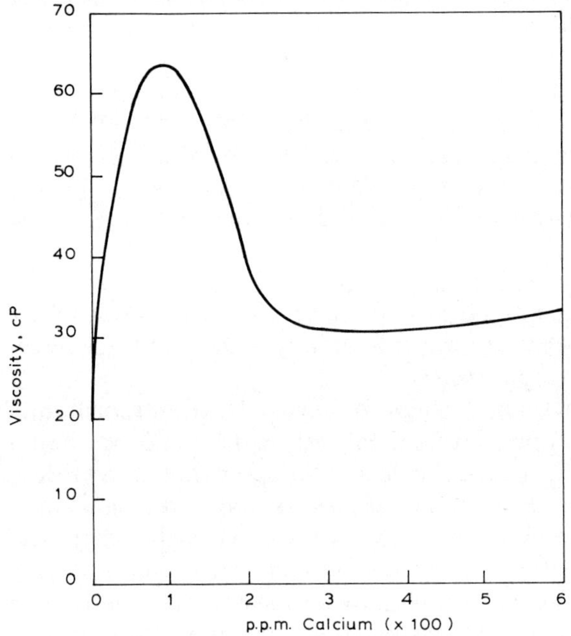

Fig. 8-13. Relationship between apparent viscosity and calcium ion concentration for a prehydrated bentonite slurry. (Courtesy of Magcobar, 1968, fig. 6-4, p. 81.)

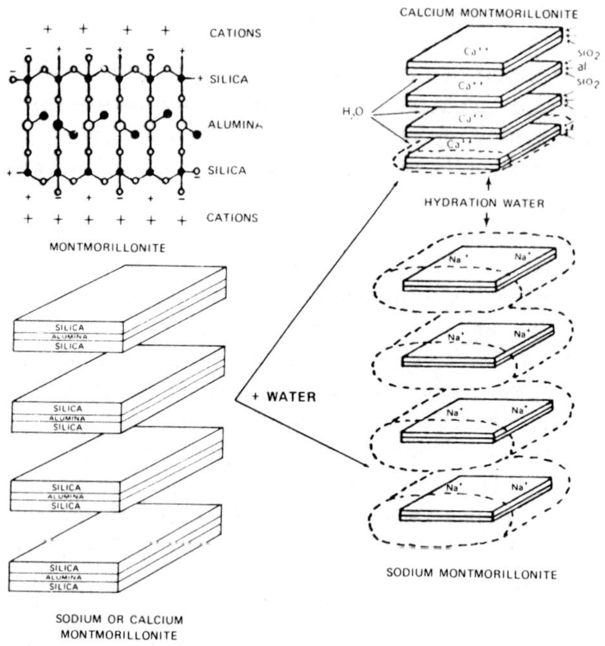

Fig. 8-14. Comparison between hydration of Na- and Ca-montmorillonite clay platelets. (Courtesy of Magcobar, 1977, fig. 2, p. 2.)

rium states shown in Fig. 8-3. The assumed equilibrium states and the degree of conversion of Na- to Ca-montmorillonite control the extent and magnitude of changes in properties.

Possibly, the attractive forces giving rise to edge-to-surface contacts also increase, resulting in stronger brush-heap structures. This may increase apparent viscosity, yield point, and gel strength. Conversion of Na- to Ca-montmorillonite increases the water loss because of increase in the permeability of filter cake. This is probably due to a considerable reduction in thicknesses of the water envelopes around clay platelets.

As shown in Fig. 8-14, the water envelopes around Na-montmorillonite platelets are thicker because Na^+ (monovalent cation), which has lower bonding energy than Ca^{2+} (divalent cation), allows greater water penetration between the clay platelets. Upon completion of base exchange reaction (Ca^{2+} for Na^+), calcium cations, bonding energy of which is much higher than that of sodium cations, impart greater attraction between planar faces of clay platelets. Consequently, less water can penetrate in between platelets. As a result of this conversion, clay volume is reduced and the free water created may easily move into formation as mud filtrate, before a low-permeability mud cake may form due to deposition of clay bundles ("flocs"). Thus, Ca-

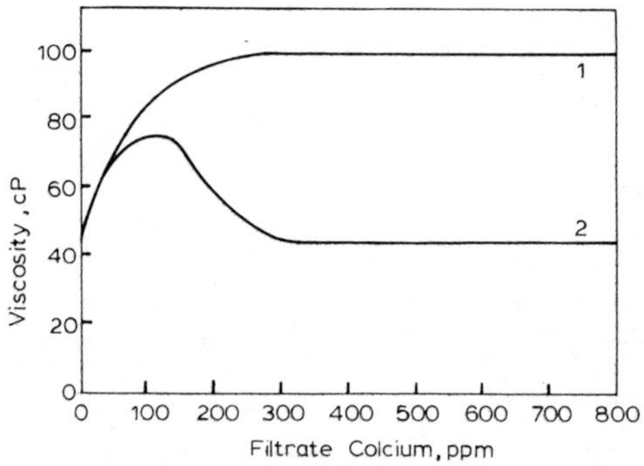

Fig. 8-15. Relationship between viscosity and calcium ion concentration in the filtrate for prehydrated alkaline muds. (Courtesy of Magcobar Services, 1968, fig. 8-3, p. 94). *1* = high-solids mud; *2* = low-solids mud.

montmorillonite forms a thick mud cake rather than a thin one forming in the case of Na-montmorillonite, when individual clay platelets are deposited to form a filter cake having a very low permeability.

Because of base exchange, aggregated state may also result in a well-dispersed mud system. In such a case, gel strength is reduced due to the non-existence of brush-heap structures. Water loss increases as the clay volume decreases with generation of additional free water. As concentration of Ca^{2+} cations increases, degree of aggregation may increase simply as a result of neutralization of negative charges on clay platelets by cations. In this state, a reduction in plastic viscosity may occur. As a result of aggregation of clay particles, there is an overall reduction of solid volume, enabling the clay "clumps" or "flocs" to move more freely in the water. Plastic viscosity decreases with decreasing internal friction. Eventually, at very high concentrations of Ca^{2+} cations, the solid (clays) and liquid (water) phases may completely separate giving rise to a very high water loss. Some specialists refer to this state as complete "flocculation" of clay.

Figure 8-15 shows increase in viscosity resulting from flocculation by Ca^{2+} cations of high-solids and low-solids lightly-treated native muds used in top-hole drilling. Subsequent decrease in viscosity is due to the base exchange reaction. As noted by Rogers (1963), another effect of Ca^{2+} cations is that of decreasing pH by displacing the H^+ cations from the clay platelets by mass action reaction. This effect is pronounced when $CaCl_2$ is added to the clay—water mixture. In the case of $Ca(OH)_2$ addition, the pH reduction is masked or overshadowed by the increased concentration of OH^- ions.

Calcium cations have different effects on properties of pre- and post-hydrated (dry) montmorillonite-in-water suspensions (see Figs. 8-13 and 8-16). Post-hydration refers to hydration of dry clay particles in water containing various ions, e.g., Ca^{2+} cations. When Ca^{2+} cation concentration is less than

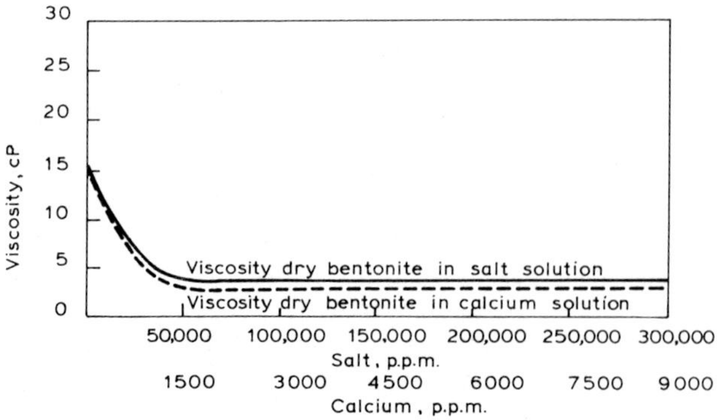

Fig. 8-16. Relationship between Ca^{2+} cation (and salt) concentration and viscosity of post-hydrated (dry) montmorillonite suspensions in water. (Courtesy of Magcobar, 1968, fig. 3-6, p. 53.)

50 ppm, there is an ordinary swelling of the bentonite clay particles. As the Ca^{2+} cation concentration increases, however, the viscosity decreases. Above Ca^{2+} concentration of 1500 ppm, no further swelling occurs and viscosity development stabilizes at 5 cP. Relationship between Ca^{2+} cation (and salt) concentration and viscosity of post-hydrated (dry) montmorillonite suspensions in water is presented in Fig. 8-16.

Gypsum and anhydrite contaminations

Gypsum ($CaSO_4 \cdot 2\,H_2O$) and anhydrite ($CaSO_4$) are commonly encountered contaminants during drilling operations. Gypsum and anhydrite are found (1) as thick beds or stringers, (2) in makeup water, (3) embedded in silts, and (4) sometimes, as salt dome caprocks. The effects of gypsum and anhydrite on mud properties are similar to those of cement. Contamination results in a marked increase in apparent viscosity, gel strength, yield point, and water loss, which is due mainly to the aggregating and flocculating effects of doubly-charged Ca^{2+} cations.

In order to distinguish gypsum and anhydrite contamination from cement contamination, tests for SO_4^{2-} anion may be performed to determine its presence. Proper treatment may then be prescribed. Tests for Ca^{2+} and SO_4^{2-} ions are discussed in Chapter 4.

Gypsum and anhydrite contaminations may also be differentiated from cement contamination by measuring pH of the mud, because cement contamination gives rise to high pH. Inasmuch as pH of the mud may be a function of several factors, however, final conclusion should be based on Ca^{2+} and SO_4^{2-} contents. As a result of gypsum and anhydrite contamination, pH reduction may occur due to base exchange (mass action) of the Ca^{2+} for the

H^+ cations attached to the clay particles.

Severe contamination may result if gypsum and anhydrite are encountered unexpectedly. Contamination is due to the combined effects of Ca^{2+} cations and SO_4^{2-} anions, and the severity of contamination is dependent upon the amount of contaminants incorporated into the mud stream.

Thicknesses of anhydrite beds may range from about 800 ft in some areas to 1—25 ft (stringers) in others. As a result, treatment problems are great. One method is to convert the mud being used to *lime* or *gyp* muds which are inert to $CaSO_4$ contamination.

The Ca^{2+} and SO_4^{2-} ions can be precipitated out as insoluble $CaCO_3$ and Na_2SO_4 by some chemicals, which may be applied in two ways: (1) Drilling fluid can be pretreated with organic thinners to prevent aggregation and flocculation of clay particles. It becomes uneconomical, however, when thick sections of gypsum or anhydrite are being penetrated. (2) Chemicals, such as Na_2CO_3 or $BaCO_3$, may be added to the contaminated mud.

The reactions involving treatment with Na_2CO_3 (soda ash) or $BaCO_3$ can be presented as follows:

$$Na_2CO_3 + CaSO_4 \rightarrow CaCO_3\downarrow + Na_2SO_4\downarrow$$
$$BaCO_3 + CaSO_4 \rightarrow CaCO_3\downarrow + BaSO_4\downarrow$$

Inasmuch as there is a 1 : 1 reaction between Na_2CO_3 and $CaSO_4$, the amount of soda ash required may be calculated from the amounts of Ca^{2+} and SO_4^{2-} ions present (see Chapter 4 for methods of analyses). Detrimental effects of Ca^{2+} and Na^+ cations are minimized as a result of precipitation of Na_2SO_4 and $CaCO_3$.

Cement contamination and treatments

Cement is utilized at various stages of drilling a well, e.g., cementing of casing, squeeze cementing, and plugging back. Cement is composed in various proportions of calcium, silica, aluminum, and iron compounds, which upon hydration and subsequent setting form a strong material. Most of the dry oil-well cements contain about 60% $Ca(OH)_2$ and varying proportions of other constituents (Rogers, 1963, p. 448). Upon hydration of cement, certain portion of this $Ca(OH)_2$ is released to the aqueous phase. Inasmuch as free lime [$Ca(OH)_2$)] dissociates in the aqueous phase to Ca^{2+} and OH^- ions, the cement contamination is due to these two ions. The degree of cement contamination depends on: (1) concentration of solids, particularly that of reactive solids, (2) condition of cement when drilling out the plug, (3) amount of cement which must be drilled out, (4) types of cement, and (5) ratio of the mud volume to the volume of drilled-out cement, and (6) chemicals used in previous treatments of the mud.

Inasmuch as a high solids content is not desirable from the standpoint of achieving reasonable rheological property control, the additional presence of cement chips aggravates the problem. In general, only a small amount of chip-sized cement is generated if cement is properly set and allowed to dry before drilling out. In such a case, cement chips contaminate only a small volume of contacted mud, which forms thick mud gel around the cement chips. Further spreading of reaction throughout the mud system is thus prevented. Cement chips can then be discarded after removal at the surface by the shale shakers or in settling pits.

Drilling out wet (or "green") cement, which has not been allowed to dry and set properly gives rise to a much greater amount of $Ca(OH)_2$ than dry and well-cured cement. Finely-ground, readily-dispersible chips of cement, having large surface area, are thus incorporated into the mud and greatly thicken it. This necessitates expensive chemical treatment, because mechanical removal of the mud-coated cement chips becomes impractical.

If the drilled-out plug has a small size, only minor portion of the mud is contaminated with cement. In such a case, it is possible to circulate this portion of the mud to the surface and discard it after separation. The effects of contamination may be considerably reduced in the presence of thinners remaining from prior chemical treatments of the mud.

The commonly used methods available for treating cement contamination, which increases apparent viscosity, yield point, gel strength and water loss, are as follows: (1) pretreatment of the mud, (2) precipitation of the contaminant and restoration of the mud properties, and (3) toleration of the effects of contamination.

As a result of cement contamination, Ca^{2+} and OH^- ions give rise to aggregated—deflocculated or aggregated—flocculated state of clay particles depending on degree of the contamination. Thinning agents may redisperse the clay platelets. An accepted practice in preventing contamination is to pretreat the mud with chemicals at the surface and circulate until return reaches the surface pit, before starting to drill out the cement, because it is always known when cement contamination will occur (see excellent discussion in Magcobar, 1977). Organic types of the phosphate and tannic families such as sodium-acid-pyrophosphate (SAPP) and quebracho, may be used as thinners.

Small degree of cement contamination may be tolerated, especially in the upper portion of the hole. Excessive pump pressure may be required, while drilling at great depths, to circulate the cement chips to the surface. Consequently, cement contamination may not be tolerated.

In the case of slight contamination resulting in slight increase in viscosity, yield point, gel strength, and water loss, the addition of a small amount of water may correct the problem. If addition of water causes a reduction in mud weight, the weighting materials must be added to control the formation pressure.

A chemical treatment, which involves removal of Ca^{2+} cations and neutralization of OH^- anions, must be prescribed if cement contamination cannot be tolerated. Depending on prior treatment of the prehydrated bentonite mud, phosphates, sodium bicarbonate ($NaHCO_3$), or soda ash (Na_2CO_3) may be used to precipitate Ca^{2+} cations out of solution (see Magcobar, 1977). The calcium cations can be removed as calcium carbonate ($CaCO_3$), which is inert and does not have detrimental effects on the mud properties.

The following phosphates have been used successfully to treat cement contamination: (1) sodium-acid-pyrophosphate (SAPP), (2) sodium-tetraphosphate, (3) sodium-hexametaphosphate, and (4) tetrasodium-pyrophosphate. Sodium-acid-pyrophosphate (SAPP) and the sodium-tetraphosphate, however, are used more commonly. The pH of the mud should be maintained in the range of 8.0—9.5, which is readily attainable in the case of cement contamination, in order to obtain good results. A sharp reduction of pH, however, may occur on continued treatment with these phosphates, which could thicken the mud. In addition, in the presence of high salt concentrations and resulting strong flocculating effects on clays, the dispersing ability of phosphates is minimized.

Some of the contaminating calcium cations are removed by the phosphates in the aqueous phase through formation of complexes of the polyphosphates (sequestering effect) or insoluble orthophosphate (Baroid, 1975, p. 32).

In treating cement contamination, phosphates are introduced through the chemical barrel directly into the contaminated mud system. The Ca^{2+} cations can be precipitated by monosodium phosphate as follows:

$$3\ Ca(OH)_2 + 2\ NaH_2PO_4 \rightarrow Ca_3(PO_4)_2\downarrow + 2\ NaOH + 4\ H_2O$$

Sodium bicarbonate ($NaHCO_3$) and soda ash (Na_2CO_3) are commonly used to treat cement contamination by removing the Ca^{2+} cations from solution in the form of calcium carbonate ($CaCO_3$). Caustic soda (NaOH), however, forms during the reaction. The following reaction occurs between sodium bicarbonate ($NaHCO_3$) and lime [$Ca(OH)_2$] in solution:

$$Ca(OH)_2 + NaHCO_3 \rightarrow CaCO_3\downarrow + NaOH + H_2O$$

The effectiveness of sodium bicarbonate is greater when the mud has undergone little or no prior tannin or phosphate treatment (Magcobar, 1965, p. 81).

Inasmuch as only carbonate anions (CO_3^{2-}) and not bicarbonate anions (HCO_3^-), precipitate the Ca^{2+} cations as $CaCO_3$, it is very important to maintain pH of the mud at approximately 8.3 or higher in order to achieve complete dissociation of $NaHCO_3$. Bicarbonate anions (HCO_3^-) cannot precipitate Ca^{2+} cations. Cement contamination, however, usually raises the pH of the mud sufficiently (>8.3). At all times, however, petroleum engineers should be aware of the deleterious effect of Na^+ cation. In order to treat 74 g of calcium hydroxide (1 mole of lime), 84 g of sodium bicarbonate are required to produce 100 g of $CaCO_3$ and 40 g of NaOH. In the case of overtreatment, the following secondary reaction may occur:

$$NaOH + NaHCO_3 \rightarrow Na_2CO_3 + H_2O$$

According to this reaction, 40 g of NaOH react with 84 g of $NaHCO_3$ to form 106 g of soda ash (Na_2CO_3). The pH adjustment of the drilling fluid can be accomplished by adding sodium tannate thinners in small quantities prior to cement contamination (Rogers, 1963, p. 451). Caustic soda (NaOH), however, is the more traditional additive:

$$Na^+ + OH^- + Na^+ + HCO_3^- \rightarrow CO_3^{2-} + 2\,Na^+ + H_2O$$

Overtreatment with $NaHCO_3$, with introduction of excessive amounts of Na^+ cations, should be avoided in order to prevent thickening of mud by Na^+ cations. Even in the absence of overtreatment, the presence of Na^+ cations prevents restoration of original viscosity, yield point, and gel strength of the mud.

Soda ash (Na_2CO_3) sometimes is used to treat the cement contamination instead of sodium bicarbonate ($NaHCO_3$). In this case, however, for every carbonate (CO_3^{2-}) anion necessary to precipitate a Ca^{2+} cation, two Na^+ cations, which may have deleterious effects on drilling fluid properties, are added instead of one. Sodium carbonate (soda ash) precipitates the Ca^{2+} cations in the form of insoluble $CaCO_3$ as follows:

$$Ca(OH)_2 + Na_2CO_3 \rightarrow CaCO_3\downarrow + 2\,NaOH$$

The very high pH of mud resulting from the high pH of soda ash ($\approx$11.0) combined with generated NaOH leads to increased viscosity, yield point, and gel strength. Consequently, expensive organic dispersants must be added to offset the "thickening" (flocculating) effects of high pH (IMCO, 1973, p. 44).

One advantage of soda ash (Na_2CO_3) over sodium bicarbonate is that Na_2CO_3 is less sensitive to (1) pH of the mud and (2) previous chemical treatments (Magcobar, 1977, p. 81).

Organic thinners (e.g., quebracho), which are used to treat cement contaminations, may be introduced through the chemical barrel directly into the contaminated mud rather than through the hopper during pretreatment (IMCO, 1977, p. 44). [Approximately 50 lb of chemical are added per bbl of fresh water to constitute the chemical bbl (Magcobar, 1965, p. 81)]. The effectiveness of quebracho in treating cement contamination depends on: (1) salt concentration, (2) severity of contamination, and (3) hardness of the water.

Dispersion by quebracho is ineffective at large salt concentrations, because of strong flocculating effect on clays. In addition, solubility of quebracho decreases with increasing salt concentration. Hard water has a similar effect

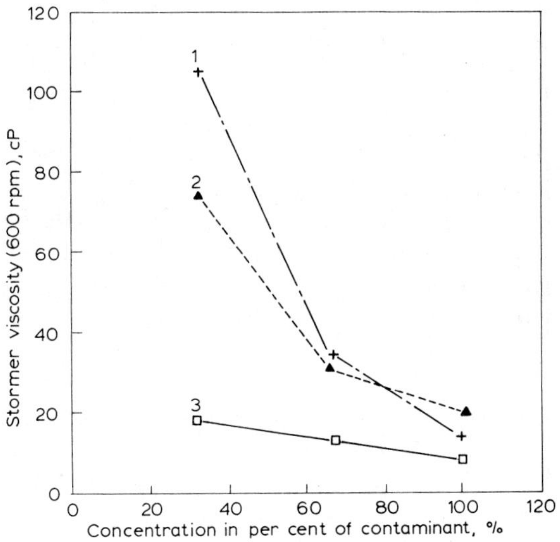

Fig. 8-17. Relationship between concentration of treating agent (sodium-acid-pyrophosphate, quebracho, and sodium bicarbonate) and apparent viscosity of cement-contaminated mud (2 g cement/100 cm^3 mud, initially composed of 6 g bentonite/100 cm^3 of water). 100% treatment = 2 g treating agent/100 cm^3 contaminated mud. (Data after Rogers, 1963, table 8-9, p. 452.) *1* = sodium bicarbonate, *2* = sodium-acid-pyrophosphate, and *3* = quebracho.

on solubility of quebracho (Baroid, 1975, p. 33). Lignosulfonates, on the other hand, are tolerant to common contaminants. Quebracho is preferred to lignosulfonates in the upper portion of the hole because of (1) its strong thinning ability, (2) lower cost, and (3) common absence of contaminants affecting it (Magcobar, 1977, p. 1). Lignosulfonates are mainly used in the lower portion of the hole, where the less refined tannic extracts, such as quebracho, become ineffective.

The relative effectiveness of some treatments for cement contamination is presented in Figs. 8-17 and 8-18. The mud was prepared by adding 6 g of bentonite to 100 cm^3 of fresh water and, then, was contaminated with 2 g of cement. As shown in Fig. 8-17, quebracho appears to be the most effective treating agent in reducing viscosity in this case. Sodium bicarbonate is not as effective as SAPP in reducing viscosity at low concentrations. Concentration of Ca^{2+} cations (ppm) remaining in the filtrate (Fig. 8-18) indicates the effectiveness of $NaHCO_3$ in precipitating out the Ca^{2+} cations. At 100% treatment, the 30-min. API water loss was 18.5, 17.5, and 15.0 cm^3 for quebracho, SAPP, and sodium bicarbonate treatment, respectively (Rogers, 1963, p. 452). Although quebracho is a very efficient treating agent for cement contamination, when large treatment must be performed, $NaHCO_3$ may be preferred because of its lower cost.

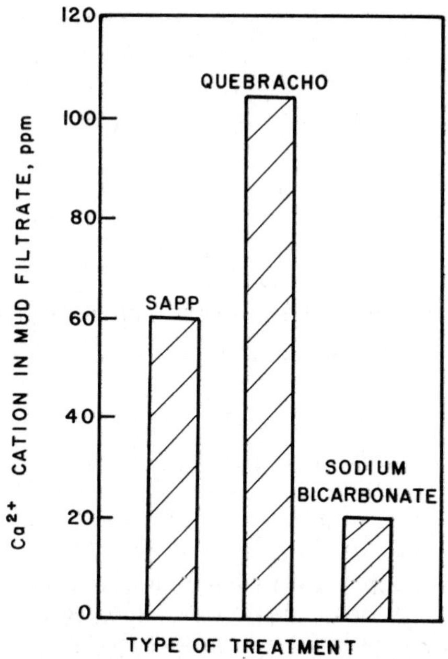

Fig. 8-18. Effects of sodium-acid-pyrophosphate, quebracho, and sodium bicarbonate treatments on concentration of Ca^{2+} cations in mud filtrate of cement-contaminated muds. (Based on data from Rogers, 1963, table 8-9, p. 452.) Initial mud (6 g bentonite/ 100 ml water) was contaminated with 2 g cement/100 cm^3 of mud and then treated with 2 g treating agent/100 cm^3 of mud. Cement contamination is equivalent to drilling 50-ft plug in 7-inch casing; pH = 11.2.

Effects of hard water on muds

Formation waters contain varying amounts of NaCl, $CaCl_2$, $MgCl_2$, $MgSO_4$, and other salts (see Degens and Chilingar, 1967). High concentrations of $MgSO_4$ are often encountered in caprocks of salt domes (Levorsen, 1967, p. 368). In oilfield waters, Mg^{2+} cations are often present in minor amounts compared to seawater, possibly due to dolomitization process in sediments and rocks: $2\,CaCO_3 + Mg^{2+} \rightarrow CaMg(CO_3)_2 + Ca^{2+}$. The hardness * of the water is determined by the relative proportions of Ca^{2+}, Mg^{2+}, CO_3^{2-}, HCO_3^-, SO_4^{2-}, Na^+, Cl^- and other ions. Total hardness is often measured in terms of ppm of $CaCO_3$ or Ca^{2+} and, sometimes, equivalents per million of Ca^{2+} cation.

* *Hard water*: $rNa^+/rCl^- < 1$; $(rCl^- - rNa^+)/(rMg^{2+}) > 1$; $(rCl^- + rSO_4^{2-}) > (rNa^+ + rK^+)$. *Alkaline water*: $rNa^+/rCl^- > 1$; $(rNa^+ - rCl^-)/(rSO_4^{2-}) > 1$; $rNa^+ + rCl^- > rSO_4^{2-}$; $rHCO_3^- > (rCa^{2+} + rMg^{2+})$; $(rNa^+ + rK^+) > (rCl^- + rSO_4^{2-})$. r = % equivalents.

If hard formation waters invade the mud stream, the degree of contamination will depend on: (1) degree of water hardness, (2) type of water-base mud, and (3) previous chemical treatments. Inasmuch as contaminating effects of Mg^{2+} and Ca^{2+} cations are similar, same treatments are applicable (Baroid, 1975, p. 46).

In addition, many oilfield waters available for preparing the mud are quite hard and bentonite clay does not swell as much in such waters. Thus, more bentonite has to be added to attain desired rheological properties, which results in additional cost. Pretreatment of the water with chemicals, however, may remove the hardness. The Mg^{2+} cations, for example, may be precipitated from solution as $Mg(OH)_2$ at a pH above 10 (Baroid, 1975, p. 46).

Temperature effects

The effects of temperature may be considered inherent to all drilling operations. Contaminating effects of temperature are twofold. First, temperature affects the viscosity of the base liquid (see Chapter 2). As a result of temperature increase, the mobility of the water molecules increases, which may have profound effects on the behavior of the clay particles and their interaction with the chemical additives. In most cases, however, the contaminating effects of temperature are reversible (Rieke, 1978, p. 80).

Secondly, and more importantly, the effects of temperature are to reduce the effectiveness of most mud-conditioning agents. For example, SAPP (sodium-acid-pyrophosphate) will perform efficiently up to a temperature of 180°F, whereas quebracho performs satisfactorily up to 225°F. Carboxymethyl-cellulose (CMC) and starch are ineffective above 300°F.

According to IMCO (1978, p. 112A), the rate of chemical reaction approximately doubles for every 15°F increase of temperature of the continuous phase, greatly shortening the effective lives of chemicals.

The temperature has severe contaminating effect on calcium-treated water-base muds. High temperature causes severe gelation or solidification of these muds. According to Magcobar (1977), the gelation is caused by the reaction between the hydroxyl anions, calcium cations, and clay in the mud to form a calcium-alumino-silicate or simple cement at temperatures exceeding 275°F.

Contamination of water-base muds by gases

Several gases (mainly O_2, CO_2, and H_2S), which are encountered during drilling as a result of (1) penetrating gaseous subterranean formations or (2) surface agitation, may have detrimental effects on the properties of water-base muds. If continuous phase of a mud is impregnated with gas bubbles, the mud is referred to as a "gas-cut" mud. According to Mercer (1976, p. 63), gas-cut mud may be formed as a result of: (1) liberation of gas by the bit as it penetrates the gaseous formations; (2) production of gas into the drilling mud from a specific zone in response to formation pressure (pore pressure) that exceeds the opposing hydrostatic pressure of a mud column; (3) cycling back of gas down the hole to appear a second time at the surface; and (4) generation of gas by degradation of mud additives. The degree of contamination depends on: (a) types and amounts of various gases encountered, (b) circulating temperature of mud, and (c) concentrations and types of various chemicals present in the mud.

TABLE 8-I

Short and long term effects of various gaseous contaminants (i.e., CO_2, H_2S, and O_2)

Short term effects	Long term effects
Carbon dioxide (CO_2)	
(1) pH reduction due to CO_2 dissolving in the aqueous phase and forming carbonic acid.	(1) CO_2 or sweet corrosion of circulating and associated drilling equipment (e.g., drillpipe).
(2) Increase in plastic viscosity due to release of dissolved CO_2 in the upper portion of the hole, where pressure is reduced forming gas bubbles and emulsion which increase plastic viscosity.	(2) High cost of replacing above equipment.
(3) Reduction in the specific weight due to entrained gas bubbles which formed as discussed above.	
(4) Foam formation at surface.	
(5) Added costs of treating and conditioning agents.	
Hydrogen sulfide (H_2S)	
(1) All of the above.	(1) H_2S or sour corrosion of circulation and associated drilling equipment.
(2) Personnel hazard.	(2) High cost of replacing such items.
(3) Hydrogen sulfide stress cracking (see corrosion chapter).	
Oxygen (O_2)	
(1) Foam formation at surface.	(1) Oxygen (O_2) corrosion of circulation and associated drilling equipment.
(2) Added cost of treating and conditioning agents.	(2) High cost of replacing such items.

Gaseous contaminants are encountered in subsurface formations in many areas of the world. The mud technologists are mostly concerned with the short-term effects, which affect the mud properties directly and instantaneously. Some of the common short-term and long-term effects of gaseous contaminants are listed in Table 8-I.

If gas bubbles are dispersed in the continuous phase of the mud, mud is lightened and, consequently, addition of weighting materials, which are expensive, is required to maintain the desired specific weight. Presence of gas bubbles increases internal friction and, thus, gives rise to increased plastic viscosity, because it is a function of internal friction resisting the flow of the fluid. The increase in viscosity, however, may be of short duration, because gas bubbles coalesce to form separate continuous phase.

Inasmuch as contaminating gases may be generated internally from degradation of some chemicals, the selected chemicals should not degrade at expected maximum bottomhole temperatures.

Corrosion of drillpipe and other tubular goods, surface connections, and all associated rubber parts is a long-term contaminating effect of gases in water-base muds. Hydrogen sulfide, oxygen, and carbon dioxide gases are highly corrosive under specified conditions. Chilingar and Beeson (1969) discussed mechanisms of oxygen corrosion, sweet or carbon dioxide corrosion, and sour or hydrogen sulfide corrosion.

Hydrogen sulfide, which occurs naturally, may also be generated internally by the bacterial action, thermal degradation, and electrochemical reactions occurring in the mud (Chesser, 1968). According to Chesser (1968), presence of even a few parts per million (ppm) of hydrogen sulfide in the mud gives rise to almost certain failure of high-strength tubular goods. Untreated lignosulfonate solutions, such as lignosulfonate muds used at the initial stage of their development, exhibit higher corrosion rate after aging (Skelly and Dieball, 1969, p. 3), because H_2S forms as a result of degradation of these compounds. Properly-maintained lignosulfonate-treated muds, however, are effective, because oxygen scavenger present in the mud minimizes oxygen corrosion of drillpipe and other tubular goods (Chesser, 1968, p. 14).

In the excellent API publication entitled "Recommended Practice-Standard Procedure for Testing Drilling Fluids, RP 13 B", a recently-developed method, which replaces the old "Hack Paper" method for sulfide determination in drilling muds, is discussed (Garrett, 1976). This method, which is referred to in the above publication as the "Garrett Gas Train Sulfide Analysis", has been found to be suitable for use at the rig site, because it is rapid, results are accurate, and the test reliability and simplicity of testing procedure are high (Garrett, 1976).

Treatment of H_2S in water-base muds requires utilization of such chemicals as H_2S-scavengers. For example, ironite sponges react on contact with the H_2S to form stable pyrite (FeS_2), which is safe (Anonymous, 1974, p. 173). This pyrite is easily removed at surface. Presence of pyrites, how-

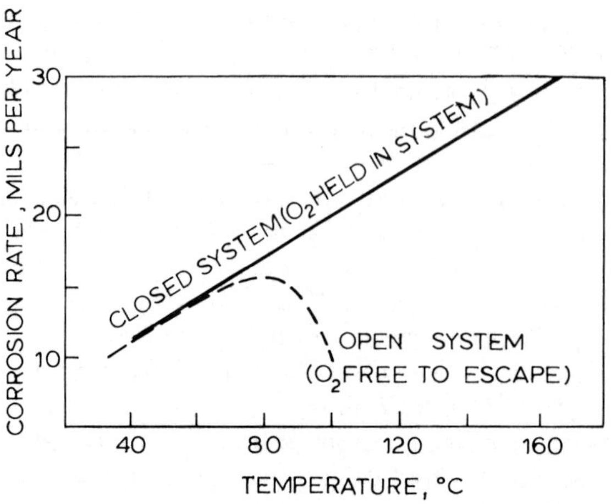

Fig. 8-19. Relationship between corrosion rate in water-base muds and temperature in the closed (O_2 is held in) and open (O_2 is allowed to escape) systems. (After Cox, 1974, fig. 2, p. 110; courtesy of World Oil.)

ever, slightly increases the plastic viscosity. Pretreatment should be done whenever possible.

Carbon dioxide, which is commonly encountered by the drilling mud, lightens the mud and, under certain conditions, becomes dissolved in the continuous phase (water) making it acidic. This leads to severe corrosion problems. Consequently, CO_2 must be removed by using some method, such as surface degasser equipment.

Inasmuch as oxygen is the leading cause of drillpipe corrosion and failure, even traces of oxygen in the mud must be avoided. As pointed out by Bradley (1970, p. 50), oxygen corrosion in water-base muds is not limited to surface atmospheric conditions. The solubility of oxygen in fresh water decreases with increasing temperature in the system open to atmosphere, which allows escape of O_2. This does not mean, however, that corrosion would decrease at higher temperatures in oil wells, because modern mud-circulating systems are mainly of the closed system where oxygen cannot escape. According to Cox (1974), corrosion rate in closed system not only does not decrease with increasing temperature, because the oxygen gas is not allowed to escape, but rather increases (Fig. 8-19). This is due to the increase in the amount of free oxygen.

Two common methods are used to remove oxygen from water-base muds: (1) use of degasser equipment, and (2) addition of oxygen scavengers, such as lignosulfonates. Simpson and Barbee (1970), Chesser (1968), and Browning et al. (1962) have discussed scavenging property of lignosulfonates. Other sulfite-base chemicals have been used extensively as oxygen scavengers

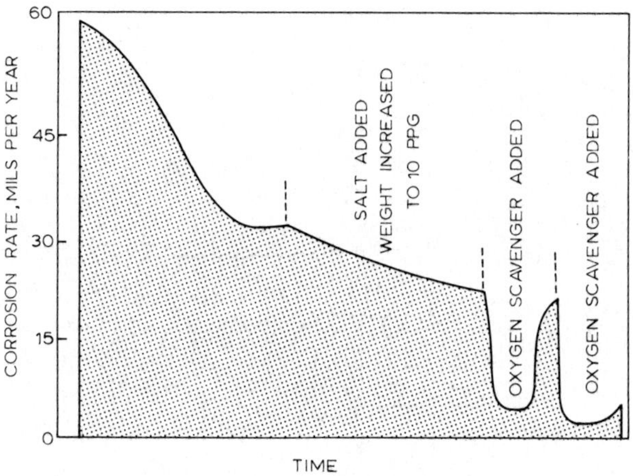

Fig. 8-20. Variation of corrosion rate with time on addition of salt and oxygen scavenger. (After Cox, 1974, fig. 6, p. 112; courtesy of World Oil.)

(Cox, 1974). As shown in Fig. 8-20, the corrosion rate may be controlled by addition of oxygen scavenger. As pointed out by Cox (1974), oxygen scavenger must be added at regular intervals in order to maintain lowest possible oxygen content, because O_2 content of the mud and the resulting corrosiveness increase with time of exposure of the mud to O_2-bearing environment.

INHIBITIVE MUDS

Since 1932, a trend was established in mud technology toward developing drilling fluids which will (1) minimize damage to the formations penetrated during drilling, and (2) increase the wellbore stability, without reducing formation productivity. Thus, *lime* and *gyp* muds were popular inhibitive muds up to 1960; *surfactant* muds, from 1956 through 1960; and lignosulfonate, polymer, and salt-water muds from 1956 to the present time.

Inhibitive mud is defined in the "Glossary of Drilling Fluid Terms" (API, 1974) as *any* drilling fluid having an aqueous phase with chemical composition that tends to retard and even prevent (inhibit) appreciable clay hydration and dispersion through chemical and/or physical means. Several mud types fit this description: (1) all calcium-treated muds, (2) surfactant muds, (3) lignosulfonate muds, (4) saturated salt-water muds, and (5) certain low-solids muds, such as polymer muds.

Inhibiting mechanisms in inhibitive muds

Inhibitive muds can be prepared by addition of (1) sufficient amount of certain selected thinners, (2) various electrolytes, (3) certain selected polymers, and (4) combination of two or more of the above.

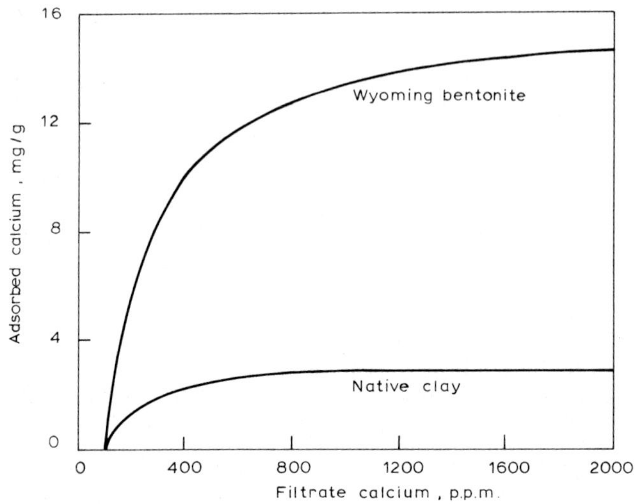

Fig. 8-21. Relationship between the amount of calcium adsorbed on clay and its concentration in the mud filtrate. (Courtesy of Magcobar, 1977, chapter 8, p. 2, fig. 2.)

Presence of high concentrations of some electrolytes in the aqueous phase may give rise to inhibiting properties of muds. Electrolytes containing doubly-charged Ca^{2+} cations and singly-charged Na^+ and K^+ cations have been used extensively. Base exchange and mass action reactions are operative in such systems and the degree of inhibition depends directly on the concentration of electrolytes in the aqueous phase and in the mud filtrate.

In calcium-treated muds, Ca^{2+} cations, having a high cation exchange capacity, will replace Na^+ cations from the clay structures and cause the clay structure to collapse, with reduction in the solid volume. This process may also occur in the formation between formation clays and calcium-containing filtrate, giving rise to wellbore stability. Shale cuttings are also similarly affected, preventing their further disintegration and dispersion.

Calcium-inhibited muds, Ca^{2+} cations of which are derived from $CaCl_2$, may be referred to as salt-inhibited muds, whereas *lime* and *gyp* muds refer to muds which derive their Ca^{2+} cations from $Ca(OH)_2$ and $CaSO_4$, respectively.

Sufficient amount of Ca^{2+} cations in the aqueous phase and mud filtrate must be present to attain completion of all the required base exchange reactions. In order to determine the concentration of Ca^{2+} cation in the mud filtrate necessary to inhibit swelling of formation clays, it is necessary to prepare curves such as the ones presented in Fig. 8-21.

The solubilities of $Ca(OH)_2$, $CaSO_4$, and $CaCl_2$ in water are greatly increased with increasing temperature. Solubility of $CaCl_2$ increases with increasing pH, whereas solubilities of $Ca(OH)_2$ and $CaSO_4$ decrease with increasing pH.

Inasmuch as Na-montmorillonite adsorbs Ca^{2+} cations to a much greater extent than do native clays (see Fig. 8-21), smaller amounts of Ca^{2+} are left in the aqueous phase to inhibit the drilled solids and formation clays.

According to Perricone and Browning (1962, p. 5), mud filtrate having a Ca^{2+} cation concentration in excess of 3,500 ppm provides a maximum degree of swelling inhibition for montmorillonite clay. Clays hydrate and disperse when the Ca^{2+} cation content in aqueous phase is depleted.

Inhibition by monovalent salts

Salt-water muds may be considered to be inhibitive muds, because at high salt concentration ($\approx$40,000 ppm) they inhibit hydration and swelling of montmorillonite and illite clays and prevent shale cuttings from disintegrating. High concentrations of Na^+ cation in the mud filtrate, however, may have detrimental effects on the formation productivity, because as a result of mass action reactions the calcium-base formation clays can be converted to more swelling sodium-base clays. As a result of solid volume increase and dispersion of clays, pores may become plugged, creating permanent reduction in permeability.

Potassium chloride (KCl) imparts inhibiting ability to the mud and,

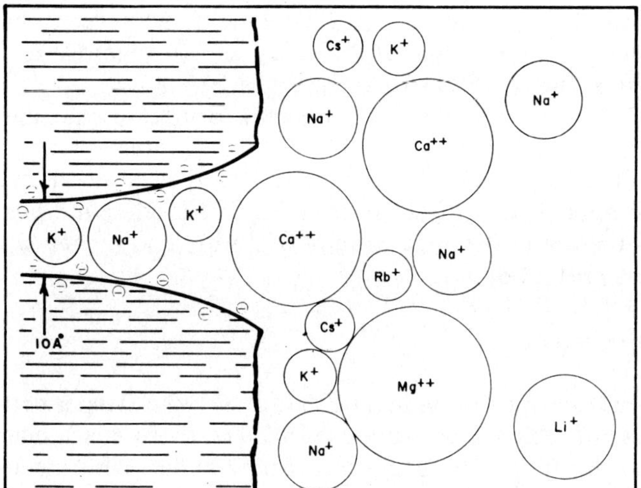

Fig. 8-22. Relative sizes of various cations (drawn to scale). The K^+ ion is able to penetrate the clay interlayer, whereas other cations cannot due to their larger sizes. (Courtesy of Baroid Petroleum Services, 1975, fig. 12, p. 21.)

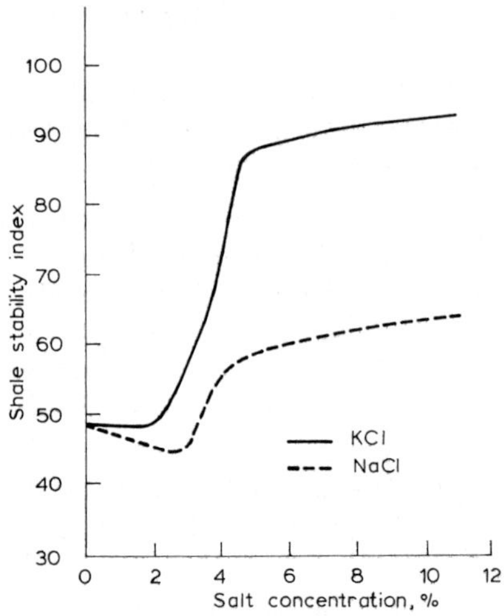

Fig. 8 23. Relationship between salt concentration (NaCl and KCl) and shale stability index. (After Mondshine, 1973, p. 14, fig. 3; courtesy of the Society of Petroleum Engineers of AIME.)

according to Mondshine (1973), has been found to be more effective than NaCl in inhibiting shale swelling. Optimum inhibition with KCl is attained at ≈5% by weight concentration. The K^+ cation acts as an inhibitor probably because of its small size and low hydrational energy (Mondshine, 1973). The K^+ cations can penetrate the clay lattices and become tightly fixed by attractive forces. This results in binding of the clay sheets together and reduction in the tendency of the clays to swell (see Fig. 8-22).

The effectiveness of KCl as an inhibitor, as compared to sodium chloride, is presented in Fig. 8-23. According to Clark et al. (1975), the concentration of KCl necessary to inhibit clay swelling varies with the clay mineralogy: hard illitic shales require approximately 10 lb/bbl of mud for adequate protection, whereas softer offshore shales may require 35 lb/bbl and higher (up to 60 lb/bbl) for adequate protection.

Inhibiting mechanisms of thinners

As pointed out by Perricone and Browning, (1962, p. 3), with deeper drilling higher bottomhole temperatures are encountered and there is a longer exposure time of mud to the formation clays and drilled solids. This places stringent requirements on the mud which must be used. An ideal water-base mud should consist of an inert or non-reactive fluid which will come in contact with formation clays (Perricone and Browning, 1962, p. 3). Consequently, new types of inhibitive muds were developed that differ from the calcium-treated inhibitive muds, effectiveness of which depends on base-exchange reactions.

According to Browning (1960, p. 3), electrokinetic and physicochemical (or mechanical) stabilizations are mainly responsible for the inhibiting ability of thinner-treated muds. Physicochemical stabilization is the more widely used process.

Physicochemical stabilization involves (1) formation of a physical barrier on the clay structural surfaces by adsorption of macromolecular substances, which possess structural viscosity due to the immobilization of dispersion media, (2) adsorption of materials that create highly viscous layers on the clay mineral surfaces (see Browning, 1960, p. 3). Certain organic polyelec-

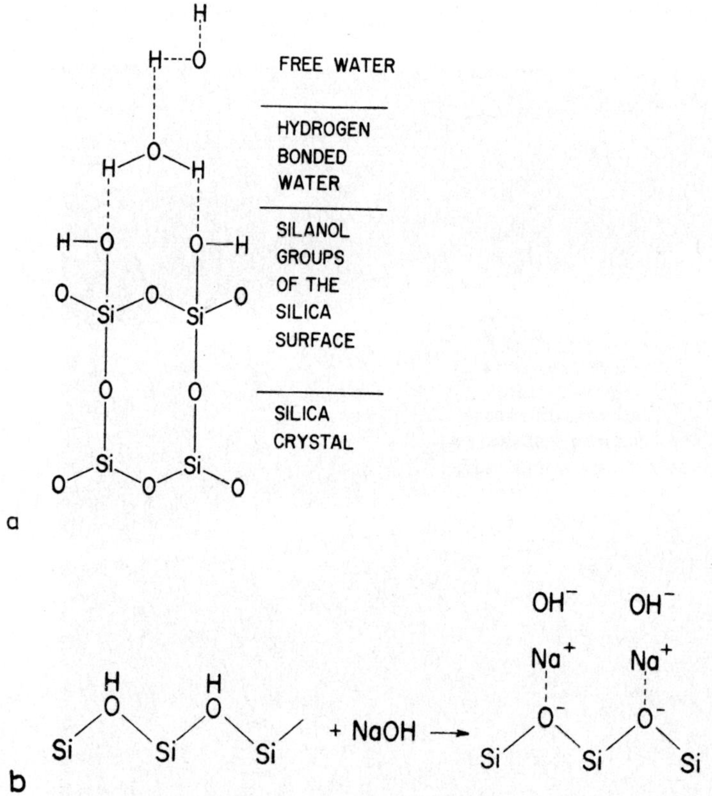

Fig. 8-24. Schematic diagram of hydrogen ion sites on clay surfaces.
a. Complex nature of a silica surface of montmorillonite in the presence of water. (After Browning, 1964, p. 4; courtesy of the Society of Petroleum Engineers of AIME.)
b. Bonding of ionized sodium ions to the silica—oxygen structural surface. (After Browning, 1964, p. 5; courtesy of the Society of Petroleum Engineers of AIME.)

trolyte agents are suitable for use in water-base muds because of their high tolerance to common contaminants encountered during drilling. Lignosulfonates of organic lignin substances have been used extensively. They have been referred to as mud conditioners, thinners, or organic polyelectrolyte agents.

Browning and Perricone (1963) have discussed the chemistry of lignosulfonates. Understanding of the adsorption mechanism of polyelectrolyte agents on clay surfaces is of utmost importance, because the degree of colloidal stabilization is a function of how well these agents are able to become adsorbed. As shown in Fig. 8-24, H^+ cations are present at different sites on the clay surfaces. Inasmuch as lignosulfonates possess natural hydrogen bonding groups, a strong hydrogen bonding is created when they come in contact with clays. Thus, lignosulfonate molecules are attached or adsorbed

Fig. 8-25. Model illustrating polymolecular adsorption of lignosulfonate molecules on clay particles. (After Browning and Perricone, 1963, p. 13, fig. 6; courtesy of the Society of Petroleum Engineers of AIME.)

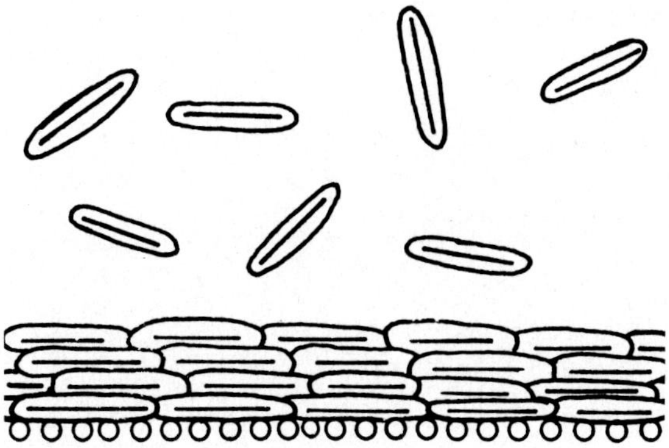

Fig. 8-26. Schematic diagram showing water loss control mechanism on addition of lignosulfonate. (After Browning and Perricone, 1962, p. 14, fig. 12; courtesy of the Society of Petroleum Engineers of AIME.)

onto the clay surfaces through hydrogen bonding. A multilayer adsorption is formed because lignosulfonates are polyelectrolytes (Fig. 8-25). Adsorption of lignosulfonate molecules on the clay surfaces creates this viscous multilayer, which stabilizes the clay particle.

According to Browning and Perricone (1963, p. 5), lignosulfonate molecules also exhibit strong lateral attraction among themselves, when adsorbed on the clay surfaces, which further stabilizes the viscous multilayer.

These viscous multilayers prevent flocculation of the clay particles by physically preventing them from coming close enough to each other for the van der Waals' attractive forces to become active. This results in the reduction of apparent viscosity, yield point, and gel strength.

Water loss is reduced because strong lateral attraction between lignosulfonate molecules results in the formation of a continuous impermeable layer, when the clay particles are deposited on the formation face. Pore spaces that may exist between deposited clay particles are also filled by large lignosulfonate molecules preventing escape of water into the formation (see Fig. 8-26).

Inasmuch as viscous multilayers also adsorb on shale cuttings, swelling and resulting dispersion of the chips are prevented. Lignosulfonate molecules present in the mud filtrate, which comes in contact with formation clays, also have a stabilizing effect. Browning and Perricone (1963) pointed out that the lignosulfonate molecules are too large to enter the clay interplanar layers and cause dispersion. Due to the plating effect, however, they prevent entrance of water molecules into the interlayer positions.

The two major hydration mechanisms of shales and formation clays, i.e., surface and osmotic hydrations, resulting in wellbore instability, are prevented from occurring by lignosulfonates. Thus, shale and formation clays maintain their original state upon entrance of mud filtrate.

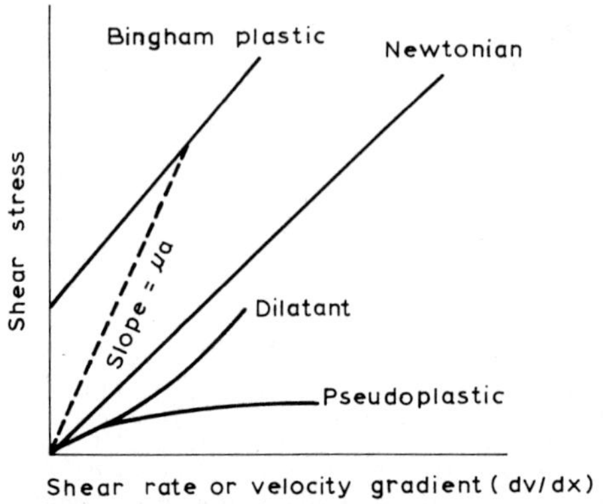

Fig. 8-27. Shear stress versus shear rate curves for Newtonian and non-Newtonian fluids.

USE OF PLASTIC VISCOSITY-VERSUS-YIELD POINT CURVES FOR EVALUATION OF DRILLING FLUID CONDITION

As shown in Fig. 8-27 shear stress τ can be related to the rate of shear, dv/dx, for Bingham fluids as follows:

$$\tau = \overline{YP} + \overline{PV}\left(\frac{dv}{dx}\right) \tag{8-2}$$

where $\overline{YP}$ is the yield point (intercept on the y-axis) and $\overline{PV}$ is the plastic viscosity (negative slope of the line). Apparent viscosity, $\overline{AV}$, on the other hand depends on the particular rate of shear:

$$\tau = \overline{AV}\left(\frac{dv}{dx}\right) \tag{8-3}$$

Thus:

$$\overline{AV} = \overline{PV} + \frac{\overline{YP}}{(dv/dx)} \tag{8-4}$$

As shown in Fig. 8-28, it is possible to plot $\overline{PV}$ versus density or $\overline{YP}$ versus density. These curves may be considered consisting of straight-line portions *AB*, *BC*, *CD*, and *DE*. The points of inflexion seem to reflect changes in certain stages in the development of the physicochemical conditions of mud. Riabchenko et al. (1968) proposed a method for evaluating stability of drilling fluids using $\overline{PV}$-versus-$\overline{YP}$ curves (see Fig. 8-29). The interval *CD*, representing the major portion of the curve, can be expressed by the following equation:

$$\overline{YP} = \tan\alpha(\overline{PV} - \overline{PV}_{min}) \tag{8-5}$$

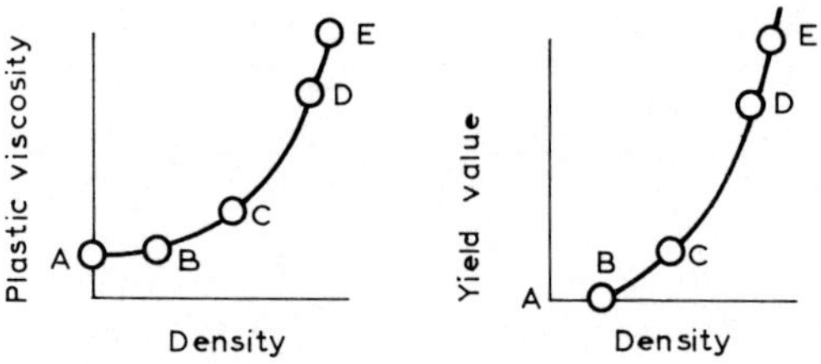

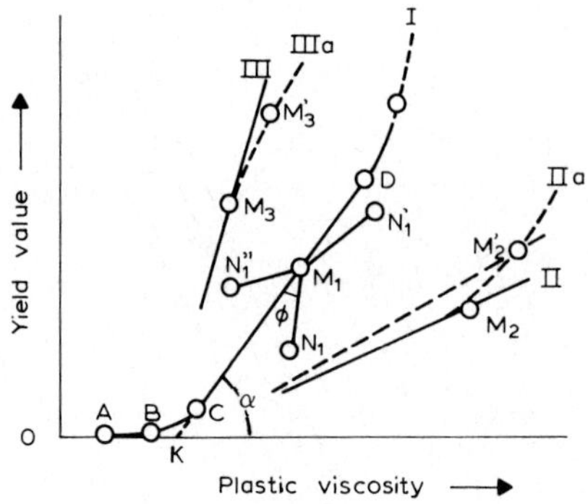

Fig. 8-28. Relationship between plastic viscosity and density, and between yield point and density. (After Riabchenco et al., 1968, fig. 1, p. 15; courtesy of the Ingenieria Petrolera, Mexico.)

Fig. 8-29. Examples of relationship between plastic viscosity and yield point for untreated fresh-water mud (curve *I*), stabilized mud system (curve *II*), and coagulated mud system (curve *III*). (After Riabchenco et al., 1968, fig. 1, p. 15; courtesy of the Ingenieria Petrolera, Mexico.)

where $\overline{PV}_{\min}$, $\tan\alpha$, and $\overline{YP}$ can be expressed in cP, mg/cm^2-cP, and mg/cm^2, respectively.

In Fig. 8-29, curve *I* represents the initial condition of the untreated mud, curve *II* represents mud stabilized with chemicals, and curve *III* (Fig. 8-29) shows behavior of coagulated mud. Angle α increases with increasing degree of coagulation and decreases with increasing stabilization of the system. The mud treating agents can be broadly divided into: (1) low-molecular weight reagents, such as soda ash; (2) high-molecular weight reagents, such as carboxy-methyl-cellulose (CMC); and (3) reagents having intermediate molecular weights, such as tannins.

On considering some point M_1 on curve *I* (Fig. 8-29), the changes in the concentration of solid phase would result in shifting of point M_1 along the curve. Addition of stabilizer will reduce the yield point without affecting the plastic viscosity (point M_1 will shift to point N_1). The length of line M_1N_1 and the angle ϕ are indicative of the degree and effectiveness of stabilization. Effectiveness of the reagents at a given concentration is lower with smaller value of angle ϕ.

Addition of substances such as soda ash, which changes the nature of ion-exchange complex and degree and nature of peptization, results in reduction of $\tan\alpha$ value, as point M_1 moves along M_1N_1' line.

Point M_1 shifts to N_1'' on addition of coagulating agent, giving rise to a negative ϕ angle.

In the case of stabilized systems (curve *II*, Fig. 8-29), which are characterized by larger values of distance $0K$, the nature of the shift of point M_2,

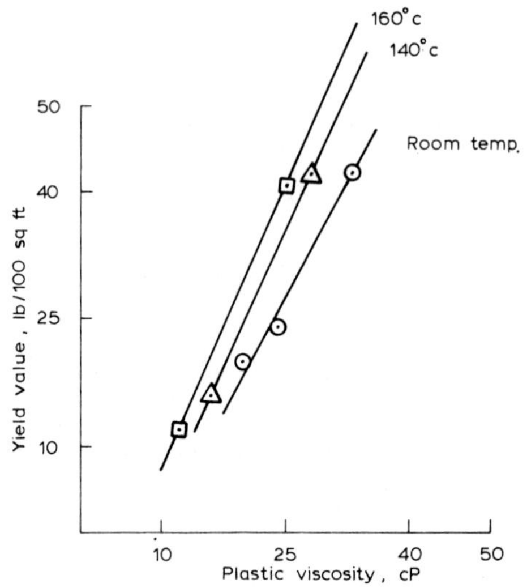

Fig. 8-30. Relationships between plastic viscosity and yield point for cutch(tannin)—dichromate treated mud at different temperatures (effect of thermal aging on stability). (After R.M. Mathur and G.V. Chilingar, personal communication, 1979.)

for example, on addition of water, stabilizer, or coagulant would not differ much from the shift of point M_1 described previously. On addition of clay, point M_2 will shift to point M_2', as a result of lower concentration of the reagent per unit volume of solid phase. On the other hand, if solid phase (without reagent) is removed from the stabilized mud, point M_2 will shift in the opposite direction indicating better mud stabilization. On addition of larger quantities of clay, point M_2' will approach the curve *I*, Fig. 8-29 (untreated mud). If muds are treated with excessive amounts of reagents, point M_2 simply slides along the curve *II* (Fig. 8-29). In the case of coagulated muds (curve *III*, Fig. 8-29), the shift of a point on the curve (e.g., point M_3) on addition of water or stabilizer is similar to the shift of point M_1 or M_2 under the same conditions. On addition of *large* quantity of clay, point M_3 will shift to the point M_3', because of lower concentration of coagulant per unit of solid phase volume.

The $\overline{PV}$—$\overline{YP}$ curves may be also used to evaluate the thermostability of the mud systems. For example, the behavior of cutch-dichromate Indian stabilizer (tannin + potassium dichromate) at three different temperatures (room temperature, 140°C, and 160°C) is presented in Fig. 8-30. The tan α value increases with increasing temperature, which indicates reduction in the stabilizing property of the system. Reverse trend is commonly true in the case of ferrochrome-lignosulfonate system.

Straight lines in Fig. 8-31 represent addition of barite to bentonite suspen-

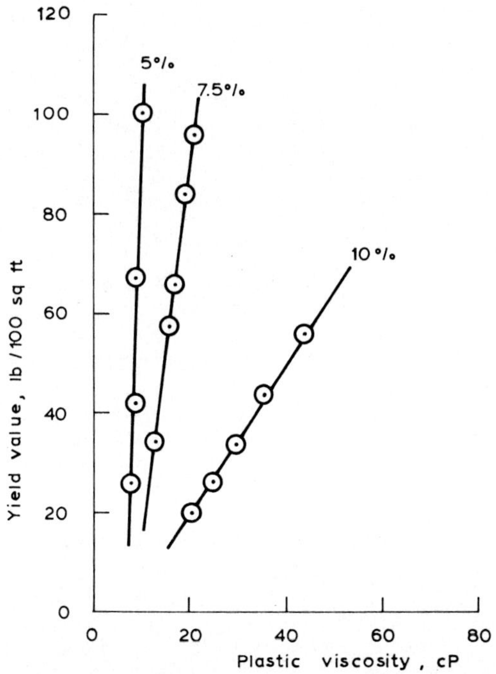

Fig. 8-31. Relationships between plastic viscosity and yield point on addition of barite to fresh-water bentonite suspensions having different concentrations (5, 7.5, and 10% by weight). Mud systems were initially treated with tannin—caustic soda in order to control the initial viscosity of the system. (After R.M. Mathur, and G.V. Chilingar, personal communication, 1979.)

sions having different concentrations. They were treated with tannin-caustic soda in order to control the initial viscosity of the system. The straight-line nature of $\overline{PV}$—$\overline{YP}$ curves indicates that for a given clay concentration, the stability of the system does not change on addition of barite. Decreasing the value of $\tan\alpha$ with increasing concentration of clay is indicative of a better stabilization of the system due to the increase in the colloidal content. The $\tan\alpha$ values for 5%, 7.5%, and 10% bentonite suspensions are 300, 39.0, and 7.8 mg/cm^2-cP, respectively.

A fresh-water mud having apparent viscosity of 50 cP was treated with (a) causticized lignite and (b) causticized lignite with dichromate. These two mud systems were weighted with barite to a specific gravity of around 2.0 (see Fig. 8-32). As shown in Fig. 8-32A, apparent viscosity increases with increasing specific gravity of the muds. Straight-line relationship exists in the case of causticized lignite, whereas in the presence of dichromate a break

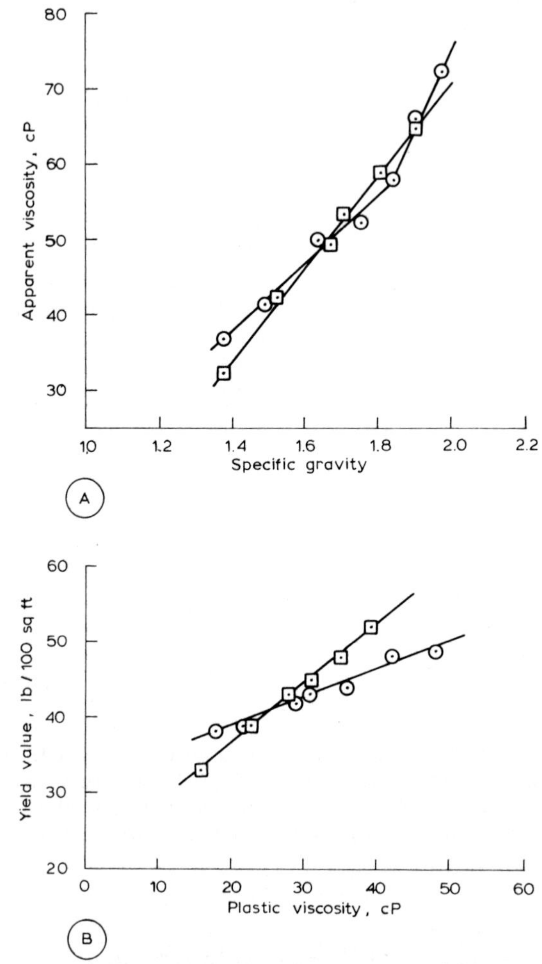

Fig. 8-32. A. Circles = relationship between apparent viscosity and specific gravity for causticized lignite plus dichromate mud system weighted with barite; squares = relationship between apparent viscosity and specific gravity for causticized lignite mud system weighted with barite.
B. Circles = relationship between plastic viscosity and yield value for causticized lignite plus dichromate mud system weighted with barite; squares = relationship between plastic viscosity and yield point for causticized lignite mud system weighted with barite. (After R.M. Mathur and G.V. Chilingar, personal communication, 1979.)

in the line is present at a specific gravity of 1.85, above which there is an abrupt increase in the apparent viscosity.

Figure 8-32B, showing relationship between $\overline{PV}$ and $\overline{YP}$, indicates that stability of the system remains unaffected at high mud weights with absence

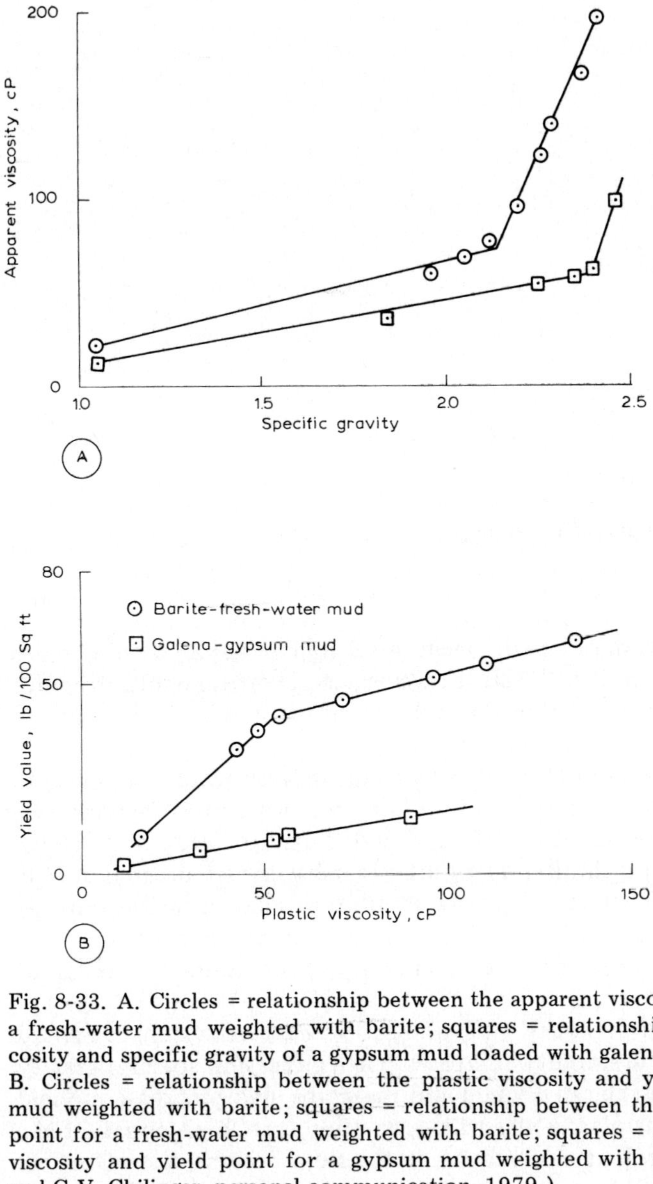

Fig. 8-33. A. Circles = relationship between the apparent viscosity and specific gravity for a fresh-water mud weighted with barite; squares = relationship between the apparent viscosity and specific gravity of a gypsum mud loaded with galena.
B. Circles = relationship between the plastic viscosity and yield point for a fresh-water mud weighted with barite; squares = relationship between the plastic viscosity and yield point for a fresh-water mud weighted with barite; squares = relationship between plastic viscosity and yield point for a gypsum mud weighted with galena. (After R.M. Mathur and G.V. Chilingar, personal communication, 1979.)

of break in the line in the presence of dichromate. The tan α values for causticized lignite and causticized lignite plus dichromate muds are 4.1 and 1.9 mg/cm^2-cP, respectively.

Figure 8-32A may give an impression that causticized lignite system is

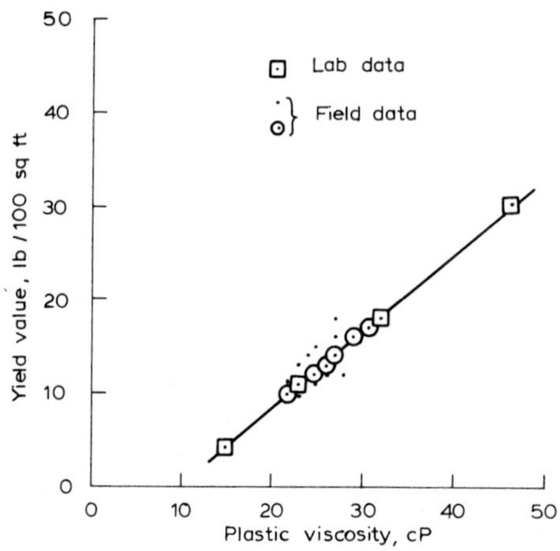

Fig. 8-34. Relationship between plastic viscosity and yield point (stability curve) determined initially (squares) in the laboratory (control curve) with field data points added (circles). (After R.M. Mathur and G.V. Chilingar, personal communication, 1979.)

superior to lignite-dichromate mud, because of higher apparent viscosities of the latter. As shown in Fig. 8-32B, however, tan α of the lignite—dichromate system is lower indicating that the stability of causticized lignite mud is enhanced in the presence of dichromate.

Figures 8-33A,B show the effect of weighting of fresh-water and gypsum muds with barite and galena, respectively. Apparent viscosities increase systematically up to specific gravities of 2.14 and 2.40 for barite and galena, respectively. Beyond these limits, apparent viscosity shows a rapid rise in both cases. The $\overline{PV}$—$\overline{YP}$ curve for barite exhibits a change in the slope at a specific gravity of about 2.15 at which an abrupt increase in apparent viscosity is observed. In the case of galena, linearity is maintained throughout the experimental limits.

The decrease in the slope of $\overline{PV}$—$\overline{YP}$ line in the case of system weighted with barite indicates that stability increases with increasing specific gravity (Fig. 8-33b). Decrease in the tan α value also indicates lower increase in yield point per unit increase in plastic viscosity and, therefore, high apparent viscosities cannot be controlled easily because thinners become less effective. This may suggest that galena is superior to barite for weighting the mud in the specific gravity range of 2.15—2.50 (and higher).

It appears that the values of tan α used in conjunction with fluid loss data may be useful tools in understanding and predicting the behavior of different mud systems. Plotting of $\overline{PV}$ versus $\overline{YP}$ stability graphs (control graphs) can be used as guidelines in the field control of the drilling mud systems and in taking timely actions (see Fig. 8-34).

REFERENCES

Anonymous, 1974. New mud additives will neutralize H_2S. *World Oil,* 179 (1, July): 173—179.

API, 1974. *Principles of Oil Well Drilling Fluid Control.* 12th ed. University of Texas Press, Lubbock, Texas, 300 pp.

Baroid Services, 1975. *Baroid Mud Handbook,* 200 pp.

Beeson, C.M., Chilingar, G.V. and Larsen, D.H., 1958. *Drilling Fluids Laboratory,* University of Southern California Publ., 73 pp.

Bradley, B.W., 1970. Oxygen causes drill pipe corrosion. *Pet. Eng.,* 42 (13, Dec.): 50—57.

Browning, W.C., 1960. Tanning agents control flow properties of mud. *World Oil,* 150 (5, Feb.): 6 pp.

Browning, W.C., 1964. *Hydroxyl Factors in Shale Control.* Spring Symp. Soc. Pet. Eng. AIME, Ft. Worth, Texas, March 23—24, SPE Pap. No. 795: 20 pp.

Browning, W.C. and Perricone, A.C., 1962. *Lignosulfonate Drilling Mud Conditioning Agents.* 37th Annu. Fall Meet. Soc. Pet. Eng. AIME, Los Angeles, Calif., Oct. 7—10, SPE Pap. No. 432: 20 pp.

Browning, W.C. and Perricone, A.C., 1963. *Clay Chemistry and Drilling Fluids.* First Conf. on Drilling and Rock Mechanics, Univ. of Texas, Austin, Texas, Jan. 23—24, SPE Pap. No. 540: 16 pp.

Carney, L.L. and Myer, R.L., 1976. *A New Approach to High Temperature Drilling Fluids.* 51st Annu. Fall Tech. Conf. and Exhibition Soc. Pet. Eng., New Orleans, Oct. 3—6, SPE Pap. No. 6025: 8 pp.

Chesser, B.G., 1968. *Corrosivity of Chrome-Treated Sodium Lignosulfonate Packer Fluid Systems.* Spring Meet. of Southwestern Distr., March 20—22, Tyler, Texas, Pap. No. 906-13-Ca: 15 pp.

Chesser, B.G. and Perricone, A.C., 1973. *A Physicochemical Approach to the Prevention of Balling of Gumbo Shales.* 48th Annu. Fall Meet. Soc. Pet. Eng. AIME, Las Vegas, Nev., Sept. 30—Oct. 3, SPE Pap. No. 4515: 12 pp.

Chilingar, G.V. and Beeson, C.M., 1969. *Surface Operations in Petroleum Production.* American Elsevier, New York, N.Y., 397 pp.

Clark, R.K., Scheuerman, R.F., Rath, H. and Van Laar, H., 1975. *Polyacrylamide-Potassium Chloride Mud for Drilling Water Sensitive Shales.* 50th Annu. Fall Meet. Soc. Pet. Eng. AIME, Dallas, Texas, Sept. 28—Oct. 1, SPE Pap. No. 5514: 16 pp.

Cox, T.E., 1974. Even traces of oxygen in muds can cause corrosion. *World Oil,* 179 (5, Oct.): 110—112.

Degens, E.T. and Chilingar, G.V., 1967. Diagenesis of subsurface waters. In: G. Larsen and G.V. Chilingar (Editors), *Diagenesis in Sediments.* Elsevier, Amsterdam, pp. 477—502.

Garrett, R.L., 1976. *New Field Method for the Quantitative Determination of Sulfides in Water-Base Drilling Fluids.* 51st Annu. Fall Tech. Conf. and Exhibition Soc. Pet. Eng. AIME, New Orleans, La., Oct. 3—6, SPE Pap. No. 6146: 6 pp.

Gatlin, C., 1960. *Petroleum Engineering — Drilling and Well Completions.* Prentice-Hall, Englewood Cliffs, N.J., 341 pp.

IMCO, 1978 (and earlier editions). *Applied Mud Technology.* 250 pp.

Levorsen, A.J., 1967. *Geology of Petroleum.* 2nd ed. Freeman, San Francisco, Calif., 724 pp.

Magcobar Services, 1977 (and earlier editions). *Drilling Fluid Engineering Manual.* 200 pp.

McCray, A.W. and Cole, F.W., 1973. *Oil Well Drilling Technology,* 4th printing. University of Oklahoma Press, Norman, Okla., 492 pp.

Mercer, R.F., 1976. Gas-cut mud: causes, effects, and cures. *World Oil,* 183 (4, Sept.): 63—72.

Mondshine, T.C., 1973. *A New Potassium Based Mud System.* 48th Annu. Fall Meet. Soc. Pet. Eng. AIME, Las Vegas, Nev., Sept. 30—Oct. 3, SPE Pap. No. 4516: 16 pp.

Olson, R.M., 1973. *Essentials of Engineering Fluid Mechanics.* 3d ed. Intext Educational Publ., New York, N.Y., 637 pp.

Perricone, A.C. and Browning, W.C., 1962. *Formation Clay Chemistry.* 37th Annu. Fall Meet. Soc. Pet. Eng. AIME, Los Angeles, Calif., Oct. 7—10, SPE Pap. No. 433: 11 pp.

Riabchenco, V.I., Mathur, R.M. and Chilingar, G.V., 1968. Un nuevo intento para evaluación de la estabilidad de los lodos de perforación empleando los parámetros reológicos. *Ingenieria Petrolera (Mexico)*, 8 (5): 14—19.

Rieke, H.H., 1978. *Drilling Fluids Laboratory Experiments.* Vandalia Publishers, Morgantown, W.Va., 218 pp.

Rogers, W.F., 1953. *Composition and Properties of Oil Well Drilling Fluids.* 1st ed. Gulf Publ. Co., Houston, Texas, 650 pp.

Rogers, W.F., 1963. *Composition and Properties of Oil Well Drilling Fluids*, 2nd ed. Gulf Publ. Co., Houston, Texas, 818 pp.

Simpson, J.P. and Barbee, R.D., 1970. How corrosive are water-base completion muds. *Pet. Eng.*, 13 (42, Dec.): 42—48.

Skelly, W.G. and Dieball, D.E., 1969. *Behavior of Chromate in Drilling Fluids Containing Chromate.* 44th Annu. Fall Meet. Soc. Pet. Eng. AIME, Denver, Colo., Sept. 28—Oct. 1, SPE Pap. No. 2530: 6 pp.

Tchilingarian, G.V., 1952. Study of dispersing agents. *J. Sediment. Petrol.*, 22 (4): 229—233.

Zwicker, D.A., 1979. Barite situation now points at serious crunch. *Drilling Contractor*, 35 (6, June): 46—85.

Chapter 9

EMULSION DRILLING FLUIDS *

DANIEL ACOSTA

INTRODUCTION TO THE THEORY OF EMULSIONS

Emulsions are a dispersion of one immiscible liquid in another immiscible liquid (e.g., oil-in-water or water-in-oil). Research in the area has shown that if small equidimensional spheres are closely packed together into a given space, they will occupy approximately 74% of the total available volume (this figure is constant, independent of sphere size). According to this theory, known as the phase-volume theory, it is not possible to have two different types of emulsions over a particular range of concentration. For example, 8% oil and 80% water could form only an oil-in-water emulsion.

When oil is continually added to a volume of water (or oil-in-water emulsion), the distance between the oil globules decreases until the globules touch each other and coalesce. An invert, or water-in-oil, emulsion forms at this concentration of oil.

Further research has shown that droplets which form an emulsion are not all of the same size and that these globules are plastic and can deform. The 74% limit of compaction of globules, therefore, could be exceeded. It was also found that the surface film present around the globules would not necessarily aid coalescence, but instead would aid compression so that only a minute space separates the globules. Inasmuch as different interfacial tensions are present in each fluid phase, this difference will determine which phase will be continuous and which will be dispersed.

Modern-day investigation into the theory of emulsion formation has dealt mainly with the works of Bancroft (1867), focusing on the fact that the interfacial film is the main contributor to the emulsion. Research work based on the nature of the interfacial film, however, has shown that the characteristics of the interfacial film with respect to strength and compactability are of prime importance in the development of a stable emulsion. The relative importance of all other factors is dependent upon their effects on the film. The amount of emulsifiers present in a system, therefore, would greatly affect the emulsion stability of a water-in-oil fluid having a high oil concentration. For example, an insufficient amount of emulsifier would allow the electrolytes to hinder the system's stability because of failure of coherent

* Edited by George V. Chilingarian

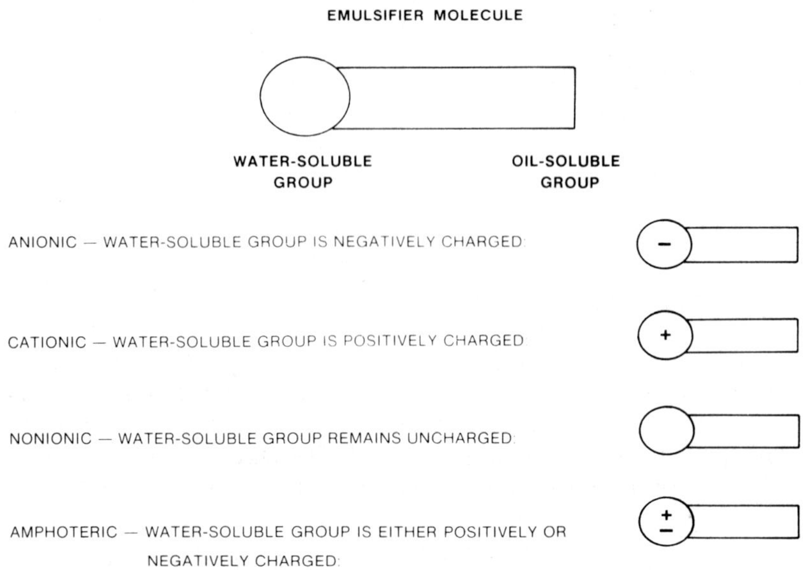

Fig. 9-1. Type of emulsifiers.

film formation around the globules of oil. It was then felt that the stability of the emulsions was not dependent on the interfacial film, but on the ability of that film to absorb the emulsifiers present in the system.

Another theory of emulsion formation is that the stability of an emulsion is due to the ionic charges of the internal phase. These charges can be obtained by ionization, droplet contact, or absorption. As shown earlier, the presence of an emulsifier alone at the interface may be sufficient to form a stable emulsion. The development of a surface charge from the ionization of the emulsifier, therefore, can be easily understood. For example, in an oil-in-water emulsion using a soap as an emulsifier, one can expect to see the head of the carbolic ion penetrating the interface and imparting a coat of negative charges around the droplet. The positive ions, which are present in the water phase, will then surround the negative charges attached to the droplets and thereby prevent coalescence through electrical repulsion. This will contribute to the stability of an emulsion.

Charges may also develop as a result of contact of droplets. As the continuous phase comes in contact with droplets of the dispersed phase, charges may develop in a manner similar to generation of charges by rubbing a plastic rod with fur.

When stabilization of oil-in-water systems are attained by the use of soaps, as described above, the charges imparted to the droplet will be negative. When cationic emulsifiers (see Fig. 9-1) are used, the droplets will develop a positive charge. The type of charge imparted to the droplet can be determined easily by the emulsifier used.

Determination of the charge in reference to the contact of droplets should be based on the following rule of thumb: When two liquids are in contact with one another, the liquid having the highest dielectric constant will be positively charged. Applying this rule to the oil-in-water emulsions, the oil will have a negative and the water a positive charge in the water-in-oil emulsion.

EMULSIFIERS AND EMULSION MUDS

The value of an emulsifier depends only upon what is being emulsified and the procedure by which the emulsion is being produced.

There are four distinguishing factors of an emulsifier: it may be either (1) molecularly soluble or (2) colloidally dispersible in water, and is either (3) hydrophobic or (4) hydrophilic. If an oil-soluble hydrophilic colloid is dissolved in oil, it is impossible to attain a stable emulsion in a water-in-oil dispersion. The same would hold true if a water-soluble hydrophobic colloid is dissolved in an oil-in-water dispersion. A hydrophilic emulsifier will favor the formation of an oil-in-water emulsion, whereas a hydrophobic emulsifier will favor the formation of a water-in-oil emulsion. In other words, the phase in which the emulsifier is more soluble will become the external phase; however, any emulsifier used must have the ability to come in contact (interface) with the oil. Consequently, a specific type of emulsifier is required to prepare oil-in-water or water-in-oil emulsion. In water-in-oil emulsion muds, a calcium-type soap is used as the emulsifying agent. The stability of the emulsion formed by a calcium soap is enhanced by the presence of electrolytes in the water (either a calcium or sodium salt, i.e., $CaCl_2$ or NaCl). Brine also prevents the hydration of any hydrophilic solids that may tend to weaken the emulsion.

The means by which an emulsion is produced depends greatly upon the capabilities of the mechanical equipment used. Inasmuch as the desired goal is the dispersion of two immiscible fluids, mechanical shear forces are required. The higher the degree of agitation, the finer the droplets of the dispersed phase. The fine dispersion may approach the 74% oil-in-water limit, i.e., percentage of oil globules as described by Becher (1966), and result in a more stable emulsion.

An emulsion mud is simply a dispersion of oil in water or water in oil. In oil-in-water emulsions, water is the continuous or external phase, whereas oil is the discontinuous or internal phase. In water-in-oil emulsions, on the other hand, water is the discontinuous phase, whereas oil is the continuous phase (inverted emulsion) (see Fig. 9-2).

Fluid systems which contain 50% or less water are generally classified as water-in-oil dispersions, whereas systems containing 80% or more water form oil-in-water dispersions. These percentages are arbitrary, however,

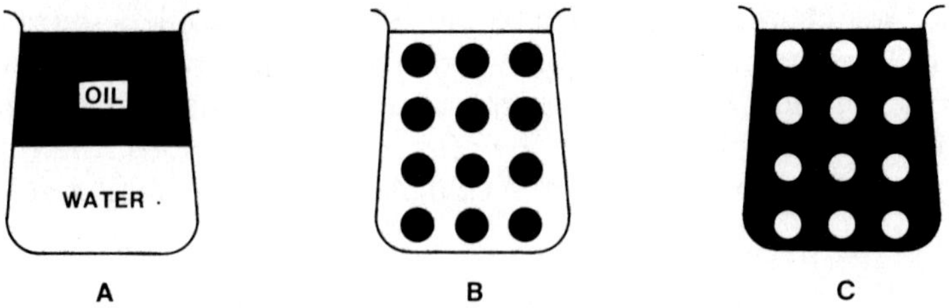

Fig. 9-2. Schematic representation of non-dispersed system (A), oil-in-water emulsion (B), and water-in-oil emulsion (C).

because under certain conditions, water-in-oil emulsions may be formed with more than 50% water, and oil-in-water emulsions may exist with more than 20% water. Determining factors include (Bennett, 1947, pp. 41—57):

(1) *Relative specific gravities of oil and water.* Large differences in specific gravity would enhance creaming of the emulsion, i.e., the formation of a more concentrated emulsion layer. The speed of creaming does not only depend on the specific quantity of the phases, but also on viscosity, size of globules of dispersed phase, and other variables.

(2) *Dielectric constants of the internal phase.* Similar charges on the dispersed phase globules aid in emulsion stability because of repulsion of the globules.

(3) *The pH of water.* Many emulsifiers are designed for various pH ranges. If pH falls outside these ranges, changes in surface tension and viscosity of the emulsifier may occur.

(4) *Type of emulsifying agents.* Many emulsifiers are sensitive to various environments, i.e., temperature, calcium content, etc. Care must be taken to use proper emulsifier for the formation to be drilled.

Determination of the type of emulsion formed can be made using a number of different methods. One of the simplest tests is based on the fact that an emulsion can only be mixed with a liquid that forms the continuous phase. When a small amount of emulsion is added to water, if the sample does not mix with water, then the continuous phase of the emulsion must be oil, i.e., water-in-oil emulsion. On the other hand, if the sample mixes with water, the continuous phase of the emulsion is water, i.e., an oil-in-water emulsion.

In addition to the above-described test, the most common field test is the voltage breakdown (V.B.) test. This test simply measures the electrical conductivity of the emulsion. An emulsion which exhibits a very high conductivity (or low V.B.) will normally have water as the continuous phase. A low conductivity (or high V.B.) would normally indicate the water-in-oil emulsion, as oil does not readily conduct electric current. This test also measures

TABLE 9-I

Comparisons of field emulsion muds showing weak and stable emulsions

Emulsion mud No.:	1	2
Diesel (ml)	175	280
H_2O (ml) (10 ppg NaCl)	175	70
Kenol-S * (ppb or lb/bbl)	9	9
Kenox ** (ppb)	10	10
VR *** (ppb)	10	10
O/W ratio	50/50	80/20
Test temp. (°F)	75	75
600 rpm VG meter reading	106	65
300 rpm VG meter reading	75	36
AV (cP)	53	325
PV (cP)	31	29
YP (lb/100 sq ft)	44	7
Gel strengths (init./10-min)	11/16	3/6
VB (Volts)	30	460

* Kenol-S = a liquid blend of selected emulsifiers, gelling agents, and fluid-loss control materials.
** Kenox = $CaCO_3$.
*** VR = powdered asphaltic-type material which is soluble in diesel oil.

the emulsion stability. Whereas a low V.B. (0—20 V) indicates a weak emulsion and possible coalescence of water, a high V.B. reading (400+ V) indicates a tight and stable emulsion (see Table 9-I).

PHYSICAL PROPERTIES OF EMULSION MUDS

The physical properties of an emulsion indicate a system's stability. These properties are dependent on the composition and the manner in which the emulsion was prepared. The properties of oil-in-water emulsions (i.e., density, plastic viscosity, yield point, etc.) basically do not differ from those of the fluids from which they were formed. The same daily treatments are necessary for the maintenance of rheological and chemical properties.

Addition of oil to a water-base mud would result in an increase in viscosity. This may be due to the oil attachment to the clay particles that draws them together. Though not detrimental to the system, the oil can be tolerated or reduced by additions of water or organic thinner, which will neutralize the charges and reduce viscosity.

The fluid density also is reduced when oil is added to the system. This is basically due to the low specific gravity of oil of about 0.85. The reduction in density is particularly beneficial in areas with lost circulation problems.

The formula used in calculating the volume of oil necessary to reduce the density of a system is as follows:

$$X = \frac{V(W_2 - W_1)}{W_1 - 7.0} \tag{9-1}$$

where: V = volume of system (bbl); W_1 = desired specific weight (lb/gal); W_2 = present specific weight (lb/gal); 7.0 = weight of one gallon of oil (lb); and X = number of barrels of oil to be added to the system.

Sample problem:

What would be the number of barrels of oil required to dilute 500 bbl of 12.0-lb/gal system to 10.0-lb/gal?

$$X = \frac{500(12.0 - 10.0)}{10.0 - 7.0} = \frac{1000}{3} = 333 \text{ bbl}$$

An equivalent amount of original fluid must be removed before the oil is added.

Emulsifying oil into a mud (clay-in-water dispersion) tends to lower the API fluid loss. This is probably due to the plugging action of the oil droplets as they are forced into the mud cake, reducing its permeability. It should be pointed out, however, that this effect is not noted at high-temperature, high-pressure test, indicating that it is only a surface effect.

Droplet size (diameter) within an emulsion is found to exceed 0.25 microns, with the maximum diameter about a hundred times greater (Becher, 1966, p. 49). As mentioned earlier, therefore, the tighter the emulsion or the more uniform the droplet size of the order of 0.25 μm, the more stable the system. Changes in the system, which would allow formation of larger diameter droplets, also enable one to measure the instability of a system (by coalescence), i.e., low voltage breakdown (V.B.) readings, which are a measure of the conductivity of an emulsion (see Fig. 9-3).

Although the above trend is found to be true in laboratory tests, the inverse occurs in oil field muds. It is found that with aging, the droplet diameters become larger and more uniform, giving rise to a stable emulsion. This phenomenon, termed "limited coalescence" (Wiley, 1954, pp. 427—437), occurs in the initial emulsion and may be carried throughout its use. Consequently, the idea that a certain droplet size is necessary in the formation of an emulsion drilling fluid is discounted. This, however, does not apply to emulsions which use finely-divided solids as a stabilizing agent (Wiley, 1954, pp. 427—437).

Concentrations of the dispersed phase and the emulsifying agent are two other factors which affect the physical properties of the emulsion. The volumes of the internal and external phases, which make up the emulsion, can be measured in volume percent. This is necessary because any changes in the

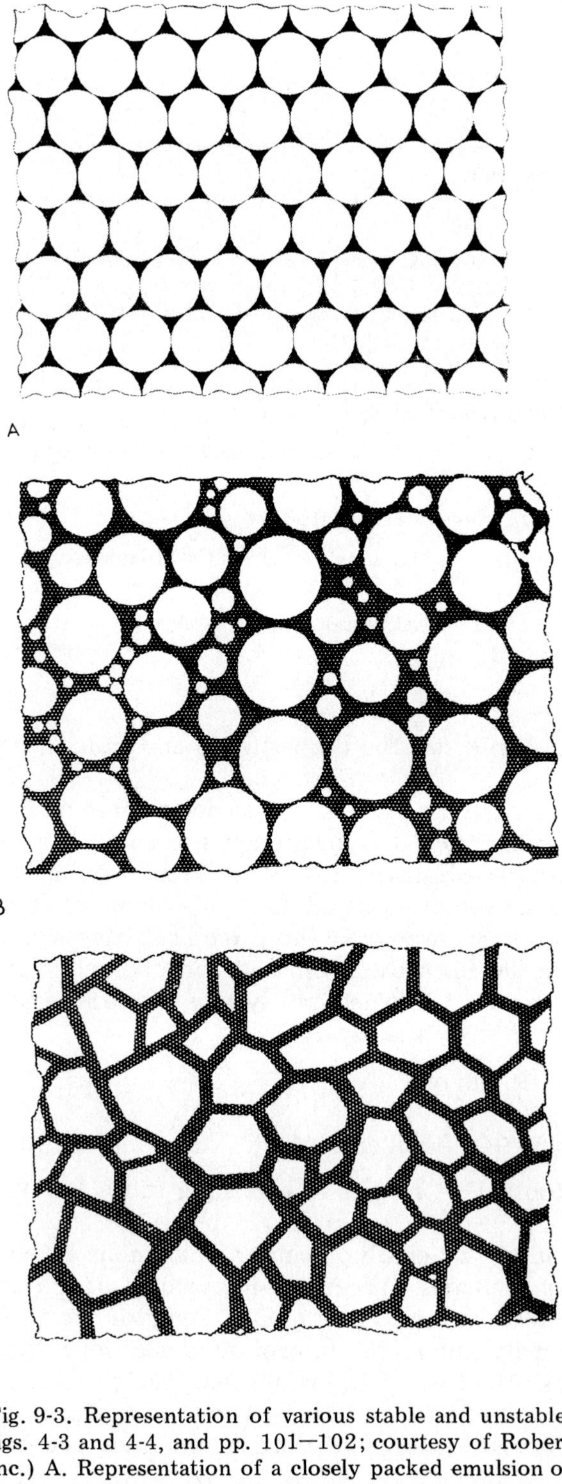

Fig. 9-3. Representation of various stable and unstable emulsions. (After Becher, 1966, figs. 4-3 and 4-4, and pp. 101—102; courtesy of Robert E. Krieger Publishing Company, Inc.) A. Representation of a closely packed emulsion of uniform spherical droplets (very stable). B. A closely-packed emulsion of non-uniform droplets (unstable). C. A closely-packed emulsion of polyhedral droplets (extremely unstable).

TABLE 9-II

Trouble shooting emulsion muds

Problem	Treatment
High viscosity	
water contamination	Add oil and emulsifiers to proper concentrations along with CaO.
high solids content	Use wetting agent; reduce size of shaker screen; dilute with oil and emulsifiers.
Low viscosity	
no gel strength	Add organophilic clay.
High fluid loss	Add CaO to tighten emulsion; check emulsifier concentration.
Emulsion breaking	Check emulsifier concentration; add CaO.
Oil separation	Provide sufficient agitation; check CaO to make sure it is hot.
Reduction in V.B.	Agitate; add CaO; insure proper chemical concentration.

volume percent of either phase will alter the properties of an emulsion (see Table 9-II).

The concentration of the emulsifying agent is the second term in need of clarification. As stated earlier, insufficient concentrations of emulsifier will prevent the formation of a stable emulsion. The concentration of needed emulsifying agent must be expressed in terms of the total volume of emulsion. Concentrations of emulsifier lower than those required may simply result from the failure to base the calculations on the total volume of emulsion. For example if 10% oil is added to a 100-bbl system with 4% emulsifier:

$$V_o = \%\ \text{oil} \times V_s = 0.10 \times 100 = 10\ \text{bbl oil} \tag{9-2}$$

$$V_e = V_s \times \%\ \text{emulsifier} = 100 \times 0.04 = 4\ \text{bbl emulsifier} \tag{9-3}$$

where: V_o = volume of oil (bbl); V_e = volume of emulsifier (bbl); and V_s = volume of system (bbl).

An important physical property of emulsion muds, which must be considered, is viscosity. Viscosity increases with increasing volume of the dispersed (or internal) phase. Increase in the concentration of emulsifying agent also causes an increase in viscosity due to the adsorption of additional films around the dispersed droplets. Addition of colloidal solids, such as organophilic clays, also changes viscosity, because of their hydration or because of tying up of additional emulsifiers. Viscosity of the final emulsion is determined by the viscosity of the external (continuous) phase, as shown by many researchers.

Oil wetting of steel occurs on using both oil-in-water and water-in-oil emulsion muds. This is not directly due to the oil, which cannot displace water from metal, but to the presence of emulsifiers. Tall oils and some sulfonate oils can be termed as "wetting agents" because of their ability to allow steel to become oil wet. This does not necessarily mean that all emulsion muds have this characteristic or that wetting agents are used as emulsifiers. Oil wetting of steel is due only to *some* emulsifiers used in formulating oil emulsion systems.

PREPARATION AND MAINTENANCE OF EMULSION MUDS

There are two types of emulsion muds: the oil-in-water and the water-in-oil (invert). These two systems are similar only to the extent that they are an emulsification of two immiscible fluids (i.e., water and oil) and that they may, if necessary, incorporate similar types of emulsifying agents. The preparation and maintenance of these systems are discussed separately.

Oil-in-water emulsions

In an oil-in-water system, oil (diesel or crude oil) is added to a water-base drilling fluid. The oil becomes the internal phase, whereas the water remains as the external phase (see Table 9-III).

Before preparing this system under field conditions, one must be certain that the initial water-base mud is in good condition and that clay is well-dispersed. This is necessary because the addition of oil is not a corrective measure for poor mud properties, but is used to impart properties not inherent to the original system (Rogers, 1963, p. 546). It must be noted that oil will reduce the weight of the water-base mud and increase its volume. Precautions must then be taken so that the starting volume of water-base mud is such that, when the calculated amount of oil is added, the final volume will be what is desired. In addition, if the density of the system is critical, barite should be added at a sufficient rate to prevent reduction in specific weight. The amount of oil added to a system may vary between 5 and 30%, which is dependent on the type of final properties desired. The majority of emulsion muds used, however, contain from 10 to 20% by volume of oil.

Oil is usually added through the hopper along with an emulsifying agent, if necessary. If the mud contains lignosulfonate or lignin compounds as the dispersing agents, however, the emulsifying agent may be eliminated. The emulsifying agent may be added through the hopper or chemical barrel. In addition, the system will require agitation to break up the oil into fine droplets. This can be accomplished by mud guns or pit stirrers (see Fig. 9-4).

TABLE 9-III

Recommended specifications of diesel or crude oil used in emulsion muds

Properties of oil	Typical specifications preferred	Acceptable range
Gravity, A.P.I.	36.0	33.0—39.0
Color, Robinson	25	20—30
Flash, PM (°F)	152	130 minimum
Pour point (°F)	—5	+10 to —15 *
Viscosity at 100°F, SSU	35.5	**
Carbon (10% bottoms)	0.01	**
Ash (%)	nil	**
Bomb sulfur (%)	0.16	0.15—0.32
Corrosion, 3-hr, at 212°F	J3	**
Cetane number	50.8	**
Suspended sediment	0.1	**
Dissolved sediment	0.9	**
Bromine number	1.43	**
Aniline point (°F)	149	135 minimum
Diesel index	54.5	**
% Aromatics (+ olefins)	21.8	18—30

* Lower pour point material may be suitable for use but should be checked in the laboratory prior to use. Oils with pour points lower than —15°F usually indicate that a pour point depressant has been added. These depressants cause a weakening of the emulsion with consequent increase in viscosity and gel strengths.
** Not critical.

Maintenance of this type of system is for the most part dependent on the type of initial water-base mud used; however, the stability of the emulsion and the amount (percent) of oil added do influence the system. Stability of the system can be determined simply by observing the filtrate of the mud, because a stable system produces a filtrate free of oil. The presence of oil in the filtrate may indicate: (1) improper dispersion of solids, and/or (2) improper concentration of emulsifying agents, e.g., due to the initial presence of such agents which may alter the emulsifying properties of the emulsifier added.

The volume percent of oil is determined with a distillation apparatus: the oil and water, boiled off from a sample of mud, are collected and their volumes are measured. The percentage by volume of oil is calculated by using the following formula:

$$\%\ \text{oil} = \frac{\text{volume of oil}}{\text{volume of mud sample}} \times 100 \qquad (9\text{-}4)$$

It must be noted that for any additional oil added to the system, the proper amount of emulsifier must also be added. Inasmuch as adding oil to a

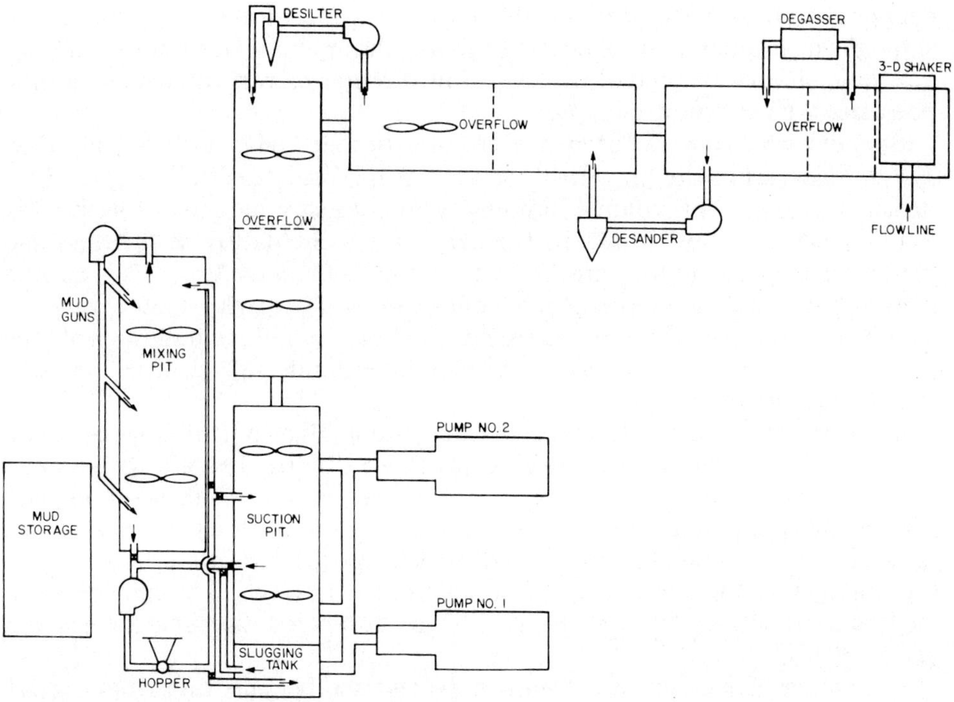

Fig. 9-4. Drilling fluids surface system.

system is similar to adding solids, water is usually added along with the oil to minimize the increase in viscosity.

Water-in-oil emulsion (invert)

The invert systems are designed to be formulated in either NaCl or $CaCl_2$ brines. The principal components include oil, water, primary concentrates, dispersants, calcium oxide, and fluid loss agents in varying concentrations, depending on the temperature limitations imposed upon the system. It is, therefore, essential that each ingredient be added in the correct proportion.

The exact details of mixing and the concentration of various chemical components vary with the type of brine used, but the general procedure is the same. First, diesel oil is placed in the pits, the amount of which is dependent on the oil/water ratio desired and the quantity of primary concentrates required. Then, vigorous agitation is applied to the oil, while the concentrates and solid components are being added. Agitation ensures proper dispersion of these materials in the liquid. Next, the required amount of salt brine is added at the point of highest agitation. Finally, after a thorough emulsification is obtained, weight material (if required) may be added through the hopper, with continued mixing. After about one hour of mixing, a sample should be tested to determine whether or not the emulsion mud possesses the desired properties.

Stability and easy maintenance are two primary advantages of an invert system. Normally, the only field treatments necessary are those required for density increases and volume building. The necessary amount of each additive is calculated and added to the mud over a circulation. Salt brines and diesel oil may be added directly with a small diameter hose. The correct amount of solid components and concentrates along with CaO should be added to obtain the final (increased) desired volume. When building volume, it is desirable to add diesel and salt brine in amounts sufficient to maintain the desired oil/water ratio.

Temperature greatly affects an invert system. High bottomhole temperatures tend to lower the viscosity of invert emulsions. These high temperatures also cause the mud to expand and weigh less. Inasmuch as oils expand much more than water, these effects are more noticeable in invert systems. High flowline temperatures may reduce the specific weight up to 0.5 ppg (pounds per gallon) as compared to ambient conditions. To enhance accuracy, it is recommended that specific weight of invert systems be determined at 75—80° F.

The tendencies of invert systems to expand and become thinner are offset by downhole pressures in the well. Oil, being more compressible than water, becomes thicker and more dense under high pressures. Density and Marsh funnel viscosity are measured at ambient temperatures, whereas viscometer should be run at 150° F. Fluid loss is determined at a pressure of 500 psi and temperature of 300° F.

Deviation of the oil/water ratios from optimum levels may require additions of diesel oil or salt brine. A distilling unit must be used to determine the volumetric percentages of oil, water, and solids in an invert system. The oil/water ratio may then be calculated as follows:

$$\frac{\%\ \text{oil}}{\text{total}\ \%\ \text{of liquids (oil + water)}} \times 100 = \%\ \text{oil in liquid phase or oil ratio} \qquad (9\text{-}5)$$

$$\frac{\%\ \text{water}}{\text{total}\ \%\ \text{of liquids (oil + water)}} \times 100 = \%\ \text{water in liquid phase or water ratio} \qquad (9\text{-}6)$$

$$\text{oil/water ratio} = \text{eq. (9-5) : eq. (9-6)} \qquad (9\text{-}7)$$

Example:

Given retort measurements: water = 30%, oil = 45% and solids = 25%.

Find oil/water ratio.

$$\frac{\%\text{ oil}}{\%\text{ total liquid}} \times 100 = \frac{45}{75} \times 100 = 60\% \text{ oil in liquid phase}$$

$$\frac{\%\text{ water}}{\%\text{ total liquid}} \times 100 = \frac{30}{75} \times 100 = 40\% \text{ water in liquid phase}$$

oil/water ratio = 60/40.

When it is desired to change an oil/water ratio, the corrected retort readings and the total circulating volume must be known. First, determine the present oil/water ratio and decide whether oil or water should be added. Then, calculate the amounts of each phase present in the system as follows:

% oil (retort) × total volume (bbl) = volume oil (bbl) (9-8)

% water × total volume (bbl) = volume water (bbl) (9-9)

% solids × total volume (bbl) = volume solids (bbl) (9-10)

If oil is to be added,

$$\frac{\text{volume oil} + X}{\text{volume water}} \times 100 = \frac{\text{desired oil ratio}}{\text{desired water ratio}} \tag{9-11}$$

If water is to be added:

$$\frac{\text{volume oil}}{\text{volume water} + X} = \frac{\text{desired oil ratio}}{\text{desired water ratio}} \tag{9-12}$$

where X = the number of barrels required.

Example:
Maintain a 70/30 oil/water ratio in an 800-bbl circulating system. Corrected retort readings are: oil = 50%, water = 25% and solids = 25%.

$$\frac{\%\text{ oil}}{\%\text{ oil and water}} \times 100 = \frac{50}{75} \times 100 = 67 \text{ oil ratio}$$

oil/water ratio = 67/33 .

Inasmuch as the ratio is too low, oil must be added to bring it to 70/30. First, calculate volume of oil, water, and solids in present system:

Volume oil = 50% × 800 bbl = 400 bbl.
Volume water = 25% × 800 bbl = 200 bbl.
Volume solids = 25% × 600 bbl = 200 bbl.

Find additional amount of oil necessary as follows:

$$\frac{400 + X}{200} = \frac{70}{30}$$

Thus, X is equal to 67 bbl, i.e., 67 barrels of oil should be added along with the appropriate amounts of weight material, emulsifier, and CaO. The final volume may be calculated and the final oil/water ratio can be checked:

$$\% \text{ oil} = \frac{467 \text{ bbl oil}}{467 \text{ bbl oil} + 200 \text{ bbl water}} = \frac{467}{667} = 70\%$$

Final oil/water ratio = 70/30.

EMULSION MUD EFFECTS ON DRILLING PARAMETERS

Penetration rate

Rate of penetration is based on the following parameters: efficiency of the drilling crew, weight on bit, hole size, rotary speed, and hydraulic program. Any change in these parameters, no matter how slight, can lengthen the days on location; with this in mind, one can see the difficulty in producing accurate data on the subject (see Chapters 2 and 3).

Nonetheless, researchers have found some striking differences on using emulsion muds as compared with regular muds. In the case of oil emulsion muds, penetration rates were greatly increased with additions of oil to the system (Table 9-IV). Although the exact cause is not known, it has been postulated that this may be due (1) to the added lubricity to the bit bearings, which would allow the cones to revolve continuously, and/or (2) to the drastic reduction of torque and drag, which would also allow a more accurate determination of weight on the bit.

Although addition of oil to a system results in dramatic increases in penetration rate, Eckel (1954, pp. 119—125) has shown that the increase in drilling rates, with all other factors being constant, reaches a peak with about 12% oil. Amounts of oil less or more than 12% produced lower drilling rates.

Although one could expect the same results with an oil-emulsion system as those for an invert system, this is not the case. Research work has shown that an invert emulsion drilling fluid produces drilling rates lower than those obtained on using an oil-emulsion mud. It must be noted, however, that an invert-emulsion system is a low-fluid loss system and is usually used at high density ranges. This would definitely decrease the drilling rate due to the hold-down pressure exerted on the formation and to the absence of a spurt loss, which is characteristic of a water-base mud. The spurt loss would tend to equalize the pressure between the cuttings and formation.

Unlike the water + clay or oil-emulsion muds, the invert-emulsion systems have an advantage in their ability to suppress the hydration of shales and reduce wall sticking and other well problems. These are important problems in drilling deep wells having troublesome shales. Although penetration

TABLE 9-IV

Penetration rates on using water-base mud versus oil emulsion muds

(After Rogers, 1963, table 9-5, p. 549; courtesy of Gulf Publishing Company, Houston, Texas)

Location	Water-base mud			Oil-emulsion mud			
	type of mud	depth drilled (ft)	penetration rate (ft/hr)	type of mud	depth drilled (ft)	penetration rate (ft/hr)	increase in drilling rate (%)
1. Sec. 28, La.	Fresh-water-phosphate	9,428—10,423	2.07	11—17% oil	9,420—10,429	2.43	17.4
2. Bully Camp, La.	Lime-starch	8,503— 9,000	2.49	12—15% oil	9,008— 9,500	5.66	127.2
3. Hull, Tex.	Lime-starch	5,067— 6,722	11.1	9—11% oil	5,067— 6,722	14.0	26.1
4. S. Gillock, Tex.	Lime-starch	9,632—11,467	4.77	10% oil	9,586—11,439	5.44	21.4
5. West Texas	Fresh-water-quebracho	8,011— 9,157	2.72	5% oil	8,022— 9,217	3.28	20.6
6. Pegasus, Tex.	Fresh-water-quebracho	9,995—11,281	4.38	10% oil	10,001—11,280	5.94	31.1
7. Pegasus, Tex.	Fresh-water-quebracho	10,481—11,574	4.58	5% oil	10,480—11,578	4.84	5.7
8. Krotz Springs, La.	Fresh-water	6,994— 9,400	6.13	10% oil	7,021— 9,400	10.8	76.2

TABLE 9-V

Comparison between bit life on using water-base muds and that on using oil-emulsion muds

(After Rogers, 1963, table 9-7, p. 542; courtesy of Gulf Publishing Company, Houston, Texas)

Rock type and location	Water-base mud			Oil-emulsion mud				
	mud type	ft per bit	bit life (hr)	mud type	ft per bit	increase in ft/bit (%)	bit life (hr)	increase in bit life (%)
1. Shale, Sec. 28, La.	Fresh-water-phosphate	66.3	32	11—17% oil	126.1	90.5	52.1	62.8
2. Shale, Bully Camp, La.	Lime-starch	71	28.6	12—15% oil	246	246	43.5	52.0
3. Shale, Hull, Tex.	Lime-starch	331	29.8	9—11% oil	414	25	29.5	—0.1
4. Shale, So. Gillock, Tex.	Lime-starch	53.3	13.26	10% oil	97.6	54.3	17.96	35.5
5. Chert and hard shale, W. Tex.	Fresh-water-quebracho	27.3	10.02	5% oil	23.9	—12.5	7.3	—37.0
6. Hard shale-lime, Pegasus, Tex.	Fresh-water-quebracho	46.0	10.5	10% oil	53.3	16.0	8.95	—15.0
7. Hard lime-shale, Pegasus, Tex.	Fresh-water-quebracho	54.6	11.95	5% oil	61.0	5.4	12.60	11.7
8. Shale, Krotz Springs, La.	Fresh-water	153	24.9	10% oil	238	55.5	22.8	—8.4

rate may be equal to, or slightly greater than, that obtained on using oil-emulsion muds, the overall days on location have been greatly reduced.

Effect on bit wear

Oil-emulsion and invert-emulsion muds possess the ability to increase the bit life. This is due to the lubricating properties of the oil, which result in better operation of the bearings over longer periods of time (Table 9-V). This inadvertently decreases (1) the number of bits required for a well, (2) the round trips necessary to change bits, and (3) the duration of drilling (days on location). All these improvements result in an overall decrease in the drilling cost.

Effect on hole size

Hole washout is a major problem encountered in using water-base muds. Control of this problem will minimize the side effects, such as stuck pipe and reaming back to bottom, resulting in better and speedier well completion.

Oil-emulsion and invert-emulsion muds have shown the ability to maintain a gauged hole with minimal washout, which is due to the low fluid loss characteristic of these systems. The ability to prevent the hydration of shales in the case of invert-emulsion systems and the "droplet plugging" of the oil-emulsion systems, prevents (a) water from entering microfissures and (b) water-wetting of the formation shales (Fig. 9-5).

The above-described protection becomes more apparent, especially on using invert-emulsion systems, in drilling soluble formations such as massive salt, anhydrite, and gypsum sections. For the same reason, dry salt is not added to oil in the formulation of an invert-emulsion system used in drilling of these sections. Inasmuch as oil does not allow these soluble salts to go into solution, a gauged hole is produced.

Torque and drag

Oil-emulsion and invert-emulsion systems greatly reduce torque and drag, due to the lubricating qualities of oil. Field tests have shown that in areas where high torque and drag were observed, the use of an oil-emulsion mud (10% oil by volume) reduced the torque from 40 to 50%, whereas drag was reduced by about 20%. Although these results are desirable in all types of drilling environments, the use of oil-emulsion drilling fluid is highly recommended in directional holes to prevent pipe-sticking.

Bit balling

Bit balling, a common occurrence in areas of gumbo shales, can cause many detrimental effects in the course of drilling. These include (1) exces-

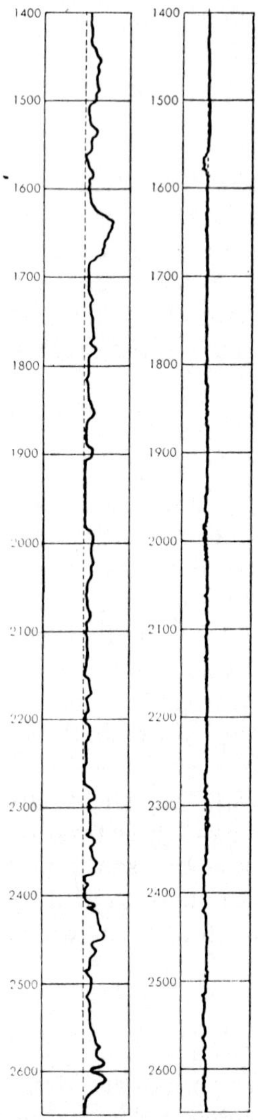

Fig. 9-5. Caliper logs of wells drilled through identical sections with fresh-water and oil-emulsion muds. (After Rogers, 1963, fig. 9-5, p. 554; courtesy of Gulf Publishing Company, Houston, Texas.) Left: water-base mud; right: oil-emulsion mud. Depth in ft; horizontal scale — hole diameter in inches (left = 0—20 inches, right = 4—20 inches).

sive surge and swab pressures while tripping, (2) excessive pump pressures, (3) low drilling rates, and (4) lost circulation. Research and field data have shown that the presence of oil in a system greatly reduces the chances of bit balling. The use of tight emulsions and wetting agents, to ensure oil-wetting of the drillstring, is also important in preventing bit balling.

Well completions

Although oil-emulsion and invert-emulsion systems have been used to increase the speed of completion, their use to increase production has not been sufficiently well recorded. Researchers have, however, recorded a few cases where an increase in production was observed. It must be noted, however, that under no circumstances were wells damaged.

EFFECT OF EMULSION DRILLING FLUIDS ON LOGGING RESULTS AND THEIR INTERPRETATIONS

The logging programs for wells drilled in oil-emulsion muds are the same as those used for water-base muds. Maintenance and interpretation are performed as if the oil were not present. This is mainly due to the fact that oil present in an emulsified state allows the mud to retain its conductivity.

Programs available for wells drilled with invert-emulsion systems are slightly different. Many of these logs, however, are obtained with the same tools used in water-base drilling fluid environments. They are as follows:

Gamma ray—induction log

Gamma ray—induction log is the basic log run in an invert-emulsion system. Its use is simple for correlation, depth control, resistivity measurements in defining shale beds, and for the detection of radioactive minerals. The gamma ray curve is used for correlation in lieu of the specific gravity curve, whereas the induction conductivity curve is used instead of the short normal curve.

Acoustic velocity log

Acoustic velocity log is helpful in correlation of formations where other logs give poor results. It can be used in conjunction with other logs to evaluate reservoir rocks and to determine lithology and porosity. This logging tool is equally effective in invert-emulsion systems as in water-base muds, because its operation does not depend upon the conductivity of the system.

Formation density log

Formation density log is not affected by either oil-emulsion or invert-emulsion systems. It is used to determine porosity and water saturation.

Neutron log

Neutron log is useful in determining gas—liquid contact, and is used in conjunction with other logs for determining lithology. It is not affected by either oil-emulsion or invert-emulsion systems.

Temperature log

Temperature log is used to determine temperature gradient in the well and to help delineate gas-bearing sands. It is not affected by the type of fluid used in the wellbore.

Other wireline services

The other wireline services, which can be used in the oil-emulsion or invert-emulsion drilling fluid environment and interpreted in the same manner as in the water-base mud environment, are as follows: caliper log, dipmeter log, formation tests, and obtaining sidewall cores.

EFFECT OF EMULSION MUDS ON CORING AND CORE DATA INTERPRETATIONS

Oil-emulsion muds are considered inferior by many petroleum engineers and geologists, mainly because inadequate core samples are obtained on using these muds. This is especially true where determination and delineation of target sands is dependent on correct core evaluation. On the other hand, the invert emulsion system allows complete retrieval of cores and their proper examination. Connate water and soluble salts remain intact in an oil-emulsion mud environment, allowing for maximum data gathering. Only refined diesel oil, however, must be used in preparing the mud. This is due to the great differences in "show" exhibited by refined diesel oil and crude oil, which enable the experienced investigator to easily distinguish between the two.

REFERENCES

Acosta, D.T., 1978. *Oil Mud Systems.* Unpublished Technical Paper, Houston, Texas, 35 pp.

Bancroft, W.D., 1867. *Applied Colloid Chemistry.* McGraw-Hill, New York, N.Y., 544 pp.

Becher, P., 1955. *Principles of Emulsion Technology.* Reinhold, New York, N.Y., 149 pp.

Becher, P., 1966. *Emulsions, Theory and Practice.* Reinhold, New York, N.Y., 440 pp.

Bennett, H., 1947. *Practical Emulsions.* Chemical Publ. Co., Brooklyn, New York, N.Y., 568 pp.

Berkman, S. and Egloff, G., 1941. *Emulsions and Foams.* Reinhold, New York, N.Y., 591 pp.

Clayton, W., 1928. *Theory of Emulsions and Their Technical Treatment.* J. and A. Churchill, London, 283 pp.

Clayton, W., 1923. *Theory of Emulsions and Emulsification.* J. and A. Churchill, London, 160 pp.

Eckel, J.R., 1954. Effect of mud properties on drilling rate. *API Dril. Prod. Prac.*, 2: 119—125.

Rogers, W.F., 1963. *Compositions and Properties of Oil Well Drilling Fluids.* Gulf Publ. Co., Houston, Texas, 818 pp.

Wiley R.M., 1954. Limited coalescence of oil droplets in coarse oil-in-water emulsions. *J. Colloid Sci.*, 9: 427—437.

Chapter 10

OIL-BASE DRILLING FLUIDS

PAUL VORABUTR and GEORGE V. CHILINGARIAN

INTRODUCTION

Although the first patent on oil-base drilling fluids was issued in 1923, they were used in actual drilling for the first time seven years later (Baroid, 1965, p. 85). Because of advances that have been made in the field of oil-base muds over the past 57 years, at the present time the performance of oil-base muds is competitive with that of water-base muds and is superior under certain adverse drilling conditions encountered on drilling for hydrocarbons at great depth. At present, oil-base muds account for 5—10% of the total volume of drilling fluids used, depending on geographic location, type of drilling (exploratory or development), and type of available equipment (see McMordie, 1975, p. 505).

In order to understand the behavior and, therefore, performance of oil-base muds, it is necessary to examine first the internal structure of this sys-

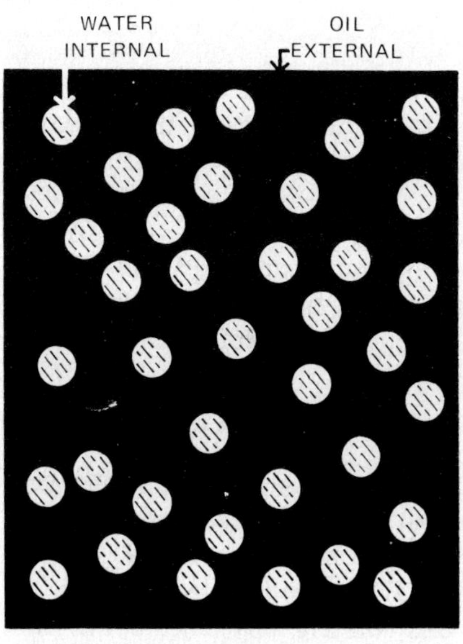

Fig. 10-1. Schematic diagram of water-in-oil emulsion. (After Magcobar Drilling Fluid Engineering Manual, 1972, p. 11-1, fig. 1; courtesy of Magcobar, Dresser Industries, Inc.).

tem. In general, oil-base mud has been defined as a system the continuous or external phase of which is any suitable oil. Oil muds may contain some water in the form of small, homogeneous in size, and uniformly-dispersed droplets commonly called the internal or discontinuous phase. The water content is governed by the desired properties of the muds. Generally, ready mixed oil muds supplied by various companies contain 2—7% water by volume. In many occasions, however, oil muds containing much larger volumes of water have been used successfully (Magcobar, 1975, p. 10-1).

At the present time there are two mud systems the external phase of which is oil, i.e., true oil muds and invert emulsion muds. Methven and Baumann (1972, p. 3) differentiated between these two systems by considering the function of water. Any water present in true oil muds does not function unlike that present in the invert emulsion muds, in which water imparts certain rheologic properties and filtration control characteristics. Rogers (1963, p. 564) distinguished oil muds from invert emulsion muds on the basis of their water contents: oil mud becomes invert emulsion mud when its water content exceeds 15% by volume. A large overlap exists in various classifications of oil muds and invert emulsion muds. A schematic diagram of water-in-oil emulsion is shown in Fig. 10-1. According to Magcobar (1975, p. 11-1), oil muds are generally very dark in color, whereas invert emulsion muds have a light tan to coffee brown color depending on the type and amount of oil used to make up the system.

COMPONENTS OF OIL MUDS

According to Methven and Baumann (1972, p. 3), true oil mud systems consist of the following components: (1) suitable oil, (2) asphalt, (3) water, (4) emulsifiers, (5) surfactants, (6) calcium hydroxide, (7) weighting materials, and (8) other chemical additives.

Among all of these components only oil and asphalt are necessary for the proper functioning of oil muds. The other components are used only for the purpose of enhancing and stabilizing rheologic properties and plastering characteristics. Each one of these components is discussed separately below.

Continuous phase — oil

Different types of oils have been used as the continuous phase in oil muds. Some of these oils, however, have exhibited superior performances over others. Performance, availability, and price are the main criteria in oil selection. The following commonly available oils have gained widespread acceptance: (1) lease crude oil, and (2) refined oils (diesel, kerosene, heavy fuel oils, etc.).

Of these two types, diesel oil is the most widely used in preparing oil mud systems because of its superior performance. Inasmuch as diesel oil is a refined product, it possesses highly uniform properties. It is also readily available and may be more economical to use than lease crude oil (IMCO Oil Muds, 1972, p. 3).

Lease crude oils have many serious disadvantages, which preclude their usage in most cases. According to Magcobar (1975, p. 11-2), in using crude oil one must make certain that it does not contain any natural or added emulsion breakers, because they may give rise to a separate continuous water phase. In addition, most crude oils contain light components which may pose a fire hazard. Consequently, light components should be removed, which leads to an increase in the costs of equipment, additives, and operation (Magcobar, 1975, p. 11-2).

Several slightly different specifications for diesel oils, which can be used as an external phase, have been developed by various companies. The specification proposed by Magcobar (1975) and presented below is intended to serve as an example of guidelines for the selection of oil:

(1) Specific weight — The specific weight should correspond to the API gravity of 30° to 40°. This spread in specific weight gives rise to a wide range of viscosity. In general, viscosity increases with increasing specific weight of the oil.

(2) Aniline point * — The aniline point, which is a measure of the aromatic content of oil, should be 140°F or higher. Aromatics are highly detrimental to rubber parts in the circulating system. The aniline point increases with increasing content of aromatics.

(3) Flash point — The flash point should be 180°F or higher. Flash point of oil is the temperature at which oil vapor ignites upon passing flame over the hot oil. In order to reduce fire hazard, the oil selected should have as high a flash point as possible.

(4) Fire point — The fire point should be 200°F or higher. This is the temperature at which continuous fire is sustained over the oil surface when flame is passed over it.

A typical specification for the oil used as a continuous phase in the oil mud system is presented in Table 10-I. For safety purposes, oil mud temperatures must be measured continuously. Thermostats should be placed in the flow line and in the active and reserve pits to monitor the temperatures. If oil mud is stored in closed, metal circulating and reserve pits, then proper ventilation is required. Oil vapor, which may accumulate over the oil surface, can be easily ignited posing an extreme fire hazard. Consequently, "fire hazard" and "no smoking" signs must be posted.

---

* See Appendix I of this chapter.

TABLE 10-I

Recommended specifications for diesel oil for use in oil-base muds (also presented in Chapter 9)

(After Imco Oil Muds, 1972, p. 3; courtesy of Imco Drilling Mud Inc.)

	Typical specifications preferred	Acceptable range
Gravity, API	36.0	33.0—39.0
Color, Robinson	25	20—30
Flash, PM (°F)	152	130 minimum
Pour point (°F)	—5	+10-—15 *
Viscosity @ 100°F, SSU	35.5	— **
Carbon (10% bottoms)	0.01	— **
Ash (%)	nil	— **
Bomb sulfur (%)	0.16	0.15—0.32
Corrosion, 3 hr @ 212°F	J3	— **
Cetane number	50.8	— **
Suspended sediment	0.1	— **
Dissolved sediment	0.9	— **
Bromine number	1.43	— **
Aniline point (°F)	149	135 minimum
Diesel index	54.5	— **
% Aromatics (+ olefins)	21.8	18—30

* Lower pour point material may be suitable for use but should be checked in the laboratory. Pour points lower than —15°F usually indicate that a pour point depressant has been added to oil. These depressants cause a weakening of the emulsion with increased viscosity and gel strengths.
** Not critical.

Discontinuous phase — water

Although the presence of water is not required in oil muds, some water is generally added to react with chemical additives in order to enhance the rheologic properties and plastering characteristic of oil mud. The water-in-oil mud exists in the form of small, homogeneous in size, and uniformly dispersed droplets in the continuous oil phase, i.e., water-in-oil emulsion. The properties of oil mud are affected to some extent by the stability and uniformity of these water droplets. The distances between droplets increase as they become smaller and more uniform in size. As a result of being farther apart, droplets require longer time to coalesce and to give rise to a separate free water phase in the oil mud. Consequently, the emulsion is more stable with greater degree of dispersion. On the other hand, if water droplets are large and nonuniform, the distances between the droplets decrease, assuming same volume of water. As a result, droplets coalesce faster, giving rise to a less stable emulsion system, with rapid separation of the water phase. The emulsion stability decreases with increasing water content in the oil mud sys-

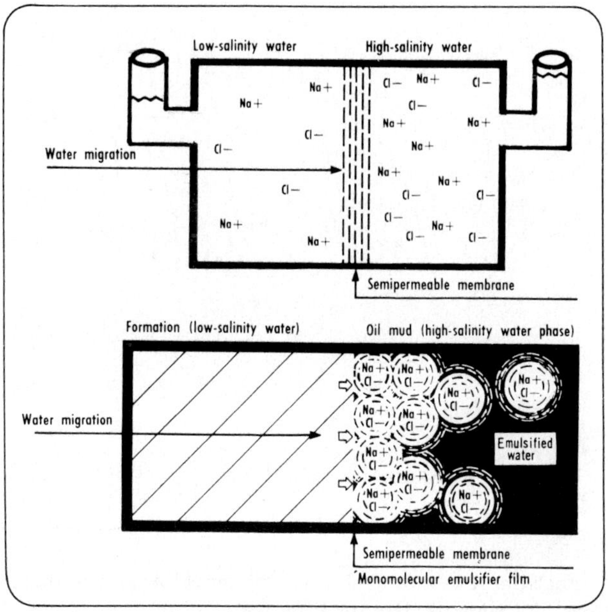

Fig. 10-2. Schematic diagram of osmotic mechanism responsible for the migration of water, which may either hydrate or dehydrate the shale clays depending on the types and concentrations of chemicals present (chemical potential). (After Baroid, 1977, II(II), p. 16, fig. 6; courtesy of the Baroid Petroleum Services.)

tem. With increasing number of water droplets, there is a greater chance of more rapid coalescence. Greater emulsion stability results on addition of continuous phase which separates the droplets. The theory of emulsion is discussed in detail in Chapter 9.

Certain compositions of water present in oil muds have been found to give rise to superior performance. According to Magcobar (1975, p. 11-4), the term "activity" is used to define the unit of measure of the chemical potential of water present in mud or shale. Water migrates toward the area having a higher chemical potential. Generally, water present in shales, which are composed of different clay types, possesses high activity. When drill bit penetrates a shale section, osmosis occurs, i.e., there is a partial hydration of shale by the flow of water in the direction of higher chemical potential or greater concentration of ions (see Fig. 10-2). This results in shale swelling with resulting sloughing, which causes wellbore enlargement and other associated problems. Mechanisms of shale swelling are discussed later in this chapter and in Chapters 2, 5, and 8.

With addition of salt to the aqueous phase, the activity of mud can be increased. Consequently, oil mud, the aqueous phase of which is saturated

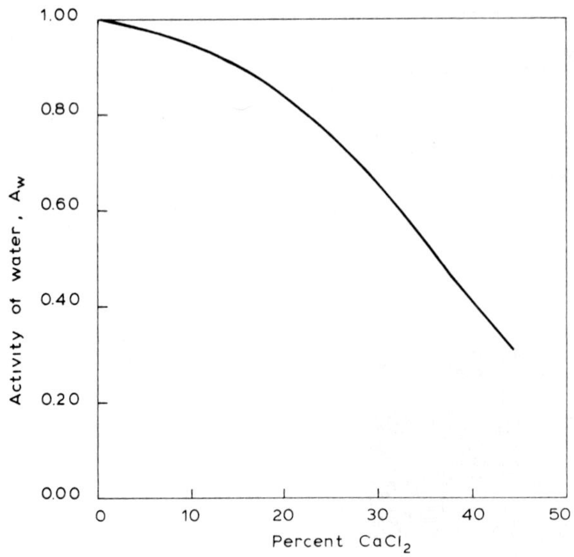

Fig. 10-3. Activity of water as a function of calcium chloride concentration. (After Magcobar, 1972, p. 11-9, fig. 7; courtesy of Magcobar, Dresser Industries, Inc.)

with salt, has a higher activity than the surrounding shale, resulting in water migration out of the shale. This loss of water hardens the shale section and stabilizes the wellbore. Inasmuch as it may also cause spalling of shale into the wellbore, the activity of the mud must be controlled, so that shale is hardened only to the point at which its spalling into the wellbore may not occur.

The electroosmotic migration of water may also occur because of difference in the chemical potential of water in the shale cutting and that present in mud (aqueous phase). If the aqueous phase is more saline, water is removed from the shale chips. As a result, the shale structure becomes more compact and does not break easily allowing better removal of shale cuttings at the surface. Relationship between the activity of water and calcium chloride concentration in the aqueous phase is presented in Fig. 10-3. The activity of water decreases with increasing concentration of salt, suggesting that mud activity would increase with addition of salt. Although either sodium chloride or calcium chloride may be used for the purpose of increasing the mud activity, the latter can be controlled to a much greater extent with calcium chloride. This is due to the fact that calcium salt lowers the activity of the aqueous phase to a much greater degree with a consequent greater increase in the activity of the mud.

The type of shale which is expected to be encountered during drilling, dictates the required mud activity (see Fig. 10-4). Consequently, geologists

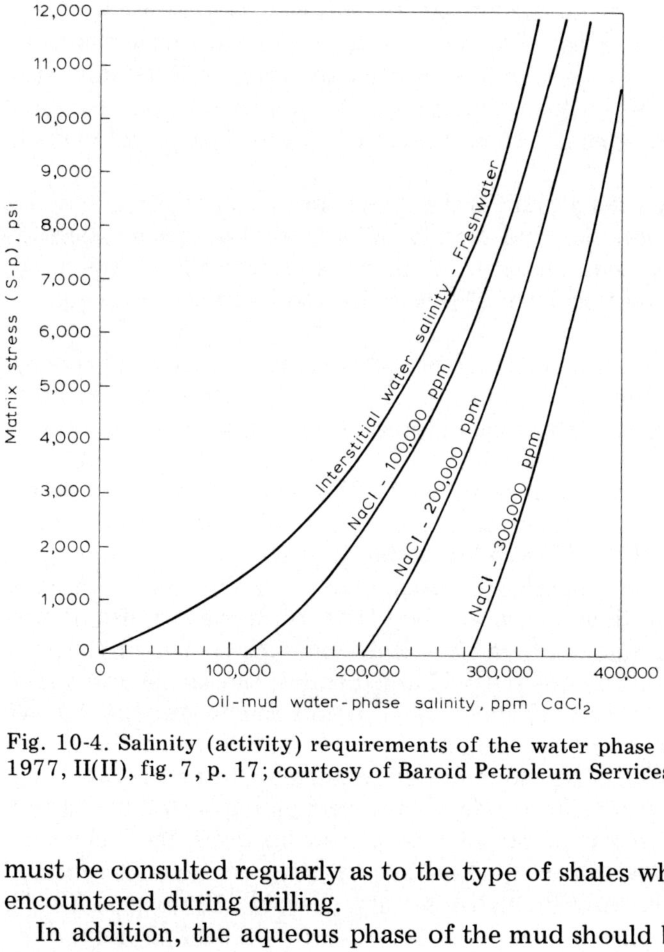

Fig. 10-4. Salinity (activity) requirements of the water phase of oil muds. (After Baroid, 1977, II(II), fig. 7, p. 17; courtesy of Baroid Petroleum Services.) (See eq. 10-3.)

must be consulted regularly as to the type of shales which are expected to be encountered during drilling.

In addition, the aqueous phase of the mud should be basic, because emulsions break if pH decreases below 7.0 (Imco Oil Muds, 1972, p. 4). Emulsion breakers present in the water and oil phases must be disabled in order to insure emulsion stability.

Bodying agents

A number of bodying agents have been used in oil muds to achieve the desired rheologic and filtration loss characteristics. Bodying agents can be classified into two groups: (1) colloidal-size materials and (2) high-molecular-weight metal soaps.

Colloidal-size materials

Asphalts which are colloidal-size organophilic materials are used in oil muds to impart required properties and control fluid loss, mainly through their adsorptive characteristic. According to Methven and Kemick (1968, p. 5), each asphalt particle has tendency to aggregate with other particles forming micelles. These micelles, in the presence of certain refined oils, associate to form network of micellar structures in the continuous oil phase and impart desired rheologic properties and control filtrate loss as do clays in water-base muds.

The most commonly used asphalt is the special air-blown asphalt, which is organophilic, hydrocarbon-like, and readily oil wettable. It does not degrade appreciably at elevated temperatures. Methven and Kemick (1968, p. 5) attributed the superior performance of this asphalt to its tremendous adsorptive characteristics.

The fact that air-blown asphalt is readily oil wettable, which is a property common to all organophilic substances, reduces the pretreatment costs necessary to make other water-wet clay materials oil wet prior to their use in oil muds.

Special oil-wettable clays have been used as bodying agents more extensively in oil muds than in invert emulsion muds. According to Methven (1971, p. 2), these treated clays are oil-wettable montmorillonite, attapulgite, and other clays. The term "synthetic" organophilic colloids has been commonly used to describe them, because these water-wet clays must be pretreated, mostly with a quaternary amine (cationic surfactant), in order to make them oil wettable. The positively charged end of the surfactant, which is attracted to the negatively charged clay particles, changes the surface chemistry of the clays and makes them oil wettable (Methven and Baumann, 1972, p. 2). These clays can be made either gel-forming or non-gel-forming, and have good barite suspension property in invert emulsion muds (Baroid, 1954, p. 300-10). As a result of pretreatment cost, however, these clays are expensive and are only used in cases where good barite suspension is required without resorting to excessively high viscosity (Baroid, 1954, p. 300-11). Such situations mostly arise in the case of invert emulsion drilling fluids.

Heavy metal soaps of fatty acids

The functions of emulsifiers in oil muds are as follows (Rogers, 1963, p. 568):

(1) Imparting weak gel strength to oil mud because gel strength is necessary for suspension of weighting materials.

(2) Emulsification of any water picked up during drilling operation (water contamination).

(3) Controlling the tightness of any water emulsion resulting from water contamination and, therefore, controlling fluid loss.

Heavy metal soaps of fatty acids (emulsifiers) are added to the oil muds in order to emulsify the water in oil. Their presence also results in emulsification of additional water which is picked up during drilling operations.

Calcium soap of fatty acid is used to emulsify the water in oil-base mud.

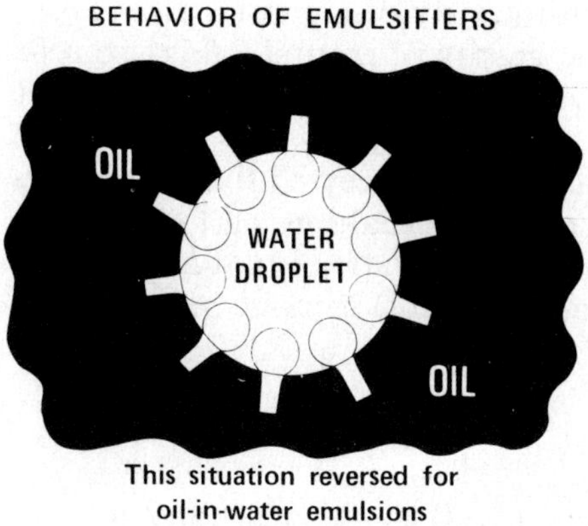

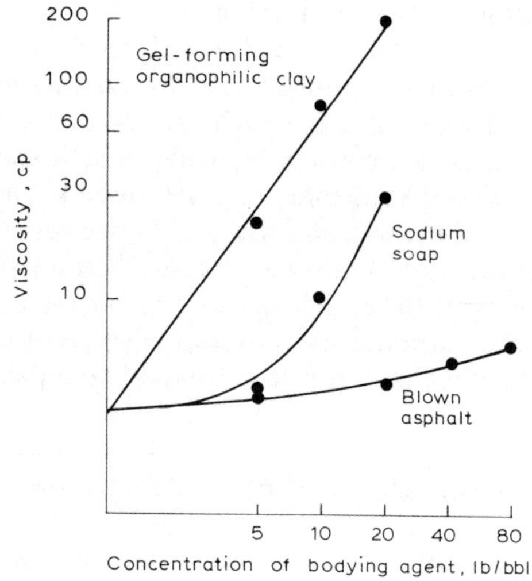

Fig. 10-5. Molecular orientation of an emulsifying agent in water-in-oil emulsion. (After Magcobar, 1972, p. 11-3, fig. 4; courtesy of Magcobar, Dresser Industries, Inc.)

Fig. 10-6. Relationship between the concentration of bodying agents and the viscosity of oil muds. (After Simpson et al., 1961, in: Rogers, 1963, p. 566.)

The molecules of this calcium soap have two large organic group ends, which are attached to a water-soluble central ion. These ends bend across the oil and water interfaces and lead to a stable water-in-oil emulsion (Imco Oil Muds, 1972, p. 4).

A monovalent soap used as an emulsifier is sodium soap. Sodium soap molecules consist of two parts: the water-soluble or inorganic part and the oil-soluble or organic part. These molecules concentrate at the oil and water interfaces and thus promote stable emulsion. Fig. 10-5 illustrates the molecular configuration at the oil—water interface. According to Rogers (1963, p. 556), sodium soap has a greater effect on viscosity through the formation of stable water emulsion than it does on oil as a high-viscosity, oil-insoluble material.

Rogers (1963, p. 561) presented graphically the effect of the bodying agents discussed above, using data obtained by Simpson et al. (1961) (see Fig. 10-6). As shown in this figure, at the same concentration, the organophilic clays are the most effective bodying agents. At low concentration (less than 5 lb/bbl), heavy metal soaps and asphalts have similar effects on viscosity.

It must be pointed out, however, that the performance of heavy metal soaps as emulsifiers is unreliable at elevated temperatures and pressures.

WEIGHTING MATERIALS

Barium sulfate (barite) is used as weighting material both in oil muds and water-base muds. It is inert and has a specific gravity of around 4.2.

Calcium carbonate, which has a specific gravity of around 2.7 is also used as a weighting material. Although it has a lower specific gravity than barite, it is readily wettable by oil. Consequently, its usage may result in reduction of surfactant treatment costs which must otherwise be incurred to make barites oil wet. Inasmuch as calcium carbonate is oil wet, it soaks up part of the continuous phase and, consequently, if optimum concentration is not maintained, its usage may result in unstable emulsion and a notable increase in viscosity.

RHEOLOGIC PROPERTIES OF OIL-BASE MUDS

In the past, the most severe drawback of oil muds, was their lack of rheologic properties and high filtration loss. Oil muds, which consist of oil alone, lack (1) the wide viscosity range to facilitate efficient transportation of cuttings to the surface, and (2) yield point and gel strength (thixotropic

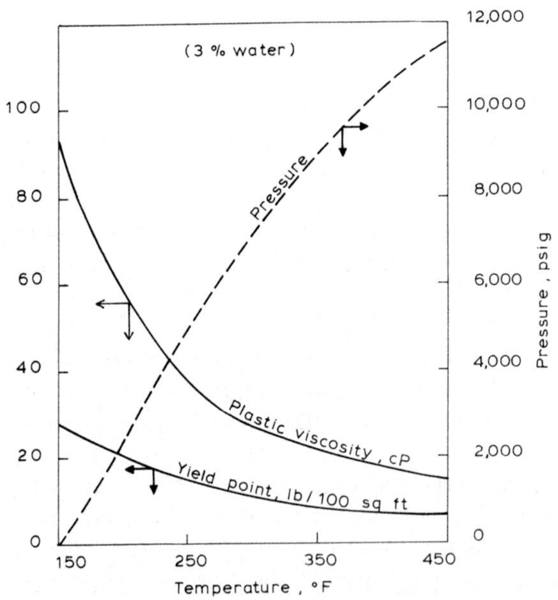

Fig. 10-7. Typical plastic viscosity and yield point of a 18.5 ppg (lb/gal) oil mud (suspension of air-blown asphalt in diesel oil) having 3% water content by volume as a function of temperature and pressure. (After McMordie, 1969, p. 84, fig. 9; courtesy of Oil and Gas Journal.)

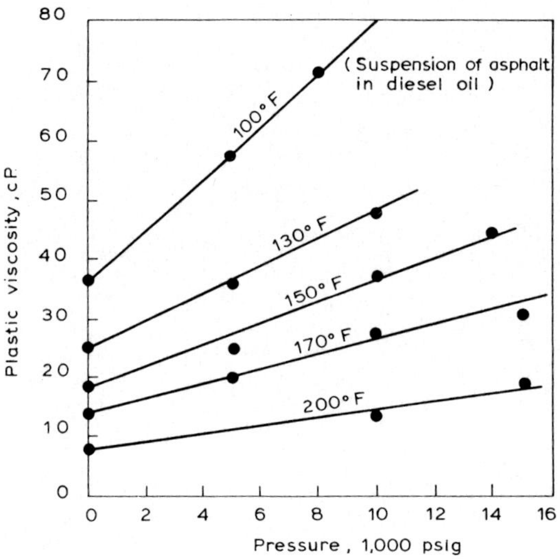

Fig. 10-8. Effects of temperature and pressure on plastic viscosity of an oil-base drilling fluid (suspension of air-blown asphalt in diesel oil). (After McMordie, 1969, p. 85, fig. 3; courtesy of Oil and Gas Journal.)

property) in order to suspend weighting material and cuttings when the circulation stops. At the present time, however, oil muds possess all the properties of good drilling fluids.

Generally, the viscosity of the base fluid (oil) provides the viscosity for the oil-base mud system. Base oils, however, have a limited viscosity range which hamper their uses. Incorporation of water in a form of water emulsion provides additional control of viscosity. A system of small water droplets dispersed in a continuous oil phase offers greater resistance to flow than the continuous phase alone, and the viscosity increases with increasing water content of the oil mud. Addition of the external continuous phase decreases the viscosity.

Oil mud viscosity, yield point, and gel strength are controlled and maintained to a great extent through addition of colloidal-size materials, particularly air-blown asphalt. As discussed previously, the network of micellar structures formed by association of micelles are entirely responsible for increasing viscosity, yield point, and gel strength of oil muds. Increase in the concentration of colloidal-size materials in the mud system raises the viscosity, gel strength, and yield point of the mud.

Oil mud containing air-blown asphalt maintains its reliability up to a bottomhole temperature of 450° F (McMordie, 1969, p. 84). As shown in Figs. 10-7, 10-8, and 10-9, oil mud containing air-blown asphalt shows a regular thinning effect (decrease in viscosity and yield point) with increasing temperature (up to 450° F) and decreasing pressure.

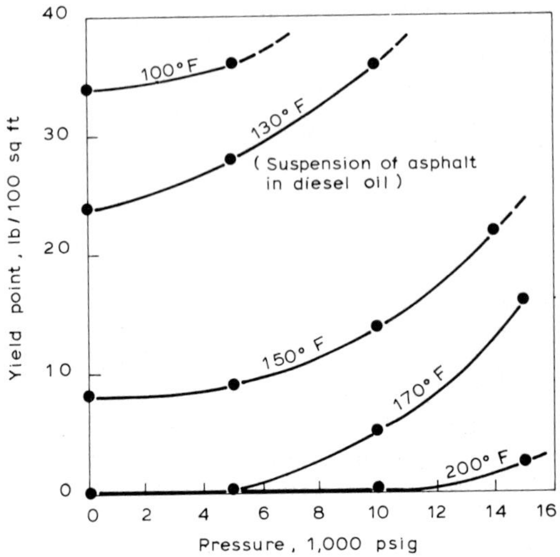

Fig. 10-9. Effects of temperature and pressure on yield point of an oil-base mud (suspension of asphalt in diesel oil). (After McMordie, 1969, p. 85, fig. 4; courtesy of the Oil and Gas Journal.)

FILTRATION PROPERTIES

Wall plastering properties of oil muds are due to the extensive criss-crossing network of micellar structures in the oil phase. These micellar structures play the same role in oil as do clays in water-base muds. After very small initial "spurt loss", these structures form a nearly impermeable cake on the face of the permeable formation and prevent any further invasion of any fluid into this formation.

According to Methven and Kemick (1968, p. 5), the air-blown asphalt consists of asphaltene molecules (dispersed phase) in a dispersion medium consisting of certain resins and oil. The aromatic portion of the dispersion medium is adsorbed on the asphaltene molecules which constitute micelles. The buildup of these micelles and their association to form micellar structures and networks are influenced by temperature, type of hydrocarbons used as base fluid, and the chemical constituents of the solvent and asphaltenes. Apparently, adsorption as well as desorption of these micelles and micellar structures may be controlled. This special feature of asphalt, particularly air-blown asphalt, enables mud engineers and chemists to control the filter cake thickness and corresponding filtrate loss.

The desorption property of these asphaltene micellar structures gives rise to trouble-free drilling and completion operations. The desorption process has been found to occur in most crude oils (Methven and Kemick, 1968, p. 6). Formation damage due to the mud cake invasion, therefore, can be

TABLE 10-II

Completion efficiency of various drilling fluids

(After Methven, 1971, pp. 6—8; courtesy of the Society of Petroleum Engineers of AIME)

Mud type	Completion efficiency (%)
(A) *Oil or gas-condensate flowing wells*	
Water-base drilling fluids	44—69
Invert emulsion drilling fluids	74
Oil-base drilling fluids	100
(B) *Dry-gas flowing wells*	
Invert emulsion drilling fluids	88
Oil-base drilling fluids	100

eliminated on using oil muds just after the initial production of the permeable zone. The range of completion efficiency (in percent) achieved on using various drilling fluids (oil muds and other types of muds) in oil, condensate, and dry-gas wells is presented in Table 10-II.

Generally, the filtration loss of oil mud is low and the filtrate is the oil (continuous phase). Oil muds, the water contents of which are negligible, have higher filtration loss than oil muds having some water content (e.g., in the range of 3—7% water by volume). This is due to the reasons discussed below.

According to Baroid (1975b, p. 18), reduction in the filtrate loss is achieved by plugging up of the pores on the face of a permeable formation by micelles and micellar structures of various sizes. Bridging and plugging of pores on the filtering medium face continue until permeability is reduced to a degree that the continuous phase molecules cannot penetrate the filter cake any further.

Water droplets emulsified in oil probably further augment this plugging and bridging process (water blocking effect). Probably, individual water droplets, surrounded by films of emulsifiers, first are deposited on the face of permeable zone and on the newly-formed mud cake. Then they adjust themselves to the irregular shape of the pores, through which filtration is occurring, and seal the pores (Baroid, 1975b, p. 18).

Two types of filtration occur downhole, i.e., static and dynamic. Static filtration loss occurs when the drilling fluid is at rest. In such a case, filtrate invades the formation as a result of difference between the hydrostatic pressure of the mud column and the formation pore pressure. Although static filtration can be controlled in all mud systems, this can be achieved to a much greater extent in the case of oil-base muds.

Dynamic filtration, which occurs during the mud circulation, is a function of static filtration. At the present time, there is no standard laboratory test method to determine the amount of dynamic fluid loss. The performance of oily filter cake, which is created and destroyed continuously during dynamic filtration test, should be thoroughly investigated and compared with performance of filter cakes formed by other types of drilling fluids.

SPECIFIC WEIGHT ("DENSITY")

The commonly used specific weights of oil muds range from 7 to 22 lb/gal. Similar to water-base muds, the base fluid in oil-base muds provides the

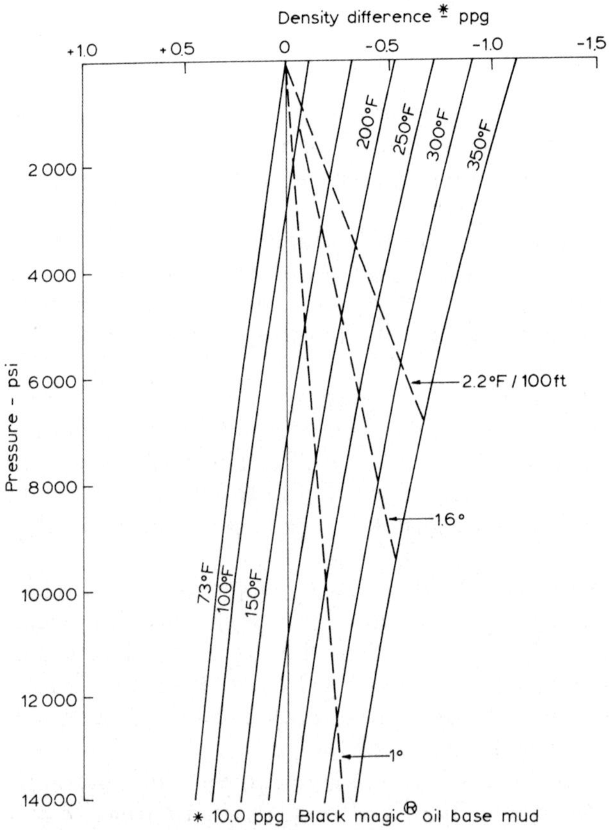

Fig. 10-10. Interrelationship among density difference (deviation from density at standard conditions), temperature, and pressure of a 10.0-lb/gal oil-base mud. (After Methven and Baumann, 1972, p. 10, fig. 1; courtesy of the Society of Petroleum Engineers of AIME.)

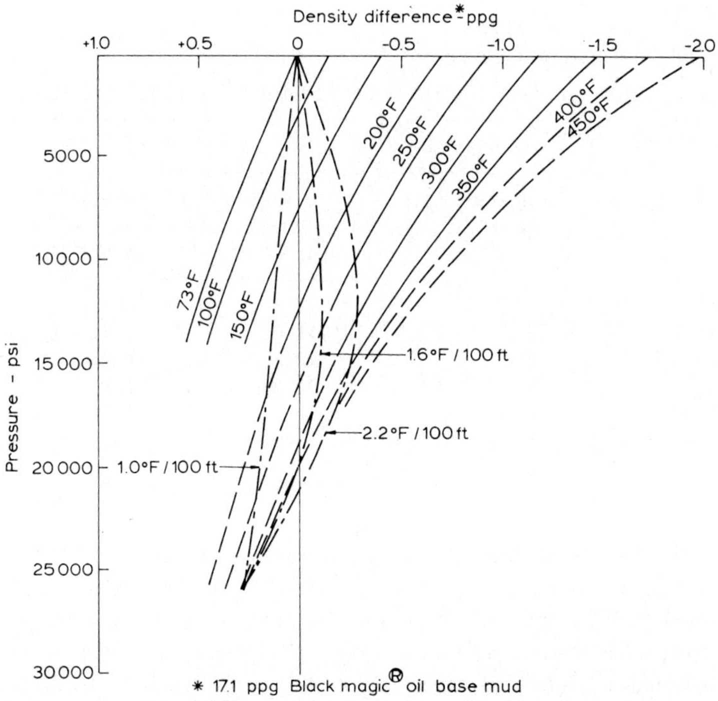

Fig. 10-11. Interrelationship among density difference (deviation from density at standard conditions), temperature, and pressure of a 17.1-lb/gal oil-base mud. (After Methven and Baumann, 1972, p. 10, fig. 2; courtesy of the Society of Petroleum Engineers of AIME.)

basic density, and weighting materials are added to it in order to raise the specific weight of mud to the desired "mud weight". Inasmuch as oil is much more compressible than water, and in oil muds oil is the external phase, the specific weight of oil mud is a function of temperature and pressure. Density varies directly with pressure and inversely with temperature (Methven and Baumann, 1972, p. 5). As shown in Figs. 10-10 and 10-11, the opposite effects of pressure and temperature on density of oil mud do not cancel out. The curves in these two figures also indicate that the initial mud weights at standard conditions have little influence on the effects of pressure and temperature on density. Finite differences, which may exist due to the mud weight differences, are mainly caused by the amount of oil-wet weighting materials in the systems. There are more oil-wet solids in the system having higher mud weight, which soak up part of the continuous phase, and, therefore, reduce part of the oil that is subjected to the effects of temperature and pressure.

TABLE 10-III

Specific weights of various saturated salt solutions

(After Baroid, 1975a, p. 5, Table 2; courtesy of the Baroid Division, NL Industries, Inc.)

Salt type	Saturated at 77°F		
	concentration (wt %)	specific weight (lb/gal)	
		at 77°F	at 212°F
Potassium chloride, KCl	26.5	9.8	9.5
Sodium chloride, NaCl	26.5	10.0	9.7
Sodium carbonate, Na_2CO_3	22.5	10.3	10.0
Calcium chloride, $CaCl_2$	46.1	12.2	11.8
Potassium carbonate, K_2CO_3	52.9	13.0	12.7

The solids primarily used to obtain the desired mud weight are barite (sp. gr. $\approx$ 4.2) and calcium carbonate (sp. gr. $\approx$ 2.7). Barite can be used to raise the weight of the mud to 22 lb/gal, whereas mud weight of 12 lb/gal may be obtained with addition of calcium carbonate (Baroid, 1975a, p. 4). In order not to impair the suspension ability of oil muds, which is due to the presence of asphalt and high-molecular weight metal soaps, higher mud weights cannot be tolerated.

By altering the specific weight of the internal phase, the density of oil muds may be controlled to some extent without using the weighting materials. The extent of this control varies directly with the amount of aqueous phase present in the oil-mud system. In the case of high water content, the weight increase by using this method may be appreciable.

Salts are the primary water-soluble materials used for increasing the weight of aqueous phase. Specific weights of some saturated salt solutions are presented in Table 10-III. For example, potassium carbonate (K_2CO_3) nearly doubles the specific weight of water at a concentration of 52.9% by weight. A typical relationship between specific weight of oil mud and total solids content of base fluids for various salt contents of water phase is presented in Fig. 10-12.

The specific weight of oil muds can be reduced by using the following methods:

(1) *Addition of the external phase.* The volume of the external phase increases and the solids content per unit volume decreases on adding additional oil; this reduces the density. The overall properties of the oil mud are least affected through addition of oil. Viscosity, yield point, and gel strength all decrease as a result.

(2) *Mechanical removal of weighting materials.* Solids control devices may

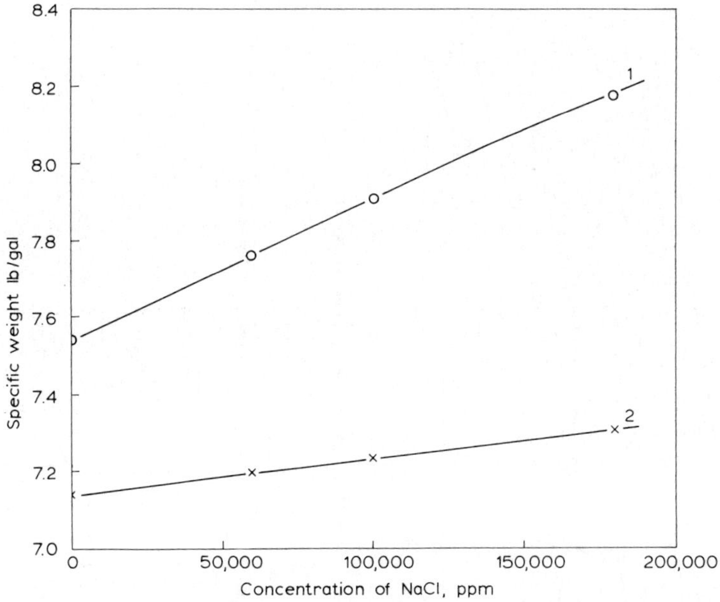

Fig. 10-12. Relationship between the specific weight of oil mud and total solids content of base fluid for various salt contents of water phase. *1* = 40% diesel oil (sp. gr. = 0.86 at 25°C); *2* = 90% diesel oil. $\gamma_m = [(\%\ \text{oil})(\gamma_o) + (\%\ \text{water})(\gamma_w)]/100$, where γ_m, γ_o, and γ_w are specific weights of oil mud, diesel oil, and water, respectively.

be used to remove excess solids, with a consequent reduction in specific weight.

(3) *Introduction of gas.* Gas which may be encountered during drilling operation, moves into the wellbore in the form of gas bubbles and lightens the mud system.

(4) *Addition of plastic microballoons.* According to Minkhailov et al. (1971, pp. 26—29), the plastic microballoons, which are filled with nitrogen and are made from phenolformaldehyde and from urea formaldehyde resins, can be used for the reduction of specific weight. The diameter of these balloons ranges from 50 to 300 microns, whereas their density ranges from 0.1 to 0.25 g/cm^3. As a result of addition of these balloons, viscosity, yield point, and gel strength increase, whereas filtration rate decreases.

Two types of specific weight must be considered, i.e., static and circulating. Static specific weight is a function of the amount of materials present per unit volume of mud when the mud is at rest. According to Methven and Baumann (1972, p. 4), the circulating specific weight of an oil mud is a function of static specific weight and the circulation friction loss which, in turn,

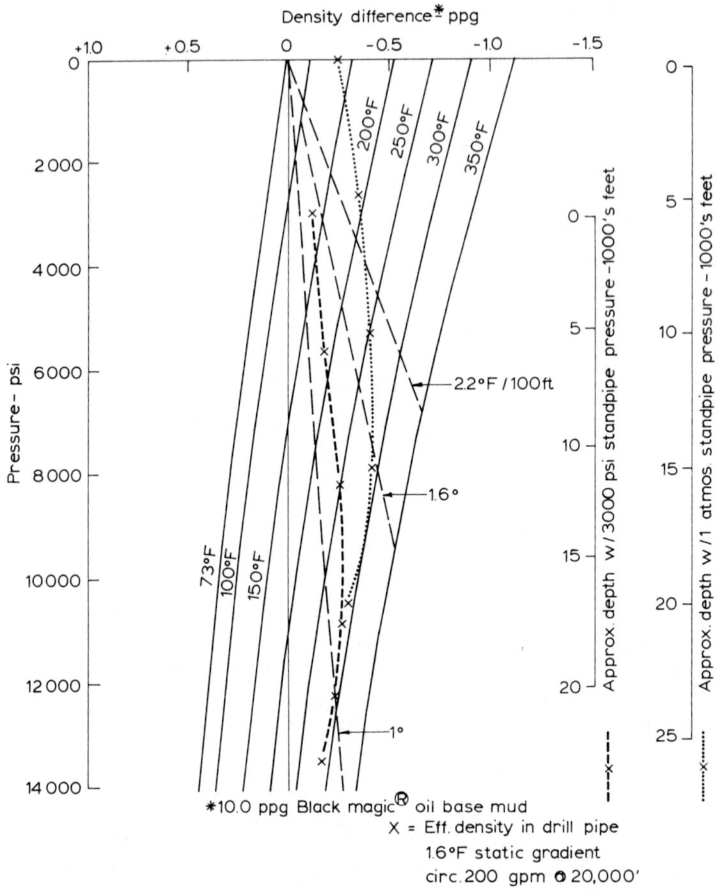

Fig. 10-13. Interrelationship among circulating density difference (deviation from standard conditions), temperature, and pressure of a 10.0-lb/gal oil-base mud. (After Methven and Baumann, 1972, p. 12, fig. 5; courtesy of the Society of Petroleum Engineers of AIME.)

is a function of temperature and pressure. Furthermore, the bottomhole temperature and pressure of circulating oil mud is a function of time, circulating rate, hole size, etc. Excluding the friction losses, bottomhole temperature and pressure alone may cause a loss equivalent to a pressure drop of 50—100 psi in the annulus (Methven and Baumann, 1972, p. 4). The pressure drop is higher on considering the effect of friction. The effects of temperature and pressure on the circulating density of two oil muds the specific weights of which are different, are presented in Figs. 10-13 and 10-14.

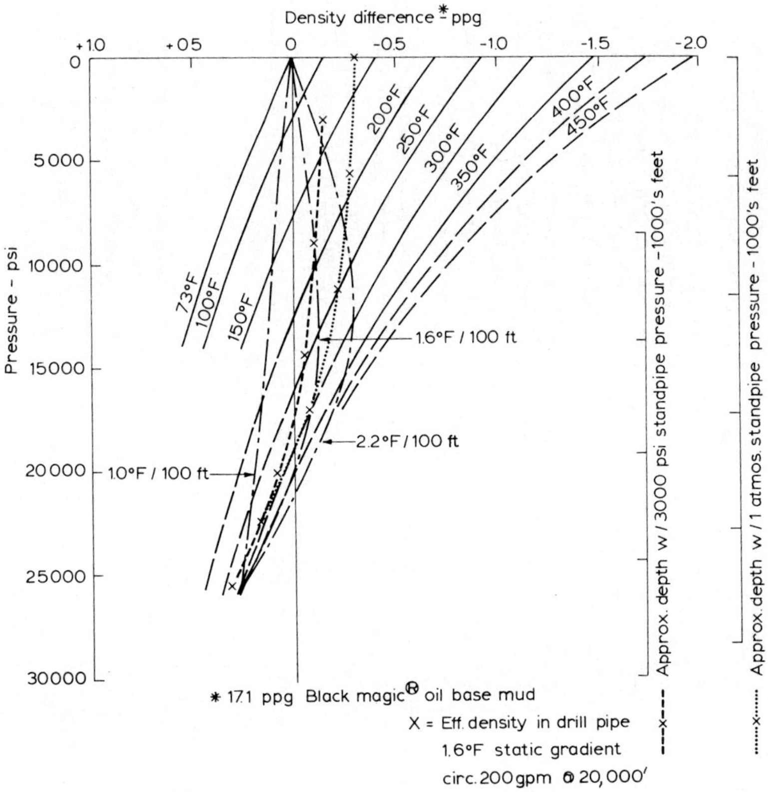

Fig. 10-14. Interrelationship among circulating density difference (deviation from standard conditions), temperature and pressure of a 17.1-lb/gal oil-base mud. (After Methven and Baumann, 1972, p. 12, fig. 6; courtesy of the Society of Petroleum Engineers of AIME.)

OIL MUD CONTAMINATIONS

Oil muds are generally quite resistant to contaminants commonly encountered during drilling, and water-base mud contaminants have very small effect on the properties of oil-base muds. Water intrusion into the system has the greatest effect on the oil mud properties. Oil mud which contains an excess of emulsifiers, however, is highly water tolerable, because water picked up during the drilling operation is immediately emulsified and becomes part of the system. Normal thickening effects result from incorporation of additional water, for which various treatments may be prescribed to maintain the desired properties.

Numerous treatments may be prescribed when the contaminant is water

TABLE 10-IV

General guidelines for treatment of oil-base mud systems

(Courtesy of the Oil Base, Inc.)

Problem	Cause	Weight	$\overline{PV}$—$\overline{YP}$ *	Solids	Water	*ES* **	Recommended treatment
Viscosity too high	Drill solids	Normal	High	High	Normal	Normal	Check solids and oil removal; dilute with diesel oil or new mud; maintain weight.
	Water-base mud	Normal	High	High	High	Low	Check solids removal; add emulsifier, dispersant, lime, and naphthenic acid; dilute with diesel oil or new mud; maintain weight.
	Water-wet barite or water flow	Low	High	Normal	High	Low	Determine water source; add emulsifier, dispersant, naphthenic acid, gelling agent, diesel oil or new mud, and barite as required.
	Weighting material or drill solids	High	High	High	Normal	Normal	Check solids removal; dilute with diesel oil; add emulsifier and dispersant to disperse drilled solids as required.
Viscosity too low	High-pressure, high-temperature filtrate is high	High	Low	High	Normal	Normal	Add gelling agent and diesel oil or new mud.
	Crude or distillate	Low	Low	Low	Normal	Normal	Utilize degasser; add naphthenic acid, gelling agent, and barite to stop flow; add new mud.

* $\overline{PV}$ = plastic viscosity; $\overline{YP}$ = yield point.
** *ES* = electrical stability.

containing water-soluble alkaline salts such as calcium chloride. All treating methods have the same aim, i.e., to maintain the oil wettability of solids; however, the methods of application and chemical additives used may differ. Bennett and McMordie (1969) proposed to use a system of water-soluble acid-blocked alkanolamide nonionic surfactant, a sulfonated acid anionic surfactant, and an oil-soluble nonionic ethoxylate-alkyl-phenol. This system maintains the oil wettability of solids in the presence of alkaline earth salt-water contamination in the oil mud.

Oil-wet solids in the mud system may also contaminate the mud if optimum concentration is not maintained for the existing specific weight. As a result of soaking up part of the continuous phase, oil-wet solids cause increase in viscosity and reduction in water-in-oil emulsion stability. Some of the problems, causes, property changes, and treatment recommendations, which may be used as guidelines for treating oil mud contaminations, are listed in Table 10-IV.

Pilots tests must be run prior to each treatment using the contaminated mud in order to determine the effects of treatment and the required amount of chemicals, etc.

FUNCTIONS OF VARIOUS COMPONENTS IN AN OIL MUD SYSTEM

Although numerous oil-base mud systems have been formulated, they all have the same basic components. Diesel oil is used as the base fluid; asphalts and high-molecular-weight metal soaps, as bodying agents; and barite and calcium carbonate, as weighting materials. Certain chemical additives are also incorporated in order to enhance various properties, because each system is designed for a particular set of drilling conditions expected to be present in the area. As an example, one system is discussed here (after Oil Base, Inc.).

This system consists of refined oil (usually diesel) as the external phase, air-blown asphalt, lime [$Ca(OH)_2$], a small amount of water, naphthenic acid, fatty acid, and a mixture of emulsifiers and dispersants.

The oil serves as the base fluid, whereas air-blown asphalt provides the rheologic properties and controls filtrate loss. Lime, which aids in stabilizing the rheologic properties and filtrate loss characteristics imparted by asphalt, also reacts with naphthenic and fatty acids to form oil-soluble calcium soap in the presence of water. This soap is the water-in-oil emulsifier and helps maintain stable water emulsion. Generally, excess of lime is added, because the unreacted lime acts as biocide and corrosion control agent when hydrogen sulfide and carbon dioxide are encountered during drilling operations. Addition of a mixture of emulsifiers and dispersants together with the internally-formed calcium soap increases the gel strength of the oil mud by dispersing the oil-wet solids.

As mentioned previously, an excess amount of lime is added to this sys-

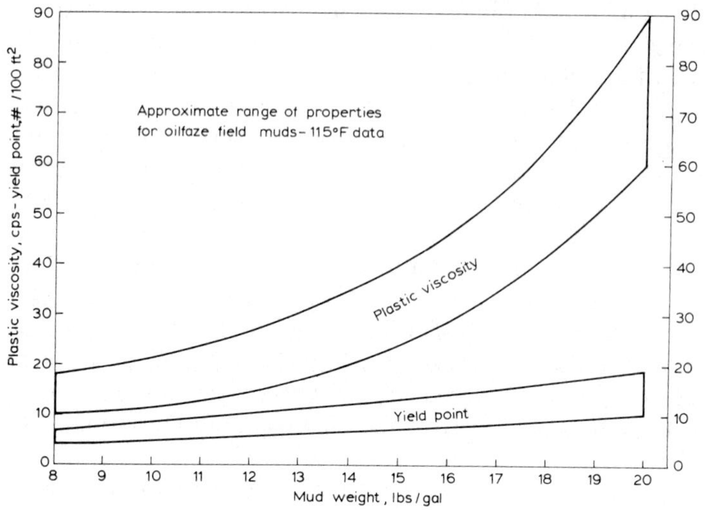

Fig. 10-15. Relationship between specific weight of mud (lb/gal) and (1) plastic viscosity (cP) and (2) yield point (lb/100 sq ft) for an oil-base mud at 115°F. (After Magcobar, 1972, p. 11-9, fig. 8; courtesy of Magcobar, Dresser Industries, Inc.)

tem to control corrosion caused by hydrogen sulfide and carbon dioxide contamination. The following reaction occurs in the presence of hydrogen sulfide:

$$Ca(OH)_2 + H_2S \rightarrow CaS + 2\,H_2O \tag{10-1}$$

One mole of lime is required to neutralize every mole of hydrogen sulfide.

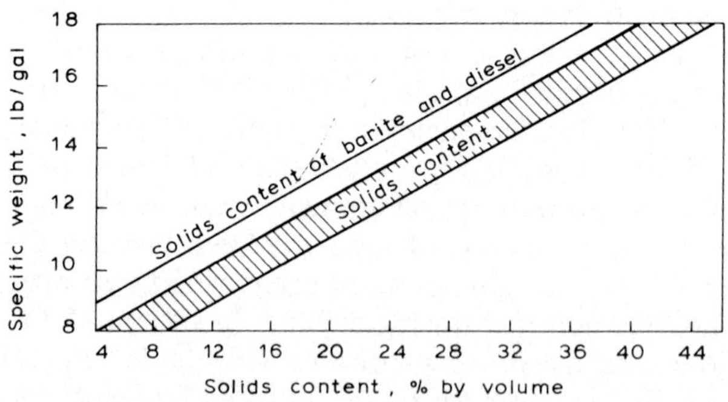

Fig. 10-16. Relationship between the specific weight of an oil mud and solids content. (Courtesy of Oil Base, Inc.) Sp. gr. of barite = 4.25.

Carbon dioxide also reacts with lime according to the following reaction:

$$Ca(OH)_2 + CO_2 \rightarrow CaCO_3 + H_2O \qquad (10\text{-}2)$$

The reaction products of these two reactions are in the form of solid precipitates of submicron size. Their presence increases the solids content of the mud and, therefore, results in an increase in viscosity. If the viscosity becomes too high, dilution with the continuous phase may become necessary. Typical properties of this mud system are illustrated in Figs. 10-15 and 10-16, with the viscosity increasing drastically at high specific weights of muds.

APPLICATION OF OIL MUDS

Oil muds are used mainly in special drilling conditions such as:

(1) Drilling through thick water-sensitive shale sections.

(2) Drilling through producing zones.

(3) Drilling through water-soluble formations, such as rock salt (NaCl), anhydrite, and gypsum.

(4) Preventing and relieving stuck pipe.

(5) Drilling high-temperature deep wells.

Wellbore instability is a problem related to drilling through water-sensitive shale sections. Darley (1969, p. 186) used the term "instability" to represent all borehole problems related to shales during and after completion of drilling operations, i.e., heaving, sloughing, caving, tight hole, and hole enlargement. Borehole problems associated with shales are due mainly to (1) swelling and (2) shrinking.

According to Darley (1969, p. 18), two swelling mechanisms may be operative in shales:

(1) *Surface hydration.* Inasmuch as water molecules are polar, they attach themselves to the unsatisfied sites on the clay surfaces. Because of large surface areas and nature of lattice structures of clay particles, large volumes of water are adsorbed on the clays.

According to Tschirley et al. (1977), the shale matrix stress may be utilized to express the magnitude of surface hydration. It is also referred to as shale hydration force, which has been defined as:

$$\text{shale hydration force (matrix stress, psi)} = \text{overburden pressure (psi)} - \text{pore pressure (psi)} \qquad (10\text{-}3)$$

This hydration force is equal to the degree of relief of stress around the wellbore when the bit penetrates the shale section (Baroid, 1977, p. 6). Consequently, a greater degree of hydration occurs with increasing hydration force.

(2) *Osmotic swelling.* Inasmuch as water migrates from zones having lower

ion concentration to areas with higher ion concentration (see Fig. 10-2), in many clays (particularly sodium-montmorillonite), in which the sodium is instrumental in the expansion of lattice structures, water may migrate into these layers depending on the ion concentration differential. Shrinking of shale, on the other hand, is the reverse of this osmotic swelling. Inasmuch as the ion concentration in the aqueous phase of oil muds is higher than that in the interstial water of shales, reverse osmotic migration takes place, i.e., water migrates out of the shale lattice layers resulting in a decrease of shale volume.

The osmotic pressure may be defined as follows (Baroid, 1977, p. 6):

$$p_{os} = \text{osmotic pressure (atm)} = RT(\theta_1 m_1 u_1 - \theta_2 m_2 u_2) \tag{10-4}$$

whereas R = gas constant; T = absolute temperature (°K); θ = osmotic coefficient of the salt solution; m = salt concentration of solution expressed as molality; and u = number of solute ions per mole. This pressure (p_{os}), which exists across the membrane between two solutions having different salinities, can be considered as either adsorption or desorption force depending upon the relative salinities of solutions across the membrane. It has been shown that this force is sufficiently high enough to cause the water movement to or from the clay matrix (see Table 10-V). The membrane referred to here is that which forms the water-in-oil emulsion in oil-base drilling fluid (see Figs. 10-1 and 10-2).

TABLE 10-V

Osmotic pressures of oil muds

(After Baroid, 1977, table 2; courtesy of Baroid Petroleum Services)

Salt	Dissolved salt concentration (ppm)	Osmotic pressure * (psi)
Calcium chloride	52,600	500
	100,000	1,100
	182,000	3,000
	250,000	5,800
	307,000	9,400
	357,000	13,900
	400,000	16,100
	456,000 (saturation)	24,400
Sodium chloride	55,000	670
	105,000	1,400
	149,000	2,200
	189,000	3,200
	226,000	4,300
	268,000 (saturation)	5,800

* Calculated for oil mud opposite a fresh-water shale at 25°C.

If the salinity of interstitial solutions in the formation to be penetrated is known, the salinity of the discontinuous phase (i.e., water) may be adjusted accordingly to obtain optimal osmotic pressure effects (see Table 10-V).

In deep, hot wells surface hydration of clays dominates the shale swelling and consequent wellbore stability. Methven and Barthel (1971, p. 4) pointed out that clays, which adsorb and/or desorb water to some extent through the above-described processes, occur in most rocks. Thus, in most drilling operations either one of these processes gives rise to borehole problems. In the case of swelling, an excess volume of shale is introduced into the wellbore. In the case of shrinking, the spalling of shale into the wellbore occurs. Methven and Barthel (1971, p. 8) discussed some special features of oil muds that minimize and/or prevent the shale problems.

Oil mud must provide enough hydrostatic pressure to hold back the flow of formation fluid into the wellbore and prevent formation collapse. The oil mud filtrate (oil) isolates the water-sensitive formation clays from direct contact with the aqueous phase present in the oil mud system. Additional isolation is achieved through use of surfactants that make clays oil wet.

Salt concentration in the aqueous phase of oil muds can be adjusted so that the vapor pressure of oil mud becomes equivalent to that of the formation. Surface hydration of clays is a function of the vapor pressure. If the water content of oil mud is low and its vapor pressure has been adjusted to equal that of the shale, clay hydration is prevented. Even if exact vapor pressure adjustment is not achieved by controlling a salt concentration of aqueous phase, the low water content alone may minimize the hydration.

Drilling through producing zones

Using oil muds to drill producing zones have proven highly effective. Oil muds do not cause clays, which are present in most producing formations, to swell; hence, little formation damage occurs. Methven (1971, pp. 2—3) discussed different types of formation damages which may be avoided on using oil muds:

(1) Shallow formation damage, which is caused by swelling of formation clays, can be avoided. Inasmuch as clay particles contacted by the oil filtrate around the wellbore do not swell, there is no permeability reduction.

(2) Clay migration, which may cause deep formation damage, can be prevented. Inasmuch as oil filtrate is a nonconductive medium, clays under reservoir equilibrium condition do not disperse. Consequently, they cannot migrate with producing fluid toward the wellbore. Such a migration, if allowed to occur, may cause severe in situ permeability reduction. Even in clean sandstones permeability may be reduced to less than 1% of the original one (Gray and Rex, in: Methven, 1971, p. 3). Reverse clay migration or invasion may also occur if drilling fluids containing clays are used. In oil muds, asphalts are used instead of clays and form micellar structures which provide rheologic properties and filtrate loss control. These micellar structures do not become implanted in the formation face permanently as do clays.

(3) Water blocking (see Appendix I, Chapter 8) is another deep formation damage which may be avoided on using oil muds. Water blocking is a consequence of water saturation increase in the oil and/or gas-bearing formation (Methven, 1971, p. 3). Partial or complete reduction of relative permeability to oil or to gas may occur and the connate water may move in the form of large globules. In the case of low capillary pressure, these water globules may plug-up pore channels between sand grains. Although oil filtrate alters the relative permeability, the water saturation does not change and, therefore, water blocking is minimized on drilling with oil muds.

As pointed out by Methven (1971, p. 5), any relative permeability reduction, particularly to gas in "dry" gas-bearing formations, may have a devastating effect on the productivity. Fortunately, oil filtrate does reduce the relative permeability to gas. Inasmuch as most dry gas reservoirs exist in high-temperature and high-pressure environments, complete vaporization of oil filtrate may be achieved by selecting oil of suitable vaporization point. Using this approach, relative permeability to gas may be restored upon commencement of production.

Drilling through water-soluble formations

Inasmuch as filtrate of an oil mud is oil, oil muds can be used to drill through water-soluble formations. For example, only minimal hole enlargement results on drilling through salt domes. Only minor property changes occur when oil muds are used in drilling through gypsum ($CaSO_4 \cdot 2\,H_2O$) or anhydrite ($CaSO_4$). Consequently, only minor treating costs are borne by the operators. On the other hand, on using water-base drilling fluids properties change drastically, necessitating high treating costs.

Minimizing stuck drillpipe

Reduction in stuck drillpipe problem on using oil muds is attributed by most investigators to the thin filter cake formed by oil mud. Oil mud filter cake gives rise to low coefficient of friction between the cake and the preferentially oil-wet steel drillpipe (Anonymous, 1976, p. 109).

LOGGING IN OIL MUDS

As a result of progress made in the field of logging technology, logging in nonconductive medium is as simple as logging in conductive medium. Although the two most basic logging tools used in water-base muds, i.e., spontaneous potential (SP) and normal resistivity, cannot be run in oil muds, other logging tools operate in both systems with similar data interpretation. Suite of logs may be designed and run in oil-base muds to obtain information

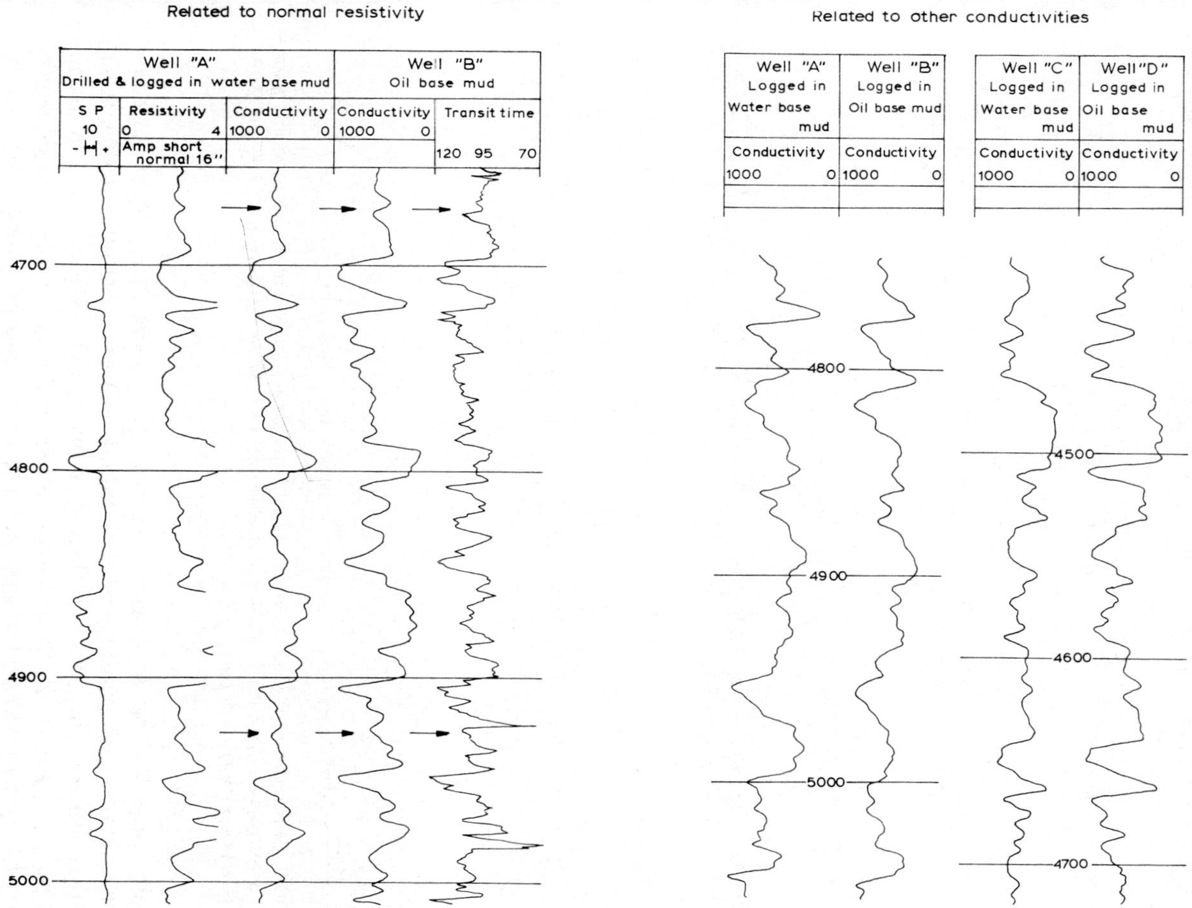

Fig. 10-17. Comparison between logs (resistivity and induction) obtained in water-base and oil-base muds. There is a good correlation. (After Wilshusen and Long, 1964, figs. 1 and 2, p. 55; courtesy of Oil and Gas Journal.)

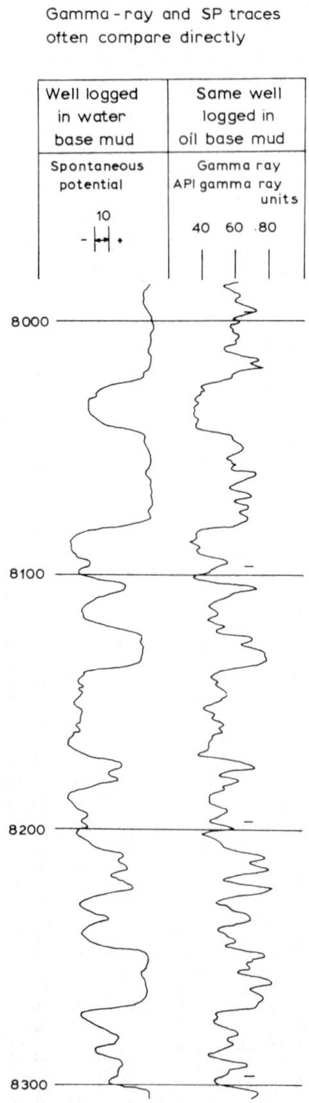

Fig. 10-18. Comparison of logs (gamma ray and spontaneous potential) obtained in water-base and oil-base muds. They show good correlation. (After Wilshusen and Long, 1964, fig. 4, p. 56; courtesy of Oil and Gas Journal.)

normally provided by spontaneous potential and resistivity logs: for example, combination of gamma ray and induction logs. As shown in Figs. 10-17 and 10-18, there is a good correlation between (1) resistivity and induction log conductivity and (2) gamma ray and spontaneous potential logs, respectively, obtained in water-base and oil-base muds (also see Chapter 9).

EQUIPMENT FOR SAFE AND ECONOMICAL HANDLING OF OIL MUDS

Rayburn (1977, p. 2) credited safe and economical handling of oil muds at wellsites to advanced planning, proper equipment selection, and coordinated applications.

Handling facilities and equipment for oil-base muds differ from those used for handling water-base muds to some extent, because of (1) economics and (2) ecology considerations. Inasmuch as oil muds are expensive, an attempt should be made to minimize their unnecessary loss. In addition, oil muds, which have oil as the external phase, have damaging effects on the environment around the wellsite. Additional clean-up costs are incurred if spill is allowed to occur.

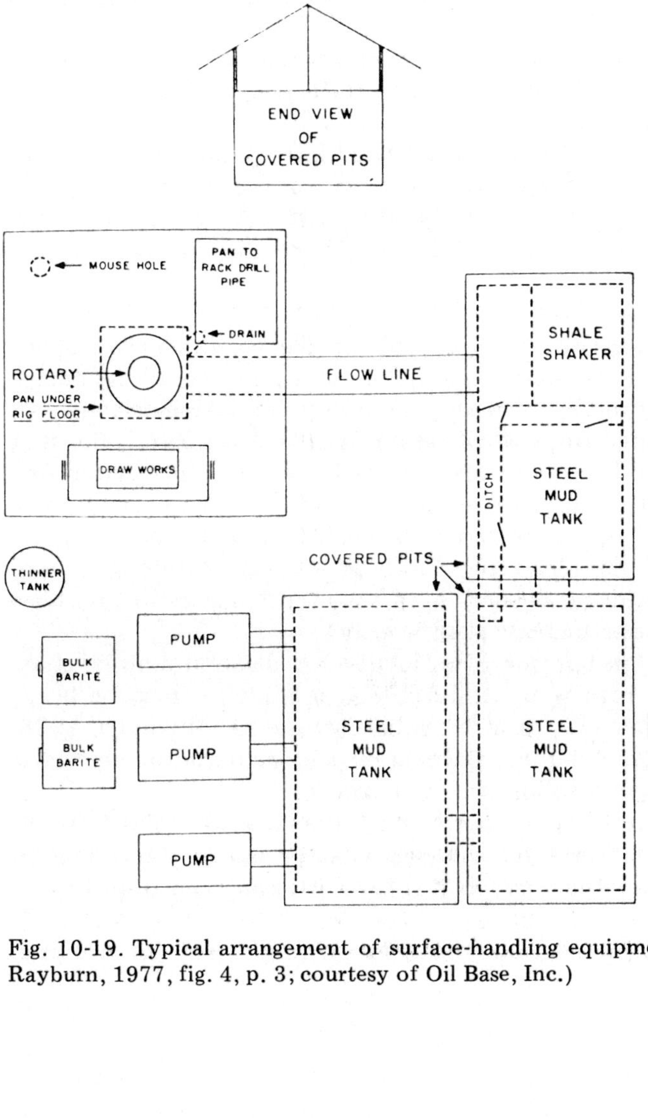

Fig. 10-19. Typical arrangement of surface-handling equipment for oil-base muds. (After Rayburn, 1977, fig. 4, p. 3; courtesy of Oil Base, Inc.)

According to Rayburn (1977, pp. 2—3), the following oil mud handling equipment is commonly used to handle oil muds. Only the equipment which differs from the one used for handling water-base muds is discussed. It may be loosely categorized into the following two overlapping groups: (1) mud handling equipment and (2) mud saving equipment. The handling equipment includes (Rayburn, 1977, p. 3):

(1) *Steel mud tanks* — Steel mud tanks offer an advantage in lowering the daily mud treating costs, and should be covered with roofs of proper height in order to prevent water addition into the system from rainfalls. Enough headroom must be provided, however, for fast and easy personnel handling of the mud during drilling operation.

(2) *Shaker screens* — Screen sizing in the case of oil-base mud is dictated by the flowline temperature and the rate of penetration; 10- to 30-mesh screen sizes are most commonly used. Flowline temperature affects the oil mud viscosity on the screens.

Mud saving equipment may consist of the following: (1) pipe and kelly wipers, (2) mud-saver sub and kelly cock, (3) drillpipe pan, (4) rotary table catch pan, and (5) mud-saver box.

A typical arrangement of oil mud surface handling equipment is shown in Fig. 10-19.

DISPOSAL OF OIL MUDS

Disposal of oil muds presents some difficult problems, with a consequent increase in operating costs. Improper disposal methods may result in damage to the environment, followed by a lengthy and costly recovering period.

McMordie (1975) divided disposal of oil muds into two phases. The first phase applies to the liquid phase of oil mud itself, whereas the second one involves disposal of solid cuttings which are brought to the surface. Oil-base mud, which is generally containerized in a closed system during drilling operation, is returned to the plant for proper disposal after drilling is completed and the mud is no longer needed. In offshore and inaccessible onshore locations operating cost may increase significantly.

Results of many tests on burning of oil muds as a disposal method show that oil mud, even with high water and drill solids contents, may be burnt almost completely, leaving only charred solids or ash (Anonymous, 1976, pp. 109—113). Even more complete burning may be achieved if oil mud is centrifuged to remove all drill solids prior to burning.

Inasmuch as all oil muds contain surfactants, which make solids oil-wet, conventional washing of the solid cuttings is impractical. According to McMordie (1975), three methods are available for treating and disposing of solid cuttings:

(1) Cuttings are containerized and returned to mud plant for disposal.

This may be costly for offshore and inaccessible onshore locations.

(2) Cuttings are washed with proper chemicals on location and then disposed of in the same manner as water-base mud cuttings.

(3) Cuttings may be buried, if local regulations and environment allow, after oil is removed from cuttings by burning. Burning of oil from cuttings, however, is highly impractical for offshore locations.

Selection of treating and handling methods is governed by local regulations, logistics, availability of treating methods, and economic considerations.

ACKNOWLEDGEMENTS

The help extended by A.J. Long and R.H. Williams is greatly appreciated by the authors.

APPENDIX 10.I — ANILINE POINT (Chilingar, 1955)

The aniline point is the temperature in degrees F at which an equivolume mixture of oil and aniline are no longer soluble in one another. Method 3601.3 in VV-L-791 describes three different aniline point test methods. Although the basic principle is the same in all cases, each test utilizes a different apparatus for mixing and for reading the separation point. The principle involved in this text is the fact that when two immiscible liquids are heated together, they become increasingly soluble in one another until a point is reached where they are completely miscible.

Figure 10.I-1 shows the solubility temperature of different mixtures of aniline and oil.

At any given composition of oil and aniline there is one temperature at which complete solubility first occurs on heating. For a 50—50 mixture this temperature, shown as T_c on Fig. 10.I-1, is defined as the aniline point. Should the composition, however, be changed even slightly, for example to 49—51 through careless measurement, then the temperature at which mixing occurs would differ from the true aniline point. Accurate measurements of aniline and oil, therefore, are essential for obtaining good results in this test.

Inasmuch as different hydrocarbons differ in the ease with which they mix with aniline, the aniline point varies with the composition of the oil. A very general and simplified rule is that the compounds which are most like aniline in structure will tend to mix most readily with aniline and, therefore, will have lower aniline points. Inasmuch as aniline is an aromatic amine, the aromatic compounds generally can be expected to have low aniline points. These compounds also attack rubber parts and have low heating values. In general, the less saturated hydrocarbons, which will mix more readily with

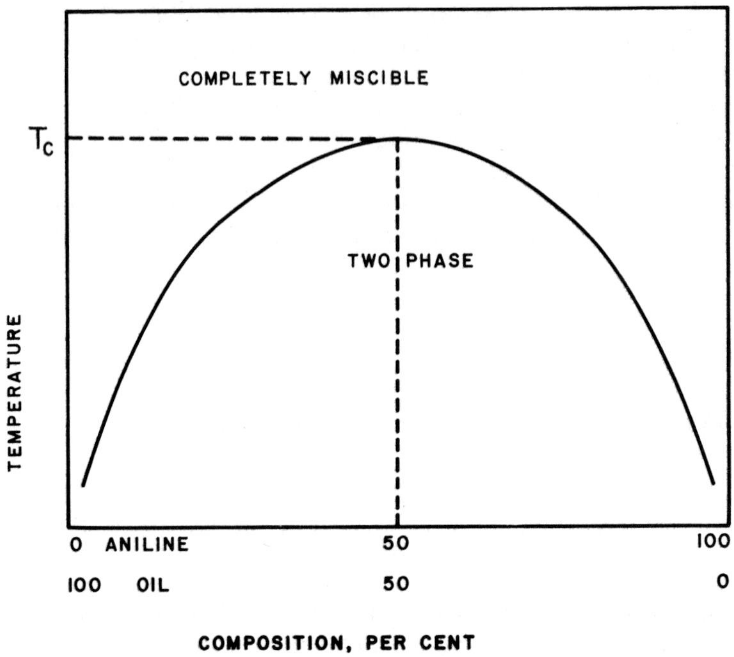

Fig. 10.I-1. Relationship between composition and solubility temperature of various mixtures of aniline and oil.

aniline (low aniline points), usually have lower heating values. Figure 10.I-2 shows the heating values and aniline points of C_8 aromatic, olefin, and paraffin. This comparison illustrates to some extent why the aniline point can be used as an empirical measure of the heat of combustion of fuels.

The aniline point is determined by heating an oil-aniline mixture to a point where a clear solution is obtained and then letting it cool to the temperature at which the mixture just becomes cloudy throughout. This is more satisfactory than trying to read the temperature at which the mixture just clears on heating, because the clear homogeneous solution gives a very sharp cloud point when the aniline point is reached.

Key points

(1) Accurate measurements of aniline and oil are essential for obtaining accurate results.

(2) The difficulty of maintaining an absolutely uniform temperature throughout the solution results in stringy clouds of separating material in cold spots as the aniline point is approached. They should be ignored, how-

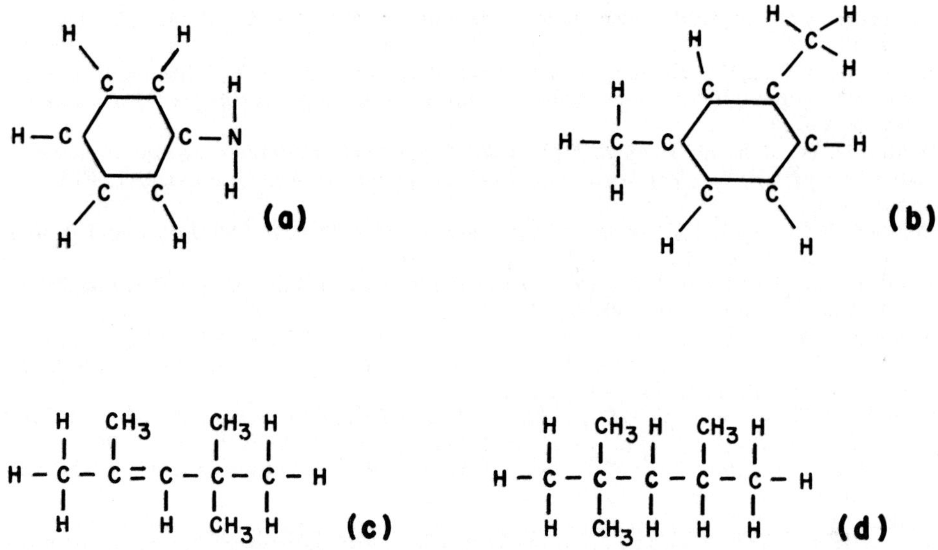

Fig. 10.I-2. Aniline (a), meta-xylene (aromatic) (b), di-isobutylene (olefin) (c), and iso-octane (paraffin) (d). The latter three have aniline points of —22°, 108.5°, and 175.6°F, respectively.

ever, and only the temperature at which the mixture is uniformly cloudy throughout is taken as the true aniline point.

REFERENCES

Anonymous, 1976. Burning oil muds may solve problems. *World Oil*, 182 (1, Jan.): 109, 112—113.

Baroid Petroleum Services, 1954. *Drilling Mud Data Book*. Sect. 300: 11 pp.

Baroid Petroleum Services, 1965. *Mud Technology Handbook*. Sect. 12: 85—101.

Baroid Petroleum Services, 1975a. Oil base drilling fluid. In: *Manual of Drilling Fluid Technology*. 26 pp.

Baroid Petroleum Services, 1975b. Borehole instability. In: *Manual of Drilling Fluid Technology*. 23 pp.

Baroid Petroleum Services, 1977. *Manual of Drilling Fluids Technology — Oil Base Drilling Fluids*. 30 pp.

Bennett, R.B. and McMordie, W.C., Jr. 1969. *Oil Base Well Drilling Fluid Composition and Method*. U.S. Patent 3,642,623, C. 2/15/72, f. 10/10/69, Appl. 865,492.

Chilingar, G.V., 1955. *Technology of Testing Petroleum Products*. USAF Publ., Wright-Patterson AFB, 76 pp.

Darley, H.C.H., 1969. *Physical and Chemical Factors Affecting Wellbore Stability*. Soc. Pet. Eng. Meet., SPE Pap. No. 2400.

IMCO Services, 1972. *IMCO Oil Muds Technical Publication*. 98 pp.

Magcobar Petroleum Services, 1975. Oil phase drilling fluid. In: *Drilling Fluid Engineering Manual*. Chapter 11: 19 pp.

McMordie, W.C., Jr., 1969. Viscometer tests mud to 650°F. *Oil Gas J.*, 67 (20, May 19): 81—84.

McMordie, W.C., 1975. Environmental aspect of chemical use in well drilling operations. In: *Environmental Protection Agency Symposium.* EPA-56051-75-004, May, Houston, Texas, 200 pp.

McMordie, W.C., Jr. and Bennett, R.B., 1974. *The Effect of Temperature and Pressure on the Viscosity of Oil-base Muds.* Soc. Pet. Eng. Meet., Houston, Texas, Oct., SPE Pap. No. 4974.

Methven, N.E., 1971. *Effects of Drilling Fluids on Gas Well Productivity.* Soc. Pet. Eng. Meet., New Orleans, La., Oct., SPE Pap. No. 3504.

Methven, N.E. and Barthel, H., 1971. *Shale Drilling with Oil Muds.* Soc. Pet. Eng. Meet., Los Angeles, Calif., Nov., SPE Pap. No. 3679.

Methven, N.E. and Baumann, R., 1972. *Performance of Oil Muds at High Temperatures.* SPE — European Spring Meet, Soc. Pet. Eng. of AIME, Amsterdam, The Netherlands, May 16—18, SPE Pap. No. 3743: 8 pp.

Methven, N.E. and Kemick, J.G., 1968. *Drilling and Gravel Packing with Oil-base Fluid System.* Soc. Pet. Eng. Meet., Houston, Texas, Sept.—Oct., SPE Pap. No. 2190.

Minkhailov, K.L., Bochkarev, G.P. and Babalyan, G.A., 1971. Low-density oil-base drilling fluids. *Neft. Khoz.*, 1971 (9, Sept.): 26—29.

Rayburn, C.A., 1977. *Rigup Tips for Handling Oil Muds.* Oil Base, Inc., Publ., 3 pp.

Rogers, W.F., 1963. *Compositions and Properties of Oil Well Drilling Fluids.* Gulf Publ. Co., Houston, Texas, 818 pp.

Wilshusen, R.C. and Long, J., 1964. How to handle logs in nonconductive muds. *Oil Gas J.*, 62 (48, Nov. 30): 55—58.

Simpson, J.P., Cowan, J.C. and Beasley, A.E., 1961. *Recent Advances in Oil Mud Technology.* Fall Meet., SPE of AIME.

Tschirley, N.K., 1977. *Manual of Drilling Fluids Technology — Oil Base Drilling Fluids.* 30 pp.

Chapter 11

POLYMER DRILLING FLUIDS

HERMAN H. RIEKE and GEORGE V. CHILINGARIAN

INTRODUCTION

A polymer can be defined as a very large molecule made up by joined repeating blocks (chains) called monomers. Polymers have high molecular weights and when added in low concentrations increase the viscosity of water and the resulting fluid's ability to suspend particles. The main advantage of using organic polymers in drilling fluids is that these substances can impart maximum viscosity at a minimum solids content. This maximum viscosity per unit weight of solids is also important at remote drilling sites where costs of delivery of viscosifiers may constitute a large part of the overall cost of the material. Although most polymers start to break down at a temperature of about 300° F, they are nevertheless used successfully in wells with bottomhole temperatures appreciably higher than this temperature. Natural gums (see Tchillingarian and Beeson, 1950, 1952) are more complex materials and commonly are more stable at higher temperatures and acidic environment than are the synthetic materials.

Polymers that are currently being used in the industry include: guar gum, xanthan gum, starch, polyacrylates, cellulosics, polyacrylamides, and maleic anhydride derivatives (Carico and Bagshaw, 1978). Over 260 trade names for polymers and polymer blends were listed in 1978—1979 in the "Guide to Drilling, Workover and Completion Fluids", published by the World Oil Journal.

Darley (1965) stated that fast rates of penetration can be maintained in hard formations by using "clear" water as a drilling fluid. In some cases, however, water cannot be used because the very high downhole filtrate loss creates various difficulties in drilling, in logging, or in producing the well upon completion. In such cases, it becomes necessary to "mud up" (increase the solids content), which results in a decrease in drilling rate. The reason that drilling rates decrease with increasing viscosity is a complex subject discussed elsewhere in this book (see Chapters 2 and 3). The principal reason for the reduction in drilling rate appears to be the increasing tendency of the drilling fluid to form a filter cake on the bottom of the hole. The pressure differential which develops across this filter cake opposes the action of the bit in dislodging chips from the rock. This pressure differential is referred to by Garnier and van Lingen (1959) as the chip hold-down pressure, which is referred to as CHDP.

Initial research was undertaken to develop a drilling fluid which would minimize the chip hold-down pressure (CHDP), but which would still have a low filtrate loss, i.e., a fluid which would deposit a filter cake on the sides of the borehole but would lay little or no filter cake on the bottom of the hole. Darley (1965) showed that the initial filtration rate and lower CHDP is dependent on the concentration of solids in the drilling mud. This was tested in the laboratory by utilizing four drilling fluids, solids of which had different particle-size distributions. These muds were filtered against Berea Sandstone cores in a dynamic test cell and the initial filtration rates were measured. This was a measure of effectiveness of the mud particles to form a filter cake and, therefore, reduce the filtrate loss. Low initial filtration rates were obtained only when a substantial concentration of solids in the 2—10 μm size range was present (Darley, 1965). Evidently this size range was critical for bridging the surface of the Berea Sandstone.

Darley's test results suggested that to get a low CHDP it was not necessary to maintain a low concentration of all solids. Inasmuch as the filtration process never gets beyond the initial stage beneath the bit, high filtration rates and, hence, low CHDP's could be maintained there if the concentration of bridging solids is kept low. A concentration of colloids sufficient to give a low filtrate loss on the sides of the borehole could then be used without affecting the CHDP.

The achievement of fast drilling rates is based on the concept of lessening the tendency of the mud to form a filter cake on the bottom of the hole. One would, therefore, expect faster drilling rates only when drilling more permeable formations. Improved drilling rates, however, have been obtained in all types of formations, including shales, permeability of which is several orders of magnitude less than that of filter cakes. It is clear, therefore, that more complex mechanisms than that of static chip hold-down are involved. The fast drilling characteristics of a drilling fluid is probably due, fundamentally, to faster pressure equalization around the chips. In the more permeable rocks, this is achieved by penetration of filtrate through the pores of the rock, and, thus, by lowering dynamic chip hold-down pressure, whereas in the less permeable rocks this is due to the easier access of fluid to the fractures created ahead of the bit.

The use of asphalt colloid emulsion gave rise to considerable savings in drilling hard, stable formations. Although fast drilling rates can also be achieved in other types of formations, the emulsion drilling fluids commonly cannot be used because of economic or technical limitations.

Some of these limitations can be overcome by modifying the formulation or by developing and using other types of fluids having low viscosities and low contents of solids. Other limitations, however, are inherent in any drilling fluid having a low viscosity and low solids content. These limitations must be overcome by mechanical means or by the development of new engineering techniques.

OTHER USES OF POLYMERS IN THE PETROLEUM INDUSTRY

It has been known for many years that the efficiency of a waterflood can be improved by lowering the water/oil mobility ratio in the system. Such a change leads to better sweep efficiency and to more efficient oil displacement in the swept zone.

Data from various laboratory waterflood tests of both small cores and long sand packs show that water mobility may be reduced and oil recovery increased by the addition of certain polymers to flood water. Reduction in water mobility, in many cases, is greater than would be expected from conventional viscosity measurements. These polymer solutions, however, do not cause significant reductions in oil mobility. The over-all effect of these mobility changes may be increased waterflood oil recovery.

Field studies have been conducted utilizing partially hydrolyzed polyacrylamide polymer in waterflooding (Sandiford, 1964). Laboratory data show that low-concentration solutions of this polymer will increase the recovery of oil with viscosity ranging from very low to very high in a nonlinear or in a nonhomogeneous system and high-viscosity oil in any system. Thus, this process may be applicable to many reservoirs. Field tests also indicate that flooding with low-concentration, partially hydrolyzed polyacrylamide solutions has possibilities of commercial application to a variety of oilfields. This and other processes are being further evaluated in large-scale field tests.

TYPES OF POLYMERS

A polymer was defined earlier as a very long-chain molecule. This is true, however, only for the linear polymers, which belong to the class of thermoplastics that resemble supercooled liquids.

Arridge (1975) stated that it is possible to have large networks of chains connected by cross-links to make what is effectively a three-dimensionally infinite molecule. Examples of this class of polymer are cured rubber and the thermosets. Resins such as the phenolics or epoxies cannot change their form after being cross-linked. Figure 11-1 illustrates these concepts schematically. Clearly there is no real distinction between thermoplastics and thermosets on the molecular level, except for the increased molecular weight caused by cross-links. If a polymer has a large number of monomer units between cross-links, it is effectively thermoplastic; however, it will not be infinitely deformable, i.e., polymer will not flow or be extrudable like polyvinyl chloride (PVC) or nylon. In addition, whereas polymer with cross-links can be swollen, it will not dissolve. It would dissolve only if there are no cross-links (Arridge, 1975). As the cross-link density increases, the polymer becomes less like a gum or liquid and more like a rubber.

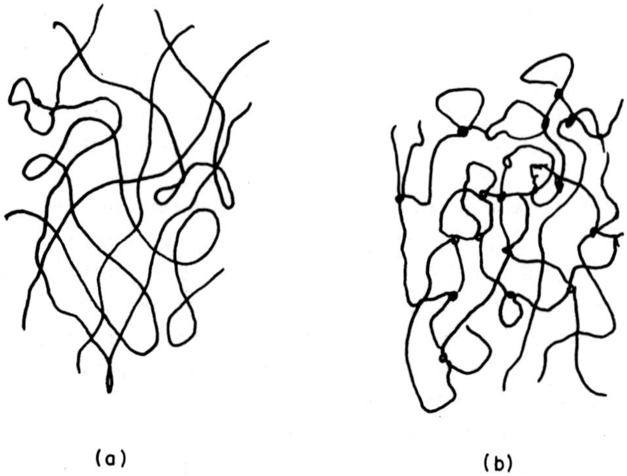

Fig. 11-1. Schematic configurations of (a) long chain and (b) cross-linked polymer molecules. (After Arridge, 1975, figs. 1.2a and 1.2b, p. 5; courtesy of Clarendon Press.)

In petroleum engineering applications, soluble polymers are employed in secondary recovery, drilling, workover, and completion operations. Solubility relations in polymer systems are more complex than those among low-molecular-weight compounds, owing to the size of the molecules, viscosity of their solutions, and the possible presence of a crystalline phase. The most common types of polymers, which are grown in the fermentation process, are polysaccharides, biopolymers and partially hydrolyzed polyacrylamide.

Both polyacrylamide and biopolymers are employed in petroleum engineering projects. The partially hydrolyzed acrylic polymers consist of long

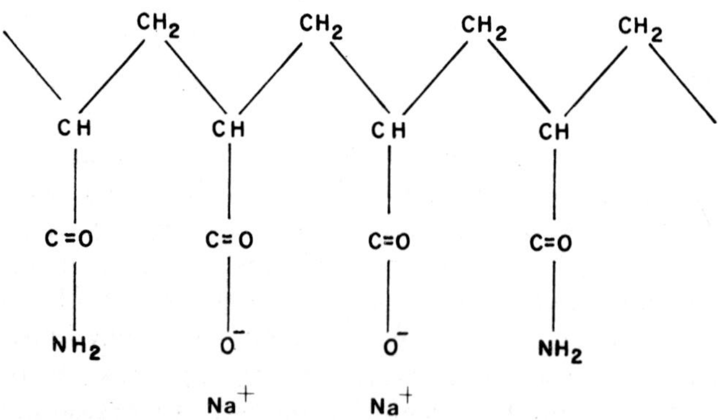

Fig. 11-2. Structure of a typical partially hydrolyzed polyacrylamide (After Burcik, 1965, fig. 1, p. 14; courtesy of Producers Monthly.)

TABLE 11-I

Comparison of friction-reducing and rheological properties of various polymers and bentonite

(After Darley and Vickers, 1975, p. 5, table 1)

Expt. No.	Fluid	Pressure drop in 10,000-ft, 2-inch tbg., 250 gpm (psi)	Fann viscometer			Pipe viscometer exponent n'
			$\overline{PV}$ (cP)	$\overline{YP}$ (lb/100 ft^2)	$\overline{YP}/\overline{PV}$ ratio	
1.	water	1278	1	0	0	1
2	21 lb/bbl Wyoming bentonite	2981	6.5	5.5	0.85	0.69
3	1.2 lb/bbl Guar gum	335	6	4	0.67	1.0
4	1 lb/bbl hydroxy-ethyl-cellulose (HEC)	384	6	3	0.5	0.85
5	1 lb/bbl carboxy-methyl-cellulose (CMC)	432	6.5	1	0.15	0.95
6	1 lb/bbl poly-saccharide biopolymer	664	4	8.5	2.1	0.46
7	0.6 lb/bbl, 30% hydrolyzed polyacrylamide	573	4.5	7	1.6	0.62
8	1.9 lb/bbl polyacrylamide	897	6	4	0.67	0.82

1975). In comparing the various polymers, however, it is important to realize that the results depend on the size of pipe, the rate of circulation, and the concentration of the polymers. Prospective fluids, therefore, should be tested in a pipe viscometer and the data converted to the specific well conditions.

If the hole cleaning problems are anticipated, a polymer which provides a high $\overline{YP}/\overline{PV}$ ratio or a low n' exponent should be used. If the hole fillup is anticipated during trips, a crosslinked polysaccharide or its mixture with another polymer should be used. Crosslinked polysaccharide suspensions have thixotropic properties, i.e., gel strength increases with time of standing, which prevents settling of cuttings. It should be remembered, however, that high gel strengths increase the tendency to swab-in gas and to promote loss of circulation through pressure surges. The polysaccharide, therefore, should be used with caution in small quantities.

Stabilized hydroxy-ethyl-cellulose (HEC) or a polymer which is stable in concentrated $CaCl_2$ solution should be used when specific weights of about 10 lb/gal are required.

Carico (1976) pointed out that the "*n*" factor from the power law model does not correlate well with the particle settling rate. It is logical to assume that the ability of drilling fluids to keep the settling rate of particles low depends upon the fluid viscosity. Most drilling fluids are non-Newtonian and can have different viscosities depending on the rate to which the fluid is subjected. Inasmuch as a settling particle encounters a very low shear rate, a low-shear-rate viscosity should be employed to describe the suspending ability of these fluids (Carico, 1976). In testing various polymer suspensions, Carico (1976) discovered that the API yield point correlation was marginal and could provide inaccurate suspension information. A new comparative suspension test was devised which utilized a standard API sand content tube (Carico, 1976). This method was used to test and rank the suspension properties of various polymer fluids.

The test consists of filling the tube to the mark labeled "mud level" ($\approx$100 cm^3) with the polymer fluid to be tested. One gram of sand is placed into the tube and the mixture is shaken vigorously. The tube is then placed upright and the settling rate is timed. The time required for the sand to reach selected graduated marks at the bottom of the tube is recorded. This proce-

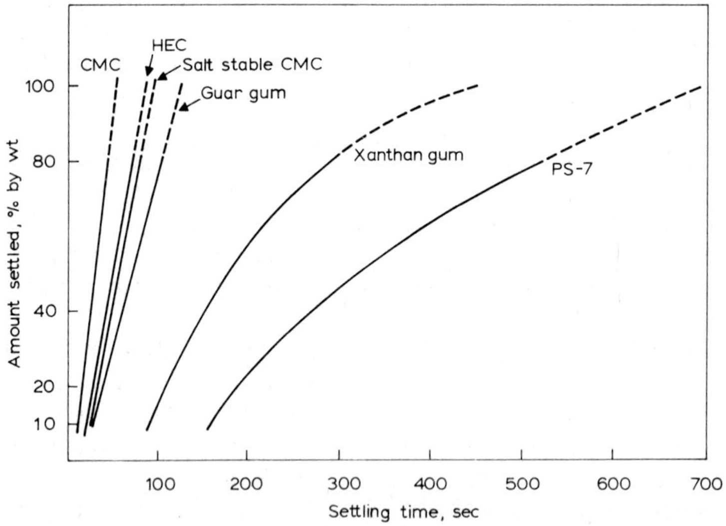

Fig. 11-3. Examples of settling times of 20—35-mesh sand in various polymer fluids (Carico, 1976, fig. 1; courtesy of the Society of Petroleum Engineers.) *PS-7* = experimental biopolymer; xanthan gum = naturally produced heteropolysaccharide (*M.W.* > 10^6); *HEC* = synthetic cellulosic derivative; *salt-stable CMC* = synthetic cellulosic derivative; *guar gum* = naturally-produced nonionic polysaccharide (*M.W.* ≈ 200,000).

dure is repeated several times without removing the sand. The computed average time is compared with average times of other fluids to determine the relative suspension ability (Carico, 1976). An example of such a comparison is presented in Fig. 11-3.

In the case of unstable shales, some polymers have a marked stabilizing influence; however, the mechanism is not well understood. It is not due to the filtration control because the stabilizing action depends on the chemistry (salinity, types of anions and cations, pH, etc.) of drilling fluid and formation water, and the types of salt, shale, and polymer. In order to determine the best formula, it is necessary to run tests on shale samples under simulated subsurface conditions (Darley, 1969). Satisfactory results have been obtained both in the laboratory and in the field with partially substituted polyacrylamide and stabilized hydroxy-ethyl-cellulose (HEC).

LABORATORY STUDIES OF POLYMER DRILLING FLUIDS

After choosing the best polymer for a particular drilling condition, a drilling fluid is prepared which has the necessary specific weight and proper rheological and filtration properties. Darley and Hartfiel (1974) discussed various polymer fluids which have been developed for the purposes of (1) increasing hole stability, (2) better hole cleaning, (3) faster drilling rates, and/or (4) prevention of formation damage (productivity impairment).

Polymers also have some unusual rheological properties, notably their ability to reduce friction losses in turbulent flow. Consequently, pressure losses in the drillpipe with flowing polymer fluids are often considerably lower than those in the case of water-base muds or water alone. For a given surface input, therefore, more hydraulic horsepower will be available at the bit, with resulting faster drilling rate. Another reason for the faster drilling rates obtained with polymer fluids is lower apparent viscosity beneath the bit (Eckel, 1968). The friction reducing property of polymer fluids is also an advantage in certain completion and production repair operations in the case of small-diameter tubing.

Results of laboratory experiments by Darley and Hartfiel (1974) indicate that the use of polymer fluids will result in considerably lower pressure losses in the drillpipe (tubing) than when conventional drilling fluids are used. Amount of the pressure drop is largely a function of pipe diameter, volumetric pumping rate, and polymer concentration (Table 11-II).

Additives such as bentonite or barite cannot be used without impairing the fast drilling characteristic of clay-free fluids. Table 11-III shows how the required properties are best achieved when stabilized hydroxy-ethyl-cellulose (HEC) is used as the base polymer. As shown in Table 11-III the effect of changes in composition on the drilling rate may be evaluated from the calculated pressure drops in 10,000 ft of $3\frac{1}{2}$-inch, 11.2-lb/ft drillpipe. Compari-

TABLE 11-II

Rheological properties of pure polymer fluids

(After Darley and Hartfiel, 1974, p. 71, table 2; courtesy of Petroleum Publ. Co.)

Expt. no.	Polymer	API filter loss (cm^3/30 min)	Fann viscometer			K' * (poundals/sq ft)
			$\overline{AV}$ (cP)	$\overline{PV}$ (cP)	$\overline{YP}$ (lb/100 sq ft)	
1	1.2 lb/bbl guar gum	66	8	6	4	0.00641
2	1 lb/bbl stabilized HEC (hydroxy-ethyl-cellulose)	55	7.5	6	3	0.0170
3	1 lb/bbl CMC	122	7	6.5		0.00826
4	1 lb/bbl polysaccharide	114	8.2	4	8.5	0.240
5	0.6 lb/bbl, 30% hydrolyzed polyacrylamide	192	8	4.5	7	0.0985
6	1.5 lb/bbl polyacrylamide		8	6	4	0.0028
7	1.5 lb/bbl guar gum	140	15	9	13	0.2404
8	2 lb/bbl stabilized HEC	17.8	20	13	15	0.312
9	water			1	0	0.00066
10	21 lb/bbl Wyoming bentonite	15.5	9.2	6.5	5.5	0.0547

		n' *	Critical Reynolds number	Pipe viscometer $\overline{AV}$ @ Re' crit. (cP)	Fanning friction factor		Pressure drop 10,000 ft (psi)	
					@ Re' crit.	@ Re' = 25,000	50 gpm 1-inch ID tbg	250 gpm 2-inch ID tbg
1	1.2 lb/bbl guar gum	1.00	3,560	9.5	0.0048	0.00152	679	335
2	1 lb/bbl stabilized HEC (hydroxy-ethyl-cellulose)	0.848	1,850	8.4	0.0091	0.00192	797	384
3	1 lb/bbl CMC	0.947	2,440	8.2	0.0068	0.00202	822	432
4	1 lb/bbl polysaccharide	0.46	1,550	8.1	0.0103	0.00270	965	664
5	0.6 lb/bbl, 30% hydrolyzed polyacrylamide	0.617	2,420	7.7	0.0068	0.00255	969	573
6	1.5 lb/bbl polyacrylamide	0.817	1,960	8.8	0.0085	0.00321	1,160	897
7	1.5 lb/bbl guar gum	0.591	1,410	14.2	0.0103	0.00104	653	303
8	2 lb/bbl stabilized HEC	0.567	1,460	16.7	0.0117	0.00123	956	425
9	water	1.0				0.0081	2,012	1,278
10	21 lb/bbl Wyoming bentonite	0.69	1,438	8.7	0.0109	0.00988	3,479	2,981

* Metzner and Reed equations (modified Power Law): (1) $d\Delta p/4l = K'(8v/d)^{n'}$, where: d = internal diameter of pipe; Δp = pressure drop; l = length of pipe; K' = consistency index; v = mean linear velocity; and n' = flow index. (2) Generalized Reynolds Number = $Re' = (d^{n'}v^{2-n'}\rho)/(gK'8^{n'-1})$, where: ρ = density and g = conversion factor = 32.17 (ft/sec/sec)(lb-mass/lb-force). (3) Relationship with conventional Power Law: $K' = K(3n+1)^n/(4n)$, where $n = n'$ (see Chapter 7).

For turbulent flow in pipes, $\Delta p = (1/d)(0.000108\, flv^2\rho)$, where Δp = pressure drop (psi); f = Fanning friction factor; l = length (ft); ρ = density (lb/cu ft); d = internal diameter of pipe (ft); and v = velocity (ft/sec).

TABLE 11-III

Properties of HEC clay-free drilling fluids

(After Darley and Vickers, 1975, p. 6, table 2)

Expt. no.	Composition	Weight (lb/gal)	API filtrate loss (cm^3)	Pressure drop, 10,000 ft ($3\frac{1}{2}$-inch × 11.2 lb/ft DP)	$\overline{PV}$ (cP)	$\overline{YP}$ (lb/100 ft^2)
1	1 lb/bbl stabilized HEC	8.3	55	135	6	3
2	6 lb/bbl stabilized HEC + lignosulphonate + ground carbonates	8.3	8.5	143	3	1
3	As for Expt. 2 but in sat. NaCl	10.1	4.3	220	6	1.5
4	Sat. NaCl	10.0	—	635	2	—
5	As for Expt. 2 but in conc. $CaCl_2$	11.0	11.0	265	3.5	2.5
6	Conc. $CaCl_2$	11.0	—	706	4.5	—
7	21 lb/bbl bentonite + barite	10.8	15	1,730	10.5	9
8	2 lb/bbl stabilized HEC	8.3	17.8	165	13	15
9	1 lb/bbl stabilized HEC + 8 lb/bbl bentonite	8.5	130	548	8.5	8.5

son of Experiments 1 and 2 shows that the filtration loss may be sharply reduced with practically no effect on the pressure drop. The properties of weighted fluids are compared in Experiments 2—7. Particularly notable are the lower pressure drops obtained on using the HEC brine fluids, e.g., for 11-lb/gal fluids, the use of HEC in concentrated $CaCl_2$ solution results in a pressure drop of 706 psi (Exp. No. 6), whereas addition of bentonite and barite gives rise to a pressure drop of 1730 psi (Exp. No. 7). A comparison of Experiments 8 and 9 shows that it is better to obtain higher viscosities by increasing the HEC concentration (pressure drop = 165 psi) rather than by adding bentonite (pressure drop = 548 psi).

Critical Reynolds number

Darley and Hartfiel (1974) plotted Fanning friction factor, *f*, versus Reynolds number, *Re'*, for several representative fluids on a log-log graph (Fig.

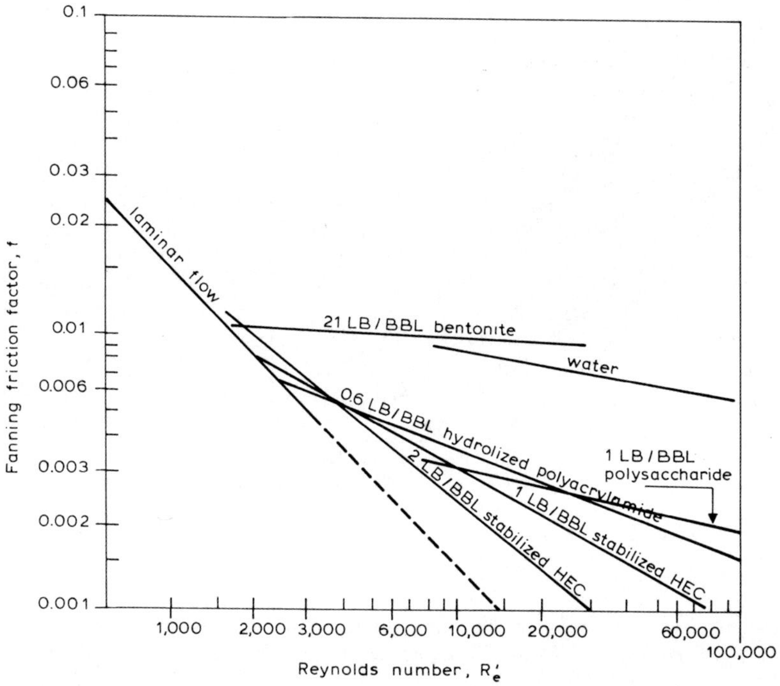

Fig. 11-4. Fanning friction factor versus Reynolds number for various fluids. (After Darley and Hartfiel, 1974, p. 73, fig. 2; courtesy of the Oil and Gas Journal.)

11-4), which shows that at high Re' numbers the polymer suspensions exhibit much lower f values than do water or bentonite suspensions. With increasing polymer concentration, friction factor, f, decreases at high Re' numbers; however, this is not necessarily the case at low Re' values. Values of f at Re' of 25,000 are presented in Table 11-II.

Pressure drop in pipes on using polymer fluids

Figure 11-5 shows pressure drop per 10,000 ft of pipe versus internal pipe diameter (ID) for several fluids at a pumping rate of 250 gpm. Pressure drop versus pumping rate (gpm) in 2-inch (ID) tubing is shown in Fig. 11-6. The major factors which govern the pressure drop are pipe diameter and pumping rate. For any given set of conditions, however, substantial decrease in pressure drop can be achieved on using polymer suspensions.

The pressure drops in a 1-inch tubing at a pumping rate of 50 gpm and in a 2-inch tubing at 250 gpm, that are listed in Table 11-II for all polymer suspensions, show reductions in pressure drop down to about one-fourth compared to that of water. Figure 11-5 shows that the relative performance of

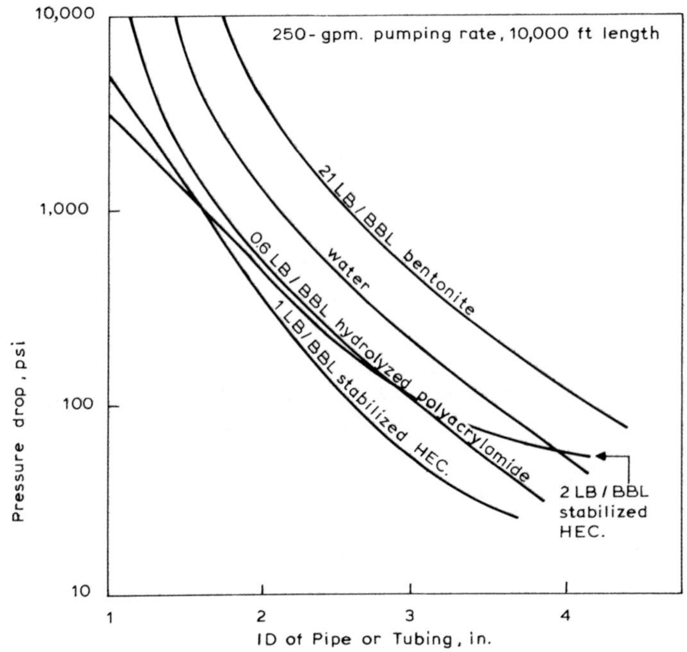

Fig. 11-5. Relationship between pressure drop and pipe (or tubing) diameter for various fluids. (After Darley and Hartfiel, 1974, p. 73, fig 3; courtesy of the Oil and Gas Journal.)

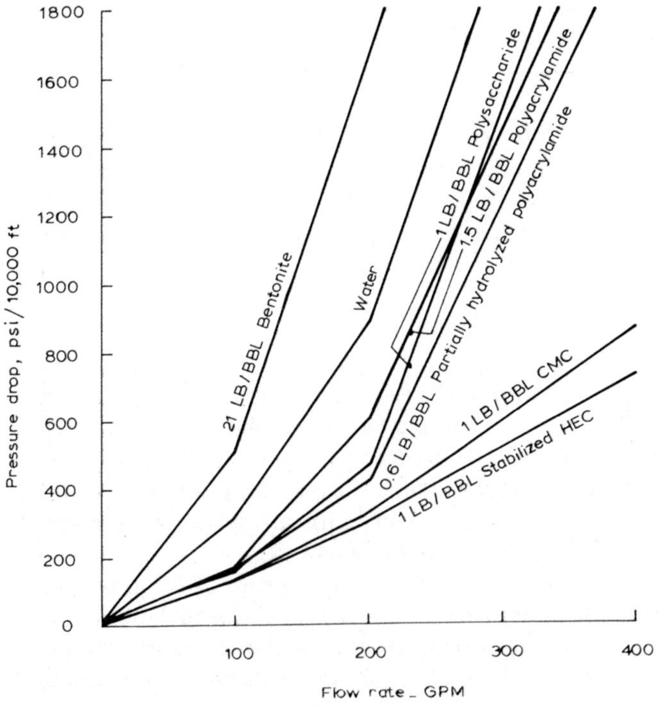

Fig. 11-6. Relationship between pressure drop in 10,000 ft of 2-inch tubing and flow rate for various fluids. (After Darley, 1976; courtesy of Brinadd Co.)

TABLE 11-IV

Rheological properties of composite fluids

(After Darley and Hartfiel, 1974, p. 72, table 4; courtesy of Petroleum Publ. Co.)

Expt. no.	Composition	Weight (lb/gal)	API filter loss (cm^3/30 min)	Fann viscometer			K' (poundals/sq ft)
				$\overline{AV}$ (cP)	$\overline{PV}$ (cP)	$\overline{YP}$ (lb/100 sq ft)	
1	1 lb/bbl stabilized HEC	8.34	55	7.5	6	3	0.0170
2	1 lb/bbl stabilized HEC and 8 lb/bbl bentonite	8.5	130	13.7	8.5	8.5	0.210
3	6 lb/bbl slugheal	8.4	8.5	3.5	3	1	0.0034
4	6 lb/bbl slugheal and 9 lb/bbl asbestos	8.5	11.4	8.7	6.0	5.5	0.06473
5	6 lb/bbl slugheal, 9 lb/bbl asbestos and 214 lb/bbl carbonate particles	10.5	8	9.5	8	3	0.0699
6	6 lb/bbl slugheal in sat'd NaCl	10.0	4.3	7.7	6	1.5	0.0263
7	5 lb/bbl hydrolyzed polyacrylamide and barite	10.3	47	19	10	10	0.538
8	10 lb/bbl attapulgite in sat'd NaCl	10.1	165	5.2	3.5	3.5	0.00556
9	Sat'd NaCl	10.0			2		0.00134
10	6 lb/bbl slugheal in $CaCl_2$	11.0	11.0	9.7	8.5	2.5	0.00853
11	5 lb/bbl hydrolyzed polyacrylamide and barite	11.6	35	22	12	19	0.545
12	21 lb/bbl bentonite and barite	10.8	15	15	10.5	9	0.239
13	$CaCl_2$	11.2			4.5		0.0302
14	2 lb/bbl stabilized HEC	8.34	17.8	20.7	13	15.5	0.3121

TABLE 11-IV (continued)

Expt. no.	Composition	n'	Critical Reynolds number	Fanning friction factor		Press drop/10,000 ft of $3\frac{1}{2}$-inch by 11.2-lb/ft D.P. (psi)	
				@ Re' crit.	@ Re' 25,000	100 gpm	400 gpm
1	1 lb/bbl stabilized HEC	0.848	1,855	0.0091	0.00192	21	135
2	1 lb/bbl stabilized HEC and 8 lb/bbl bentonite	0.562	1,117	0.0141	0.00521	67	548
3	6 lb/bbl slugheal	1.0	1,680	0.00923	0.00264	20	143
4	6 lb/bbl slugheal and 9 lb/bbl asbestos	0.662	1,340	0.0120	0.0050	47	450
5	6 lb/bbl slugheal, 9 lb/bbl asbestos and 214 lb/bbl carbonate particles	0.386	1,810	0.00830	0.00490	98	611
6	6 lb/bbl slugheal in sat'd NaCl	0.75	1,650	0.00982	0.00317	36	220
7	5 lb/bbl hydrolyzed polyacrylamide and barite	0.478	2,380	0.00720	0.00232	51	319
8	10 lb/bbl attapulgite in sat'd NaCl	0.61	1,570	0.0102	0.00782	73	905
9	Sat'd NaCl	1			0.00785	54	635
10	6 lb/bbl slugheal in $CaCl_2$	0.996	2,070	0.00781	0.00252	31	265
11	5 lb/bbl hydrolyzed polyacrylamide and barite	0.489	2,430	0.00655	0.00243	59	370
12	21 lb/bbl bentonite and barite	0.531	1,870	0.0083	0.00795	80	1,730
13	$CaCl_2$	1			0.00763	66	706
14	2 lb/bbl stabilized HEC	0.567	1,460	0.00117	0.00123	50	165

different fluids may change with pumping rate and pipe diameter (Darley and Hartfiel, 1974). For example, polysaccharide is better than polyacrylamide at low flow rates, whereas the reverse is true at high flow rates. Similarly, on increasing the concentration of stabilized HEC from 1 to 2 lb/bbl, pumping pressure decreased in a 1-inch pipe, but not in larger-diameter pipes.

Pressure drop on using composite fluids

The pressure drops obtained on using various composite fluids flowing through a 10,000-ft, $3\frac{1}{2}$-inch, 11.2-lb/ft drillpipe at rates of 100 and 400 gpm are presented in Table 11-IV. Many of the additives significantly increase the pressure drop at a flow rate of 400 gpm (Darley and Hartfiel, 1974).

FIELD STUDIES OF POLYMER FLUIDS

Field experience and the results of laboratory investigations have established that the accumulation of solids in the drilling fluid reduces the drilling rate (Gray and Kellogg, 1953; and Eckel, 1954). According to Gallus et al. (1958), fresh-water or brine drilling fluids may be kept free of suspended drilled solids by adding a water-soluble acrylamide-carboxylic acid copolymer at the flowline. If 0.01 to 0.2 lb/bbl of polymers are added to drilling fluids containing $<5\%$ clay solids by weight, solids flocculate and settle rapidly to the pit bottom. Water loss to permeable formations is reduced when solutions are injected at the pump suction and, thus, the permeability to oil is not reduced. Use of polymers in drilling fluids results in (1) higher drilling rates, (2) longer bit life, and (3) lower drilling fluid costs.

In the field test by Gallus et al. (1958), 1% aqueous solution (3.5 lb/bbl) of the polymer was mixed in a large-volume tank (or in manifolded drums) suitably positioned to permit gravity feed to the mud system. Mechanical or air agitators and aspirator-type solids dispersers were used in dissolving the polymer.

In order to clarify the solids-laden fluid flowing from the wells, a continuous stream of the polymer solution was added just below the shale shaker, where the fluid from the flowline discharged into the pits. Thus, sufficient agitation was provided for adequate dispersion of the polymer solution in the drilling fluid.

The polymer solution was added at a rate sufficient for maintaining clear fluid at the pump suction. Rate of injection, which was dependent on the nature of solids and their quantity in the fluid, was varied from $\frac{1}{2}$ to 4 lb of polymer in solution per hour. High-yield solids such as bentonite, which disperse readily in the mud, require higher quantities of polymer. Low-yield solids such as kaolinite and shale clays which did not disperse well required

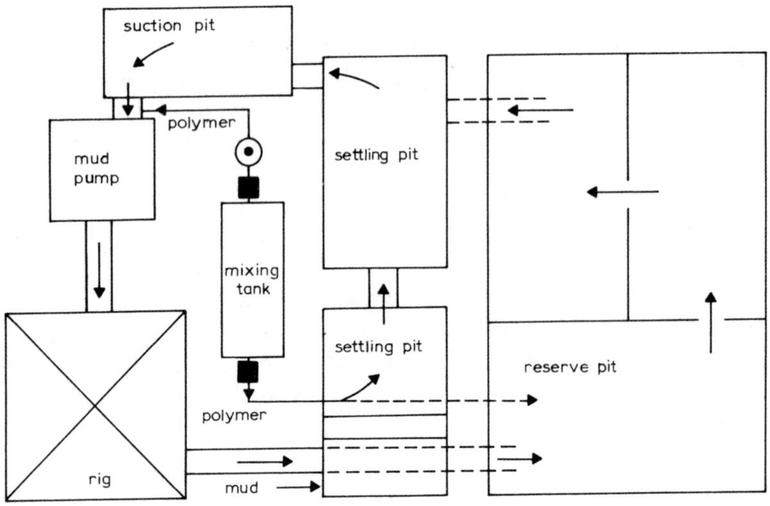

Fig. 11-7. Flow system when using flocculating agent. (Gallus et al., 1958, p. 71, fig. 2; courtesy of the Society of Petroleum Engineers of AIME.)

the lower quantities of polymer (Gallus et al., 1958).

The fluid discharging from wells was circulated through earthen reserve pits, when practical, because of the large settling area and to avoid frequent jetting of the steel circulating pits for removal of settled solids. A typical fluid circulating system used during field tests is shown in Fig. 11-7. When circulation through a reserve pit was not possible, the settling pit was jetted as solids accumulated in sufficient amounts to begin flowing into the suction pit.

Polymer solutions were injected directly into the pump suction line in order to reduce fluid loss to permeable formations. From 1 to 6 lb of polymer were injected depending on the permeability of formations. Electric logs of adjacent wells were consulted whenever possible to determine depths at which permeable strata might be encountered. Gallus et al. (1958) conducted 17 field tests, as presented in Table 11-V. Polymers were used to settle out solids from fresh water and brine and to prevent fluid loss to permeable formations in wells which ranged in depth from 234 to 11,160 ft.

Drilling rates obtained on a well in Andrews County, Texas, where the drilling water was clarified with the polymer, was compared with the drilling rates of two other wells drilled in the same area without the polymer treatment (Fig. 11-8, field test No. 2). All three wells were drilled with the same rig operated by the same crew. The drilling rate of a well in Midland County, Texas (field test No. 14), drilled with water clarified with the polymer from a depth of 8535—9800 ft, was compared with the drilling rates for the same

TABLE 11-V

Results of field tests using an acrylamide carboxylic acid copolymer

(After Gallus et al., 1958, table 2, p. 73; courtesy of the Society of Petroleum Engineers of AIME.)

Test no.	Area	Depth (ft)	Results
1	Winkler County, Tex.	459—3,559	Trip time reduced six hours. Drilling time reduced 35 percent. Bit life increased 25 percent.
2	Andrews County, Tex.	352—4,050	Drilling time reduced 39 percent. Bit life increased 18 percent. Trouble-free trips.
3	Scurry County, Tex.	4,995—7,560	Reduced fluid loss to formations: 80 percent. Reduced mud costs: 50 percent
4	Winkler County, Tex.	550—8,747	Reduced drilling time: 41 percent Increased bit and pump liner life: 26 percent.
5	Scurry County, Tex.	2,730—6,750	Reduced fluid loss to formations: 95 percent. Increased penetration rate: 34 percent. Increased bit life: 19 percent.
6	Scurry County, Tex.	4,460—7,000	Reduced fluid loss to formations: 50 percent. Increased penetration rate: 31 percent. Increased bit life: 22 percent.
7	Andrews County, Tex.	5,500—11,160	Drilled 1000 ft deeper than adjacent wells with water.
8	Lipscomb County, Tex.	3,460—6,135	Saved $2000 on mud costs. Regained circulation by slugging 20 lb of polymer.
9	Lipscomb County, Tex.	365—3,493	Reduced seepage 66 percent. Increased penetration rates and bit life.
10	San Juan County, N.M.	263—5,020	Drilled to 5020 ft in 11 days (fastest drilling rate recorded in area). Stopped water loss to formations.
11	San Juan County, N.M.	234—4,700	Prevented water seepage. Reduced drilling time. Increased bit and pump liner life 26 percent.
12	Lipscomb County, Tex.	1,719—3,115	Saved nine bits. Saved 128 rotating hours.

TABLE 11-V (continued)

Test no.	Area	Depth (ft)	Results
13	Andrews County, Tex.	5,435—10,710	Saved 13 bits. Saved 85 rotating hours over 5000-ft interval.
14	Andrews County, Tex.	8,550—9,800	Saved $2\frac{1}{8}$ bits, 22 hours drilling time, over 1250-ft interval. Chemical cost \$25. Savings: \$ 2000.
15	Leo County, N.M.	4,250—4,770	Reduced fluid seepage from 40 to 0 bbl/hr. Reduced drilling time. Increased bit and pump liner life.
16	Cleveland County, Ohio	925—2,735	Inadequate fluid volume and settling area prevented maintenance of clear water at pump. Highly bentonitic solids required reserve pit circulation.
17	Alberta, Canada	2,100—7,800	Reduced drilling time nine days to 7800 ft. Good bore conditions.

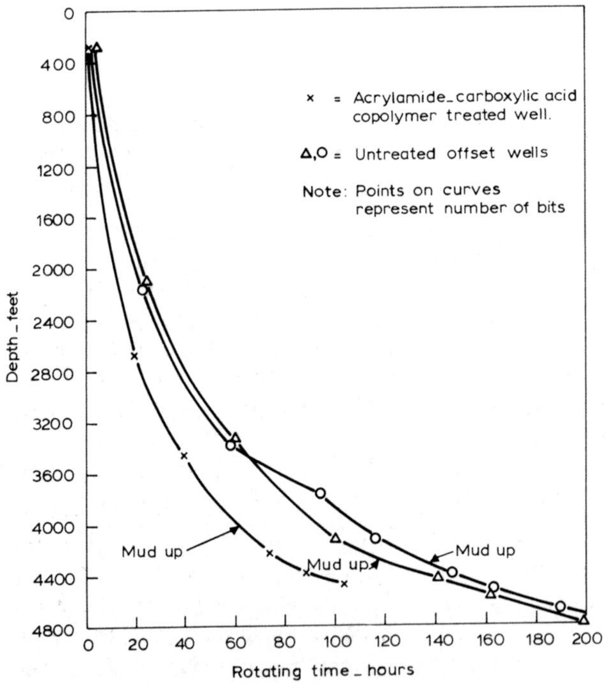

Fig. 11-8. Effect of treatment with acrylamide-carboxylic acid copolymer on drilling rates in Andrews County, Texas. (After Gallus et al., 1958, p. 73, fig. 8; courtesy of the Society of Petroleum Engineers of AIME.)

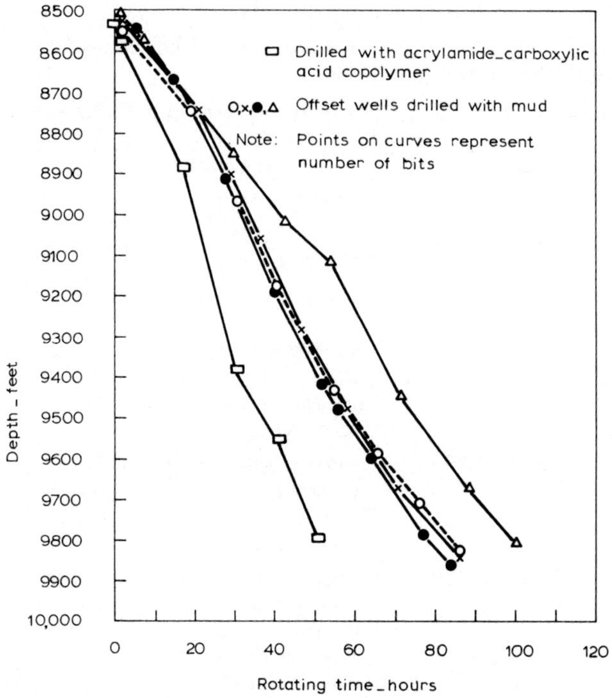

Fig. 11-9. Effect of treatment with acrylamide-carboxylic acid copolymer on drilling rates in Midland County, Texas. (After Gallus et al., 1958, p. 74, fig. 9; courtesy of the Society of Petroleum Engineers of AIME.)

interval in four adjacent wells, drilled with conventional drilling mud (Fig. 11-9). A 43% increase in the drilling rate and a 37% decrease in drill bit requirements was observed on using polymer treated system. All compared wells were drilled with the same rig and crew. In addition, water requirements, which are high in some areas such as New Mexico and West Texas, have been reduced by as much as 50%.

Some typical savings in cost achieved by drilling with water treated with polymers are presented in Table 11-VI (Gallus et al., 1958). According to Gallus et al. (1958), the 17 field tests conducted using the polymer for the purpose of removing suspended solids from the drilling water and preventing fluid loss to permeable formations, in general, proved to be successful. Mud costs were reduced by as much as 50%, penetration rates were increased by an average of 50%, and bit life was extended by approximately 25% (see examples in Tables 11-IV and 11-V, and Figs. 11-8 and 11-9).

Typical savings obtained through the use of a polymer are shown in Table 11-VI. The total savings realized on the five wells selected varied from

TABLE 11-VI

Typical cost savings realized in 1958 through use of acrylamide-carboxylic acid co-polymer. Rig time charge = \$1,000/day; bit cost = \$200/bit

(After Gallus et al., 1958, p. 74, table 3; courtesy of Society of Petroleum Engineers of AIME.)

Field test no.	Area	Depth (ft)	Principal effects	Total cost savings (\$)
1	Winkler County, Tex.	459—3,559	Reduced trip time — six hours Reduced drilling time — 35 percent Increase bit life — 25 percent	2,500.00
5	Scurry County, Tex.	4,995—7,560	Reduced water loss — 80 percent Reduced total mud costs — 50 percent	4,000.00
7	Scurry County, Tex.	2,730—6,750	Reduced water loss — 95 percent Increased penetration rate — 34 percent Increased bit life — 19 percent	4,100.00
9	Andrews County, Tex.	5,500—11,160	Obtained \$0.50/ft footage price reduction from contractor	2,500.00
20	Alberta, Canada	2,100—7,800	Reduced drilling time — nine days	9,000.00

\$ 2,500 to \$ 9,000 (constant 1958 dollars), and the cost of the polymer ranged from \$ 300 to \$ 500/day. These field results indicated that the use of polymer with clear water for drilling, increased both penetration rates and bit life and sharply reduced mud costs (Gallus et al., 1958).

In field test No. 1 in the Winkler County, Texas, the first round trip at 1750 ft required only 2 hours, whereas the first trip in the redbeds usually requires from 4 to 9 hours, because of the pipe dragging in the borehole. This fast trip time can be attributed to the low fluid loss of the polymer solution with consequent thin filter cake that did not hinder the movement of the drillstring. There was less sloughing of the redbeds during this and subsequent tests in the same area as compared to results obtained with conventional drilling fluids. This indicates that the polymer retards the hydration of the redbeds.

Field test No. 3 was conducted in the Scurry County to confirm the plugging effect of the polymer which was observed in the laboratory. Gel—chemical mud and lost-circulation materials are usually required in this area because of very high fluid loss to permeable formations from the surface down, which becomes excessive below a depth of 5000 ft. Polymer solution was injected into the pump suction line at a rate of 1.3—4 lb of polymer per hour on the test well. Fluid loss to permeable formations was reduced by 80% as a result of this treatment, and the generally troublesome shale section between 6100 and 7500 ft was penetrated without difficulty. The mud cost was reduced to approximately half of the average for 13 other wells which were drilled in the area with conventional drilling fluids (Gallus et al., 1958).

On using polymer in drilling another well (field test No. 5) in the same field, the drilling fluid loss to permeable zones was reduced to 5% of the normal fluid loss.

The use of polymer for sharply reducing fluid loss was also demonstrated in other wells drilled in Texas, Oklahoma, and New Mexico in U.S.A. For example, in test No. 8 on a well in the Texas Panhandle, where circulation was lost completely, full returns were regained immediately upon injection of a slug of concentrated polymer solution (approximately 20 lb polymer/10 bbl water).

In the case of field test No. 4 in the Winkler County, Texas, the borehole, which was drilled with clear water plus polymer, had a very thin filter cake and was devoid of bit cuttings and cavings to permit setting of casing without mudding up.

Because of insufficient pit volume for settling of the flocculated solids, field test No. 16 was unsuccessful. Circulation was possible only through two 250-bbl capacity steel mud pits because the reserve pit, which would have provided sufficient settling space, was below the level of the steel pits. If sufficient pit volume is not available for settling the flocculated solids, the frequency of jetting for the purpose of removing solids is too high for efficient operation.

When the use of a large reserve pit is impractical or impossible, wet cyclone separators or other mechanical separating devices, that are capable of reducing the solids content of the drilling fluid, may aid in reducing the frequency of jetting (Gallus et al., 1958).

In the case of occurrence of hole washouts, the low cuttings carrying capacity of the solids-free drilling fluid is a disadvantage. In the case of washouts, the bit cuttings and flocculated solids tend to accumulate at the bottom of borehole when circulation is stopped. Inasmuch as these solids are easily removed upon resumption of circulation, however, there is no tendency for the drillstring to stick.

Conversion from clear water to mud in drilling these test wells was necessary at various drilling depths for such reasons as: (1) improvement of the bit

TABLE 11-VII

Rheological properties of XC polymer solutions (KELZAN XC). The polymer solutions exhibit pseudoplasticity

(Courtesy of Kelco Oil Field Products.)

KELZAN XC polymer concentration	Apparent viscosity (cP)	Plastic viscosity (cP)	Yield point (lb/100 sq ft)	n *	K *	Viscosity at 50 sec^{-1} ** (cP)
Fresh water						
0.5 lb/bbl	6.9	3.8	6.2	0.41	3.9	39
1.0 lb/bbl	9.9	4.7	10.4	0.35	8.6	68
1.5 lb/bbl	16.0	6.5	19.0	0.32	17.5	135
Seawater						
0.5 lb/bbl	5.6	3.1	5.0	0.44	2.7	30
1.0 lb/bbl	9.0	4.0	8.0	0.36	7.5	64
1.5 lb/bbl	15.0	6.0	18.0	0.31	17.5	120
Saturated salt water						
0.5 lb/bbl	6.8	3.9	5.8	0.43	3.4	35
1.0 lb/bbl	11.6	6.5	10.2	0.38	6.0	68
1.5 lb/bbl	17.25	6.5	21.5	0.28	24.0	140

* n and K at 100 and 200 rpm; K is in dynes-sec/cm^2.
** Typical annular shear rate.

cuttings carrying capacity to keep the hole clean, (2) to provide samples of adequate size for inspection, (3) to increase drilling fluid density in drilling through abnormally-pressured formations, and (4) to combat loss of circulation with bridging materials.

In order to convert back to a clay-base mud after using the polymer, injection of polymer solution must be discontinued 1—2 hours (or 100—200 ft of drilling depth) before adding clay. Because of adsorption on the clays, the polymer completely disappears in a relatively short period of time.

The polymer has been found to be both practical and economical in areas where clear-water drilling is feasible (Gallus et al., 1958).

Deily et al. (1967) developed a drilling fluid that exhibits a viscosity close to that of water at the bit and, yet, provides good hole cleaning action and cuttings carrying ability. They used a water-soluble biopolymer produced by action of bacteria on carbohydrates (a high-molecular-weight linear polysaccharide which is sometimes called an XC polymer). Table 11-VII presents the rheological properties of Kelzan * XC polymer solutions. These solutions

* Tradename, Kelco Oil Field Products.

Field test results with XC polymer mud
(After Deily et al., 1967, table 3, p. 80; courtesy of the World Oil)

Test Well	Objectives	Results	Remarks
1	Use XC polymer.	Successful.	Anhydrite stringers flocculated drilled solids, but had no effect on polymer. Water loss increase controlled by additional polymer.
2	Observe crosslinking with chromic chloride.	Successful. Crosslinking decreased polymer consumption about 50%.	Crude oil was readily emulsified. Maintenance procedures were simple. Mud showed good carrying and suspending ability. Water loss control was augmented with strach and bentonite. Anhydrite mildly flocculated drilled solids.
3, 4	Allow rig crew to run polymer mud without assistance.	Achieved.	Concentrated chromic chloride solution exposed to —20° F caused salt crystallization and settling. Motivated change to dry crystalline form.
5, 6	A. Combat severe calcium contamination.	XC polymer built viscosity in gyp-saturated system with no difficulty.	
	B. Operate in salt environment up to 65,000 ppm.	Salt additions caused fluid loss increases which additions of polymer did not reduce; high salt concentrations did not affect final viscosities, but did decrease rate of polymer solubility.	With apparent viscosities in the range of 10-20 cp, XC polymer provides fluid loss control about 10-12 cc API. Solution rate of XC polymer may be increased by exposure in mixing hopper to high, efficient shear.
7	Determine polymer behavior in an environment of saturated calcium and salt.	Polymer built viscosity satisfactorily in saturated calcium system. Fluid loss control was not achieved to degree desired, and starch was used to augment. Salt retarded rate of hydration of XC polymer markedly.	Paraformaldehyde was used to stop and prevent polymer fermentation.
8	A. Observe XC polymer system behavior in well with thick shale sections that are potentially troublesome.	Crooked hole problems masked true evaluation of XC in thick shale sections.	Paraformaldehyde prevented polymer fermentation. Gyp contamination did not disturb system.
	B. Observe polymer behavior over a long period of time in which bottomhole temperature was 220° F.	Mud remained stable for the 25 day test period. Performance was not affected.	
9	A. Study ability to suspend barite with mud weight up to 13 ppg.	Barite suspension was adequate with mud weights to 13 ppg.	
	B. Observe system stability for more than 30 days at temperatures above 250° F.	XC polymer was stable for building viscosity for more than 50 days. Bottom hole temperatures of 255° F were encountered with no observable viscosity degradation.	Polymer fermentation was controlled by sodium trichlorophenate in concentrations of about 300 ppm.
	C. Determine degasser ability to separate entrained gas bubbles in XC polymer system.	Degassing was accomplished over more than 50 days with no problem.	After bentonite was added at 14,056 feet to improve fluid loss, both viscosity and gel strength increased.

illustrate the polymer's ability to viscosify and provide necessary rheological characteristics over a wide salinity range and at low concentrations. This type of mud system appears to present important means of reducing well costs. During the field tests, penetration rates were increased and material transportation costs were reduced. In addition, on using desanders, bit life was significantly prolonged and water consumption was reduced up to 40% (Deily et al., 1967).

Gallus et al. (1958) have found some limitations of polymer muds: (1) they cannot be used at temperatures above 300° F, (2) they are subject to bacterial attack, and (3) fluid loss control agents may be required to reduce fluid loss values to below about 12 cm^3.

Many wells have been drilled with biopolymer muds in the United States, Canada, Libya, and other countries. During initial testing, operating and technical strengths and weaknesses were closely studied (e.g., see Table 11-VIII). Cost reduction to varying degrees was observed in the case of some of the first eight test wells.

The following important field observations were made by Deily et al. (1967) concerning the biopolymer muds:

(1) For viscosity building, the amount of biopolymer needed is 1/10—1/20 of the total mud treating material formerly used.

(2) Water consumption is reduced.

(3) On using drilling fluids of nearly equivalent properties, bit life is longer with biopolymer drilling fluid, primarily as a result of reduced bearing wear.

(4) There was some improvement in the penetration rate in the case of wells drilled with biopolymer muds.

PRODUCTIVITY REDUCTION AS A RESULT OF MUD DAMAGE

Keelan and Koepf (1977) studied the productivity reduction of laboratory cores by drilling fluids. Montmorillonite, as discussed in previous chapters, is used most frequently in water-base clay muds to provide proper rheological properties and filtration control. Mud solids which form on the wellbore wall (filter or mud cake) restrict filtrate flow into the formation; however, some fines do move with the filtrate into the pore channels, especially prior to the formation of filter cake. These fines block flow channels and reduce productivity (e.g., see Glenn and Slusser, 1957). Various effects of reaction of the filtrate with clays in the formation may further reduce the flow of hydrocarbons. When both of these impairment mechanisms are present, their deleterious effects are additive. Reservoir productivity can be further reduced as a result of the increase in reservoir water saturation, because of water block and reduction in the relative permeability to oil (Keelan and Koepf, 1977).

Flow charts of laboratory tests available for evaluating prospective muds

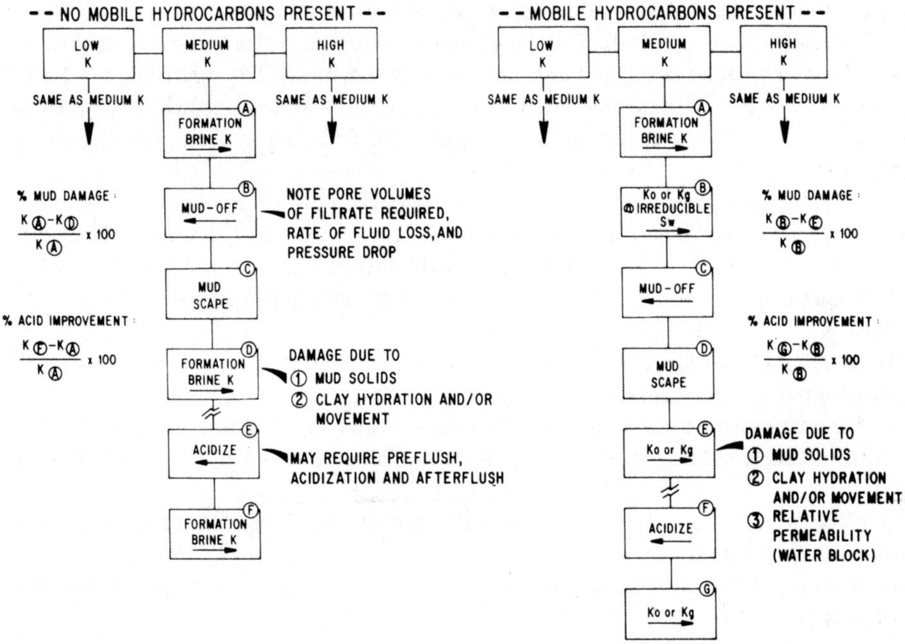

Fig. 11-10. Flow chart of tests for evaluating proposed drilling fluids for potential damage and remedial acidization results. (Keelan and Koepf, 1977, fig. 2, p. 484; courtesy of the Society of Petroleum Engineers of AIME.)

are presented in Fig. 11-10. Mud is exposed to the coreface under pressure differential. No mobile hydrocarbons are present in the most common test sequence. Inasmuch as water saturation was kept constant, the water-block effect was not present (Keelan and Koepf, 1977). In Figure 11-10, arrows indicate the direction of fluid flow, which is selected to simulate reservoir conditions.

The test samples must cover the expected range of permeabilities and porosities of formations, because some mud systems are suitable in only certain permeability and porosity ranges. Pore sizes, pore shapes, and types of pore geometry are interrelated with permeability and porosity.

Equations commonly used to indicate the magnitude of formation damage by drilling fluids are presented on the flow chart (Fig. 11-10). The rate of fluid loss and pore volumes of filtrate that enter the test plug before mud-off should be recorded. The pressure differential used in the laboratory test should be scaled to correspond to that expected in the field during drilling (Keelan and Koepf, 1977).

A test sequence that incorporates mobile hydrocarbons and the possibility of water-blocking in a formation, is also illustrated in Fig. 11-10. In order to calculate water saturation, the filtrate recovered during hydrocarbon flow in Step E should be recorded. If all extraneous filtrate water introduced before mud-off is produced in this step, remaining permeability reduction can be attributed to clay hydration and/or penetration of mud solids into formation. Commonly used mud-damage equations for this system are shown in Fig. 11-10.

Keelan and Koepf (1977) reached the following conclusions:

(1) Remedial work is required to restore formation productivity after contact with the mud, because all drilling fluids damage the formation.

(2) Return permeability decreases with increasing filtrate loss to achieve mud-off.

(3) Return permeability increases with an increase in the illite content of the clays used in preparation of muds.

(4) Return permeability increases with increasing content of diesel oil in the mud. Diesel, which was a discontinuous phase, was not lost to the pore spaces.

(5) Return permeability increases with increasing pressure differential imposed during cleanup.

(6) Return permeability decreases with addition of chemicals to control shale sloughing.

SUMMARY AND CONCLUSIONS

Penetration rates

Darley (1976) has shown that laboratory microbit penetration rates approach those obtained with water alone if solids content is below about 2% (Fig. 11-11). If filter cake forms on the bottom of the hole, it creates a chip-hold-down pressure. Thus an ideal situation would be formation of normal filter cake on the sides of the borehole and prevention of its formation beneath the bit. This can be achieved if the concentration of solids in the bridging size range (which corresponds to the inert solids fraction) is kept to a very low value.

Darley (1976) suggested the following in order to increase the bit penetration rates:

(1) Viscosity should not be raised by the filtration control additives (whenever required). Lignosulfonates proved to be satisfactory in this respect.

(2) Pressure losses in drillpipe should be minimized and, thus, for a given input provide more hydraulic horsepower at the bit for bottomhole cleaning (see Figs. 11-6 and 11-12). In addition, the use of friction-reducing polymers may also result in minimum viscosities beneath the bit.

(3) Solids content must be kept low and disintegration of cuttings must be

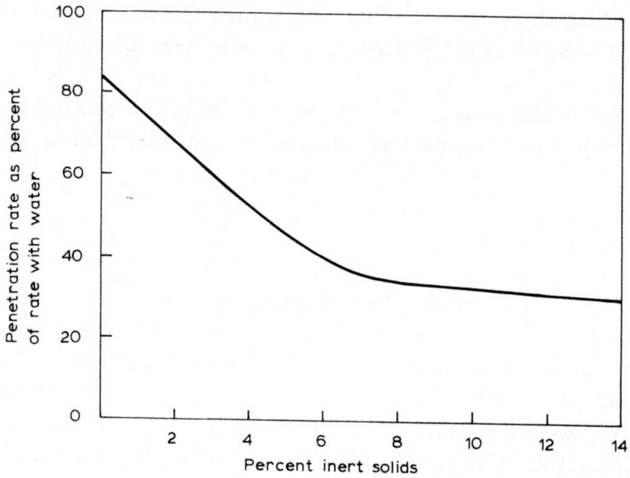

Fig. 11-11. Relationship between penetration rate and inert solids content (microbit drilling rates). (After Darley, 1976, fig.1; courtesy of Petrol. Engr. Int.)

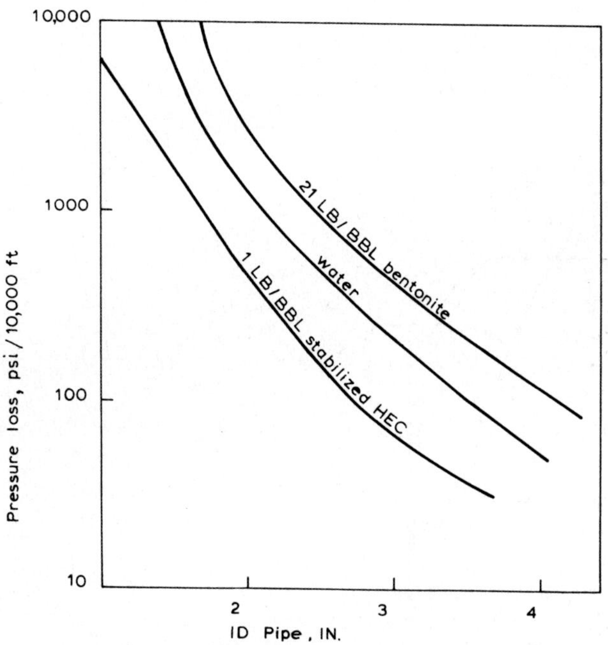

Fig. 11-12. Relationship between pressure loss (per 10,000 ft of pipe) and internal diameter of pipe for various fluids at 250 gal/min. (Courtesy of Brinadd Co.)

prevented. A polymer mud having a low $\overline{YP}/\overline{PV}$ ratio is best for settling fines in pits. Settling can be improved by the addition of a polyacrylamide copolymer flocculant.

(4) The borehole must be maintained close to gage in order to be able to clean it with a low-viscosity fluid. Consequently, shale stabilization is of utmost importance.

Hole cleaning

Drilling fluids having low $\overline{YP}/\overline{PV}$ ratio prevent gas-cutting and swabbing pressures and enhance settling. If, however, gage hole is not maintained and its diameter enlarges, then a fluid having a high $\overline{YP}/\overline{PV}$ ratio is desirable.

In the case of polymer mud, which is a shear-thinning fluid, a decrease in the upward velocity in the enlarged portion of borehole is automatically compensated for by an increase in viscosity (Darley, 1976). On the other hand, in the case of the Newtonian (or near Newtonian) fluids either the pumping rate must be increased or more viscosifier must be added, which increases the pressure drop in the annulus.

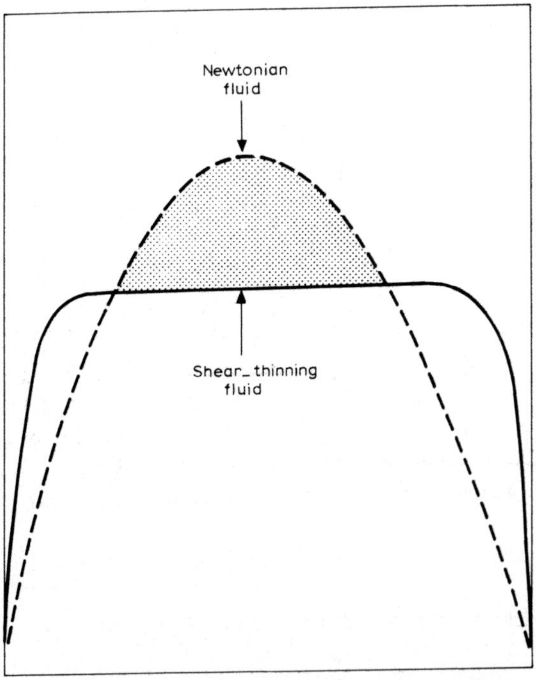

Fig. 11-13. Flow velocity profiles for shear-thinning fluid and the Newtonian fluid. (After Darley, 1976; courtesy of Petrol. Engr. Int.)

For the same pressure drop, a better hole cleaning is achieved with shear-thinning fluid because of a flatter velocity profile in the annulus (Fig. 11-13).

Stabilization of water-sensitive shales

According to Darley (1976), partially hydrolyzed polyacrylamide and stabilized HEC stabilize water-sensitive shales. This action is greatly increased in the presence of soluble salts. Potassium chloride reduces clay hydration to a greater extent than do equivalent amounts of sodium and calcium chloride (Clark et al., 1975).

In the foothills of Alberta, the use of 0.5 lb/bbl polyacrylamide plus 10 lb/bbl of potassium chloride kept the boreholes closer to gage and greatly reduced time previously spent drilling out bridges and fill. Over 1,000,000 ft of hole encountering hard, brittle and, often, mechanically incompetent shales have been successfully drilled in 40 wells. Occasionally, prehydrated bentonite was added later if additional hole-cleaning capacity was required (Darley, 1976).

In offshore wells in Spain, Portugal, Brunei, and the North Sea, where severe tight hole and wash-out problems have been prevalent in gumbo shales, the polyacrylamide-plus-potassium chloride muds were used successfully. Polyacrylamide was used in 0.25 to 0.5 lb/bbl concentration, with up to 60 lb/bbl of KCl to prevent shale swelling (Darley, 1976).

Cost comparison and necessary precautions

Inasmuch as polymers and KCl are expensive, polymer mud costs are higher than those of conventional muds. Fewer hole problems and faster drilling rates, however, compensate for these higher costs. As an example, the findings of Cornett (1977) are presented in Fig. 11-14. He reported on a third South Texas Austin-Edwards well in Dimmit County drilled with a brine-based stabilized HEC polymer drilling fluid. It resulted in saving of 250 rotating hours and $7\frac{1}{2}$ days from the time of spudding to the time of completing drilling to a total depth. Although HEC polymer system cost was about \$ 18,000 compared to \$ 11,000 for a gel system (bentonite water-base mud), the total well costs for the same depths indicated savings of \$ 25,000. This was due to the lack of formation damage, lower testing costs, and better completions.

According to Darley (1976), the following precautions should be observed for best results:

(1) Frequent maintenance checks to ensure proper level of polymer concentration.

(2) Prevention of drilled solids buildup in the mud, by having adequate solid separating equipment (combinations of desanders, desilters, and fine screens) available.

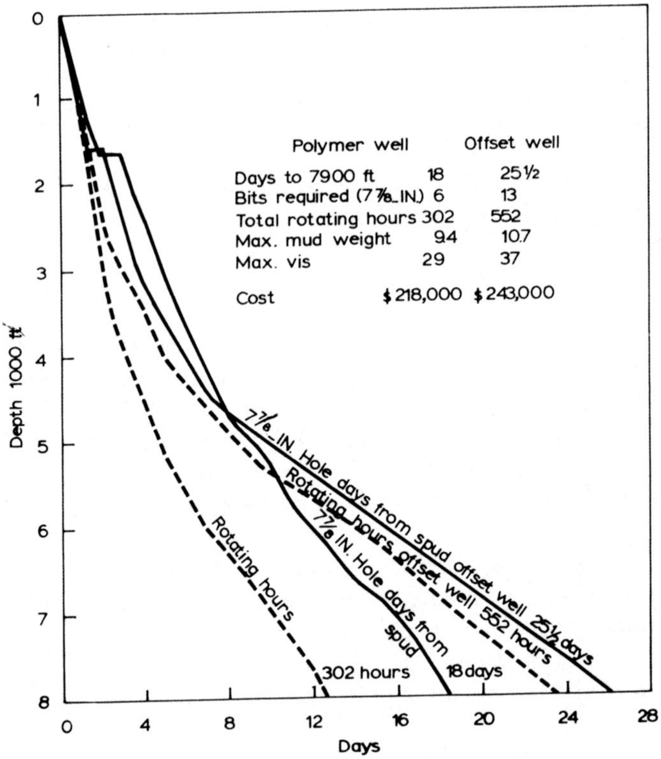

Fig. 11-14. Drilling curves for South Texas wells using polymer mud and water-base bentonite mud (cost comparison). (After Cornett, 1977, fig. 1; courtesy of Petrol. Engr. Int.)

(3) Efficient mixing by equipment available at the well to insure complete and rapid dispersion of polymer into the mud.

(4) Performance of laboratory tests prior to drilling in order to determine the optimum mud composition.

REFERENCES

Arridge, R.G.C., 1975. *Mechanics of Polymers.* Clarendon Press, Oxford, 246 pp.

Billmeyer, F.W., 1962. *Textbook of Polymer Science.* Interscience, New York, N.Y., 601 pp.

Burcik, E.J., 1965. A note on the flow behavior of polyacrylamide solutions in porous media. *Prod. Mon.*, 29(6): 14—16.

Carico, R.D., 1976. *Suspension Properties of Polymer Fluids Used in Drilling, Workover and Completion Operations.* 46th Annu. Calif. Reg. Meet. Soc. Pet. Eng., Long Beach, Calif., SPE Pap. 5870: 6 pp.

Carico, R.D. and Bagshaw, F.R., 1978. *Description and Use of Polymers Used in Drilling, Workovers, and Completions.* AIME Prod. Tech. Symp., Hobbs, N.M., SPE Pap. 7747: 6 pp.

Clark, R.K., Scheuerman, R.F., Rath, H. and VanLaar, H., 1975. *Polyacrylamide-Potassium Chloride Mud for Drilling Water-Sensitive Shales.* 50th Annu. Fall Meet. SPE—AIME, Dallas, Texas, Sept. 28—Oct. 1, SPE Pap. 5514.

Cornett, J., 1977. Polymer-brine mud speeds Austin well. *Pet. Eng. Int.*, 49 (11, Oct.): 118.

Darley, H.C.H., 1965. Designing fast drilling fluids. *J. Pet. Technol.*, 17(4): 465—470.

Darley, H.C.H., 1969. A laboratory investigation of borehole stability. *J. Pet. Technol.*, 21(7): 883—892.

Darley, H.C.H., 1976. Advantages of polymer muds. *Pet. Eng. Int.*, 48 (11, Sept.): 46—52.

Darley, H.C.H. and Hartfiel, A., 1974. Tests show that polymer fluids cut drill-pipe pressure losses. *Oil Gas J.*, 72 (24, June 17): 70—73.

Darley, H.C.H., and Vickers, R.C., 1975. Clay-free aqueous drilling fluids. *Proc. Southwestern Petroleum Short Course, Lubbock, Texas,* 4 pp.

Deily, F.H., Holman, W.E., Lindblom, G.P. and Patton, J.T., 1967. New low-solids polymer mud designed to cut well costs. *World Oil,* 165 (1, Jul.): 77—80.

Eckel, J.R., 1968. How mud and hydraulics affect drilling rate. *Oil Gas J.*, 66(25): 69—73.

Gallus, J.P., Lummus, J.L. and Fox, J.E., Jr., 1958. Use of chemicals to maintain clear water for drilling. *Trans. A.I.M.E.*, 213: 70—75.

Garnier, A.J. and Van Lingen, N.H., 1959. Phenomena affecting drilling rates at depth. *Trans. AIME,* 216: 232—239.

Glenn, E.E. and Slusser, M.L., 1957. Factors affecting well productivity drilling fluid particle invasion into porous media. *Trans. AIME,* 210: 132—139.

Gray, G.R. and Kellogg, W.C., 1955. A cross-section of typical mud problems. *World Oil,* 14(2): 102—117.

Keelan, D.K. and Koepf, E.H., 1977. The role of cores and core analysis in evaluation of formation damage. *J. Pet. Technol.*, 29(5): 482—490.

Mungan, N., 1968. *Programmed Mobility Control in Polymer Floods.* Paper, 3rd Symp. Assoc. de Recherche sur les Techniques de Forage et de Production, Pau, France.

Mungan, N., Smith, F.W. and Thompson, J.L., 1966. Some aspects of polymer floods. *J. Pet. Technol.*, 18(8): 1143—1150.

Sandiford, B.B., 1964. Laboratory and field studies of water flood using polymer solution to increase oil recovery. *J. Pet. Technol.*, 16(8): 917—925.

Smith, F.W., 1970. The behavior of partially hydrolyzed polyacrylamide solution in porous media. *J. Pet. Technol.*, 22(2): 148—156.

Snell, G.W. and Schutz, G.F., 1966. Polymer chemicals aid in unique recovery process. *Pet. Eng.*, 38(2): 53—59.

Sparlin, D.D., 1974. An evaluation of polyacrylamides for reducing water production. *J. Pet. Technol.*, 28(8): 906—914.

Tchillingarian (also Chilingar), G.V. and Beeson, C.M., 1950. *Iranian Gums as Treating Agents for Drilling Fluids.* Pet. Branch, Am. Inst. Min. Metall. Eng., Los Angeles Meet., Oct., No. 78G: 8 pp.

Tchillingarian, G.V. and Beeson, C.M., 1952. Natural organic colloids of Iranian origin as mud treating agents. *Pet. Eng.*, 24(4): B45—52.

Chapter 12

DRILLING FLUID SOLIDS REMOVAL *

GEORGE S. ORMSBY **

INTRODUCTION

For many years drill solids were removed mostly using settling pits and vibrating shaker screens, few of which had openings less than one-sixteenth of an inch across. The mechanical processing of rotary drilling fluids entered a new phase with field-practical decanting centrifuges in 1952 (see Bobo and Hoch, 1953), efficient 6-inch hydrocyclones in 1954 (O'Shields and Wuth, 1955), much more efficient 4-inch hydrocyclones in 1962 (Ormsby, 1965), and very fine-mesh shale shaker screens in 1966 (Mondshine, 1966). Spin-offs and special equipment are being developed constantly.

The available approaches to solids removal for better mud control and at reduced mud cost are presented in this chapter.

SYSTEM MATERIAL BALANCE RELATED TO MUD COST

Many active mud systems tend to increase in surface volume. Consequently, the excess must be disposed of by some means after being hauled away or jetted out. The mud system material balance is presented in Fig. 12-1.

The subsurface mud system consists only of the hole. This subsurface system volume increases at the rate of drilling new hole, plus caving or sloughing rate. Drilling the hole does not change the level of drilling fluid in the surface tanks, as every 100 bbl of hole drilled, for example, adds 100 bbl of cuttings, resulting in 100 bbl of hole being filled with mud. The same is true of caved volume and whether the drilled formation is bentonite or granite.

Filtration to the formation tends to reduce the mud pit level at rates varying from hundreds of gallons per minute while drilling sands and silts with "clear water" in rare cases, to small losses using filtration control, or to no

---

* This chapter is freely adapted from Chapter 6, "Drilling Practices Manual", Preston L. Moore, First Edition, The Petroleum Publishing Co. Professor George V. Chilingarian edited and rewrote some portions of this chapter.

** George S. Ormsby was the founder of Pioneer Centrifuging Company (now a part of Geosource Solids Control), which sponsored the preparation of this chapter.

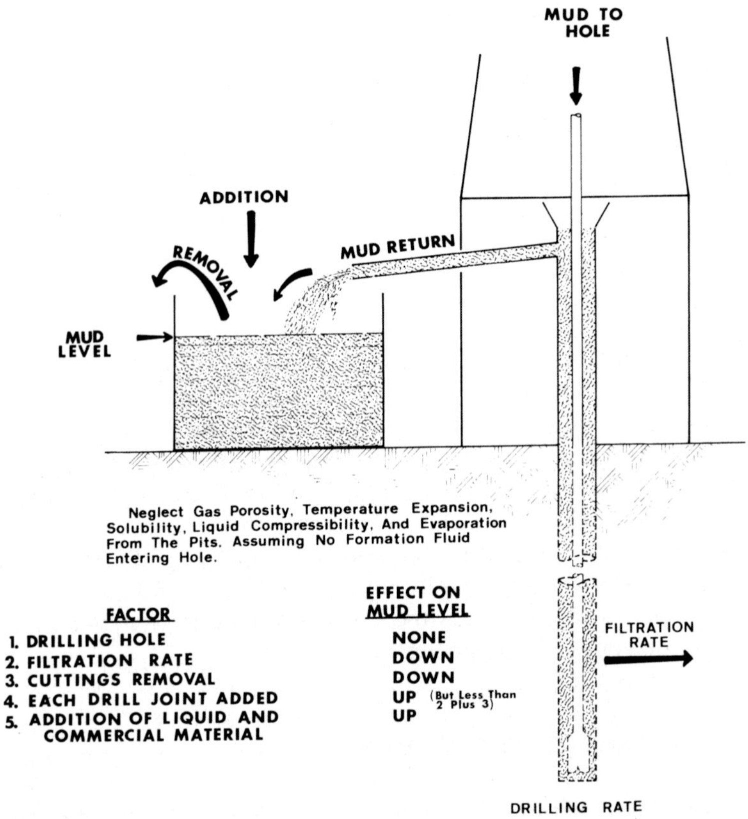

Fig. 12-1. Mud system material balance.

filtration loss in impermeable zones. Removal of solids tends to reduce system level. This reduction can be more than the drilling rate, even though the net solids removal rate is always less than the drilling rate, because solids are not removed nearly as dry as they exist in the formation. The volume rate of removal of solids with associated water may be much more than the one indicated by the hole drilling rate.

Drill-joint addition slightly increases the level in the mud tanks; each 31 ft of drill-joint displaces approximately 1 cu ft of mud (= 7.5 gal or 28 l). During the drilling of equivalent 9.4 m of hole, the volume of cuttings formed is much more than the volume of steel added (in a $7\frac{7}{8}$-inch hole it is roughly 10 times as much as the steel volume of $4\frac{1}{2}$-inch I.F. pipe).

The drilling of a gas sand displaces some liquid from the hole by expansion of the gas rising in the annulus. Thus, the liquid level in the surface tanks will increase temporarily. If a gas sand is continuously drilled, the level in the tanks begins to decrease slightly as part of the pore volume occupied by the formation gas is replaced with drilling fluid in the new portion of the hole.

The breaking out of gas at the surface reduces total mud system volume. After the gas zone is past and the gas is circulated out, the surface mud system level returns to a point slightly lower than when the gas zone was first encountered. The fact that the gas expansion in the annulus is a major temporary effect should not mask the underlying and continuing minor reducing effect on the surface volume and level.

There are other *very* minor effects that are sometimes misinterpreted. Solubility is a negligible factor causing a decreasing trend in surface level, too small to be noticed in the field. Temperature expansion should increase the mud system volume in the early stages of shutdown, assuming the subsurface temperature is higher than the surface temperature. The reverse is true if subsurface temperature is lower than the ambient temperature; however, in any normal field case the change would be very difficult to detect. Excluding entrained gas, liquid compressibility does not appreciably affect the circulating system.

The natural trend of a surface mud system is to decrease in volume as the hole is drilled, formation fluid entry being the only natural factor capable of reversing this trend. In normal overpressure drilling, with mud hydrostatic head being greater than the formation (pore) pressure, any excess surface system volume, which must be drained off, jetted out, or trucked away, can be caused only by factor No. 5 in Fig. 12-1, i.e., addition of liquid and commercial solids materials. Water is a major cost item in some areas due to scarcity.

Inasmuch as addition of various materials to increase the mud level in the pit is made doubly expensive by the problem of disposing of the excess volume of mud, there should be valid reasons for doing it. Before the advent of modern solids removal equipment, the valid reason for "dumping" or "watering back" was to control mud properties. At the present time, such courses of action may be due to failure either to understand the problem or the solids removal equipment available.

MUD TREATMENTS

There are three basic ways of conditioning drilling fluids, all relating to solids: (1) adding solids or their equivalent, (2) removing solids, and (3) treating solids (chemically). Removing gases from the drilling fluid becomes very important when drilling fluid has high gel strength or high yield point.

The major *addition* of solids is from the drill bit. Commercial colloidal and soluble solids are added to increase yield point, gel strengths, and plastic viscosity, and to decrease filtration rate with minimum increase in solids content. Heavier and coarser commercial solids are added to increase the density of the slurry with minimum effect on the mud properties and solids volume.

The effect of emulsified oil droplets in water-base muds is similar to that of colloidal solid particles. The same is true of water droplets in continuous oil phase (oil-base muds).

Removing solids is necessary before a drilling fluid is pumped (returned) to the drill bit. The properties of recirculated drilling fluids must be changed (returned) to as near the optimum properties as is economically practical. The more effective the mechanical solids removal, the less dilution and chemical treatment is required. If the total annular return fluid is being discarded after one pass through the bit, no separation effort is necessary. In most situations, however, this is neither possible nor desirable because of either economic or ecologic considerations, or both.

Treating solids ("mud chemistry" or mud "treating") is the science and art of adding specific soluble materials to a drilling fluid to alter the behavior, directly or indirectly, of some specific solids in the slurry. The chemicals act on the clay particles and hydratable shales, and not on the larger inert solid particles. Even though chemicals such as a salt may be soluble, they do add to the total solids content. They should not be added, therefore, unless needed for a specific reason on a specific job.

DYNAMIC FILTRATION AND WATER DILUTION

The addition of water does not necessarily dilute a drilling mud in a drilling situation. In some situations, the dynamic fluid loss can be very high. Downhole filtration as high as 400 gpm (gal/min) has been reported (Gassett, 1964) in long stretches of open hole in massive sand and silt sections and with very little selective solids control. Failure to replace the filtrate loss would cause a decrease in liquid volume and resultant increase in per cent solids, weight, and viscosity, independent of the drilling rate. If the filtration loss is constantly replaced, however, the properties of a mud are continuously restored, much like drinking water restores body fluids being lost by sweating. Whether this "pass-through water" amounts to most or a minor part of the total water requirement, depends upon the ratio of the dynamic downhole filtration rate to the solids removal rate.

With no addition of solids and if the amount of water added is above that necessary to balance filtration loss, the mud is diluted, i.e., the solids content (%) decreases, and the properties change. Solids added by the bit increase the total solids content, without increasing the surface system level. If the surface level remains constant, water being added is replacing the solids being removed and the liquid being lost with the solids, and is restoring filtrate lost downhole. The total amount of solids can be altered only by the addition and removal of drilled and commercial solids regardless of the amount of water being added.

The API Committee on Standardization of Drilling Fluids Materials became concerned about the inherent inaccuracy of mud stills, as compared to the margins of solids control many operators were attempting to hold. A "Task Group on Determination of Liquids and Solids" was set up at the annual 1968 meeting. A later report (API Task Group, 1969) of this group seems to confirm that dependable accuracy of field mud retorts does not justify the actions and expenditures many operators make based on mud solids tests. There is a good reason to believe that field readings of per cent solids can be in error by an amount ranging from minus two percentage points to plus four percentage points from the true non-evaporating solids in a mud. Perhaps more significant, from a remedial point of view, is the fact that retorting affords no clue to the size or shape of the undesirable solids, whether they are indeed undesirable, and how they might best be removed.

PARTICLE-SIZE REFERENCES AND TERMINOLOGY

To appreciate any discussion of small particles, some appreciation of size is necessary. The micron is the most convenient unit to use in discussing small particles. Table 12-I shows the size of common materials in microns (Perry, 1950; Stanford Research Institute, 1961). The finger tip sensitivity can be used to determine the effectiveness of coarse solids removal by rubbing a mud filter cake between the fingers. Inasmuch as the red blood corpuscles are very small, they constitute a good mental size reference when looking at tabulated sizes. Particle-size distributions of the solids in drilling fluids show that most single particles are invisible.

TABLE 12-I

Sizes of some materials (and senses) in microns

Material and/or sense	Diameter (size) (microns)	
Human hair	30—200	
Pollen	10—100	
Portland cement dust	3—100	
Milled flour	1—80	
Talcum powder	5—50	
Red blood corpuscles	$7\frac{1}{2}$	
Human eye resolution, normal	40	(min.)
Cosmetic preparations	35	(max.)
Face and lip skin sensitivity	35	(min.)
Finger tip sensitivity	20	(min.)
Between-teeth sensitivity	6—8	(min.)
Human eye resolution, absolute	6—8	(min.)

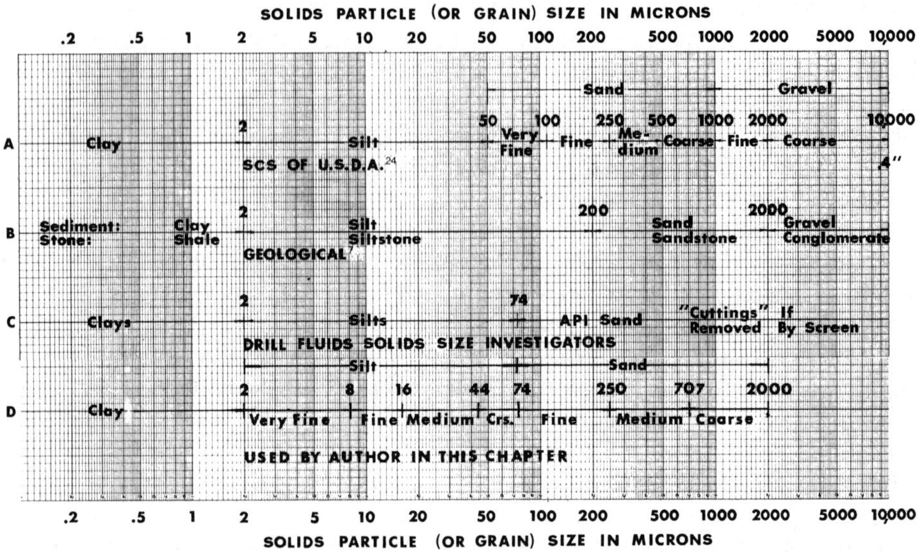

Fig. 12-3. Some particle-size terminology and usage.

Inasmuch as it is sometimes inconvenient to refer to all sizes in microns, various disciplines have adopted some standard terminology for convenience. When an industry or group lacks a standard, individual investigators must use improvised or borrowed terms. Figure 12-3 illustrates the terminology problem and compares some scales often used in the field (Longwell et al., 1941; U.S. Dept. of Agriculture, 1961).

The word "clay", used in this chapter, consistently refers to particles less than two microns in size. "Silt" as used here covers the 2—74 micron-size range. "Sand" and "API Sand" are used interchangeably and refer to particles larger than 74 microns in size, as determined by an API Sand Test, or by any test designed to determine size irrespective of grain composition.

Any further divisions in size, such as fine silt, coarse silt, coarse sand, etc. refer to subdivisions of the silt and sand sizes as indicated in Scale D of Fig. 12-3 and used by the author. These size scales do not change with the nature of particles, whether they are drilled limestone, quartz, dolomite, basalt, or chert particles, commercial bentonite, barite, or steel particles from the drilling tools.

A source of confusion in the application of solids removal equipment has been the use of the word "clay" by most mud-treating engineers referring to all low-specific-gravity commercial and drilled solids in the system, i.e., to all solids excluding commercial weighting material (Petroleum Extension Service, 1969). This usage is expressed in the following equation:

$$\frac{100\%}{\text{avg. sp. gr. of solids}} = \frac{\%\text{ clay by wt.}}{2.6} + \frac{\%\text{ barite by wt.}}{4.2} \tag{12-1}$$

Although this is one of the most basic expressions used in drilling fluid solids calculations, and some of the equations presented in this chapter are derived using it, the terminology shown here is indeed unfortunate. The above relationship concerns the specific gravities of two mixed solid materials, their weight percentages in the mixture, and the average specific gravity of the mixture. The confusion arises because the term "clay", which is used in this mathematical expression for materials having specific gravity of 2.6, has been recognized and used by many investigators including earth scientists as meaning a size range, whether or not the size ranges are in exact agreement (see Fig. 12-3).

For example, it is correct to state that "decanting centrifuges can separate liquid and clay from the larger solid particles in the drilling muds", using the term "clay" strictly in the size sense. On the other hand, it is wrong to say that decanting centrifuges can separate low-specific-gravity solids (or drilled solids) from barite in the drilling fluids. Yet the latter statement has been widely accepted because of the ambiguous usage of the term "clay" in many mud treatment studies. Thus, the above equation should be changed to a much more meaningful and accurate expression when applied to mud being pumped into the hole:

$$\frac{100\%}{\text{avg. sp. gr. of solids}} = \frac{\%\ \text{low-sp.-gr. solids}}{2.5} + \frac{\%\ \text{barite}}{4.2} \tag{12-1a}$$

RELATION OF SPECIFIC SURFACE AREA OF SOLID PARTICLES TO MUD PROPERTIES

The effect of solid particles size on mud properties is illustrated in Fig. 12-4 (Ritchey, 1959). Although another "regular" barite might exhibit same variation, whereas barite of 3 microns or less in size in another test would not follow the same line in Fig. 12-4, a similar basic difference would exist. The difference in results when particle size is varied in a mud slurry is primarily due to magnitude of surface area, which determines the degree of adsorption (tying-up) of water. More water is adsorbed with increasing surface area. Table 12-II shows how the surface area increases as particle size decreases, even though the shape and quantity (net volume and weight) remain exactly the same for one pound of glass balls (Orr and Dallavalle, 1959).

Table 12-II also shows that if the shape is changed, the surface area is affected. Spherical shape gives rise to the least surface area possible for any given volume. Particles of crushed quartz, having the same volume as quartz balls, have more area per unit of weight and consequently would adsorb more water per pound of solids. When dry solids are added to a mud system, water is adsorbed from the available liquid. This adsorption has the effect of

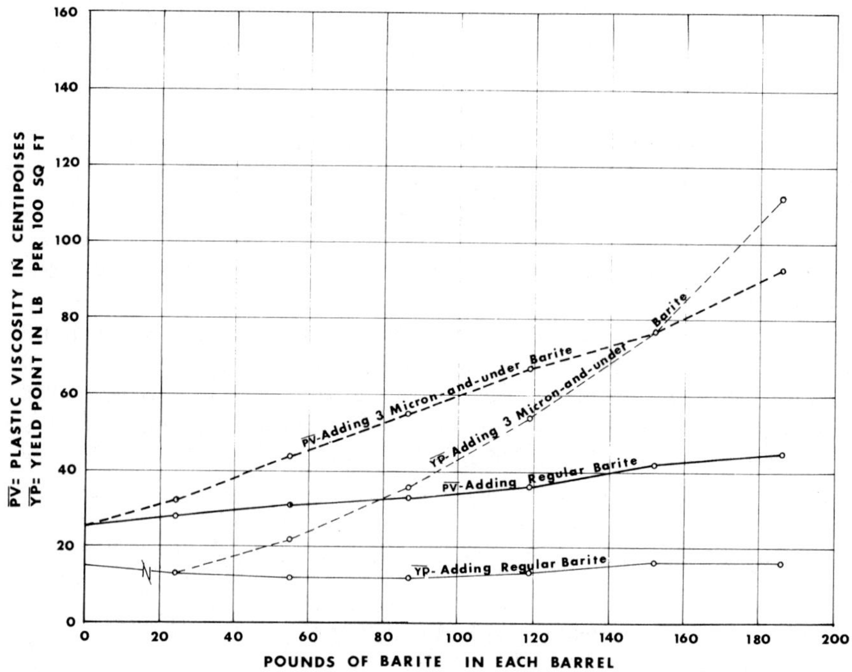

Fig. 12-4. The effect of particle size of solid particles on mud flow properties.

TABLE 12-II

Effect of particle size and shape on surface area *

Particle diameter or equivalent diameter (microns)	Type of particles	Surface area (sq ft/lb)	Surface area (m^2/kg)
50.0	Glass spheres	234.5	48.0
5.0	Glass spheres	2,345	480.4
1.0	Glass spheres	11,725	2,402
1.0	Crushed quartz	17,160	3,515
0.5	Glass spheres	23,450	4,804
0.5	Crushed quartz	34,320	7,030
0.1	Glass spheres	117,250	24,018
0.1	Crushed quartz	171,600	35,151
0.003	Silica gel	2,930,000 *	600,192 *

* Includes interior area of sponge-type pore openings.

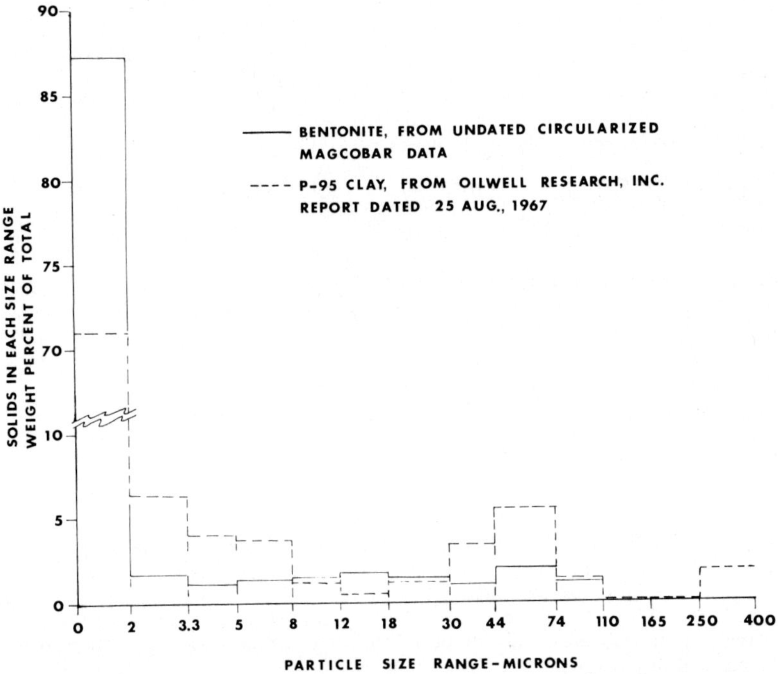

Fig. 12-5. Distribution of dry commercial bulk bentonite and P-95 clays.

"concentrating" the solids (including bentonite) already in the system. The effect (an increase in plastic viscosity and yield point) can be the same as adding bentonite, even though the solids added may not be efficient viscosifiers like bentonite and increase in these properties may not be desired or even expected.

This also explains why it may be necessary in weighted muds to remove fine solids to reduce viscosity and, at the same time, to add bentonite or some other high quality commercial product to control fluid loss. Viscosity consideration often will not permit the addition of any more of the colloidal solids necessary to control filtration, unless total solids surface area is first reduced by removing a portion of the existing clays, most of which may be of poor quality (i.e., having low degree of swelling). One should be aware, however, that some drilled formations contain clays very beneficial to the mud system.

Soft flaky minerals reduce most readily to clay size, and most clays are of a flatter shape than crushed quartz, ground barites, etc. Behavior of bentonite, which may be finely ground and sized (Fig. 12-5), after the first two hours in fresh water indicates that it undergoes a large increase in available

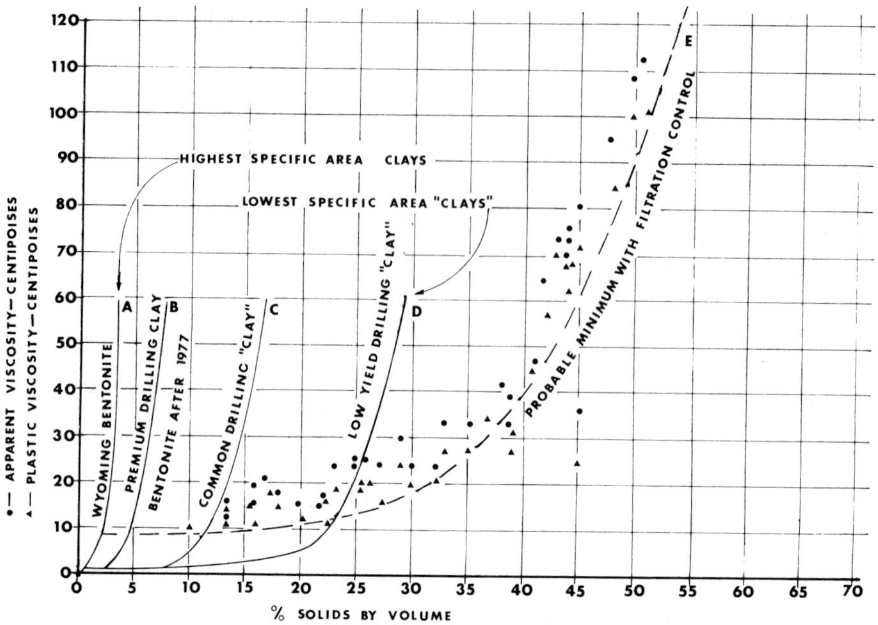

Fig. 12-6. The effect of particle size and shape of solids (surface area) on relationship between viscosity and solids content.

surface area, manifested by increase in viscosity and reduction in water loss.

Curves *A*, *B*, *C*, and *D* of Fig. 12-6 are adapted directly from the "Typical Clay Yield Curve" found in "Principles of Drilling Fluids Control" (Petroleum Extension Service, 1969). Curve *A* shows that high-quality bentonite has more surface area than can be hydrated by (adsorb) 95% water by volume. Curves *C* and *D* show that some materials marketed as "clay" have questionable value as viscosifiers (viscosity builders), apparently having a low specific surface area.

Since around 1977, published clay yield curves have omitted the "Wyoming Bentonite" curve shown in the older curves, because high-quality bentonite is very scarce at the present time. The first curve now is the one labeled in Fig. 12-6 as "premium drilling clay". It is now commonly labeled "attapulgite and bentonite". "Bentonite" or "gel", as now purchased, is sometimes a mixture of a low-grade bentonite and a polymer, or even polymers mixed with attapulgite or sepiolite. Consequently, results are not always predictable, due to polymer reactions to hole conditions.

Curve *E* of Fig. 12-6 probably shows approximately minimum plastic viscosity possible at any given percent solids content in a water-base mud, having filtration rate of less than 15 cm^3 as determined by API standard low-pressure, low-temperature test. It is based on the field data points which are

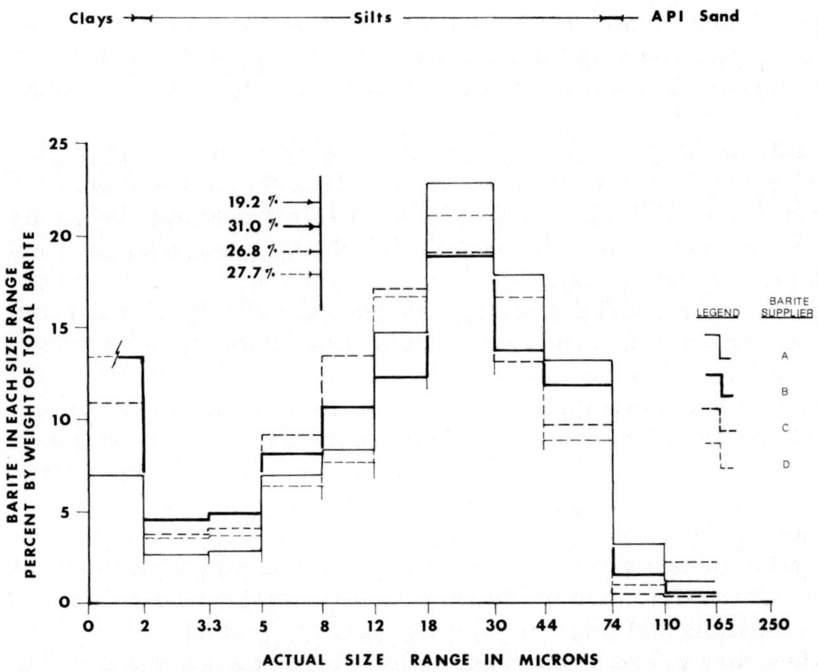

Fig. 12-7. Typical particle-size distribution of API-grade barite.

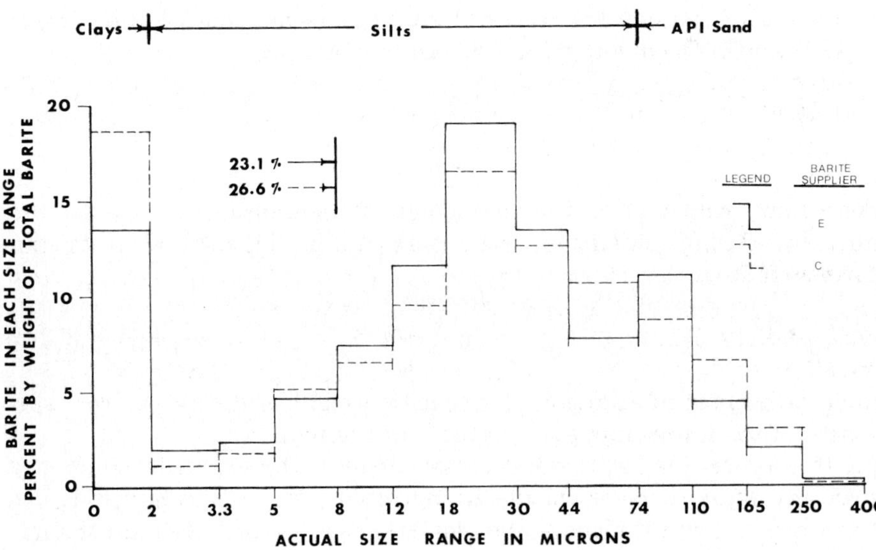

Fig. 12-8. Particle-size distribution of special grinds of barite from Iran.

the minimum found in over 1000 field tests screened for validity and covering all mud types. The only points below the curve represent muds which were mixed quickly to control a "kick", with little or no control of filtration.

If the plastic viscosity of a water-base mud is significantly higher than values of Curve E, it can be reduced by reducing the colloidal content, whether or not the total solids content is reduced. If the viscosity values are near Curve E, the plastic viscosity can be reduced only by reducing the total solids content. This latter may or may not be possible, depending on the necessary slurry density and the low-specific-gravity solids (light solids) content. The size distribution of the light solids present determines the economics of reducing total solids content.

The particle-size distribution of barite, milled to meet API specifications, is somewhat similar all over the world. A maximum of 3% by weight is allowed to be of API sand size and at least 5% must be too large to pass through a U.S. Standard No. 325 screen, i.e., greater than 44 microns. The latter requirement does not mean that large particles are desirable, but rather insures that the average grind is coarse enough to minimize the content of colloidal particles. Variation in the amounts of barite in the colloidal range is sometimes noticeable and sometimes creates viscosity problems, due to the specific surface area effect. Typical variations in barite particle size distribution from suppliers of API-grade barites are shown in Fig. 12-7. Usually barite suppliers add phosphates, tannins, lignins, etc., either at the bulk plants or at the mill, to counteract the viscosity-increasing effect of the water adsorption by the colloidal barite fraction. Special particle-size distributions of barite ground for use in Iran to achieve some of the highest drilling mud weights in the world are shown in Fig. 12-8.

STOKES' LAW AND SEPARATION OF SOLIDS

Stokes' Law, which gives the mathematical relationship of the factors governing the settling velocity v_s of a sphere in a liquid is stated in its simplest form as follows:

$$v_s = \frac{d_s^2(\gamma_s - \gamma_l)}{18\mu_l} \qquad (12\text{-}2)$$

where: d_s = diameter of a sphere, γ_s = specific weight of the sphere, γ_l = specific weight of the liquid, and μ_l = viscosity of the liquid.

Using this expression, or a modification thereof, the settling velocity of a sphere in any liquid can be calculated and predicted. In addition, one can allow any object to settle in a liquid, measure its rate of fall, and calculate the size of a particle which would settle at that rate. This gives an "equiva-

tain them it is weighted. It is important to note here that as the change is made from an unweighted to a weighted mud system, the philosophy (or logic) of solids treatment also must be changed. As shown in Table 12-III, the problems change almost completely.

FACTORS GOVERNING TOTAL SOLIDS CONTROL IN WEIGHTED MUDS

Although great improvement has been made in solids removal techniques in the last twenty years, in weighted muds there are overriding factors that cause total solids to vary. If the reasons are not understood, the blame (or credit) may fall on the type of mud, supervision, removal equipment, etc., when variation in total solids content actually may be caused by some other factor listed in Table 12-IV.

A popular concept is that the total solids content in weighted muds usually will be high, unless large sums of money are spent to keep it low. The lowest total solids contents in weighted muds usually occur when medium-hard formations are being drilled with long-tooth bits. If large cuttings can be brought out of the hole without being broken up, any shale shaker that is not bypassed will remove a high percentage of the cuttings, resulting in low total solids content.

Unconsolidated formations drilled with weighted muds usually have a large content of silt and fine sand. Inasmuch as these particles cannot be separated from the bulk of the barite, a relatively higher total solids content will be the result.

Slow drilling in holes of normal size, whether in hard formations or not, gives rise to fine cuttings. Total solids content when drilling slowly with weighted muds will be higher than when consolidated formations are drilled at faster rates with long-tooth bits. The amount of money spent on the mud system will not basically alter the facts stated in this and the preceding two paragraphs.

BASIC MUD SLURRY

The following basic equations relate contents of solid components of mud slurries. Literally countless mathematical expressions can be derived from these relations. Various published nomographs are based on these and similar equations:

$$\frac{100}{S_t} = \frac{F_s}{2.5} + \frac{F_b}{4.2} \qquad (12\text{-}1a)$$

$$F_t = F_s + F_b = 100 \qquad (12\text{-}4)$$

TABLE 12-IV

Factors which control total solids content in weighted muds (numbered in order of decreasing influence)

Item	Contributes to low total solids	Contributes to high total solids	
	First order of importance		
1a Type of formation	Medium hard Unconsolidated sand and gravel	Unconsolidated silt	Very hard
1b Bit cutter type	Long teeth	All Small diamond	Slow drilling
2 Mud density	Minimum	Above minimum	
3 Bit jet horsepower	Adequate	Inadequate	
4 Annular lift	Adequate	Inadequate (rare if hole is near gauge size)	
5 Rig shale shaker	Constant efficient operation	Bypassed	
	Second order of importance		
6 Full-flow hydrocyclone removal of sand	Effectiveness varies with formation and bit type	Not applicable	

7	Rig screen mesh	Fine *	Coarse *
8	Fine screen used to return to system part of the liquid, clays, and silts from hydrocyclone	Secondary separation does not further reduce solids	Variable increase in total solids content, but more than centrifuge salvage (see item 10 below); no viscosity reduction
9	Removing directly from system a fraction of clay and liquid with centrifuge, while maintaining weight and volume	Variable effect on total solids content, but good viscosity reduction	Primary separation cannot increase total solids content
10	Centrifuge used to return to system the liquid and clay from the hydrocyclone underflow (see item 6 above)	Secondary separation cannot reduce solids content; not normally recommended on water-base weighted muds	Variable increase in total solids content but less than screen salvage (see item 8 above); no viscosity reduction
11	Chemical treatment to prevent shale cuttings dispersion	Variable (but may help screen removal?)	Normally does not cause, but decreases viscosity to total-solids-content ratio
12	Chemical treatment to disperse shale to clay-size particles	Variable, but can help centrifuge removal of shale as clays	Normally does not cause, but increases viscosity to total-solids-content ratio

* Whether or not a finer screen will help noticeably in this primary separation depends upon the comparative size relation between the cuttings reaching the surface and the screen mesh, and whether or not the finer mesh can be maintained in proper operation. If a "fine" screen cannot operate properly at full flow, a coarser screen will maintain lower total solids content than a finer screen which is bypassed.

$$V_h + V_w + V_t = 100 \tag{12-5}$$

$$(100)(S_m) = (V_t)(S_t) + (V_w)(S_w) + (V_h)(S_h) \tag{12-6}$$

$$P_t = P_b + P_s \tag{12-7}$$

$$P_t = \frac{(V_t)(S_t)(8.34)(42)}{100} \tag{12-8}$$

$$P_b = \frac{(V_t)(F_b)(S_t)(8.34)(42)}{(100)(100)} \tag{12-9}$$

$$P_s = \frac{(V_t)(F_s)(S_t)(8.34)(42)}{(100)(100)} \tag{12-10}$$

$$S_m = \frac{\gamma_m}{8.34} \tag{12-11}$$

where: γ = specific weight of slurry (lb/gal, U.S.); F = weight percent of total solids; S = specific gravity with respect to water; V = percent of total volume; and P = pounds of solids in one API bbl (1 bbl = 42 U.S. gal). Subscripts: b = barite or other heavy material suspended in the mud to increase weight; s = low-specific-gravity (all non-heavy-weighting) solids, including dissolved salts; m = slurry of total mud components; h = oil; t = total dissolved and suspended solids (net dry); and w = water.

On assuming that $S_b = 4.2$; $S_s = 2.5$; $S_h = 0.85$; and $S_w = 1.0$, it becomes much easier to derive extremely useful equations from the above relationships, e.g.:

$$S_t = \frac{(11.99)(\gamma_m) - 0.15V_w - 85}{V_t} + 0.85 \tag{12-12}$$

$$P_t = (3.5028)(V_t)(S_t) \tag{12-13}$$

$$F_s = \frac{617.4}{S_t} - 146.9 \tag{12-14}$$

The variations between graphs and nomographs due to assumption of different specific gravities for the low-specific-gravity solids and for the barite are not important. The margin of error in solids determination far outweighs these minor differences. Great caution must be exercised, however, in using graphs that attempt to classify muds as "good" and "bad" on the basis of specific gravity relationships, because such graphs are marketing tools for selling mud products.

SHALE SHAKERS

The primary solids removal devices in their best order of series operation can be listed as follows: (1) shale shaker, (2) sand trap (or settling tank, or shale tank, etc.), (3) hydrocyclone desander, (4) hydrocyclone desilter, and (5) centrifuge. Regardless of improvement of drilling mud economics with the other mechanical devices, at present none of them operate as well without the protection of the shale shakers. The sand trap serves mainly to protect the downstream equipment temporarily if something happens to a shale shaker screen.

The term "shale shaker" is used in drilling mud work (API, 1973) to cover all the devices that in another industry might be differentiated as "shaking screen", "vibrating screen", and "oscillating screen" (Perry, 1950). Although all of these types are in use, most would probably fall in the "vibrating screen" classification. The old cylindrical rotating (or "squirrel cage") screen is only rarely in use at the present time.

The volume of liquid discharge that can pass through a vibrating screen depends upon: percent open area, mesh opening size, type of screen (weave), speed and amplitude of vibration, type of motion, mud flow properties, solids content of liquid discharge, feed rate of oversize solids, and type of oversize solids. Of these, the speed and amplitude, type of motion, and screen itself depend on equipment choice, whereas others are a function of the formation types penetrated and drilling practices. The order of importance of these variables depends on a particular situation.

The volumetric rate of solids discharge that can be handled depends on: type of screen motion, speed and amplitude, horsepower, type of solids, mesh design, and wire strength. The size range of solids that can be separated ("cut") is determined by: mesh opening size, type of screen motion, speed and amplitude, and screen mesh weave.

Type of motion is determined both by the eccentric weight or reciprocator ("vibrator") arrangement and by the suspension system. If either one is not in good condition, the shaker cannot perform properly.

Research on fine-screen separation and promotion of fine screens caused a general trend to much finer screens (from 12×12 and 16×16 to 50×50, 80×80, and 100×100 mesh) in the late sixties and early seventies (see Fig. 12-9). Although results were not always satisfactory, there was a great advance in knowledge (API Committee, 1974; Brandt, 1973). There are indications that a considerable advance in technology and application may take place in the late seventies. Under most conditions, finer separation can be accomplished by a square mesh than with the "equivalent" oblong mesh (see Fig. 12-9). Brandt (1973) cited an example of a square 40-mesh and an "equivalent" 40-mesh oblong screen, both of which separate approximately 420-micron dry material. In a 45-sec Funnel viscosity drilling mud, the number 40 oblong cloth separated particles down to approximately 300 microns, whereas the number 40 square cloth separated particles down to about 200 microns.

The new screen designation system adopted by both the IADC and API is presented in Fig. 12-10 and Table 12-V. These designations, which are

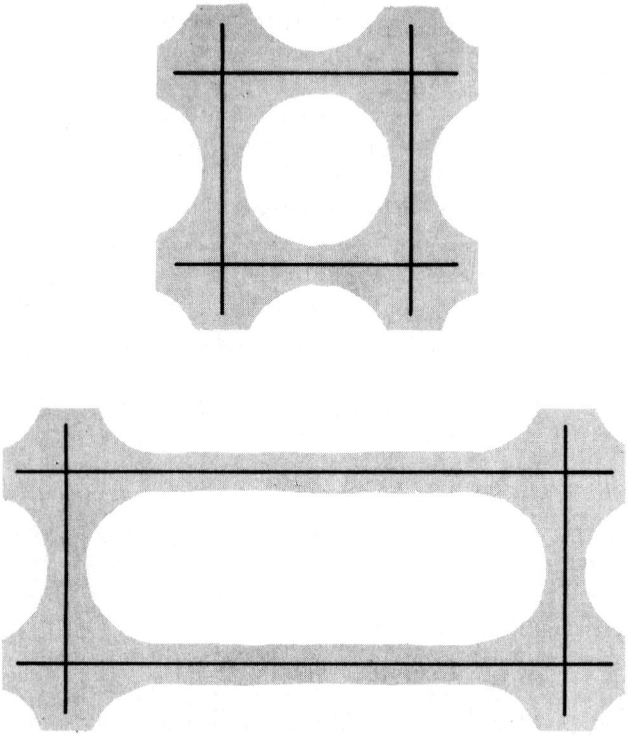

Fig. 12-9. Square and oblong "equivalent" screen meshes, with effect of wire coating. (After Brandt, 1973.)

described in API RP 13E, are much more informative than the simple mesh counts used before. Unequal mesh designations have been completely meaningless. In one part of the United States, an "oblong 40" screen customarily has been a 60 mesh by 20 mesh screen. Another shaker manufac-

TABLE 12-V

IADC—API designations of representative test screens

Test screen	API—IADC designation	Wire diameter calculated (microns)
U.S.S. No. 50	API 50 × 50 (295, 33.7)	213
U.S.S. No. 80	API 80 × 80 (177, 31)	140.5
U.S.S. No. 100	API 100 × 100 (149, 35.2)	102
Tyler No. 150	API 150 × 150 (104, 37)	65.2
U.S.S. No. 200	API 200 × 200 (74, 34.0)	53
U.S.S. No. 325	API 325 × 325 (44, 31.7)	34

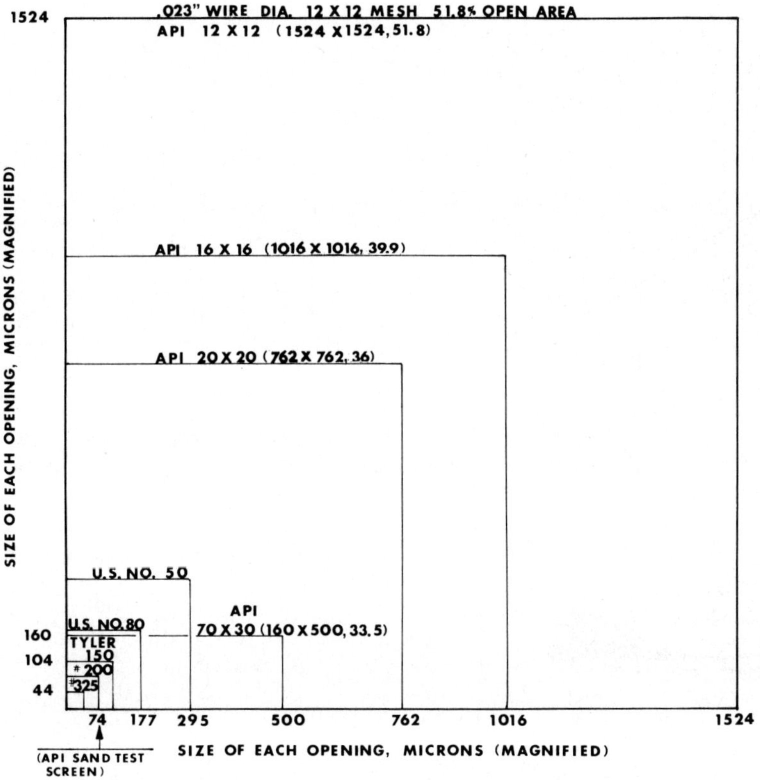

Fig. 12-10. Scale comparison of common screen cloth openings.

turer furnishes "B-100" which is indicated on his chart as an oblong mesh having 60 × 40 mesh count. It is highly recommended to use the API—IADC standard designation, which shows mesh count, opening in microns, and percent open area in that order.

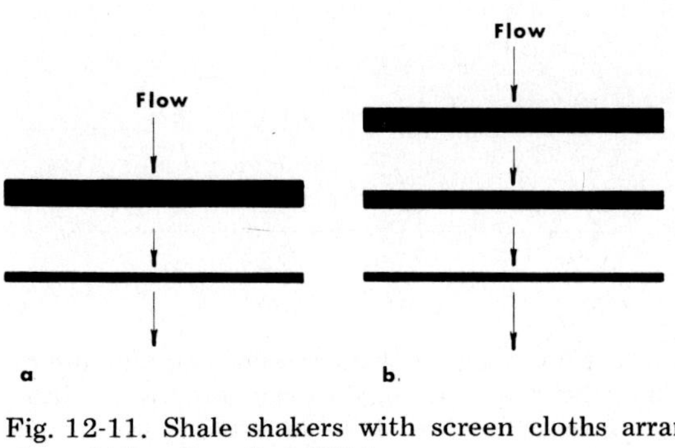

Fig. 12-11. Shale shakers with screen cloths arranged in series. a. Dual-stage. b. Triple-stage. (After Brandt, 1973.)

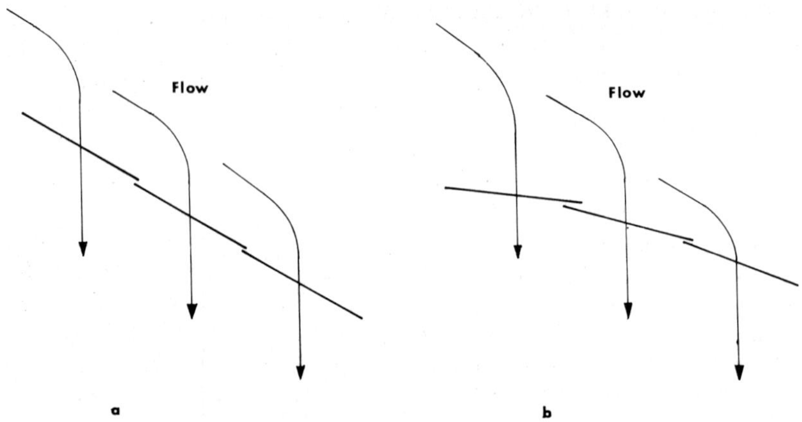

Fig. 12-12. Shale shakers with screen cloths arranged in parallel. a. Constant-angle deck. b. Differential-angle single deck. (After Brandt, 1973.)

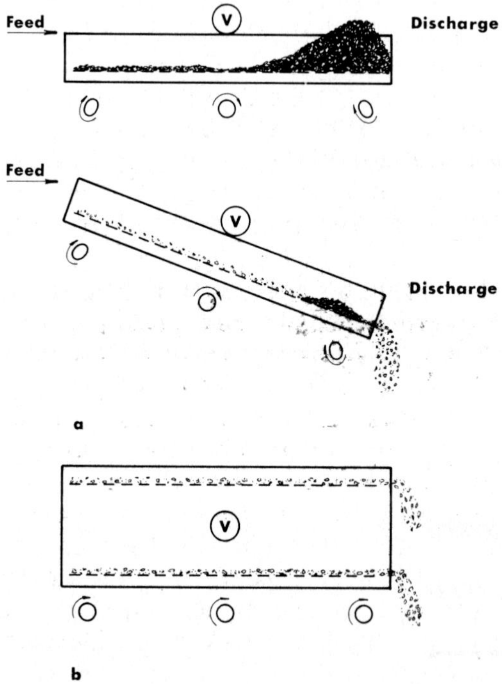

Fig. 12-13. Effect of shale shaker motion. a. Unbalanced motion creating combination of circular and elliptical motion and showing the deck inclined to overcome the resulting solids pileup. b. Balanced motion, with uniform circular motion and resultant even solids flow irrespective of deck angle. (After Brandt, 1973.)

If a shale shaker unit (one vibrating mechanism with screen cloth and all appurtenances) has multiple screens in a series arrangement, as shown in Figs. 12-11a and 12-11b, the particle-size separation will be determined by the finest mesh screen, which should be the bottom screen. If a shale shaker unit has multiple screens in a "parallel" arrangement (Figs. 12-12a and 12-12b), the size separation is determined by the coarsest screen in the unit through which liquid passes.

Although less effective motion can be overcome by sloping the deck towards the solids discharge end (see Fig. 12-13a), this may be at the expense of wetter, oversize solids carrying more drilling fluid over. The balanced motion, with the vibrator being located in the center of gravity of the vibrating mass, imparts a uniform circular motion capable of moving the oversize solids even slightly uphill, the transport being relatively independent of the angle (see Fig. 12-13b).

If mud properties and formation solids in the feed remain equal, liquid throughput capacity depends mainly on the transport of oversize solids from the screen cloth. If the mesh openings plug up with near-size particles, or if the mesh openings are coated over with sticky, undersize particles in a viscous mud, liquid throughput capacity becomes negligible and the screen cloth is plugged ("blinded"). For this reason selection of screen cloth mesh can be critical to the entire solids removal installation in a mud system.

If the screen installed creates a blinding problem for the shale shaker unit, a change to a higher percent open area is recommended, which may require a coarser mesh, if the problem is wire coating. If the problem is near-size plugging, mesh opening must be changed, and a smaller opening may solve the problem better than a larger one. The screen should never be bypassed, even if a 10-mesh cloth is necessary to pass the mud.

A shale shaker can be overloaded with solids even if it is properly designed for solids transport and fitted with a screen of proper mesh selection for the particle-size distribution. The amount of solids to be removed always varies with the drilling rate (hole volume) and the sloughing rate if the latter is occurring. If a solids overload occurs, it is also reflected in a reduction in liquid throughput capacity. A two-stage screen can help in such cases if properly used.

Two-stage screens in the field often are not used properly. One common practice is to keep a screen in only one of the two decks. The two-stage equipment is then used as a single-stage screen. Another poor practice is to use the same type of fine screen on both the upper and lower decks. The reason given is that the screens have such a short life that the bottom one is simply a "spare" for the top one.

Another questionable practice is to use a fine screen on the top deck and a coarse screen on the lower deck. Various reasons have been given, a typical one being: "When the fine screen breaks, at least the coarse screen is there". A more logical but insufficient excuse is: "The fine screen can be checked and changed more easily on top".

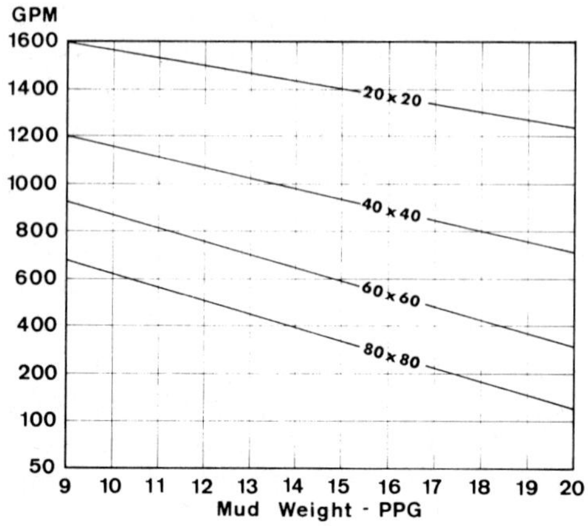

Fig. 12-14. Effect of mud weight on flow rate through screen cloth. Single screen or bottom screen of a series. (After Brandt, 1973.) PPG = lb/gal, GPM = gal/min.

A more serious problem exists when the top screen is so coarse that it does not remove any solids. The lower fine screen then performs as a single-stage screen. The upper screen should be fine enough to remove a significant fraction of the solids load. Field practice indicates that if approximately 75% of the surface area of a screen is needed to pass the slurry through, a finer screen can not be used on that deck.

Figure 12-14 shows the effect of slurry weight on the liquid throughput capacity of a screen. It is general for the finest mesh cloth of screen in a series, or of a single screen, because the effect of cuttings load is omitted. The effect portrayed here is a combination of that of viscosity, fine particles, and the reduction in effective screening area for some typical meshes (Brandt, 1973).

The frequency and severity of screen overload problems (bypassing) is due in great measure to the fact that the effect shown in Fig. 12-14 and the drilling effect are additive. Screen liquid capacity decreases with increasing drilling rate and mud weight and with decreasing mesh size. In addition, as cuttings size decreases toward mesh size and as viscosity increases, the liquid capacity decreases. If the choice of the shale shaker screens for an operation is left to someone familiar only with the usual "catalog specifications" and the most adverse combination of shale shaker conditions possible are not considered, the shale shakers may be inadequate. They must be bypassed. As a result, mud costs go up and drilling efficiency suffers. Before a person unfamiliar with a particular rig operation specifies a certain screen mesh, he should seek the advice of the toolpusher or the mud engineer, but preferably both.

From casual observation of conventional shale shakers operating with screen cloths having mesh coarser than 30 × 30, it has been assumed for years that shale shakers remove solids with relatively small loss of mud. Some preliminary investigations (Roper, 1972), however, have strongly indicated that mud loss ranges from one gallon of mud per gallon of net dry fresh cuttings removed under the best conditions and using coarser screens, to four gallons of mud lost per gallon of net dry fresh cuttings removed under severe conditions and with very fine screens. This range of mud loss was found with primary removal screens operating normally and not blinded by coating or plugging.

Some shale shaker designs no doubt are superior to others in removing excess liquid from oversize solids. For example, a screen cloth over which the solids move uphill should logically salvage more liquid mud (from the solids surface area) through the screen than the one moving the oversize solids downhill; however, no published results of any investigation of the difference are available. In the case of heavy solid loads any difference would tend to be lessened.

At the present time, various new models of shale shakers are being introduced. Evaluation, however, is not complete, because field testing and proving takes a long time. SWACO, in the U.S.A., has introduced a full-flow, primary shale shaker having circular shape. Montello Inc. has introduced (Cagle and Wilder, 1978) a differentially-angled deck shale shaker with a "sandwich screen" in three layers, designed to reduce near-size plugging. The Thule United shale shaker with variable-speed, hydraulically-driven vibrator is being tested in the North Sea, whereas a variable-speed, air-driven shale shaker is reported to be undergoing testing in the U.S.A.

Economic limits exist for all methods of separation. As conditions become more severe, the shale shaker design principles and construction details become more critical, and maintenance costs climb rapidly. Individuals responsible for specifying shale shakers must familiarize themselves with the field operating characteristics and economic limits of the various separation equipment available.

OPERATING PRINCIPLES OF SAND TRAPS

If the shaker screen never develops a tear that could pass the oversize solids through and is never bypassed during drilling, the major justification for a sand trap ("shale trap" or "settling tank") would disappear. Inasmuch as such perfect conditions are not attainable at the present time, the sand trap is extremely important (see Ormsby, 1965, 1973). The major operating principles of sand traps can be summarized as follows:

(1) The sand trap, which usually receives the liquid slurry passing through the shale shaker, should receive all liquid slurry bypassing the shale shaker and flowing into the active mud tanks.

(2) Inasmuch as a sand trap is a gravity-settling compartment, it must not be stirred and/or be used as a suction compartment for any removal process.

(3) A sand trap must have a discharge control assembly, which can be easily and quickly opened and closed, so that the settled solids can be dumped with a minimum loss of mud.

(4) The sand trap should be only dumped and not "washed out". If the bottom is not sloped to the solids pile angle, the settled solids should be allowed to form their own sloped sides. "Cleaning the bottom", other than possibly at moving time, only increases mud loss and mud cost.

(5) Inasmuch as Stokes' Law applies in a sand trap, large quantities of barite (as well as API sand) may be settled from weighted drilling fluids. Provisions for bypassing the undersize screen discharge slurry from the carrying pan directly to the next processing compartment is advisable when using weighted muds. All compartments except the sand trap are stirred in a well-designed active system to prevent settling out of barite. The sand trap must not be bypassed if there is any problem with the shale shaker.

(6) The mud exit from the sand trap should be over a retaining weir to a stirred compartment.

Incorporation of a proper sand trap into a solids removal system increases the utility of the other removal equipment and often enables continuation of drilling under unexpectedly severe solids loading conditions.

DEGASSERS

Degassers sometimes are essential to the solids removal process, even though shale shakers remove a good portion of the gas from a badly gas-cut mud, especially if the yield point is as low as 10 lb/100 ft^2. Generally, special degassing equipment is not necessary if the yield point of the mud is $\leqslant$10 lb/100 ft^2.

Slurry-handling centrifugal pumps feeding abrasive oil-field muds to hydrocyclones lose efficiency when pumping gas-cut muds. They will "gas-lock" when gas content exceeds 3% by volume, according to Mission Manufacturing Co. Hydrocyclones do not function properly if pumping (feed) head is not constant or if there is gas or air in the feed. Provision for degassing equipment, therefore, should be made between the sand trap and the first hydrocyclones (Ormsby, 1973).

A relatively new and widely used degasser patented and sold by DRILCO is quite efficient in most cases. Based on tests by the author on a similar equipment, and on reports from users of the DRILCO degasser, the following statements can be made: If the yield point and gel strengths of the mud are low ($\overline{YP}$ below 10 lb/100 ft^2), the DRILCO degasser leaves very little gas or air in the mud. If, on the other hand, the yield point and gel strengths are high ($\overline{YP}$ above 15—20 lb/100 ft^2), the DRILCO degasser leaves some gas and/or air that is objectionable to some operators and not to others.

If the yield point is extremely high, it may not be possible to reduce the gas content sufficiently with the DRILCO machine and, in fact, it may entrain significant quantities of air in a mud which was initially air- and gas-free.

HYDROCYCLONES

Hydrocyclones can accomplish the finest separation of any primary separation equipment normally used on the full-flow circulating rate of an unweighted mud system. The hydrocyclone (see Fig. 12-15) has a conical-shaped portion in which most of the settling takes place. Usually, a cylindrical feed chamber, sometimes called the "barrel", is located at the large end of the conical section. At the apex of the conical section is the underflow opening for the solids discharge. In operation, the underflow opening is usually facing downward for convenience, and "bottom" or "top" as used in this chapter concerning hydrocyclones refers to this type of arrangement (position).

Near the top end of the feed chamber, the inlet nozzle is situated tangential to the inside circumference and on a plane perpendicular, or nearly so,

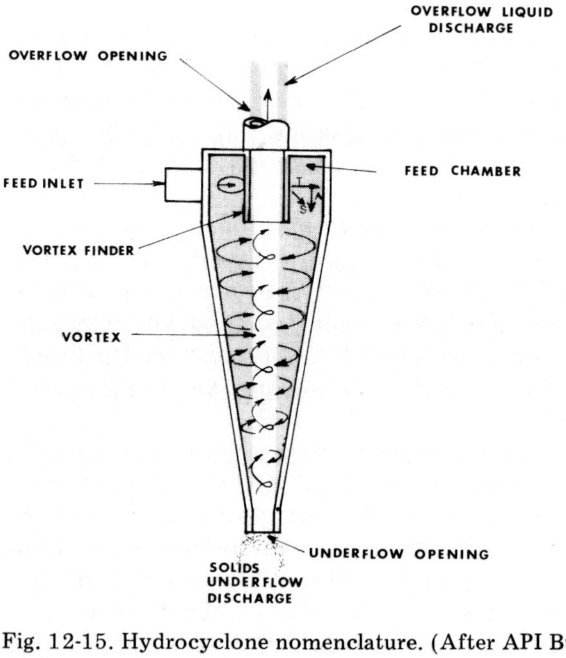

Fig. 12-15. Hydrocyclone nomenclature. (After API Bull. 13C.)

to the top-to-bottom central axis of the hydrocyclone. A hollow cylinder, called the "vortex finder", extends axially from the top into the barrel of the hydrocyclone past the inlet. The inside of the vortex finder forms the overflow outlet for the liquid discharge or effluent. The overflow outlet is much larger than the underflow opening. The "size" of a hydrocyclone is determined by the largest inside diameter of the conical portion. All dimensions are critical to the operating characteristics of a hydrocyclone having any specific design and size.

The hydrocyclone owes its centrifugal force to the tangential velocity of the slurry entering the feed chamber. An axial velocity component is created by the axial thrust of the feed stream leaving the blind annular space of the feed chamber. As a result, there is a downward spiraling velocity represented by S in the velocity diagram in Fig. 12-15, in which T represents the tangential and A the axial velocity components.

Balanced design of hydrocyclones

In a hydrocyclone, the incoming stream spirals down along the wall of the conical section to the underflow opening, reverses axial direction, and then spirals upward to exit at the overflow opening. A properly-designed hydrocyclone can be adjusted in such a manner that when a clean liquid is fed into it, there is no underflow discharge. When separable solid particles are fed into such a hydrocyclone, solids are discharged at the underflow opening, each particle carrying a liquid film covering its surface area.

The solid particles in the turbulent, downward-spiralling feed stream are settled outward toward the wall, following modifications of Stokes' Law. When the liquid stream reverses to spiral upward, the solid particles continue spiralling downward through the underflow opening due to their greater mass and higher inertia forces (Fig. 12-16). They discharge at high velocity.

When a balance-design hydrocyclone is adjusted with the underflow opening too small in relation to the balance-point opening, a "beach" of dry cone wall exists between the edge of the actual underflow opening and the balance point, or the point of liquid turnback. If the hydrocyclone is capable of settling silts, this adjustment or design will cause sticking and plugging, because fine silts stick as they lose their liquid film on traversing the beach. This undesirable adjustment is described as "dry bottom" and the trouble it causes is known as "dry plugging".

If a balance-design hydrocyclone is opened to an underflow opening greater than the balance point diameter, a cylindrical shell of spiralling liquid will discharge unless there are enough separable solids to occupy the unbalanced discharge space. This adjustment is termed "wet bottom" and, unless excessive, it usually does not, contrary to general belief, result in high mud losses under normal drilling conditions. Actually, a wet bottom adjustment results in cheaper and better drilling fluid, having better hole-conditioning

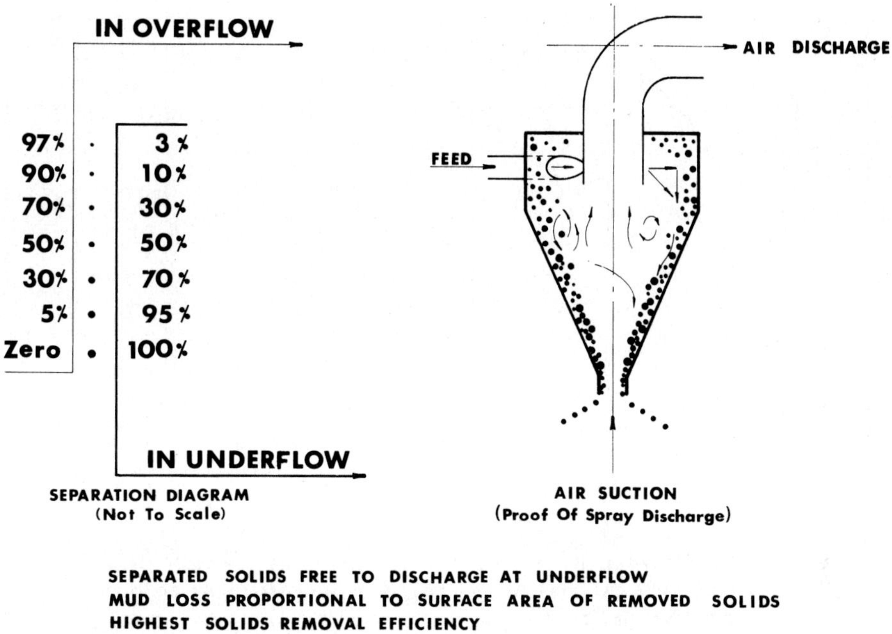

Fig. 12-16. Balanced-design hydrocyclone in spray discharge.

properties as compared to a dry-bottom adjustment or dry-bottom design. The ideal balance point adjustment is not always known and is not absolutely necessary. Environmental problems may dictate a dryer discharge which can be obtained with a secondary separator, discussed later in this chapter.

Figure 12-16 depicts a balanced-design hydrocyclone operating in spray discharge. If the air is being sucked in at the underflow opening, there is a spray discharge; this can be felt with the end of a finger for absolute verification. The air is simply entering the low-pressure zone created by removal of the air from the vortex by the friction of the high-velocity, upward-spiralling inner stream, which takes the vortex air out with it through the overflow opening. The flow rates in mud hydrocyclones are so high that total retention time, from inlet to either exit, is less than one-third of a second.

The opposing flows of two fluids through the underflow opening in spray discharge [air moving (entering) upward and a liquid—solids slurry moving (leaving) downward], indicate that the underflow opening is not acting as a choke. The underflow opening acts as a circular weir (ring dam), used in liquid—liquid separation such as in a cream separator. As long as the underflow opening acts as a circular weir, the hydrocyclone has capacity to discharge all solids that can settle to the wall and move to the bottom. Only in this manner can a hydrocyclone remove solids at its highest possible effi-

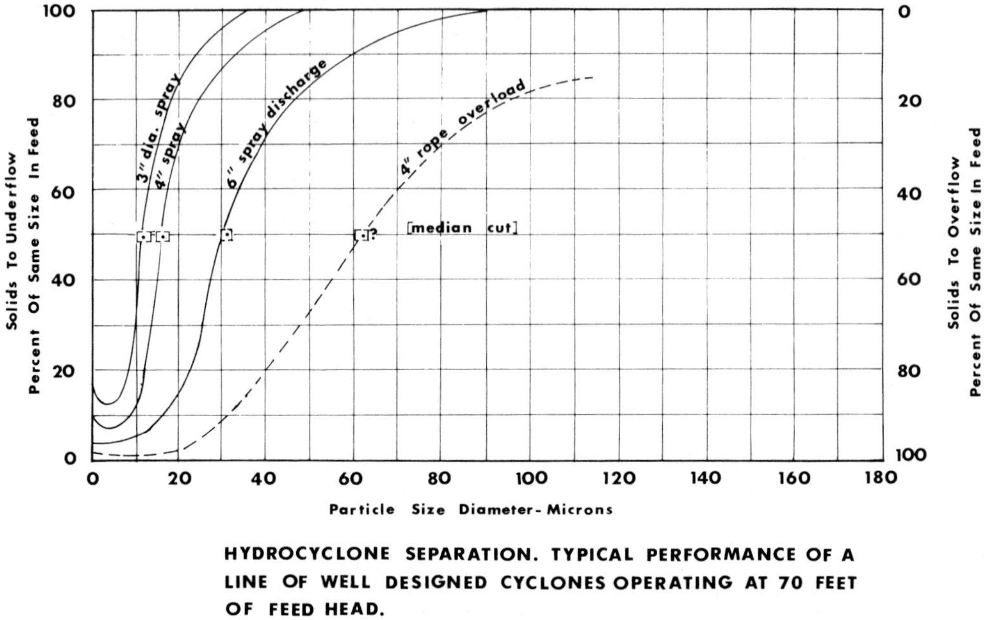

Fig. 12-17. Typical separation performance of a hydrocyclone.

ciency while saving maximum volume of liquid.

There are two spiraling countercurrent turbulent streams, one moving downward near the cyclone wall and the other moving upward near the vortex. Eddy currents are created at the boundaries of these two streams as shown in Fig. 12-16. These currents sweep up some solid particles from near the wall and move them towards the center (upward moving stream). Some particles are too massive to be swept up, whereas others are so small that they seldom get out the bottom. Particles too small to be separated are found in the free liquid phase of the feed, underflow, and overflow in about the same ratio. The separation curve for a hydrocyclone varies with design, size, feed head, specific gravity and shape of solids, solids content in the feed, feed viscosity, and adjustment. Separation curves shown in Fig. 12-17 are presented for illustration purposes only. These curves are valid only for one particular design and one set of test conditions.

When a cyclone delivers more solids to the underflow than can pass through its opening, the cyclone is "overloaded" and a balanced-design underflow will assume a shape of a rope or of link sausages. Thus, the terms "rope discharge" and "sausage discharge" have been coined for the overload condition. There is no air suction at the underflow opening in rope discharge and, in a balanced design, this proves the existence of overload (see Fig.

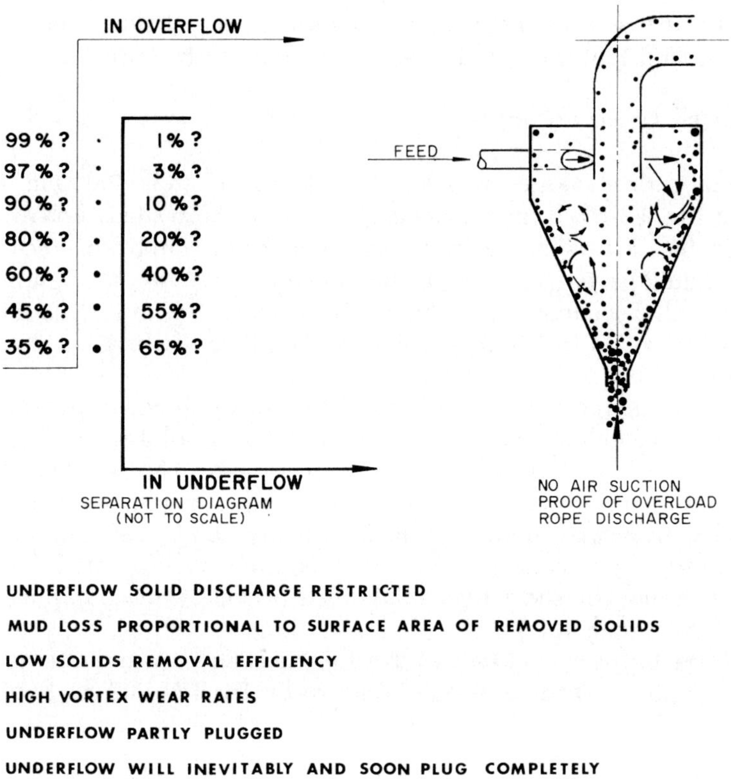

Fig. 12-18. Hydrocyclone operated in rope discharge (balanced design).

12-18). In the overload condition, the underflow opening is no longer a weir, but acts as a choke.

In the rope discharge, not all of the particles arriving at the bottom of the hydrocyclone are discharged. The smallest ones, which have the least chance of getting out of the overloaded bottom, return instantly to the top and leave with the overflow. This increases the overflow slurry density, causing the weight of the mud system to increase continuously and more rapidly than when the underflow discharge is of a spray type.

Smaller particles have more surface area per unit weight than coarse ones. Inasmuch as in rope discharge the smaller solids leave at the top, the rope underflow has less surface area per unit weight of solids than does a spray discharge. The rope underflow has less liquid per unit weight of solids and has a higher specific weight than a spray type underflow, because the liquid film accompanying the removed solids is a function of their surface area.

In the above discussion of rope and spray discharge, it was assumed that the solids in the feed of both cyclones are of the same general type, shape, and size distribution, and that the liquid phase of the mud is the same.

Choke-bottom design of hydrocyclones

Another design of hydrocyclones, which is described as a "choke bottom" design, is shown in Fig. 12-19. It has no balance point and both the overflow and underflow openings act as chokes at all times under all conditions. For example, if clear liquid is fed to a choke-bottom cyclone with a 10-mm bottom opening, it will discharge a forced underflow stream 10 mm in diameter. Adding separable solids to the feed will cause solids to discharge in the underflow, with little effect on the discharge volume rate.

Liquid underflow discharge is not a function of solids discharge unless an overload condition is reached so that not all of the arriving solids can leave with the underflow. The best adjustment when operating a choke-bottom hydrocyclone will be a compromise between the minimum opening required to prevent underflow discharge of all mud during the low-solids feed periods (during making drillpipe connections), and the maximum opening required to remove cuttings during the short high-solids feed period (while pumping the kelly down in fast-drilling hole).

Many of the choke-bottom cyclones in the field actually can operate as balanced cyclones if the overflow headers are reconstructed. When two

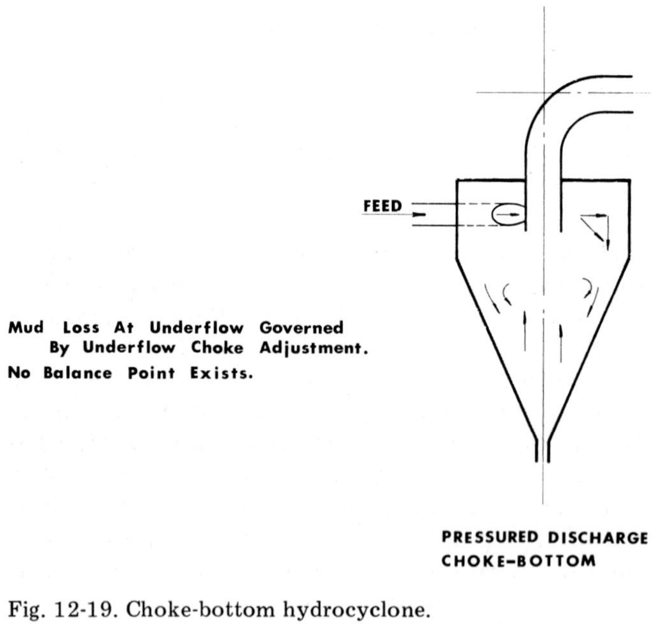

Fig. 12-19. Choke-bottom hydrocyclone.

hydrocyclones on opposite sides of the header have their overflow pipes pointing directly and straight at each other, a high shear and back-pressure situation is created. This can cause choke-bottom performance of even-balanced-design cyclones. Although expensive, it is economically advantageous to rework the overflow header so that each cyclone overflow slants downstream into the header at a 45° angle. This reduces the back-pressure and will permit balanced cyclone operation in most instances.

Hydrocyclone capacity

As with the shale shakers and all other mechanical separation devices, the hydrocyclones have dual capacity: *feed* capacity and underflow *solids discharge* capacity. Although the two capacities are related as far as design is concerned, the equipment available commercially has extreme variations in comparison to each other.

"Feed capacity" is the volume rate of feed slurry that a specific hydrocyclone will accept at a specific feed head, which may be expressed in feet or meters. Head can be expressed as pressure only at a specific feed slurry weight.

As an example, a hydrocyclone of certain size and make is rated as having a feed capacity of 100 gpm (gal/min) at 75 ft of head. If the feed were water, which has a hydrostatic gradient of 0.433 psi/ft, the ideal feed pressure would be 32.48 psi. If the feed slurry weight were to increase to 16.68 lb/gal (specific gravity = 2.0), the hydrostatic head gradient would be doubled to 0.866 psi/ft, and 75 ft of feed head would be equivalent to a pressure of 64.95 psi. The volume feed rate of a 16.68-lb/gal slurry at 64.95 psi would be close to that of a 8.32-lb/gal slurry at 32.48 spi, because the head is the same.

Centrifugal pumps, operating at constant speed (electric motor or closely governed engines) and feeding the same equipment, automatically maintain a constant head as slurry weight changes. Thorough knowledge of the subject is strongly recommended to anyone operating or installing hydrocyclones (see Ormsby, 1973).

Inasmuch as pumping mud through a hydrocyclone is expensive and is of no economic or physical benefit, the *solids removal ability* and not the feed capacity is the real key to the value of the equipment. Solids removal ability consists of (1) the high separation forces and internal flow pattern to settle finer particles, which are the most difficult ones to handle, to the underflow opening; and (2) a large balanced underflow opening so that the cyclone can discharge the settled solids at a maximum rate before reaching solids overload. If a balanced cyclone does reach the overload condition, the underflow will assume a rope shape and removal ability is minimized as shown in Fig. 12-17. Obviously a hydrocyclone with a balanced design can be operated economically with a much larger opening than the one having a choke-bottom design.

TABLE 12-VI

Mohole hydrocyclone evaluation based on timed and measured data taken during tests

Cone		Feed (gal/min)	Rating A	Rating B	A × B
make	diameter (inches)		UF ** solids, as vol. rate % of feed solids	UF solids, % by wt of total UF solids	overall rating
W	3	16.76	11.72	56.25	670
X	2	22.03	11.69	49.30	575
W	4	57.12	8.94	56.83	507
X	6	108.39	5.95	47.26	281
Y	4	62.65	4.55	47.96	218
W *	6 *	88.73 *	3.28	59.31	194
Z	4	44.90	1.55	65.25	101
X	12	407.00	1.80	54.46	98
Z	8	167.70	0.746	66.70	49.8

* Company "W" requested, prior to any evaluation work, that their 6-inch cone test be deleted as it was not taking the proper feed volume due to an undetermined malfunction; however, request was denied.
** UF = underflow.

A series of hydrocyclone performance tests were run in 1964 under the auspices of the National Science Foundation using facilities of the "Mohole Project", with the actual designer or head engineer responsible for field performance, adjusting his own hydrocyclone in each test. Since that test series, at least one of these cyclones has been discontinued in the oilfield market, at least three have been modified, and at least three new models have been introduced by manufacturers. The differences between designs and operations of various hydrocyclones and the dilemmas pointed out by these tests, however, still provide valuable insights.

Table 12-VI presents summaries of performance results, in order of decreasing solids removal capacity (Mohole, 1964). Rating "A" gives top rating for removing the highest percentage of solids entering, whereas rating "B" penalizes for liquid in the underflow. These two contrary ratings multiplied by each other give an overall ("total") performance rating, which surprisingly was in almost exactly the same order as rating "A" (i.e., vol. rate % of feed solids removal only). This arrangement is in the approximate order of practical assistance to a field engineer desiring to maintain minimum solids content as economically as possible in his unweighted mud system. The cyclones having higher ratings will maintain a low mud weight more economically than those having lower ratings, given the same feed materials.

In Table 12-VII the data have been rearranged in the approximate order of increasing purchase cost per volume rate of feed. In specifying new equip-

TABLE 12-VII

Mohole hydrocyclone evaluation, arranged in order of increasing cost per volume rate of feed

Cone make	Cone diameter (inches)	Feed (gal/min)	UF solids, as vol. rate % of feed solids	Total rating
X	12	407.00	1.80	98
Z	8	167.70	0.746	49.8
W *	6 *	88.73 *	3.28	194
X	6	108.39	5.95	281
Y	4	62.65	4.55	218
W	4	57.12	8.94	507
Z	4	44.90	1.55	101
X	2	22.03	11.69	575
W	3	16.76	11.72	670

* Malfunctioning.

ment, this "initial cost" often determines the choice. The planner often does not have field experience and usually is forced to keep the initial equipment cost down for total processing (feed rate) at some specific maximum rig pump capacity. The cost per unit of feed rate capacity is inversely proportional to the cost per unit of solids discharge ability.

There are big differences in performance between hydrocyclones of the same size, as shown in Table 12-VI. In Table 12-VIII the cyclones are pre-

TABLE 12-VIII

Mohole hydrocyclone evaluation presented in original order (see Table 12-VI), but arranged in columns to show the effect of design

Cone manufacturer and size				Feed head (ft)	Total rating = ($A \times B$)
W	*X*	*Y*	*Z*		
3				76.1	670
	2			105.7	575
4				76.1	507
	6			76.8	281
		4		76.8	218
6 *				76.1	194 *
			4	67.1	101
	12			67.1	98
			8	80.7	49.8

* Malfunctioning.

sented in the same order of rating as in Table 12-VI, but they are tabulated for different manufacturers (different designs). As shown in Table 12-VIII, smaller cyclones of the same design have greater separation ability, but different design can cause even more difference than size. There is no substitute for observing equipment performance in the field and specifying equipment based on that performance.

Hydrocyclone plugging effects

Although the results of underflow plugging in hydrocyclones are as disastrous as blinding of shale shakers, the effects sometimes are not as obvious. Figure 12-20 illustrates the change in balanced hydrocyclone operation from normal to plugging condition at the underflow. A cyclone plugged in this manner, however, offers no clue to the damage it is inflicting. It is nullifying the cleaning work of all other cyclones on the unit by introducing unprocessed mud directly into the cleaned overflow mud from other cyclones on the same header. At the same time, the inner parts of the cyclone and the overflow pipe are rapidly being destroyed by the particles that should have

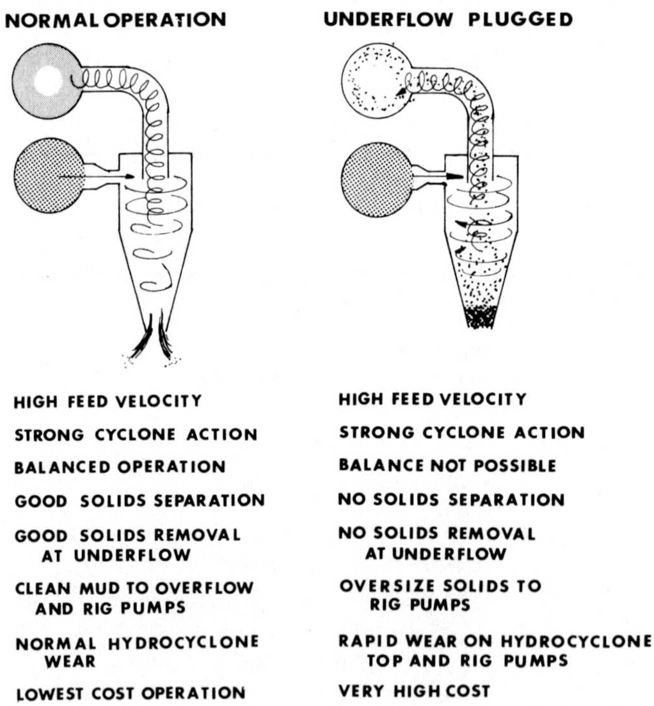

Fig. 12-20. The effects of plugging the underflow discharge of hydrocyclones.

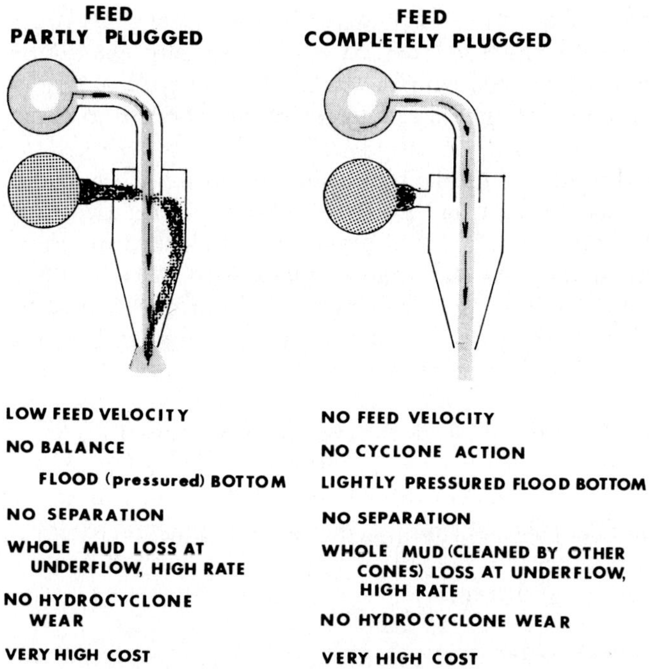

Fig. 12-21. The effects of plugging the feed inlet of hydrocyclones.

been discharged at the bottom. Thus, a plugged cyclone must be unplugged or removed from the manifold immediately.

Underflow plugging is usually caused either by the presence of a dry beach or solids overload. The dry beach effect can be due either to improper design (dry bottom) or too small an adjustment at the underflow. Solids overload can be overcome by (1) adjusting to a larger underflow opening, if possible, (2) installing more cyclones in the stage, (3) removing more solids in a stage ahead of the overloaded hydrocyclones, or (4) by purchasing cyclones with larger balanced underflow openings.

Figure 12-21 shows the effects of feed plugging. With partial plugging, although there is a mud inflow, the velocity is low in the feed chamber. The cyclone body acts as a swirling funnel.

The discharge itself has the appearance of a smooth inverted funnel (flared skirt). This flaring shows that the mud is entering tangentially through the feed inlet. The underflow, which is pit mud, may be discharging at a volume rate from 5 to 100 times the volume rate of a normal balanced underflow discharge. The underflow may even contain some cleaned mud from the overflow headers. The cost of this mud loss and the aggravation of the end-of-location cleanup problems cannot be tolerated.

If *all* cyclones on a header are flooding out the bottom with smooth flare characteristics, the trouble is not due to a plugged cyclone inlet. The problem is located upstream from the feed header and is usually caused by replacement of a 1800-rpm by a 1200-rpm motor on the centrifugal pump. Sometimes it is due to an improper impeller change, or a suction problem at the pump.

Complete plugging of the inlet will result occasionally in no discharge at the underflow, with air suction occurring through the apex into the overflow manifold. More often, it results in a high discharge rate of cleaned overflow mud to the waste pit. This underflow discharges straight down with no flare. Regardless of the reason, any hydrocyclone not performing properly must be removed or corrected. If a hydrocyclone is operating properly, it is *saving* money, whereas if it is not functioning properly, it is *costing* a great deal of money.

Feed plugging is caused by poor supervision and maintenance of the mud system. Bypassed shale shakers or torn screen cloths are the most common offenders. Special large-open-area suction screens installed at the end of the centrifugal pump suction, inside the suction pit, can reduce plugging by 90%; however, even this is not enough to cope with a bypassed shale shaker, especially in the absence of a good sand trap.

DECANTING CENTRIFUGES

The decanting centrifuge is the only liquid—solids separation device used on drilling fluids that can remove (decant) all free liquid from the separated solid particles, leaving only adsorbed or bound water on the surfaces of particles. This adsorbed liquid does not contain soluble salts, such as clorides, nor clays, such as bentonite. Dissolved salts and colloidal clay solids are present in the continuous, free liquid phase from which the decanting centrifuge separates the inert solids. The adsorbed water can be removed from the separated solids only by evaporation, which has been neither desirable nor practical for drilling fluids.

Although centrifuges of other types, notably the solids-discharge-nozzle type with vertical axis, have been tested periodically but unsuccessfully in the U.S.A. since the early thirties for use on weighted water-base muds, the decanting centrifuge was first tested and proved adaptable as a practical field tool in 1953 (Bobo and Hoch, 1953). It moved rapidly into the rental field and is now recognized universally as a useful tool, though not as well understood as it should be.

The standard conical centrifuge, which has proved so dependable for separation purposes in weighted water-base muds, is most accurately described as a “continuously-conveying, conical-bowl, decanting centrifuge”. A schematic cutaway of the separating chamber is shown in Fig. 12-22. The size of the machines is given in maximum bowl inside-diameter and length. The best

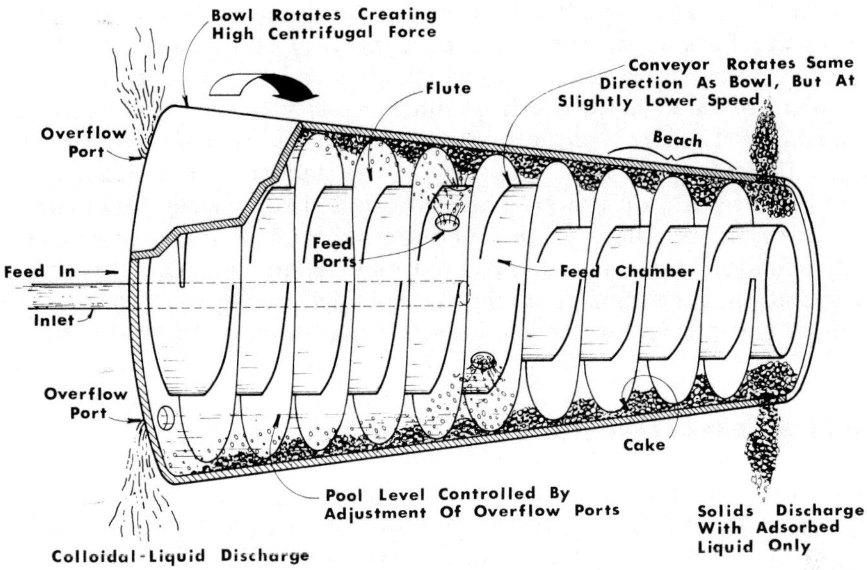

Fig. 12-22. Sectional view of decanting centrifuge.

known centrifuges have been the Sharples 14 inches × 20 inches, Bird 18 inches × 28 inches, Pioneer Centrifuging Co. (now Geosource) 18 inches × 28 inches, Baroid 18 inches × 28 inches, and Gilreath 14 inches × 22 inches (the latter probably is no longer manufactured).

The principle of operation is the same for all these different centrifuges. Larger machines usually have higher load capacities at equal separation ability and finer separation ability at the same loads. Departures from conventional design in the case of Gilreath centrifuge resulted in an increase in feed capacity at a very high sacrifice of separation ability.

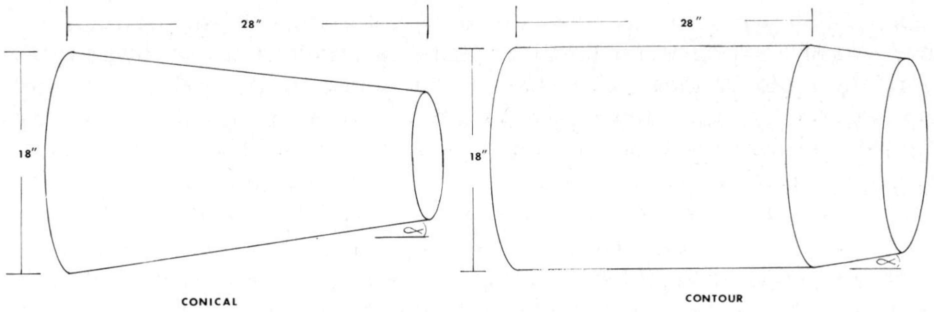

Fig. 12-23. Comparison of "conical" and "contour" bowl shapes of decanting centrifuges.

About 1972, Pioneer Centrifuging Company evaluated a "contour bowl" machine in the field as an environmental control machine (see Fig. 12-23) and for those very rare situations in which standard conical bowl machine capacity was not enough for weighted mud viscosity control. About 1974, Baroid introduced their "contour bowl" centrifuges, which have two to three times the solids and liquid capacity of the standard conical designs, at about the same separation power. Their application is primarily for environmental control, as secondary separators discussed later. It is not possible to justify their added cost on most weighted mud separation work. The principle of operation is the same as for the conical bowl machine.

Still larger and higher speed machines are being tested in the field periodically.

Operating principles of centrifuges

As shown in Fig. 12-22, the separation occurs inside the bowl, which is rotated at speeds ranging from 1500 to 3500 rpm, depending on the bowl size, conveyor design, application, and preventive maintenance program. Inside the bowl usually there is a double-lead conveyor connected to the bowl through an 80 : 1 gear box so that the conveyor will lose one revolution for every 80 revolutions of the bowl. Thus, there is a relative rotation of 22.5 rpm when the bowl speed is 1800 rpm. This relative rotation is the conveying speed, which can be varied by using other gear ratios.

The mud slurry to be treated is metered to the centrifuge through a feed tube centered in the hollow axle to the feed chamber. From there the feed passes through feed ports into the leads in the separation chamber. The scrolls (blades) form the walls of the leads or channels. As the slurry is slung outward and joins the annular ring called the "pool", it accelerates to the approximate rotational speed of the conveyor and bowl. The depth of the pool is determined by the adjustment of the eight to twelve overflow openings (ports).

The slurry flows through the leads toward the overflow ports under very high centrifugal force and the flow is laminar in most machines and applications. Following Stokes' Law, the solids settle outward to the walls, the most massive particles settling first and the smaller ones settling further along the leads. Particles, which are not committed to the solids cake against the wall by the time the slurry stream is increasing in velocity to exit through the overflow ports, do not settle and pass out of the overflow ports with the liquid. In the normally recommended operating ranges, retention time in the machines varies from 10 to 80 sec, with the most common range being between 30 and 50 sec.

As the solids settle outward against the wall, the conveyor continuously plows or scrapes them toward the small end of the bowl, out of the pool, and across that portion of the wall between the pool and the solids underflow discharge openings called the "beach". As the solids cake crosses the beach, the free liquid is removed by both centrifugal squeezing and centrifugal draining of the cake. The free liquid drains back to the pool, carrying the colloidal and dissolved solids. As the separated solids discharge out the underflow, they contain only adsorbed water, i.e., the solids are *damp*.

The only purpose of diluting the feed to a decanter is to reduce the viscosity of the feed as an aid in maintaining separation ability. The best rule of dilution is to add only enough water to reduce viscosity of the effluent to 37 Marsh funnel seconds; water should not be added to reduce the viscosity below 35 sec. Greater dilution will decrease the settling time out of proportion to the increase in settling rate.

Dilution is applicable only to water-base muds. Dilution with oil to facilitate centrifuging is not advisable in the case of continuous oil-phase muds, which respond better to heat. Heating above 90°F (32°C) for centrifuging usually is neither economical nor necessary, but separation does improve at a decreasing rate up to at least 140°F (60°C).

Capacities of centrifuges

The solids discharge capacity of a decanting centrifuge is limited by the volume of solids cake that can be moved by the conveyor and discharged by the underflow openings per unit of time. The liquid discharge capacity of a decanter is limited by the capacity of the overflow ports to discharge liquid at a particular pool depth adjustment. The feed rate capacity of the decanting centrifuge is determined by either the volume of separable solids in the feed or by the free liquid and colloids content of the feed, depending on whether discharge limit is being approached for liquid or solids. Thus, feed rate of a mud low in solids content is limited by liquid discharge capacity, whereas the feed limit of a mud very high in separable solids content (a hydrocyclone underflow or a weighted mud) is determined by the solids discharge capacity.

If a very fine separation is desired, it may be necessary to hold down feed volume rate in low solids feeds to maintain (1) laminar flow in the settling chamber and (2) proper retention time, to achieve fine separation ability.

Separation limits of centrifuges

An effective decanting centrifuge for weighted water-base mud should be capable of making a separation approximately at the 2-micron barite particle size. Settling according to Stokes' Law, light solids will be separated down to 3 microns ($1\frac{1}{2} \times 2$ microns) in the same machine under the same conditions. In a continuous oil-phase mud, the effective viscosity of the oil is greater than that of the water in a water-base mud. Thus, the separation in the centrifuge is not as fine, the cut increasing approximately by the square root of

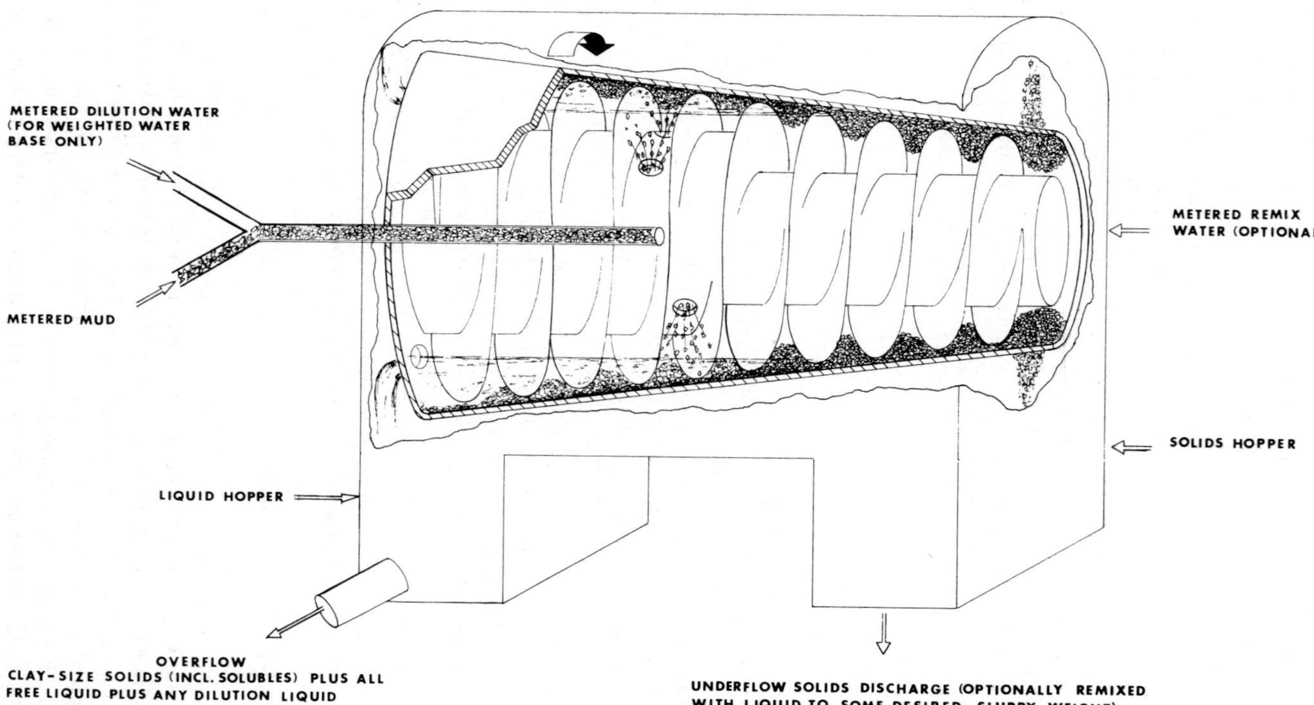

Fig. 12-24. Schematic diagram of a decanting centrifuge.

the ratio of the effective viscosities in the overflow (Burdyn et al., 1965).

In field operation, a decanting centrifuge is fitted with (1) a housing over the bowl, (2) liquid and solids "hoppers" to receive the liquid and the solid discharges separately, (3) slurry feed pump, (4) dilution water connection, and (5) remix water connection if desired (Fig. 12-24). The feed mud is calibrated with a meter of one kind or another. Although the bowl speed usually can be varied, it should be done by the centrifuge technician, as it is seldom either necessary or desirable. In order to reduce damage to expensive gears, shearpins, trips, limit switches, etc. are installed to shut down the equipment automatically in the case of a solids overload.

Due to the lack of free liquid, the underflow solids discharge from the bowl of a decanting centrifuge is very heavy and very thick. Its consistency is like that of glazier's putty, and it cannot flow or be pumped. Remix water, added between the bowl and housing to provide some free liquid, quickly thins the underflow. It is better, however, to drop the underflow straight down or at an angle not more than 45° from the horizontal. Addition of two quarts or two liters of remix water per minute makes the underflow much easier to handle if it returns to the system. Addition of one gallon or four liters of water per minute imparts to the underflow a thick-soup consistency.

Dilution water connections should be equipped with a flow gauge or meter and a flow adjusting valve. Remix connections are usually provided. Remix water can be added in the housing to aid in softening the underflow solids for easier mixing into the system, or for reducing the underflow solids slurry density to the range of system density desired.

THE PERFORATED ROTOR CENTRIFUGAL SEPARATOR

The perforated rotor centrifugal separator was developed in the middle 1960's (see Burdyn et al., 1965). The machine was expected to control viscosity in weighted water-base muds, by discarding clay-size solids while salvaging silt-size barite, in competition with the decanting centrifuge. Its advantage is portability in that the solids discharge can flow through a pipe or hose. Thus, the machine is relatively independent of location in relation to the mud tanks. It can be brought at any time and be placed almost anywhere on the drilling location.

This separator is not a decanter, because part of the liquid-clay phase must pass out with the solids discharge, as otherwise it cannot flow. For this reason the dilution water must be completely compatible with the drilling mud. Uniformly, this dilution requirement has been 70% of the undiluted mud feed rate.

The "perforated rotor centrifugal separator" or "perforated rotor mud separator" is not a centrifuge, because the member containing the processing slurry does not rotate. For the sake of simplicity, however, this machine is often referred to as a "centrifuge". The author is promoting the term "holey rotor", which is short, accurate, and easy to remember.

Operation of the perforated rotor centrifugal separator

The perforated rotor separator consists of a 6-inch outside-diameter cylindrical-shell rotor 40 inches long (with 474 one-half inch holes drilled in the shell and evenly spaced) spinning in a horizontal position inside an 8-inch inside-diameter, stationary pipe housing, fitted with fluid seals at each end through which the axles of the rotor protrude (Fig. 12-25). A slurry to be treated is fed by a positive-displacement metering pump and is mixed with 70% dilution water (mud/water volumetric ratio of 10/7) from a second metering feed pump. Part of the dilution water is used to flush the rotating seals. The diluted feed enters the annular separation chamber through a circumferential opening at one end of the annular space. (See Nelson, 1970.)

Under centrifugal force, larger solids are concentrated against the outer annular wall (Stokes' Law). An underflow solids discharge opening is located at the opposite downstream end of the annular space, on the periphery. All feed material, both liquid and solid, would discharge out through this opening if not restrained. A positive-displacement pump operating as a mud-

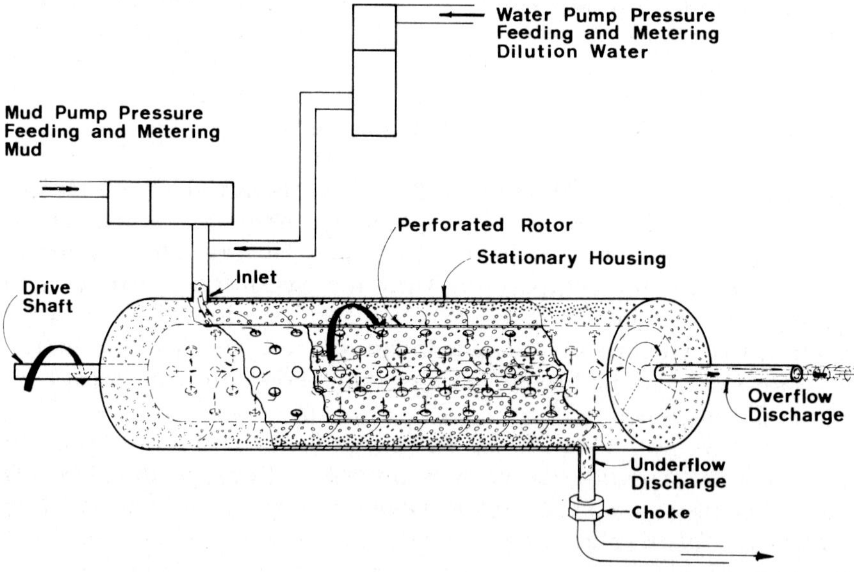

Fig. 12-25. Schematic diagram of a perforated rotor centrifugal separator, available from Milchem exclusively.

driven motor was originally arranged to hold back-pressure on this underflow opening and to permit the underflow solids slurry to discharge at a metered and controllable rate. A special adjustable choke is now being used with good results.

Separation parameters

Depending on the ratio of the rate setting of the underflow discharge metering pump to the rate setting of the feed metering pump, a certain volume fraction of the feed is forced out the overflow; i.e., that volume portion of the feed which is not permitted to leave at the underflow discharge can only leave as the overflow. When rotating at constant, near-design speed, the particle size of the median cut ranges from 3.5 to 5 microns when specific gravity of particles is around 2.6 (equivalent to 2.4 to 3.5 microns median cut for barite).

For the "holey rotor" to maintain its positional flexibility, it must discharge the underflow as a very fluid stream that can flow from some convenient operating location to the active mud system without line plugging problems. In the most recent operating literature, Milchem Inc. recommends that the underflow solids slurry weight should be approximately 3 lb/gal higher than the undiluted feed (active mud system) weight.

If the perforated rotor machine were operated at the condition of underflow specific weight being equal to the feed specific weight at zero dilution, there would be 100% underflow discharge of all feed and no overflow. At the recommended operating dilution of 70% of the feed volume rate, the liquid—colloidal phase of a 15-lb/gal (sp. gr. = 1.8) feed mud would be diluted at an average ratio of approximately one-to-one. If the underflow slurry specific weight were equal to undiluted feed mud specific weight, the amount of undersize material (clays) removed with the overflow discharge would constitute approximately 50% by volume of the clays in the feed. If dilution were doubled to two-to-one, the clay fraction removed as overflow would be approximately 67% by volume; whereas at three-to-one dilution ratio, the clay fraction removed in overflow would consitute approximately 75% by volume of the clays in the feed.

Instead of regulating the underflow slurry at a specific weight equal to that of the undiluted feed mud, according to modern instructions it is necessary to operate at an underflow specific weight of 3 lb/gal (sp. gr. = 0.36) heavier than the undiluted feed mud. This forces more liquid-clay phase out of the overflow, increasing the removal of clay and, thereby, improving the operation. The underflow discharge weight of 20 lb/gal must not be exceeded, because sooner or later plugging occurs with high specific weights.

Cut point decreases as feed rate increases, but the machine makes an acceptable cut within its recommended operating parameters. The greatest confusion related to this machine has arisen from the fact that total feed capacity and total separating capacity are not necessarily the same. Only that portion of the liquid phase which is forced out as the overflow removes clays from the system. The rest of the free liquid phase must be discharged in the underflow in order to keep the latter flowing. Part of the clays return to the system with the free liquid in the underflow.

EQUIPMENT APPLICATIONS – PRIMARY SEPARATION

The proliferation of machines and ideas concerning liquid—solids separation in the last decade necessitates some classification of processes for any ordered discussion. Primary separation, as used here, refers to the use of liquid—solids separators to remove either a liquid-clay fraction or a solids fraction from the active mud system. Any separator accepting feed *directly* from an active system and returning either the liquid discharge (having lower solids content than the feed stream) or the solids discharge (with higher solids content than the feed stream) back to the active system, in effect removes a selective fraction from the system, and thus performs a primary separation. Figure 12-26 is a schematic diagram of typical primary separation equipment as related to a drilling fluid system.

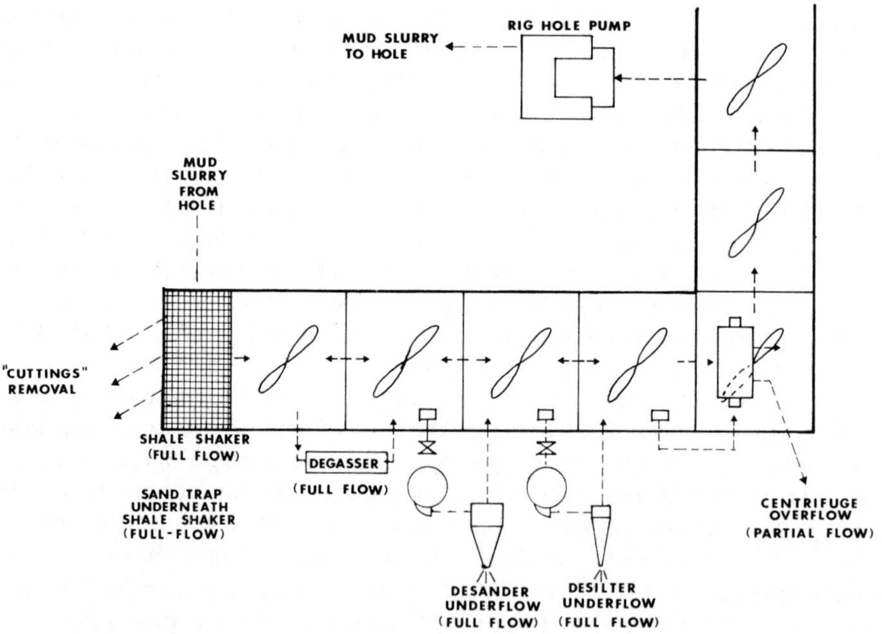

Fig. 12-26. Schematic diagram of a primary separation equipment as related to a drilling fluid system.

Understanding and use of good and adequate solids removal equipment can reduce mud costs drastically as compared to the cost of removing solids by discarding a portion of the whole mud (dilution). Dilution or "watering-back" was once the mainstay of solids control after mud had passed the conventional shale shaker. The technique of dilution involves increasing mud pit level with needed addition of water and then discarding a fraction of the whole mud. These steps can then be repeated. Obviously this procedure does not change the particle size distribution, and it is the most expensive and wasteful way to control mud properties, i.e., solids.

Mechanical devices, on the other hand, selectively remove solids and, thus, alter the particle-size distribution. They permit the removal of more of the unwanted solids with lower loss of needed solids (mud property control) than is possible by dilution. In addition, mud pit level is maintained. Whole mud should never be discarded, except in the case of real emergencies, which are very rare.

FULL FLOW PROCESSING OF UNWEIGHTED MUD

Full-flow processing of all drilling fluid from the borehole is done for the purpose of formation solids removal, specifically those in the inert particle size range (larger than clay size). In this application, the best 4-inch or smaller hydrocyclones or desilters can achieve the optimum removal. For the desilting to be performed economically and efficiently, balanced-design hydrocyclones must be operated with spray-type underflow discharge. Solids discharge overload or rope discharge must be avoided.

Underflow discard contains less liquid per unit volume than system (feed) mud that would be discarded to remove the same weight or volume of solids. The underflow also contains selectively the coarsest and the most troublesome of the inert solids in the system, with far less of any beneficial solids, than an equal volume of feed. This statement is valid for any type of primary, full-flow solids removal device used for unweighted muds.

A knowledgeable drilling supervisor with an adequate removal system at his disposal does not permit discarding of whole mud from the active system under normal drilling conditions. In legitimate surplus mud situations (e.g., rapid weight-up and cementing) he will attempt to save as much drilling fluid as possible in steel storage tanks equipped with adequate stirrers and then will salvage it back into the active system as quickly as practical when the emergency is over.

Effect of oil on specific weight

If oil is added to a water-base mud system and emulsified, specific weight of the mud is temporarily decreased due to the lower specific weight of oil.

Within a very short time, however, the specific weight of the system returns to the original weight and then increases above it. This is due to: (1) the increased viscosity caused by the oil, (2) the oil particles acting as solid particles and interfering with separation, and (3) the increased underflow waste problem due to the presence of oil. Consequently, oil should not be added to water-base muds for the purpose of reducing specific weight for drilling.

Oversize and overload protection

At slow drilling rates, regardless of bit type, cuttings are small and their volume cannot possibly overload a full-flow installation of effective desilters. If it could be assured that no formation cavings could enter the feed, no other full-flow separation device would be necessary for this situation. Inasmuch as some part of every hole is usually drilled at a rapid penetration rate, protection against both plugging with oversize particles and overload with solids volume usually is necessary for the desilters.

The results of various full-flow solids removal procedures are generalized in Fig. 12-27. Optimum solids removal in unweighted muds is obtained by continuous spray discharge underflow from desilters having adequate capac-

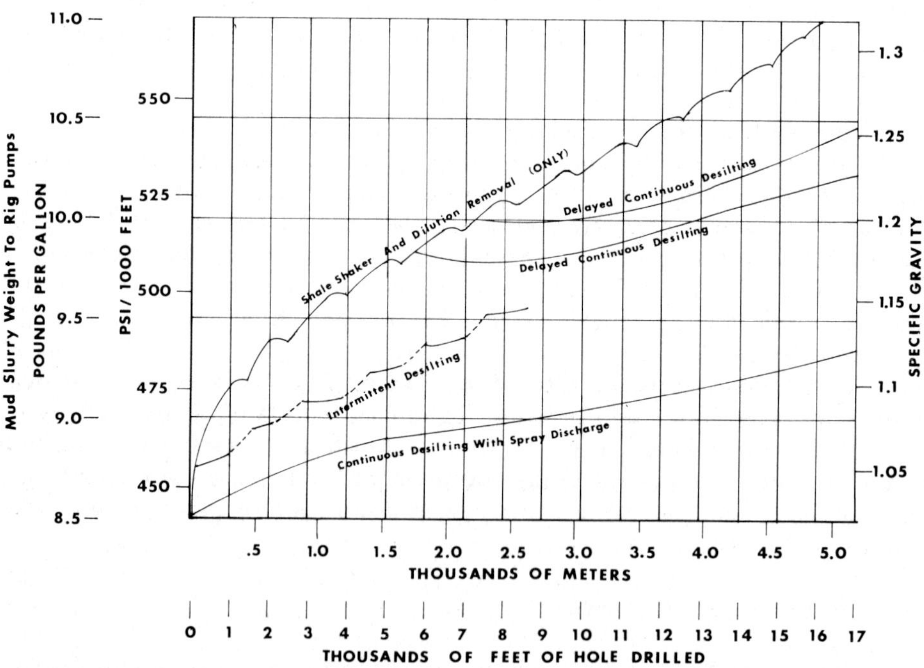

Fig. 12-27. Effect of various solids removal techniques on mud weight.

ity to handle the full circulation rate as a minimum. It is assumed in this case that a shale shaker upstream is adequate to protect against oversize solids particles that could plug the desilter inlets, and that whatever other solids removal equipment (such as desanders), which may be necessary to prevent solids discharge overload of the desilters, are installed.

Delayed and intermittent desilting

The upper line in Fig. 12-27 represents the "shale shaker-plus-dilution-and-discard" technique, which was necessary before efficient hydrocyclones were available. For economical considerations, operators often wait until mud weight is fairly high before beginning to desilt. The result of this delayed desilting is also shown. The additional solids in the system, as evidenced by the increased mud weight above that of a continuously desilted mud, decrease in size due to mechanical degradation and most of them no longer can be removed by the desilters.

Delayed desilting has another adverse effect besides increased mud weight. The specific surface area of solid particles (per unit weight or volume) increases with decreasing size of particles (see Table 12-II). The volume of liquid film associated with the surface area of all solid particles removed from a mud by hydrocyclones and by screens must vary directly with the surface area per unit weight of solids removed, i.e., more liquid phase is lost with increasing content of finer solids. Delay in removing the solids will always result in greater liquid loss, making it more expensive to remove those that can still be removed.

Periodic desilting is another false "money saver", because it never results in minimum mud weight or best mud performance. With effective cyclones, the cost per unit weight of solids removal is minimized when desilting is continuous, and the mud level in mud pits is kept nearly constant while drilling. Mud weight increases constantly, but very gradually.

Particle sizes in desilted muds

Figure 12-28 shows the particle-size distribution of the solids from two desilted (low silt content or small silt size) drilling fluids from very different situations. Curve *A* shows the distribution of solids in a drilling fluid (at the rig pump suction) that has been processed first through a 12 × 12 mesh shaker screen, and then through well-designed balanced desilters at a very low underflow solids load, always in spray discharge. This is probably as "clean" as a desilted mud could be. Content of particles that are larger than eight microns is small. There were no desanders on the job, but even their presence would not have changed the solids content in the mud, because the light load supplied to the desilters permitted them to operate at a maximum efficiency.

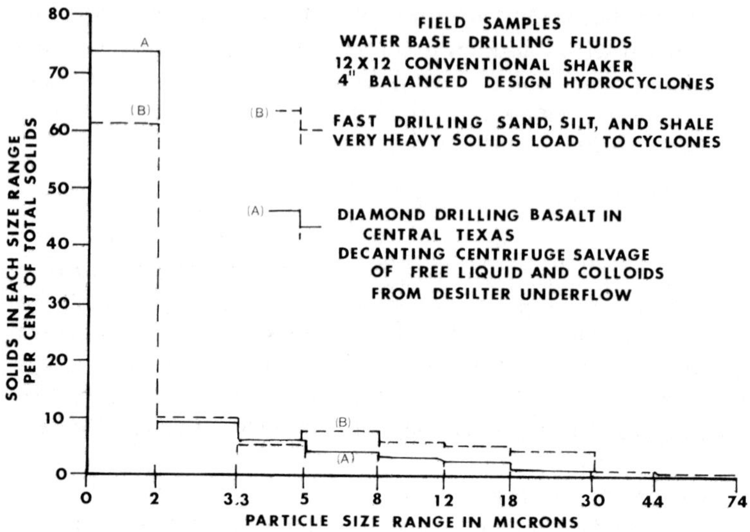

Fig. 12-28. Comparison of solids removal (screened and desilted slurries).

Curve *B* represents the distribution of solids in a drilling fluid that has been processed through the same type of well-designed desilters, but at very high underflow solids loads. During most of the drilling time (until shortly before the sample was taken) the underflow discharge was roping periodically. There were no desanders and the conventional shale shaker, with 12 × 12 mesh cloth, separated very little material because the formation was composed of poorly consolidated sand, silt, and shale. The shale shaker prevented any feed plugging of the cyclones, which were operated with full underflow openings (slightly wet bottom). Had there been a stage of good desanders ahead of the desilters or more desilters were provided for greater underflow discharge capacity, the solids distribution might have been as fine as that of curve *A*. In this drilling condition (fast drilling of sand, silt, and sticky shale with a water-base mud), very fine shale shaker screens could not have functioned well enough to have kept the desilters in spray discharge without blinding and losing mud. Properly-designed 6-inch desanders certainly could have helped. Nevertheless, the mud was desilted sufficiently so that low-silt mud drilling benefits were obtained (Ormsby, 1965).

Most of the 73% colloidal material in mud (*A*) of Fig. 12-28 was purchased to increase viscosity and gel strengths and to reduce filtration loss. There were no expandable clays in the formation which would add any desirable properties to the mud. No commercial clays were added to mud (*B*). The shale formation in that coastal area contains sufficient quantity of bentonitic clay to provide filtration control, viscosity, and hole stabilization, but only if these clays are constantly kept in the system (saved) by good desilting practice. In drilling any formation where bentonitic clays are not present, they must be added judiciously in order to attain proper filtration control for borehole stabilization.

Particle-size distribution in desanded muds

Size distributions of solids in two desanded muds are shown in Fig. 12-29. Both of these muds first have passed through conventional 12 × 12 mesh shale shakers and then through a full-flow set of 8-inch hydrocyclones. Curve *B* represents a mud which was used to drill a very hard formation in central part of southern Oklahoma, using a flat milled-tooth bit. There is a small amount of API sand (sand-size particles) which these particular hydrocyclones could not remove, even at the low solids load. A much higher percentage of solids above eight microns in size are present compared to either of the desilted muds in Fig. 12-29.

Curve *A* of Fig. 12-29 represents an unusual drilling condition encountered in the first few thousand feet of drilling in the delta area of Nigeria, where formation consists of coarse sand and fine gravel, with almost no formation particles less than 74 microns in size. Curve *A* shows that the desanders have left some of the API sand (sand-size particles). All the colloidal material in the mud is commercial bentonite, added at the rate of one 50-lb sack per ft of hole primarily to increase gel strengths and reduce mud flow (loss) through the coarse formation pores.

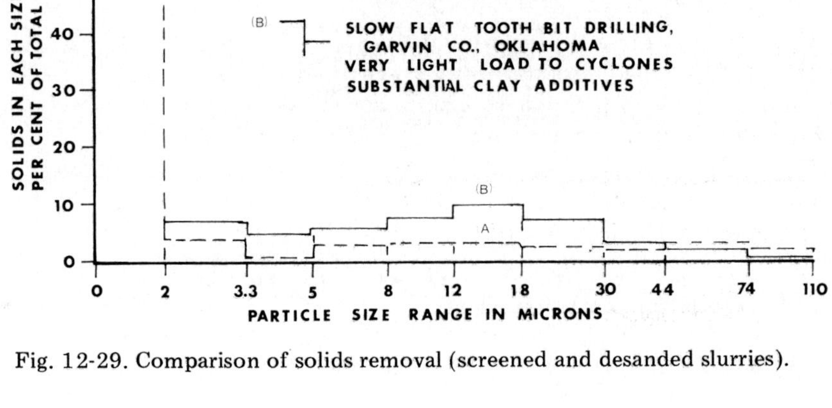

Fig. 12-29. Comparison of solids removal (screened and desanded slurries).

If the Nigerian mud (Fig. 12-29, curve *A*) was more thoroughly desanded, it would have had the characteristics of a desilted mud, simply because there is such a small amount of formation particles between 8 and 74 microns in size. Better desanders, with more underflow solids capacity, would have removed all the sand-size solids. On the other hand, in the situation shown by curve *B* in Fig. 12-29, the solids load did not present a problem. Properly-designed desilters would have produced a low-silt mud in conjunction with only a shale shaker to remove oversized particles. The slow drilling rate would not overload desilters properly sized for circulating rate.

Particle-size distribution of coarse-screened mud

Figure 12-30 shows two extremes of solid-size distributions in unweighted muds. Curve *A* of Fig. 12-30 represents a mud from the Niger delta area that has been passed through the 12 × 12 mesh shale shaker screen only. While not from the same drilling rig, this mud is from the same area and is very similar to the mud represented by curve *A* of Fig. 12-29 before desanding with hydrocyclone desanders. The small amount of solid particles 2—74 microns in size below the shaker screen is striking and is due to their near-absence in the unconsolidated and clean gravel and sand formations. The commercial clay particles are distinguished from the API sand (sand-size solids) not removed by the screen.

Curve *B* of Fig. 12-30 represents a very good example of the cuttings-size distribution from a diamond (core) bit. A considerable amount of commercial viscosifiers and filtration loss reducers were present in the mud. Ob-

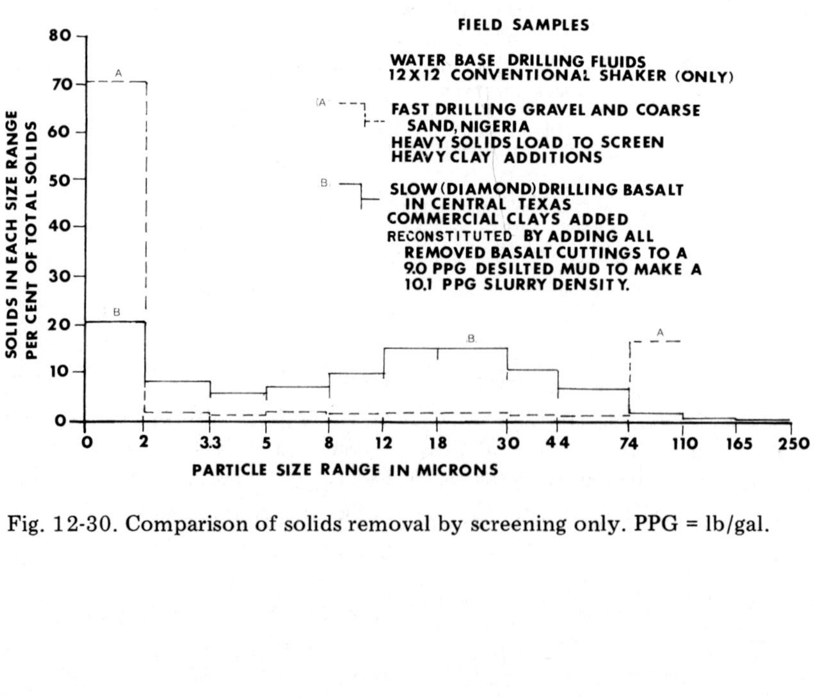

Fig. 12-30. Comparison of solids removal by screening only. PPG = lb/gal.

viously, most of the clay particles and at least half of the particles between 2 and 3.3 microns in size are not diamond bit cuttings, whereas probably all of the particles above 5 microns in size are diamond bit cuttings. None of these particles were removed by the 12 × 12 mesh shale shaker cloth. The finest of practical full-flow rig screens would remove only a minor fraction of these diamond bit cuttings.

The grain-size distribution in Fig. 12-30 illustrates the variations that exist in the case of solids removal from unweighted muds. Figures 12-29 and 12-30 illustrate the change that occurs in particle-size distributions as the full-flow primary separation is improved by adding desanding and then desilting stages. Very fine screening would appear similar to desanding if illustrated on graphs such as Figs. 12-29 and 12-30.

In muds weighted with barite, the problem of solids removal is more complicated (see Figs. 12-7 and 12-8). Comparison of the API-barite grain-size distributions (which are consistently similar) with the light-solids size distributions in Figs. 12-29 and 12-30, shows that removing light solids coarser than barite with fine screen cloths, without removing any barite, should be fairly effective in some cases and impossible in others, with actual results varying between the extremes.

Loss of barite

As mentioned before, reduction in liquid flow capacity of screen cloth, due to weighted mud effect, can result in a serious problem. Operators, therefore, use primary shale shaker screens coarse enough to take total flow without creating blinding problems and for many years have processed the weighted mud through hydrocyclones in order to remove the coarsest drill solids with acceptable losses of coarse barite. Inasmuch as according to Stokes' Law the low-specific-gravity particle must have $1\frac{1}{2}$ times the equivalent spherical diameter of the barite particle to fall at the same velocity, Figs. 12-7 and 12-8 indicate how difficult it would be not to lose some of the coarser fraction of the barite. At the same time, these operators have reported their losses of barite at the underflow to be negligible under most conditions.

Several facts must be pointed out here: (1) The API barite standards limit the API sand-size barite fraction and a loss of barite (with other solids) in this size range is beneficial to reduction of equipment erosion and filter cake problems. The same is true of that portion of barite above 44 microns in actual size, (66-micron Stokes' Law equivalent of light solids). (2) Desilters are not as efficient in separation ability for heavy, weighted muds and, therefore, their removal is not as severe as would otherwise be expected. The median cut point increases proportionally to the square root of the plastic viscosity, and hydrocyclones are normally rated using water. (3) If there is a moderate load of coarse solids in the API sand-size range which must be removed, it is very easy to adjust equipment in order to lose virtually no barite below 44 microns in actual size. In rare situations it may be permissible to adjust the cyclone underflow *slightly* smaller than balance, provided it will operate without plugging. (4) It is easy to test the hydrocyclone underflows to determine the cost of replacing the solids which are being removed.

Benefits from size refinement

If some fraction of the particles in the 44—74 micron size range and all those larger than 74 microns in size can be removed continuously from a weighted mud, any one or a combination of the following several possible benefits may be obtained: (1) abrasion in the rig pumps, manifolds, etc. is reduced; (2) mud filter cake becomes smoother in character and may become thinner; (3) differential pressure sticking problem is usually reduced if it has been present; (4) the rate of buildup of low-grade colloidal material, both light solids and barite, is reduced, easing (but usually not eliminating) the viscosity control problems; (5) total solids content may be reduced in some cases; and (6) overall drilling progress may be speeded up. These possible benefits are apparently related to the desilting of unweighted mud (Ormsby, 1965). Such benefits in unweighted muds have been the result of changes in particle-size distribution of solids and not changes in their average specific gravity or total solids content. It is reasonable to assume that benefits obtained in the case of weighted mud are also related to size and character of solids, and only indirectly to their specific gravity.

Replacement of light solids and barite

Improving the full-flow primary separation (removal of solids) in weighted muds using hydrocyclones downstream from the shale shakers does involve an added cost of barite makeup that has been commonly misunderstood. For example, every 100 lb of light solids removed must be replaced by about 80 lb of barite and 20 lb of water (actually, 78.75 and 21.25 lb, respectively, assuming that barite has a specific gravity of 4.2 and that light solids have a specific gravity of 2.5). In this process the fraction of larger particles will be reduced, whereas the average specific gravity of solids will be increased slightly. Under no circumstances would the specific gravity of total solids be decreased.

A decrease in the content of coarsest material will improve the drilling operation. It is necessary, therefore, to remove the coarser fraction of the solid particles in the mud. Economic considerations do not involve relative proportions of light solids and barite removed, with two extreme removal situations being possible (see Table 12-IX).

If the object is to remove most of the particles larger than 44 microns in

TABLE 12-IX

Cost of restoring weight and volume of a barite-weighted mud after a removal process

Type of solids removed and removal rate (lb/hr or kg/hr)	Rebuilding (adding back), neglecting chemicals (lb/hr or kg/hr)	
	barite	water
Low-specific-gravity, 1000 lb/hr	800	Equivalent to water removed with solids, plus 200 lb/hr
Barite, 1000 lb/hr	1000	Equivalent to water removed with solids
Low-specific-gravity, 500 lb/hr Barite, 500 lb/hr	900	Equivalent to water removed with solids, plus 100 lb/hr
Any and all solids, 1000 lb/hr	900 ± 100	Equivalent to water removed, ∓ 100 lb/hr

size in a weighted mud, replacement cost is not much more for the removed barite than it is for the removed light solids, and the benefits are about the same. Also it must be realized that initial barite loss on a "new" system will be higher than subsequent losses as the content of larger size barite is reduced. Further barite loss will be almost completely restricted to makeup barite. This operation should be performed only at the full-flow rate and never on the partial-flow rate.

PRIMARY SEPARATION AT PARTIAL FLOW

Partial flow processing in primary removal of solids should be used only to reduce the clay, or sometimes the soluble solids, content of the mud. Obviously, any attempt to remove all the colloidal phase from the total circulating rate volume of a weighted drilling mud may lead to immediate disaster. It is dangerous, expensive, and futile to remove the total liquid phase whether it is replaced without clays or an attempt is made to replace the total liquid phase with needed properties at the full rig circulating rate.

The decanting centrifuge

The decanting centrifuge has been a very popular clay removal machine because it assures a complete removal of clays from the small fraction of the volumetric circulating rate of mud it was intentionally selected to process.

Although the decanting centrifuge is most often referred to as a "barite salvage" machine, this name is used as an economic justification and does not signify the machine function. All primary separation machines have a removal function, and barite can be "saved" simply by not pumping mud from the pit at all. The primary problem in a water-base weighted mud is viscosity increase, which is due to the natural and constant increase in content of clay particles, usually not of high quality. Their presence in large quantity prevents addition of better filtration control agents and, in combination with the silt and sand present, they may severely retard drilling progress.

Although the dilution-and-discard approach to remove these excess low-quality colloids is effective, it is expensive due to the high cost of barite necessary to replace the total weighting material (including silt-sized, low-specific-gravity particles) in the discard. Unpublished data from several parts of the world indicate the *minimum* dilution (or equivalent mechanical treatment) amounts to an average of 5% of the total system each day a weighted mud is in use. A maximum apparently exists only rarely and is equal to the maximum dilution rate at which a clay-colloidal phase can be rebuilt and controlled successfully. It is around 10% of total circulation rate. The 18-inch conical-type decanting centrifuges, operated at a barite cut point of 2-micron particles, have a feed capacity ranging from an extreme of about 3—12% of the circulating rate, depending upon volumetric circulating rate and mud weight.

An example is presented here to illustrate the economic comparative evaluation of "dilution-and-discard" versus "decanting" technique for a weighted water-base mud in order to keep the colloidal phase under control.

Figure 12-31 is a diagram that combines certain, non-variable, mathematical facts (on the left side) concerning a barrel of 17.0-lb/gal of barite-weighted, water-base mud containing 5% oil and 37% solids, the assumed specific gravities being 4.2, 2.5, and 0.85 for the barite, light solids, and oil, respectively. On the right side, some of the important functions, facts, and trends concerning the clay and the inert solids, and the liquid fractions are presented.

To complete the information on the right-hand side, two basic assumptions were made: (1) the light solids clay fraction was assumed to be 50% of the total light solids, and (2) the clay-sized barite fraction was assumed to be 15% of total barite. Once these assumptions, which are within a realistic range, were made, equations presented earlier were used to calculate the content of all fractions, whether by weight or by volume.

If a 10% dilution-and-discard (or "discard-and-make-up") technique is used, the make-up additions include 10 lb (1.14% by volume) of low-specific-gravity filtration control material, 472.4 lb of barite, 25.93 gal of water, and 2.1 gal of oil per makeup bbl of mud (see Fig. 12-32). When added to the remaining mud (90% of the original system), the reduction in total solids content is not appreciable. The decrease in clay-colloidal phase ranges from

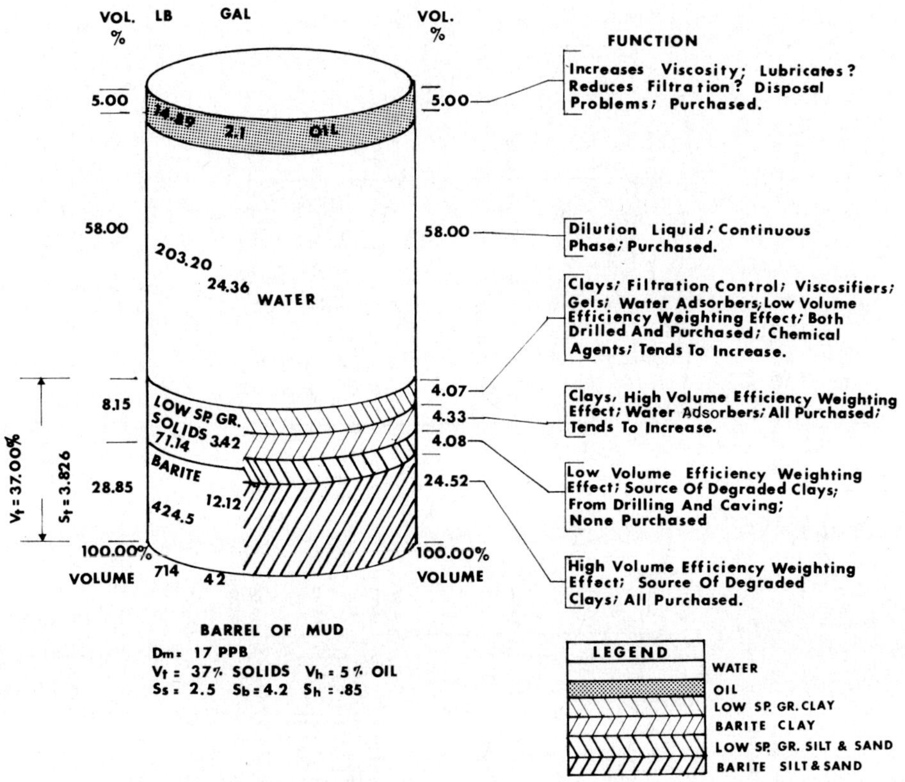

Fig. 12-31. Evaluation of a 17-lb/gal barite-weighted, water-base mud containing 5% oil and 37% solids, using 1 bbl mud as a basis.

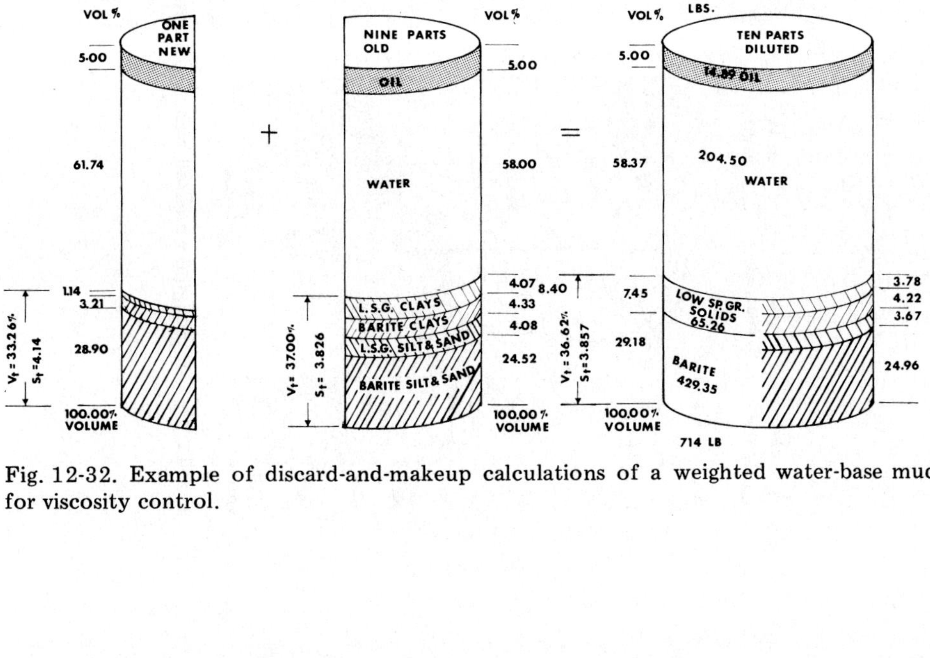

Fig. 12-32. Example of discard-and-makeup calculations of a weighted water-base mud for viscosity control.

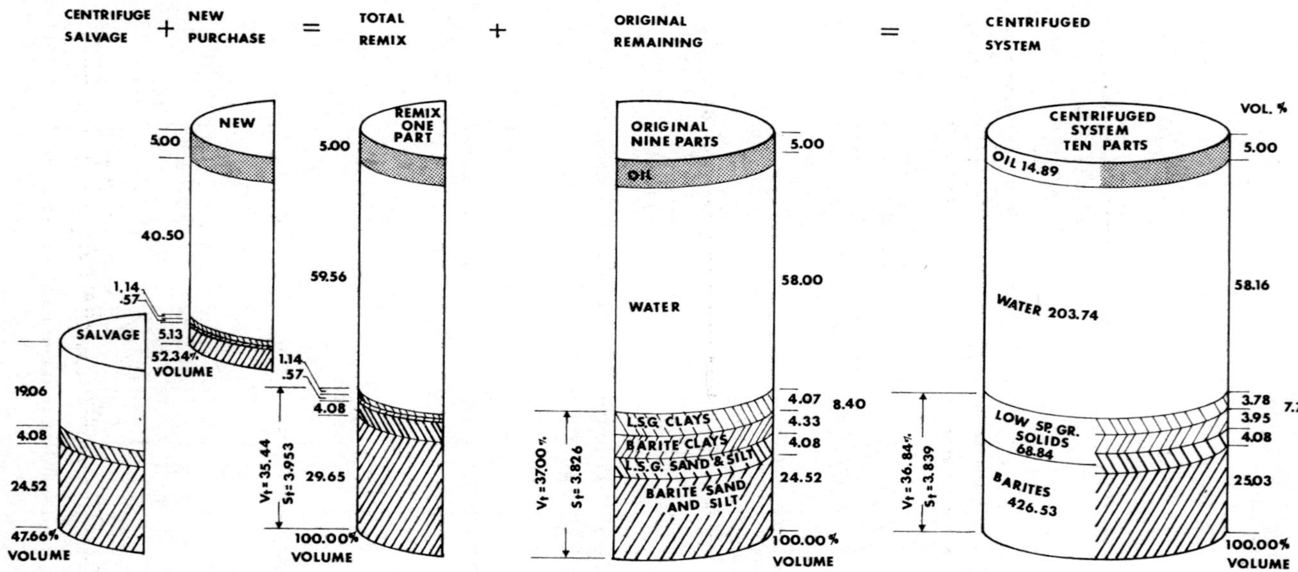

Fig. 12-33. Centrifuging weighted water-base mud.

TABLE 12-X

Comparison of discard-and-makeup method with centrifuging of a 17-lb/gal weighted water-base mud

Materials *	Original mud system	Discard 10% and dilute		Dilute 10% by centrifuging			
				Makeup			New system per barrel
		makeup purchase	new system	underflow salvage	purchase	total remix	
Oil (gal/bbl)	2.1	2.1	2.1	—	2.1	2.1	2.1
Water (gal/bbl)	24.36	25.93	24.51	8.01	17.00	25.01	24.43
Low-sp-gr clay (vol %)	4.07	1.14	3.78	—	1.14	1.14	3.78
Clay-sized barite (vol %)	4.33	3.21	4.22	—	0.57	0.57	3.95
Total clay (vol %)	8.40	4.35	8.00	—	1.71	1.71	7.73
Low-sp-gr silt and sand (vol %)	4.08	—	3.67	4.08	—	4.08	4.08
Barite silt and sand (vol %)	24.52	28.90	24.96	24.52	5.13	29.65	25.03
Total low-sp-gr solids (vol %)	8.15	1.14	7.45	4.08	1.14	5.22	7.86
Total barite (vol %)	28.85	32.11	29.18	24.52	5.70	30.22	28.98
Total solids (vol %)	37.00	33.26	36.62	—	—	35.42	36.84
Total low-sp-gr solids (lb/bbl)	71.41	10.00	65.26	35.73	10.00	45.73	68.84
Total barite (lb/bbl)	424.50	472.39	429.35	360.73	83.86	444.59	426.53
Total solids (lb/bbl)	495.91	482.39	494.61	396.46	93.86	490.32	495.37
Average sp. gr. of total solids	3.826	4.14	3.857	3.958	3.916	3.953	3.839

* 42 U.S. gal = 1 API barrel.

8.40 to 8.00% by volume. It would be lower if the clay content of the newly-added barite had not been assumed to be 10%, which is conservative.

As shown in Fig. 12-33, all clays from the 10% of the system centrifuged are removed. As compared to discard makeup in Fig. 12-32, the same amount of fluid loss control agent (10 lb/bbl), the same amount of oil (2.1 gal/bbl), and much smaller amount of new barite are required. Reduction in total solids content is lower, whereas the viscosity reduction is greater, because while total light solids removal is only 50%, removal of clay-colloidal material is complete. The small makeup requirement of barite reduces the amount of clay-sized barite addition. Table 12-X summarizes these and other details, including the difference in the amount of barite purchased (472.39 lb/bbl compared to 83.85 lb/bbl, for discard makeup and centrifuge makeup, respectively). Economic justification of the process for centrifuging is thus demonstrated.

In cases where excessive viscosity presents a problem, on a net volume basis, the clay-size barite can be as troublesome as low-grade light clay solids and, therefore, must be removed. The cost of replacing the clay-size fraction removed is approximately the same, regardless of whether the removed clay-size fraction is barite, light solids, or a mixture of the two as in this case. The key to economical clay-phase removal is to make a fine cut, to remove the free liquid and clay from the processed fraction, and to use as little dilution (mud and cleanup cost) as possible.

Perforated rotor centrifugal separators

The holey rotor machine, as already discussed, removes the major fraction of the clays from the feed slurry. This is true if the underflow weight is 3 lb/gal heavier than the feed mud before dilution. A much greater fraction of clays will be removed from a 17-lb/gal (sp. gr. = 2.04) mud than from a 12-lb/gal (sp. gr. = 1.44) mud, because the liquid-clay phase is larger in the lighter mud, causing dilution of the liquid-clay phase to be lower in the lighter mud. At feed mud weights above 17 lb/gal, this advantage is lessened, because the specific weight of solids discharge slurry should not exceed 20 lb/gal. This limit on underflow weight is based on the author's discussions with individuals operating this equipment in the field in various operating areas.

The perforated rotor separator requires two to three times the dilution water required by a decanting centrifuge, both being operated properly, to do the same job (see Table 12-XI). When dilution water compatible with the mud is available in sufficient quantity and the overflow volume can be utilized or disposed of easily, and a good decanting centrifuge is not available or cannot be set up properly on the system, the perforated rotor machine can be of great assistance in controlling viscosity in a weighted water-base mud system, with considerable reduction in expenses.

Viscosity reduction by hydrocyclones

If the average weighted mud were fed to a balanced hydrocyclone capable of separating even a sizeable fraction of the inert solids, the underflow opening cannot handle the solids load and would rope. Inasmuch as a very high separation ability is necessary to separate the major part of an API barite grind and efficient separation requires spray discharge, weighted muds must be greatly diluted in order to reduce the total solids content before entering special barite hydrocyclone units.

Dilution must be sufficient to reduce solids content to 6—12% volume, i.e., between 9.0 lb/gal (sp. gr. = 1.08) and 9.8 lb/gal (sp. gr. = 1.18) feed slurry density. Such a dilution involves large quantities of water, much greater than those required by a holey rotor, for example. At high dilution, proper head, and spray discharge conditions, the best barite hydrocyclones for field work under laboratory test conditions can separate to the underflow about 80% of the barite having API grain-size distribution and greater than 2 microns in size. With such a good separation, the free liquid film on the barite particle surfaces keeps underflow specific weight down to 14—17 lb/gal (sp. gr. = 1.7 to 2.05) or lower.

Hydrocyclones do a fairly good job of separating the colloidal phase to the overflow in spite of the amount of liquid film on particles in the underflow, because of the high dilution ratio. The high dilution ratio results in the reduction of the colloidal content per unit of liquid phase to such a degree that only a small fraction of clay-size material is found in the underflow.

In heavier muds, the low specific weight of the underflow at maximum separation creates a problem. Either (1) care must be exercised to mix new barite all the time the cyclones are operating in order to maintain the mud weight, or (2) the hydrocyclones must be operated with a rope underflow to raise the underflow weight up to or above pit mud weight, with a resulting heavy loss of very fine to medium silt-size barite in the cyclone overflow. Although special coarse barite grinds can reduce this loss, other problems are created, i.e., abrasion, differential sticking, and hole instability.

The cost of hydrocyclones designed for partial-flow removal of clays from weighted muds is lower and they are cheaper to operate than decanting centrifuges and perforated rotor; however, their use cannot be justified over decanting centrifuges, nor over perforated rotor machines, on the total mud cost basis. Hydrocyclones can be considered for barite salvage if (1) good decanting centrifuges are not available or cannot be serviced, (2) perforated rotor separators are not available, (3) water for dilution is available in large quantities and at low cost, and (4) the overflow disposal does not present a problem. If mud weights are above 17 lb/gal (sp. gr. = 2.05), the effect of hydrocyclones on drilling rate and hole conditions must be watched carefully.

TABLE 12-XI

Comparative study of various solids separators [a]

	Variations by Feed Mud Weight					
sp. gr.:	1.44	1.68	1.92	2.16	2.40	2.64
lb/gal:	12	14	16	18	20	22
Feed mud capacity (not including dilution water), (gal/min):						
decanting centrifuge, 18 in. × 28 in. conical	21.7	15.5	11.9	9.7	8.5	8.0
perforated rotor separator	17.3	14.7	13.2	12.3	11.8	[b] NR
hydrocyclone units	25.0	16.7	12.9	10.8 [b]	[c]	[c]
Recommended dilution water at above feed mud rates, (gal/min):						
decanting centrifuge	3.0	3.0	3.0	3.0	3.0	3.0
perforated rotor separator	12.1	10.3	9.2	8.6	8.3	8.1
hydrocyclone units (average)	53.7	64.3	67.1	68.6 [b]	[c]	[c]
Underflow density (lb/gal):						
decanting centrifuge, bowl	21—25 ppg, depending upon feed content					
decanting centrifuge, hopper	10—25 ppg, depending upon remix adjustment and feed					
perforated rotor separator	15	17	19	20	20	[b] NR
hydrocyclones, spray discharge	14—17, depending upon feed content					
hydrocyclones, rope discharge	17—22 [b], depending upon the cut point					
Separation of colloidal size particles to overflow as percent of colloidals feeding						
decanting centrifuge	95—98%± at all feed and underflow hopper weights					
perforated rotor separator	75—50%± as feed weight varies					
hydrocyclone units	95%± at all feed and underflow weights					
Recovery of API grind barite larger than two microns as a percent of the same size feeding						
decanting centrifuge	98%± at all weights					
perforated rotor separator	92%± at all feed weights					
hydrocyclones, spray discharge	75% maximum sustainable					
rope discharge	65% downward to 20% or plugged					

[a] Primary separation, partial flow, and special clay removal units of the best designs available in 1973, operated according to manufacturers' instructions and extrapolated, whenever necessary, according to published data or actual tests. Variations are possible by "pushing", but at sacrifice of some result.

[b] Hydrocyclones are not recommended for normal muds which are so heavily weighted with API barite.

[c] These hydrocyclone underflow weights should not be attempted without especially coarse ground barites. One should be aware of differential-sticking problems.

Hydrocyclones capable of removing clay-sized solids from weighted muds have small dimensions (Pioneer DSC-300C 3-inch SOLIDSMASTER used in their CLAYMASTER©, and the Dorr-designed SWACO 50-mm cyclone used in their CLAYJECTOR©) and are subject to plugging, which (whether feed or underflow) cannot be tolerated at all. When in use, therefore, hydrocyclones on this service must be checked frequently. To facilitate quick, frequent checking, the underflow of each individual cyclone should be visible at all times, without dismantling any part of the unit.

Table 12-XI summarizes the critical separation characteristics of three types of centrifugal separators of best design available for removal of colloids, with consequent viscosity reduction, from weighted muds. It may not be possible to obtain or install always the one that is functionally best suited, but it is good practice to be aware of the compromise being made. In any mud solids separation problem, the benefits and savings to be obtained are more important than the rent or amortization differences between various types (and designs) of machines.

The practice of misnaming solids separation equipment for the obvious purpose of confusing the customer should be condemned.

SECONDARY SEPARATION OF UNWEIGHTED MUDS (EQUIPMENT APPLICATION)

Although full-flow primary solids removal (always selectively of the larger size particles) reduces the disposal problem compared to dilution, it does not eliminate it.

In desilting unweighted muds, the liquid films associated with the separated silt particles create a waste disposal problem and may cause a very expensive liquid-phase loss. Disposal problems remain when desilting any drilling fluid in highly-populated areas, or desilting an unweighted oil mud in any area. Desilting in the Arctic also involves serious problems, because water necessary for desilting is extremely expensive, and liquid waste cannot be left in the area.

In such cases, one solution to the problem would be to confine the removal of solids to coarse material that does not have a high specific surface area and which can be removed by full-flow shale shaker screens. With properly-designed screens having sufficient capacity, the free liquid draining from the large cuttings should not present a major problem. Mud will have a higher solids content, however, due to the remaining particles that could have been removed by desilting hydrocyclones. This gives rise to (1) higher mud weight and hydrostatic head on the formation, (2) thicker and coarser filter cake with increased tendency for differential sticking in permeable formations, (3) slower drilling progress, (4) shorter bit footage life, (5) increased pump and drillstring abrasion, and (6) hole instability in the presence of weak formations. Economic remedy lies in secondary separation discussed below.

Hydrocyclone—decanting centrifuge combination

Desilting the mud with full-flow hydrocyclone desilters (primary separation) and then using a decanting centrifuge in secondary separation, to separate the liquid-clay phase and return it to the active system, is very effective (see Fig. 12-34). The underflow of the decanting centrifuge, with no free liquid present, contains silt and sand removed by the hydrocyclone desilters. This decanter underflow can be handled by a dirt scoop or blade, but cannot be sucked into a vacuum truck because of the lack of free liquid. Its disposal is a very simple problem compared to other alternatives.

In expensive liquid-phase muds, e.g., water-base muds in the Arctic and in the desert, or oil-base muds anywhere, reduction in liquid cost alone is a sufficient economic justification for using the decanting centrifuge for secondary separation. In some areas such as the City of Los Angeles and Europe, reduction in disposal cost also pays for the decanter. In all cases the reduction in mud cost has proved to be more than the added expense of running the machine on unweighted muds.

In the early 1970's, a Canadian environmental agency stopped drilling in

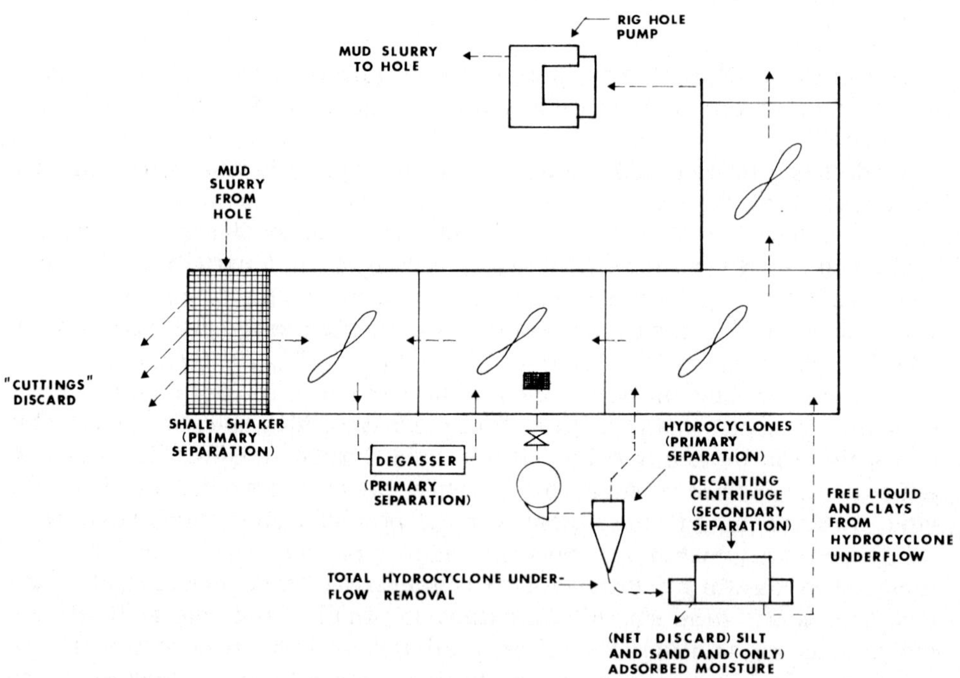

Fig. 12-34. Decanting centrifuge performing secondary separation after primary separation by desilters.

the McKenzie Delta area until liquid mud waste could be eliminated. Drilling soon started again using the decanter, which returned the liquid-clay phase from the hydrocyclone underflows back to the system. Mud cost reductions were two to five times higher than the cost of amortization and maintenance of the decanting centrifuges. Drilling benefits were reported to be at least as good as when the total underflow of the desilters had been discarded. This quickly became standard procedure for unweighted muds in most areas of Canada, with probably as many installations of this type as in all other parts of the world together.

Compared with the method of discarding total desilter underflow, the reported specific weights of systems in which the liquid-clay is returned by decanters are only about 0.1 lb/gal (sp. gr. = 0.012) heavier. The explanation may lie in the fact that when the total underflow is discarded, there is a tendency to add commercial solids with the water makeup. When the liquid circuit is closed with a decanter, no materials of any kind should be added to the mud system unless there is concrete evidence that it is needed. The mud remains extremely stable, the hole tends to stay in very good condition, and penetration rates remain very high.

Perforated rotor machines

The free-flowing underflow requirement of the perforated rotor centrifuge prevents its use as a secondary separation unit on the underflow of hydrocyclones. The inherent characteristics of separation mechanism of these machines prevent their modification to provide an underflow which is devoid of free liquid. In addition, they would lose their installation flexibility advantage.

Hydrocyclone—fine-screen combination

Special fine shale shaker screens for secondary recovery from primary separation hydrocyclones (Robinson and Heilhecker, 1975) are normally not recommended for unweighted water-base muds, because silt particles removed by the primary separation desilters are returned to the system. This decreases the drilling benefits achieved through removal of silt-size solids. In addition, there is an increase in active system specific weight, compared to desilting without secondary separation of the underflow, by approximately 0.5 lb/gal (sp. gr. = 0.06) and higher.

The secondary-separation screening process diminishes the volume of discard by returning the silt with associated free liquid and clays to the system. In a situation where the increased mud weight, total solids content, and abrasiveness of the returned solids do not harm the drilling operation, the secondary-separation screens can be used to advantage.

If there are adverse drilling effects, mud cost savings from secondary separation cannot justify its use. If in such a case the underflow liquid loss from the primary separation desilters cannot be tolerated, installation of a decanting centrifuge is the sensible alternative to stopping the desilting operation.

SECONDARY SEPARATION OF WEIGHTED MUDS

Although recorded test data are very scarce and no articles on the subject have been published, hydrocyclone desilters have been run full-flow on weighted muds at least as far back as 1976. Larger desanders, some "homemade", have been used on weighted muds routinely since about 1957. Although there were common reports of good results, no scientific report on which to base cost analyses has been published. Most operators recently have been reluctant to attempt to improve the condition of weighted mud using full-flow hydrocyclone fearing that barite replacement costs would be prohibitive, in spite of the fact that most drilling rigs are equipped with desilters and/or desanders and it is not difficult to evaluate the cost and the drilling benefits in the field.

Referring to Fig. 12-35, the underflow of the primary hydrocyclones removes only a small fraction of the total solids in the cyclone feed. This underflow contains, selectively, the coarsest fraction in the mud. It is the

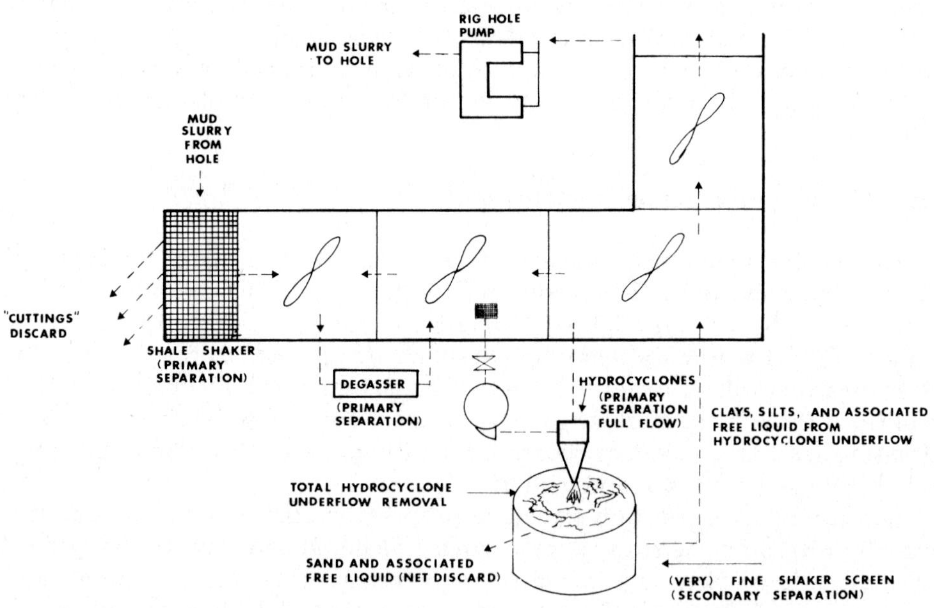

Fig. 12-35. Special shale shakers for secondary separation of weighted muds following primary-separation desilters.

most detrimental to drilling, drilling equipment, and the hole, whether barite or drilled solids are involved. On an equal weight basis, it costs 80% as much (or more) to replace light solids (weight effect) as it does to replace barite. In order to reduce drilling cost it is better to remove the coarser fraction rather than to attempt to distinguish closely between barite and light solids.

As shown in Table 12-II, the larger-size solids in the mud, both barite and low-specific-gravity solids, furnish degraded particles to the clay-size fraction. The bigger the particles, the more they contribute to the colloidal phase by degradation, because of more forceful collisions due to their greater mass (Ormsby, 1965).

Inasmuch as the clay-size materials increase the viscosity that plague most weighted muds, it is reasonable to assume that constant removal of the larger particles, including barite, might reduce the rate of viscosity buildup in a weighted mud by removing part of the potential supply of degraded colloids.

Hydrocyclone—fine-screen combination

Hydrocyclone—fine-screen combinations called "mud cleaners" (Robinson and Heilhecker, 1975), which were developed for weighted muds, were discussed briefly under the Unweighted Mud section. Logically, their use is normally confined to weighted muds (muds containing heavy commercial suspended solids, such as barite) (see Fig. 12-35).

In using the hydrocyclone—fine-screen combination, it is of primary importance to remember that the whole purpose of operating any mechanical separator is to remove material that is detrimental to the drilling operation and, thereby, improve the operation. Cost is one of the variables that must be improved.

The second fact to remember is that in the mud cleaner combination all removal from the system is performed by the cyclones. Drilling benefits which may accrue are due to this cyclone removal of solids. In a weighted mud, the cyclones remove mostly sand, some coarse silt, a minute amount of medium silt, and a large amount of free liquid film, which is loaded with clay and a small amount of fine silt particles.

All solids of every size that are discarded must be replaced by barite and a small amount of water. The fine screen, placed under the hydrocyclone to process the solids underflow discharge, will return most of the medium silt and a major part of the clay-sized material and fine silt back to the system. Whether or not it returns much of the coarse silt depends upon the screen mesh and its operation. This process will reduce barite usage compared to operating the cyclones without the screens. Two factors, however, tend to counteract this saving:

(1) The benefit of the removal of solids by cyclone may be reduced by the screen return. This can be checked by observation and, possibly, avoided by proper screen choice.

(2) If installation of a decanting centrifuge becomes necessary in order to control viscosity in the weighted mud, it would seem logical to stop returning the clay-liquid phase back to the system with the secondary screen.

Inasmuch as "mud cleaners" have become quite controversial, a supervisor must evaluate his own situation prior to making a decision to use them.

Hydrocyclone—decanting centrifuge combination

In 1969, there were preliminary reports of attempts to apply centrifuges on secondary separation in Europe for processing the desilting hydrocyclone underflow on weighted muds in order to reduce disposal hauling and to investigate effects on the drilling. Excellent data were obtained in 1971 on both weighted and unweighted muds (Maikranz, 1972).

In order to use a decanting centrifuge successfully on the underflow of primary cyclones for weighted muds and to close the liquid circuit, one or more of the following conditions must exist:

(1) The formations drilled must have low "mud-making" tendency, i.e., they must not tend to break up easily into clay-size particles.

(2) Sufficient volume of cuttings must be of sand size so that a major portion of solids can be removed with the hydrocyclones.

(3) The drilling must be able to tolerate higher mud flow properties than the norm for most drilling operations.

Obviously, application of the secondary centrifuge will probably remain rare in the case of weighted muds.

In Table 12-XII, items 1, 3, and 7 pertain to the screening, desilting by hydrocyclone, and secondary separation by the decanting centrifuge, respectively. It is extremely important to realize that in the case of the existence of a viscosity problem, a second centrifuge to remove colloids (item 4 or 5) or special cyclones (item 6) are not necessary. The secondary-separation centrifuge already in operation is returning colloidal material to the system that would be removed by another centrifuge in primary separation. Thus, they will be working against each other. The first remedial step would be to stop the secondary separation and divert the desilter underflow to waste. If this cannot be tolerated, secondary separation must be continued, but the centrifuge liquid overflow must be diverted from the system. It can be used on an unweighted mud job. In either case, a liquid-clay phase will be removed from the system, with consequent reduction in viscosity when it is replaced with water.

If the first step of allowing total desilter underflow to leave the system fails to give sufficient viscosity control, then the decanting centrifuge should be converted to primary separation (item 4, Table 12-XII). This will give rise to two effective avenues for colloidal removal simultaneously: (1) the colloids in the free liquid film in the desilter solids underflow, and (2) the colloids in the centrifuge liquid overflow. In addition, the larger particles are

TABLE 12-XII

Effects on total solids content, mud weight, and plastic viscosity in the case of primary and secondary separations for full and partial flow processing (assuming constant mud level in the pits and that specific weight of weighted mud remains constant)

Item	Type of equipment [1]	Separation	Type of flow	Character of the discard	Effect on:				
					unweighted mud			weighted mud	
					total solids	mud weight	plastic viscosity	total solids	plastic viscosity
1	Shale shaker screens	Primary	Full	Wet to damp	Holding down	Holding down	Negligible	Holding down	Negligible
2	Desanding [2] hydrocyclones	Primary	Full	Wet	Holding down	Holding down	Negligible	Holding down	Negligible
3	Desilting hydrocyclones	Primary	Full	Very wet	Holding down	Holding down	Negligible	Holding down	Negligible to slightly down
4	Decanting centrifuges	Primary [3]	Partial	Low vol. — liquid	M.A. [4]	M.A. [4]	M.A. [4]	Slightly down	Down
5	Perf. rotor separator	Primary	Partial [5]	Medium vol. — liquid	M.A. [4]	M.A. [4]	M.A. [4]	Slightly down	Down
6	Special barite cyclones	Primary	Partial [2]	High vol. — liquid	M.A.	M.A.	M.A.	Slightly down	Down
7	Decanting centrifuge	Secondary	(Desilter underflow)	Damp	Slightly up	Slightly up	Negligible	M.A.	M.A.
8	Special screens	Secondary	(Desilter underflow)	Wet to damp	Up (M.A.)	Up (M.A.)	Up (M.A.)	Up, but [6] variable	Up [3]

[1] Best designs available.
[2] On oil muds, use only in conjunction with item 7 or item 8.
[3] On oil muds use only as item 7 for active system work.
[4] Misapplication. Items 1, 2, 3 are the effective primaries on unweighted mud.
[5] On oil muds use only to split into heavy and light phases for special purposes.
[6] Compared to operation of the cyclones *without* a screen.

still being removed by the desilter. Maintaining weight and volume with barite and water will then resolve the most severe viscosity problems in weighted-mud drilling.

It should be kept in mind that returning a liquid-clay fraction back into the system from which it came cannot increase or decrease viscosity. Discarding this liquid-clay fraction, however, will reduce viscosity if it is replaced with water.

Two-stage centrifuging

Occasionally reference is made to "double stage" centrifuging in the case of weighted muds (see Fig. 12-36). This process began on the Gulf Coast of Texas and Louisiana, U.S.A., in the late 1950's, fell into disfavor in the early 1960's, and was not used for several years. Several misconceptions were involved in using this process: (1) Assumption that drill solids, or at least the detrimental ones, are all in a size range just above clay size; examination of Figs. 12-28, 12-29, and 12-30 indicates the error of this assumption. (2) An assumption that there are no barites in this range (see Figs. 12-7 and 12-8). (3) Assumption that colloidal materials (including excessive amounts of chemicals) do not cause problems and do not need to be discarded.

A modern revival of centrifuge—centrifuge combinations has a quite different objective and approach. The first machine is operated as a normal clay-rejection, silt-salvage machine for weighted mud. The secondary ma-

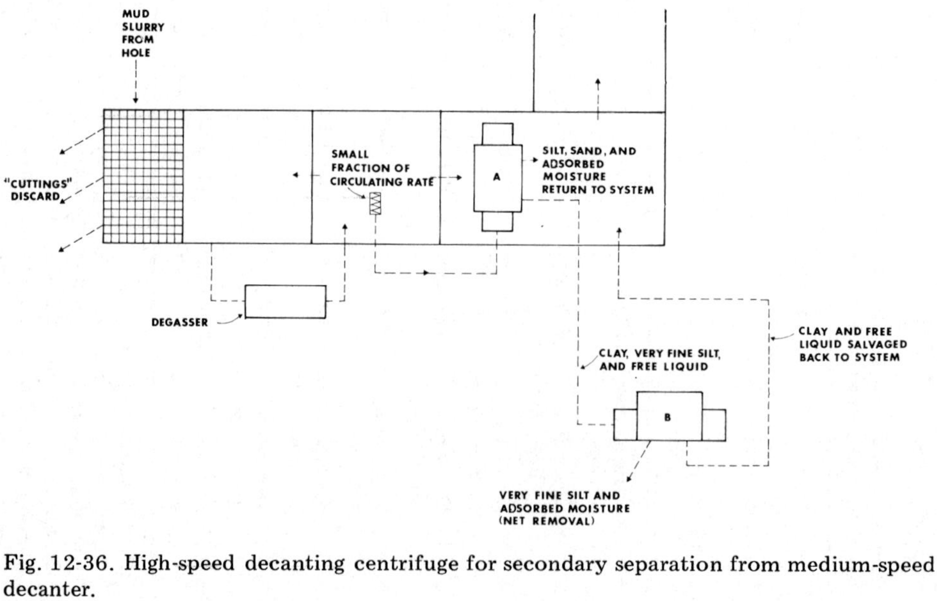

Fig. 12-36. High-speed decanting centrifuge for secondary separation from medium-speed decanter.

chine receives the liquid overflow from the primary centrifuge, containing barite particles less than 2 microns in size and low-specific-gravity solids below 3 microns in size.

The general idea was that in the case of separation down to a fraction of a micron ($\frac{1}{2}$ to $\frac{1}{4}$ micron, for example), only the colloidal material would remain in the liquid phase. It was assumed that only a very small portion of drilled solids can contribute purely colloidal material. Thus, the latter would not build up in the system and viscosity can be controlled in a weighted mud without discarding any liquid phase.

This approach, however, has not been successful, because formations in many areas can contribute some high-grade colloidal material. In addition, formation solids can become extremely fine, whether they are of a high-grade or not. Although the field operation of centrifuges with the proper geometry at the necessary *g*-forces to produce nearly clean water requires sophisticated maintenance and costs a great deal, development efforts continue. There are some areas where the liquid-clay disposal from weighted mud is expensive enough to justify drastic means.

SOLIDS CONTROL IN THE TRANSITION ZONE BETWEEN WEIGHTED AND UNWEIGHTED MUDS

The problem of controlling solids in weighted muds is quite different from that for unweighted muds. For example, an ordinary unweighted fresh-water mud will usually contain approximately 50 lb of solids per bbl (143 kg/m^3). If clay constitutes a very high percentage of these solids, viscosity can be easily controlled with a small amount of thinner.

On the other hand, a 15-lb/gal (sp. gr. = 1.8) mud, for example, will contain approximately 380 lb of solids per bbl (1085 kg/m^3). Obviously, the solids content cannot be significantly reduced if specific weight has to be maintained. Even if half of this material were in the clay range, viscosity could not be controlled. The content of clay-size material will tend to increase, but clay content must be held to a minimum. This has long posed a dilemma to operators who reach a depth where the formation pore pressure requires only slightly higher hydrostatic head than the desilted unweighted mud in the hole can provide, and the operator does not want to experience the drilling disadvantages of stopping the desilters. This problem, which occurs mainly in the mud weight range of 9 lb/gal (sp. gr. = 1.08) to 10.5 lb/gal, has been resolved in three basic ways in the past:

(1) Desilting is continued, while barite is added as higher specific weight is required. This appears to be the best method. If the liquid circuit has been closed with a decanting centrifuge, it may be necessary to stop the liquid salvage due to viscosity buildup. If the underflow discard from the desilters then presents a problem, a secondary screen might be used under the

cyclones (mud cleaner combination), *provided* (a) the drilling benefits of the solids removal are still obtained when the silt and liquid-clay are returned to the system, and (b) the viscosity buildup does not continue.

Usually, hydrocycloning with desilters can be continued up to mud weights of 13 lb/gal (sp. gr. = 1.55) without a decanting centrifuge if the cyclone underflow is discarded, i.e., without the mud cleaner screened return. At about 13 lb/gal and, sometimes at lower weights, regular weighted-mud treatment is necessary.

(2) A much quicker method often has been used in faster-drilled holes with thin, rapidly-changing transition zones. Intensive desilting using minimum mud weight is continued until the transition zone is reached. Then, simultaneously, desilting is stopped, decanting centrifuging for removal of colloidal material is started, and rapid weighting with barite is commenced, while fast drilling is continued. Several operators have reported increasing mud weight by 3 lb/gal (sp. gr. = 0.36) in one circulation. According to them, this can be done only with a thoroughly desilted mud. This method has proved to be very successful under proper supervision. It is a field procedure, however, and should not be used in exploration wells.

(3) The most common method involves stopping the desilting and allowing mud weight to increase due to the incorporation of drilled solids. Barite is added as higher weight is needed. For viscosity control, a centrifuge is used or the discard-and-makeup method is resorted to. Using this technique, the uncontrolled coarse solids cause problems such as abrasion and sidewall sticking. Usually, the sidewall sticking zone starts at a depth where desilting stops whenever this particular weight-up technique is used.

The secondary separation screens (mud cleaners) may be of benefit in the "twilight zone" of neither weighted nor unweighted muds. As shown in Table 12-XII, the return of the liquid phase from the screen to the unweighted system causes some weight increase (about 0.5 lb/gal) compared to the technique of discarding the desilter underflow. Barite can be added if a more rapid increase in mud weight is needed. As the transition to a weighted mud is completed, additional measures can be taken if needed. Each operator must design the best system for the specific conditions which exist in his well.

INSTALLATION AND MAINTENANCE OF EQUIPMENT (CONCLUDING REMARKS)

The subjects of proper mud system design, equipment installation engineering, and maintenance programs have been discussed by various authors (e.g., Brandt, 1973; Ormsby, 1965, 1973).

In the case of any drilling operation in the world, poorly maintained solids removal equipment usually reflects the attitude of the management.

Equipment will not be kept in operating condition if management does not make it clear that failure to maintain and operate it properly will not be tolerated. If an installation is not properly designed, most likely it will be changed soon if proper maintenance is insisted upon.

Equipment inadequate for the job cannot operate properly. Higher solids removal capacities are needed in the case of faster drilling rates. The potential drilling benefits justify installation of the best equipment available.

In addition, training for supervisors and key men on the drilling rig (both operator and contractor) is absolutely necessary.

REFERENCES

API Committee on Standardization of Drilling Fluid Materials, Task Group on Solids Removal Devices, Luther Bartlett, Task Group Head, June 1974. *Am. Pet. Inst., API RP 13-C,* 1st ed., 22 pp.

API Task Group on Determination of Liquids and Solids, D.B. Anderson, Chairman, June 9, 1969. (Set up in 1968 by API Committee on Standardization of Drilling Fluid Materials.) Official minutes, Meeting, Denver, CO., 4 pp.

Bobo, R.A. and Hoch, R.S., 1953. *The Mechanical Treatment of Weighted Drilling Muds.* AIME Pap. No. 290-G, Dallas, Texas, 18—21 Oct.; *Trans. AIME,* 1954, 201: 93—96.

Brandt, L.K., 1973. Remarks on fine screen shakers. *Proc. Annu. Rotary Drill. Conf. IADC,* Houston, Texas, Feb. 28—Mar. 2, p. C-1—6.

Burdyn, R.F., Hawk, D.E. and Patchen, F.D., 1965. New device for field recovery of barite: II scale-up and design. *Proc. 2nd Conf. Drill. Rock Mech.,* Univ. Texas, 20—21 Jan., pp. 31—40.

Cagle, W.S. and Wilder, L.B., 1978. Layered shale shaker screens improve mud-solids control. *World Oil,* 186 (6, April): 6 pp.

Gassett, P., 1964. *Drilling Tests Conducted on a Drilling Platform Offshore Cameron Parish, Louisiana.* Data related to the author's work circa 1966.

Gill, J.A., 1966. Drilling mud solids control — a new look at techniques. *World Oil,* 163 (5, Oct.): 121—124.

Longwell, C.R., Knopf, A. and Flint, R.F., 1941. *Outlines of Physical Geology,* 2nd ed. Wiley, New York, N.Y., 381 pp.

Maikranz, E., 1972. *Experience with a Centrifuge installed to Recover Mud from the Desilter Underflow.* Brigitta-Elwerath Betriebsführungsgesellschaft, unpublished Intracompany report, 24 pp. plus 46 pp. of appendices.

MOHOLE Project, 1964. *MOHOLE-calculated data.* MOHOLE test work on hydrocyclones conducted near Uvalde, Texas, sponsored by Natl. Sci. Foundation, Stonewall Associates library.

Mondshine, T.C., 1966. New fast-drilling muds also provide hole stability. *Oil Gas J.,* March 21: 84—90.

Nelson, M.D., 1970. *Removal of Fine Solids from Weighted Muds.* API Pap. No. 926-15-H, presented before the Southern District Division of Production, Houston, Texas, 4—6 March, 4 pp.

Ormsby, G.S., 1965. *Desilting Muds with Hydrocyclones.* Trans. Rotary Drill. Conf., Am. Assoc. Oilwell Drill. Contractors, Dallas, Texas, 8—9 Feb., pp. 78—83.

Ormsby, G.S., 1966. *The Results of an Analysis of a Dual Centrifuging Process on a Weighted Inverted Oil Emulsion Mud Containing Asphalt Additives.* 10 Mar. (unpublished), Stonewall Associates, Inc. library, 21 pp.

Ormsby, G.S., 1973. *Why Your Removal System Won't Remove.* Trans. Rotary Drill. Conf., Int. Assoc. Drill. Contractors, Houston, Texas, 28 Feb.—2 Mar., pp. B-1—B-25.
Orr, Jr., C. and Dallavalle, J.M., 1959. *Fine Particle Measurement,* 1st ed. MacMillan, New York, N.Y., 353 pp.
O'Shields, R.L. and Wuth, Don E., 1955. New mud desander cuts drilling cost. AAODC Annual Meeting, Sept., Houston, Texas. *Drilling Contractor,* Sept.—Oct.: 4 pp.
Perry, J.H., 1950. *Chemical Engineers Handbook,* 3rd ed. McGraw Hill, New York, N.Y., 1942 pp.
Petroleum Extension Service, 1969. *Principles of Drilling Fluid Control,* 12th ed. University of Texas Press, Austin, 215 pp.
Pioneer Centrifuging Company, unpublished data files.
Ritchey, R.L., 1959. *Effect of Particle size on $\overline{YP}$ and Viscosity of Drilling Muds.* Salt Water Control Co. (SWACO) Publ., Feb., folded brochure.
Robinson, L.H. and Heilhecker, J.K., 1975. *Solids Control in Weighted Drilling Fluids.* 48th Annu. Fall Meet. Soc. Pet. Eng. AIME, Las Vegas, SPE Pap. 4644; *J. Petrol. Technol.,* 1975 (Sept.): 1141—1144.
Roper, W.F., 1972. Leader of screen section of Solids Removal Task Group, verbal statements in Task Group Meetings leading to API RP-13C Publ.
Stanford Research Institute, 1961. Characteristics of particles and particle dispersoids. *Bull. from Stanford Res. Inst. J.,* 3rd quarter, 1 p. (Dept. 300, Stanford Res. Institute, Menlo Park, Calif.)
U.S. Dept. of Agriculture, S.C.S., 1971. *A Guide for Interpreting Engineering Uses of Soils,* 1971 Manual. Director of Documents, Washington, D.C. (Also see *International Standard Soil Classification System,* adopted by International Society of Soil Scientists in 1934.)

Chapter 13

LOW DENSITY FLUIDS; PACKER FLUIDS AND CASING PACKS; AND LUBRICITY AND FRICTION REDUCERS

B.S. GUPTA, GEORGE V. CHILINGARIAN and PAUL VORABUTR

LOW-DENSITY FLUIDS

Introduction

Low-density drilling fluids used in balanced or underbalanced drilling consist of air, mist, stable foam, and aerated mud foam with back pressure. "Air drilling" implies the use of air or natural gas as the circulating medium instead of the regular mud, whereas the term "aerated mud" implies the simultaneous introduction of air and mud together into the standpipe in order to drill special types of formations.

The aerated drilling fluids have certain advantages and disadvantages. The main advantage of air as a circulating fluid is that being the lowest density fluid, it imposes minimum pressure on the formation to be drilled. High penetration rates have been achieved in hard and dry formations with the use of air as a circulating fluid. In addition to high penetration rate, longer bit life results through the use of this medium as compared to mud (see Chapter 2). According to Angel (1957, p. 1), drilling rates as high as 90 ft/hr have been attained in shales. Air drilling, however, is restricted to areas where high-volume water sands are not present ahead of the producing zone. The rate of water influx that can be handled in the case of air drilling is also not well known.

Other inherent disadvantages on using air or natural gas (particularly the latter) as drilling fluids include (1) possibility of downhole fires and explosions, and (2) sloughing of formations due to underbalance of stresses around the wellbore (Cooper et al., 1977, p. 53). Possibility of downhole explosions are of particular concern in air drilling operations. Small dust-like particles are generated as a result of (1) rock cuttings (chips) being ground and pulverized by the drillstring in the annulus, and (2) collision of cuttings with each other, the tool joints, and the wall of the borehole due to the high-velocity forces (Cooper et al., 1977, p. 54). In the presence of moisture, seal rings may form at tight places in the annulus, which create pressure chambers. With additional influx of natural gas from gas-bearing zones being penetrated by the bit, an explosion may easily occur (see Fig. 13-1).

Figure 13-1 illustrates the inflammable range of natural gas. Even at small

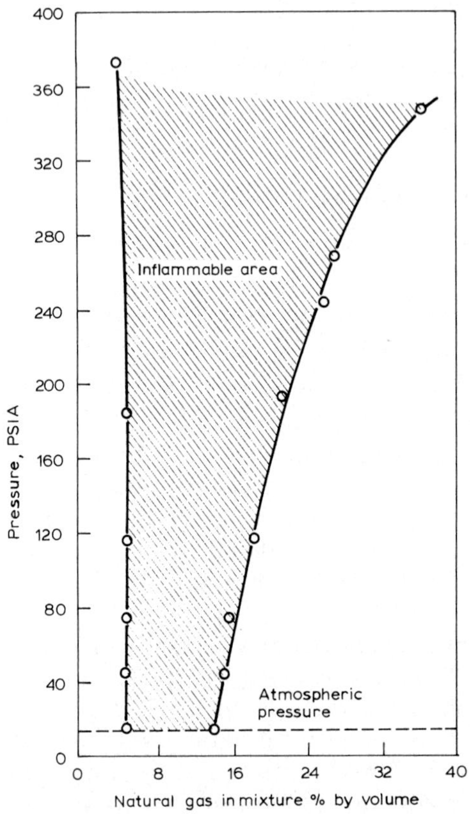

Fig. 13-1. Effects of pressure on the limits of inflammability of Pittsburgh natural gas. (After Bureau of Mines Report of Investigations 3798; courtesy of Bureau of Mines.)

concentrations of natural gas in the air present in the annulus, either added intentionally or resulting from the gas influx on penetrating gas-bearing zones, the air—gas mixtures can be highly inflammable at low pressures.

Aerated drilling fluids

Besides having formations suitable for air drilling, the most important consideration in drilling with air is the volume of air required. According to Cooper et al. (1977, p. 58), air drilling often fails because of insufficient volume of air necessary to clean the hole efficiently under certain conditions, e.g., wet hole, sloughing shales, and influx of formation water.

As pointed out by Cooper et al. (1977, p. 58), a practical rule of thumb for determining adequate air volume is that time required to clean the hole should not greatly exceed one minute per 1000 ft of depth. Table 13-I shows

TABLE 13-I

Minimum air requirements for achieving efficient hole cleaning for various hole and drill-pipe sizes; estimated penetration rate = 50 ft/hr.

(After Cooper et al., 1977, p. 65, table I; courtesy of the Society of Petroleum Engineers of AIME)

Hole size (inches)	Drill-pipe (inches)	Depth and volume required, Scf/M						
		1000	2000	4000	6000	8000	10000	12000
$17^1/_2$	$6^5/_8$	6150	6600	7150	7700	8400	8950	9500
	$5^1/_2$	6450	6700	7400	8000	8550	9100	9650
	$4^1/_2$	6600	6850	7550	8100	8700	9250	9800
$12^1/_4$	$6^5/_8$	2600	2800	3000	3400	3700	4150	4500
	$5^1/_2$	3000	3100	3350	3900	4300	4700	5000
	$4^1/_2$	3100	3350	3700	4200	4500	4900	5400
11	$6^5/_8$	2000	2150	2350	2650	3200	3550	3900
	$5^1/_2$	2300	2400	2700	3000	3300	3700	4000
	$4^1/_2$	2500	2600	2900	3200	3400	3800	4100
$9^7/_8$	$5^1/_2$	1700	1950	2150	2400	2600	2850	3100
	5	1900	2000	2300	2600	3000	3300	3600
	$4^1/_2$	1950	2100	2400	2700	3200	3500	4000
9	5	1450	1550	1850	2150	2350	2550	2850
	$4^1/_2$	1550	1700	2000	2300	2500	2700	3000
	$3^1/_2$	1700	1850	2200	2600	2700	3000	3300
$8^3/_4$	5	1350	1500	1700	2000	2200	2400	2700
	$4^1/_2$	1450	1650	1850	2140	2400	2600	2850
	$3^1/_2$	1600	1750	2000	2280	2600	2800	3000
$7^7/_8$	$4^1/_2$	1100	1200	1500	1800	2100	2350	2650
	$3^1/_2$	1300	1400	1650	1900	2120	2400	2700
$7^3/_8$	$3^1/_2$	1070	1150	1300	1550	1800	1950	2150
$6^3/_4$	$3^1/_2$	850	1000	1200	1420	1650	1850	2100
$6^1/_4$	$3^1/_2$	710	790	930	1030	1280	1420	1650
	$2^7/_8$	790	850	970	1150	1420	1700	1800
$4^3/_4$	$2^7/_8$	430	500	650	790	930	1070	1200

minimum air volume requirements for various hole and drillpipe sizes (i.e., various annular volumes). Inasmuch as there seems to be no upper limit on the air volume requirement, the volume of air generally used is slightly larger than that listed in Table 13-I.

Air drilling is also used to prevent formation damage, and severe lost circulation can be prevented by reducing the drilling fluid pressure gradient that exists in the low-pressure zones. The use of aerated muds proves to be most successful in drilling vugular and cavernous limestones and dolomites. In certain areas of the Gulf of Mexico, there are coral formations several thousand feet in thickness which are filled with seawater to the sea level. Lost circulation occurs when these formations are drilled with regular mud, because the drilling mud pressure gradient is higher than the formation fracture gradient. Even on drilling with seawater, lost circulation occurs because of the annulus circulating pressure and the weight of the cuttings in the fluids. When air is mixed with the circulating seawater, however, the low density of the resulting aerated fluid permits introduction of fluids from the formation. This influx of formation water into the circulating fluid helps to carry cuttings to the surface, as a result of increase in the annular velocity.

At shallow depths, limestones and dolomites frequently contain some cavernous intervals, which are generally hard and compact. On drilling such zones with aerated fluid, the annulus pressure becomes less than that of the formation, resulting in the influx of formation fluids. According to Rogers (1963, p. 661), amounts of fluids produced from such formations have been of the order of thousands of barrels per hour.

Advantages of aerated drilling fluids

The advantages of aerated drilling fluids can be summarized as follows:

(a) Economy.

(b) Lifting of cuttings to the surface, and bit and drillstring cooling are adequately handled.

(c) No pipe sticking occurs in the case of low-permeability and low-pressure zones.

(d) Prevention of lost circulation.

(e) Reduction of fire possibilities because of the presence of water in the drilling fluid.

(f) Weight buildup can be quickly achieved by shutting off the source of injected air.

Use of aerated fluids

Only muds of good quality and having low gel strength can be aerated. After each cycle the mud must be de-aerated for its proper handling by the pumps.

In order to avoid corrosion, maintenance of high pH and/or use of corrosion inhibitors is an absolute necessity. In a few days, the drillpipe string may be damaged, if proper precautions against corrosion are not taken.

Inasmuch as the gas expands on rising in the annular space, the velocities are high, i.e., of the order of 2000—4000 ft/min. This condition could result in the damage of the open hole and, therefore, sufficient casing protection must be provided. According to H.B. Woods (in: Angel, 1957, p. 1), in quarry drilling an annular velocity of 3000 ft/min of standard air is required for best results. Angel (1957) derived the following equation for determining the circulation rates necessary to produce annular velocities which are equivalent in lifting power to some velocity of standard-density air, v_e (specific gravity of the drilled solids was assumed to be 2.70):

$$\frac{6.61S(T_s + Gh)\,Q^2}{(d_h^2 - d_p^2)^2 v_e^2} = \sqrt{[p_s^2 + bT_{av}^2]\,\exp(2ah/T_{av}) - bT_{av}^2} \tag{13-1}$$

where:

$$a = \frac{Sq + 28.8Kd_h^2}{53.3q}$$

It is a function of gas gravity, hole size, solids specific gravity, drilling rate, and circulation rate.

$$b = \frac{1.625 \cdot 10^{-6} q^2}{(d_h - d_p)^{1.333}(d_h^2 - d_p^2)^2}$$

It is a function of circulation rate, hole diameter, and pipe outside diameter.
d_h = hole diameter (ft).

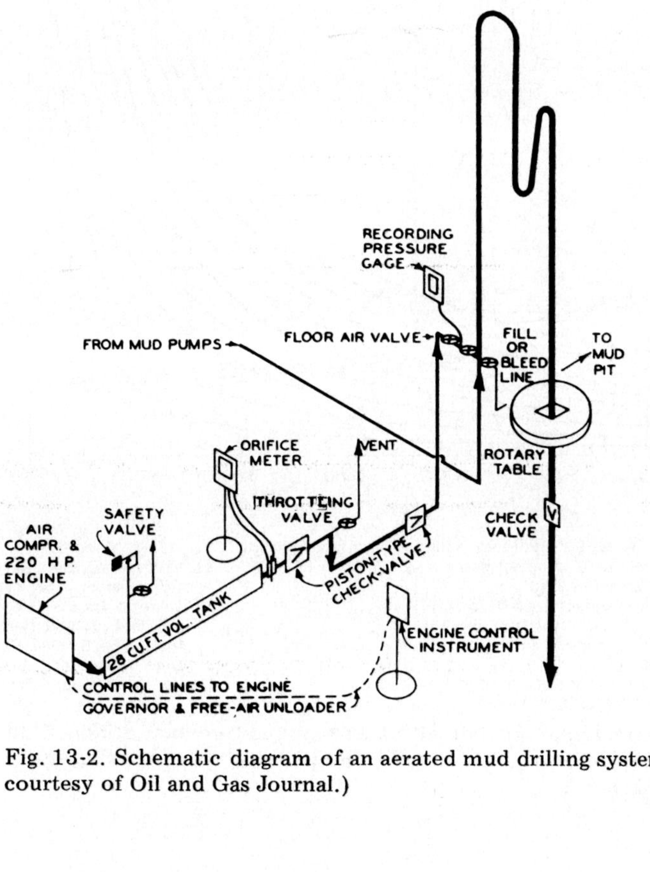

Fig. 13-2. Schematic diagram of an aerated mud drilling system. (After Bobo et al., 1955; courtesy of Oil and Gas Journal.)

d_p = pipe outside diameter (ft).
e = 2.71828, base of natural logarithms.
G = annular temperature gradient (°F/ft).
h = depth below the surface to any point under consideration (ft).
K = drilling rate (ft/hr).
p_s = pressure in the annulus at the surface (lb/sq ft abs.).
q = circulation rate (standard cu ft/min).
S = specific gravity of gas related to air (dimensionless).
T_s = absolute temperature in the annulus at the surface (°R).
T_{av} = average absolute temperature of a flow stream (°R).
v_e = velocity of standard density air (ft/min).

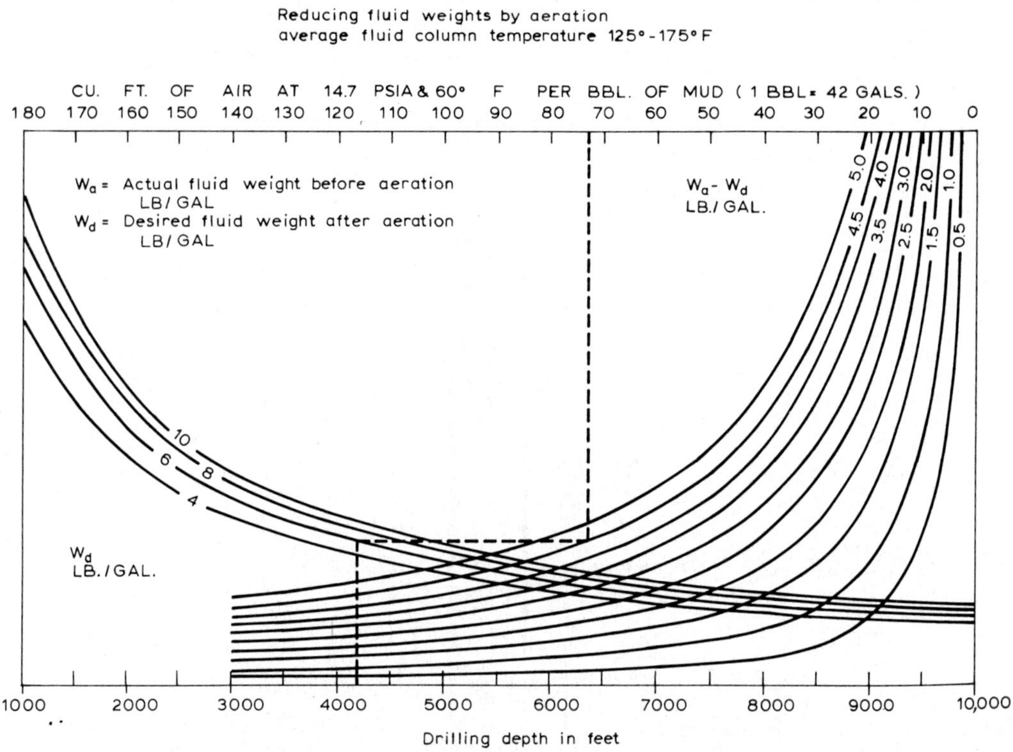

To find air requirements
1) Find drilling depth in feet on bottom scale.
2) Go up to appropriate curve of desired fluid weight after aeration (W_d)
3) Go across to the appropriate curve of the difference between the actual fluid weight before aeration and the desired fluid weight after aeration (W_a - W_d)
4) Go up to the cubic feet of air needed per barrel of mud on top scale

Example (follow dotted lines)
1) 4200' deep hole
2) Up to 6 LB./GAL. desired fluid weight after aeration
3) Over to 4.5 LB./GAL. (10.5 LB./GAL. actual fluid weig minus 6 LB./GAL. Desired fluid weight)
4) Up to 73 cubic feet of ai

Fig. 13-3. Nomogram for determining the amount of air required to reduce drilling fluid weight by aeration. (After Cooper et al., 1977, p. 68, fig. 13; courtesy of the Society of Petroleum Engineers of AIME.)

In cases where the drilling fluid pressure is considerably less than the formation pressure and the influx of foreign fluids becomes excessive, pumps and fresh mud should be available so that the well can be brought under control easily.

Bobo et al. (1955) have presented a schematic diagram of an aerated mud drilling system as shown in Fig. 13-2. Air is injected into the system between the mud pumps and the standpipe. The air/mud ratio is controlled by regulating the inputs of each fluid.

Figure 13-3 may be used to determine the air requirement to reduce fluid weights by aeration.

Air and preformed stable foam, which constitute the true "no-solids" drilling fluids, initially do not contain any solids. The entrained drilled formation solids are discharged directly to a waste sump. The lifting capacity of foam depends more on its viscosity rather than on velocity, which is the controlling factor with air or mist. With appropriate back pressure and rotating head pack-off equipment, circulating fluids with specific weights of 3.75—70 lb/cu ft can be attained with preformed stable foam fluids.

Stable foam

Bikerman (1973) calssified states of aggregation of phases as follows: (1) liquids dispersed in gases (mists, fogs); (2) gases dispersed in liquids (foams, gas emulsions); (3) solids dispersed in gases (fumes, smokes); (4) gases dispersed in solids (solid foams, foam rubber, etc.); (5) liquids dispersed in liquids (emulsions); (6) liquids dispersed in solids (gels); (7) solids dispersed in liquids (suspensions, sols); and (8) solids dispersed in solids.

The "stiff" foam, i.e., the AEC-type foam developed by the Atomic Energy Commision, consists of a gel-base mud containing compatible foaming surfactant. Very viscous foam forms on injecting this fluid into an air stream. This foam has excellent wall building and hole cleaning characteristics. Contamination with crude oil, salt water, or calcium cation, however, cannot be tolerated, because foamers compatible with montmorillonite clays are not usually compatible with crude oil, and clays tend to ball up in the presence of salt water or calcium cation.

The original AEC foam consisted of a gel-detergent base mix consisting of $12\frac{1}{2}$ lb/bbl of bentonite, 1 lb/bbl soda ash, and $\frac{1}{2}$ lb/bbl of gum guar. Although this foam worked well in fresh water, salt water flocculated the bentonite. A foam which could be used both in fresh and salt water consists of 1 lb/bbl of Cypan, $\frac{1}{2}$ lb/bbl of Drispac, and 1% anionic foamer.

Injection of fresh or salt water with foaming surfactant into a gaseous phase in a foam generator gives rise to a preformed stable foam. This foam, which has excellent hole cleaning ability at very low annular velocities, is resistant to contamination by crude oil, salt water, solids, and calcium cation. It has desirable thermal properties for drilling in permafrost areas and at high-temperature conditions.

The surfactants do not exceed 10% by volume of the fluid to be foamed (Rogers, 1963) and may be lower than 1% of the total volume (Hutchison et al., 1969a, b). Most foaming agents are non-ionic and organic; however, they may cause foams to be susceptible to dissolution under the effects of various hydrocarbons and ions. Such foams can be stabilized by (1) carboxy-methyl-celullose (CMC) in highly-concentrated brines (Zore et al., 1970), (2) a mixture of decomposed proteins and inorganic salts (Kapca et al., 1971), and (3) minor amount of unneutralized ammonia (Hutchison et al., 1969b).

Foam can be used as an alternative in an air-drilled well on encountering a water-saturated formation, with a water influx in excess of approximately 20—50 bbl/hr.

Foam is also used in clean-out and remedial work in already drilled wells. Foamed hydrocarbons have been tested successfully both in the laboratory and in the field (in several areas) in cleanout and remedial work (Smith and Holcomb, 1979, p. 76). Foam has been successfully used as the primary drilling fluid in West Texas, California, and limestones and dolomites in Colorado (Hutchison and Anderson, 1972).

According to Kennedy (1974), foam drilling has speeded up the completion of drilling wells in West Virginia. Drilling with foam in the Carpathian region of U.S.S.R. reduced the problems of cavern formation, caving, and cuttings removal encountered on using air and liquid drilling (Zore et al., 1970).

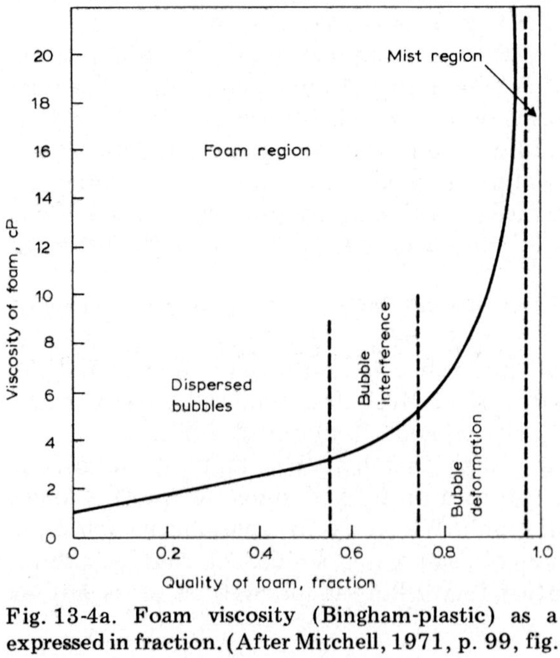

Fig. 13-4a. Foam viscosity (Bingham-plastic) as a function of the quality of foam expressed in fraction. (After Mitchell, 1971, p. 99, fig. 5; courtesy of Oil and Gas Journal.)

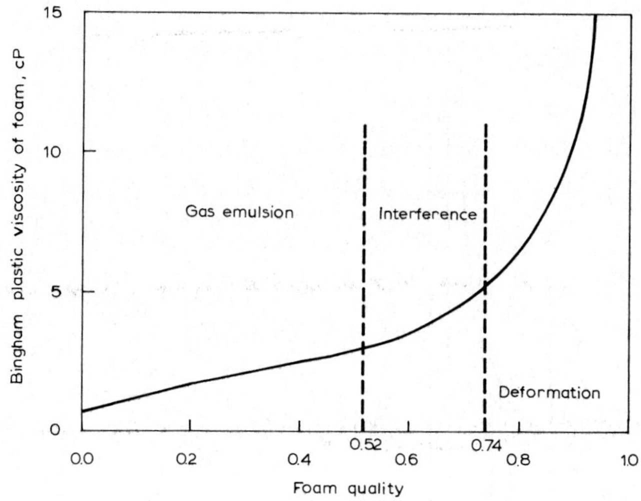

Fig. 13-4b. Relationship between Bingham-plastic viscosity of foam and foam quality. (After Smith and Holcomb, 1979, fig. 1, p. 76; courtesy of Petroleum Engineer International.)

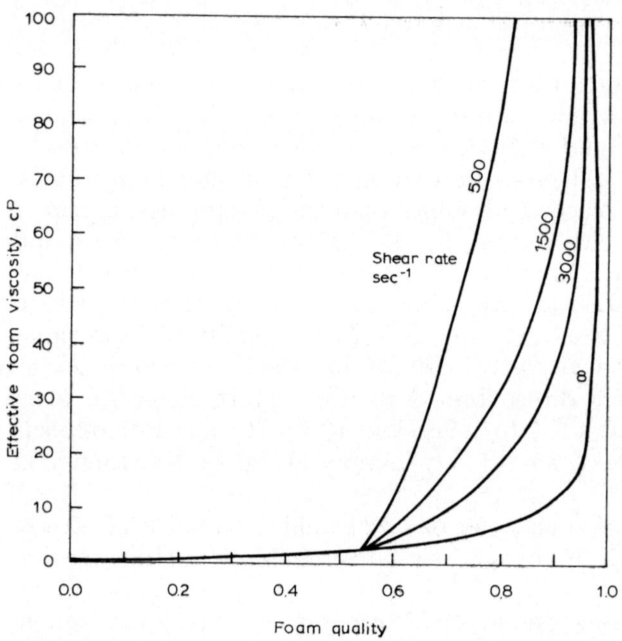

Fig. 13-4c. Relationship between effective viscosity and foam quality at various shear rates. (After Smith and Holcomb, 1979, fig. 3, p. 78; courtesy of Petroleum Engineer International.)

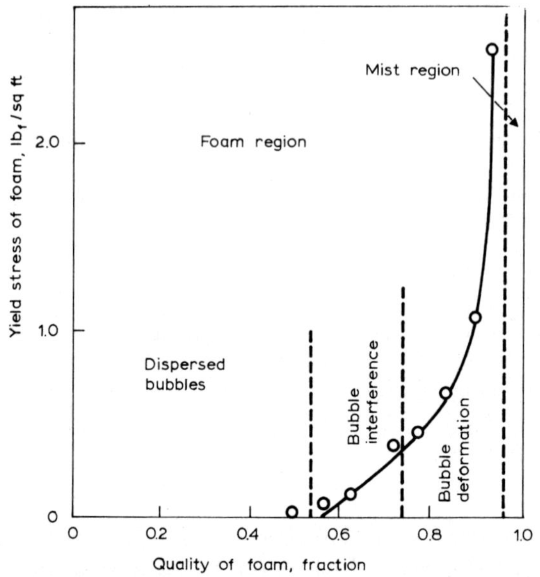

Fig. 13-5. Relationship between yield stress of foam and foam quality expressed as fraction. (After Mitchell, 1971, p. 99, fig. 6; courtesy of Oil and Gas Journal.)

Foams could exhibit non-Newtonian flow, inasmuch as they are colloidal systems in which individual units are much larger than ordinary molecular sizes. With increasing gas/liquid ratio, the foam tends to be less of a mass of semi-independent bubbles and more of a mass of liquid membranes. Inasmuch as it is easier to displace the bubbles than the membranes, apparent viscosity should increase with an increase in the foam quality (Raza and Marsden, 1967). Mitchell (1971) proved this to be true and found that foams having a quality less than 0.54 are Newtonian in nature, whereas those with a quality greater than 0.54 are Bingham fluids. As the quality of foam varies from 0 to about 0.96, the viscosity varies from about 1 to over 22 cP (Fig. 13-4). The yield stress varies from 0 to over 4 lb/sq ft as the foam quality changes from about 0.54 to 0.96 (Fig. 13-5). If shear rate exceeds 20,000 sec^{-1}, however, the foam of any quality becomes Newtonian in nature.

Even though viscosities of foams can be comparable to mud fluid viscosities, depending on quality of foam, their bottomhole circulating specific weights are much lower.

Other investigators (see Beyer et al., 1972) prefer to use the liquid-volume fraction, LVF, at temperature T and pressure p:

$$LVF_{(T,p)} = \frac{V_L}{V_L + V_{G(T,p)}} \tag{13-2}$$

Foam quality, QF, at a particular temperature, T, and pressure, p, is related to LVF as follows:

$$LVF_{(T,p)} + QF_{(T,p)} = 1 \tag{13-3}$$

Several equations have been developed for determining parameters for use of foam in fracturing and flow (NOWSCO, 1975):

$$\mu_e = \mu_p + \frac{38.7559(T_y)(d^3)}{Q_t} \tag{13-4}$$

where μ_e = effective viscosity (cP); μ_p = plastic viscosity (cP); T_y = yield stress (lb/sq ft); d = inside diameter of tube (inches); and Q_t = total flow rate (bbl/min).

$$N_R = \frac{15{,}916.2(Q_t)(\gamma_f)}{d\mu_e} \tag{13-5}$$

where N_R = Reynold's number, and γ_f = specific weight of foam at a particular temperature, T, and pressure, p, in lb/gal.

$$\Delta p/L = \frac{11.408(f)(\gamma_f)(Q_t^2)}{d^5} \tag{13-6}$$

where $\Delta p/L$ = pressure drop (psi/ft); f = Fanning friction factor; and L = depth (ft).

$$\gamma_f = 8.331\ (LVF) \tag{13-7}$$

An example of interrelationship among liquid volume fraction (LVF), plastic viscosity (μ_p), and yield stress (T_y) is presented in Table 13-II.

TABLE 13-II

Interrelationship among liquid volume fraction (LVF), plastic viscosity (μ_p), and yield stress (T_y)

(After NOWSCO, 1975)

Liquid volume (LVF)	Plastic viscosity, μ_p (cP)	Yield stress, T_y (lb/sq ft)
0.35	4.0	0.15
0.30	4.6	0.25
0.25	5.4	0.40
0.20	6.45	0.60
0.15	8.1	0.82
0.10 (est.)	11.4	1.03
0.05 (est.)	20.7	2+

Mist

Injection of small volumes of mud or water containing a foaming surfactant into a high-velocity gaseous stream results in a "mist" drilling fluid. The foaming surfactant is carried in the gaseous stream to the bottom of the hole where it mixes with the formation water. This increases the lift efficiency of the high-velocity gaseous stream and permits removal of water from the hole.

Rheological properties of foams

The principal advantage of low-solids (less than 2%) fluids (air, mist and foam) is the low viscosity existing at the bit, which is one of the main factors affecting the penetration rate (Kennedy and Meyer, 1968).

According to Sherman (1950), the following physical characteristics may affect the rheological properties of foams: (1) viscosity of the gas in the internal phase, (2) viscosity of the fluid in the external phase, (3) relative volume concentrations of the gas and liquid phases, (4) type of surface-active agent (surfactant) and the nature of the interfacial film formed, (5) electrical forces, and (6) bubble-size distribution. Although in most cases only the first three characteristics are important as far as rotary drilling is concerned, the bubble size may be of great importance if pressure is lowered to the extent that bubbles can expand and coalesce. The foam half-life is also of great importance (see Table 13-III). The use of foam may be restricted if its half-life is too short.

Mitchell (1971) defined the "quality" of the foam (QF) as: (volume of gas)/(total volume of gas plus liquid). He conducted experiments with foam by varying its quality, but keeping the viscosity of both the internal and external phases constant, i.e., using air and water, which are the most common foam constituents. He minimized the bubble effect by conducting the tests under a constant pressure of 1000 psi.

Krug and Mitchell (1972) developed charts and equations needed to determine the volume and pressure requirements for air and water at minimum hydraulic horsepower for various size holes, drilling rates, and depths (see Figs. 13-6 and 13-7).

Drilling in permafrost areas

Ferrians et al. (1969) defined permafrost as soil or rock, with or without included moisture or organic matter, which has been exposed to a temperature below 0°C continuously for two or more years. The total area covered by permafrost is estimated to be as high as 20% of the earth's surface, i.e., land masses north or south of 65°N or S latitudes. Permafrost is found from 8 to 18 inches below the surface to a depth of between 800 to 2000 ft. The

TABLE 13-III

Foam half-life with various concentrations of typical foamer in various hydrocarbons

(After Smith and Holcomb, 1979, table 3, p. 84; courtesy of Petroleum Engineer International)

Hydrocarbon	Concentration of foamer (vol %)	half-life (min)
Kerosene	0.1	3.0
	0.2	3.2
	0.3	4.6
	0.4	4.7
	0.5	7.2
Xylene	0.1	1.1
	0.2	5.0
	0.3	8.4
	0.4	17.8
	0.5	19.3
Diesel	0.1	4.5
	0.2	5.0
	0.3	6.2
	0.4	7.3
	0.5	7.0
San Andres crude oil	0.1	7.3
	0.2	7.8
	0.3	10.4
	0.4	14.8
	0.5	16.9
Condensate	0.1	6.2
	0.2	8.3
	0.3	8.9
	0.4	9.4
	0.5	9.7

ice content varies from up to 90% in the upper 20% layer of permafrost to no ice in the deepest portions. This ice occurs in lenses from a few inches to a few feet in thickness (Staff, 1969).

Air drilling has been used for drilling in permafrost areas for very shallow boreholes, such as those needed for the initial surface casing (Dier, 1969).

Foam drilling in permafrost areas has solved many of the problems encountered on using conventional muds, including polymer muds, and air drilling. It optimized the economics related to penetration rate and equipment transportation and reduced the problems of cuttings removal and hole deviation and enlargement. Other advantages of foam include maximum cuttings size, minimum wetting, no recycling, and good identification of

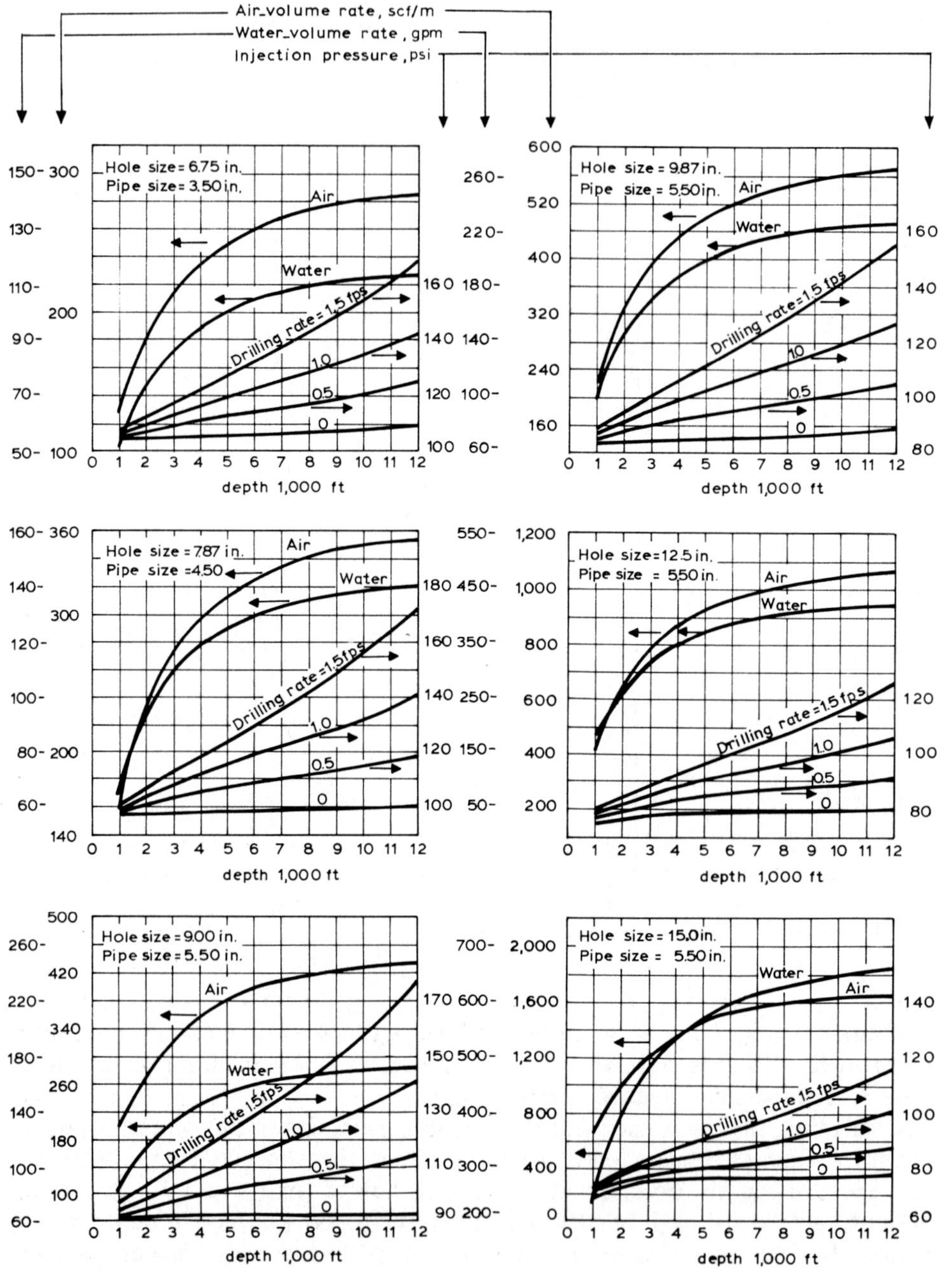

Fig. 13-6. Required volumetric rates of air and water at minimum hydraulic horsepower for various (1) size holes, (2) drilling rates, and (3) depths. (After Krug and Mitchell, 1972, p. 63, fig. 2; courtesy of Oil and Gas Journal.)

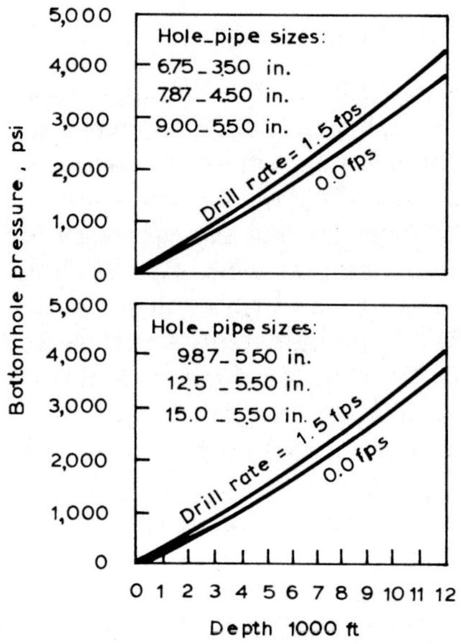

Fig. 13-7. Required air circulating pressure at minimum hydraulic horsepower for various (1) circulating volumes, (2) hole sizes, (3) drilling rates, and (4) depths (use in conjunction with Fig. 13-6). (After Krug and Mitchell, 1972, p. 63, fig. 3; courtesy of Oil and Gas Journal.)

hydrocarbon shows. Inasmuch as the essential constituents of foam are air, water, and a small content of organic foaming agent, which is biodegradable, foam is not harmful to the ecology and is easily destroyed. Formation samples are obtained from the collapsed foam.

Low heat conductivity and low heat capacity decrease the heat transfer that could melt permafrost and result in hole enlargement. Preformed foam can be kept only a few degrees above freezing ($\approx$35—45° F) and can be used without any freezing point suppressants. One of the serious problems encountered on drilling with regular muds in the permafrost areas is caused by the freezing of fluids left between the drillhole and the casing string, especially in the borehole irregularities caused even by small washouts. Expansion caused by freezing can result in the collapse of casing string (McGhee, 1973). Foam, on the other hand, (1) minimizes washouts, (2) is much more easily displaced than muds by either oil-base pack or cement, and (3) does not present a danger to the casing by the freezing expansion, even if left behind the casing, because it contains only 2% water by volume (see Anderson, 1971).

PACKER FLUIDS AND CASING PACKS

Clay-laden or solid-free liquids, which are left in the annular space between the tubing and casing, are known as packer fluids. The packer fluid provides a hydrostatic head for pressure differential reduction across the packer in order to prevent development of potentially dangerous pressure on casing, casing interior, and tubing exterior. This is particularly necessary for high gas/oil ratio wells, condensate wells, and dry gas wells. Packer fluids must remain thermally stable for long periods of time. At shallow depths, the temperature effect is negligible, whereas at greater depths the packer fluids have a tendency to deteriorate, due to the high temperature and pressure. In addition, the packer fluids have the following functions:

(1) Prevention of settling of solids, while remaining fluid for easy placement and removal of tubing and casing.

(2) Prevention of corrosion of the outside of tubing and the inside of casing.

At present, high-grade steels are used for casing and tubing strings, and because they are more subject to damage from corrosion than ordinary steels, it is necessary to use packer fluids pretreated with chemical agents.

Casing packs are special fluids which are placed between the casing and the borehole walls. They protect casing against corrosion, contain formation fluid pressures, and, in some cases, enable recovery of the casing when it is not needed any longer. The long-term stability is an important consideration in the preparation of casing packs, because they are usually placed for long periods of time without periodic checking.

Types of packer fluids

Water-base muds

Water-base drilling fluids in general are not good packer fluids. Original drilling fluid, which has been used for drilling the well, is an ideal packer fluid, however, as far as *economy* is concerned. Water-base muds prove to be quite satisfactory in case of shallow wells, where the temperatures do not increase above 250° F. In order to prevent corrosion, a high pH of the order of 12—13 and above is desirable. The pH drops below 9.5 at temperatures higher than 200—220° F. A decrease in pH results in the loss of protection against corrosion and loss of other desirable mud properties. The danger inherent in sulfur-containing organic additives is evident. For example, the thermal degradation of lignosulphonates at temperatures of 300° F and higher gives rise to both H_2S and CO_2 (see Chapter 8).

Bacteria (especially sulfate-reducing bacteria) are usually active in water-base muds, unless (1) bactericides are added in adequate concentrations, (2) pH of mud is higher than 12, and (3) mud is saturated with NaCl.

Corrosion inhibitors have been used in the past. Due to the large surface

TABLE 13-IV

Possible combinations of caustic soda, carboxy-methyl-cellulose, bentonite and barite used as packer fluids

(After Tschirley, 1979, vol. II, table V. Courtesy of Baroid Petroleum Services)

Mud specific weight (lb/gal)	Caustic soda (lb/bbl)	CMC (lb/bbl)	Bentonite (lb/bbl)	Barite (lb/bbl)
11.0	1.0	2.0	10.0	148
12.0	1.0	2.0	10.0	234
13.0	1.0	2.0	10.0	310
14.0	1.0	2.0	9.0	393
15.0	1.0	2.0	8.0	485
16.0	1.0	2.0	7.0	585
17.0	1.0	2.0	6.0	696
18.0	1.0	2.0	3.0	820
19.0	1.0	2.0	4.0	958

area of the solids in the drilling fluid, however, the efficiency of corrosion inhibitors is reduced because they form a film around the surfaces of the solid particles. Proper protection can be obtained by mixing the inhibitor with a calculated quantity of oil.

A sweet crude or diesel oil may be mixed with the corrosion inhibitor to increase its efficiency and economy.

Mixtures of bentonite, carboxy-methyl-cellulose (CMC) and caustic soda have been used where the downhole temperatures do not exceed 275° F.

Several combinations of caustic soda, bentonite, barite, and CMC can be used as shown in Table 13-IV.

Although lime muds are essentially noncorrosive, they tend to solidify and become unpumpable with time and at high temperatures. Muds contaminated with cement [$Ca(OH)_2$] may behave similarly.

Oil-base muds

Oil-base muds can be prepared having specific weight of 7.8—22.0 lb/gal. They are considered to be noncorrosive, because of being nonconductive and the fact that they do not support the growth of microorganisms. Oil-base muds are much more stable than water-base muds and do not tend to solidify as a result of high temperatures when left in the borehole for long periods of time. At very high temperatures, additional gelling agents are added to prevent settling of solids (weighting material) due to the reduction in viscosity and gel strength. In general, long-term suspension of weighting material is more difficult to achieve in the case of oil-base muds as compared to water-base muds. An oil-base mud may contain 5—10% of water and is known as "oil mud". Invert emulsion muds containing 20—30% of water can also be used as packer fluids.

A careful planning is very essential in the selection of a particular packer fluid and depends upon a number of factors and local conditions.

Solids-free brine solutions

Brine solutions have been successfully used as packer fluids for several reasons: (1) clear brine solution minimizes formation damage (see Fig. 13-8); (2) solids settling problems are avoided; (3) solutions do not solidify, which allows easy circulation; and (4) clay swelling inside the perforations is avoided. Obviously, the specific weight of brine solutions is limited by the (1) type of salts and (2) solution temperature, which controls salt solubility in the aqueous phase. Common problem encountered with clear brine solutions, however, is corrosion.

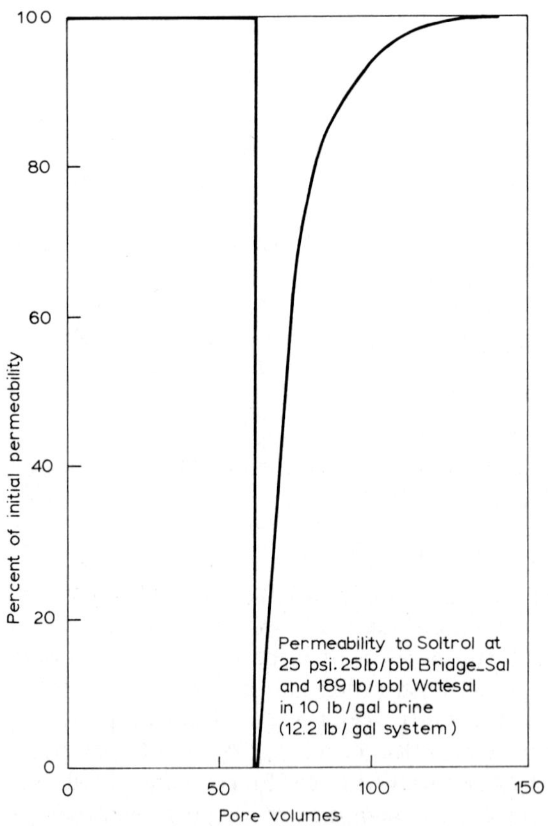

Fig. 13-8. Desirable return permeability for completion fluids (Packer Fluids, Casing Packs, etc.). (After Mondshine, 1977, p. 126, fig. 6; courtesy of Oil and Gas Journal.)

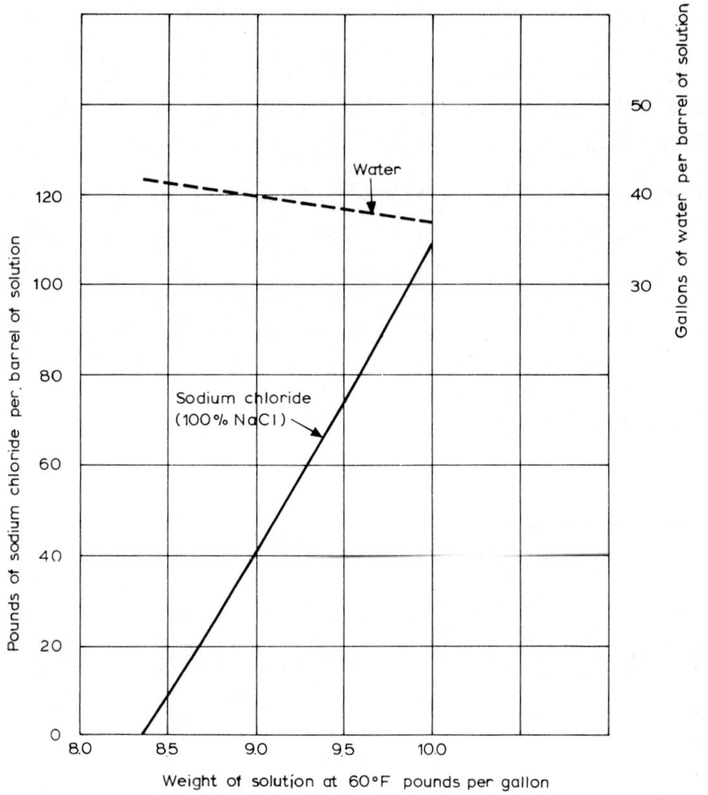

Fig. 13-9. Requirements of various components for preparing sodium chloride solutions. (Courtesy of Columbia Southern Chemical Corp.)

Figures 13-9 and 13-10 may be used to determine the amount of commonly used salts necessary to add in order to obtain a required specific weight of brine solution. More than one type of salts may be used to achieve desired specific weight. Reference temperatures used in the above calculations are 60°F, and the following equation may be used to calculate the change in density of these solutions (IMCO, 1978, 18-2):

$$\text{Density change (lb/gal)} = 0.003\ (T_1 - T_2) \tag{13-8}$$

where T_1 = existing temperature (°F); and T_2 = desired temperature (°F).

Common clear brine solutions which are used today include (1) sodium chloride, (2) sodium chloride—sodium carbonate, (3) sodium chloride—calcium chloride, (4) calcium chloride, (5) calcium chloride—zinc chloride, and (6) calcium bromide—calcium chloride.

It is of utmost importance to maintain these brine solutions at a pH of 11

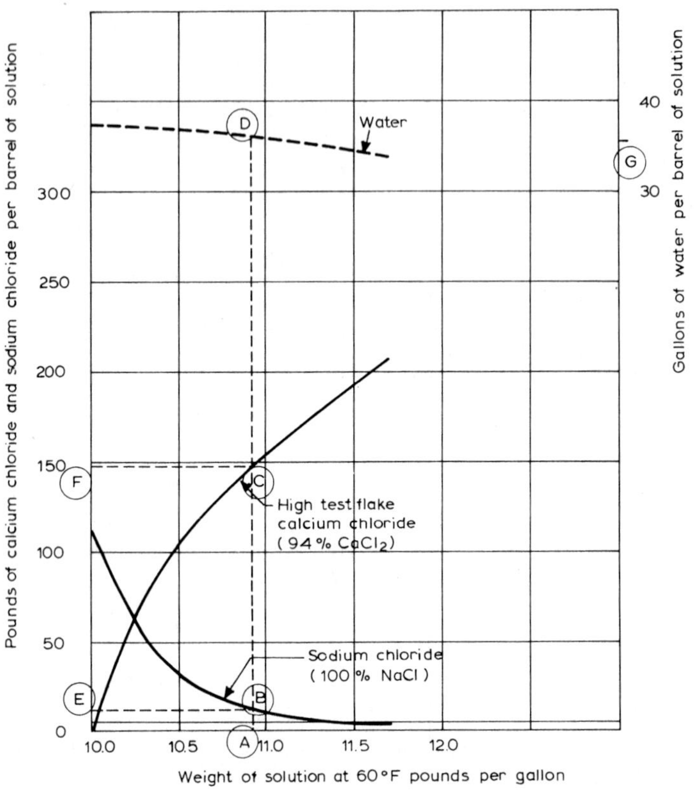

Fig. 13-10. Requirements of various components for preparing calcium chloride—sodium chloride solutions. (Courtesy of Columbia Chemical Corp.) Example: enter at *A*; proceed to *B*, *C*, and *D*; read required amounts of NaCl, $CaCl_2$, and water at *E*, *F*, and *G*, respectively.

and greater in order to avoid severe corrosion of tubular goods with caustic soda.

According to IMCO (1978, 18-3), there are several inherent problems (drawbacks) in using these various brine solutions, some of which have been mentioned previously: (1) the magnitude of specific weight is limited, (2) there is possibility of severe corrosion, (3) specific equipment may be required to handle certain salts such as zinc chloride, (4) some salts are hazardous to personnel involved, and (5) high cost of chemicals.

It is possible to increase the weight of saturated salt solutions by adding weighting materials if viscosifiers and gelling agents (e.g., CMC and numerous other polymers) are added to increase the carrying capacity of the solution.

Casing packs

Casing packs, which are placed between the walls of the borehole and casing, are usually oil-base or invert emulsion drilling fluids. They are used to (1) prevent external well-casing corrosion and (2) aid in casing recovery.

Casing packs are especially useful in wells drilled through permafrost and in steam-injection wells.

Casing packs must (1) have low fluid loss into formation over long periods of time, (2) limit thermal convection currents in certain cases, and (3) act as a noncorrosive barrier between the casing and the formation fluids.

Rotation and reciprocation of the casing are necessary for proper placement of casing pack, i.e., to cover the pipe completely. Uninsulated areas are especially susceptible to corrosion.

Recovery of casing

On plugging or abandoning of a well, it is desirable to salvage as much as possible of the casing from the borehole. Both protection string and intermediate casing are salvageable from the cement top to the surface. A proper casing pack can facilitate recovery of casing by (1) reducing friction between the casing and the borehole walls and (2) maintaining hole stability during the period of time between recovery of casing and its setting.

In designing a casing pack, one should mainly consider the formation pore pressure and water adsorption by shales from the casing packs. The pore pressures must be balanced by hydrostatic fluid pack pressure, whereas proper salinity control of the fluid will prevent destabilization (swelling and water-wetting) of shales. Obviously, the flow properties (rheology) should be properly adjusted in the same manner as for packer fluids. In addition, high-temperature, high-pressure fluid loss from the fluid to the formation should be kept at a minimum. It is advisable to place a specially-formulated spacer between the casing pack and cement.

Casing packs for steam-injection wells and wells in permafrost areas

In steam-injection wells, casing failure is a serious problem and initially occurs when temperature-generated compressive stresses exceed the yield strength of the casing. The compressive stresses are relieved upon subsequent cooling when the well is producing or is shut-in. The deformation produced during steam injection, however, causes a tensile stress as the casing temperature returns to normal. According to Wilhite and Dietrich (1966), this tensile force sometimes results in joint failure by fracture or pull-out. Thus, if casing pack would permit stress-free expansion and contraction during alternate heating and cooling, deformation and failure could be avoided. Excellent oil mud packs for steam injection wells have been designed by Baroid (1977a, p. 13) and other companies.

Because the temperature of the produced oil in the permafrost areas is much higher than the temperature of the frozen ground, the casing must be insulated from the ground in order to prevent melting of the ice. Provisions are made to allow casing to expand and contract with the temperature changes caused by shutting in and producing the well.

The gel strength of oil mud casing packs containing organophilic clay can be inhibited from yielding initially by mixing the mud at temperatures below 50°F (Baroid, 1977a, p. 13). This fluid is pumped behind the casing. Having a grease-like consistency, an oil-base casing pack, which serves to retard heat transfer, is formed in place. It also acts as an expansion—contraction medium. As an example, the gel strength of an oil mud casing pack (containing 0.475 bbl diesel oil, 0.475 bbl still bottoms, 0.05 bbl of 25% solution of $CaCl_2$, 4 lb/bbl of EZ MUL, and 45 lb/bbl of GELTONE) rises to about 240 lb/100 sq ft at a temperature of 170°F.

LUBRICITY AND FRICTION REDUCERS

One of very important functions of drilling fluids is to provide adequate lubrication to the bit-bearings and drilling string. If the drilling mud is properly treated by the addition of lubricants, one can expect a longer bit life because of lubrication of bit-bearings and reduction in their rate of wear.

Lubricating properties of drilling fluids depend on their composition (e.g., type of continuous phase) and the subsequently added additives. Thus, different drilling fluids possess different degrees of lubricity. This is evidenced by different lubricity coefficients of the fluids used in drilling fluid formations: 0.50 for air, 0.35 for water, and 0.07 for diesel oil (Baroid, 1977a, p. 5). Thus, oil-base drilling fluids may be expected to provide better lubrication than water-base drilling fluids and air (or foams). As pointed out

TABLE 13-V

Effect of oils on lubricity

(After Tschirley, 1979, vol. II, table V; courtesy of Baroid Petroleum Services)

Hydrocarbon	Concentration (bbl/bbl)	Oil spotted directly without mixing into mud	Oil incorporated into mud with high-shear mixing
None	0	0.13	0.23
Diesel oil	0.1	0.07	0.23
Crude oil	0.1	0.07	0.23
Fuel oil No. 6	0.1	0.07	0.23
Asphalt-kerosine	0.1	0.07	0.23
Kerosine	0.1	0.13	0.23

TABLE 13-VI

Lubricity of oil-base and water-base drilling fluids

(Courtesy of Baroid Petroleum Services, 1977a, vol. II, section I, p. 21, table IX)

Mud type	Lubricity coefficient
Unweighted water-base drilling fluid (fresh water was organically thinned)	0.24
Unweighted oil mud	0.09
High-weight, water-base drilling fluid (fresh water was organically thinned)	0.22
High-weight oil mud	0.08

by Tschirley (1977, p. 6), lubricity with oil is best obtained when the oil forms a continuous phase or when emulsification is minimized. Oil, when present as a continuous phase, wets the metal surfaces and forms a thin film of lubricant (Magcobar, 1977, p. 4). Table 13-V shows the effectiveness of oil as a lubricant. Oil spotted directly without mixing into drilling fluid (i.e., oil is a continuous phase) provides greater lubricity to the mud than when incorporated into the drilling fluid with high-shear mixing.

Lubricity coefficients of typical drilling fluids are presented in Tables 13-VI and 13-VII. Data obtained in the laboratory are also presented for comparison.

Lubricity of drilling fluids not only minimizes the aforementioned hole problems, but also affects the drilling rate which is discussed.

Frictional resistance to rotation of the drillstring, which is called torque, and frictional resistance to hoisting and lowering of the drillstring, which is called drag, are reduced on using lubricants.

According to Browning (1960, p. 100), the basic properties of the lubricants can be summarized as follows:

(a) They must react with or adsorb on the metal surfaces.

(b) The adsorbed material must be either plastic or pseudoplastic with a high yield point.

(c) The softening or melting point should not be too low.

(d) They should be insoluble in water as well as in oil.

In the past, bentonite, graphite, asphalt, diesel and crude oils, fine mica, ground nut shells, linseed oil, castor oil, fatty acids, etc. have been added to the mud in order to achieve this purpose. The industry, however, lacked the proper instrumentation in order to compare lubricity of one mud with another. Recently, Baroid Petroleum Services have designed an E.P. Mud Tester, for measuring and comparing lubricities of various types of muds.

In this apparatus, the speed of rotation of the drillpipe and the pressure

TABLE 13-VII

Lubricity coefficients of various drilling fluids

(After Tschirley, 1979, Manual of Drilling Fluid Technology, vol. II, section VII, p. 9, table VIII; courtesy of Baroid Petroleum Services)

Muds	Coefficient
Lab prepared muds	
Water	0.36
Low-solids mud	0.42
Low-solids mud treated with special lubricant	0.06
Low-solids polymer mud	0.42
Low-solids asbestos mud	0.47
Unweighted organic thinned mud	0.24
Unweighted organic thinned mud treated with special lubricant	0.16
Medium density organic thinned mud, 13 lb/gal	0.24
High-density organic thinned mud, 18 lb/gal	0.22
Seawater mud, unweighted	0.20
Sat. salt-water mud, unweighted	0.24
Oil mud, unweighted	0.09
Oil mud, 18 lb/gal	0.08
Field muds	
Organic thinned mud, 14.5 lb/gal	0.28
Organic thinned mud, 12.6 lb/gal	0.23
Organic thinned mud, 12.0 lb/gal	0.26
Organic thinned mud, 9.8 lb/gal	0.31
Organic thinned mud, 10.1 lb/gal	0.25
Brackish-water mud, 16.3 lb/gal	0.25
Brackish-water mud, 18.0 lb/gal	0.24
Seawater mud, 12.0 lb/gal	0.23
Oil-mud, 18.5 lb/gal	0.13
Brackish-water mud, 10.8 lb/gal	0.38
Brackish-water mud, 10.8 lb/gal treated with special lubricant	0.17
Sat. salt-water mud	0.27
Sat. salt-water mud with special lubricant	0.18

with which the pipe bears against the wall of the hole can be more-or-less simulated. A ring and a block (steel) are used in this apparatus to simulate drillpipe and borehole wall. The coefficient of friction, u, between two solids is equal to the frictional force, F, divided by the load or force, W, perpendicular to the surface, i.e., $u = F/W$. According to Amontons' law, the coefficient of friction is independent of the apparent area of contact. In the Baroid tester, the load W is the force with which the test block is pressed against the test ring through the torque arm. The dial on the torque arm indicates the magnitude of this force in inch—lb. On the other hand, the force, F, required to slide the ring surfaces and the block across one another, is measured by the ammeter in amperes required to turn the test ring at a prescribed rpm (number of revolutions per minute).

Rosenberg and Tailleur (1959) developed high-pressure lubricants in order to increase the life of bit bearings. These extreme-pressure ($\approx$30,000 psi) lubricants are able to reduce interfacial tension between liquids and solids, and possess the ability to lubricate steel surfaces. Extreme-pressure lubricants have also been found useful, when used alone or with oil, in reducing the torque and drag of the drillpipe in the drilling operations and solving problems associated with stuck pipe and other similar problems. One of the main problems of using extreme-pressure lubricants was their dissolution in oil, which retards its efficiency; however, this problem was overcome partially by sulfurization with elemental sulfur.

Browning (1960, p. 98), has pointed out the difference between normal and high-pressure lubrication. In normal lubrication, a highly viscous oil forms a film between two metal surfaces and, therefore, prevents direct metal-to-metal contact. Under heavy loading when the steel plates collide, however, metal galling and wear of bearings occur.

In such cases, special lubricants, such as iron sulfide, are used. When the bearing load becomes sufficiently great, metal-to-metal contacts occur between the "hills" of the moving surfaces. This gives rise to high temperatures at the points of contact, with consequent melting of the lubricant and formation of a metallic compound, which still acts as a lubricant.

Commonly used lubricants

Diesel oil lubricates very well when loosely emulsified. On the other hand, when tightly emulsified, it lubricates poorly. In the case of *conditioned* drilling fluids, very few of the commonly used lubricants are effective. Only sulfurized fatty acid with diesel oil, and blended triglycerides and alcohols lower the lubricity coefficient of fresh-water and salt-water muds (Baroid, 1977b, p. 6).

Sulfurized fatty acid is added in concentrations of 2—6 lb/bbl. It should not be added to muds containing calcium cation, however, because of formation of grease-like soap. In case the calcium cation cannot be treated out, oil should also be added to keep the grease-like reaction product from being removed by the shaker screen (Baroid, 1977b, p. 12).

Some commonly used polymers, such as Impermex, Cellex, Drispac, Cypan, and Dextrid, significantly improve the lubricity of drilling fluids, in concentrations of 1—4 lb/bbl.

The lower lubricity coefficient exhibited by mud containing barite is due to the lubricating mechanism, variously described as a "graphite" or a "ball-bearing" effect between surfaces rather tightly pressed together. It is analogous to the use of corn meal in place of wax for lubrication of the floors in barn dancing places.

REFERENCES

Anderson, G.W., 1971. Near gauge holes through permafrost. *Oil Gas. J.*, 69(38): 128—140.

Angel, R.R., 1957. *Volume Requirement for Air or Gas Drilling.* 32nd Annu. Fall Meet. Soc. Pet. Eng., SPE No. 8736: 8 pp.; *J. Pet. Technol.*, (Dec.): 325—330.

Baroid Petroleum Services, 1977a. Lubricity. In: *Manual of Drilling Fluids Technology*, 2(7): 12 pp.

Baroid Petroleum Services, 1977b. Packer Fluids and Casing Packs. In: *Manual of Drilling Fluids Technology*, 2(5): 14 pp.

Beyer, A.H., Millhone, R.S. and Foote, R.W., 1972. *Flow Behavior of Foam as a Well Circulating Fluid.* Pet. Eng. AIME Meet., Dallas, Texas, SPE No. 3986.

Bikerman, J.J., 1973. *Foams.* Springer, New York—Heidelberg—Berlin, 337 pp.

Bobo, R.A., Ormsby, G.S. and Hoch, R.S., 1955. Phillips tests air-mud drilling. *Oil Gas J.*, (Jan. 24).

Browning, W.C., 1960. Thinning agents control flow properties of mud. *World Oil*, 150 (2, Feb. 1): 94—103.

Cooper, L.W., Hook, R.A. and Payne, B.R., 1977. *Air Drilling Techniques. Deep Drilling and Production Symposium.* Soc. Pet. Eng. AIME, Amarillo, Texas, April 17—19, SPE Pap. No. 6435.

Dier, J.S., 1969. Techniques for setting drilling rig piling and surface casing under permafrost conditions. *Oilweek*, 19(49): 19—20.

Ferrians, O.J., Kachadoorian, R. and Green, G.W., 1969. Permafrost and related engineering problems in Alaska. *U.S. Geol. Surv. Prof. Pap.*, 678: 37 pp.

Hutchison, S.O., 1969. Foam workovers cut costs 50%. *World Oil*, 169 (6, Nov.): 73—91.

Hutchison, S.O. and Anderson, G.W., 1972. Preformed stable foam aids workover, drilling. *Oil Gas J.*, 70(20): 74—79.

Hutchison, S.O., Anderson, G.W. and McKindell, J.C., 1969a. *Generation and Use of Foamed Well Circulating Fluids* (Patent). Chevron Res. Co., Pat. No. 3,463,231.

Hutchison, S.O., Anderson, G.W. and McKindell, J.C., 1969b. *Ammoniated Foamed Well Circulation Fluids and Uses Thereof* (Patent). Chevron Res. Co., Pat. No. 3,486,560.

IMCO Services, 1978. *Applied Mud Technology.* 250 pp.

Kapca, F., Jaluvka, S. and Fiala, F., 1971. *Improvements in or Relating to Foam-Flush Drilling.* Patent Great Britain, 1,250,186, c 10/20/71, F 11/6/68; Pr Czech. 11/13/67; USSR Agency.

Kennedy, J.L., 1974. Air drilling may play role in deep Appalachian play. *Oil Gas J.*, 73(38): 136—137.

Kennedy, J.L. and Meyer, R.L., 1968. Low solids fluids reduce costs. *Oil Gas J.*, 66(29): 132—137.

Krug, J.L. and Mitchell, B.J., 1972. Charts help find volume, pressure needed for foam drilling. *Oil Gas J.*, 70(6, Feb. 7): 61—64.

McGhee, E., 1973. MacKenzie Delta: Getting the most from a short drilling season. *Oil Gas J.*, 71 (37): 71—76.

Magcobar, 1977. Lubricity. In: *Drilling Fluids Engineering Manual.* 14 pp.

Mitchell, B.J., 1971. Test data fill the gap on using foam as drilling fluids. *Oil Gas J.*, 69(36, Sept. 6): 96—100.

Mondshine, T.C., 1977. Completion fluid uses salt for bridging, weighting. *Oil Gas J.*, 75(34, Aug. 22): 124—128.

NOWSCO, 1975. *Stable Foam Fracture.* Nitrogen Oil Well Service Company.

Raza, S.H. and Marsden, S.S., 1967. The streaming potential and rheology of foam. *Soc. Petrol. Eng. J.*, 7(4): 359—368.

Rogers, W.F., 1963. *Properties and Composition of Oil Well Drilling Fluids.* Gulf Publ. Co., Houston, Texas, 818 pp.

Rosenberg, M. and Tailleur, R.J., 1958. *Increase Drill Bit Life Through Use of Extreme Pressure Lubricant Drilling Fluid.* Soc. Pet. Eng. Meet., Dallas, Texas, Oct., SPE No. 1152G: 8 pp.

Sherman, P., 1950a. Water-in-oil emulsions. I. Influence of dispersed-phase concentration on emulsion viscosity. *J. Soc. Chem. Ind. (London)*, 69 (Suppl. No. 2): S70—74.

Sherman, P., 1950b. Water-in-oil emulsions. II. Effect of pH on the reciprocal mobility. *J. Soc. Chem. Ind. (London)*, 69(2): S74—75.

Smith, M.A. and Holcomb, D.L., 1979. Foamed hydrocarbons: Compatible stimulation and cleanout agents. *Pet. Eng. Int.*, 51(11, Sept.): 76—84.

Staff, 1969. Slope drillers keep operations simple. *Oil Gas J.*, 67(33): 33—34.

Tschirley, N.R., 1977. Lubricity. In: *Manual of Drilling Fluids Technology*. Baroid Petroleum Services, 26 pp.

Tschirley, N.R., 1979. *Manual of Drilling Fluids Technology*. Baroid Petroleum Services.

Wilhite, G.P. and Dietrich, W.R., 1966. *Design Criteria for Completion of Steam Injection Well.* Soc. Pet. Eng. Meet., Dallas, Texas, Oct., SPE No. 1560: 7 pp.

Zore, O.U., Vasyuk, V.M., Reukin, V.T., Bronzou, A.S., Muradyan, I.M. and Shanovich, L.P., 1970. Use of flowing foam as a circulating agent in well drilling. *Nauch.-Tekh. Sb. Ser. Burenie*, (1): 16—18.

Chapter 14

LOST CIRCULATION

B.S. GUPTA, GEORGE V. CHILINGARIAN and PAUL VORABUTR

INTRODUCTION

Lost circulation is defined as the loss of an appreciable part or the entire volume of drilling fluid through the borehole into the cavernous, vuggy, fractured, and/or highly porous formations (unconsolidated formations

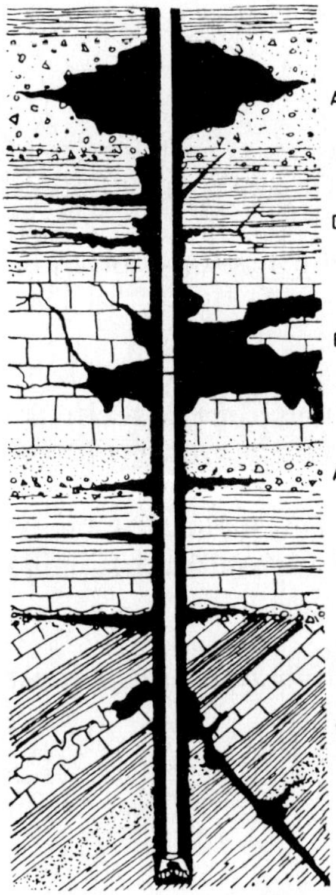

Fig. 14-1. Various types of lost circulation: (*A*) very permeable unconsolidated formations; (*B*) vugular and cavernous formations; (*C*) faulted, jointed, and fissured (natural) formations; and (*D*) induced faulted, jointed, and fissured formations. (After Magcobar, 1968, fig. 1, p. 158; courtesy of Magcobar.)

having a permeability in excess of 14 darcys). This has presented one of the greatest problems to the drilling industry. Lost circulation can take place while drilling is in progress or during the "trips", when pressure surges occur due to the lowering of drillpipe or casing into the hole. After the occurrence of lost circulation, the level of the drilling fluid in the annulus may drop and stabilize at a particular level depending upon the formation pressure.

The cavernous, vuggy, fractured, or simply highly porous formations have generally high permeabilities and large openings which are capable of accepting considerable volumes of drilling fluids (see Fig. 14-1). If voids and fractures are large enough, they cannot be plugged by the solids present in drilling fluids, such as clays, cuttings, etc.

According to Rogers (1963, p. 634), the maximum allowable drilling fluid loss is of the order of one barrel per hour. This loss can be measured in the pits at the surface. Preventive measures must be adopted in case the mud loss exceeds 1 bbl/hr.

FAVORABLE CONDITIONS FOR LOST CIRCULATION

The following conditions favor lost circulation:

(1) If the size of the pore openings exceeds the maximum particle-size of the screened mud. The large size of the mud screen openings is 20 mesh, which corresponds to about 0.83 mm.

(2) Unconsolidated formations which can be easily fractured or those having a permeability in excess of 14 darcys.

(3) Presence of vugular and/or cavernous formations.

(4) Presence of induced fractures owing to the circulating pressure of high-density drilling fluids. Static drilling fluid pressures rarely cause lost circulation; however, the pressure surges due to the downward movement of the pipe may create problems.

(5) Presence of natural fractures and large displacement faults. The breakdown pressure of such formations is considerably lower than that of similar formations without fractures.

Howard and Scott (1951) discussed several identifying features of lost circulation zones, which are summarized in Table 14-I.

UNDESIRABLE EFFECTS OF LOST CIRCULATION

As a result of lost circulation, the level of the drilling fluid in the annulus is lowered, causing the hydrostatic pressure in the annulus to become lower than the formation pore pressure. This can lead to a disastrous blowout. In addition, due to the loss of drilling fluids into the fractured, vuggy, and/or

TABLE 14-I

Identifying features of lost circulation zones

(After Howard and Scott, 1951, table I, p. 172; courtesy of AIME)

Natural fractures	Induced fractures	Cavernous	Unconsolidated formations
1. May occur in any type rock. 2. Loss is evidenced by gradual lowering of mud in pits. If drilling is continued and more fractures are exposed, complete loss of returns may be experienced.	1. May occur in any type rock, but would be expected in formations with characteristically weak planes such as shale. 2. Loss is usually sudden and accompanied by complete loss of returns. Conditions are conducive to the forming of induced fractures when mud weight exceeds 10.5 lb/gal. 3. Loss may follow any sudden surge in pressure. 4. When loss of circulation occurs and adjacent wells have not experienced lost circulation, induced fractures should be suspected.	1. Normally confined to limestone. 2. Loss of returns may be sudden and complete. 3. Bit may drop from a few inches to several feet just preceding loss. 4. Drilling may be "rough" before loss.	1. Gradual lowering of mud level in pits. 2. Loss may become complete if drilling is continued. 3. Since it is known that the rock permeability must exceed 14 darcys before mud can penetrate and that oil or gas sand permeability seldom exceeds 3.5 darcys, it is improbable that loose sands are ever the cause of mud loss to an oil or gas sand unless the loss can be attributed to the ease with which this type of formation fractures.

cavernous formations, mud cost would increase greatly.

Inasmuch as drilling fluid does not return to the surface together with the cuttings, no information is obtained about the formation being drilled.

In the case of lost circulation, there are more chances of stuck pipe, owing to the differential pressure between the mud column and the formation. This eventually leads to expensive fishing jobs.

In the case of lost circulation into a potential pay zone, productivity is usually lowered considerably.

FORCES REQUIRED FOR FRACTURING FORMATIONS

Hydraulic fracturing is commonly used to stimulate production. Consequently, numerous studies have been made in order to determine forces required to fracture underground formations. The fracture orientation depends on the relative magnitude of three principal stresses (Fig. 14-2), and is usually normal to the direction of the least compressive stress. Some rocks, such as those composed of salt, behave as plastic bodies at depths greater than 10,000 ft because of high stresses and temperatures. It is commonly believed that formations are plastic near the borehole and elastic farther out.

Some investigators believe that vertical fractures are created by the hydraulic pressures at greater depths (>3000 ft) where rocks are less perme-

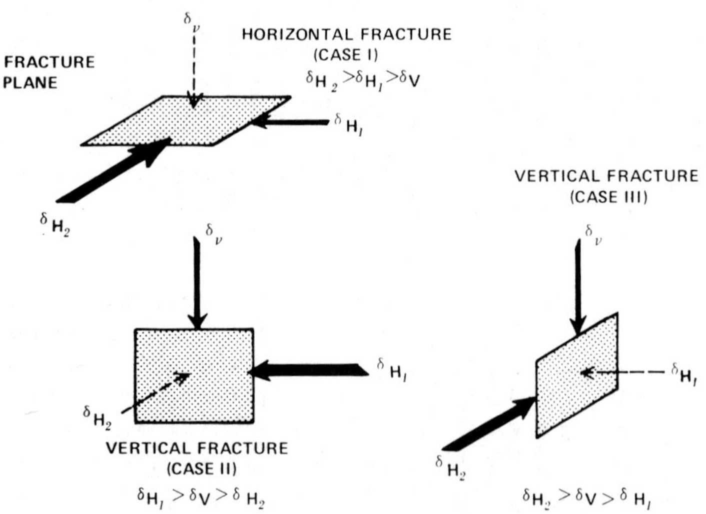

Fig. 14-2. Fracture orientation as a function of relative magnitudes of three principal stresses. (After Fertl and Chilingarian, 1978, p. 301, fig. A-4.)

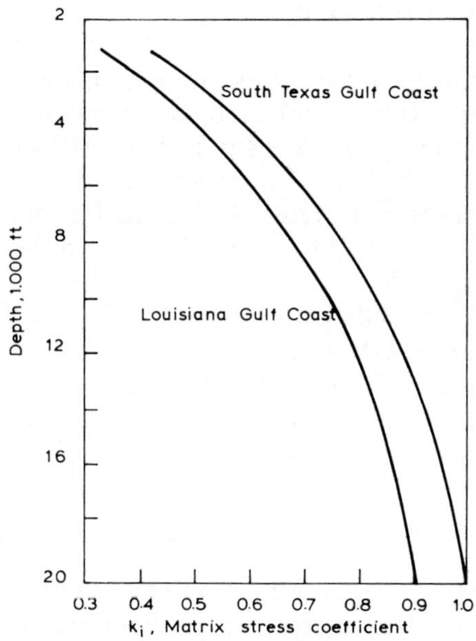

Fig. 14-3. Matrix stress coefficient for Louisiana and South Texas Gulf Coast areas. (After Matthews and Kelly, 1967, fig. 2, p. 93; courtesy of Oil and Gas Journal.)

able*, whereas the horizontal fractures form at shallower depths where the fluid easily penetrates the formation. Pre-existing fractures are widened if they form an angle of 45° or less with the vertical. Other investigators believe that *only* vertical fractures could form as a result of hydraulic fracturing.

The hydrostatic pressure at which the exposed formation will fracture depends upon the pore pressures of the fluids within the rock, with an additional increment attributable to the strength of the rock. Matthews and Kelly (1967, pp. 92—106) have proposed the following relation:

$$F = p_p/D + k_i\sigma/D \qquad (14\text{-}1)$$

where: F = fracture gradient, i.e., drilling fluid pressure at a particular depth at which fracturing is induced (psi/ft); p_p = pore pressure at depth of interest (psi); D = depth (ft); σ = matrix stress at depth, D, (psi); k_i = dimensionless matrix coefficient for the depth at which the value of σ would be the normal matrix stress. The matrix stress coefficient, k_i, should be determined empirically for a given drilling area. For example, Fig. 14-3 indicates k_i values for two Gulf Coast areas.

Fracture gradient is greater than the pore pressure gradient, but may approach it at greater depth (e.g., 16,000 ft).

---

*A better explanation is the increase in vertical stress due to the added overburden.

DEVICES FOR LOCATING THE LOST CIRCULATION ZONE

If proper steps are taken to identify and locate the lost circulation zone and the exact depth of loss, serious operational problems can be avoided during drilling. At shallow depths, the exact location of the depth of lost circulation into naturally-permeable zones, which can be sealed with normal sealing agents, is not important. At depths greater than 5000 ft and if the lost circulation interval constitutes more than 500 ft of open hole, where the normal sealing agents are ineffective, it is important to locate the exact point of loss and seal the hole adequately before drilling is continued. The lost circulation zone may not be located at the bottom of the hole. In fact, it could occur anywhere in the hole and may show its effect at any time. Several commercial methods are available for finding the point of loss, some of which are discussed below.

Temperature survey

A number of service companies perform temperature surveys, which proved to be quite successful in locating the thief zones, where temperature gradients are approximately of the order of 1°F/100 ft.

After running the first temperature survey to establish the gradient in the hole, a quantity of the mud is pumped into the hole and second survey is

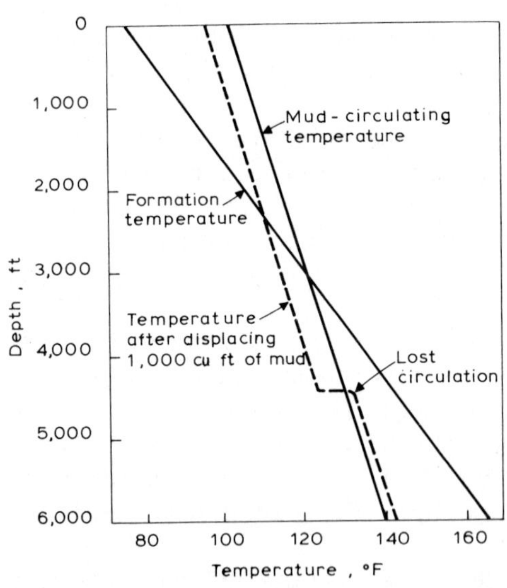

Fig. 14-4. Lost circulation zone located by temperature survey (idealized conditions). (After Goins and Dawson, 1953, fig. 1, p. 170; courtesy of Oil and Gas Journal.)

taken. A rather sudden change in the difference between these two surveys indicates presence of the lost circulation zone. One of the advantages of temperature survey is the fact that it can be made in drilling fluids heavily loaded with sealing agents. In addition, only small quantities of drilling fluid are required.

Initial temperature survey shows an approximately constant temperature gradient (increase of temperature with depth). A temperature survey of the well can be made while going in and coming out of the borehole. As the cool surface drilling fluid is pumped into the borehole, the mud temperature at the point of loss is decreased to that of mud which has been pumped in. The second temperature survey should be made immediately after pumping the mud in, so that temperature equilibrium is not reached. An idealized temperature survey showing the lost circulation zone is presented in Fig. 14-4. A sharp change in temperature gradient is usually observed at the point of loss, as shown in Fig. 14-5, for example.

In Fig. 14-5, curves *1* and *2* represent the downward and upward runs

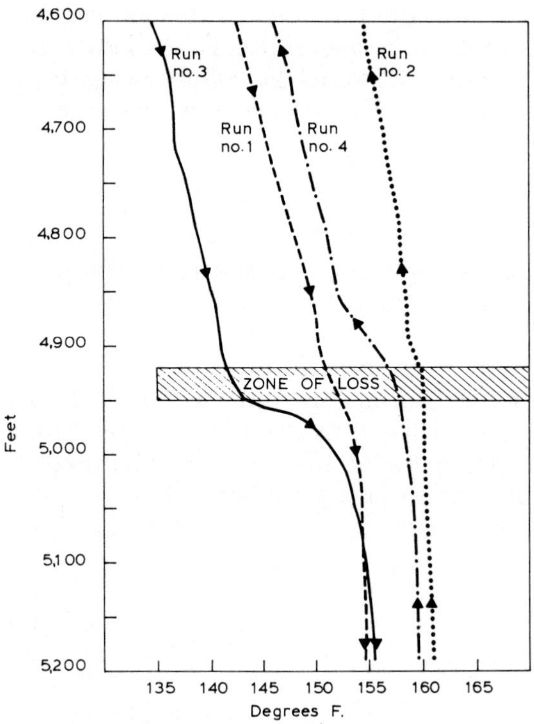

Fig. 14-5. Example of locating lost circulation zone by temperature survey. (After Goins and Dawson, 1953, fig. 3, p. 170; courtesy of Oil and Gas Journal.)

made prior to pumping mud. Curves *3* and *4* represent the downward and upward runs made immediately after pumping approximately 1000 cu ft of mud into the borehole. Interpretation of this temperature survey shows that the point of loss occurs at a depth of 4920—4950 ft.

Spinner survey

Spinner survey is not suitable in cases where large quantities of sealing materials are present in the mud. In using this method, a small spinner attachment is run into the borehole on a conductor cable in such a way that movement of mud either spins or turns the propeller. The speed (revolutions per minute) of the rotor is recorded on film as series of dashes and spaces (see Chilingar and Beeson, 1968).

Radioactive tracer survey

Radioactive tracer survey method is based on the measurement of the radioactivity of the formations. After running a gamma-ray log, a slug of radioactive material (e.g., carnotite) is displaced into the borehole above the point of expected loss and squeezed into the formation. A second gamma-ray log is then run and compared with the first one. A decrease in radioactivity is shown at the point of loss as a result of loss of the radioactive material into the formation. Although this method is quite expensive, the point of loss can be located exactly.

Hot wire survey

Hot wire method is based on the measurement of temperature changes in the borehole by a precalibrated resistance wire which is very sensitive to the temperature variations. Initially, hot wire tool is run in the borehole and the resistance is noted. Fresh mud is then pumped into the borehole. If the point of loss is below the tool, flowing mud will change the wire resistance. On the other hand, if the resistance does not change, then the tool is located below the point of loss. This type of tool can be used in any kind of mud; however, it requires a large amount of mud in order to survey the borehole.

Transducer method

Bardeen and Teplitz (1956, p. 36) designed an instrument which is composed of a hollow metal cylinder with an entrance at the top larger in cross-section than the exit at the bottom. This restricts the flow of mud through the cylinder. There is a window fitted with a neoprene diaphragm on one side of the cylinder. A movable electrode on the diaphragm moves back and forth between two fixed electrodes. When the pressure differential varies across the diaphragm, the potential varies in the electric circuit which, in turn, indicates the rate of flow of the mud. The potential difference between the movable and fixed electrode is measured at the surface.

When the instrument is above the point of loss, the readings at the surface are normal. On the other hand, when the instrument is below the point of loss, no signal is generated by the instrument because fluid does not flow through. In order to locate the point of loss, the instrument is moved successively to different portions of the borehole until the signal obtained at full flow is reduced to that at no flow.

This instrument operates in all types of muds containing sealing agents. The only disadvantage is that it requires considerable mud flow and the equipment may not be readily available.

PREVENTIVE METHODS FOR LOST CIRCULATION

Because most circulation losses result from improper drilling techniques, drilling programs should be conducted in such a way that loss of drilling fluid will not occur. Some of the preventive methods which have provided good results are summarized below:

(1) Reduction of mud specific weight; however, it should be high enough to control the formation pressure.

(2) Lowering mud viscosity and gel strength.

(3) Avoiding pressure surges due to the fast movement of pipes into the borehole.

(4) Avoiding to spud kelly rapidly with pumps on.

(5) Prevention of balling of the bit and collars.

(6) Practical utilization of proper estimates of fracture (both vertical and horizontal) and pore pressure gradients.

Sometimes circulation losses do occur, however, even if these simple precautions are taken.

It is important to differentiate between the specific weight of the drilling fluid as measured at the surface and that actually existing downhole when drilling in a weak formation. Equivalent and annular specific weights are very important and can be significantly greater than those measured at the surface, particularly when the penetration rate and flow properties (mainly viscosities and gel strengths) are high (see Chapter 8).

According to Fertl (1976, p. 233), the optimum downhole equivalent circulating specific weight should exceed the formation pore pressure only by about 0.2—0.4 lb/gal. Such close tolerances between the equivalent formation pore pressure and drilling fluid specific weight allow the drillstring to be pulled with little or no swabbing, give rise to good penetration, and reduce the possibility of differential sticking. (See Chapter 16.)

Of great concern in any drilling operation is the occurrence of lost circula-

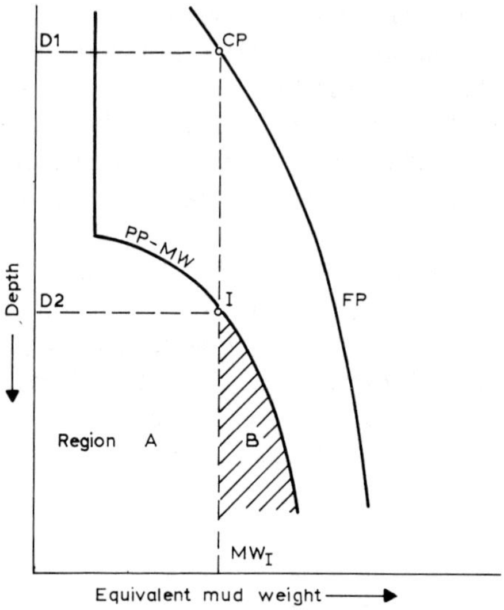

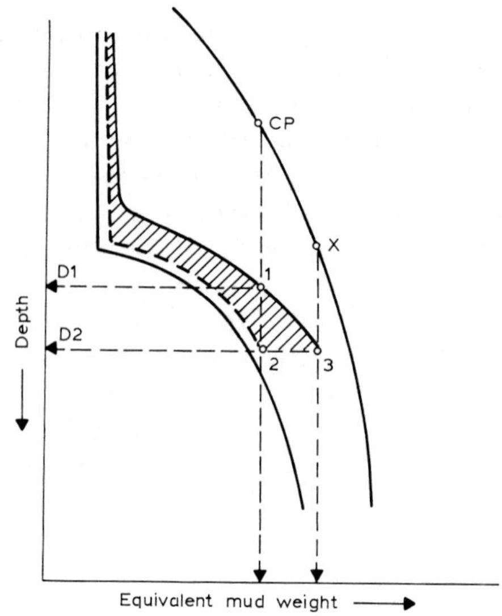

Fig. 14-6. Casing point selection under perfectly balanced conditions. "Safe" drilling and well killing conditions are located in the region *A*, whereas in the region *B* well kick cannot be handled and circulation is lost. *CP* = casing point; *FP* = fracture pressure; *PP* = pore pressure; and *MW* = mud weight. (After Fertl, 1976, fig. 6.4, p. 234.)

Fig. 14-7. Schematic diagram of lost circulation due to an overbalanced drilling condition. (After Fertl, 1976, fig. 6.8, p. 237.)

tion below the casing shoe when the wellbore is experiencing gas, oil, or water "kick". Such occurrences are not uncommon and are very hazardous. The results are very costly in terms of personnel expenses and equipment losses, environmental damages, and public distrust. Complete shutdown of Santa Barbara Channel (California, U.S.A.) drilling for a long period of time after an accident serves as a good reminder of the importance of correct determination of casing string setting depth, so that drilling may be safely conducted without induced lost circulation.

Figure 14-6 illustrates the concept of casing setting depth and induced lost circulation. For the sake of simplicity, it is assumed that pore pressure is equal to downhole equivalent circulating specific weight. The casing depth corresponding to the depth D_1 will allow drilling to continue to the depth D_2. Region *A* shows the depth interval where on occurrence of kick, the wellbore can still be safely circulated and killed if necessary (Fertl, 1976, p. 234).

As previously mentioned, small overbalance is maintained while drilling; however, large overbalance may lead to lost circulation, as shown in Fig. 14-7. The shaded area represents an unnecessary overbalance. With this heavy drilling fluid present in the hole, drilling may proceed and kick may still be handled safely to a depth D_1, whereas the excessive drilling fluid weight at a depth D_2 would cause stuck pipe or even formation failure (i.e., fracturing), inducing lost circulation at point below casing shoe (Fertl, 1976, p. 238).

Once it has been established that there is a lost circulation zone, which is also called a thief zone, the point of loss should be accurately determined by one or more of the methods previously discussed.

Lost circulation occurs both in soft and hard formations. In the case of lost circulation, reliable means of properly healing the hole and regaining circulation must be used. In most cases, a lost circulation material is spotted in the zone of loss. Sometimes, it is used as a part of the mud system. The lost circulation material used must satisfy the following requirements:

(1) The material must be capable of forming a mud-tight seal for the remainder of the drilling operation.

(2) The seal must be strong and not easily broken or removed.

(3) The material should not cause sticking of the drillpipe.

(4) The material should not damage the productive zone.

Not all of the materials used satisfy the above requirements. A few polymer substances, however, do possess these characteristics.

MATERIALS USED FOR PREVENTION OF LOST CIRCULATION

There are three types of materials used for correcting lost circulation:

(1) *Fibrous materials.* Fibrous materials include raw cotton, bagasse, flax shive, wood fiber, bark fiber, textile fiber, mineral fiber, leather, glass fiber, peat moss, feathers, and beet pulp.

(2) *Flaky materials.* Flaky materials include cellophane, cork, mica, corn cobs, cotton-seed hulls, and vermiculite.

(3) *Granular materials.* Granular materials include perlite, coarse bentonite, ground plastic, nut shells, nut hulls, ground tires, asphalt, wood, corn cobs, and coke.

Tests conducted to evaluate the effectiveness of these materials have shown that the best sealing agents are granular (see Fig. 14-8). It has been established, however, that even the granular particles will not seal fractures greater than 0.22 inches in width. Sometimes, these materials are easily dislodged due to the formation of a bridge across the zone, which results in renewed fluid loss.

There are two types of seals that are formed by these lost circulation materials: (1) fracture seal at the face of wellbore and (2) fracture seal within the wellbore (see Fig. 14-9). The latter type of sealing is preferred because (1) a more permanent bridge forms within the formation, and (2) lost circulation materials do not become dislodged easily as a result of pipe

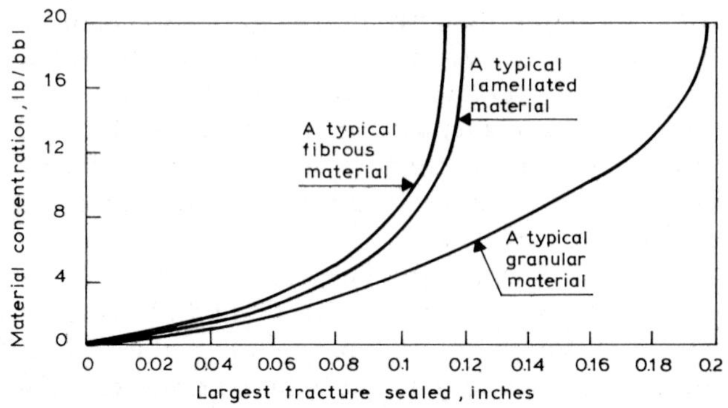

Fig. 14-8. Performance of lost circulation materials in sealing fissures. (After Howard and Scott, 1951, fig. 14, p. 177; courtesy of AIME.)

movement in the wellbore and are not subjected to as much erosion as a result of fluid movement (downhole hydraulics).

Inasmuch as proper bridging must be formed within the formation for sealing to be effective, the particle-size distribution of lost circulation materials is of utmost importance. A mixture of large and small particles is most commonly used. It is believed that the larger materials are transformed into the formation where bridging occurs. Smaller-size particles then plug the smaller openings until formation of an impermeable bridge (see Fig. 14-9) (Magcobar, 1968, p. 160).

Because the size of openings and width of fractures causing lost circulation are rarely known, no acceptable mathematical model has been developed to calculate the optimum sizes and size distributions of lost circulation materials to stop lost circulation. Field experience in a particular area, therefore, must be heavily relied upon to make the final decision.

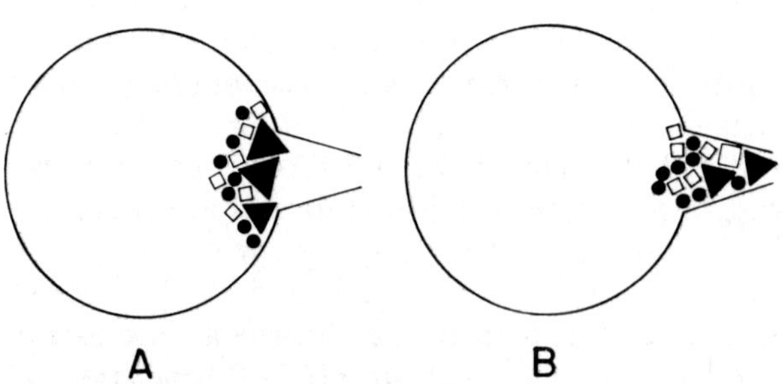

Fig. 14-9. Schematic diagram of fracture sealing with lost circulation materials at the face of wellbore (A) and within the formation (B). (After Magcobar, 1968, fig. 2, p. 160; courtesy of Magcobar.)

At the present time, several companies supply a packaged mixture of various lost circulation materials. For severe loss problems, however, dehydratable plugging pills are used. The use of gunk plugs (diesel oil + bentonite, or diesel oil + bentonite + cement) proved to be successful. On mixing water or water-base clay mud with diesel oil + bentonite mixture, rapid hydration of bentonite results in formation of a strong solid material. The following is an example of the steps which can be carried out:

(1) Approximately 350 lb of bentonite are mixed per barrel of diesel oil, which acts as a carrying fluid; or 200 lb of bentonite plus 200 lb of cement are used per barrel of diesel oil.

(2) As this slurry is pumped down the drillpipe, simultaneously the drilling fluid is pumped down the annulus. The intermingling of bentonite and water results in a very viscous "gunk" mixture which seals off the zone of loss. The drillpipe must be placed just above the zone of loss. About 10 bbl of diesel oil should precede the bentonite, in order to prevent the slurry from setting up in the drillpipe.

A squeeze pressure of around 100—300 psi should be applied while spotting the slurry.

A high-yield point soft plug forms as a result of reaction of bentonite with water. In the presence of cement, the slurry eventually sets to a rigid mass which has considerable compressive strength. The initial yield point of the plug should not exceed 5 psi, in order to pump the plug into the lost-circulation zone.

It should be kept in mind, however, that if (1) the plug sets up too soon, (2) the plug is spotted too far away from the zone, or (3) the zone of loss is not properly located, then the plug will set up in the open hole and not across the lost-circulation interval, which is a common problem. In addition, a permanent zone damage occurs when the gunk plug is set.

A few polymer plugs, which have been introduced during the past 10—15 years, consist of self-complexing blend of polymers with catalysts that forms a rubber-like gel in situ. These solid-free systems, which initially have a very low viscosity ($\approx$1 cP), change to tough plastics after a predetermined time ranging from 1 minute to 8 hours. Certain chemicals control the setting time, whereas others may enable plugs to revert back to water after a predetermined time.

Attapulgite plus barite mixtures, together with a combination of lost circulation materials, establish good bridge and seal upon dehydration. A waiting period of 4—8 hours (or longer) is recommended.

The viscosity of the slurry should be as high as possible and still be pumpable. For example, if the final specific weight desired is 15 lb/gal, it is necessary to add 15 lb of attapulgite, 1 lb of gypsum, and 485 lb of barite to a barrel of water. The final volume would constitute 1.323 bbl, and 0.756 bbl of water should be added per each bbl of finished slurry.

(1) Pressure generated by breaking the drilling fluid gel can be determined using the following formulas developed by Melrose et al. (1958) and Burkhardt (1961):

(a) *inside the pipe:*

$$p = 4\,L\,G/d_{ip} \tag{14-2}$$

where p = pressure required to break the mud gel in psi; L = pipe length in ft; G = gel strength in lb/100 sq ft; and d_{ip} = inside diameter of pipe in inches; and:

(b) *in the annulus:*

$$p = 4\,L\,G/(d_h - d_{op}) \tag{14-3}$$

where d_h = diameter of borehole in inches, and d_{op} = outside diameter of pipe in inches.

(2) Pressure generated by inertia of the mud column can be determined as follows:

(a) *for closed pipe strings:*

$$p = L\,\gamma_f\,a_p\,d_p^2/g\,(d_h^2 - d_{op}^2) \tag{14-4}$$

where γ_f = specific weight of mud in lb/gal; a_p = acceleration of pipe in ft/sec/sec; and g = gravity acceleration (ft/sec/sec); and:

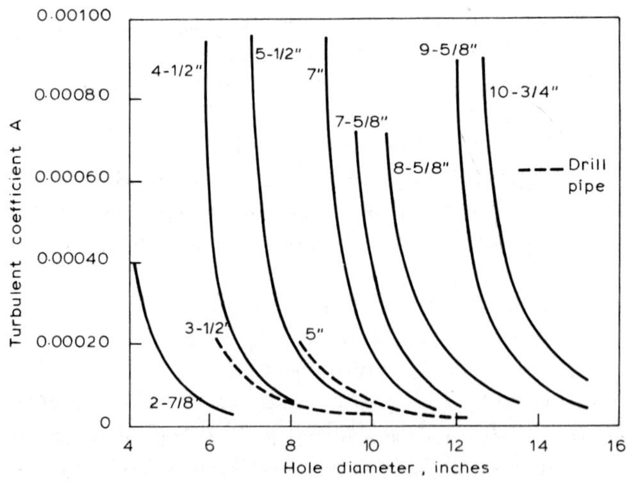

Fig. 14-10. Turbulent flow geometry coefficient A as a function of wellbore diameter and nominal pipe-string diameter. (After Burkhardt, 1961, fig. 9, p. 603; courtesy of the Society of Petroleum Engineers of AIME.)

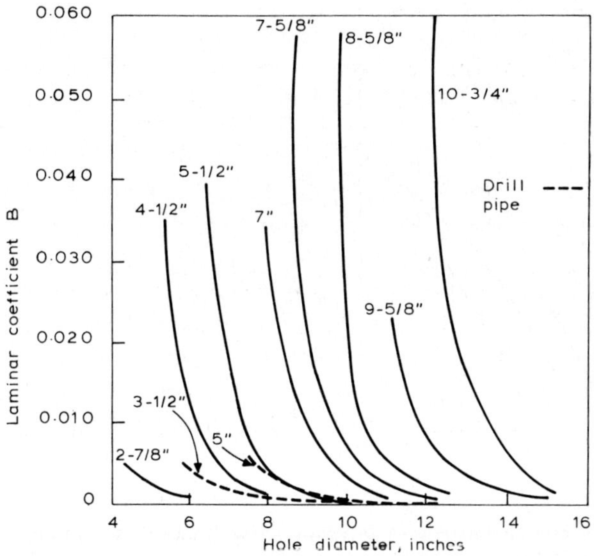

Fig. 14-11. Laminar flow geometry coefficient B as a function of wellbore diameter and nominal pipe-string diameter. (After Burkhardt, 1961, fig. 10, p. 603; courtesy of the Society of Petroleum Engineers of AIME.)

(b) *for open pipe strings:*

$$p = L\,\gamma_f\,a_p\,(d_{op}^2 - d_{ip}^2)/g(d_h^2 - d_p^2 + d_{ip}^2) \tag{14-5}$$

(3) The following approximate formulas can be used in computing viscous drag pressure surges for closed pipe-strings:

(a) *laminar flow:*

$$p_s = B\,\mu_p\,v_p + \overline{YP}/0.3(d_h - d_{oe}) \tag{14-6}$$

where p_s = pressure surge in psi/1000 ft of pipe; B = geometry coefficient for laminar flow; v_p = pipe velocity in ft/min; μ_p = plastic viscosity in cP; $\overline{YP}$ = yield point in lb/100 sq ft; and d_{oe} = effective pipe outside diameter in inches, and:

(b) *turbulent flow:*

$$p_s = A\,\mu_p^{0.21}\,q^{0.806}\,v_p^{1.8} \tag{14-7}$$

where q = volumetric mud flow rate in ft^3/min. Figures 14-10, 14-11, 14-12, and 14-13 can be used for determining the values of A, B, α, and β. Pressure surges should be determined for both laminar and turbulent cases and the higher of the two values must be used. For open pipe-strings, β is added in the first part of eq. 14-6 (right-side) and α is added in the right side of eq. 14-7.

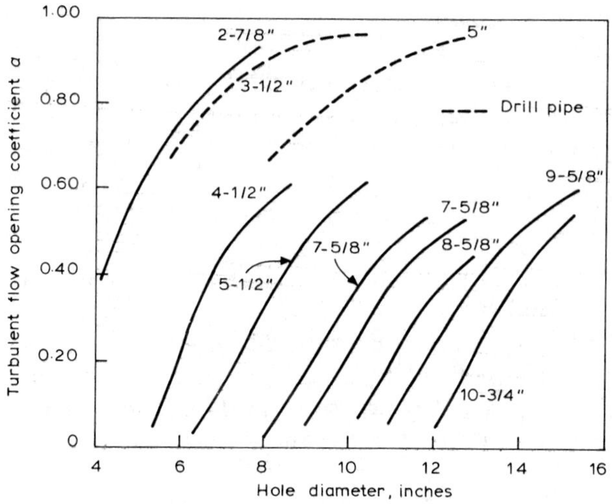

Fig. 14-12. Turbulent flow opening coefficient α as a function of wellbore diameter and nominal pipe-string diameter. (After Burkhardt, 1961, fig. 12, p. 604; courtesy of the Society of Petroleum Engineers of AIME.)

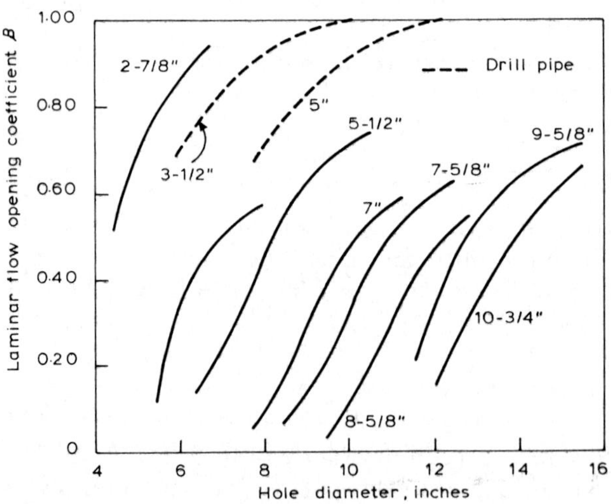

Fig. 14-13. Laminar flow opening coefficient β as a function of wellbore diameter and nominal pipe-string diameter. (After Burkhardt, 1961, fig. 13, p. 604; courtesy of Society of Petroleum Engineers of AIME.)

The equations presented above show that two phenomena may result from tripping in and out of the hole while drilling: (1) suction effect or swabbing and (2) surge effect. Both are functions of (1) drilling fluid properties, namely, gel strength (see equations), (2) clearance between pipe and hole (Fig. 14-14), (3) actual pipe size, (4) length of drillstring, and (5) tripping speed.

As the drilling string is being pulled out of the hole, frictional drag on the entire length of the drillstring and the piston effect of the bit and drillcollar create a suction on the bottom of the hole. This may literally pull formation fluids, which have been confined by the pressure of the drilling fluid hydrostatic column, into the wellbore. Strong swabbing effect creates sudden change in the stress distribution around the wellbore, which may lead to a stress distribution that causes fracturing.

Conversely, pressure surge (increase in bottomhole pressure) may result from running pipe in the hole. The amount of surge depends on the previously mentioned factors. Figures 14-15, 14-16, and 14-17 illustrate the effects of various factors on pressure surge. Pressure surge increases with increasing gel strength and increasing frictional drag or adherence of drilling fluid to the drillpipe and/or casing surfaces. Most likely in any well the clearance between the hole (or casing) and pipe is determined by the economics and type of well (exploratory, development, etc.), which are beyond control. Drilling fluid properties (i.e., yield point and gel strength) and pipe running speed, however, may be controlled. The effect of pipe running speed

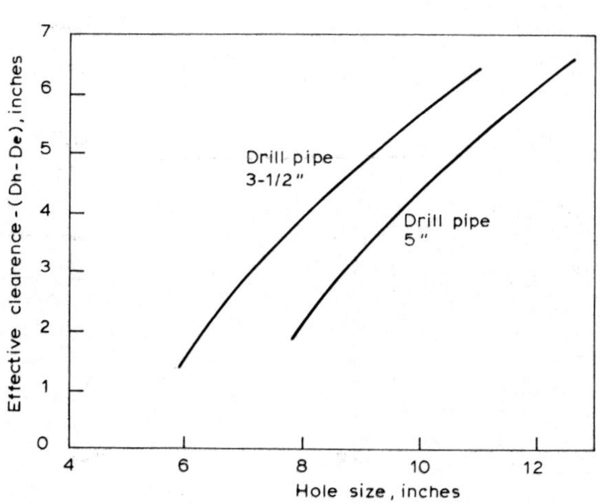

Fig. 14-14. Effective annular clearance ($d_h - d_{oe}$) as a function of hole diameter and nominal drillpipe diameter. (After Burkhardt, 1961, fig. 11, p. 604; courtesy of the Society of Petroleum Engineers of AIME.)

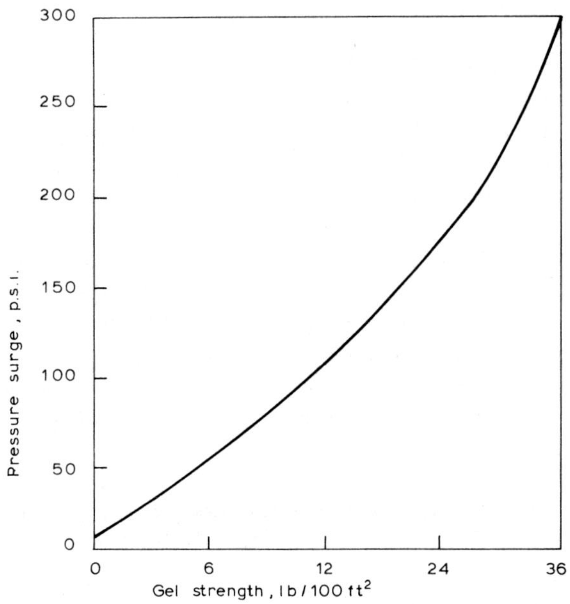

Fig. 14-15. Relationship between gel strength and surge pressure, which is proportional to depth, for a $4\frac{1}{2}$-inch drillpipe in $10\frac{3}{4}$-inch casing. Depth = 7000 ft. (After IMCO, 1978, fig. 20-3, p. 20-2; courtesy of IMCO Services.)

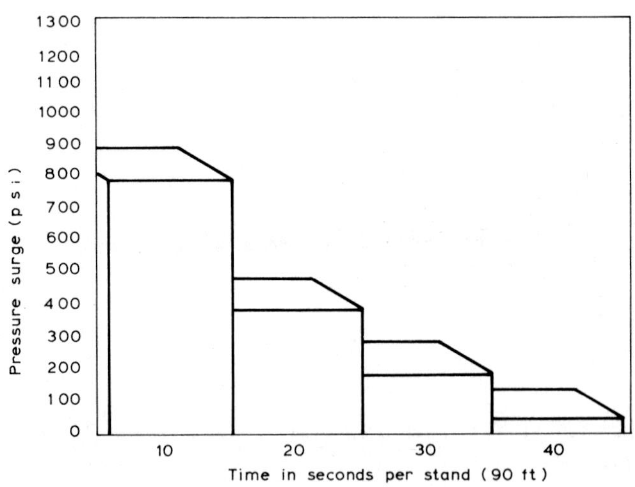

Fig. 14-16. Pressure surge in a 15,000-ft well reflecting speed of running pipe and hole clearance; $3\frac{1}{2}$-inch drillpipe in 8-inch borehole. (After IMCO, 1978, fig. 20-5, p. 20-3; courtesy of IMCO Services.)

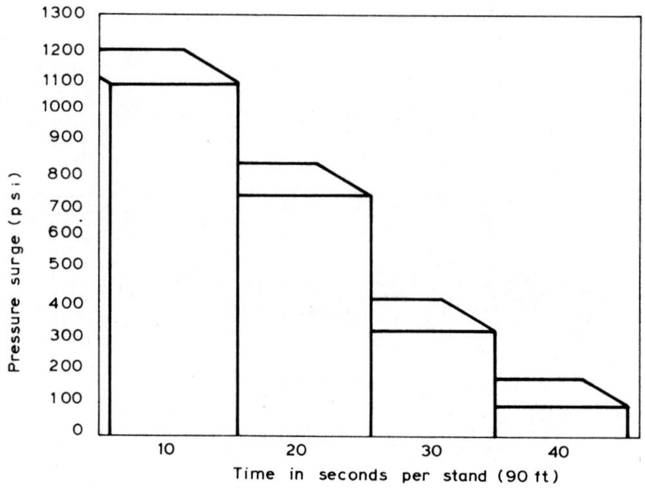

Fig. 14-17. Pressure surges in a 15,000-ft well reflecting speed of running pipe and hole clearance; $4\frac{1}{2}$-inch drillpipe in 8-inch borehole. (After IMCO, 1978, fig. 20-6, p. 20-3; courtesy of IMCO Services.)

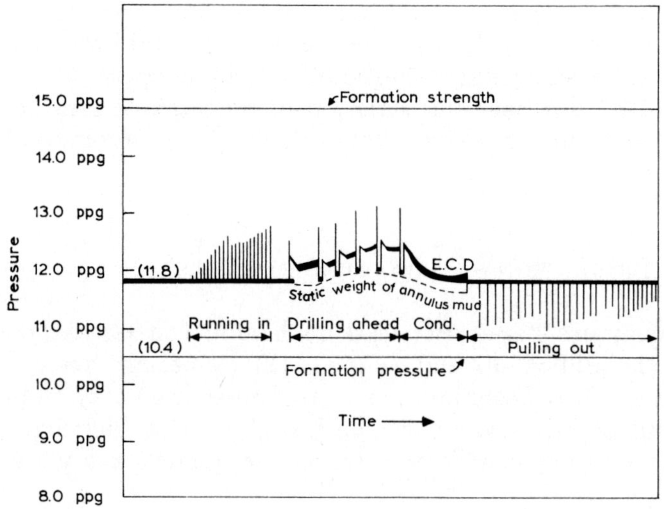

Fig. 14-18. Variation of bottomhole pressure with time for normal drilling cycle of a low-gel-strength drilling fluid. (After IMCO, 1978, fig. 20-1, p. 20-2; courtesy of IMCO Services.)

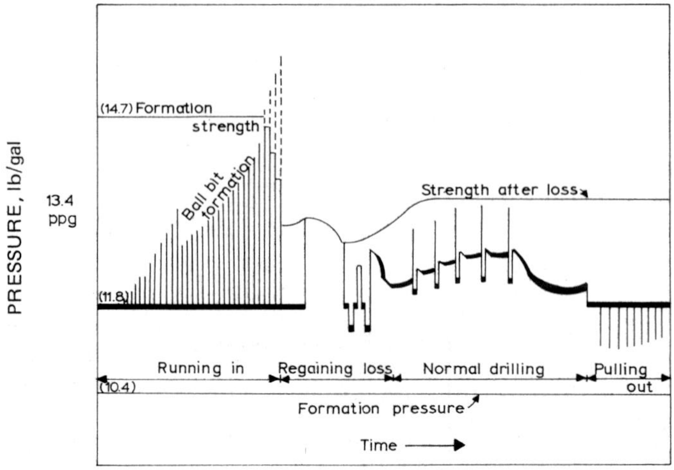

Fig. 14-19. Variation of bottomhole pressure with time for a normal drilling cycle for high-gel-strength drilling fluid. (After IMCO, 1978, fig. 20-2, p. 20-2; courtesy of IMCO Services.)

is shown in Figs. 14-16 and 14-17; the faster the movement, the higher the surge. The additional bottomhole pressure increase due to surge may be sufficient to break down the weak and unconsolidated formations causing lost circulation. Hence, when planning a well, it is of utmost importance to know the fracture (vertical and horizontal) gradients of the formations to be penetrated. The personnel in charge of drilling (tool-pushers and drilling engineers) must be fully aware of the significance of surge and swabbing effects during the pre-spud meeting and throughout the drilling operation.

Figures 14-18 and 14-19 illustrate the variation of bottomhole pressure during normal drilling cycle (pressure versus time) due to the swabbing and surge effects of low- and high-gel strength drilling fluids.

PREVENTION OF LOST CIRCULATION — CONCLUDING REMARKS

Several remedial techniques have been used to handle lost circulation upon its occurrence: (1) pulling up and waiting, (2) squeezing, and (3) cementing. Application of these techniques and procedures used vary from area to area and depend on the severity of the lost circulation, personnel preference based on prior experiences in an area, and equipment and material availability.

It must be stressed that lost circulation must not be taken lightly, because loss of drilling fluid in the wellbore often leads to loss of control of formation pressure and a possible kick and blowout situation. Many lost circulations that occur may be avoided with good drilling practices.

REFERENCES

Anonymous, 1977. New lost circulation remedy is easy to use. *World Oil*, 184(4, March): 76—77.

Bardeen, T. and Teplitz, A.J., 1956. Lost circulation information with a new tool for detecting zones of loss. *Trans. Am. Inst. Min. Met. Pet. Eng.*, 207: 36—41.

Burkhardt, J.A., 1961. Wellbore pressure surges produced by pipe movement. *Trans. Am. Inst. Min. Met. Pet. Eng.*, 222: 595—605.

Chilingar, G.V. and Beeson, C.M., 1968. *Surface Operations in Petroleum Production.* Am. Elsevier, New York, N.Y., 397 pp.

Dawson, D.D. and Goins, W.C., 1953. Bentonite-diesel oil squeeze. *World Oil*, 137(5, Oct.): 222—223.

Fertl, W.H., 1976. *Abnormal Formation Pressures.* Elsevier, Amsterdam, 382 pp.

Fertl, W.H. and Chilingarian, G.V., 1978. Fracturing of reservoir rocks. In: G.V. Chilingarian and T.F. Yen (Editors), *Bitumens, Asphalts and Tar Sands.* Elsevier, Amsterdam, pp. 296—306.

Goins, W.C. and Dawson, D.D., 1953. Temperature survey to locate zone of lost circulation. *Oil Gas J.*, 52(7, June 22): 170—171.

Howard, G.C. and Scott, P.P., 1951. An analysis and control of lost circulation. *Trans. AIME*, 192: 171—182.

IMCO Services, 1978. *Applied Mud Technology.* 210 pp.

Magcobar Services, 1968. *Mud Engineering.* 224 pp.

Magcobar Services, 1972. *Drilling Fluids Engineering Manual.* 200 pp.

Matthews, W.R. and Kelly, J., 1967. Formation pressure and fracture gradient from logs. *Oil Gas J.*, 65(8, Feb. 20): 92—106.

Melrose, J.C., Savins, J.G., Foster, W.R. and Parrish, E.R., 1958. A practical utilization of the theory of Bingham plastic flow in stationary pipe and annuli. *Trans. AIME*, 213: 316—322.

Messenger, J.U., 1973. Common rig materials combat severe lost circulation. *Oil Gas J.*, 71(25, June 18): 57—64.

Rehm, B., 1976. Lost circulation complicates well-killing operations. *Oil Gas J.*, 74(8, Feb. 23): 121—126.

Rogers, W.F., 1963. *Composition and Properties of Oil Well Drilling Fluids*, 2nd ed. Gulf Oil Publ. Co., Houston, Texas, 690 pp.

Chapter 15

CORROSION IN DRILLING OPERATIONS

T.A. BERTNESS and GEORGE V. CHILINGARIAN

INTRODUCTION

The National Bureau of Standards (1978) reported to the U.S. Congress that the cost of corrosion in the U.S.A. in 1975 was 70 billion dollars plus or minus 30%, and that 10—45% of this loss was avoidable. This economic loss represented about 4% of the gross national product.

The principles of corrosion must be understood in order to effectively select materials and to design, fabricate, and utilize metal structures for the optimum economic life of facilities and safety in operation.

Corrosion in various forms is the major cause of drillpipe failures which add significantly to drilling costs. The trend toward (1) drilling of deeper wells, (2) use of higher strength steels, (3) presence of higher stresses, and (4) use of lower pH drilling fluids contribute to increased susceptibility of metals to failure due to corrosion.

Definition of corrosion

Corrosion is defined as the chemical degradation of metals by reaction with the environment. The destruction of metals by corrosion occurs by: (a) direct chemical attack at elevated temperatures (500+° F) in a dry environment, and (b) by electrochemical processes at lower temperatures in a water-wet or moist environment.

Corrosion attacks metals in drilling operations through electrochemical processes in the presence of electrolytes in drilling, completion, and packer fluids.

CORROSIVE AGENTS IN DRILLING OPERATIONS

The components which promote the corrosion of steel in drilling operations are oxygen, carbon dioxide, hydrogen sulfide, salts, and organic acids. Destruction of metals is influenced by various physical and chemical factors which localize and increase corrosion damage.

The conditions which promote corrosion include:

(a) Energy differences in the form of stress gradients or chemical reactivities across the metal surface in contact with a corrosive solution.

(b) Differences in concentration of salts or other corrodants in the electrolytic solutions.

(c) Differences in the amount of deposits, either solid or liquid, on the metal surfaces, which are insoluble in the electrolytic solutions.

(d) Temperature gradients over the surface of the metal in contact with a corrosive solution.

(e) Compositional differences in the metal surface.

REQUIREMENTS FOR ELECTROCHEMICAL CORROSION

Corrosion continues provided electrically conductive metal and solution circuits are available to bring corrodants to the anodic and cathodic sites.

Four conditions must be present to complete the electrochemical reactions and corrosion circuit:

(1) *Presence of a driving force or electrical potential.* Difference in reaction potentials at two sites on the metal surface must be sufficient to drive electrons through the metal, surface films, and liquid components of the corrosion circuit.

(2) *Presence of an electrolyte.* Corrosion occurs only when the circuit between anodic and cathodic sites is completed by an electrolyte present in water.

(3) *Presence of both anodic and cathodic sites.* Anodic and cathodic areas must be present to support the simultaneous oxidation and reduction reactions at the metal—liquid interface. Metal at the anode ionizes.

(4) *Presence of an external conductor.* A complete electron—electrolytic circuit between anodes and cathodes of the metal through (a) the metal surface films, (b) surrounding environment, and (c) fluid—solid interfaces is necessary for the continuance of corrosion.

In the environment surrounding the metal, the presence of water provides conducting paths for both corrodants and corrosion products. The corrodant may be a dissolved gas, liquid, or solid. The corrosion products may be ions in solution, which are removed from the metal surface, ions precipitated as various salts on metal surfaces, and hydrogen gas.

ELECTROCHEMICAL CORROSION

The conditions needed to promote many types of corrosion can be found in most industrial facilities. The basic electrochemical reactions, which occur simultaneously at the cathodic and anodic areas of metal causing many forms of corrosion damage, are as follows:

(1) At the cathode, the hydrogen (or acid) ion (H^+) removes electrons

from the cathodic surfaces to form hydrogen gas (H_2): $2\,e^- + 2\,H^+ \rightarrow 2\,H^0 \rightarrow H_2$ (in acidic solution). If oxygen is present, electrons are removed from the metal by reduction of oxygen: $4\,e^- + O_2 + 4\,H^+ \rightarrow 2\,H_2O$ (in acidic solution); $4\,e^- + 2\,H_2O + O_2 \rightarrow 4\,OH^-$ (in neutral or alkaline solution).

(2) At the anode, a metal ion (e.g., Fe^{2+}) is released from its structural position in the metal through loss of the bonding electrons and passes into solution in the water as soluble iron, or reacts with another component of the environment to form scale. The principal reaction is:

$$Fe - 2\,e^- \rightarrow Fe^{2+} \tag{15-1}$$

Thermodynamic data indicate that the corrosion process in many environments of interest should proceed at very high rates of reaction. Fortunately, experience shows that the corrosion process behaves differently. Studies have shown that as the process proceeds, an increase in concentration of the corrosion products develops rapidly at the cathodic and anodic areas. These products at the metal surfaces serve as barriers that tend to retard the corrosion rate. The reacting components of the environment may be depleted locally, which further tends to reduce the total corrosion rate.

The potential differences between the cathodic and anodic areas decrease as corrosion proceeds. This reduction in potential difference between the electrodes upon current flow is termed polarization. The potential of the anodic reaction approaches that of the cathode, and the potential of the cathodic reaction approaches that of the anode. Electrode polarization by corrosion is caused by (1) changing the surface concentration of metal ions, (2) adsorption of hydrogen at cathodic areas, (3) discharge of hydroxyl ions at anodes, or (4) increasing the resistance of the electrolyte and films of metal-reaction products on the metal surface.

Changes (increase or decrease) in the amount of these resistances by the introduction of materials or electrical energy into the system will change the corrosion currents and corrosion rate.

A practical method to control corrosion is through cathodic protection, whereby polarization of the structure to be protected is accomplished by supplying an external current to the corroding metal. Polarization of the cathode is forced beyond the corrosion potential. The effect of the external current is to eliminate the potential differences between anodic and cathodic areas on the corroding metal. Removal of the potential differences stops local corrosion action. Cathodic protection operates most efficiently in systems under cathodic control, i.e., where cathodic reactions control the corrosion rate.

Materials may cause an increase in polarization and retard corrosion by adsorbing on the surface of the metals and thereby changing the nature of the surface. Such materials act as inhibitors to the corrosion process. On the other hand, some materials may reduce polarization and assist corrosion. These materials, called depolarizers, either assist or replace the original reactions and prevent the buildup of original reaction products.

Oxygen is the principal depolarizer which aids corrosion in the destruction of metal. Oxygen tends to reduce the polarization or resistance, which normally develops at the cathodic areas, with the accumulation of hydrogen at these electrodes. The cathodic reaction with hydrogen ion is replaced by a reaction in which electrons at the cathodic areas are removed by oxygen and water to form hydroxyl ions (OH^-) or water:

$$O_2 + 2\,H_2O + 4\,e^- \rightarrow 4\,OH^- \text{ (in neutral and alkaline solutions)} \tag{15-2}$$

$$O_2 + 4\,H^+ + 4\,e^- \rightarrow 2\,H_2O \text{ (in acid solution)} \tag{15-3}$$

Polarization of an electrode surface will reduce the total current and corrosion rate. Though the rate of metal loss is reduced by polarization, equipment failures may increase if incomplete polarization occurs at the anodes. For example, inadequate anodic corrosion inhibitor will reduce the effective areas of the anodic surfaces and thus localize the loss of metal at the remaining anodes. This will result in severe pitting and destruction of metal.

Resistances to the corrosion process generally do not develop to the same degree at the anodic and cathodic areas. These resistances reduce the corrosion rate, which is controlled by the slowest step in the corrosion process. Electrochemical corrosion comprises a series of reactions and material transport to and from the metal surfaces. Complete understanding of corrosion and corrosion control in a particular environment requires knowledge of each reaction which occurs at the anodic and cathodic areas.

Components of electrochemical corrosion

The various components which are involved in the process of corrosion of metal are: (1) the metal, (2) the films of hydrogen gas and metal corrosion products, (3) liquid and gaseous environment, and (4) the several interfaces between these components. Metal is a composite of atoms which are arranged in a symmetrical lattice structure. These atoms may be considered as particles which are held in an ordered arrangement in a lattice structure by bonding electrons. These electrons, which are in constant movement about the charged particles, move readily throughout the lattice structure of metal when an electric potential is applied to the system. If bonding electrons are removed from their orbit about the particle center, the resulting cation will no longer be held in the metal's crystalline structure and can enter the electrolyte solution. Electrochemical corrosion is simply the process of freeing these cations from their organized lattice structure by the removal of the bonding electrons. Inasmuch as certain of the lattice electrons move readily within the metal under the influence of electrical potentials, the locations on the surface of the metal from which the cations escape and the locations from which the electrons are removed from the metal need not be and generally are not the same. Corrosion will not occur unless electrons are removed from some portion of a metal structure.

All metals are polycrystalline with each crystal having random orientation with respect to the next crystal. The metal atoms in each crystal are oriented in a crystal lattice in a consistent pattern. The pattern gives rise to differences in spacing and, therefore, differences in cohesive energy between the particles, which may cause preferred corrosion attack. At the crystal boundaries the lattices are distorted, giving rise to preferred corrosion attack. In the manufacture and processing of metals, in order to gain desirable physical properties both the composition and shape of the crystals may be made non-uniform, distorted, or preferably oriented. This may increase the susceptibility of the metal to corrosion attack. Undistorted single crystals of metals experience comparatively little or no corrosion under the same conditions which may destroy commercial pieces of the same metal. Compositional changes in metal alloy crystals and crystal boundaries, which are present in steels and alloys, can promote highly localized corrosion.

CHEMISTRY OF CORROSION AND ELECTROMOTIVE FORCE SERIES

Oxidation takes place when a given substance loses electrons or a share of its electrons. On the other hand, reduction occurs when there is a gain in electrons by a substance. A substance that yields electrons to something else is called a reducing agent, whereas the substance which gains electrons is termed an oxidizing agent. Thus, electrons are always transferred from the reducing agent to the oxidizing agent. In the example below, two electrons are transferred from metallic iron to cupric ion:

$$\underset{\text{metallic iron}}{Fe^0} + \underset{\text{cupric ion}}{Cu^{2+}} \rightarrow \underset{\text{ferrous ion}}{Fe^{2+}} + \underset{\text{metallic copper}}{Cu^0} \qquad (15\text{-}4)$$

The emf series is presented in Table 15-I; potentials given are those between the elements in their standard state at 25°C and their ions at unit activity in the solution at 25°C. A plus (+) sign for E^0 shows that, for the above conditions, the reduced form of the reactant is a better reducing agent than H_2. On the other hand, a negative (—) sign indicates that the oxidized form of the reactant is a better oxidizing agent than H^+. Thus, in general, any ion is a better oxidizing agent than the ions above it.

Actual electrode potentials

In the emf series, each metal will reduce (or displace from solution) the ion of any metal below it in the series, providing all of the materials have

TABLE 15-I

Electromotive force series

Electrode reaction	Standard electrode potential, E^0 in Volts, 25°C
Li = $Li^+ + e^-$	+3.05
K = $K^+ + e^-$	+2.922
Ca = $Ca^{2+} + 2e^-$	+2.87
Na = $Na^+ + e^-$	+2.712
Mg = $Mg^{2+} + 2e^-$	+2.375
Be = $Be^{2+} + 2e^-$	+1.85
Al = $Al^{3+} + 3e^-$	+1.67
Mn = $Mn^{2+} + 2e^-$	+1.029
Zn = $Zn^{2+} + 2e^-$	+0.762
Cr = $Cr^{3+} + 3e^-$	+0.74
Ga = $Ga^{3+} + 3e^-$	+0.53
Fe = $Fe^{2+} + 2e^-$	+0.440
Cd = $Cd^{2+} + 2e^-$	+0.402
In = $In^{3+} + 3e^-$	+0.340
Tl = $Tl^+ + e^-$	+0.336
Co = $Co^{2+} + 2e^-$	+0.277
Ni = $Ni^{2+} + 2e^-$	+0.250
Sn = $Sn^{2+} + 2e^-$	+0.136
Pb = $Pb^{2+} + 2e^-$	+0.126
H_2 = $2\ H^+ + 2e^-$	0.000
Cu = $Cu^{2+} + 2e^-$	−0.345
Cu = $Cu^+ + e^-$	−0.522
2 Hg = $Hg_2^{2+} + 2e^-$	−0.789
Ag = $Ag^+ + e^-$	−0.800
Pd = $Pd^{2+} + 2e^-$	−0.987
Hg = $Hg^{2+} + 2e^-$	−0.854
Pt = $Pt^{2+} + 2e^-$	ca. −1.2
Au = $Au^{3+} + 3e^-$	−1.50
Au = $Au^+ + e^-$	−1.68

unit activities. The activity of a pure metal in contact with a solution does not change with the environment. The activity of an ion, however, changes with concentration and the activity of a gas changes with partial pressure.

An electrode reaction, in which a metal M is oxidized to its ion M^{n+}, liberating n electrons, may be represented by the relation: $M = M^{n+} + ne^-$. The actual electrode potential of this reaction may be calculated from the standard electrode potential by use of the following expression:

$$E = E^0 - \frac{RT}{nF} \ln(M^{n+}) \qquad (15\text{-}5)$$

where E = actual electrode potential at the given concentration (Volts); E^0 = standard electrode potential (Volts); R = universal gas constant, 8.315 Volt

TABLE 15-II

Variation in actual electrode potentials of iron and cadmium with change in concentration of the ions

Reaction	Activity (moles/kg water)			
	1	0.1	0.01	0.001
	Actual electrode potential (Volts)			
$Fe = Fe^{2+} + 2e^-$	+0.440	+0.470	+0.499	+0.529
$Cd = Cd^{2+} + 2e^-$	+0.402	+0.431	+0.461	+0.490

Coulombs/°K; T = absolute temperature (°K); n = number of electrons transferred; F = the Faraday, 96,500 Coulombs; and M^{n+} = concentration of metal ions.

At 25°C, 2.303 RT/F = 0.05915 and the formula becomes:

$$E = E^0 - \frac{0.05915}{n} \log_{10}(M^{n+}) \quad (15\text{-}6)$$

The actual electrode potential for a given environment may be computed from the above relation. Table 15-II shows how the actual electrode potentials of iron and cadmium vary with change in concentration of the ions.

It is apparent from Table 15-II that iron will reduce cadmium when their ion concentrations are equal, but the reverse holds true when the concentration of cadmium ion becomes sufficiently lower than that of the ferrous ion.

It is well to note that the standard electrode potentials are a part of the more general standard oxidation—reduction potentials. Textbooks on physical chemistry also contain a general expression for calculating the actual oxidation—reduction potential from the standard oxidation—reduction potential.

GALVANIC SERIES

Dissimilar metals exposed to electrolytes exhibit different potentials or tendencies to go into solution or react with the environment. This behavior is recorded in tabulations in which metals and alloys are listed in order of increasing resistance to corrosion in a particular environment. Coupling of dissimilar metals in an electrolyte will cause destruction of the more reactive metal which acts as an anode and provides protection for the less reactive metal which acts as a cathode.

CORROSION OF STEEL

In most corrosion problems, the important differences in reaction potentials are not those between dissimilar metals, but are those which exist between separate areas interspersed over all the surface of a single metal. These potential differences result from local chemical or physical differences within or on the metal, such as variations in grain structure, stresses and scale, inclusions in the metal, grain boundaries, and scratches or surface condition. Steel is an alloy of pure iron and small amounts of carbon present as Fe_3C with trace amounts of other elements. Iron carbide (Fe_3C) is cathodic with respect to iron.

Inasmuch as in typical corrosion of steel anodic and cathodic areas lie side by side on the metal surface, in effect it is covered with both positive and negative sites. During corrosion, the anodes and cathodes of metals may interchange frequently.

TYPES OF CORROSION

Numerous types of metal destruction can result from the corrosion process, which are listed under the following classes of corrosion:

(1) *Uniform attack.* The entire area of the metal corrodes uniformly resulting in thinning of the metal. This often occurs to drillpipe, but usually is the least damaging of different types of corrosive attacks. Uniform rusting of iron and tarnishing of silver are examples of this form of corrosion attack.

(2) *Crevice corrosion.* This is an example of localized attack in the shielded areas of metal assemblies, such as pipes and collars, rod pins and boxes, and drillpipe joints. Crevice corrosion is caused by concentration differences of a corrodant over a metal surface. Electrochemical potential differences result in selective crevice or pitting corrosion attack.

Oxygen dissolved in drilling fluid promotes crevice and pitting attack of metal in the shielded areas of a drillstring and is the common cause of washouts and destruction under rubber pipe protectors.

(3) *Pitting corrosion.* Pitting is often localized in a crevice but can also occur on clean metal surfaces in a corrosive environment. An example of this type of corrosion attack is the corrosion of steel in high-velocity sea water, low-pH aerated brines, or drilling fluids. Upon formation of a pit, corrosion continues as in a crevice but, usually, at an accelerated rate.

(4) *Galvanic or two-metal corrosion.* Galvanic corrosion may occur when two different metals are in contact in a corrosive environment. The attack is usually localized near the point of contact.

(5) *Intergranular corrosion.* Metal is preferentially attacked along the grain boundaries. Improper heat treatment of alloys or high-temperature exposure may cause precipitation of materials or non-homogeneity of the metal struc-

ture at the grain boundaries, which results in preferential attack.

Weld decay is a form of intergranular attack. The attack occurs in a narrow band on each side of the weld due to sensitizing or changes in the grain structure due to welding. Appropriate heat treating or metal selection can prevent the weld decay.

Ring worm corrosion is a selective attack which forms a groove around the pipe near the box or the external upset end. This type of selective attack is avoided by annealing the entire pipe after the upset is formed.

(6) *Selective leaching.* One component of an alloy is removed by the corrosion process. An example of this type of corrosion is the selective corrosion of zinc in brass.

(7) *Erosion—corrosion.* The combination of erosion and corrosion results in severe localized attack of metal. Damage appears as a smooth groove or hole in the metal, such as in a washout of drillpipe. The washout is initiated by pitting in a crevice which penetrates the steel. The erosion—corrosion process completes the metal destruction.

The erosion process removes protective films from the metal and exposes clean metal surface to the corrosive environment. This accelerates the corrosion process.

Impingement attack is a form of erosion—corrosion process which occurs after the breakdown of protective films. High velocities and presence of abrasive suspended material and the corrodants in drilling fluids contribute to this destructive process.

The combination of wear and corrosion may also remove protective surface films and accelerate localized attack by corrosion. This form of corrosion is often overlooked or recognized as being due to the wear. The use of inhibitors can often control this form of metal destruction. For example, inhibitors are used extensively for protection of downhole pumping equipment in oil wells.

(8) *Cavitation corrosion.* Cavitation damage results in a sponge-like appearance with deep pits in the metal surface. The destruction may be caused by purely mechanical effects in which pulsating pressures cause vaporization with formation and collapse of the bubbles at the metal surface. The mechanical working of the metal surface causes destruction, which is amplified in a corrosive environment. This type of corrosion attack, examples of which are found in pumps, may be prevented by increasing the suction head on the pumping equipment. A net positive suction head should always be maintained not only to prevent cavitation damage, but also to prevent possible suction of air into the flow stream. The latter can aggravate corrosion in many environments.

(9) *Corrosion due to variation in fluid flow.* Velocity differences and turbulence of fluid flow over the metal surface cause localized corrosion. In addition to the combined effects of erosion and corrosion, variation in fluid flow can cause differences in concentrations of corrodants and depolarizers,

which may result in selective attack of metals.

(10) *Stress corrosion.* Stress-corrosion cracking of metals is produced by the combined action of corrosion and tensile stress. The term stress-corrosion cracking often is applied to all cracking that is related to stress and corrosion: (1) stress-corrosion cracking, (2) hydrogen embrittlement and blistering, (3) sulfide cracking, (4) corrosion fatigue, (5) stress alloying, and (6) caustic embrittlement.

Although stress-corrosion cracking can occur in most alloys, the corrodants which promote stress cracking may differ and be few in number for each alloy. Cracking can occur in both acidic and alkaline environments, usually in the presence of chlorides and/or oxygen.

Stress cracking of metals can occur quickly with exposure to specific corrodants, and the time to failure is shortened by: (1) concentration of specific corrodants, (2) increase in stress, and (3) increase in strength and hardness of steel.

Stress-corrosion cracking slowly produces a network of cracks that penetrate metal at right angles to the tensile stress. The anodic reaction ($M \rightarrow M^{2+} + 2\,e$) extends the cracks across the lines of tensile stress. The effect is to reduce the cross-sectional area of the metal and to raise the tensile stress to a point of failure.

FORMS OF CRACKING IN DRILLING ENVIRONMENTS

Hydrogen embrittlement (sulfide cracking) and corrosion fatigue are two forms of cracking which are associated with drilling and producing environments.

Hydrogen embrittlement (sulfide cracking)

Hydrogen embrittlement occurs as a sudden cracking of metal, caused by the entrapment of hydrogen within the lattice structure. The cracking of the metal may proceed in a stepwise rupturing manner.

The corrosion process in an acid environment produces atomic hydrogen:

$$Fe + 2\,H^+ \rightarrow Fe^{2+} + 2\,H^0 \text{ (atoms)} \tag{15-7}$$

$$Fe - 2\,e \rightarrow Fe^{2+} \text{ (anode)} \tag{15-8}$$

$$2\,H^+ + 2\,e \rightarrow 2\,H^0 \text{ (cathode)} \tag{15-9}$$

Some of the hydrogen atoms, which are formed in the cathodic reaction, penetrate the metal to form molecular hydrogen. The remaining hydrogen atoms combine to form molecules of hydrogen gas at the metal surface. The adsorption of hydrogen atoms by the metal causes a loss in ductility and cracking of high-strength steels.

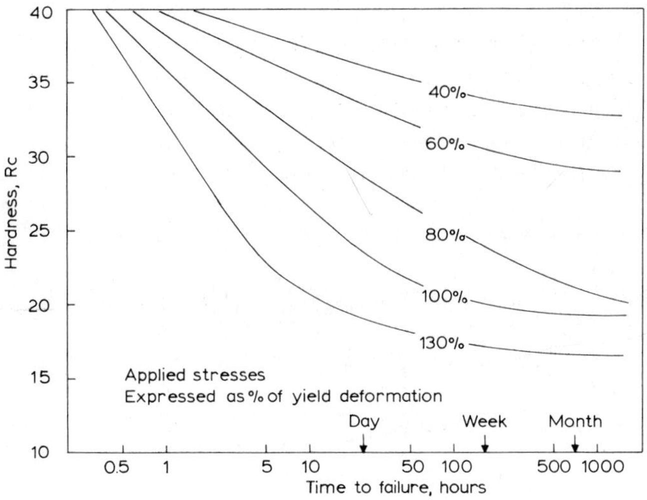

Fig. 15-1. Relationship (approximate) between hardness and time to failure at different applied stresses, expressed as percentage of yield deformation, for carbon steels in 5% NaCl solutions containing 3000 ppm H_2S. (After Hudgins, 1969, p. 42, fig. 1; courtesy of Materials Protection.)

Materials which interfere with the pairing of atoms of hydrogen to form hydrogen gas at cathodic areas of metal, enhance the penetration of atomic hydrogen into the steel. Hydrogen sulfide in drilling fluids supplies sulfide ions which prevent the pairing of hydrogen atoms to form hydrogen gas. Thus, the penetration of atomic hydrogen into steels is promoted by the presence of hydrogen sulfide.

The time to failure by hydrogen embrittlement is shortened by increasing (1) concentration of hydrogen sulfide, (2) stress, and (3) strength and hardness of steel. This behavior is illustrated by Hudgins et al. (1966) in Figs. 15-1 and 15-2.

The embrittlement (sulfide stress cracking) occurs generally in steels with yield strengths above 90,000 psi or above Rockwell C-20-22 hardness. Lower-strength steels are not subject to sulfide stress cracking; however, they are subject to hydrogen blistering.

Hydrogen blistering

Hydrogen penetration of low-strength steels may cause blistering, which appears as bumps on the metal surface. The hydrogen atoms form hydrogen molecules at points of defect in the metal. The hydrogen gas cannot penetrate into nor escape from the crystalline structure. The pressure of hydrogen gas increases sufficiently to part the metal, create a void, and raise a blister on the steel surface.

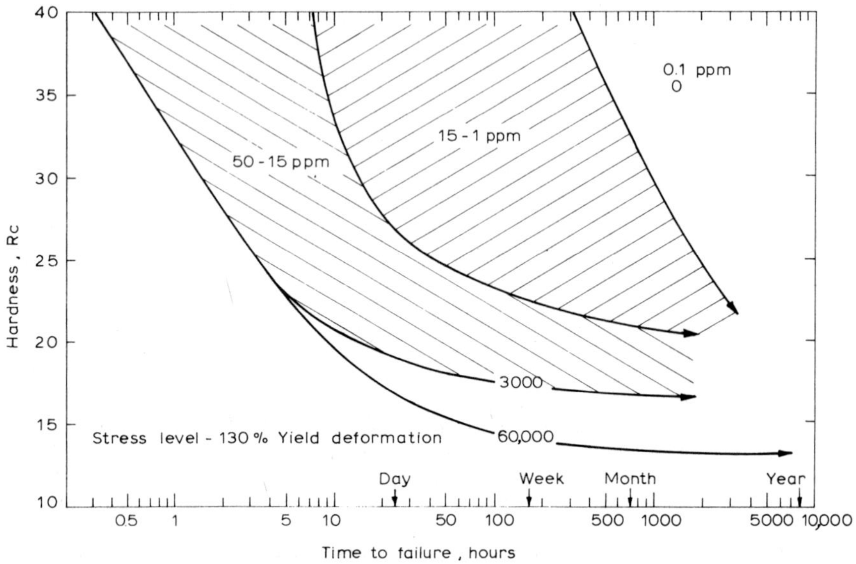

Fig. 15-2. Relationship (approximate) between hardness and time to failure for carbon steels in 5% NaCl solution containing various concentrations of H_2S. Stress level = 130% of yield deformation. (After Hudgins, 1969, p. 43, fig. 2; courtesy of Materials Protection.)

Corrosion fatigue

Cyclic stresses at high levels cause fatigue failures of metal. In a corrosive environment, the progress of fatigue failure is accelerated by electrochemical corrosion. Corrosion fatigue is the combined action of corrosion and fatigue (cyclic stressing), which results in early fracture of metal. Corrosion fatigue is the principal cause of drillpipe failures. By establishing limits to the stresses repeatedly applied to metal, failure by fatigue in a non-corrosive environment can be avoided. The *endurance limit* is the maximum cyclic stress level which can be applied to a metal without a fatigue failure.

In an environment with continued corrosion, instead of exhibiting an endurance limit, metal will fail due to the growth of corrosion fatigue cracks. The time to failure or number of stress cycles necessary to cause failure is decreased with increasing severity of corrosive environment and level of stress as illustrated in Fig. 15-3.

Corrosion fatigue is enhanced by corrodants in a corrosive environment that cause pitting of steel, such as oxygen and the acid gases (H_2S and CO_2), which are often present in drilling fluids. Increasing the strength of steels will not improve the resistance to corrosion fatigue. Instead it may shorten

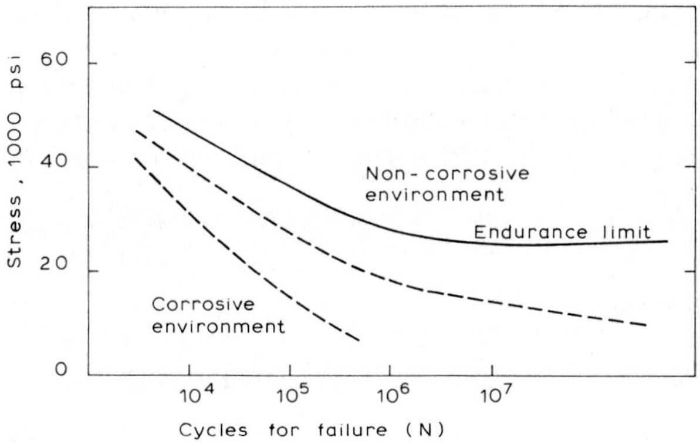

Fig. 15-3. Relationship between endurance strength of steel stress and number of cycles needed for the occurrence of failure in corrosive and non-corrosive environments. (After Bertness, 1957, p. 131, fig. 1; courtesy of Division of Production, American Petroleum Institute.)

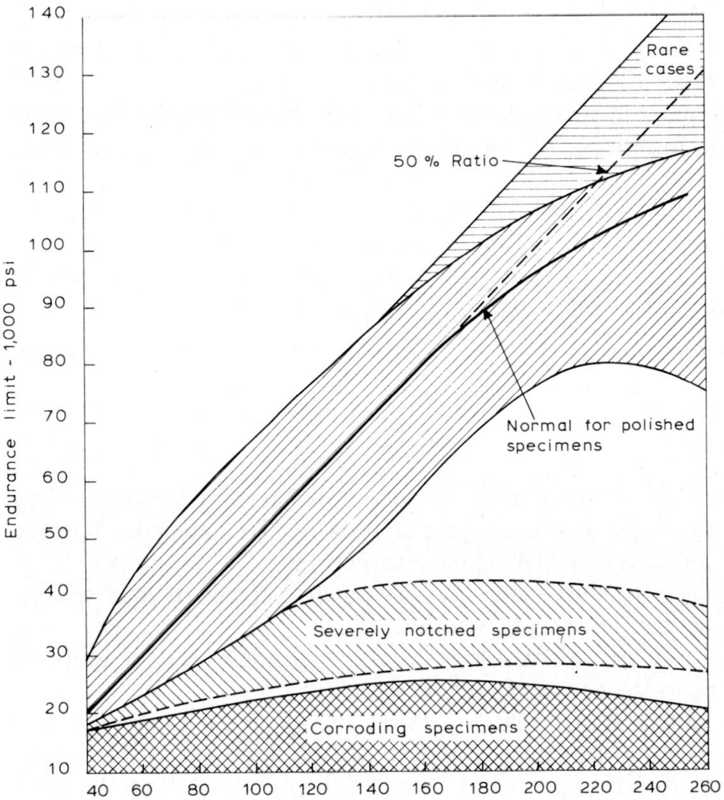

Fig. 15-4. Relationship between endurance limit and tensile strength for polished, notched, and corroding specimens (applied to ordinary corrodible steels). (After Battelle Memorial Institute, 1949, p. 78, fig. 93; courtesy of John Wiley and Sons, Inc.)

the cycles to failure. High-strength steels are more susceptible to pitting than the low-strength steels.

Notches, such as the ones caused by tongs or slips, and corrosion under protectors will accelerate corrosion fatigue failure. The relationship of fatigue to environment, tensile strength, and surface condition of steel is illustrated in Fig. 15-4.

CORRODANTS IN DRILLING FLUIDS

Corrodants in drilling fluids include oxygen, hydrogen sulfide, and carbon dioxide.

Oxygen

Oxygen dissolved in drilling fluids is the major cause of drillpipe corrosion. As a depolarizer and electron acceptor in cathodic reactions, oxygen accelerates the anodic destruction of metal. The high-velocity flow of drilling fluids over the surfaces of drillpipe continues to supply oxygen to the metal and is destructive at concentrations as low as 5 ppb.

The presence of oxygen magnifies the corrosive effects of the acid gases (H_2S and CO_2). The inhibition of corrosion which is promoted by oxygen is difficult to achieve and is not practical in the drilling fluid system.

Removal of oxygen from the drilling fluid by physical deaeration, followed by chemical removal of residual oxygen, is recommended.

Chemical scavengers for the removal of oxygen are sodium sulfite, bisulfites, hydrozine, and sulfur dioxide:

$$Na_2SO_3 + \frac{1}{2}\,O_2 \rightarrow Na_2SO_4 \tag{15-10}$$

$$SO_2 + \frac{1}{2}\,O_2 + H_2O \rightarrow H_2SO_4 \tag{15-11}$$

$$N_2H_4 + O_2 \rightarrow N_2 + 2\,H_2O \tag{15-12}$$

The reaction rates are complex in many water systems and are affected by temperature, pH, hydrogen sulfide, and the presence of catalysts. Snavely and Blount (1969) and Snavely (1971) have shown that:

(a) SO_2 or Na_2SO_3 is not effective for scavenging oxygen in sour waters, and H_2S must be removed prior to treatment.

(b) Hydrozine is not sufficiently reactive for scavenging O_2 at ambient temperatures, except in the presence of Cu^{2+}.

It is concluded that tests of reaction rates and chemical scavenger and catalyst requirements should be made for each drilling fluid.

Oxygen corrosion of drillpipe occurs while the pipe is out of the hole. Pitting can develop rapidly under particles of mud solids which are left on the pipe. Pits provide the sites for further local attack of the drillpipe while it is in service. Proper cleaning with fresh water for the removal of salts and mud solids is recommended.

Cleaned drillpipe should be sprayed with a protective coating prior to storage.

Hydrogen sulfide

Hydrogen sulfide is most damaging to drillpipe by promoting sulfide cracking or embrittlement as discussed in the Stress Corrosion section. General corrosion attack by hydrogen sulfide is also significant and is influenced by the presence of carbon dioxide, oxygen and salts. The nature of the attack on metal is related to the alloy composition and strength of steel.

The corrosion of mild steel in distilled water containing hydrogen sulfide is illustrated by Watkins and Wright (1953). (See Fig. 15-5.)

The data in Fig. 15-5 indicates that high concentrations of hydrogen sulfide may act to inhibit corrosion of *mild* steel. High concentrations of hydrogen sulfide are catastrophic, however, in the case of high-strength steels, producing rapid embrittlement.

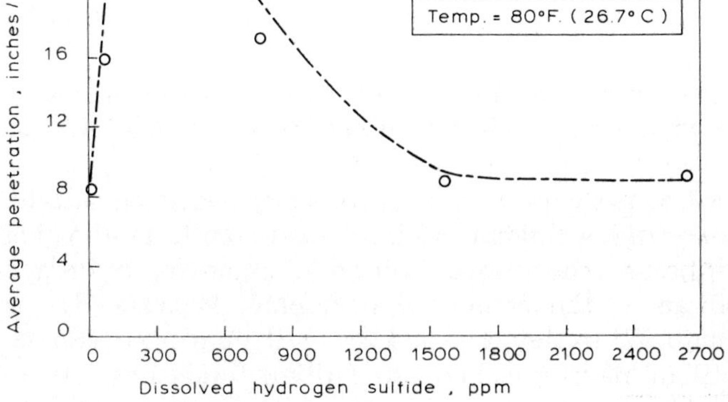

Fig. 15-5. Corrosive action of hydrogen sulfide on steel in distilled water at 80°F. (After Watkins and Wright, 1953, p. B-55, fig. 5; courtesy of the Petroleum Engineer.) mpy = mils per year = [wt loss (mg)] : [SG (or g/cm^3) × 16.387 (cm^3/cu inch) × area (sq inches) × yr (days/365)]; in the case of steel coupons having SG of 7.86, the formula can be simplified: mpy = [wt loss (mg) × 68.33] : [area (sq inches) × hours exposed].

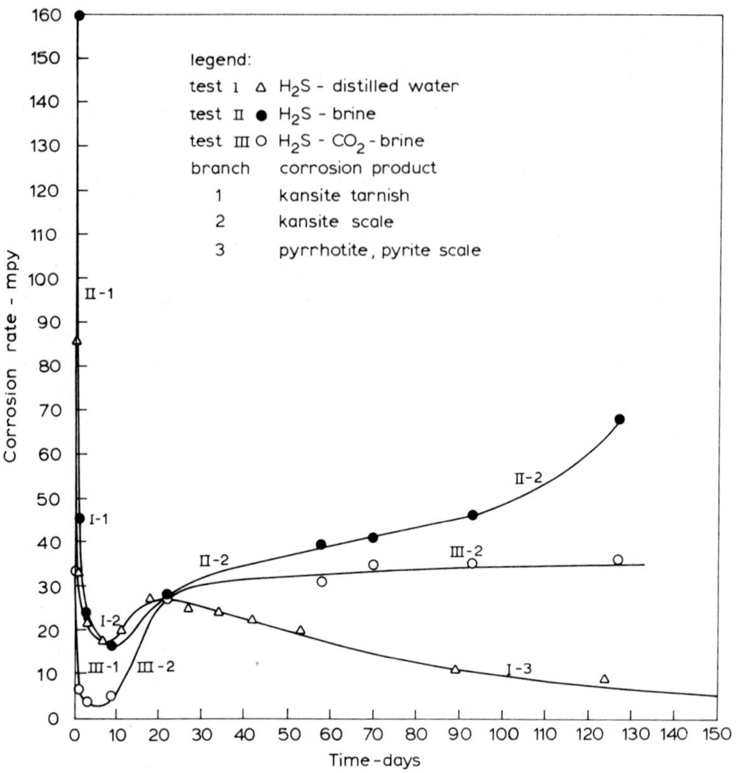

Fig. 15-6. Corrosion rates of 1020 mild steel from tests I, II, and III in mixtures of hydrogen sulfide, carbon dioxide, and brine. (Modified after Meyer et al., 1958, p. 113t, fig. 7; courtesy of "Corrosion".)

The influence of hydrogen sulfide, brine, and carbon dioxide mixtures upon corrosion rates of mild steel is illustrated by Meyer et al. (1958) in Fig. 15-6.

Obviously, the removal of the dissolved gases (oxygen, hydrogen sulfide, and carbon dioxide) from drilling fluids is an important step in minimizing corrosion damage to drillpipe. The primary object of removing hydrogen sulfide from drilling fluids is the safety of personnel, because H_2S is extremely toxic. The limit for repeated exposure is 10 ppm. Exposure to concentrations of 800+ ppm may result in death. Drilling fluids must, therefore, be treated to neutralize hydrogen sulfide gas as it enters the fluid by flow from the formation or from the drilled cuttings.

If presence of hydrogen sulfide is expected, the pH of drilling fluid should be held above 10. The reactions with caustic soda are as follows:

$$pH = 7.0: H_2S + NaOH \rightarrow NaHS + H_2O \qquad (15\text{-}13)$$

$$pH = 9.5: NaHS + NaOH \rightarrow Na_2S + H_2O \quad (15\text{-}14)$$

Hydrogen sulfide scavengers are also added to drilling fluids for the purpose of pretreatment or removal of the gas. These materials include (1) zinc carbonate, chromate, and oxides, (2) iron oxide, and (3) copper carbonate. Copper carbonate should not be used for the purpose of pretreatment nor in excess of the sulfide requirement due to the possible corrosive effects.

Ironite ® Sponge ®, which is the product of a reaction (controlled oxidation) using highly reactive, specially formulated chemical-grade iron powder as the raw material, can be used as an H_2S scavenger. It consists mainly of Fe_3O_4 and is characterized by a high surface area (10 m^2/kg or approximately 50,000 sq ft/lb). The specific gravity of the dry material is around 4.5—4.6 g/cm^3, and particle size ranges from 1.5 to 50.0 μm, with 90% being between 2 and 20 μm. Inasmuch as the material retains very little magnetism, it is not attracted to drillpipe or casing.

Ironite Sponge reacts with H_2S according to the following equations:

$$Fe_3O_4 + 4\,H_2S \rightarrow 3\,FeS + 4\,H_2O + S \quad (15\text{-}15)$$

$$FeS + S \rightarrow FeS_2 \quad (15\text{-}16)$$

$$Fe_3O_4 + 6\,H_2S \rightarrow 3\,FeS_2 + 4\,H_2O + 2\,H_2 \quad (15\text{-}17)$$

Reactions 15-15 and 15-16 predominate in basic environment, whereas reaction 15-17 occurs in acidic environment. One pound of material (0.453 kg) reacts with 0.7 lb (0.318 kg) of H_2S.

The speed of reaction can be expressed by the following equation:

$$dS_t/dt = -3000 \times (S_t)^2 \times (H^+)^{1.06} \times (I) \quad (15\text{-}18)$$

where S_t = total dissolved sulfides in filtrate, ppm; t = time, min; H^+ = hydrogen ion concentration, moles/l; and I = Ironite Sponge concentration, lb/bbl (0.351 × kg/m^3).

Replacement of water-base drilling fluids with oil-base systems, provides protection to the drillpipe by eliminating the electrolyte which is essential to corrosion. The oil-base systems contain some emulsified water, alkalinity of which must be maintained.

The H_2S gas, which is carried by the oil-base drilling fluid, must be removed by gas separators and vacuum degassers. The removed gases must then be neutralized for the protection of personnel.

Carbon dioxide

Carbon dioxide is present in most formation fluids as a component of formation gases and in solution in water and oil. Carbon dioxide dissolves in water and forms carbonic acid:

$$CO_2 + H_2O \leftrightharpoons H_2CO_3 \quad (15\text{-}19)$$

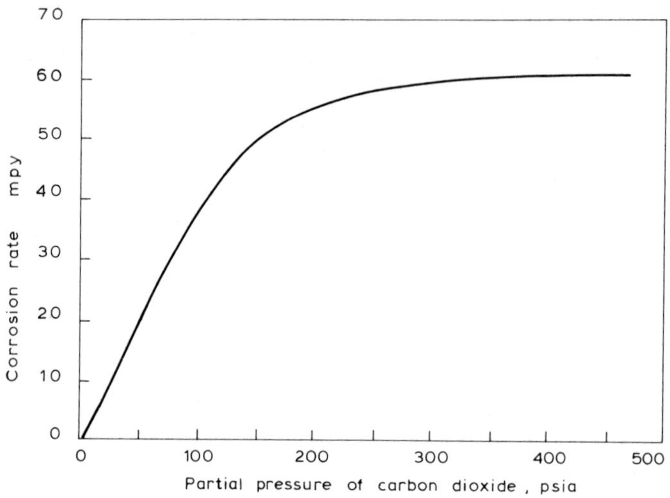

Fig. 15-7. Relationship between corrosion rate of steel and partial pressure of CO_2. (After Rhodes and Clark, 1936; courtesy of Industrial and Engineering Chemistry.)

$$H_2CO_3 \rightleftharpoons H^+ + HCO_3^- \tag{15-20}$$

$$HCO_3^- \rightleftharpoons H^+ + CO_3^{2-} \tag{15-21}$$

As the partial pressure of carbon dioxide increases, more acid ions are formed and the water becomes more corrosive. The relationship of the partial pressure of carbon dioxide and the corrosion rate of steel is illustrated in Fig. 15-7.

Carbon dioxide is usually removed from drilling fluids by degassers. The pH and alkalinity is adjusted, preferably with caustic soda, to avoid scaling or precipitation of carbonates.

ALKALINITY (pH) OF ENVIRONMENT

The pH or hydrogen activity of water influences the corrosion rate of steel. The effect of pH on the corrosion of steel is dependent upon metal composition, stresses, oxygen concentration, and the type of acid which controls the pH.

The effect of pH on corrosion of steel in water containing 5 ppm of oxygen is shown in Fig. 15-8.

In a high alkaline range the corrosion reaction is under anodic control and proceeds at high rates:

$$O_2 + Fe - 2\,e \rightarrow FeO_2^{2-} \tag{15-22}$$

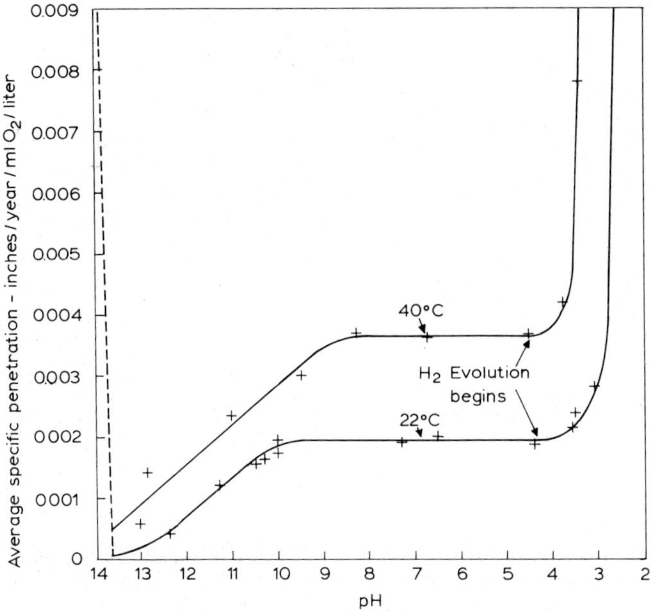

Fig. 15-8. Effect of pH on corrosion of mild steel. (Modified after Whitman et al., 1924; also see Uhlig, 1948, p. 129, fig. 2; courtesy of Industrial and Engineering Chemistry.)

In the neutral range and in mild alkaline solutions the corrosion rate is under cathodic control, which provides some corrosion protection. Ferrous hydroxide, which provides a protective layer on the metal surface, forms in this environment. The actual corrosion rate is dependent upon the diffusion of oxygen to the metal surface. Corrosion increases with increasing oxygen concentration and abrasion and in the presence of turbulent, high-velocity flow, which is often imposed on drillpipe.

In the acid pH range, corrosion is under anodic control and the metal composition influences the corrosion rate extensively. Trace elements in steel and stresses affect the corrosion damage.

The type of acid in solution determines the pH at which corrosion increases rapidly with the evolution of hydrogen gas. As shown in Fig. 15-8, hydrogen evolution starts at a pH near 4, with corrosion accelerating as pH is reduced. Hydrogen evolution at a pH of about 4 occurs in the presence of strongly dissociated acids.

Carbonic acid from the solution of carbon dioxide reacts with iron at a pH of 6 with the evolution of hydrogen gas from severe corrosion. This illustrates the importance of pH control throughout the circulating mud system where carbon dioxide is present.

Relationship of failure by embrittlement (sulfide cracking) and pH is illustrated in Fig. 15-9.

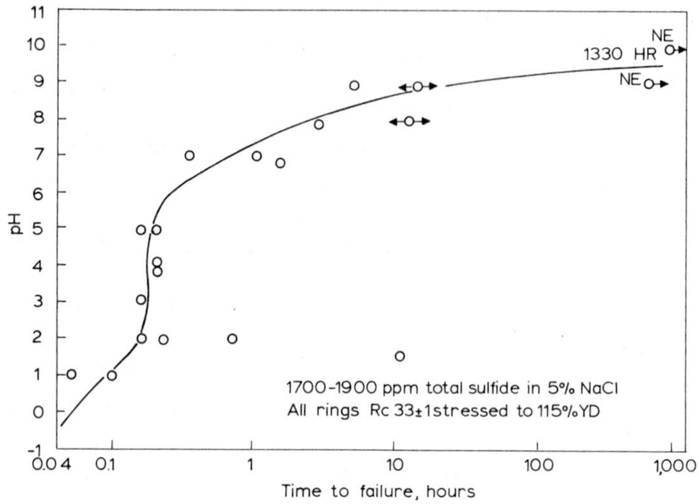

Fig. 15-9. Relationship between pH and time to failure. (After Hudgins, 1969, p. 43, fig. 3; courtesy of Materials Protection, National Association of Corrosion Engineers.)

Figures 15-8 and 15-9 illustrate the importance of pH influence on the rate of corrosion. The most important corrosion control measure is to remove oxygen, carbon dioxide, and hydrogen sulfide (soluble gases) from the drilling fluids. The effectiveness of both corrosion inhibitors and pH is enhanced by the removal of soluble gases.

CATHODIC PROTECTION

Cathodic protection is not a practical method for the control of corrosion of drillpipe or the internal surfaces of well casings. Cathodic protection, however, is applied successfully for the corrosion control of external casing surfaces.

The first step in the control of external casing corrosion is to provide a complete cement sheath and bond between the pipe and the formation over all external areas of the casing strings. This practice is essential to the successful well completions.

ROLE OF BACTERIA IN CORROSION

The influence of sulfate-reducing bacteria in the corrosion process has been the subject of extensive investigations. The subject is complex and the reader is referred to the comprehensive treatment of the subject by Davis (1967). (Also see Chilingar and Beeson, 1969.)

Microbial corrosion has not been significant in the corrosion of drillpipe. Sulfate-reducing bacteria have produced serious corrosion to well casings as reported by Doig and Wachter (1951). Pitting occurred on the external surfaces of the casing where drilling fluid was present between the casing and the wall of the hole. The low pH of the mud and the presence of organic nutrients were favorable for the growth of sulfate-reducing bacteria.

The pH favorable for bacterial growth ranges from 5 to 9. Bacteria will thrive in areas of stagnant flow even in high-pH mud systems, however, provided other requirements, i.e., temperature and contents of organic nutrients, salts, and oxygen, are satisfied.

Sulfate-reducing bacteria are anaerobic and thrive only in the absence of oxygen. In an aerated mud system, oxygen is depleted in stagnant areas, e.g., along the walls of mud pits and behind casing. This allows the sulfate-reducing bacteria to grow. Equation can be presented as follows:

$$H_2SO_4 + 8\ H + \text{bacteria} \rightarrow H_2S + 4\ H_2O \qquad (15\text{-}23)$$

Thus, hydrogen atoms formed at cathodic areas of metals are removed and are utilized to reduce sulfates to sulfides. The end products are live bacteria and corroded iron.

The bacterial attack of organic additives of drilling fluids may result in excessive use of mud chemicals and a rapid reduction of pH. The mud will become more corrosive under these conditions. Thus, bacteria must be controlled by the use of high pH (e.g., 10.5) or bactericides.

For excellent and comprehensive treatment of corrosion, the reader is referred to Uhlig (1965), Fontana and Greene (1967), and NACE TPC Publication No. 5 (1979).

REFERENCES

Battelle Memorial Institute, 1949. *Prevention of the Failure of Metals under Repeated Stress.* Wiley, New York, N.Y., 295 pp.

Bertness, T.A., 1957. Reduction of failures caused by corrosion in pumping wells. *API Dril. Prod. Prac.*, 37: 129—135.

Chilingar, G.V. and Beeson, C.M., 1969. *Surface Operations in Petroleum Production.* Am. Elsevier, New York, N.Y., 397 pp.

Davis, J.B., 1967. *Petroleum Microbiology.* Elsevier, Amsterdam, 604 pp.

Doig, K. and Wachter, A.P., 1951. Bacterial casing corrosion in the Ventura Field. *Corrosion,* 7: pp. 221—224.

En Dean, H.J., 1977. Avoiding drilling and completion corrosion. *Pet. Eng.*, 10(9): 23—28.

Fontana, M.G. and Greene, N.D., 1967. *Corrosion Engineering.* McGraw-Hill, New York, N.Y., 391 pp.

Hudgins, C.M., 1969. A review of corrosion problems in the petroleum industry. *Mater. Prot.*, 8(1): 41—47.

Hudgins, C.M., McGlasson, R.L., Mehdizadeh, P. and Rosborough, W.M., 1966. Hydrogen sulfide cracking of carbon and alloy steels. *Corrosion*, 22(8): 238—251.
Ironite Products Co., 1979. *Hydrogen Sulfide Control.* 41 pp.
May, P.D., 1978. Hydrogen sulfide control. *Drilling-DCW*, April.
Meyer, F.H., Riggs, O.L., McGlasson, R.L. and Sudbury, J.D., 1958. Corrosion products of mild steel in hydrogen sulfide environments. *Corrosion*, 14(2): 109t—115t.
National Association of Corrosion Engineers, 1979. *Corrosion Control in Petroleum Production.* NACE TPC Publ. No. 5, 101 pp.
Ray, J.D., Randall, B.V. and Parker, J.C., 1978. *Use of Reactive Iron Oxide to Remove H_2S from Drilling Fluid.* 53rd Annu. Fall Tech. Conf. Soc. Pet. Eng. AIME, Oct. 1—3, 1978, Houston, Texas, 4 pp.
Rhodes, F.H. and Clark, J.M., 1936. Corrosion of metals by water and carbon dioxide under pressure. *Ind. Eng. Chem.*, 28(9): 1078—1079.
Simpson, J.P., 1979. A new approach to oil-base muds for lower-cost drilling. *J. Pet. Technol.*, 31(5): 643—650.
Snavely, E.S., 1971. Chemical removal of oxygen from natural waters. *J. Pet. Technol.*, 23(4): 443—446.
Snavely, E.S. and Blount, F.E., 1969. Rates of reaction of dissolved oxygen with scavengers in sweet and sour brines. *Corrosion*, 25(10): 397—404.
Staehle, R.W., 1978. $ 70 billion plus or minus $ 21 billion, editorial. *Corrosion*, 34(6): i—iii.
Uhlig, H.H. (Editor), 1948. *The Corrosion Handbook.* Wiley, New York, N.Y., 1188 pp.
Uhlig, H.H., 1965. *Corrosion and Corrosion Control*, 3rd ed. Wiley, New York, N.Y., 371 pp.
Watkins, J.W. and Wright, J., 1953. Corrosive action on steel by gases dissolved in water. *Pet. Eng.*, 25 (12, Nov.): B50—B57.
Wendt, R.P., 1979. *The Kinetics of Ironite ®Sponge ®H_2S Reactions.* Pet. Div. Am. Soc. Mech. Eng., Energy Technol. Conf., Houston, Texas, Nov. 5—9, 1978, 7 pp.
Wendt, R.P., 1979. Control of hydrogen sulfide by alkalinity may be dangerous to your health. *Pet. Eng. Int.*, 51(6, May): 66—74.
Wendt, R.P., 1979. Alkalinity control of H_2S in muds is not always safe. *World Oil*, 188(2, Aug.): 60—61.
Whitman, W., Russell, R. and Altieri, V., 1924. *Ind. Eng. Chem.*, 16: 665.

Chapter 16

DRILLING THROUGH OVERPRESSURED FORMATIONS

WALTER H. FERTL and GEORGE V. CHILINGARIAN

INTRODUCTION

Many factors can cause abnormally high formation pressures, that is, pressures higher than hydrostatic. In some areas, a combination of several causes prevails (Table 16-I). Because conditions can vary widely, special care should be taken not to assume that the cause of overpressures established from experience in a well-known area is necessarily the cause of similar condition in a nearby basin, which may not yet have been adequately tested by drilling.

Abnormally high formation pressures have been encountered worldwide in clastic and carbonate intervals ranging in age from the Cenozoic Era (Pleistocene) to as old as the Paleozoic Era (Cambrian), in formations as shallow as a few hundred feet and at depths exceeding 25,000 ft (7620 m).

Because detection and quantitative evaluation of overpressured formations is critical to exploration, drilling, and production operations involving hydrocarbon and geothermal sources, an interdisciplinary technical team approach was recommended to optimize the safety, engineering, and financial aspects of operating in such hostile subsurface environments (Fertl, 1976).

Proper drilling operations and mud and casing programs constitute major cost factors, particularly in high-pressure areas. In order to design an effective and economical mud program, the fracture pressure gradient as well as the pore-fluid pressure variations must be known. If pressure exerted by the drilling fluid exceeds the strength limit of the formation, fractures form resulting in lost circulation. Subsequent drop in the hydrostatic pressure head in the wellbore creates a potential blowout condition.

Experience indicates that costly misinterpretations are best avoided by studying a combination of several measured formation-pressure indicators, including seismic, drilling, and well logging data (Fig. 16-1). Although several such indicators should be monitored at the wellsite, not all will always turn out to be usable or necessarily needed in any one drilling operation. Drilling and well logging overpressure detection and evaluation techniques are summarized in Table 16-II, whereas parameters affecting such practical formation pressure detection techniques are compiled in Table 16-III. The best plan for drilling operations in overpressured environments incorporates (1) gathering and interpreting all known information (i.e., planning the well), (2)

TABLE 16-I

Possible causes for abnormal formation pressures

Origin	Comments
Piezometric fluid level	Effect of regional potentiometric surface. Artesian water system, such as Artesian Basin, Florida, U.S.A., Great Artesian Basin, Australia, and North Dakota Basin, U.S.A.
Structure of permeable reservoir	Pressure transmission to shallower part of reservoir. Large anticlines, steeply-dipping beds, etc.
Rate of sedimentation and depositional environment	High depositional rates in clastic sequence and high shale/sand ratios.
Tectonic activities	Local and regional faulting, folding, lateral sliding and slipping; squeezing caused by downdropping of fault blocks; diapiric salt, sand, or shale movements; earthquakes; etc.
Osmotic and salt-filtering phenomena	On regional basis, e.g., San Juan Basin, New Mexico; Western Sedimentary Basin, Canada; San Joachim Valley, California; U.S.A. Gulf Coast; and Paradox Permian Basin, Illinois, U.S.A.
Diagenetic phenomena	Postdepositional alterations: (1) montmorillonite → mixed layer clays → illite, with resulting water release; (2) gypsum → anhydrite + water; volcanic ash → clay minerals + carbon dioxide.
Secondary precipitation of cementing materials	Calcium sulphates, sodium chloride, dolomite, siderite, calcite, silica, etc. may act as (1) sealing barriers ("pressure caps"), and (2) directly cause overpressure by decreasing pore space due to crystal growth in closed reservoirs (e.g., NaCl in Markovo oil pool in the Osinskiy Series, U.S.S.R.).
Repressuring of reservoir rocks	Caused by hydrocarbon production, drilling operations, and massive water injection programs (secondary recovery). May occur across faults, behind casing, or may be due directly to planned injection.
Massive areal rock salt deposition	Presence of salt (NaCl) beds which are impermeable. Examples present in U.S.A., U.S.S.R., North Africa, Middle East, North Germany, etc.
Paleopressures	Old, completely sealed-off reservoir rocks. Depth change due to uplifting or erosion.
Thermodynamic effects	Formation temperature changes cause fluid pressure variations.
Biochemical effects	About 2- to 3-fold volume increase caused by breakdown of hydrocarbon molecules.
Permafrost environment	Formation of giant frost heaves (pingos). Freeze-back pressure around shut-in arctic oil wells. Gas hydrate reservoirs (e.g., in Mackenzie Delta, Canada).

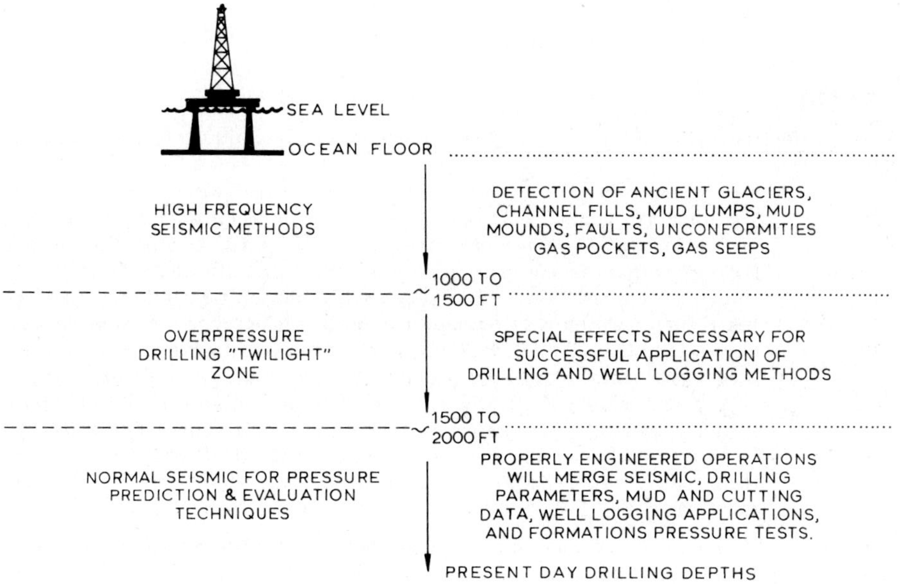

Fig. 16-1. Offshore exploration scene. (After Aud and Fertl, 1976, fig. 1, p. 122.)

TABLE 16-II

Drilling and well logging overpressure detection and evaluation techniques (after Fertl and Chilingar, 1976, table 1, p. 5; courtesy of Society of Petroleum Engineers of AIME)

Drilling operations	Drilling parameters (instantaneous)	Drilling rate, torque, drag, hole fill (→ reaming), modified *d*-exponent, analytical drilling model concepts.
	Logging-while-drilling (instantaneous or semi-instantaneous)	Electrical and acoustic-type transmission concepts (using cables, drill string, mud system); analysis of drill string vibrations; downhole gas detection tool.
	Drilling mud parameters (while drilling, but delayed by circulation lag-time)	Mud weight, gas content, temperature, flow rate, hole fill-up, pit level and total pit volume, salinity (resistivity, conductivity), well kicks.
	Drill cuttings analyses (while drilling, but delayed by time required for sample return)	Density, shape, size, color, volume over shale shaker, and moisture content, "lithofunction" plots, shale factor of cuttings (cation exchange capacity). Cuttings slurry and/or filtrate: resistivity, color, pH, redox potential, bicarbonate content, specific anion and cation concentrations, and filtration rate.
Well logging	Formation parameters from geophysical well logs	Electrical surveys: resistivity, conductivity, salinity, shale formation factor; acoustic (sonic) surveys: interval transit time and wave train presentations (VDL, signature log, etc.); bulk density surveys: density log, downhole gravity meter; hydrogen index (neutron-type logs); thermal neutron capture cross section (pulsed neutron logging); nuclear magnetic resonance; gamma ray spectral analysis logs.

TABLE 16-III

Parameters affecting practical formation pressure evaluation techniques (after Fertl and Chilingar, 1976)

Geologic factors	Geologic age changes. Compaction effects on regional (e.g., basin edge vs. basin center) and local scales and differential compaction across structures. Sand/shale ratio in clastic sediments. Lithology effects: pure shales (soft, hard); limey and silty shales; bentonitic markers; gas-bearing ("shale gas"), organic-rich, bituminous shales; types and amounts of clay minerals in shales (depending on depositional environment and/or diagenesis). Heavy minerals (siderite, pyrite, mica, etc.). Drastic formation water salinity variations across overpressure discontinuities (or barriers), including unconformities, pinchouts, and faults; proximity to large salt masses, mud volcanoes, geothermal "hot" spots, etc. Steep, thin, overturned beds. Pore pressure gradients within single thick shale interval.
Borehole environment	Borehole size, shape and deviation. Shale alteration and hydration (exposure time of open hole to drilling mud). Type of drilling mud (freshwater, salt-water, oil-base). Type and amount of weighting material (barite, etc.) and/or lost circulation material (mica, etc.). Degree of "gas-cutting" in mud.
Drilling conditions	Hole size, shape, and deviation. Mud programs and mud hydraulics (circulation rate). Rotary speed. Bit type (button, diamond, insert, etc.) Bit weight/bit diameter ratio. Bit wear (sharp, new bits vs. dull, old bits). Degree of overbalance. Floater ("heave" action) vs. fixed on- or offshore rig.
Sample selection	Type and size of sample (avoid sand, cavings, recirculated shales). Sampling technique. Sampling frequency. Analysis methods: for example, cuttings density — variable density column, multiple density solution technique (float and sink method), mercury pump technique, and mud balance technique. (Proper calibration is of utmost importance.)
Geophysical well logging	Difference in basic measuring principles. Shales properties are anisotropic. Sonde spacing. Depth of tool investigations. Temperature ratings. Proper tool calibration. Tool malfunction (overlapping repeats or reruns).
Parameter plotting techniques	Interval (or sample) selection. Sampling frequency. Linear and logarithmic plots. Plot comparable data (not compatible are: bulk density from logs vs. those of cuttings, short normal vs. induction log resistivities). Proper selection of "normal" compaction trendlines (discrepancies become enhanced with increasing depth of wells). Use all information available. Experienced and properly trained personnel is required.

educated guesses about the unknown (i.e., preliminary drilling program), and (3) the flexibility to deviate from the plan whenever necessary (i.e., revising the drilling program).

Drilling parameters

Drilling rate (penetration rate)

Drilling rate is a function of weight on the bit, rotary speed, bit type and size, hydraulics, bottom-hole cleaning properties of drilling fluid, and formation characteristics. Under controlled conditions of constant bit weight, rotary speed, bit type, and hydraulics, the drilling rate in shales decreases uniformly with depth. This is due to increase in degree of compaction of shales with depth; however, in pressure transition zones and highly overpressured zones the penetration rate increases. Slower penetration rate is frequently observed in the sealing pressure barrier (cap rock) overlying this transition zone. Any other major lithological changes in the shales (silty and/or limey shales, mudstones, etc.) will also cause penetration rate variations.

Penetration rate should be plotted manually at proper depth increments (5- to 10-ft increments in slow-drilling formations or in 30- to 50-ft increments in fast-drilling intervals). Plotting such data points, however, should not lag too much behind the total drilling depth (not more than twice the plotted depth increment behind the total well depth reached by the bit). At the present time, drilling rate recorders automatically plot rate in feet per hour versus depth.

Simple rules of thumb, such as the one proposed by Forgotson (1969) that a twofold penetration rate increase indicates the onset of overpressures, do not always apply. For example, an increase in mud weight to 12 lb/gal upon encountering the transition zone, may partially mask any further pressure increase with depth. It is also of interest to note that the first unit of mud weight (lb/gal) in excess of formation pore pressure will reduce the drilling rate more than each subsequent unit of mud weight (lb/gal) increase (Moore, 1974).

Complications may also arise during bit drilling, which may mask any penetration rate change due to overpressure. Penetration rate even may decrease due to fluctuating rotary torque and erratic action of the drill bit on the bottom of the borehole.

Normalized rate of penetration (d-exponent)

Inasmuch as it is not always possible and/or feasible to maintain the bit weight and rotary speed constant, the concept of the d-exponent was developed by Jorden and Shirley (1966). Data required to calculate the d-exponent, a dimensionless number, are the penetration rate (R, in ft/hr), bit size (diameter D, in inches), weight on bit (W, in lb), and rotary speed (N, in rpm):

$$d = \frac{\log(R/60N)}{\log(12W/10^6 D)} \tag{16-1}$$

Basically, plots of d-exponent versus depth show a decreasing trend with depth. In transition zones and overpressured environments, the calculated values diverge from the normal trend to lower than normal values. Quantitative pressure evaluation can then be made on using the equivalent depth method (Fertl, 1976, p. 123) or specially constructed, transparent overlay of parallel, equivalent mud weight lines for the specific depth scale used for the d-exponent (see Fig. 16-5).

The values of d-exponent are affected by any change in the basic input parameters in eq. 16-1. Furthermore, it is often difficult to establish reasonable values for bit weight in *soft* formations. Any major lithologic change in the shale section (e.g., limey or silty shales, mudstones, and marls) will also affect the d-exponent. The same is true in poorly maintained drilling fluid systems and for drastic changes in mud weight.

Inasmuch as eq. 16-1 ignores the direct effect of mud weight on drilling rate, a modification of the d-exponent has been proposed by Harper (1969) to normalize the d-exponent for the effective mud weight as follows:

$$d_c = d \cdot (MW_1/MW_2) \qquad (16\text{-}2)$$

where d_c = modified (corrected) d-exponent, MW_1 = normal mud weight, and MW_2 = actual mud weight used.

Consequently, d_c-plots represent a substantial improvement over d-plots

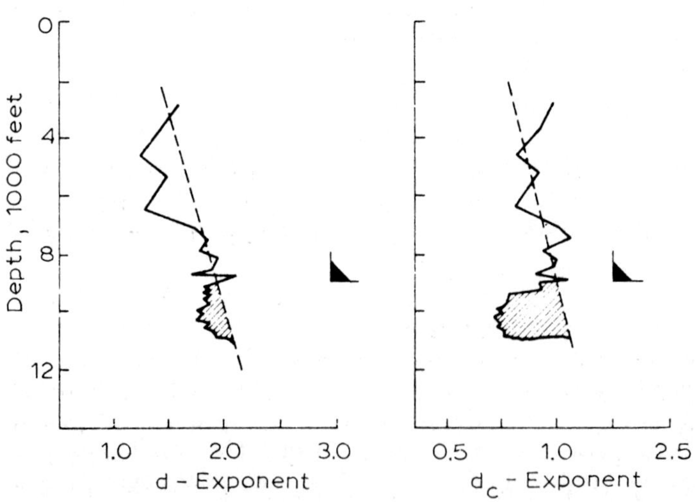

Fig. 16-2. Comparison plots of depth versus d-exponent and d_c-exponent in same well. Protective casing seat is at a depth of 8700 ft. Note that d_c-exponent defines overpressured zone more clearly. (After Fertl, 1976, fig. 4-6, p. 125.)

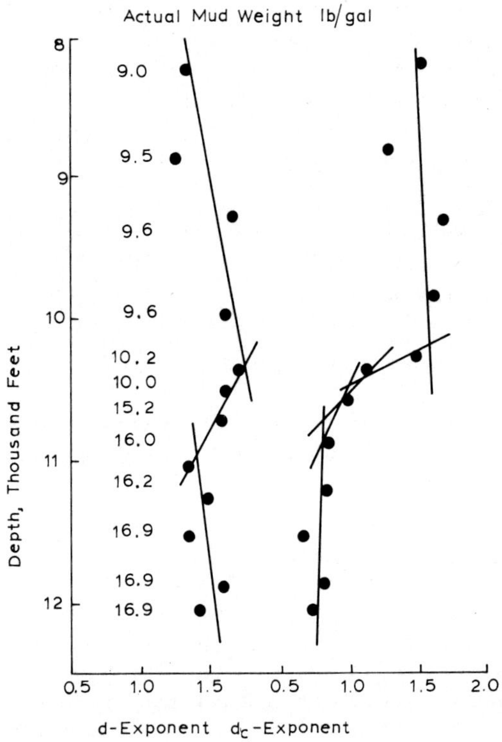

Fig. 16-3. Comparison of plots of depth versus d-exponent and d_c-exponent based on drill bit data, Louisiana well, U.S.A. (After Fontenot and Berry, 1975, fig. 3, p. 127; courtesy of Oil and Gas Journal.)

because mud weight effects are considered (Figs. 16-2, 16-3, and 16-4). They are used by the industry worldwide, both on- and offshore. Quantitative pressure methods use the (1) equivalent depth method, (2) transparent overlays of parallel equivalent trend lines for mud weight or pore fluid pressures (Fig. 16-5), or (3) a specially designed d_c-slide rule.

Drilling information is normally used in clastic sediments for calculating d_c-exponent. In several areas, in drilling through mixed lithologies (i.e., sands, shales, limestones, and dolomites) such computations often give good results. Frequency of calculating d_c-exponent depends on how fast formations are being drilled. Usually, d_c-exponent is determined for every 10 ft increment in depth. In fast drilling areas, 25-, 50-, or even 100-ft depth increments may be adequate, whereas in slow-drilling areas, i.e., hard-rock intervals, 5-ft depth intervals may become necessary.

The mathematical relationships shown in eqs. 16-1 and 16-2 clearly indicate the effects of drilling variables on d_c-values. Effects of hydraulics,

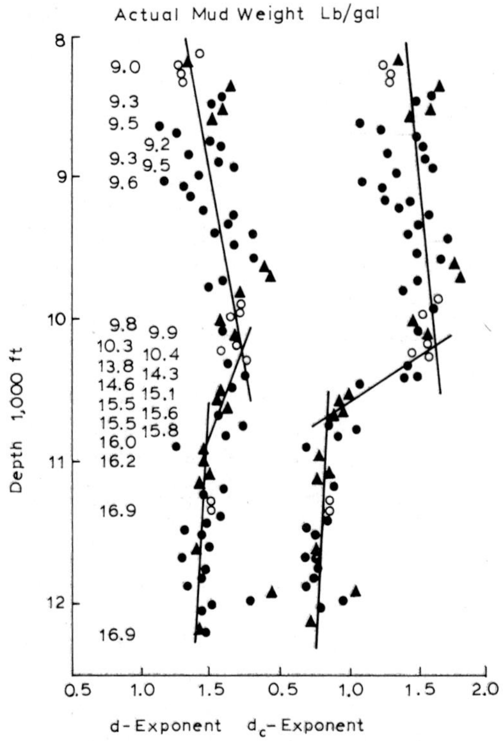

Fig. 16-4. Comparison of plots of depth versus of d-exponent and d_c-exponent based on drilling data averaged over 50-ft intervals; Louisiana well, U.S.A. (After Fontenot and Berry, 1975, fig. 6, p. 129; courtesy of Oil and Gas Journal.)

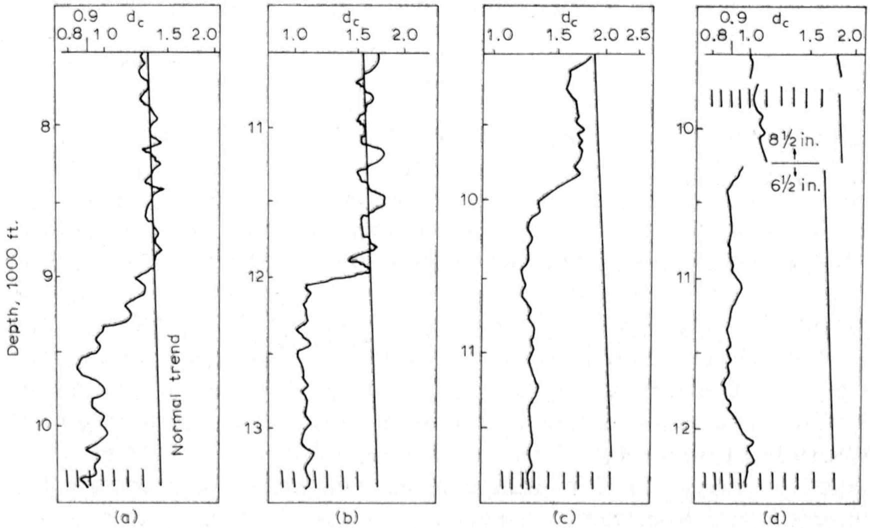

Fig. 16-5. Plotting d_c-exponent on logarithmic scale with use of mud weight overlay technique in South Texas (a, d), South Louisiana (b), and Oklahoma (c). (After Zamora, 1972, fig. 2, p. 70; courtesy of Oil and Gas Journal.) Note effect of bit size change (from $8\frac{1}{2}$ inches to $6\frac{1}{2}$ inches) on normal trendline in the South Texas well (d). Trendlines represent constant mud specific weights.

weight on bit, bit size and type, and overbalance in the case of drilling through soft, clastic formations are discussed below.

Effect of drilling hydraulics

The equations for d- and d_c-exponents are based on the assumption that drill cuttings are being effectively removed. In most wells drilled in the mid-1960's, transition zones were encountered at moderate depths, i.e., 8000 ft or slightly deeper, and hydraulics programs were usually adequate. Minor fluctuations in circulation rate did not significantly affect the penetration rate.

Recent drilling in areas where overpressured zones are encountered in shallow, unconsolidated clastic formations, indicates the increased importance of proper hydraulics programs in achieving effective penetration rates in these soft formations. The latter are drilled utilizing a combination of tooth cutting and jetting action. As a result, increased jetting action will increase penetration rate which, in turn, will result in decreasing values of the d_c-exponent. This gives an appearance of non-existent transition zone. Furthermore, maintaining a constant circulating rate when a transition zone is anticipated in unconsolidated sediments will minimize effects of hydraulics.

Transition zones at depths as shallow as 1500 ft have been sucessfully defined by controlled drilling conditions (i.e., maintaining constant bit weight, rotary speed, mud weight, and circulation rate). Frequently, the fast penetration rate is measured by timing with a stop watch the drill joints lowered during drilling.

Effect of drill bits

When button and diamond bits are used instead of standard rock bits, an empirical correction is required, i.e., a 1-inch reduction in the bit diameter is used in calculations. Reliability of this concept in actual field applications, however, is debatable.

There is a decreae in d_c-values when a button or diamond bit is used. It should be assumed that, at the point of a bit change, the first shale layer drilled with the button or diamond bit has the same pore pressure as the last shale layer drilled with a rock bit. Hence, a shift of the normal compaction trend line in the d_c-versus-depth plot is required to account for the change in bit type. This shift is similar to the one required due to the change in bit size. The effect of bit size change is shown in Fig. 16-5. Consequently, trend-line shifts become necessary when the weight-on-bit/bit size ratio is significantly changed, regardless of the type of the bit.

Bit wear will cause a decrease in penetration rate and, thus, increase in the d_c-exponent. Consequently, dull bits can mask the presence of a transition zone. Consideration of bit wear in graphical and computerized techniques will eliminate this limitation.

An alternative solution is to study the offset well data and previous bit performance in order to determine when to pull a bit *prior* to drilling through transition zones. Although this approach may result in an additional trip, it will greatly reduce the hazards of a well kick and time required to kill it, or even of a blowout, should the downhole mechanical problems occur.

Overbalance

The d_c-exponent is also affected by the difference between hydrostatic pressure of the mud column and formation pressure (i.e., amount of overbalance). A linear, instead of the required hyperbolic, correction will introduce a serious error into calculations with increasing overbalance. In other words, arbitrary and "emotional" mud weight increases should be avoided.

Knowledge of formation pressure while drilling is important for achieving safe and economic operations and in the selection of proper casing seats. The d_c-exponent, supplemented with several other pressure indicators, provides the drilling engineer with the required information for making proper decisions.

Drilling rate equations

Application of proper drilling rate equations in order to establish correlations between formation and borehole pressures require modern instrumentation at the rig site to record simultaneously several drilling parameters.

For example, a general equation for penetration rate in shales has been developed by Combs (1968) using data from six Louisiana wells in a regression-type analysis. Combs' correlation shows that penetration rate (R) is proportional to weight on the bit (W), rotary speed (N), and bit hydraulics, each raised to a fixed power:

$$R = R_0 \cdot (W/3.5)^{n_1} \cdot (N/200)^{n_2} \cdot (Q/3D_hD_n)^{n_3} \cdot f(P_d) \cdot f(T) \qquad (16\text{-}3)$$

where:

D_h = borehole diameter (inches).
D_n = bit nozzle diameter (inches).
f = function of.
n_1, n_2, n_3 = weight, speed, and hydraulics exponents; Combs' recommendations: $n_1 = 1.0$, $n_2 = 0.6$, $n_3 = 3$.
N = rotary speed (rpm).
Q = flow rate (gal/min).
P_d = differential pressure (lb/gal/1000 ft).
R = penetration rate (ft/hr).
R_0 = shale drillability with sharp bit at zero differential pressure (ft/hr).
T = bit wear index—equivalent rotating index.
W = weight on bit (in 1000 lb).

According to Combs (1968), penetration can be predicted from eq. 16-3 with a standard deviation of approximately 30%, whereas pore pressure can be predicted with a standard deviation of about 1.0 lb/gal equivalent pore pressure.

In connection with computerized drilling control, Young (1968) expressed penetration rate as follows:

$$R = G \cdot N^{m} \cdot (W - W_{int}) \cdot K \tag{16-4}$$

where a reciprocal drilling constant (K), the weight intercept value (W_{int}), and the bit rotary speed exponent (m) are determined by a five-spot drilling test. G is a function of the normalized bit tooth height.

Drilling performance optimization and identification of overpressured formations have been also investigated by Wardlaw (1969).

Several other empirical equations for drilling rate determination have been developed by different investigators. Consequently, field data from (1) previously drilled adjacent wells and/or (2) feedback data from short-interval testing in the subject well are usually needed in order to determine input parameters.

Unfortunately, drilling rates change with lithologic variations, even though the pressure gradient remains constant. Deviated holes, severe dog legs, drilling from floating vessels, and frequently occurring water-sensitive and sloughing shales can make these indicators questionable.

Development of fully-automated pressure-detection techniques is difficult, except for local geographic areas where the lithology is well known. Other pressure indicators, such as chemistry of produced water, drilling mud properties, density of formation cuttings, and wireline and related tools must still be used meticulously.

Porosity and formation pressure logs

Computerized models establish relationship between formation characteristics and drilling parameters, and provide an early indication of formation type and porosity and pore pressure variations (Zoeller, 1970; Boone, 1972). At the present time, several service companies have made similar "data units" commercially available, and many oil companies and drilling contractors have developed their own drilling program models.

For example, a field example of formation porosity and pressure logs in an offshore Louisiana wildcat (U.S.A.) is shown in Fig. 16-6 (Fertl, 1976, p. 131). There is a $9\frac{5}{8}$-inch casing seat at a depth of 14,401 ft, which is followed by a fast pressure increase over the next 600 ft.

Bourgoyne et al. (1971) proposed a general equation relating various controllable drilling variables and drilling performance, which can be expressed as follows:

$$R = K \cdot f_1(W/D) \cdot f_2(N) \cdot f_3(H) \cdot f_4(\Delta p) \tag{16-5}$$

where: R = rate of penetration (ROP); K = drillability constant or nor-

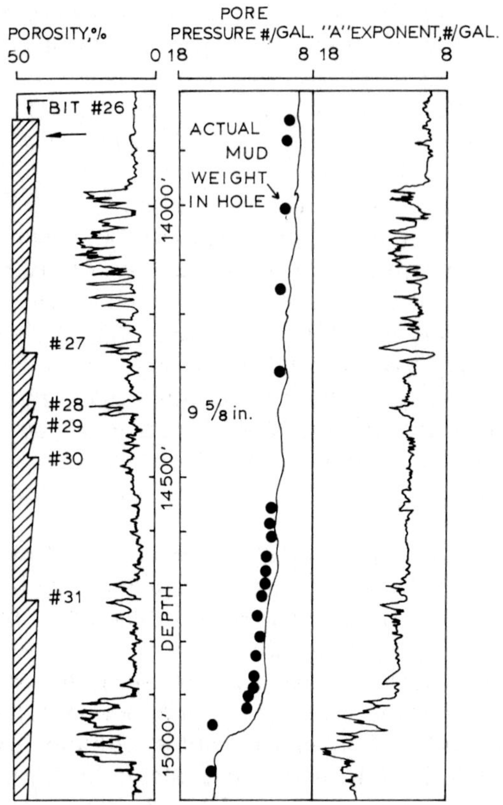

Fig. 16-6. Field example of formation porosity and pore pressure logs in an offshore Louisiana wildcat, U.S.A. The curve on the right ("A" exponent) gives the value for pore pressure taking porosity and lithology variations into account. (After Fertl, 1976, fig. 4-10, p. 131.) Pore pressure in psi is expressed in terms of equivalent mud weight in lb/gal.

malized ROP; $f_1(W/D)$ = function describing the effect of bit weight, W, per inch of bit diameter, D, on ROP; $f_2(N)$ = function defining the effect of rotary speed, N, on ROP; $f_3(H)$ = function defining the effect of tooth dullness, H, on ROP; and $f_4(\Delta p)$ = function defining the effect of the differential pressure across the hole, Δp, on ROP.

Normalized penetration rate, K, is related to the bulk density by the following equation:

$$\rho_b = 2.65 - 1.65 \frac{S_g + \log K^{-6}}{S_g} \tag{16-6}$$

where ρ_b = bulk density in g/cm^3, and S_g = rock strength parameter (e.g., a value of 5.2 in the U.S. Gulf Coast area).

These bulk density values are then converted to formation pseudo-porosity, which closely agrees with calculated porosity from geophysical well logs, such as a density log. Bourgoyne (1971) developed a graphical approach for these concepts for wellsites without digital facilities.

Belotti and Gerard (1976) discussed a similar system, which was developed in Europe and was already successfully tested in more than 40 wells. The system consists of a set of sensors which acquire necessary information and transmit it to an on-site data unit, where the information is scaled, displayed, recorded and processed by the mini-computer. Storage on magnetic tape cassettes allows playback of magnitude of overpressure, porosity, and geologic information at desired depth intervals for comparative studies with other individual pressure indicators, such as well logs.

Herbert and Young (1972), using historical field data from several Louisiana Gulf Coast wells, developed equations based on regression analysis for predicting pore pressures. When the results of this analysis are applied to drilling data, the transition from normal pore pressures to overpressures can be predicted. This, however, can be done only on a geographically-regional basis.

Correlations between well log data and rock drillability have produced encouraging results (Gstalder and Raynal, 1966; El-Hadidi, 1970). Acoustic transit time data from geophysical well logs can be used to predict rock drillability, provided lithology is independently known.

The basic concepts of the SNAP log (Lutz et al., 1972) apply the vibrations from the tricone bit, as measured at the Kelly, to give an instantaneous log of the formation characteristics while it is being drilled. Initial field tests in boreholes of the Aquitanian Basin, France, appeared to be very promising, whereas additional tests of this technique in the U.S.A. were rather disappointing.

Logging-while-drilling

Research is underway now to develop methods for recording formation, mud, and bit data at the bottom of the borehole, and then transmitting these data to the surface. The great need for the development of additional energy resources and the fact of simultaneously ever-increasing costs of drilling and exploration, incentive for methods providing downhole real-time measurements have become greater than ever. Proposed methods and granted patents are numerous. Future potential of some of these concepts appears to be great; however, a reliable, accurate, and feasible data transmission system from the bottom of the wellbore to the surface still needs to be developed.

New logging-while-drilling measurement systems essentially perform only two basic functions: (1) recording of the desired parameters at the bottom of the wellbore, and (2) data transmission to the surface. Downhole measurements comprise: (1) well control information, (2) directional drilling control, (3) drilling optimization, and (4) formation evaluation. In 1976, more than ten different logging-while-drilling systems were under development with more than twenty companies actively participating in this research work (McDonald, 1976). Basically, four different types of data telemetry are under investigation, namely, (1) mud-pressure pulses, (2) insulated conductor or cable, (3) electromagnetic methods, and (4) acoustic methods.

Mud column pressure pulsing appears to have the early lead for commercial availability as a logging-while-drilling system (McDonald, 1976). It may be quite possible, however, that the final logging-while-drilling method will be a combination of several concepts mentioned above.

Torque

Torque variations are continually monitored at the drilling rigs. Torque usually increases gradually with depth due to increased wall-to-wall contact of drillpipe and wellbore. In the presence of underbalance (i.e., negative differential pressure), overpressured shales tend to flow or heave into the borehole. Hence, a drastic increase in torque may serve as an additional pressure indicator.

Drag

Drag is defined as the excess in hook load over the free handling load. Drastic increase in drag may signal the presence of an overpressured formation. In the presence of a gradual pore pressure increase, however, as in very long transition zones or in the case of drilling from floating vessels, this pressure indicator becomes questionable. Furthermore, increase in drag may be caused by bit balling, severe dog legs, deviated holes, differential sticking, and extra volume of cuttings influx into the wellbore while drilling through transition zones.

Drilling mud parameters

Mud-gas cutting

Mud logging aids in formation evaluation and detection of overpressured zones. As early as 1945, the use of mud-gas logging was recommended as an overpressure indicator and as a warning of impending blowouts (Pixler, 1945). Similarly, Rochon (1968) proposed mud-gas anomalies as an aid in controlling drilling mud hydrostatic head—pore pressure relationships. If formation permeabilities are extremely low, the degree of gas cutting can be roughly correlated with the amount of underbalance (Goldsmith, 1972). Generally speaking, however, mud-gas cutting may or may not be directly related to increased formation pressures, because it is greatly affected by the geological environment penetrated and the drilling techniques used in the subject well (Fertl, 1973; Daw et al., 1977).

Several factors which may affect mud-gas logs and, thus, complicate their use for pressure detection include (Fertl, 1973): (1) potential pay zones, (2) connection gas, (3) Kelly air, (4) downtime, (5) gas composition, (6) presence of lignite and (7) coal seams, (8) degradation of mud additives, (9) gas flushing, (10) volcanic material, (11) deep-seated mud volcanoes, (12) faults, (13) shale gas, (14) thermodynamic processes, and (15) recycled gas.

Typical gas detection and analysis equipment incorporates one of the two standard detectors: (1) the catalytic filament detector, and (2) the flame ionization detector.

Flowline specific weight of drilling fluids

Reduction in the specific weight of drilling fluid at the mud flowline can be used as an additional indicator for gas cutting and possible presence of overpressured formations. Consequently, continuous mud weight indicators have become an integral part of many on-site sophisticated data collection and analysis units.

Issenmann and Lucon (1971) discussed a newly-developed continuous mud weight recorder, which basically consists of a constant-height column through which mud from the wellhead outlet is circulated by a special pump at a constant flow rate. The weight of the mud is measured by a pressure gauge located at the bottom of the column and then transmitted to a recorder.

Another high-resolution drilling fluid monitor system, which consists of pressure and density sensors, has been developed by Goddard et al. (1973). Field acceptance of this system has been enthusiastic. The measurements are based on the principle that the degree of absorption of gamma rays by a material is a function of the density of that material. With proper calibration, radioactive densometers provide density measurements with an accuracy of ±0.1 lb/gal for muds in the 7—20 lb/gal specific weight range.

Pressure kicks

Present-day balanced drilling techniques require a very narrow margin between effective pressure control and threatened blowout. Differential pressures are frequently reduced considerably below 500 psi. Improper pressure balance, therefore, may cause well kicks, which can be a direct indication of presence of overpressured formations. Whereas pressure kicks have occurred at pressure differentials as high as 9 lb/gal, fortunately the control of most kicks requires less than 2 lb/gal mud weight increase.

Pressure kicks in high-pressure wells are influenced by the following factors: (1) difference between pressure due to the hydrostatic mud-column weight and formation pressure; (2) thermodynamic behavior of the gas; (3) interaction of gas with drilling fluids (especially oil-base); (4) downhole pressures and temperatures; and (5) time required for the circulation of mud (which is a function of depth) and recording of the transmitted pressures. Further considerations include (a) well location, including onshore and offshore remote areas; (b) deep-water drilling; and (c) pressure kicks associated with drilling shallow surface holes and zones below the protection pipe.

Excellent discussions and detailed reviews of present-day well control methods are available in the literature. The articles by O'Brien and Goins (1960), Goins and O'Brien (1962), Goins (1968, 1969), Rehm (1969), Moore, (1974), West (1976), Bourgoyne (1976), Nance (1977) and Adams (1977) are notable examples.

In general, however, proper pressure control requires installation of blowout preventor valves, adjustable surface chokes, accurate and reliable pressure reading equipment, gas separators, and sophisticated drilling and mud analysis equipment. In addition, presence of a trained rig personnel, with a sound understanding of the basic concepts involved and having a planned control program to meet any emergency, is of utmost importance.

Flowline temperature

In a pressure transition zone, the formation pressure increases with depth at a rate above the normal one. The same appears to hold true for the rate of formation temperature increase with depth.

Inasmuch as heat conductivity varies with rock and fluid characteristics of subsurface formations, overpressured, high-porosity shales act as "thermal

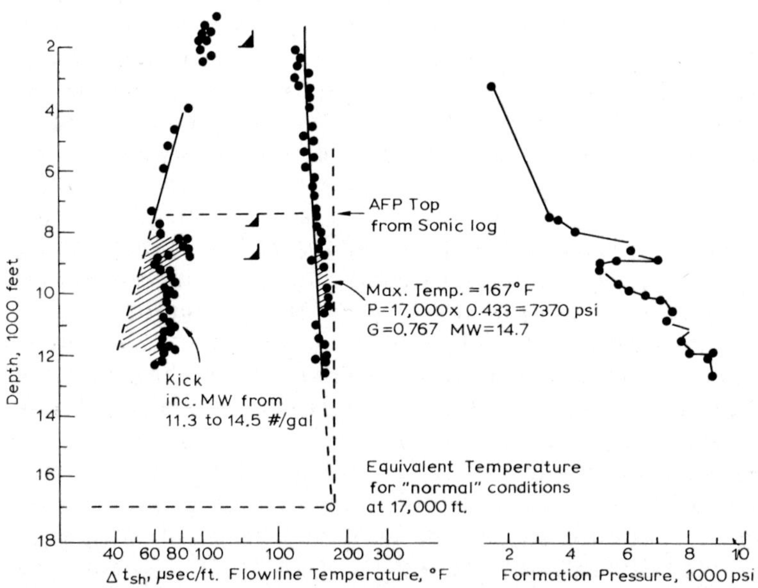

Fig. 16-7. Attempt of quantitative pressure evaluation using flowline temperature data in a Texas wildcat, U.S.A. For comparison, acoustic plot and pore pressure variations with depth are presented. Casing seats are also shown. G = geostatic pressure gradient in psi/ft; AFP = abnormal formation pressure; and MW = specific weight of drilling fluid in lb/gal. (After Fertl, 1976, fig. 4-22, p. 146.)

barriers", thereby locally increasing the geothermal gradient (Jones, 1968; Fertl and Timko, 1970). Lewis and Rose (1970) proposed a mathematical model relating overpressures and high formation temperatures, which is based upon basic heat flow considerations (Guyod, 1946).

Changes in flowline temperature gradients of up to 10°F/100 ft have been observed prior to and/or when entering overpressured intervals. This pressure indicator, however, is also affected by lithology, circulation and penetration rates, tripping the drill string for bit change, long risers in deep-water drilling, and drilling through permafrost intervals. Thus, certain precautions and refinements are required in use and interpretation of temperature gradients (Wilson and Bush, 1973) when plotting flowline temperatures, including (1) use of temperature readings (points) of each drill bit run, (2) replotting segments of individual bit runs end-to-end without regard to actual temperature values, and (3) use of inlet and outlet flowline temperatures.

The possible application of flowline temperature data as a semiquantitative pressure indicator in a Texas wildcat has been illustrated by Fertl (1976, p. 146—147). Figure 16-7 shows a similar, or at least complementary, evaluation of both acoustic log and flowline temperature data in this well.

Resistivity, chloride ion content, and other novel methods

Relationship between salinity of formation waters and formation pressure variations in consolidated and unconsolidated rocks has been discussed by Chilingarian and Rieke (1968), Chilingar et al. (1969), Overton and Timko (1969), and Fertl and Timko (1970). At the present time, several service companies measure the increase or decrease in the chloride content of the drilling mud between inlet and outlet of the mud stream in an attempt to relate apparent variations to drilling and pressure conditions. In addition, many of the modern drilling rig data gathering units present the possibility of measuring and recording both inlet and flowline drilling fluid resistivities.

Other novel concepts and techniques, such as variations in the contents of specific ions, redox and pH measurements on the drilling mud stream, and various physical and chemical analyses of drill cuttings, have been investigated by Fertl and Timko (1973). These new concepts offer possibilities of developing additional continuous pressure indicator methods. Additional research and proper equipment will be required to develop reliable interpretive wellsite techniques.

Pit level and total pit volume

Pit level indicators, which monitor variations in the total mud volume, may show mud-volume reduction caused by lost circulation or increase in mud volume due to the fluid entry into the wellbore as a result of unexpected high formation pressures. Recently, ultrasonic equipment has been developed to measure accurately drilling fluid levels in mud tanks without having any contact with the mud (Dupin de Saint Cyr, 1973). Hence, this method is particularly effective on floating, deep-water offshore drilling rigs.

Hole fill-up

If the drillstring is pulled, the mud volume needed to fill the borehole should be equal to the displaced pipe volume. Keeping the hole full is especially critical at the time when drill collars are pulled, because on pulling the same length of collars as that of drill pipe, the level of drilling mud in the bore hole will fall four to five times as fast. Furthermore, if salt water, oil, or gas from the formation enter the well, the mud volume required to fill the borehole will be less than the displaced volume of the pipe pulled out. Consequently, this gives the first indication of a pressure kick. Measurement of mud volume used to fill the borehole can be checked from changes in the pit mud level.

Mud flow rate

Flow rate measurements are superior to pit level checks, because even small flows can be detected before they become sufficiently large to show on any pit level measuring device. More time is available, therefore, to take proper control measures.

Wellsite data units

Numerous digital systems have been developed by the drilling industry for improved, more economical, and safer drilling operations.

For example, Boone (1975) presented a functional diagram of a system that provides (1) porosity and pore pressure logs at bit depth, and (2) offers all essential data for drilling optimization. The basic outline of this operational system is illustrated in Fig. 16-8.

The monitoring system recently installed on a turret-moored drillship, "Discoverer 511", has been discussed in detail by Pomrenke and Cox (1976). The locations of the various sensors on board of this drillship are schematically shown in Fig. 16-9.

Data units, which provide immediate, foot-by-foot, real-time evaluation, are also provided by several service companies. For example, the ONBOARD III system, which is offered by Magcobar, provides a highly-advanced digital well data analysis; around 64 variables can be monitored during drilling operations. The system also includes 20 customer-oriented programs to analyze and display more than 200 results. The computer system used includes a 2.5 million-word tape storage for well histories and a 2.5 million-word storage for programs. Commands, interrogations and analyzed output are executed in the Data Unit at the wellsite or through the customer's office terminal, whether the well is onshore or offshore. Some of the equipment used in measuring and recording these continuously monitored 64 surface

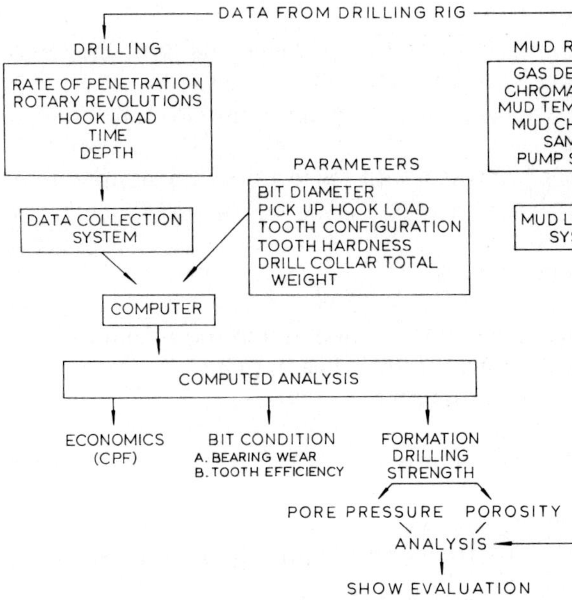

Fig. 16-8. Wellsite data collection and analysis concepts. (After Boone, 1975, fig. 1, p. 126; courtesy of Petroleum Engineer.)

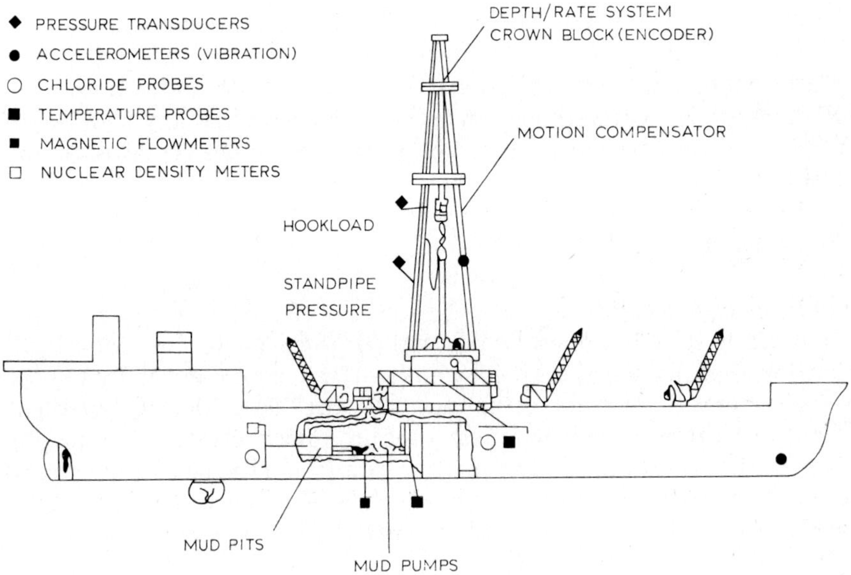

Fig. 16-9. Drilling monitor system on Discoverer 511. (After Pomrenke and Cox, 1977, fig. 4, p. 82; courtesy of World Oil.)

and downhole variables include:

(1) Automatic chromatograph recorder for hydrogen, helium, methane, ethane, carbon dioxide, propane, isobutane, and butane contents.

(2) Total gas analysis recorder for monitoring the content of total combustible gases in drilling fluids.

(3) Hydrogen sulfide analyzer for detecting H_2S content in drilling mud in the range of 0—50 ppm.

(4) Mud temperature recorder.

(5) Mud pump output volume recorder.

(6) Pit volume totalizer.

(7) Flowline sensor giving digital readout in one-percent increments.

(8) Mud weight recorder at the inflow point (suction pit).

(9) Mud weight recorder at the outflow point (flowline).

(10) Dual pump stroke counters.

(11) Rotary revolutions totalizer for providing total work turns by each bit on bottom.

(12) Trip guard system, along with drillpipe counter for determining the exact amount of fluid required to fill the borehole for any number of stands pulled.

(13) System for digital electronic display of pump speed.

(14) System for digital electronic display of pump pressure.

(15) Automatic drillpipe counter.

(16) System for digital rate of penetration display.

(17) System for depth display.

(18) System for electronic hook load and weight-on-bit readout.

(19) Continuous rate of penetration recorder.

(20) Alarm panel with warning lights and horn for critical indicators.

(21) Rig interphone communications system for enabling any member of the crew to be in contact with tool pusher, personnel on the rig floor, or any other area on the rig.

Shale cuttings parameters

Shale bulk specific weight

Bulk specific weight (commonly referred to erroneously * as "density") of drill cuttings usually increases with depth. It can be measured by (1) the high-pressure mercury pump technique, (2) the fluid density gradient column, and (3) the mud balance method. Care must be taken in selecting and preparing the drill cuttings for analysis. Multiple cuttings samples have to be tested due to the variance in sample data. The average bulk specific weight value, for a given depth, is then plotted on a linear or logarithmic

---

* See p. 47.

scale versus borehole depth, thus establishing normal compaction trend lines. Inasmuch as shale porosity increases in overpressured zones, any decrease in bulk specific weight may indicate the presence of overpressured environments. Quantitative pressure evaluation is then possible by the equivalent depth method (Fertl, 1976) or from empirical curves established for a given area (Boatman, 1967).

Major limitations are (1) examination of cavings and/or recirculated cuttings which constitute contaminating part of drill cuttings being investigated, and (2) limited care taken by rigsite personnel to collect and analyze samples. Several other factors, which may greatly affect measured bulk specific weight values of drill cuttings, include:

(1) Presence of shale gas which decreases apparent bulk specific weight values.

(2) Presence of organic-rich shales, which results in lower bulk specific weight values.

(3) Lithologic variations, e.g., presence of silty or sandy shales, mudstones, and marls. Variation in carbonate content of shales also affects bulk specific weight.

(4) Presence of heavy minerals, such as pyrite (Permian Basin, U.S.A.; offshore Cameroon, Africa; South China Sea area), siderite (South China Sea area; Mackenzie Delta, Canada), and mica (biotite and muscovite; North Sea area), will increase bulk specific weight values.

(5) Age boundaries, unconformities, differential compaction, structural effects, and position within the clastic basin may affect the normal compaction trend line (Fertl, 1977).

Shale factor

The shale factor can be successfully measured by the methylene blue test (Nevins and Weintritt, 1967). This shale formation factor method (Gill, 1968; Mondshine, 1969; Gill and Weintritt, 1970) may be equated with the cation exchange capacity (CEC) of solids carried out by drilling fluid out of the wellbore. This CEC value, in turn, can be related to the water-holding capacity of drill cuttings or montmorillonite content. The shale factor also appears to be a supplementary and useful indicator for the detection of impermeable pressure seals (cap rocks) on top of the overpressured zones.

Volume of shale cuttings

During drilling, entry into overpressured environments is characterized by an increase in penetration rate, which gives rise to an increase in volume of cuttings over the shale shaker.

Shape and size of cuttings

In pressure transition zones, the shape of drill cuttings is angular and sharp, rather than rounded as found in normal, hydrostatic pressure environments. Furthermore, cuttings from high-pressure formations are unusually large and splintery in appearance.

Novel pressure indicator methods

Several unconventional methods for detecting overpressured formations and, possibly, even predicting transition zones ahead of the drill bit have been investigated in detail by Fertl and Timko (1973a, b). These methods, some of them patented by these authors, include:

(1) Shale cuttings resistivity.
(2) Filtration rate of shale slurry.
(3) Filtrate (shale water) color.
(4) Shale cuttings moisture index.
(5) Litho-function plots.
(6) Bicarbonate content of shale slurry filtrate.
(7) Redox and pH potential measurements.
(8) Distribution of specific ions in shale slurry filtrates.

In this extensive study, Fertl and Timko (1973a, b) have made several interesting observations which may become of great practical significance, provided additional research will result in development of practical methods and their interpretation for routine application at the rigsite.

Geophysical well-logging techniques

Some of the most reliable quantitative overpressure detection and evaluation techniques are based on geophysical wireline-logging methods, even though these methods are "after-the-fact" techniques, i.e., the wellbore has to be drilled prior to logging. Hence, several short logging runs are sometimes necessary. Logs that can detect abnormally high formation pressures are summarized in Table 16-II; they have been discussed in detail by Fertl (1976, pp. 177—230). The schematic response of several well-logging parameters to (a) hydrostatic and (b) overpressured environments is illustrated in Fig. 16-10.

Except for the SP curve data, all parameters which are recorded in shale formations, rather than in sands, are plotted versus depth. Trend lines are then established for normal compaction. Interpretation of the logs depends on the magnitude of departure from the normal trend, i.e., the divergence of the formation pressure from the normal hydrostatic pressure at a specific depth.

The acoustic and short normal resistivity logs are the tools most valuable for quantitative in-situ pressure evaluation. Application of these methods, however, is not always simple and straightforward. Awareness of possible pitfalls, which are listed in Table 16-III, will allow full utilization of geophysical well logs as one of the best available methods to detect and evaluate abnormally-high formation pressures.

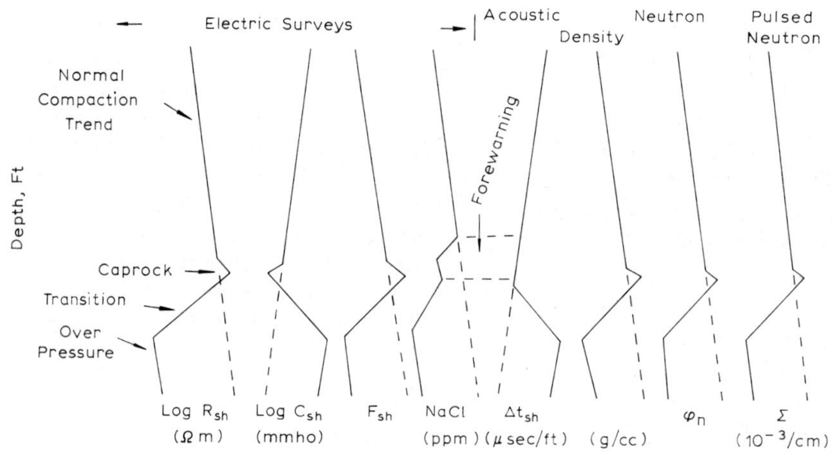

Fig. 16-10. Schematic response of well-logging parameters to normal and overpressured environments. (After Fertl and Timko, 1971, fig. 6, p. 11.) C_{sh} = shale conductivity (mmho); R_{sh} = resistivity (Ωm), resistivity log; F_{sh} = formation resistivity factor, dimensionless; Δt_{sh} = acoustic transit time (μsec/ft), acoustic log; ρ = density (g/cm^3), density log; ϕ_N = porosity index (%), neutron log; Σ_{sh} = neutron capture cross-section of shale (10^{-3} cm^{-1}), pulsed neutron log.

Newly developed wireline formation testers, which allow the recording of a large number of pressure tests over the entire length of the uncased borehole, have been rapidly accepted by the drilling industry. Such measurements can then be used to "calibrate" other drilling and/or log-derived pressure indicators.

DRILLING CONCEPTS IN OVERPRESSURED ENVIRONMENTS

Maximum well control and minimum cost are key factors in present-day drilling operations. To achieve this, a basic understanding of two key formation pressures, i.e., formation pore pressure and fracture pressure, is a prerequisite.

Formation pressure can be determined from drilling and logging parameters. Formation fracture pressures (or gradients) can be calculated using one of the several models available (Table 16-IV) or can be determined by actual tests in the well. Pressure testing below the casing shoe can (1) test the cement job, and (2) determine the integrity (fracture gradient) in the first weak (permeable) zone below the casing seat. Further drilling objectives define the possible limits for such a pressure leak-off test. Figure 16-11 shows two formation integrity (leak-off) tests in a well located in the Mackenzie Delta, Canada (Daw et al., 1977). Such tests are carried out by

TABLE 16-IV

Prediction methods for fracture pressure gradients

Authors and formulas *	Comments
Hubbert and Willis (1957) $(FP/D)_{min} = 0.5[(P_o/D) + (P_f/D)]$ $(FP/D)_{max} = P_o/3D + 2P_f/3D$	Difference between minimum and maximum fracture pressure gradients decreases with increasing pore pressure. Hence, importance of surge pressures increases with depth.
Matthews and Kelly (1967) $FP/D = P_f/D + K_i\delta/D$	Introduction of variable matrix stress coefficient (K_i), which can be empirically determined from fracture initiation (breakdown) pressure. This method is used widely.
Eaton (1969) $FP/D = P_f/D + \left(\frac{\mu}{1-\mu}\right)\delta/D$	Extension of concept proposed by Matthews and Kelly above. Eaton introduced Poisson's ratio into the mathematical model. This method is used widely.
Taylor and Smith (1970)	Empirical correlation of injection test data in normally-pressured formations only. Location: offshore of Louisiana near Mississippi delta. Minimum propagation test data correspond with fracture initiation pressure gradient given by Matthews and Kelly (1967).
MacPherson and Berry (1972)	Developed quantitative correlation between ratio of sand elastic modulus to overburden pressure and fracture pressure gradient. Based on data from 10 oil fields offshore of West Louisiana.
Christman (1973) $FP/D = K_i[(P_o/D) - (P_f/D)] + P_f/D$ where $P_o/D = \frac{0.4335}{D}(\rho_w D + \bar{\rho}_b D)$	Expanded technique of Matthews and Kelly (1967). Effect of water depth can be taken into account.
Anderson et al. (1973) $FP/D = \left(\frac{2\mu}{1-\mu}\right)(P_o/D) + \alpha\left(\frac{1-3\mu}{1-\mu}\right)(P_f/D)$	Based on Terzaghi's (1923) concept of effective stress. $\alpha = 1 - C_r/C_b$. Standard deviation = 0.083 psi/ft, between calculated and measured fracture pressure gradients based on 29 field tests in U.S.A. Gulf Coast area.
Althaus (1975) $FP/D = K_i\ (19.2 - P_f/D) + P_f/D$ where $K_i = 0.23375 + 4 \cdot 10^{-5} D$	Based on method by Matthews and Kelly (1967). K_i is determined by regression type analysis of field data.

* Where: D = depth of zone of interest (ft); FP = fracture pressure at point of interest (psi); P_o = overburden pressure (psi); P_f = formation pressure, psi; $\delta = P_o - P_f$; K_i = matrix stress coefficient, dimensionless; μ = Poisson's ratio; ρ_w = density of water (g/cm^3); $\bar{\rho}_b$ = average bulk density (g/cm^3); C_r = intrinsic compressibility of solid rock material;

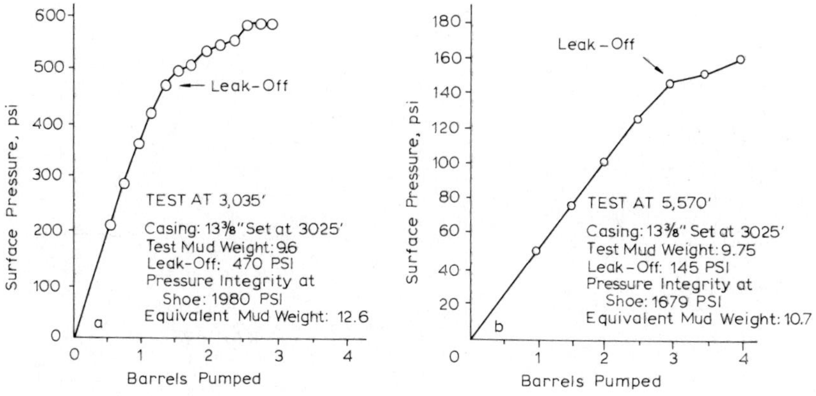

Fig. 16-11. Formation integrity (leakoff) tests in well Immerk B-48, Mackenzie Delta, Canada. a. Test at a depth of 3035 ft indicated that formation breakdown would occur at 12.6 lb/gal equivalent mud weight. b. Test at a depth of 5570 ft resulted in a leakoff at 10.7 lb/gal equivalent mud weight. (After Daw et al., 1977, fig. 3, p. 44; courtesy of World Oil.)

pumping drilling fluid into the borehole against a closed annular choke at a slow constant rate and plotting fluid volume pumped versus surface pumping pressure. The surface pressure at the point of leak-off plus the static mud column pressure, shows the maximum pressure the borehole may tolerate (contain).

As Moore (1974) candidly pointed out, however, in drilling deeper it becomes obvious that "if the operator exceeds the leak-off pressure from his last test, he is inviting problems and in all probability he will not be disappointed."

Casing point selection

Casing in the borehole protects penetrated formations and the wellbore from contaminations and drilling hazards. In both hydrostatic and overpressured zones the recommended mud weight frequently exceeds formation pressure by a margin of 0.2—0.4 lb/gal.

Casing point selection method under perfectly balanced conditions (i.e., formation pressure is equal to the hydrostatic pressure of the mud column) is shown in Fig. 16-12. Indispensible, basic requirements are (1) a good cement job, and (2) the absence of weak zones somewhere along the open hole, including lignite and other coal beds, lime streaks (in cap rocks), carbonate shell beds, and small faults or joint systems intersecting the wellbore.

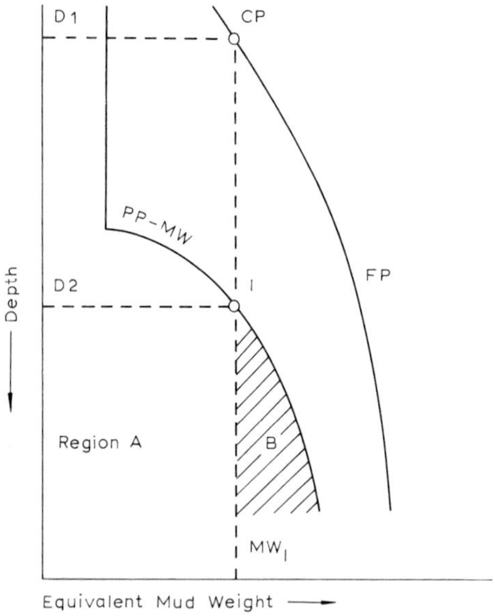

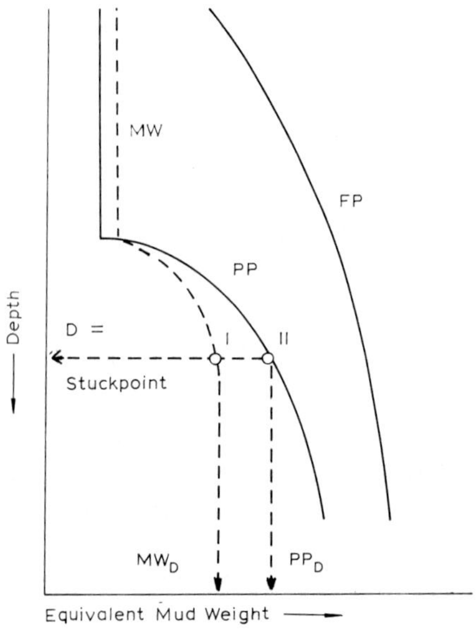

Fig. 16-12. Casing point selection under perfectly balanced conditions. *Region A* = "safe" drilling and well killing conditions; *Region B* = well kicks cannot be handled and circulation is lost. *CP* = casing point; *FP* = fracture pressure; *PP* = pore pressure; and *MW* = specific weight of drilling fluid. (After Fertl, 1976, fig. 6-4, p. 234.)

Fig. 16-13. Schematic diagram (depth versus equivalent mud weight) showing reasons for drillpipe sticking due to an underbalanced mud system. (After Fertl, 1976, fig. 6-6, p. 236.) Symbols are defined in Fig. 16-12. *D* = well depth at which pipe sticking occurs.

Drillpipe sticking

Drillpipe sticking due to underbalanced conditions is shown in Fig. 16-13. In the overpressured section, the equivalent mud weight exceeds the pore pressure down to a depth *D*, where the drillpipe will get stuck. As the drilling bit approaches this stuck-point, the borehole starts caving in and the torque increases simultaneously.

Generally speaking, pipe-sticking forces depend on (1) difference between hydrostatic drilling mud and formation pressures (differential pressure), (2) permeability of formation, (3) thickness of zone, (4) thickness and slickness of filter cake, (5) length of time the pipe remains motionless against the formation, (6) hole and pipe size, and (7) pipe shape. Inasmuch as the latter is very critical, special drillcollar configurations have been developed, including (1) spiral collars with circulation grooves, (2) square drillcollars, and (3) shouldered drillcollars. In addition, key-seating is recognized as one of the primary reasons for pipe sticking.

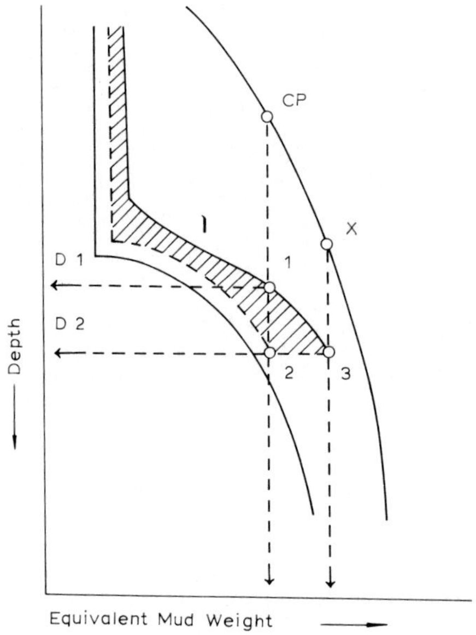

Fig. 16-14. Schematic diagram illustrating reasons for lost circulation due to an overbalanced mud system. (After Fertl, 1976, fig. 6-8, p. 237.) Shaded area shows the region of unnecessary and dangerous overbalance. *CP* = casing point.

Lost circulation

Lost circulation may occur for several reasons, including excessively weighted drilling fluid (i.e., overbalanced mud condition), as shown in Fig. 16-14. Shaded area shows region of unnecessary and dangerous overbalance. Due to the overbalanced conditions, drilling has to be stopped at a depth D_1 (point *1*), whereas proper mud weights would have allowed drilling down to a depth D_2 (point *2*) without adding additional protection pipe. Excessive mud weight at depth D_2 (point *3*) would cause pipe sticking and lost circulation.

Surge and swab pressures

Surge and swab pressures are also important parameters in well control. For example, running the drillstring into fluid-filled borehole causes a pressure surge (increase in pressure), whereas pulling the string out of a liquid-filled borehole causes a swab pressure (reduction in pressure).

According to Moore (1974), over 25% of all blowouts result from swab pressures, whereas excessive surge pressures are frequently responsible for initiating lost circulation and/or physical formation damage (i.e., solids blocking).

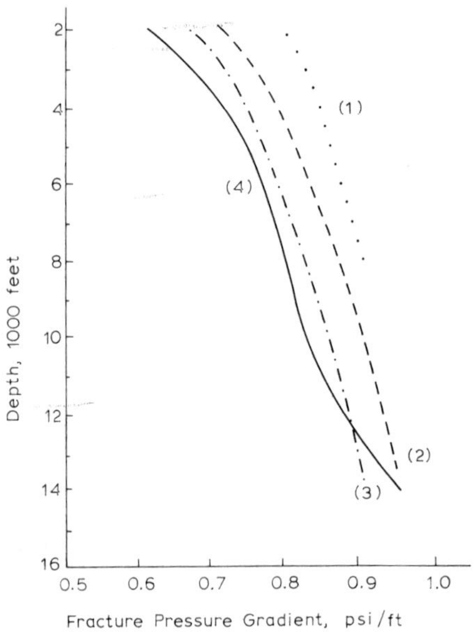

Fig. 16-15. Comparison of several fracture pressure gradient predictions in the U.S.A. Gulf Coast area. (After Fertl, 1976, fig. 6-18, p. 250.) *1* = East Louisiana (after Taylor and Smith, 1970); *2* = South Texas (after Matthews and Kelly, 1967); *3* = Louisiana Gulf Coast (after Matthews and Kelly, 1967); and *4* = average empirical trend for ten offshore fields in West Louisiana (after MacPherson and Berry, 1972).

Fracture pressure prediction

Prediction methods for fracture pressure gradients are summarized in Table 16-IV, whereas a graphical comparison of several prediction methods used in the U.S.A. Gulf Coast area is shown in Fig. 16-15.

Well planning charts providing step-by-step guidelines for safe and economic well-drilling operations are available to petroleum engineers (Eaton, 1970). In addition, recent deep-water drilling (in waters over 2000 ft in depth) with dynamically positioned drillships and from semisubmersible platforms requires special considerations for fracture gradient computations (Christman, 1973). This, in turn, affects casing point depth selection. For example, Fig. 16-16 illustrates the effect of water depth on the length of casing, design being based on fracture pressure gradients typical for Tertiary shales in the North Sea area.

Casing programs in overpressured formations

Properly engineered well design plans must be effective, safe, and soundly applied for optimum results. Whereas drilling programs are designed from the

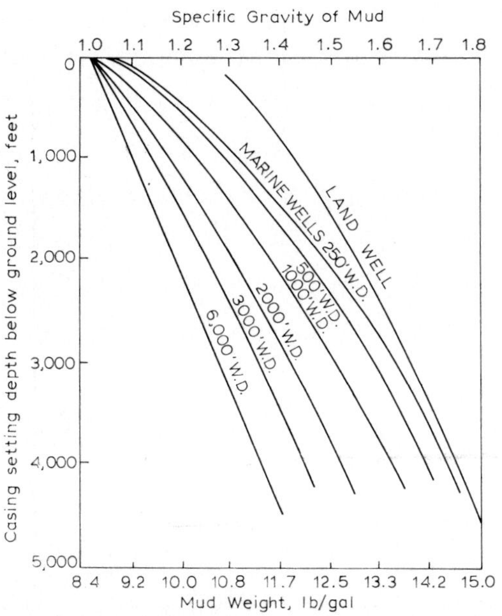

Fig. 16-16. Effect of water depth (*W.D.*, in ft) on casing setting depth based on fracture pressures in North Sea Tertiary shales. (After Koch, 1976, fig. 3, p. 48; courtesy of World Oil.) For example, 11.7-lb/gal mud in an onshore well requires about 800 ft of casing protection, whereas in 6000 ft of water a casing string of 4200 ft would be required for a 11.7-lb/gal mud.

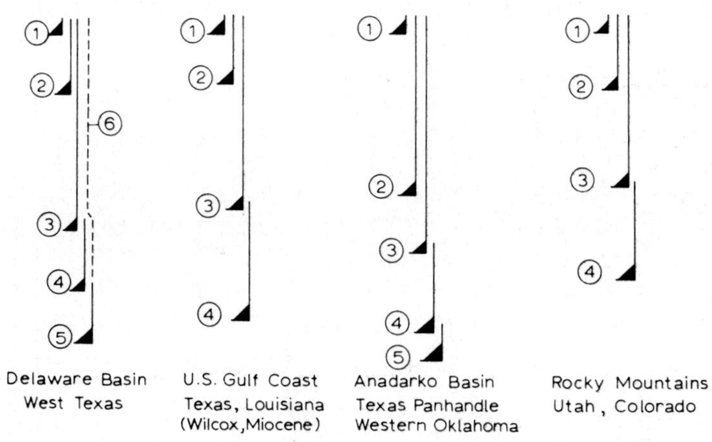

Fig. 16-17. Typical, generalized casing programs in overpressured areas in the U.S.A. The encircled numbers refer to Table 16-V. (From Fertl, 1976, fig. 6.26, p. 261.)

TABLE 16-V

Generalized typical casing programs in several abnormally-pressured U.S.A. basins (after Fertl, 1976, table 6.VI, p. 260)

Region*	Typical casing sizes (inches)	Approximate average casing point depth (ft)			
Delaware Basin (West Texas)		*Ward Co.*	*Reeves Co.*	*Pecos Co.*	*Winkler Co.*
	(1) 20" conductor	5,000	5,000	1,700	5,000
	(2) $13\frac{3}{8}$"	11,000	11,000	10,500	13,000
	(3) $9\frac{5}{8}$" or $10\frac{3}{4}$"	14,500	16,000	15,500	17,000
	(4) $7\frac{5}{8}$" liner	20,000	21,000	22,000	22,500
	(5) liner tie-back** to surface, 7"($7\frac{5}{8}$") × 5"($5\frac{1}{2}$") or $7\frac{5}{8}$" only				
U.S. Gulf Coast (Texas, Louisiana)		*South Texas*	*Upper Texas*	*Onshore Louisiana*	*Offshore Louisiana*
	(1) 20" (locally plus 30")	—	—	—	—
	(2) $13\frac{3}{8}$" or $10\frac{3}{4}$"	9,000	3,000	3,000— 4,000	4,000
	(3) $9\frac{5}{8}$" or $7\frac{5}{8}$"	12,500	12,000	13,000—15,000	13,000
	(4) 7" liner and/or $5\frac{1}{2}$" (or 5")	15,000	15,500	15,000—18,000 (plus)	20,000 (plus)
Anadarko Basin (Texas Panhandle, western Oklahoma)		*Oklohoma*	*Texas*		
	For 20,000-ft wells:				
	(1) 24" conductor	no	yes		
	(2) 20" (or 16") conductor	2,000	4,000		
	(3) $13\frac{3}{8}$" (or $10\frac{3}{4}$")	12,000	13,500		
	(4) $7\frac{5}{8}$" (or $9\frac{5}{8}$") liner	19,000	16,700		
	(5) 5" (or 7") liner	to TD	to TD		
Rocky Mountain area (Utah, Colorado)		*Uinta Basin (Utah)*			
	(1) 20" conductor				
	(2) $13\frac{3}{8}$"	1,500			
	(3) $9\frac{5}{8}$"	9,500—12,500			
	(4) 7" liner	to TD			
		Piceance Basin (Colorado)			
	(1) $13\frac{3}{8}$" conductor				
	(2) $9\frac{5}{8}$"	5,000			
	(3) 7"	14,500			
	(4) 5" liner	to TD			

* Schematic is shown in Fig. 16—17.

** Typical liner tie-back strings in deep Ellenburger wells.

surface to the bottom of borehole, well planning is a reverse process and starts at the bottom of the hole. It is important to note, however, that there is no single, unique casing program. Each one has to be tentative, flexible, and continuously reappraised for each well. Hence, optimum casing programs are the result of continuous evaluation and incorporation of additional data and new experiences, and are not the outgrowth of a one-time, single decision. Generalized typical casing programs in several abnormally-pressured U.S.A. basins are listed in Table 16-V and Fig. 16-17, as an example.

WEIGHTED MUD SYSTEMS

Drilling fluid system components required for proper maintenance and control of various types of fluids have been compiled by Hutchison and Anderson (1974), as shown in Table 16-VI.

Drilling into overpressured formations requires properly designed and controlled mud weights in order to balance formation pressures. In addition, drilling mud control must also focus on other mud properties, including gel strength, viscosity, and water loss.

As pointed out by Moore (1974), contrary to popular belief total solids content does not always (consistently) tend to be high in weighted muds. The differences in mud solids problems between weighted and unweighted

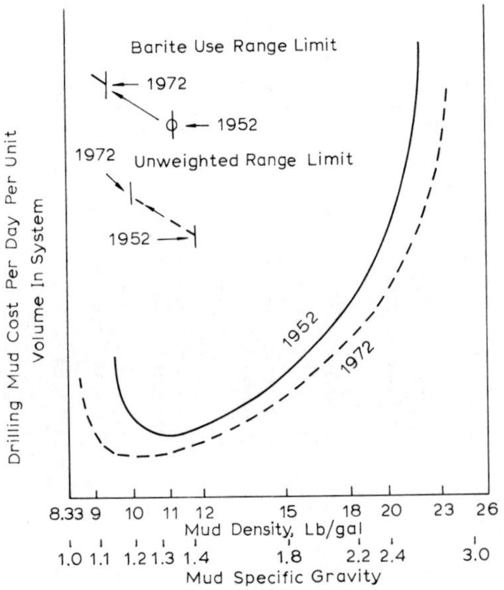

Fig. 16-18. Economic effect of improved solids control on water-base drilling fluids in normal and overpressured environments. (After Moore, 1974, fig. 6-29, p. 176; courtesy of The Petroleum Publishing Company.)

TABLE 16-VI

Drilling fluid system components required for properly engineered maintenance and control of various types of fluids (after Hutchinson and Anderson, 1974, table 3, p. 86; courtesy of World Oil)

Equipment	Water/light treated clay	Dispersed weighted	Non-dispersed	
			Unweighted	Weighted
Wellhead				
BOPs	As required for well control	As required for kick control	As required for kick control	
Rotating head or stripper	May be used for under-bal. drlg.	May be used for under-bal. drlg.	May be used for under bal. drlg.	
Variable choke	May be used for back pressure control	Essential for kick control	Essential for kick control. May be used for back pressure control.	
Surge chamber and mud separator	Not used	Essential for kick control to save gas cut mud	Essential for kick control to reclaim gas cut mud	
Discharge lines	Flow line to shakers	Flow line to shakers, choke line to surge tank and waste sump	Flow line to shakers, choke line to surge tank and to waste sump	
Solids removal				
Shaker screen	Dual w/med. screens	Dual w/med. to fine screens	Double deck w/ med. & fine screens	Double deck w med. & fine screens
Sand trap	20-30 bbl. w/45° bottom — large diam. dump	Be careful not to dump liquid mud	Important to settle coarse material ahead of desander-desilter units	
Desander	Important — run ahead of desilters	May be used to remove coarse material ahead of desilting equipment	Important to prevent overloading desilters or centrifuge	
Desilter	Important to remove fine silts for low weight	Not used	Essential for low solids muds	Not recommended
Mud cleaner	Important to remove fine silts for low weight	Excludes lt. solids on medium weight mud	Essential for drilled solids control	Useful on medium weight muds
Centrifuge	Not used	Essential for economic control of high wt. muds	May be used to reclaim liquids and dump drilled solids	Essential for high wt. mud control

Degasser	May be used ahead of treatment if gas encountered	Essential in kick control for true wts.	Essential for good kick control practice	
Treatment				
Mixing hopper	Minimum required for mixing clay, gel chemicals	Maximum for rapid wt. increase on kicks	Maximum required to mix polymers; located after solids removal	
Bulk barite	Not used	Essential in high wt. systems and kick control	Not used	Essential in kick control
Stirring & agitation	Adequate for lost circ. material, top guns best	Essential to prevent settling	Important to maintain uniform system	
Waste pit size				
Normal	May be small to large depending on depth and if used for settling large amt. solids and liquids	Medium, will have smaller amount of solids and more liquids	Medium, will have large amount of solids and smaller amount of liquids	
Town lot	Most use small steel tanks and haul all solids and liquids to an approved disposal site. Volumes of solids and liquids critical to both storage space and trucking costs.			
Offshore	All waste collected in small steel tanks for treatment before disposal.			
Solids disposal				
Normal	Solids are left in bottom of sump after liquids are removed and buried. Deflocculation treatment of liquids may help settle more solids.			
Town lot	All solids must be hauled to an approved disposal site. Disposal cost may be more than mud costs.			
Offshore	Clean washed solids may be dumped overboard or hauled to shore on waste barges.			
Liquid disposal				
Normal	Non-oily fresh water fluids may be spread on land. Oily or salty fluids must be hauled to approved disposal site. May treat liquids to neutralize and deflocculate to clarity before spreading or pumping into lakes or streams. Liquids may be solidified and mixed with soil.			
Town lot	All liquids must be hauled to approved disposal site.			
Offshore	Liquid waste is collected and pumped ashore through waste lines or barged ashore for disposal.			
Biodegradability	Not degradable	Lignosulfonates biodegradable at low pH	Most polymers, starches, CMC are biodegradable	
Ecological effects	Fresh water clay muds can be very beneficial to soils, particularly sandy soils. Sodium polyphosphates used as thinners degrade to orthophosphate fertilizer. Lignite, lignins and tannins are humic acids.	Lignosulfonate muds should not be harmful provided chromates are not used.	Polymer muds should not be harmful as long as chromic compounds and chlorophenate bactericides are excluded.	

TABLE 16-VI (*continued*)

Equipment	Water/light treated clay	Dispersed weighted	Non-dispersed	
			Unweighted	Weighted
Liquid storage				
Active	May be small to very large, depends on depth and lost circulation	Must be adequate to fill hole on trips; normally 300 to 700 bbls.	Must be adequate to fill hole on trips, normally 300 to 700 bbls.	
Reserve	May be small to very large if used for settling	Should be adequate to displace cement and for lost circulation	Should be adequate to displace cement and for lost circulation	
Waste	Must be adequate to contain drld. solids and waste liquids	Adequate to contain smaller volumes of drld. solids and liquids	Must be adequate to contain large volume of hole solids and smaller volumes of waste liquids	
Dry storage	Adequate for mud and chemicals	Proper height and size essential	Should be proper height, size and location for mud and chemicals	

Equipment	Oil muds	Preformed stable foam	Gas, air and mist
Wellhead			
BOPs	As required for well control	Generally 1 positive, 1 pipe ram & Hydril	As required for control
Rotating head or stripper	Not used	Essential to divert foam. Use rotating head w/ Kelly and a stripper w/ power swivel	Essential to divert returns to waste sump
Variable choke	Essential for kick control	Essential for back pressure control	Not used
Surge chamber and mud separator	Essential to reclaim expensive mud when gas cut	Surge chamber or cyclone type foam suppressor when foaming into a tank	Not used
Discharge lines	Flow line to shakers, choke line to surge tank and/or separator	Blooey line to waste or surge tank	Blooey line to waste sump

	careful of mud loss		
Desander	Can be used by dumping into tank of solvent	No solids removal equipment needed. Normally, foam is a one-pass system, foam and solids go to waste sump	No solids removal equipment used. Air and solids to waste sump
Desilter	Not used		
Mud cleaner	Not used		
Centrifuge	Can be used to reduce wt.		
Degasser	Can be used if gas cutting a problem	Can use gas trap to separate gas and liquids	May use low pressure trap to reclaim gas
Treatment			
Mixing hopper	Maximum required for mixing new mud	Not required unless additive used	Not needed
Bulk barite	May be used on high wt. systems	Not used	Not needed
Stirring & agitation	Essential to prevent settling and maintain viscosity	Needed if additive used but not after foamer added	Not needed
Waste pit size			
Normal	Small, will have large amount of solids, very little liquids	Small, will have large amount of solids and small amount of liquid	Small, solids very fine, very small liquid
Town lot		Can foam into 500 bbl. tank and defoam	Not used due to dust problem
Offshore		Can use existing production facilities	Not used normally
Solids disposal			
Normal		Solids are very clean and can be buried	Solids very fine but clean, can bury
Town lot	Oil-wet cuttings may require different disposal site	Solids collected in tank after defoaming and hauled to dump	Not used, severe dust problem
Offshore	Oil-wet solids must be washed w/solvent and detergent to dump	Solids carried by foam thru prod. facilities to waste	Not used normally

TABLE 16-VI (*continued*)

Equipment	Oil muds	Preformed stable foam	Gas, air and mist
Liquid disposal			
Normal	Methods of incinerating oily waste are being investigated	Small am'ts of liquids. May reclaim fluids after foam breaks on long time drilling jobs	Little to no fluids for disposal.
Town lot	Oil or oily waste may require a different disposal site or method	Small volumes of liquids can be discharged into existing prod. facilities	Little or no fluids to dispose of
Offshore			
Biodegradability	Not degradable	Foamers, additives are biodegradable	Foamers used in mist are biodegradable
Ecological effects	Oil muds are no more hazardous than oil and should present no problems with adequate controls against spillage and an adequate disposal system	Foam drilling compatible with ecology since relatively small volumes of liquids are used. Drld. cuttings are large and no dust is generated	Air drlg. can result in serious dust problem since cuttings are ground very fine and blown out with air. Addition of foamer helps, but does not cure completely
Liquid storage			
Active	Adequate to fill hole on trips	20 to 100 bbl. divided tank needed so that foamable solution can be mixed and used alternatively	10 to 20 bbl. foamer tank for misting
Reserve	May require two tanks; one for new lt. wt. mud, one for heavier mud	Not needed if water supply adequate, may be used to recycle foam solution	None
Waste	Minimal since only solids dumped, expensive mud saved	Long pit desirable to contain surging foam returns	Dry dust difficult to contain
Dry storage	Adequate for mud and chemical	Not normally needed	Not needed

water-base muds are summarized in Table 12-III, whereas the factors that control total solids content in weighted muds are listed in Table 12-IV.

Reduction in cost at any given mud weight over the period from 1952 to 1972, due to increased efficiency of mechanical solids separation equipment used in drilling fluid systems, is shown in Fig. 16-18.

Mud weight requirements in drilling through overpressured formations can be met by use of high-density solids in the fluids. Barite ($BaSO_4$), which is non-toxic, non-reactive solid, is frequently used. American Petroleum Institute specifications require a minimum specific gravity of 4.2, because the use of low-grade barite (specific gravity <4) results in mud viscosity problems, higher concentration of chemical thinner required, and excessive circulating pressures in deep, high-temperature wells.

Under special circumstances, such as drilling in superpressured environments (e.g., in Middle East) and/or fighting wells already out of control, other heavy minerals with specific gravities of about 7 have been used successfully. These include various types of iron ores and lead sulfide. They, however, are expensive, must be ground to extremely fine particle size, and necessitate higher viscosity of the mud system.

REFERENCES

Adams, N., 1977. Deep waters pose unique well kick problems. *Pet. Eng.*, 49(5): 25—36.

Althaus, V.E., 1975. A new model for fracture gradient. *5th CWLS Form. Eval. Symp.*, Calgary, Alta., May, Pap. M: 12 pp.

Anderson, R.A., Ingram, D.S. and Zanier, A.M., 1973. Determining fracture pressure gradients from well logs. *J. Pet. Technol.*, 25: 1259—1268.

Aud, B.W. and Fertl, W.H., 1976. Overpressure in the twilight zone. *Pet. Eng.*, 48(5): 122—218.

Belotti, P. and Gerard, R.E., 1976. Instantaneous log indicates porosity and pore pressure. *World Oil*, 183(5): 90—94.

Boatman, W.A., 1967. Shale density key to safer, faster drilling. *World Oil*, 165(2): 69—74; also *J. Pet. Technol.*, 19: 1423—1431.

Boone, D.E., 1972. Porosity and pressure log performs well in the North Sea. *Pet. Petrochem. Int.*, 12: 36—38.

Boone, D.E., 1975. Drilling strength analysis in evaporites. *Pet. Eng.*, 47(9): 124—131.

Bourgoyne, A.T., 1971. A graphic approach to overpressure detection while drilling. *Pet. Eng.*, 43(9): 76—78.

Bourgoyne, A.T., 1976. Graphic approach to kick severity calculations. *Pet. Eng.*, 48(9): 22—28.

Bourgoyne, A.T., Rizer, J.A. and Myers, G.M., 1971. Porosity and pore pressure logs. *Drilling Contractor*, May—June: 36—45.

Chilingar, G.V., Rieke III, H.H., Sawabini, S.T. and Ershaghi, I., 1969. *Chemistry of Interstitial Solutions in Shales versus that in Associated Sandstones.* 44th Annu. Fall Meet., Soc. Pet. Eng. AIME, Denver, Colo., SPE 2527: 18 pp.

Chilingarian, G.V. and Rieke III, H.H., 1968. Data on consolidation of fine-grained sediments. *J. Sediment. Petrol.*, 38(3): 811—816.

Christman, S.A., 1973. Offshore fracture gradients. *J. Pet. Technol.*, 25: 910—914.

Combs, G.F., 1968. *Prediction of Pore Pressure from Penetration Rate.* 43rd Annu. Fall Meet., Soc. Pet. Eng. AIME, Houston, Texas, Sept., SPE 2162: 16 pp.

Daw, R.N., Myers, D.L. and Mercer, R.F., 1977. Mud gas logs help predict occurrence of geopressures. *World Oil,* 182(2): 43—46, and 182(3): 61—75.

Dupin de Saint Cyr, P., 1973. Ultrasonic device monitors mud pit levels on floaters. *World Oil,* 177(7): 57—58.

Eaton, B.A., 1969. Fracture gradient prediction and its application in oil field operations. *J. Pet. Technol.,* 21: 1353—1360.

Eaton, B.A., 1970. How to drill offshore with maximum control. *World Oil,* 171(5): 73—77.

El-Hadidi, S., 1970. *Use of Well Logging Data for Predicting of Rock Drillability.* M.S. Thesis, Univ. California, Berkeley, Calif., 90 pp.

Fertl, W.H., 1973. What to remember when interpreting mud-gas cutting. *World Oil,* 177(4): 67—72.

Fertl, W.H., 1976. *Abnormal Formation Pressures, Implications to Exploration, Drilling, and Production of Oil and Gas Resources.* Elsevier, Amsterdam, 382 pp.

Fertl, W.H., 1977. Shale density studies and their application. In: G.D. Hobson (Editor), *Developments in Petroleum Geology.* Applied Science Publishers, Barking, Essex, pp. 293—327.

Fertl, W.H. and Chilingar, G.V., 1976. *Importance of Abnormal Formation Pressures to the Oil Industry,* Spring Meet. European Soc. Pet. Eng. AIME, Amsterdam, Apr., SPE 5946: 10 pp.

Fertl, W.H. and Timko, D.J., 1970. How abnormal pressure detection techniques are applied. *Oil Gas. J.,* 68(2): 62—71.

Fertl, W.H. and Timko, D.J., 1971. *Parameters for Identification of Overpressure Formations.* 5th Conf. on Drilling and Rock Mechanics, Soc. Pet. Eng. AIME, Univ. Texas, Austin, Texas, Jan., SPE 3223: 14 pp.

Fertl, W.H. and Timko, D.J., 1973a, b. How downhole temperatures, pressures affect drilling, Part 9 (1973a) and Part 10 (1973b). *World Oil,* 176(2): 47—51, and 176(4): 62—66.

Fontenot, J.E. and Berry, L.N., 1975. Study compares drilling rate-based pressure prediction methods. *Oil Gas J.,* Sept. 15: 123—138.

Forgotson, J.M., 1969. Indication of proximity of high pressure fluid reservoir, Louisiana and Texas Gulf Coast. *Bull. Am. Assoc. Pet. Geol.,* 53: 171—173.

Gill, J.A., 1968. Applied drilling technology, an engineered package for pressure detection and control. *Drilling Contractor,* Mar./Apr.: 128—140.

Gill, J.A. and Weintritt, D.J., 1970. *Shale Factor: a Diagnostic tool for Formation Logging.* 11th Prof. Well Log Analysts Symp., Los Angeles, Calif.: 11 pp.

Goddard, R.D., Guest, R.J. and Anderson, T.O., 1973. High resolution fluid measurements improve drilling. *Drilling Contractor,* Mar./Apr.: 55—64.

Goins, W.C., 1968. Guidelines for blowout prevention. *World Oil,* 167(4): 88—106.

Goins, W.C., 1969. *Blowout Prevention.* Gulf Publ. Co., Houston, Texas, 260 pp.

Goins, W.C. and O'Brien, T.B., 1962. How to detect and control threatened blowouts. *Oil Gas J.,* 60(42): 143—151.

Goldsmith, R.C., 1972. Why gas-cut mud is not always a serious problem. *World Oil,* 175(5): 51—54.

Gstalder, S. and Raynal, J., 1966. Measurement of some mechanical properties of rocks and their relationship to rock drillability. *Trans. AIME,* 237: 991—996.

Guyod, H., 1946. Temperature well logging. *Oil Weekly,* Oct. 28: 35—47.

Harper, D., 1969. *New Findings from Overpressure Detection Curves in Tectonically Stressed Beds.* 40th Reg. Meet. Soc. Pet. Eng. AIME, Los Angeles, Calif., Nov., SPE 2781: 9 pp.

Herbert, W.E. and Young, F.S., 1972. Estimation of formation pressure with regression models of drilling rate. *J. Pet. Technol.,* 24: 9—15.

Hubbert, M.K. and Willis, D.G., 1957. Mechanics of hydraulic fracturing. *Trans. AIME*, 210: 153—168.
Hutchison, S.O. and Anderson, G.W., 1974. What to consider when selecting drilling fluids. *World Oil*, 179(5): 83—94.
Issenmann, O. and Lucon, C., 1971. Present state of gas-logging techniques with the use of a continuous mud weight recorder. *J. Can. Pet. Technol.*, 12: 9—11.
Jones, P.H., 1968. *Hydrodynamics of Geopressures in the Northern Gulf of Mexico Basin.* 43rd Fall Meet. Soc. Pet. Eng. AIME, Houston, Texas, Sept., SPE 2207: 15 pp.
Jorden, J.R. and Shirley, O.J., 1966. Application of drilling performance data to overpressure detection. *J. Pet. Technol.*, 18: 1387—1394.
Koch, R.D., 1976. Determining casing setting depths in deep-water wells. *World Oil*, 182(3): 45—52.
Lewis, C.R. and Rose, S.C., 1970. A theory relating high temperatures and overpressures. *J. Pet. Technol.*, 22: 11—16.
Lutz, J., Raynard, M., Gstalder, S., Quic-aud, C., Raynal, J. and Muckleroy, J.A., 1972. Instantaneous logging based on a dynamic theory of drilling. *J. Pet. Technol.*, 24: 750—758.
MacPherson, L.A. and Berry, L.N., 1972. Prediction of fracture gradients from log-derived elastic moduli. *Log Analyst*, 13(5): 12—19.
Matthews, W.R. and Kelly, J., 1967. How to predict formation pressure and fracture gradient. *Oil Gas J.*, 65(8): 92—106.
McDonald, W.J., 1976. *Logging While Drilling: a Survey of Methods and Priorities.* Trans. SPWLA, Pap. U: 14 pp.
Mondshine, T.C., 1969. New technique determines oil-mud salinity needs in shale drilling. *Oil Gas J.*, 67(28): 70—75.
Moore, P.L., 1974. *Drilling Practices Manual.* The Petroleum Publ. Co., Tulsa, Okla., 448 pp.
Nance, G., 1977. How to calculate maximum surface pressure for floating drilling. *Pet. Eng.*, 49(1): 46—54.
Nevins, M.J. and Weintritt, D.J., 1967. Determination of cation-exchange capacity by methylene blue adsorption. *Bull Am. Ceram. Soc.*, 46: 587—592.
O'Brien, T.B. and Goins, W.C., 1960. The mechanics of blowouts and how to control them. *API Drill. Prod. Prac.*, 27: 41—55.
Overton, H.L. and Timko, D.J., 1969. The salinity principle, a tectonic stress indicator in marine sands. *Log Analyst*, 10(6): 34—43.
Pixler, B.O., 1945. *Some Recent Developments in Mud Analysis Logging.* Fall Meet. Soc. Pet. Eng. AIME, Houston, Texas, Oct., SPE 2026: 8 pp.
Pomrenke, R.W. and Cox, T., 1976. Latest technology to supervise drilling on Discoverer 511. *World Oil*, 182(5): 79—83.
Rehm, W.A., 1969. Pressure control in drilling. *Oil Gas J.*, 67(31)—68(7).
Rochon, R.W., 1968. *The Effect of Mud Weight in Mud Logging Gas Anomalies.* Monarch Logging Co., San Antonio, Texas, 12 pp.
Taylor, D.B. and Smith, R.K., 1970. Improving fracture gradient estimates in offshore drilling. *Oil Gas J.*, 68(15): 67—72.
Terzaghi, K.V., 1923. Die Berechnung der Durchlässigkeitsziffer des Tones aus dem Verlauf der Hydrodynamischen Spannungserscheinungen. *Sitzungsber. Akad. Wiss. Wien, Math. Naturwiss. Kl. Abts.*, 2A: 105—132.
Wardlaw, H.W.R., 1969. *Drilling Performance Optimization and Identification of Overpressure Formations.* 4th Conf. Drilling and Rock Mechanics, Soc. Pet. Eng. AIME, Univ. of Texas, Austin, Texas, Jan., SPE 2388: 12 pp.
West, E.R., 1976. Hydraulic control of deep well blowouts. *Pet. Eng.*, 48(5): 68—81.
Wilson, G.J. and Bush, R.E., 1973. Pressure prediction with flow line temperature gradient. *J. Pet. Technol.*, 25: 135—142.

Young, F.S., 1968. *Computerized Drilling Control.* 43rd Fall Meet., Soc. Pet. Eng. AIME, Houston, Texas, Sept., SPE 2241: 12 pp.
Zamora, M., 1972. Slide rule correlation aids *d*-exponent use. *Oil Gas J.*, 70(51): 68—72.
Zoeller, W.A., 1970. *The Drilling Porosity Log.* 45th Fall Meet., Soc. Pet. Eng. AIME, Houston, Texas, Oct., SPE 3066: 10 pp.

Chapter 17

BLOWOUT PREVENTION EQUIPMENT

WILLIAM L. CLARK, GEORGE V. CHILINGARIAN and PAUL VORABUTR

This chapter is dedicated to Dr. Richard Seaver, president of Hydril Company, for his important contributions to the petroleum industry and education.

INTRODUCTION

As pointed by Rehm et al. (1975), a potential blowout is manifested by several signals, starting with increased flow at the flowline. Rehm et al. (1975) outlined the following basic pressure control rules, which are used as an introduction to blowout prevention equipment described in this chapter:

(1) *Shut-in the well kick as soon as possible.* Usually well kicks of less than 15 bbl can be killed with a minimum difficulty, whereas well kicks of 15—30 bbl are rather difficult to control without losing circulation. It may be impossible to bring the bubble to the surface without causing lost circulation when a well kick exceeds 30 bbl.

(2) *Determine shut-in drillpipe pressure (SIDPP).* The amount by which the reservoir pressure exceeds that exerted by the column of drilling fluid inside the drillpipe (i.e., drillpipe pressure) must be determined. The reservoir pressure and the necessary increase in the mud specific weight can then be calculated as follows:

$$p_r = (MG \times D) + SIDPP \tag{17-1}$$

where p_r = reservoir pressure, MG = drilling fluid pressure gradient, D = depth of the reservoir, and $SIDPP$ = shut-in drillpipe pressure; and:

$$\Delta\gamma_m = (SIDPP/D) \times f \tag{17-2}$$

where $\Delta\gamma_m$ = increase in the specific weight of drilling fluid, and f = conversion factor (19.2 lb/gal, 144 lb/ft^3, 10 kg/l). It should be remembered, however, that *SIDPP* calculations assume no lost return and that pressure buildup may be slow due to skin damage or low permeability of reservoir rocks.

(3) *Circulate using the choke.* The bottomhole pressure must be maintained slightly above reservoir pressure in order to kill a well. The initial circulating pressure (*ICP*) is the sum of system pressure loss (*SPL*), i.e., circulating pressure, and the shut-in drillpipe pressure (*SIDPP*). Under certain conditions, a small pressure increment (margin) is also added.

(4) *Reduce the pump rate by one-half.* Horsepower or pressure rating must be reduced while killing a well. The circulating rate must be reduced by

one-half, a value which is obviously not absolute but is generally appropriate.

(5) *Increase the mud specific weight.* The drillpipe pressure must be gradually reduced from the time the new drilling fluid enters the drillpipe and until it reaches the bottom of the hole. The reduction in pressure is about equal to the original *SIDPP* value, and the final circulating pressure (*FCP*) is equal to:

$$FCP = [\gamma_{mi}/\gamma_{mf}] \times SPL \tag{17-3}$$

where γ_{mi} = specific weight of mud at initial circulating pressure, γ_{mf} = specific weight of mud at final circulating pressure, and *SPL* = system pressure loss. As a final check, the pump is stopped and well is shut-in. If the new *SIDPP* is equal to zero, the drillpipe contains drilling fluid having a proper specific weight.

(6) *Constant casing pressure.* Although constant casing pressure can cause changes in bottomhole pressure, use of casing pressure eliminates the necessity to calculate the system-pressure loss and the required drillpipe pressure reduction.

In deep, slim holes in which annular pressure drops are high, casing pressure techniques to kill the well also should not be used if the gas is near the surface, because the potential change in bottomhole pressure is too great.

After holding casing pressure constant by use of the choke until the pumping rate reaches some predetermined value, the drillpipe gauge pressure should be used as the initial circulating pressure.

If there is no gas expansion and no change in the length of the invading fluid column in the annulus, the casing pressure will stay constant while drilling fluid specific weight changes in the drillpipe.

The drillpipe pressure should be used as the final circulating pressure, instead of the casing pressure, when the new mud reaches the bit.

(7) *Maintain a margin value.* Inasmuch as bottomhole pressure is reduced when the drillpipe is raised, a certain minimum safety marginal value must be *added* to the calculated drilling fluid specific weight required to dominate the reservoir. A reasonable marginal value, *MV*, can be calculated using the following equation:

$$MV = \frac{\overline{YP}f}{(d_h - d_p)} \tag{17-4}$$

where $\overline{YP}$ = yield point (lb/100 ft^2); d_h = hole diameter (inches); d_p = pipe diameter (inches); f = units constant.

In an excellent publication entitled "Practical Pressure Control", Rehm et al. (1975) presented a series of problems using standard instrument panel. In the opinion of the writers this manual must be used as a guide by all petroleum engineers, because control of a well is extremely critical.

BLOWOUT PREVENTION EQUIPMENT

The blowout preventer (BOP) stack mounted on top of wellhead and casing is presented in Fig. 17-1. The equipment in the BOP stack provides the means to contain well pressures. This pressure containment equipment prevents the uncontrolled discharge of drilling fluids from the hole. After containment of detected abnormal or uncontrolled wellbore pressures, the BOP stack provides the drilling crew the time needed to evacuate the location, and then prepare and excute corrective action. The BOP stack also contains the pressures while new drilling fluids are circulated through the well.

The typical BOP stack usually consists of one or two annular type blowout preventers, two to four ram-type blowout preventers, a choke line, and choke manifold (used to circulate kicks out of a shut-in well), a kill manifold, and a hydraulic closing unit (used to operate the BOP stack and manifold components). Associated and auxiliary equipment such as diverters, drillpipe valves and marine risers (for subsea operations) are also considered part of blowout prevention equipment systems.

The primary function of the BOP stack is to confine well fluids in the wellbore, and allow controlled volumes of fluids to be bled from the wellbore through the choke manifold.

The BOP stack is mounted atop the wellhead and a casing string. The pressure rating of the BOP stack is a function of the expected downhole

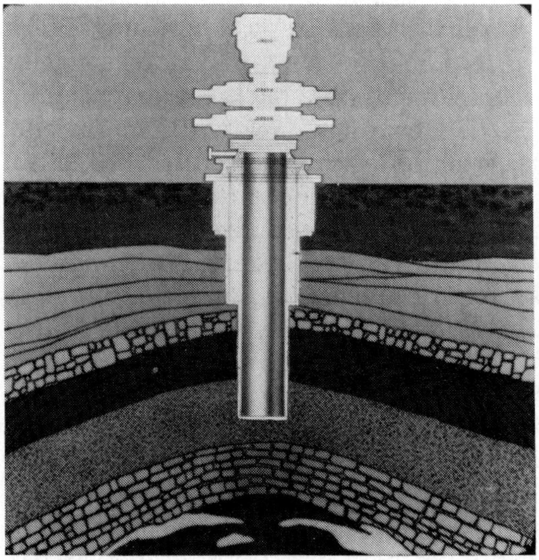

Fig. 17-1. BOP stack mounted atop well head and casing.

Fig. 17-2. Typical BOP stack on offshore platform.

pressures. The bore size of the BOP stack is a function of the tool or drill bit diameter, which must pass through the BOP in order to drill for the next size of casing. Several different bore size and pressure ratings of BOP stacks may be used to drill a well. The well completion BOP size may be determined by the production string size and the applicable running equipment. A typical BOP stack on offshore platform is shown in Fig. 17-2.

ANNULAR BLOWOUT PREVENTER

An annular blowout preventer sealed on drillpipe is presented in Fig. 17-3. Usually the first piece of equipment operated in case of a kick is the annular blowout preventer. Annular blowout preventers were designed for the purpose of providing wellbore closure under any conditions, regardless of the size or shape of objects in the wellbore. The universal feature of the annular BOP is the ability to close the wellbore and seal off anything present in the bore, i.e., drillpipe of any size, drillpipe tool joints, tubing, casing, kelly (square or hexagonal), or drillcollars. In the case of empty wellbore, the annular BOP can completely seal off the open bore. The universal sealing feature also allows drillpipe to be rotated in the shut-in well or allows drillpipe to be moved up or down (stripped) with the well shut-in. This operation is often necessary when a kick occurs with the drill bit being off the bottom of the hole. It can be stripped back to the bottom in order to begin recirculation of the drilling fluid for the entire depth of the hole.

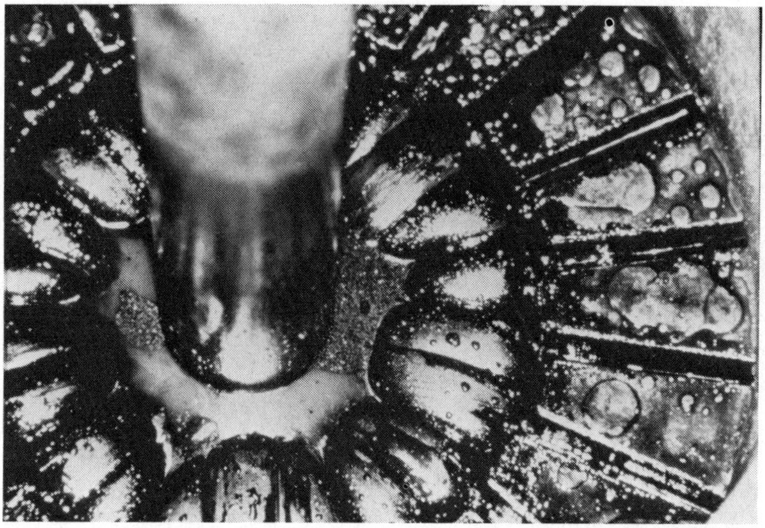

Fig. 17-3. Annular BOP sealed on drillpipe.

The design of most annular blowout preventers is similar. The principle of operation is that an operating piston squeezes a steel reinforced rubber packing unit around whatever is present in the wellbore. Hydraulic pressure, which is introduced into the BOP closing chamber, forces the piston upward. As the piston moves upward, the rubber packing unit is squeezed inward to a sealing engagement. Most annular BOP closing operators are affected by well pressure and require some adjustment of closing pressure to obtain optimum performance.

Some models of annular blowout preventers are assisted in sealing off by wellbore pressure. After a seal-off is achieved, closing pressure should be reduced in proportion to the wellbore pressure. Other models require an increase of closing pressure in proportion to the wellbore pressure. It is imperative that the manufacturer's instructions be properly followed.

To open the BOP, hydraulic pressure is applied to the opening chamber. This pressure forces the piston downward and the natural resiliency of the rubber opens the packing unit to full diameter.

Principal annular blowout preventers available today

Cameron type D

The Type D annular BOP uses a two-piece rubber packing unit. The actuating piston is raised by hydraulic pressure to achieve a seal-off. The slope on the inside of the BOP head causes the packing unit to squeeze inward. The steel reinforcing inserts of the packing unit rotate inward to support the

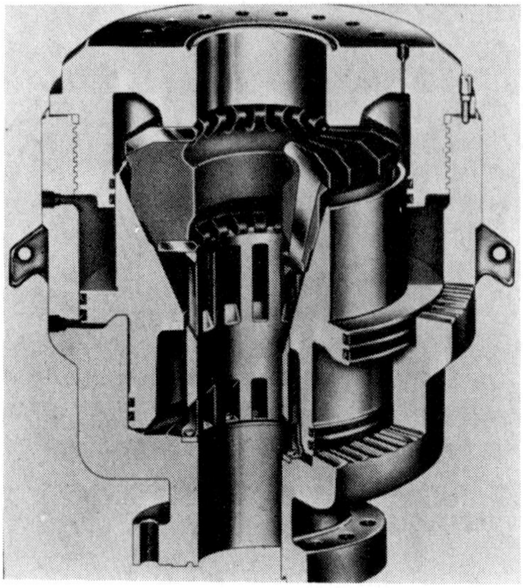

Fig. 17-4. Hydril GK type annular BOP.

rubber for seal-off. This BOP is available in the popular sizes and having different pressure ranges.

Hydril GK and MSP

The Hydril GK and MSP types of annular blowout preventers, which were the first ones available to the industry, are still available in their original design concept. The Hydril GK type annular BOP is presented in Fig. 17-4. Hydraulic closing pressure raises the piston to achieve a seal-off engagement. The slope of the piston bowl forces the packing unit inward. The cast steel inserts move inward during seal-off operation while supporting the rubber packing unit.

These two blowout preventers are assisted by well pressure after initial seal-off. Optimum control is realized by reducing closing pressure after achieving seal off. These units are presently available in bore sizes of $4\frac{1}{16}$—30 inches and pressure ratings of 500—20,000 psi.

Hydril GL

The Hydril GL BOP was designed and developed for subsea and surface drilling operations. Hydraulic pressure raises the piston to squeeze the packing unit inward toward a sealing engagement, using the same principle as the GK BOP. This blowout preventer, which was developed for long stripping life and fast field maintenance, has the standard closing and opening chambers plus a secondary chamber. Control hookup of the third chamber offers flexibility of control for different applications.

Reagan type KFL

The Reagan type KFL BOP utilizes three packer elements to seal-off on the drillstring: insert packer, main packer, and packer sleeve. Three hydraulic lines are required to operate the KFL BOP, i.e., two lines to actuate the insert packer locking dogs and one line to close or open the packing elements.

The KFL blowout preventer, equipped with surge dampeners to facilitate stripping tool joints for subsea operation, is not a self-energizing BOP. When closing the KFL BOP, it is essential to maintain pressure at 500 psi greater than the wellbore pressure.

NL-Shaffer spherical

The NL-Shaffer spherical annular BOP (Fig. 17-5) utilizes a spherical head covering the top of the packing unit to translate vertical loading on the packing unit into radial movement. Upward movement of the piston compresses the packing unit against the head. The spherical shape of the head squeezes the packing unit inward for a sealing engagement. The height of this BOP is reduced through the use of a short piston and by the spherical geometry of the head.

Well diverters

Well diverters represent a special application of large-bore, annular-type blowout preventers, normally with working pressure of 2000 psi or less. They are used while drilling for and running surface casing to provide protec-

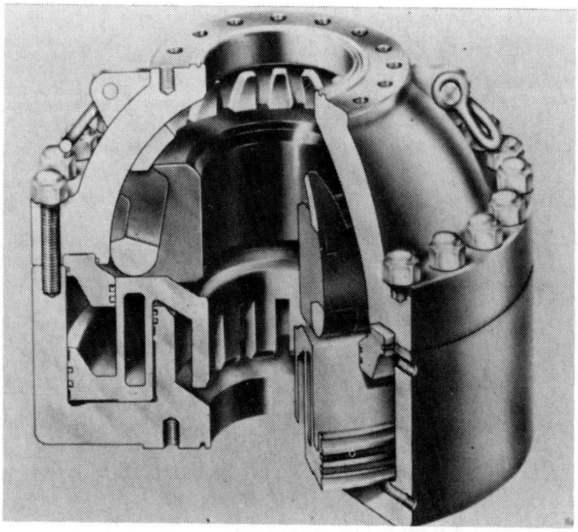

Fig. 17-5. NL-Shaffer spherical annular BOP.

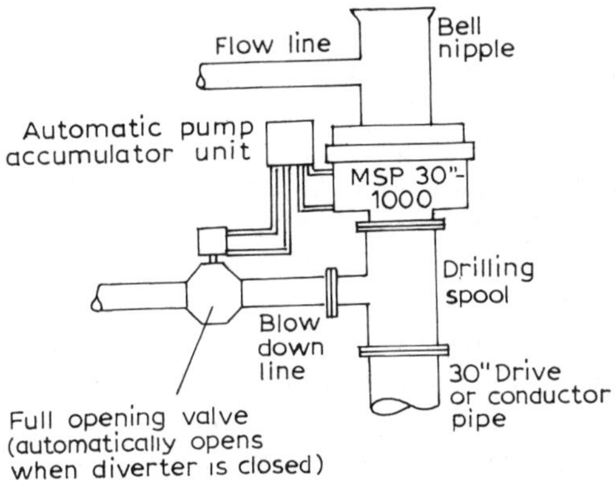

Fig. 17-6. Diverter hook-up.

tion from shallow gas pockets. If shallow gas is encountered, the diverter is closed and the gas is diverted through a flow line—blow down line away from the drill rig, while maintaining control of the escaping gas. Pressure containment at this stage of well completion (uncased and/or not cemented) is not normally feasible or advisable.

The 30-inch or $21\frac{1}{4}$-inch bore blowout preventers are used for diverters. The BOP bore is determined by the bit size required for the surface casing.

Hydril MSP $29\frac{1}{2}$-inch-500 and MSP 30-inch-1000 (Fig. 17-6)

Hydril MSP large-bore blowout preventers are used for diverters. They are closed and sealed in the same manner as the smaller Hydril MSP blowout preventers. The $29\frac{1}{2}$-inch BOP operates with only a closing chamber. These blowout preventers can close on casing, kellys, drillpipe, or open hole.

Reagan 30-inch KFL 1000

Reagan riser diverter — type KFD or KFDG

Reagan riser diverter (type KFD or KFDG) is similar to the KFL BOP except that it is adapted for use on marine riser systems. The unit is attached to, or built into, the rig floor structure on top of the riser. Misalignment is accomodated by a ball joint below the diverter in the KFD version, whereas the KFDG type is a gimbal mounted unit. Packoff is achieved down to about $4\frac{1}{2}$ inches on pipe depending on the insert used. The insert must be removed to pass tools and drill bits.

Vetco type HY diverter

Vetco type HY marine riser diverter is mounted atop the marine riser. Applying and maintaining hydraulic pressure in the closing chamber maintains seal-off on the drillpipe. Removal of the sealing unit is required to pass drill bits.

RAM BLOWOUT PREVENTERS

Ram type blowout preventers (Figs. 17-7 and 17-8) perform very important functions within the BOP stack. The ram BOP is essentially a specialized type of valve which can be closed in the wellbore. Similar in operation to a gate valve, the ram BOP has gates called rams which meet at the center of the hole. The faces of the rams are fitted with large rubber packers shaped to close around specific sizes of drillpipe, tubing, or casing. Blind (or CSO) rams completely shut-off the open hole. There are several types of rams which are described below. Hydraulic pressure applied to the operators causes the rams to close and seal.

Fig. 17-7. Outside view of ram BOP.

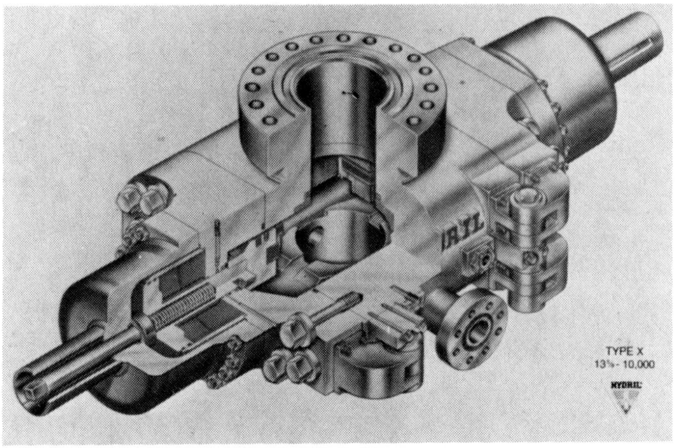

Fig. 17-8. Cutaway view of ram BOP.

Pipe rams

The inside ends of pipe rams are contoured to seal around pipe to close-off the area in the wellbore between the sides of the hole and pipe in the bore. Separate rams are necessary for each size (outside diameter) of pipe that is used in the wellbore. Thus, if drilling is begun with $4\frac{1}{2}$-inch drillpipe, a size ram block must be installed to close around $4\frac{1}{2}$-inch pipe. They will be unable to close around larger diameter pipe and would not make a seal around pipe with a smaller diameter. If the drillpipe size is changed, e.g., from $4\frac{1}{2}$-inches to $3\frac{1}{2}$-inches, the rams would have to be changed to accomodate closure on the new size pipe. When drilling with a tapered diameter drillstring, pipe rams of each drillpipe size are frequently installed in separate BOP cavities.

Blind rams

Blind rams are also called CSO (complete shut-off) rams. They will not seal around any pipe in the bore but close-off the open hole. Rubber packers provide a seal when the rams are closed. When blind rams are closed, the drillstring and bit must be raised above the BOP.

Shear—blind rams

Shear—blind rams are special blind rams with a built-in cutting blade that shears the pipe and then seals off the open hole in one motion. Shear—blind rams, which are always used on subsea drilling operations where the BOP stack is on the sea floor, are only infrequently used on surface-mounted BOP stacks. This ram is usually placed above the pipe rams in the stack so that drillstring can be hung-off on a tool joint on one of the lower rams before shearing. This technique makes the drillstring easier to retrieve at a later time and prevents it from falling into the wellbore after shearing.

Offset rams

There are single-offset and dual-offset rams, which are used when running tubing strings during well completion. The single-offset rams, which seal-off the wellbore around a string of tubing in the bore, are used in the case of multiple-string completions, causing the tubing string to be offset from the hole center.

Multiple configurations

Ram-type blowout preventers are available in multiple configurations, i.e., two or more ram-type blowout preventer bodies are joined. These multiple configurations can be accomplished when the BOP bodies are manufactured together as a unit. The arrangement of the blowout preventers in the stack varies according to well conditions, hole depth, and contractor's requirements.

Ram blowout preventers may also have side outlets. These outlets are usually placed just below the rams and are used to control annular wellbore flow through the choke manifold or from the kill manifold during well control operations. Side outlets can be placed on either or both sides of the BOP. These outlets are kept closed and sealed off by side outlet valves during normal drilling operations.

Operation of ram-type preventers

To operate the ram BOP, hydraulic fluid is introduced into one end of an operating hydraulic cylinder. The piston is connected to the ram by a rod. The pressurized hydraulic fluid moves the piston and ram block inward until a seal-off is achieved. Seal-off is accomplished at two places: (1) across the face of the ram block (around the pipe or across the bore of an open well) and (2) between the top of the ram and the body of the BOP. This sealing effectively confines pressure within the wellbore. Seal-off of most ram blowout preventers is assisted by well pressure. After initial seal-off, the well pressures make the seal-off tighter.

To open the ram BOP, hydraulic fluid pressure is applied to the opening side of the piston. This force retracts the rams into the open position.

Typical choke and kill manifold is presented in Fig. 17-9. Choke and kill lines can be connected to the wellbore (1) through drilling spool side outlets mounted between the ram blowout preventers or (2) through the side outlets of the ram BOP body. The use of the side outlets in the ram BOP is common where the overall height of the BOP stack is critical.

If a kick or gas bubble (when formation pressure is in excess of the drilling fluid hydrostatic pressure) enters the wellbore because of insufficient drilling fluid column pressure, the kick is circulated out of the well through the choke lines. After a kick is detected, the appropriate BOP is closed and the shut-in drillpipe and casing pressures are determined. Using this data, a well control technique is selected to fill the hole with a drilling fluid having a new specific weight that will permit balanced-pressure drilling.

This new drilling fluid is pumped down the drillpipe to the bottom of the hole, while a controlled volume of fluid is bled out through the choke manifold. Drillpipe and casing pressures are controlled during the pumping operation by the selected orifice size in the choke. Chokes are devices with either a fixed or variable aperture size used to control the release of well fluids under pressure (Reifel, 1976, p. 243).

When drillstring is out of the hole, the well is shut-in under pressure, *and* it is not feasible to run the drillstring back into the hole, the kill line may be utilized to pump a heavy weight mud into the wellbore to balance the formation pressure.

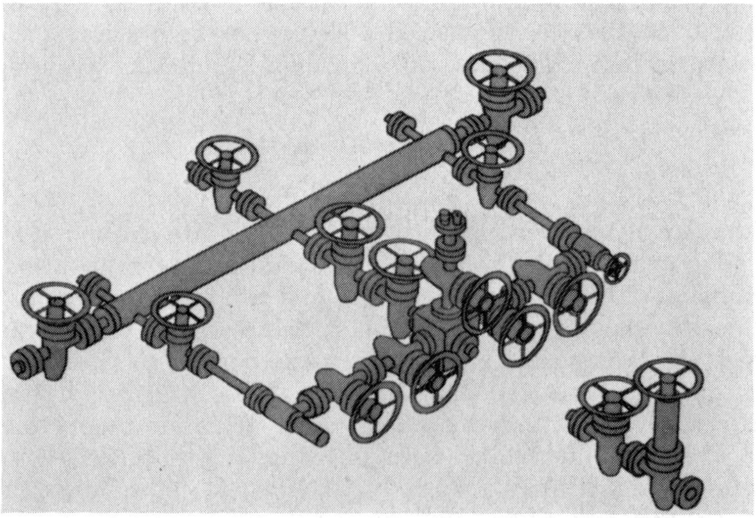

Fig. 17-9. Typical choke and kill manifold.

Fixed-type choke

Fixed-type of choke valve, often called positive choke, has a replaceable fixed-diameter orifice or flow bean. It is available in a variety of sizes, and change of the choke size is usually accomplished by switching to an alternate choke. Plugging with formation cuttings often occurs (Reifel, 1976, p. 244).

Variable-type choke

Variable-type or adjustable chokes are very popular because they provide increased flexibility. Plugging, which is likely, can sometimes be relieved by fully opening the choke temporarily. Abrasive muds wear these chokes out rather rapidly (Reifel, 1976, p. 241; Goins, 1969, p. 21). Chokes are frequently used which can be remotely controlled.

Choke and kill lines would be of a rating equal to or greater than the BOP stack pressure rating (Reifel, 1976, p. 241).

BOP CLOSING UNIT

The blowout preventer closing unit (Fig. 17-10) provides the hydraulic power to operate the elements of the BOP stack: the ram and annular blowout preventers, the choke and kill valves, and, in subsea operations, several of the connectors within the BOP stack and between the BOP stack and the sea

Fig. 17-10. Closing unit.

floor. The safety of personnel and the success or failure of a well control operation depends upon an operational control system. These units are normally designed and built to provide immediate closing power to all BOP stack equipment and to maintain the recommended pressure to assure continued safety if well pressure increases. The fast closure of the BOP under critical well conditions avoids damage to the rams and packers that seal against well pressure. There are many sources of power that could be utilized to operate the BOP stack, i.e., primary rig power, steam boilers, engine-driven pumps, remote cement pumps, manually-operated pumps, rig air, etc. Each one of these power sources has drawbacks which preclude its use as a primary power source for BOP stack operation. Thus, stored hydraulic power (hydraulic fluid under pressure in gas-charged accumulators) is used as the primary power source for the BOP closing unit.

The closing unit or pump accumulator, which is used to provide this power source, includes the components listed below. Although each one of the components can be mounted separately, they form the closing unit system:

(1) Reservoir for hydraulic fluid.

(2) Accumulator bottles where fluid is stored under pressurized gas (nitrogen) for use upon demand.

(3) Pumps to charge hydraulic fluid into the accumulator bottle.

(4) Control valves to transfer stored fluid to operating device (ram BOP, annular BOP, choke or kill valve, or connector).

(5) Pressure regulator valve to provide fluid at appropriate pressure to operating device.

(6) Remote control panels to operate BOP stack from safe, expedient, or redundant locations.

ACCUMULATOR BOTTLES

Most accumulator bottles on BOP closing units in use today are bladder-type accumulators. The accumulator consists of a metal body containing a rubber bladder. The bladder is precharged with nitrogen or other inert gas. Hydraulic fluid is pumped into an accumulator bank compressing the gas charge in the bladder. The entry port of the accumulator manifold is closed by means of a back-pressure valve (check valve) and the exit port is controlled by means of a 4-way operator valve. When the stored hydraulic fluid is required, the operator valve is opened and the fluid is directed to the end device. The pressure regulator valve controls the pressure level.

The closing unit should contain power sources to recharge the accumulator bottles as quickly and rapidly as possible when the end devices are operated. Pumps may be driven by electric motors, pneumatic motors, gasoline engines, diesel engines, or any combination thereof with appropriate back-ups. Most units have two different sources of pump power.

Closing units usually have a working pressure rating of 3000 psi. This is the maximum pressure most ram blowout preventers require for operation with normal pressure requirements at a lower value. Pressure regulator valves are used to reduce this stored high pressure for normal operations. Lower pressure is set for direction to end devices, while 3000 psi pressure is maintained in the accumulator bank.

The closing unit is usually located as far from the BOP stack and rig floor as is prudent and efficient. Long distances between closing unit and stack increase the operation time of end devices. The remoteness of the closing unit also provides additional safety by removing it from immediate dangers of blowouts. Remote control panels are designed to operate the complete BOP control systems from the rig floor or other remote locations and provide a fast response.

These panels can have complete control of the closing unit system or can perform only certain critical functions. Communication between remote control panels and the closing unit can be accomplished pneumatically, hydraulically, electrically, or using combinations thereof. In subsea operations, accumulator bottles may also be located on the subsea BOP stack with a remote or interface manifold also being located on the stack. Communications between control panels on the surface and the "closing unit" on the sea floor can be accomplished through hydraulic, electronic or acoustical transmissions.

CLOSING UNIT CAPACITY

Closing units should be designed and manufactured with a minimum volume capacity of usable fluid to close one pipe ram, the annular BOP, and to open the hydraulic choke line valve(s) (API, RP53). Usable fluid volume is defined as the volume of fluid recoverable from an accumulator between the accumulator operating pressure and a pressure which is 200 psi above the precharge pressure. Additional functional capacity is recommended.

INSIDE BLOWOUT PREVENTERS

When there is drillpipe in the hole, formation pressures can be transmitted to the surface either through the annulus or the drillpipe bore. Ram and annular blowout preventers provide means for containing these pressures in the annulus or the open hole.

Complete control is achieved only when both the annulus and the drillpipe bore are closed-off. Inside blowout preventers or drillpipe valves provide this protection.

MANUAL SHUT-OFF VALVES

Manual shut-off valves (see Fig. 17-11) are utilized above the rotary table to provide drillpipe blowout protection. These valves are called kelly cocks, kelly valves, and drillpipe safety valves. The upper kelly cock or kelly valve is placed between the swivel and the kelly. Its basic purpose is to provide a means of closing the drillstring when the kelly is down through the rotary table and the drillstring cannot be lifted off bottom.

A kelly cock is also placed between the kelly and the upper joint of the drillstring, usually separated from the upper drillstring joint with a saver sub. In this position it is called the lower kelly cock and not only can close-off the drillstring in case of a kick or blowout but also is used as a mud saver. The lower kelly valve and saver sub remain made-up with the kelly. When closed, it prevents drilling fluid from flowing out of the kelly while making up the next length of drillpipe. This promotes (1) safety by keeping the rig floor clean and (2) economy by preventing the loss of expensive drilling fluids.

Inasmuch as many kicks occur while drillpipe is being tripped out of the hole and the kelly assembly is set into the rat hole, making the kelly assembly inaccessible, another type of valve is required for drillpipe protection. The drillpipe safety valve can be stabbed into the top joint of the drillstring. The drillpipe safety valve is a kelly valve with connections matching the drillpipe tool joints that should be maintained accessible on the drilling rig floor.

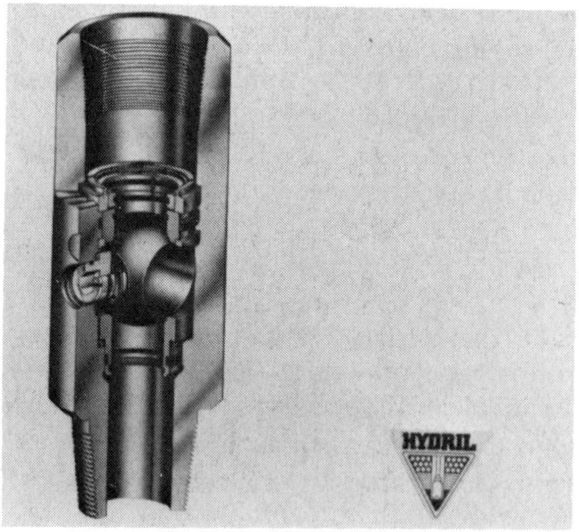

Fig. 17-11. Kelly valve (Hydril Kellyguard).

AUTOMATIC VALVES

Automatic valves are check valves that are usually located in the drillstring just above or below the drillcollars. There are several types of these valves: flapper, float, drop-in (check), and inside BOP.

The *flapper valve* is mounted in the drillstring where it remains during the drilling operation. The flapper valve opens readily with downward pressure of circulation. When formation pressures enter the drillstring with an upward force, the flapper valve closes the drillstring. Flapper valves when called *float valves* also prevent formation cuttings from clogging drill bit nozzles. Inasmuch as these valves are subjected to the cuttings and drilling fluid abrasion, they need frequent replacement.

The *drop-in* type check valve remains topside on the drilling rig floor until needed. When inside blowout protection is needed, it is dropped or pumped down the drillstring to a special landing sub. The ball check closes the bore to upward pressures, but readily opens when recirculation with downward pressure is started. This valve has the advantage of not being subjected to the abrasive action of cuttings and drilling fluid. The valve is removed from the drillstring when the bit is tripped out.

The *inside BOP*, which is check valve type, is normally held in the open position for makeup into the drillstring. After the makeup, the hold-open device is removed and the valve closes off the drillstring. Fluid can be pumped into the string.

GLOSSARY

Hydrostatic pressure (p_h) — pressure exerted on the bottom of a hole due to the weight of drilling fluid:

$$p_h = 0.052 \times \gamma_f \times D$$

where 0.052 * = constant when p_h is in psi; γ_f = specific weight of drilling fluid (lb/gal); and D = depth of hole (ft).

True Vertical Depth (*TVD*) — straight vertical distance from top to bottom of a hole.

Formation Pressure (p_f) — pressure of interstitial fluids. Normal pressure gradient = 0.433 to 0.465 psi/ft. Above-normal pressure (surpressure) may reach gradient of 1.0 psi/ft.

System Pressure Loss (*SPL*) — total pressure loss from pump throughout the circulating system, i.e., the pressure required to pump drilling fluid from the pump, up the standpipe, through the rotary hose, down the kelly and

---

* 0.052 lb/sq inch = (7.48 gal/cu ft) (1 sq ft/144 sq inch).

drillstring, through the jets at the bit, and up the annulus to the top of the hole.

Slow Circulating Rate (*SCR*) — piston mud pump speed used for well kick procedures. Generally, it is about one-half of the normal drilling speed.

Annulus Pressure Loss (*APL*) — the pressure on the annulus required to pump drilling fluid from the bottom to the top of the hole.

Shut-in Drillpipe Pressure (*SIDPP*) — pressure read from the gage on top of the drillpipe after shut-in.

Shut-in Casing Pressure (*SICP*) — pressure read from the gage on top of the casing after shut-in.

Initial Circulating Pressure (*ICP*) — pressure required in drillpipe during the well kill, and read from the drillpipe pressure gage: $ICP = SIDPP + SCR$. If $SIDPP = 600$ psi and $SCR = 750$ psi, then $ICP = 1350$ psi.

Equivalent Circulating Density (*ECD*) — additional density caused by pump pressure added to mud weight (true bottomhole pressure):

Hydrostatic pressure + Annulus pressure loss = $(p_h + APL)/(0.052 \times D)$

Example: If $p_h = 5200$ psi, $APL = 200$ psi, and $D = 10{,}000$ ft, then:

$$ECD = \frac{5200 + 200}{0.052 \times 10{,}000} = 10.38 \text{ lb/gal}$$

Final Circulating Pressure (*FCP*) — pressure required to hold on drillpipe when circulating *Kill Weight Mud* (*KWM*) from the bottom to the top of a hole:

$$FCP = SCR \times \frac{KWM}{OMW}.$$

Example: If *Original Mud Weight* (*OMW*) = 10 lb/gal; *Kill Mud Weight* (*KMW*) = 11.5 lb/gal; and *Slow Circulating Rate* (*SCR*) = 750 psi, then:

$$FPC = 750 \times \frac{11.5}{10} = 862.5 \text{ psi}.$$

Imposed pressure (p_i) — back pressure held by choke during well kill procedure (*SIDPP*). p_i is equal or greater than drillpipe pressure.

Original Mud Weight (*OMW*) — original specific weight of mud used during the drilling process.

Kill Weight Mud (*KWM*) — total mud weight required to balance formation pressure:

$$KWM = \frac{p_h + SIDPP}{0.052 \times D}$$

Example: Well is shut-in after it kicks. If $p_h = 5200$ psi (10 lb/gal mud at a depth of 10,000 ft) and $SIDPP = 600$ psi, then

$$KWM = \frac{5200 + 600}{0.052 \times 10{,}000} = 11.15 \text{ lb/gal}$$

Hydraulic Diameter (*HD*) — number of barrels per foot of depth:

$$HD = (d_h^2 - d_p^2)/1{,}029$$

where d_h = hole diameter, d_p = drillpipe outside dimension (O.D.), and 1029 = constant.

Example: If d_h = 8.5 inches and d_p (O.D.) = 4.5 inches then:

$$HD = \frac{8.5^2 - 4.5^2}{1{,}029} = 0.050 \text{ bbl/ft}$$

ACKNOWLEDGEMENTS

The help extended by Richard Churchwell and Ray Carter is greatly appreciated by the authors.

REFERENCES

American Petroleum Institute, 1978. *Recommended Practices for Blowout Prevention Equipment Systems.* 58 pp.

Goins, W.C., 1969. *Blowout Prevention.* Gulf Publ. Co., 214 pp.

Rehm, B., Hebert, R. and Hebert, J., 1975. *Practical Pressure Control.* The Petroleum Publ. Co., Tulsa, Okla., 87 pp.

Reifel, M.D., 1976. *Offshore Blowouts and Fires. The Technology of Offshore Drilling, Completion and Production.* Compiled by ETA Offshore Seminars, Inc. The Petroleum Publ. Co., Tulsa, Okla., 425 pp.

Chapter 18

DIRECTIONAL DRILLING

EASTMAN WHIPSTOCK COMPANY *, DAN MADICK, GEORGE V. CHILINGARIAN and PAUL VORABUTR

INTRODUCTION

Every well should be drilled under appropriate controls that will result in penetrating the formation at a predetermined location. The technology of deviating a wellbore along a planned course to a subsurface target is termed controlled directional drilling.

The principles of application are the same whether controlled directional drilling is used to hold a wellbore as close as possible to the vertical or to deviate it deliberately from vertical.

The first step in directional drilling is to specify the target area, i.e., the zone which must be penetrated by the well at a stated depth and location. The size and shape of the target area usually depends on geological structures and the location of the producing zones with respect to lease lines and spacings. The target area should never be reduced to an unrealistic size and, consequently, considerably increase the cost of the operation.

In selecting an optimum surface location for the rig, advantage should be taken of natural formation deviation tendencies. Formation attitudes have a marked effect on the drift of the borehole. For example, drilling through alternately hard and soft formations with a well-stabilized bit usually results in a course perpendicular to the bedding plane. If the bedded formation dips in excess of 45°, however, the bit will tend to drill parallel to the bedding.

Directional tendencies are also affected by formation attitudes (dip and strike). If, for example, the proposed direction is due updip, it follows the natural bit tendencies and drift angle can be readily built. On the other hand, if the proposed direction is left of updip direction, the bit tends to turn to the right; whereas if the direction is right of updip direction, the bit will deviate to the left. Selection of an optimum location, therefore, should be based on all subsurface information in order to take advantage of the formation attitudes and to minimize undesired borehole deviation.

Larger-diameter boreholes are easier to control directionally than smaller diameter ones, because slim holes require smaller, more flexible drillcollars

* The help of Cliff Ellis is gratefully acknowledged.

and pipes. Smaller collars also limit the range of weights available. In slim hole drilling, therefore, formation characteristics have more effect on drawing the borehole off course.

Directional drilling owes its origin to the mining industry. In 1905, a whipstock was used for deflecting diamond drillholes in the Transvaal and Rand, South Africa. Earliest application in the petroleum industry was made in the 1920's. In general, two techniques were used: (1) the selection of drilling weights, speeds, straight-hole bits, and drillcollars to keep deviation to a minimum; and (2) the corrective action for restoring deviated wellbores to vertical position. After using directional drilling for preventive and corrective purposes in California, USA, operators began to purposely deflect boreholes to reach particular objectives.

Beginning in 1927, many wells in the Signal Hill Field, the Long Beach area, and Huntington Beach area in California, were bottomed beneath tidelands of the Pacific Ocean. During this period, deflection tools such as the removable whipstock and the knuckle joint were developed.

A determining factor in the success of directional drilling project is the selection of best kick-off point (*KOP*), i.e., depth at which the directionally drilled section is started (Fig. 18-1). Usually, higher drift angles (15°—45°) should be used, because they allow greater choice in selecting the most desirable kickoff point and give greater stability of direction than smaller angles (5°—10°).

Inasmuch as the natural tendencies of the bit are to move in a curved

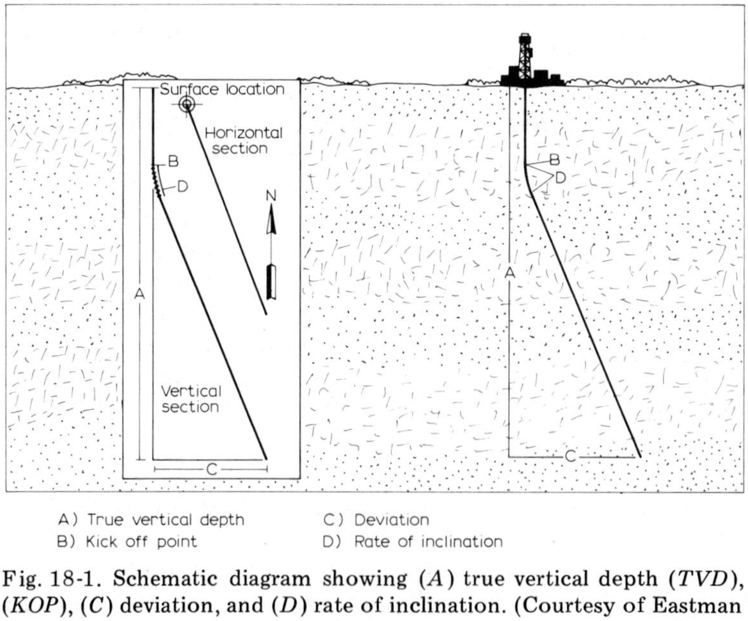

Fig. 18-1. Schematic diagram showing (*A*) true vertical depth (*TVD*), (*B*) kickoff point (*KOP*), (*C*) deviation, and (*D*) rate of inclination. (Courtesy of Eastman Whipstock, Inc.)

path, a carefully planned directional plan must permit a displacement of a few degrees on either side of a straight horizontal line connecting the surface and target location. Resisting the normal tendencies of bit's movement would only result in increased rig time and unsatisfactory performance.

The acceptable amount of lead (or lateral displacement, i.e., projected angular displacement of the bit, either to the right or left of the proposed objective) must be within certain limits. In order to avoid interference with existing wells, the borehole must be maintained within an imaginary cylinder surrounding the axis of the proposed well in high well density areas. Similar restrictions may also apply to the bottomhole target in a reservoir rock.

Most casing programs used in directional boreholes are similar to those used in straight-hole drilling projects. In deep boreholes or boreholes having high inclination, however, it becomes necessary to install rubber protectors on the drillpipe in order to reduce wear on the casing and drillpipe.

Control of frictional properties of drilling fluids is extremely important in directional boreholes in order to decrease the drag. Friction-reducing additives are usually added to drilling fluids, in addition to keeping mud weight and viscosity under control continuously.

The specifications and limitations which in the past used to rigidly restrict directional drilling are more relaxed at the present time and permit faster, more accurate, and more economical operations. Larger bottomhole targets permit selection of the best relation between the surface location and the target.

Higher angles (up to 45°) result in a greater stability of direction; even greater angles (70°–80°) are becoming quite common. Dogleg limits are kept within realistic limits, which eliminates unnecessary equipment wear and damage, resulting in monetary savings.

The effects of drillstring magnetism and those of adjacent boreholes on survey instruments should be taken into consideration. Drillstring usually becomes magnetized in a borehole. In addition, there is residual magnetism in the casing of previously drilled holes.

In order to combat the effect of magnetic forces that tend to deflect a compass, an adequate length of non-magnetic elements can be used to compensate for the known drillstring magnetism and other unknown effects. Empirical charts are available to determine the correct length of the non-magnetic sections required for surveying the hole. In addition, compass can be positioned in a non-magnetic collar. Obviously, geographic location should be taken into consideration.

TYPICAL APPLICATIONS OF CONTROLLED DIRECTIONAL DRILLING

The technique of application of controlled directional drilling was first applied to the drilling of relief wells in order to pump mud and water for the

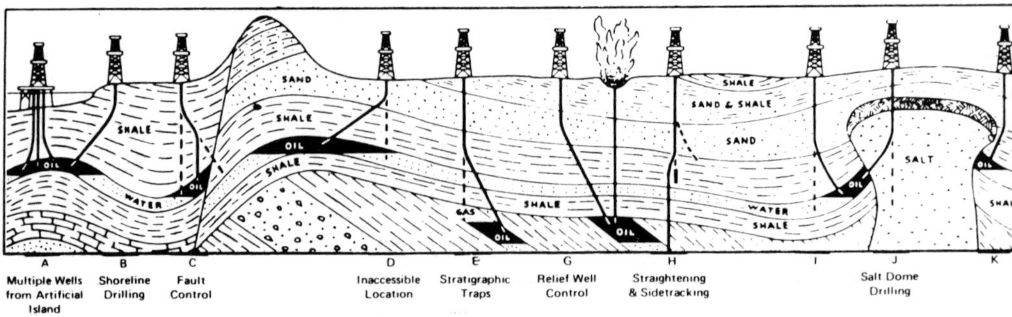

Fig. 18-2. Objectives of directional drilling. (Courtesy of Kuster Company.)

purpose of "killing" a wild or cratered well. Most common application of directional drilling technique is in offshore drilling, where an optimum number of wells can be drilled from a single platform or artificial island. A governing factor in the feasibility of the offshore industry is economics.

Another very important application of directional drilling is in fault control, where the borehole is deflected across or parallel to the fault for more efficient production. Drilling a vertical well through a steeply inclined fault plane could result in slip and shear of the casing.

Directional drilling programs are also used to overcome the problems encountered during drilling in salt dome areas. In order to reach the producing formations, which often lie underneath the overhanging cap of the dome, the well is first drilled at one side of the dome and is then deviated around and under the overhanging cap.

The objectives of directional drilling can be summarized as follows (see Fig. 18-2):

(1) To return crooked holes to a vertical course.

(2) To sidetrack obstructions such as lost tools, etc.

(3) To drill wells under surface locations where it is impossible or undesirable to place a drilling rig.

(4) To deflect in the up-structure direction.

(5) To redrill old wells for new production.

(6) To correct bottomhole spacing and obtain proper drainage areas.

(7) To kill blowouts.

(8) To drill multiple wells from a single location.

(9) Exploratory drilling.

BASIC HOLE PATTERNS

The first step in controlled directional drilling is the selection of the basic direction pattern and the calculation of the angle and direction. Three types of patterns (Fig. 18-3) are generally used and the selection is dependent

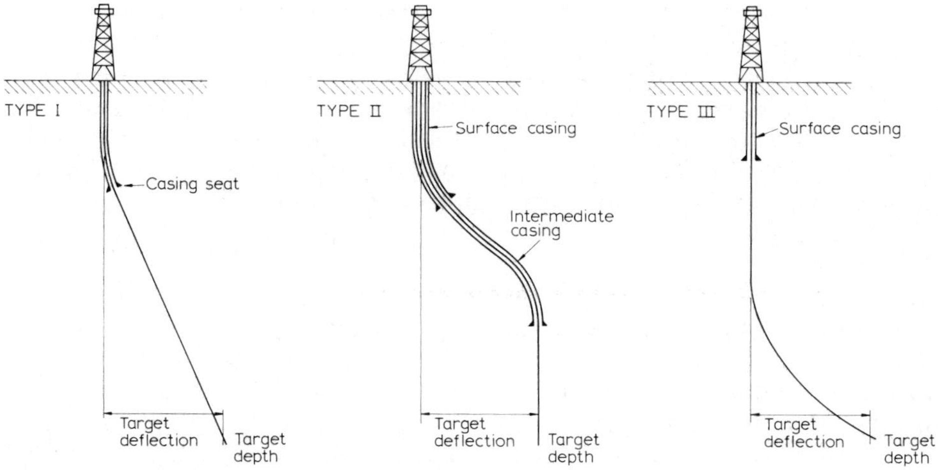

Fig. 18-3. Three basic types of hole patterns. (Courtesy of Kuster Company.)

upon the types of geologic structures, drilling mud, casing program, and spacing.

In the Type I pattern, the initial deflection angle is obtained at a relatively shallow depth. From the initial deflection, the angle is maintained (locked-in) to the target. Surface casing is set through the deviated section and cemented. Type I patterns are generally used in moderate-depth drilling and in drilling deep wells requiring a large lateral displacement.

In Type II pattern, which is also called S-curve pattern, the borehole is also deflected at a relatively shallow depth. Surface casing is set and cemented through the deviated section, and drilling is continued until the lateral displacement is reached. The hole angle is then reduced and/or returned to vertical in order to reach the target. An intermediate string of casing is set through the second vertical hole, and drilling is continued in a vertical direction until total depth is reached. Type II pattern is used where the intermediate casing is needed to control possible formation problems. It also permits more accurate bottomhole spacing in multiple pay zones.

The deflection in Type III pattern is started well below the surface. The angles are high, the lateral distance is small, and the deflected part of the borehole is seldom cased. This pattern is particularly suited to special situations such as (1) drilling through faults or salt-dome areas, and (2) redrilling or repositioning of the bottom part of the borehole.

DEFLECTION TOOLS

Inasmuch as correct initial deflection and direction are essential to a successful directional drilling operation, a prime requirement is the selection of proper deflection tool. Proper selection depends primarily on the type of formations encountered at the kickoff point depth.

Deflection tools originally used were the whipstock and knuckle-joint. The whipstock, which was initially developed in the mining industry, was later adapted to the oilfield use. A further refinement was the knuckle-joint tool for deflection. At the present time, however, they are in smaller demand than the simpler downhole motors and jet bits.

Whipstocks

Standard removable whipstock. The standard removable whipstock (Fig. 18-4A) is used to initiate the deflection and direction of the well, to sidetrack cement plugs, or straighten crooked holes. It consists of a long inverted steel wedge that is concave on one side to hold and guide a whipstock drilling assembly. It is also provided with (1) a chisel point at the bottom to prevent the tool from turning, and (2) a heavy collar at the top to withdraw the tool from the hole.

Circulating whipstock. The circulating whipstock (Fig. 18-4B) is run, set, and drilled like the standard whipstock. The drilling fluids, however, flow through a passage to the bottom of the whipstock and circulate the cuttings out of the hole, insuring a clean seat for the tool. It is most efficient for washing out bridges and bottomhole fills.

Both whipstocks are used with a drilling assembly which consists of a proper-sized whipstock drill bit, a spiral stabilizer, and an orienting sub rigidly attached to the whipstock by means of a shear pin.

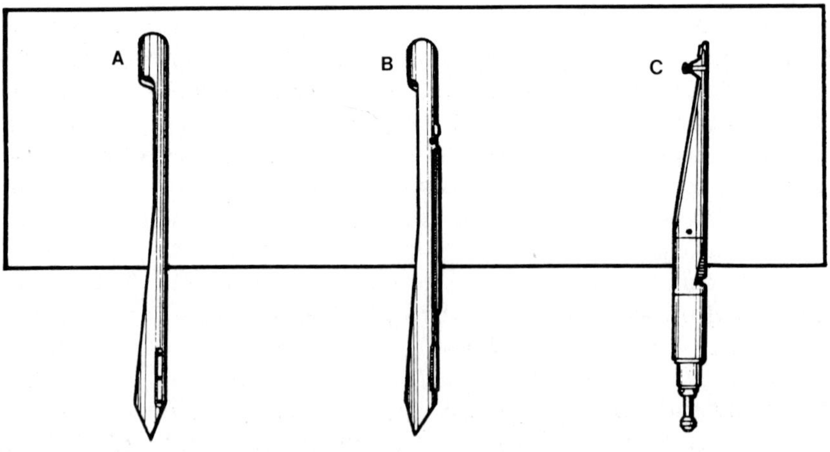

Fig. 18-4. Different types of whipstocks. A. Standard removable whipstock. B. Circulating whipstock. C. Permanent casing whipstock. (Courtesy of Eastman Whipstock, Inc.)

When the assembly is lowered and oriented to the bottom of the hole, proper weight is applied to set the whipstock and shear the pin. The bit is then drilled down and an initial hole, referred to as "rat hole", is drilled, usually 12—16 ft below the toe of the whipstock.

This hole, drilled with a bit having smaller O.D. is opened up to the regular O.D. hole with a pilot bit or hole opener, and a survey is made to record the amount of deflection. If additional follow up is required, a new bottom assembly consisting of a full gauge bit and stabilizer, a small non-magnetic pup joint, and a standard drillpipe is run. With this type of assembly, called a "limber hook up", surveys must be run at more frequent intervals than normal.

Permanent casing whipstock. The permanent casing whipstock (Fig. 18-4C) is designed to remain permanently in the well. It is mainly used to bypass collapsed casing or junk in the hole, or to re-enter and drill out old wells. By means of a setting trigger, it is set to an assembly consisting of a starting mill, an orienting sub, and the standard drill string. When the assembly is properly oriented in the cased wellbore, the shear pin is parted and permanently secures the whipstock in the well. The starting mill is then slowly rotated and guided by a sacrificial lug directly into the casing sidewall.

After a section of the casing has been milled away, a new assembly, called the "speed mill" assembly, is lowered to establish the wellbore. It consists of a full-gauge diamond or carbide flat-bottom mill, a carbide stabilizer and the standard drillstring assembly. About 4 ft of open hole is drilled before the standard drilling assembly can be run.

Knuckle joint

The knuckle joint is a special type of drill support, designed to deflect a well without using a whipstock. It includes a spring-loaded ball type universal joint, connected to the drillpipe, which allows the bit to drill at an angle to the axis of the drillstring. Its applications are limited, however, due to unpredictable characteristics in design which prevent proper orientation. The use of knuckle joint is, therefore, limited to non-directional and blind side-tracking.

Jet bits

The jet bit (Fig. 18-5) is a fast and economical deflection tool which is used to deviate wellbores in soft formations. It is a modified two- or three-cone bit, with three jet openings (two small and one large).

The bit is made up with a jetting assembly consisting of an extension sub, a full-gauge stabilizer, a non-magnetic drillcollar, two standard collars, another stabilizer, and the normal drillstring.

When the assembly is lowered and oriented, the large nozzle is oriented in

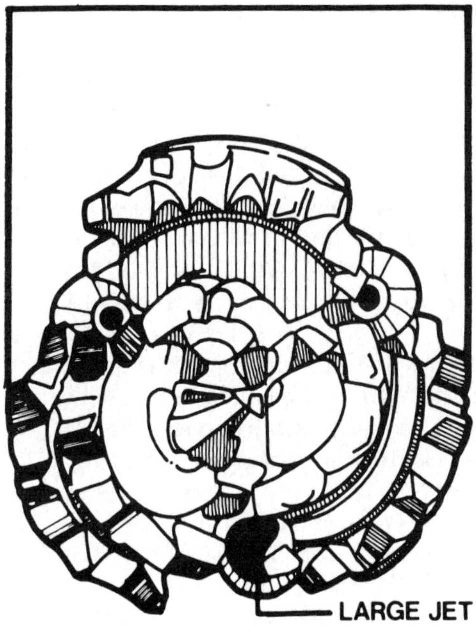

Fig. 18-5. Jet bit.

the proper direction and all but one of the jet openings are closed off, or substantially reduced in size. Maximum circulation is then established and the washing action is begun. The drillstring is set on bottom and spudded periodically, washing out a large pocket opposite the larger jet bit eye.

This jetting and spudding continues at necessary intervals until the initial curvature is established. When the desired length of hole has been drilled, normal drilling procedures are resumed to further increase the curvature. The above procedures can be repeated as many times as necessary to establish the desired borehole angle and deflection (without pulling out the jetting assembly, because surveys are taken at regular intervals to check the course of the borehole).

Jet deflection, therefore, is most economical, because it permits fast drilling rate without unnecessary trips to change assemblies and condition the hole. It also permits drilling of a full-gauge borehole with gentle changes in direction, when used in soft formations, and cuts down the number of rotating hours. In addition, the hydraulic jetting tool can also be used, which increases the depth at which deflections can be made. It is a downhole surge pump which does not require any special equipment, besides the available rig pumps and regular drilling hookup.

Downhole hydraulic motors

The downhole motor is one of the most widely used deflection tools at the present time. It is driven by drilling mud flowing down the drillstring to

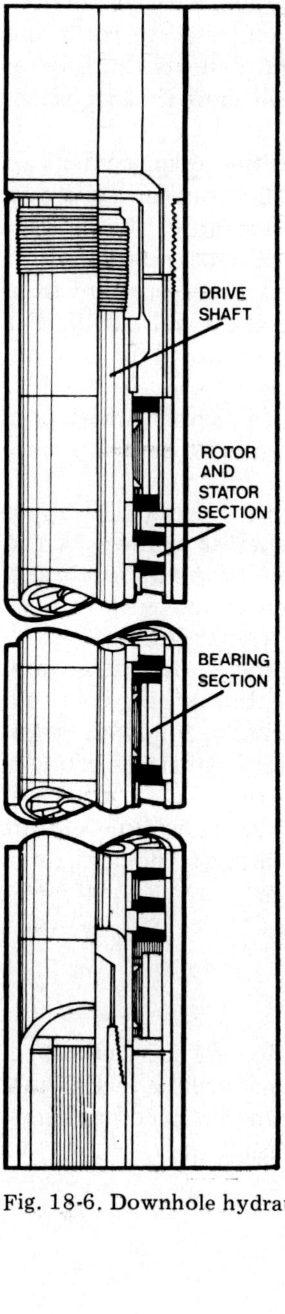

Fig. 18-6. Downhole hydraulic motor. (Courtesy of Eastman Whipstock, Inc.)

produce rotary power downhole, thus eliminating the need for rotating the drillpipe (see Fig. 18-6).

The first type of the downhole motor, referred to as the turbine-type motor or turbodrill, is a rugged multistage axial tool which has proven to be highly efficient and reliable, especially in areas of medium to hard drilling. It consists of multistage vane-type rotor and stator section, bearing section, drive shaft, and bit-rotating sub. The first stage is comprised of rotor and stator of identical profile. The stator is stationary and deflects the flow of drilling fluid to the rotor, which is locked to the drive shaft that transmits the rotary motion to turn the bit.

The second type of downhole motor is the positive displacement or helicoid hydraulic motor. It consists of a two-stage helicoid motor, dump valve, connecting rod assembly, and bearing and shaft assembly. The helicoid motor has a rubber-lined spiral cavity with an elliptical cross-section which houses a sinusoidal steel rotor. As the mud is pumped under pressure from above, it is forced downwards between the rotor and the spiral cavity. The rotor is thus displaced and turned by the pressure of the fluid column, powering the drive shaft the rotational force of which is used to turn the bit.

Both types of downhole motors can be used with an assembly, which consists of full-gauge bit, downhole motor, bent sub or hydraulic bent sub, non-magnetic drillcollar, and normal drillstring.

The bent sub is used to impart a constant deflection to the tool. Its upper thread is cut concentric to the axis of the sub body, whereas its lower thread is cut with an axis inclined 1—3° in relation to the axis of the upper thread. A hydraulic bent sub, which was developed later, can be locked into position for straight drilling or unlocked and reset for directional drilling.

Downhole motors, therefore, present many advantages over the whipstock. When jetting becomes impractical, they permit the drilling of a full-gauge hole at the kickoff point, thus eliminating costly follow-up trips. Orientation is also more accurate because the motors penetrate along a smooth gradual curve in buildup and drop-off portions. Corrections, if needed, can be made downhole without making a trip. Downhole motors also eliminate the need for clean-up trips due to bridges, doglegs, etc., because the tool can be circulated and drilled to bottom.

DRILLING THE DEVIATED SECTION OF THE HOLE

Upon establishing the initial deflection and direction of the well, directional control is accomplished mostly with conventional drilling techniques. Three basic principles are used to drop the hole angle when it becomes necessary, i.e., fulcrum, stabilization, and pendulum principles.

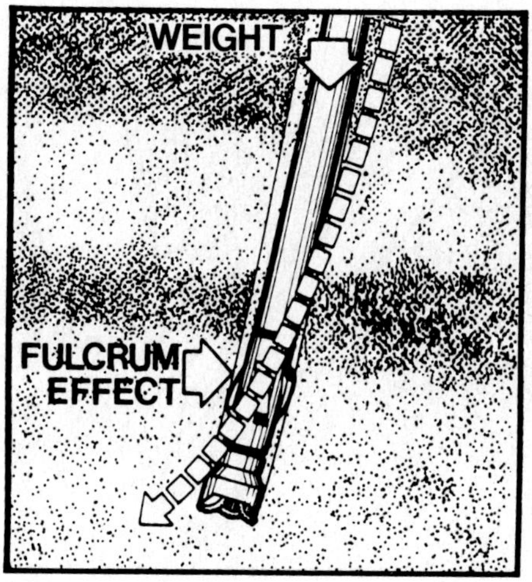

Fig. 18-7. Schematic diagram illustrating the fulcrum principle. (Courtesy of Eastman Whipstock, Inc.)

Fulcrum principle

Further increase in hole curvature to a predetermined drift angle magnitude is accomplished by the application of the fulcrum principle. A reamer or stabilizer (fulcrum) is inserted into the drillstring just above the bit. When proper weight is applied, this stabilizer has a crowbar effect on the string and causes the hole to pick up angle. A kick sub or extension can also be added above the stabilizer for extra leverage at the point of fulcrum (Fig. 18-7).

The more limber the portion of the assembly just above the fulcrum, the faster the increase of angle. Selection of a particular assembly is, therefore, determined by the buildup rate required for specific operations.

The standard buildup assembly, which is relatively limber, is used for normal buildup rates and is retrieved once the average angle is reached.

The "Gilligan angle build" and the "short hookup" assemblies are used during rapid buildup rates. Due to their extremely limber characteristics, however, they require careful monitoring by close interval surveying.

Another assembly used to buildup angle is the downhole motor assembly, which is similar to the downhole motor deflection assembly.

Stabilization principle

When a correct hole angle has been built, the angle must be maintained to total depth or until the hole is returned to vertical. This stabilization requires stiff bottomhole assemblies and strict attention to the rotary speed and

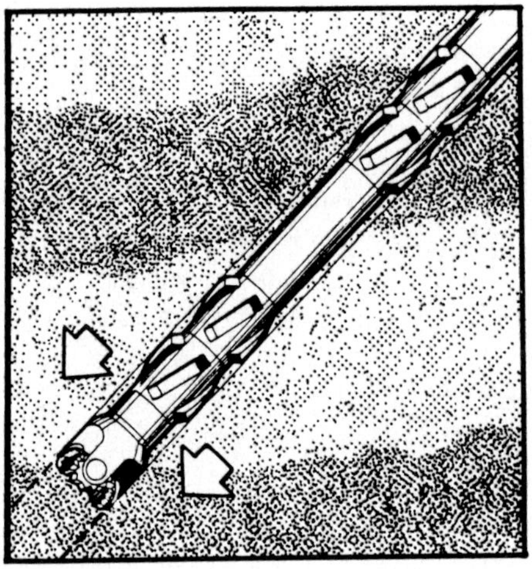

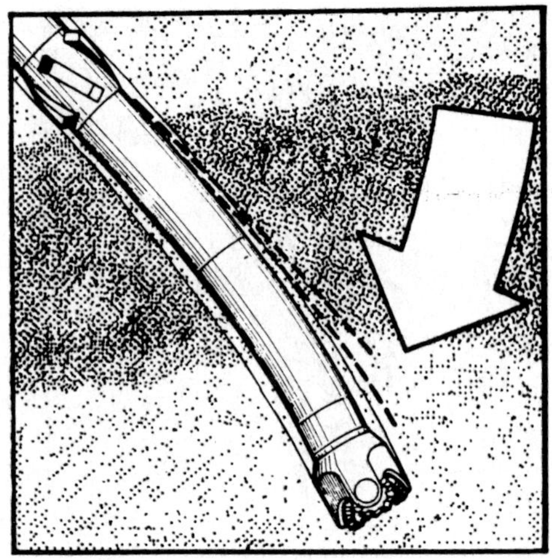

Fig. 18-8. Schematic diagram illustrating the stabilization principle. (Courtesy of Eastman Whipstock, Inc.)

Fig. 18-9. Schematic diagram illustrating the pendulum principle. (Courtesy of Eastman Whipstock, Inc.)

weight applied to the bit. A common stiff-hole assembly is the standard packed assembly, which includes one reamer or stabilizer just above the bit and another reamer above a large bottom collar (Fig. 18-8). This assembly is of rigid design. The stiff collars that are used are small enough to be fished from the hole, yet large enough to prevent the bit from wandering either to the right or left of the intended course.

The maximum packed assembly is even more rigid in design and is well suited for use in areas of extreme lateral drift. It includes two full-gauge stabilizers or a square collar-type stabilizer just above the bit, large bottom collar, and two more stabilizers above the collar. The stiffness of the stabilizers near the bit prevents it from drifting off course.

The packed undergauge near-bit assembly is used to maintain direction and drop the angle very gradually in the average-angle portion of the well. It is composed of an undergauge near-bit stabilizer just above the bit, full-gauge stabilizer, short square collar, and two full-gauge stabilizers above it. It is the stiffness of the short collar which causes the assembly to resist going around a curve, forcing the bit to drill in an inclined but straight line.

Pendulum principle

The pendulum principle (Fig. 18-9) is another basic principle which is applied to drop the hole angle, when it becomes necessary, as in a Type II pattern. The pendulum effect is produced by removing the stabilizer placed just above the bit while retaining the upper stabilizer. As a result, while the remaining stabilizer holds the bottom collar away from the low side of the wall, gravity acts upon the bit and bottom collar and tends to pull them back to vertical. Inasmuch as the bit is pressed against the low side of the hole by the weight of the bottom collar and can drill sidewise as well as forward, the hole angle decreases as the bit drills ahead and is, thus, straightened.

A common drilling assembly used to drop the angle is the gradual angle drop assembly, which is designed to maintain hole direction while gradually dropping the drift angle. Another assembly, the standard angle drop, is designed to drop the angle at a standard rate of $1\frac{1}{2}°$ per 100 ft. Finally, the Gilligan angle drop assembly is used in order to achieve a maximum angle drop.

PREVENTION AND CORRECTION OF PROBLEMS ENCOUNTERED IN DIRECTIONAL DRILLING

Lateral drift

Specially-designed bits have certain corrective properties for limiting unwanted bit deviation tendencies. These bits, referred to as true rolling bits, have a zero cone offset — as opposed to normal drill bits — which allow the bit to rotate on a true axis and drill in a fairly straightforward fashion by decreasing wobble (Fig. 18-10).

Another tool ("Rebel Tool") was specifically designed to limit lateral drift because it counteracts, without orientation, the bit tendency to deviate ("walk") from the intended course. It looks like a short drillcollar with two curved paddles attached to a free-moving shaft (Fig. 18-11), which is recessed along the side. As the top paddle rotates to the low side of the hole, it is forced by weight into the recess while the bottom paddle extends and pushes against the sidewall in the desired direction. The right or left direction, which alters the well course and maintains angle, is determined by the length of the paddle.

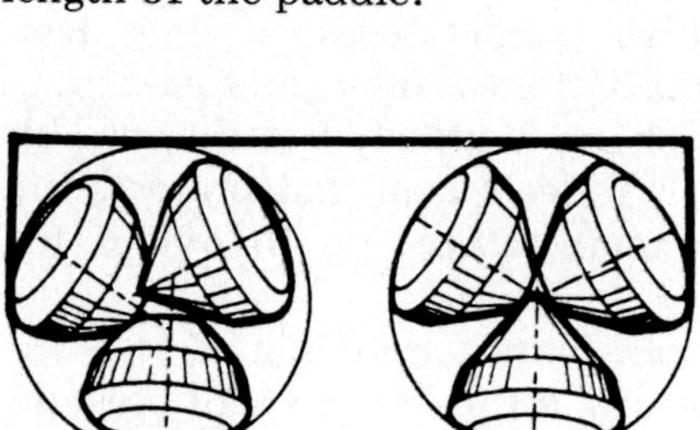

Fig. 18-10. True rolling bits.

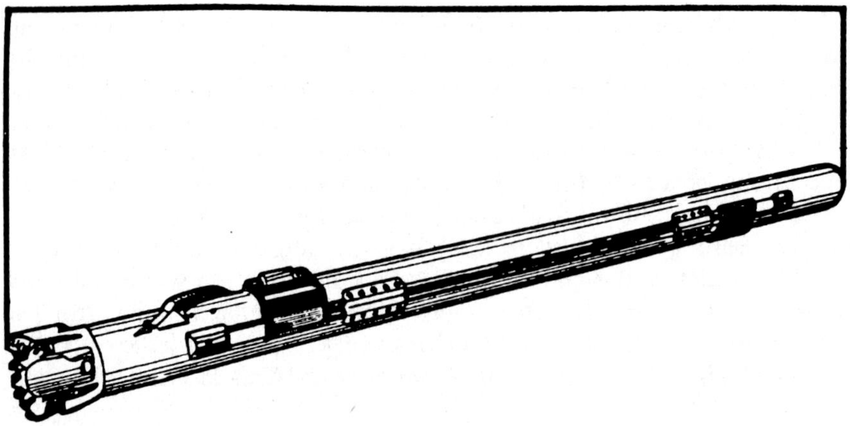

Fig. 18-11. The "Rebel Tool". (Courtesy of Eastman Whipstock, Inc.)

Dogleg severity

All directional wells are by definition doglegged, inasmuch as doglegging can be defined as the total curvature of the wellbore. Doglegging can also mean that the total curvature of the well is affected by a change in direction and in inclination. This type of dogleg should be detected early, by means of surveys at regular intervals, because severe doglegs can be responsible for many problems, with large monetary losses. These doglegs do not cause immediate problems, because the drillcollars are under compression and can easily accommodate themselves to the variations in the hole course. The detrimental effects, such as damage to the drillstring, key seats, or worn casing, appear much later. Doglegs, therefore, must be detected and corrected as early as possible. Slight doglegs must be reamed and surveyed and, if they cannot be eliminated, sidetracked.

There is, however, a "threshold" value under which limited damage is sustained by the drillstring or casing. It is termed the "permissible dogleg", which is sensitive to drilling practices and is determined by the dimensions of the tubular element and its tension at the depth of the doglegged interval.

When combined with high humidity, high temperatures in some deep boreholes may have some adverse effects on batteries, films, and electronic components of survey instruments. This problem is usually best resolved by using a high-temperature heat shield, which reflects heat and protects the instruments. These shields can withstand temperatures up to 500°F for 4 hours.

Inasmuch as most of the survey instrument barrels are designed to withstand a specific amount of hydrostatic pressure, a careful check of pressure limits minimizes the risk of equipment damage caused by high hydrostatic pressure in deep boreholes.

DIRECTIONAL SURVEYING

Directional surveying permits the determination of (1) the bottomhole location relative to the surface location at a given vertical depth, (2) location of possible doglegs or excessive hole curvatures, (3) the monitoring of the course direction and drift during the drilling process, and (4) orientation of the deflecting tools.

Various survey instruments and methods are available which provide a fast, accurate, and economical way of monitoring the location of wellbore. The drift and course direction of the wellbore at specific depths is determined by one type of survey called the single shot survey, whereas multiple shot surveys are used to record several individual readings at required depth intervals. Surveys can be either magnetic requiring a non-magnetic collar or others which do not require non-magnetic collars.

Magnetic surveys requiring a non-magnetic collar

Magnetic instruments must be run inside a non-magnetic drillcollar made of a special nickel alloy in order to position or space the compass section away from the magnetic influence of the drillstring. Non-magnetic drill-collars were discussed by Sperry-Sun Co. (1978, p. 6003), including guide for their selection. Magnetic surveys include (1) magnetic single shot, (2) magnetic drop-type survey, and (3) open hole survey.

Magnetic single shot

The first type of magnetic survey is the magnetic single shot, which is used to record simultaneously the magnetic direction of the course of an uncased wellbore and its inclination from the vertical. The instrument consists of three basic units: (1) timing device or a motion sensor unit, (2) camera section, and (3) angle-indicating unit. The timing device is used to operate the camera at a predetermined time. Inasmuch as it is quite difficult to accurately predict the time element involved in lowering the instrument, however, a new timing element, referred to as a motion sensor, was developed to eliminate needless downtime. It is an electronic device which operates an electrical light system in the camera, just seconds after motion ceases. The camera itself is pre-focused and loaded with a film disc made of a special heat-resistant material. The angle-indicating unit combines a magnetic compass and a plumb bob, which gives the angle of inclination from vertical.

In actual operation, the timing device is first set to a predetermined time and the instrument is assembled in a protective barrel and run on a wireline or dropped into the well on a dull bit. When it reaches bottom, the timing device operates the camera. A photographic record is, thus, made of the relative positions of the compass and plumb bob. The instrument is then returned to the surface where the film is developed and placed in a reading

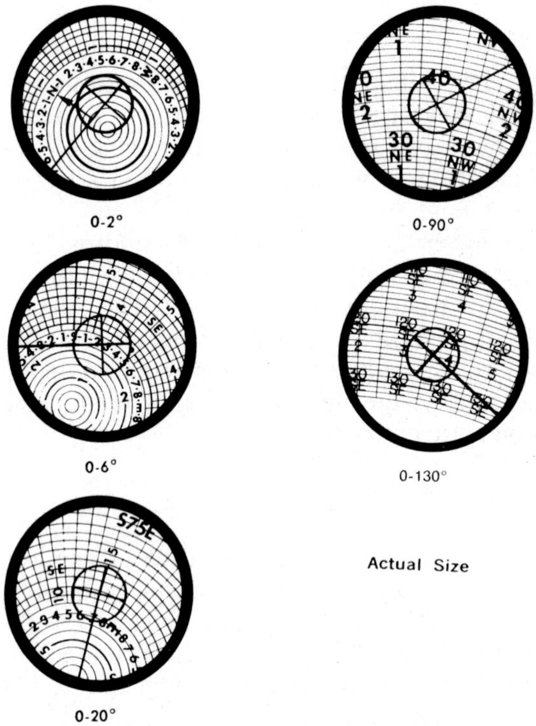

Fig. 18-12. Examples of Sperry-Sun Company film discs. (Courtesy of Sperry-Sun Co., Houston, Texas.)

device ("reader"), which shows the exact direction and inclination angle of the borehole (see Fig. 18-12).

An example of single-shot assembly is presented in Fig. 18-13.

Magnetic drop-type survey

The drop survey appears to be one of the most economical methods developed to determine the drift and course direction of an uncased wellbore. It is also run after the well has been drilled, and utilizes a 10-mm film strip to take several recordings at predetermined depths. The mechanically-operated photographic instrument is equipped with a timing device, synchronized with a surface watch. After the instrument is assembled in its protective barrel, it is dropped through the drillstring to land on the baffle plate, a round metal stand which rests on the pin end of the sub or bit. As the drillstring is retrieved, the downhole data is recorded, which is transformed into survey information when the film strip is developed and read.

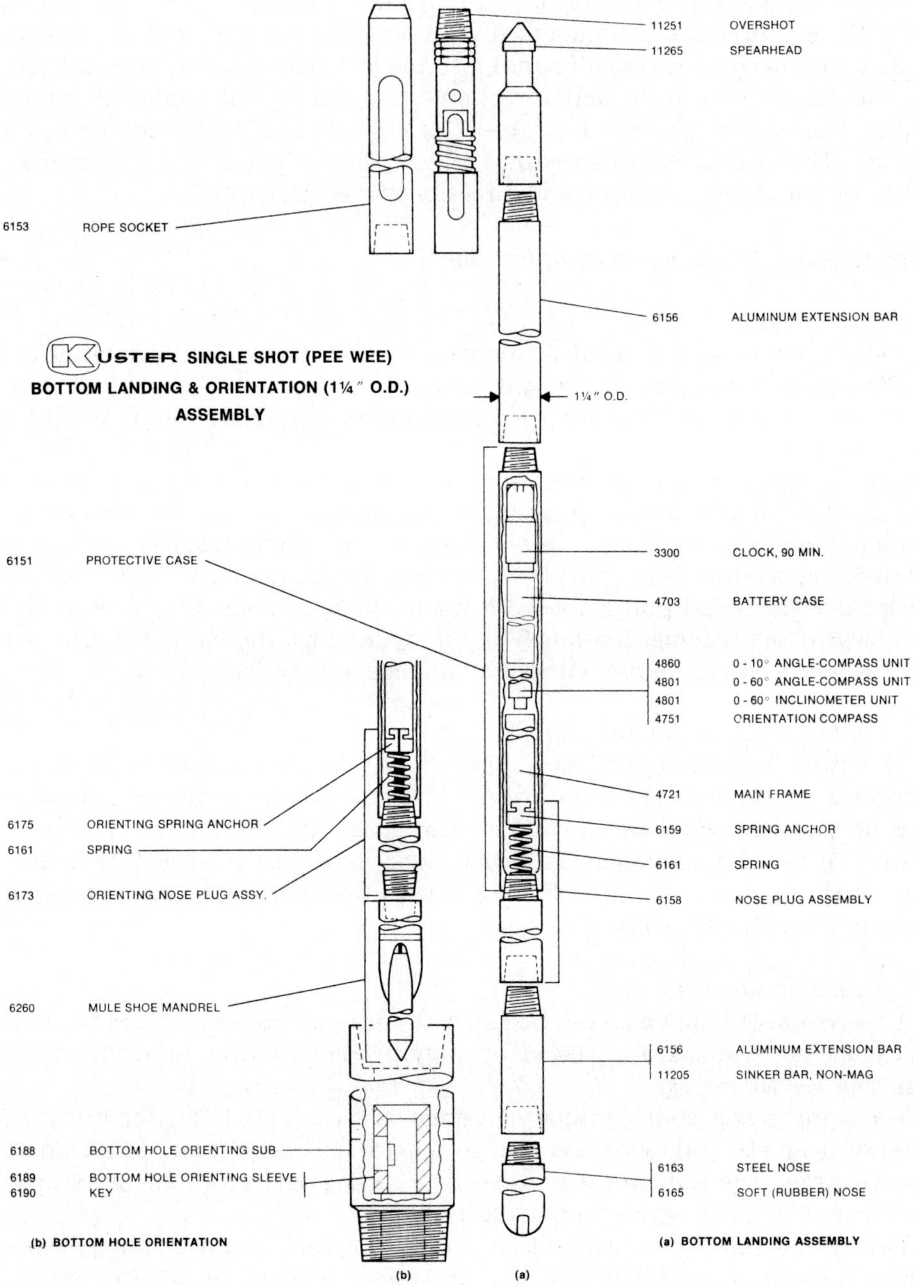

Fig. 18-13. Single-shot assembly. (Courtesy of Kuster Company.)

Open hole survey

The open-hole multiple shot type of magnetic survey is used to determine the drift and direction of an uncased wellbore after the well has been drilled. A battery-powered instrument is run into the well and operates mechanically to photograph the angle and direction of a compass at a predetermined depth. Like the single shot instrument, it consists of a timing device and a camera. The camera is loaded with a 16-mm film strip instead of a film disc, to record the changes in angle and direction at various depths.

Surveys run without non-magnetic collars

Drift indicator

The drift indicator is a mechanically-operated device used primarily for obtaining drift readings in straight holes. It is also used to compute the total displacement or deviation of a wellbore within a given radius at a total vertical depth.

The instrument is protected from shock and pressure by a snap-lock barrel. The unit, which is present at the surface, is run on wireline or dropped on a dull bit. After a given elapsed time, the instrument punches a printed paper disc once, which then rotates automatically 180°. About 45 sec later a second punch record is made allowing a positive check on the accuracy of the reading. Inasmuch as this type of instrument is not affected by magnetism, it can be run without a non-magnetic collar.

Drop multishot inclination survey

The drop multishot inclination instrument is used to record multiple surveys after the well has been drilled. Inasmuch as the inclination capabilities of the angle unit are not affected by magnetic interference, it can be spaced inside of steel collars in either cased or uncased holes. The instrument is simply dropped through the string to the baffle plate, and surveying is accomplished as the drillstring is retrieved from the well.

Gyroscopic surveys

The gyro instrument can also be run without non-magnetic collars because the magnetic compass is replaced by a gyro compass disc controlled by a high-rpm electric rotor.

The gyro single shot's primary purpose is to orient deflection tools in areas of magnetic influence and also to determine the drift and direction of the wellbore. The instrument consists of a timing device, a camera section, and the very sensitive gyro compass section.

The gyro unit is first oriented to a known direction, and the timing device is set. The instrument is encased in a protective outer barrel and lowered to the survey station where the survey is recorded. The instrument is then retrieved from the well and the film is unloaded and developed.

The gyro multiple shot instrument is used to record the drift and direction of a cased or uncased wellbore hole on a 10-mm film. The battery-powered instrument and camera section have the same characteristics as the standard drop multishot instrument. The gyro unit is again oriented to known direction, and the timing device is synchronized with a surface watch. After the instrument is assembled, it is run on wireline through the wellbore.

Inasmuch as gyro systems are very sensitive, they should be handled with extreme care. Their compasses should also be checked for drift when the assembly is retrieved.

To insure complete accuracy of the downhole data, all survey instruments must be aligned parallel to the axis of the wellbore.

A very popular method, the centralized method, is accomplished by equipping the instrument barrel with rubber fingers or, for gyro surveys, with metal belly springs called centralizers. As the instrument is lowered to and centered on the baffle plate, its upper section is centered by the centralizers which align the instrument to the wellbore axis.

BOTTOMHOLE ORIENTATION METHODS

Correct deflection and direction of the hole can only be accomplished by accurate orientation of the deflection tools. This bottomhole orientation requires a great deal of engineering data to precisely determine in what direction the tool should be facing, and to face it in that direction when it reaches bottom. Calculation of the direction in which the tool should be faced is most readily solved by graphical methods. It can be obtained by drawing a simple vector diagram or by using a chartered vector analysis called ouija board.

Several methods of bottomhole orientation, which are described below, are available to actually face the deflection tool in the desired direction.

Surface method of orientation

Surface method of orientation involves orienting the tool entirely from the surface, by aligning a derrick telescope with a sighting bar and repeating the process until the tool is set on bottom. This method, however, requires a great amount of work and may yield unreliable readings.

Direct method of orientation

The direct method of orientation provides a faster, more accurate, and more economical way of orienting deflecting tools and jet bits. Using this technique, the directional data is taken simultaneously with the other data. Two types of this method are (1) muleshoe and (2) magnetic.

The muleshoe method requires the use of a non-magnetic drillcollar or pup joint and special orienting equipment used with a single-shot instrument. The orienting sub or muleshoe sub contains a slot or key, lined up directly with the face of the tool during assembly. A non-magnetic collar is then made up in the orienting sub before the assembly is lowered into the hole. Next, a single shot instrument is run with the orientation bull plug and muleshoe. When it reaches bottom, it is set gently down on the muleshoe key where it is aligned with the tool face. The only face indicators which are not facing in the same direction are the pendulum and compass charts. The reference line of the deflecting tool is recorded in direct relation to the magnetic direction of the borehole, and the developed film shows the relationship between the tool face, drift, and magnetic north.

If the tool is not facing properly, only minor calculations are necessary to determine how much the drillstring must be rotated. Several positioning operations may be needed due to the friction of the pipe against the side of the hole. These operations must all be checked by surveys.

The applications of muleshoe method are limited when a large amount of jetting is required, because the orienting sub does not allow full circulation.

The magnetic method of orientation (direct method) was designed for drilling in areas where a large amount of jetting is required. It eliminates the orienting sub so that full circulation through the drillpipe and collars is available. A non-magnetic collar equipped with orienting magnets and a special orienting instrument are used. The non-magnetic collar is equipped with six small magnets, which are either in physical alignment with the tool face or with mathematical compensation. The orienting instrument combines a swinging pendulum, a regular compass, and a needle-type compass, which is locked in place by the magnets in the collar.

The assembly is first spaced out or matched to the non-magnetic collar so that the magnetic position indicator lands opposite the orienting magnets. After the drilling assembly is made up and run to bottom, the instrument assembly is run into the drillpipe and the data is recorded. The developed disc shows the image of the needle-type compass superimposed on the regular compass, indicating the relative positon of the magnets and the direction of the hole. A special reading equipment (reader) is used to calculate the direction in which the tool is facing. The drillstring is then rotated the required number of degrees and a check survey is run.

After the magnetic limitations of the pup joint is reached, the various instruments used in the direct method can also be utilized without modification in the indirect method of bottomhole orientation, discussed below.

Indirect method of orientation

The indirect method of orientation is based on the direction of the low side of the hole and requires prior knowledge of the hole direction. The wellbore must also have a minimum inclination of 3° for the orienting equipment to operate satisfactorily. This method involves utilization of (1) mechanical-orienting tools and (2) survey-orienting instruments.

The first type of the indirect method is based on the principle that a free-rolling ball, confined to a circular race within the tool, will seek the low side of the hole. A key, located in the stem of the assembly, is first set to a position representing the low side of the hole, and the orienting stem is locked in place. The orienting assembly is then made-up and run into the hole, where, after reaching the bottom, it may have to be raised and turned a few degrees to align the ball with the slot on the orienting stem. Only then does the tool close, which is indicated at the surface by a recognizable pressure drop.

This orienting tool, which can be used with any deviation tool, is especially useful when circulation is required during orientation.

In the second type of indirect method, i.e., utilization of survey-orienting instruments (with special magnetic sub, BHO), the south poles of the magnets are aligned with the scribe line on the outer surface of the lower sub. The angle between the scribe line and deflection tool is measured with a protractor affixed on the sub which is used to determine the relationship between the scribe line and the face of the deflecting tool. Once the sub is made-up, the assembly is lowered to the bottom. An orienting instrument equipped with a timing device, a special compass or a swiveling magnetic disc, and a plumb bob unit is run on wireline through the drillstring until it seats on the sub. The north pole of the compass is magnetically locked in place in the magnetic field created by the south poles of the two magnets in the sub, thus indicating the relative position of the tool face. As the plumb bob unit seeks the low side of the hole, the reference line of the tool is recorded in direct relation to the angle indicator, i.e., the low side of the hole.

A special chart reader is used to calculate the direction in which the tool is facing. Inasmuch as this tool contains magnets which can be positioned exactly like the magnets in the sub, a simple adjustment of the dials will show the exact relationship between the low side of the hole and the face of the tool. Several attempts may be needed to rotate the tool the same number of degrees as the pipe was turned.

Gyrosopic single-shot method of orientation

Gyroscopic single-shot bottomhole orientation method utilizes a standard single-shot instrument with an adapter to a gyro unit. The instrument is, therefore, not affected by magnetic field or by the close proximity of iron in rig substructures.

The gyro unit is first oriented to a known axis, prior to lowering into the well, and then timed to a surface watch, assembled in a steel barrel and run on wireline. After the data is recorded, the instrument is retrieved, the disc is developed and read, and computations are made and plotted immediately.

This type of instrumentation eliminates the need to orient the drillpipe while going into the hole.

Directional orientation tool (DOT) method

Directional orientation tool was developed in order to properly maintain the orientation of downhole motors while drilling. This electronic device is a downhole sensing probe which continuously monitors the torque of the drill-string. It instantly feeds this data, through a surface data processor, to a remote readout at the driller's station. The sensing probe, the top of which is connected to the wireline, is made up above a spacer bar and standard mule-shoe assembly. The assembly is then lowered to the non-magnetic collar where the muleshoe locks itself into the bent sub. Electronic high side and magnetic signals are then continuously transmitted up the wireline, enabling the driller to be constantly aware of the tool facing-direction.

METHODS OF COMPUTATION

Before any computation is made of the hole total displacement and course direction at a given depth, a correction must be made from magnetic to true bearing, because readings on all magnetic survey instruments are made with respect to magnetic north. The degree of correction varies from area to area and can be obtained from isogonic charts. (See Fig. 18-14.) If directional

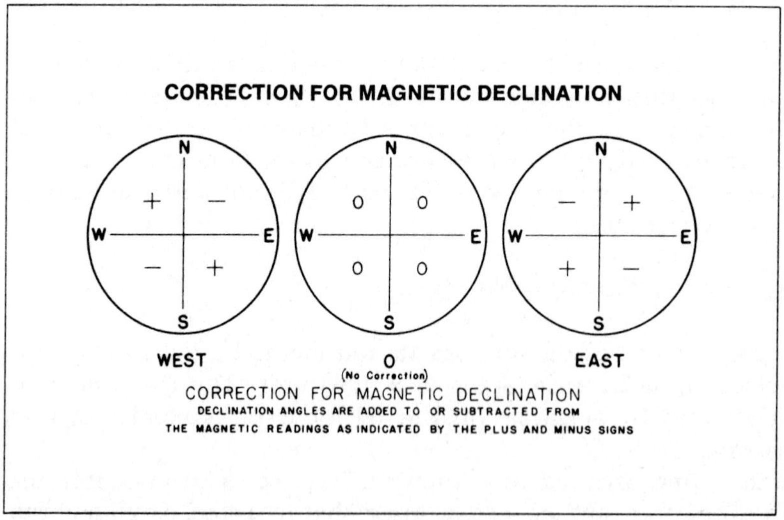

Fig. 18-14. Correction for magnetic declination. (Courtesy of Kuster Company.)

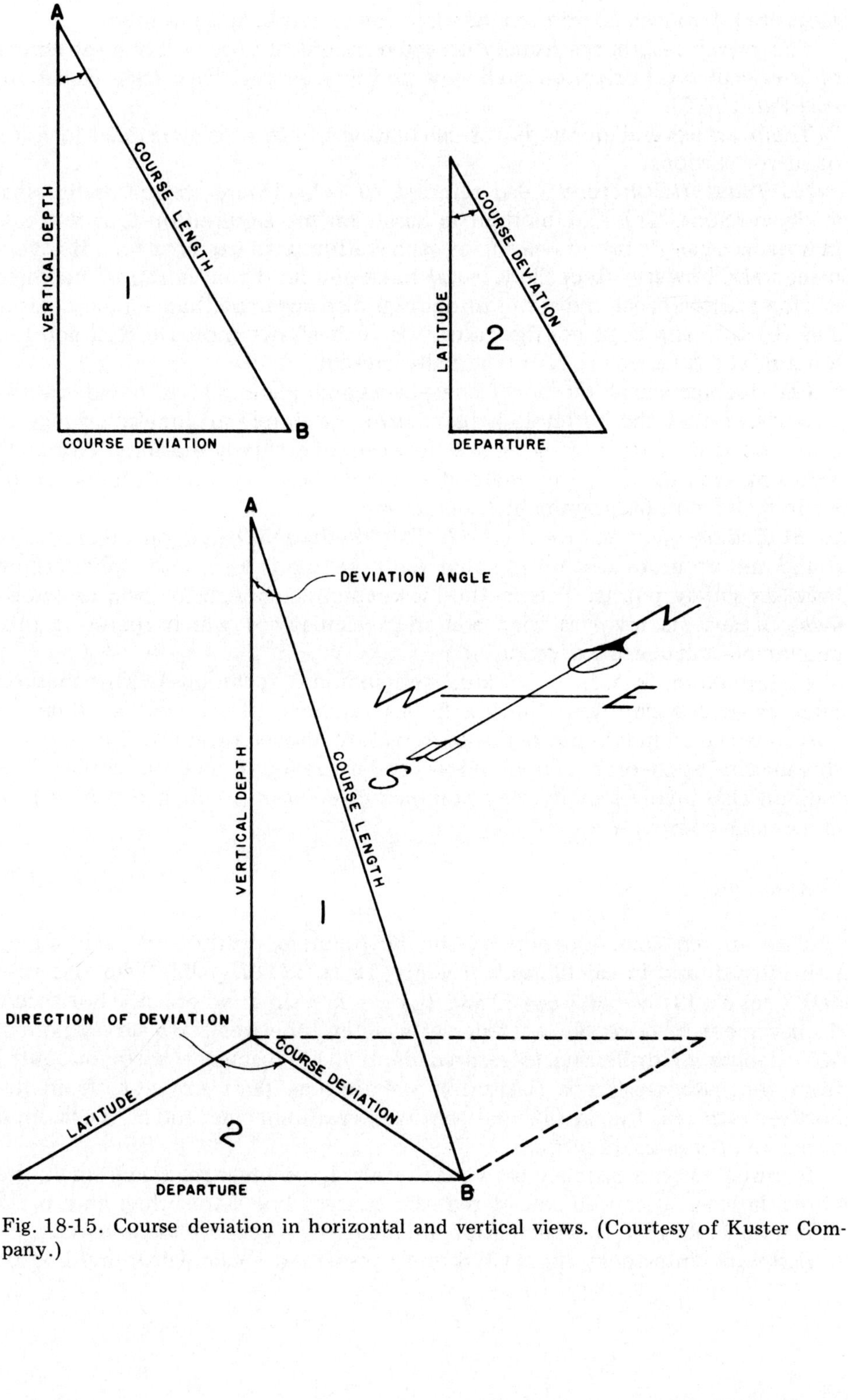

Fig. 18-15. Course deviation in horizontal and vertical views. (Courtesy of Kuster Company.)

plots are referenced to grid north, additional correction is required.

The survey results are usually depicted graphically by plotting the course of the well in a horizontal plan view and in a vertical view (side elevation) (see Fig. 18-15).

There are several methods for calculating the three-dimensional location of survey stations:

(1) *Tangential method* (also referred to as backward station or terminal angle method). This old method is based on the assumption that the hole maintains same drift and bearing of each station until the next one. It is very inaccurate, however, especially in (a) build-and-hold configuration, in which it shows less vertical and more horizontal displacement than actually exists; and (b) turn-and-hold configuration where it shows more vertical and less horizontal displacement than is actually present.

(2) *Average angle method.* Angle-averaging technique is based on the assumption that the borehole is parallel to the simple arithmetic average of both the drift and bearing angles between two survey stations. Although deficient theoretically, this method is simple because computations can be made with a non-programmable calculator.

(3) *Radius of curvature method.* This method is based on theoretically sound and accurate assumption that the borehole is a smooth, spherical arc between survey points. This method is not simple enough for field use, however, because it involves very complex calculations which require a programmable calculator or computer.

(4) *Minimum curvature method* (circular arc technique). This method involves assumption that the borehole is a spherical arc with a minimum curvature, i.e., a maximum radius of curvature between stations. Inasmuch as this method is theoretically sound, it can be applied to most boreholes. This method also involves many very complex calculations, which require a programmable computer.

Calculations

Three-dimensional location of the bottomhole position of each survey point is required in calculations (see Fig. 18-16): (1) $H = MD \times \sin\alpha$; (2) $z = MD \times \cos\alpha$; (3) $y = H \times \cos\beta$; and (4) $x = H \times \sin\beta$; where H = horizontal displacement of borehole; z = true vertical depth between two survey points; MD = length of drillstring (measured depth); y = distance (north or south) from the east—west axis (latitude); x = distance (east or west) from the north—south axis (departure); α = vertical deviation angle; and β = horizontal angle, i.e., compass direction.

Inasmuch as the distance between the survey points is relatively small, the borehole interval can be considered as a straight line rather than an arc. As shown in Table 18-I, the calculation are made for successive increments.

Eastman Whipstock, Inc. (1978, p. 5) presented a calculation method of

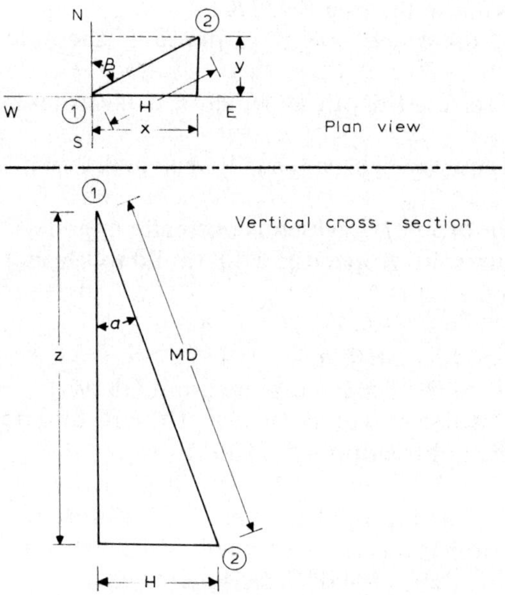

Fig. 18-16. Plan and vertical views of survey points *1* and *2*.

TABLE 18-I

Three-dimensional location of bottomhole position of each survey point

Survey point	Course length, *MD* (ft)	Vertical deviation angle, α	Vertical depth, z (ft)	Total vertical depth, *TVD* (ft)	Horizontal displacement, *H* (ft)
1	100	1°30′	99.97	99.97	2.62
2	100	1°50′	99.95	199.92	3.20
3	100	2°30′	99.90	299.82	4.36

Survey point	Compass direction,	Latitude, y (ft)		Departure, x (ft)		Total latitude, y (ft)		Total departure, x (ft)	
		N	S	E	W	N	S	E	W
1	N10°W	2.58			0.45	2.58			0.45
2	N40°W	2.45			2.06	5.03			2.51
3	N15°E	4.21		1.13		9.24			1.38

directional plans involving the following terms (see Fig. 18-1):

(1) *True vertical depth (TVD) of the objective* = total depth of the hole projected into a vertical plane.

(2) *Kickoff point (KOP) depth* = selected depth at which the deviation is started.

(3) *Deviation* = horizontal deviation, which is usually calculated with respect to the surface location.

(4) *Rate of inclination buildup (or drop-off)*, which is normally expressed in degrees per 100 ft drilled. Figures in Appendix 18.I (buildup charts) greatly simplify calculations.

Example of calculation of types I and III patterns

Given the following information, find the *TMD* (total measured depth): *KOP* (kickoff point) = 2000 ft; *TVD* (total or true vertical depth) = 10,000 ft; *Dev.* (horizontal deviation) = 2455 ft; and buildup = 2°/100 ft.

Solution:

Use standard buildup chart (see Appendix 18.I) designed to facilitate the computation of horizontal and vertical projections of wells which fall into Types I and III patterns (build-and-maintain type of directional wells).

Usable vertical depth = *KOP* — *TVD* = 10,000 — 2000 = 8000 ft.

Find intersection between usable depth (vertical scale) and the horizontal deviation (horizontal scale), i.e., 8000 ft and 2455 ft. This gives the maximum angle of deviation from the slant or degree line which is equal to 18°.

The chart inset for 18° shows that *MD* (measured depth) = 900 ft, *VD* = 885.26 ft, and *Dev.* = 141.97 ft.

Add the given *KOP* depth to the above measurements to find the actual depth at completion of buildup:

MD = 2000 + 900 = 2900 ft

TVD = 2000 + 885.3 = 2885.3 ft

Find the measurements below the buildup by solving the values of the remaining triangle:

Actual *VD* below buildup = given *TVD* — *VD* at buildup = 10,000 — 2885.3 = 7114.73 ft

$$\text{Total } MD = \frac{MD \text{ at buildup} + VD \text{ below buildup}}{\cos \text{max angle}} = 10{,}381 \text{ ft}$$

Example of calculations of type II pattern

The solution for Type II patterns varies from the previously described one, because it is necessary to calculate the inclination build and to compute the return back to vertical.

Find the total measured depth, given the following information: *KOP* = 2000 ft; *TVD* = 10,000 ft; *Dev.* = 2258 ft; buildup = 2°/100 ft; and drop-off = 1°30′/100 ft.

Solution:

The first part of calculation is the same as that described in the previous example: usable depth = 8000 ft; maximum angle at completion of buildup = 18°; *MD* at completion of buildup = 900 ft; and *VD* at completion of buildup = 885.26 ft.

From the inset of composite buildup chart for 1°30′ off to vertical, one can obtain the measured depth, vertical depth, and deviation which are necessary to bring the well back to vertical from 18° to 0°:
MD = 1200 ft; *VD* = 1180.38 ft; and *Dev.* = 188.26 ft.

The maximum angle is obtained by solving the following equation:

$$\frac{Dev. - (Dev.\text{ for buildup} + Dev.\text{ for drop-off})}{\text{Usable depth} - (VD\text{ for buildup} + VD\text{ for drop-off})} = \text{Tangent of max angle}$$

The measured length of locked-in segment =

$$\frac{\text{Usable depth} - (VD\text{ for buildup} + VD\text{ for drop-off})}{\cos\text{ max angle}} = 6238.71\text{ ft}$$

Total measured depth = *TMD* = *MD* at *KOP* + *MD* during buildup + *MD* of locked-in segment + *MD* during drop-off = 10,339 ft.

ACKNOWLEDGEMENTS

The help extended by Cliff Ellis, K. McElvain, Dick Darling, Carl Kisinger and Ray G. Wall is greatly appreciated by the authors.

REFERENCES

Eastman Whipstock, Inc., 1978. *Introduction to Directional Drilling.* 35 pp.
Gatlin, C., 1960. *Petroleum Engineering, Drilling and Well Completion.* Prentice-Hall, Englewood Cliffs, N.J., 341 pp.
Kuster Company, 1979. *Single Shot Survey Instruments.* 27 pp.
Kuster Company, 1979. *Multishot Survey Instruments.* 19 pp.
Sperry-Sun, 1979. *Catalogue 789.* pp. 5998—6015.

UNIFORM 1°00′ INCREASE IN DRIFT PER 100′ OF HOLE DRILLED

MEASURED DEPTH	VERTICAL DEPTH	DEVIATION	DRIFT	MEASURED DEPTH	VERTICAL DEPTH	DEVIATION	DRIFT
0.00	0.00	0.00	0°00′				
100.00	99.99	0.87	1°00′	3100.00	2950.95	818.37	31°00′
200.00	199.96	3.49	2°00′	3200.00	3036.21	870.62	32°00′
300.00	299.86	7.85	3°00′	3300.00	3120.55	924.35	33°00′
400.00	399.68	13.96	4°00′	3400.00	3203.94	979.54	34°00′
500.00	499.37	21.80	5°00′	3500.00	3286.35	1036.18	35°00′
600.00	598.90	31.39	6°00′	3600.00	3367.76	1094.25	36°00′
700.00	698.26	42.71	7°00′	3700.00	3448.15	1153.73	37°00′
800.00	797.40	55.76	8°00′	3800.00	3527.48	1214.61	38°00′
900.00	896.30	70.54	9°00′	3900.00	3605.74	1276.86	39°00′
1000.00	994.93	87.04	10°00′	4000.00	3682.90	1340.47	40°00′
1100.00	1093.26	105.27	11°00′	4100.00	3758.94	1405.41	41°00′
1200.00	1191.25	125.21	12°00′	4200.00	3833.84	1471.67	42°00′
1300.00	1288.87	146.85	13°00′	4300.00	3907.56	1539.23	43°00′
1400.00	1386.11	170.19	14°00′	4400.00	3980.10	1608.06	44°00′
1500.00	1482.92	195.23	15°00′	4500.00	4051.42	1678.16	45°00′
1600.00	1579.29	221.95	16°00′	4600.00	4121.51	1749.48	46°00′
1700.00	1675.17	250.36	17°00′	4700.00	4190.35	1822.02	47°00′
1800.00	1770.54	280.43	18°00′	4800.00	4257.91	1895.74	48°00′
1900.00	1865.37	312.16	19°00′	4900.00	4324.17	1970.64	49°00′
2000.00	1959.63	345.54	20°00′	5000.00	4389.11	2046.68	50°00′
2100.00	2053.30	380.56	21°00′	5100.00	4452.72	2123.84	51°00′
2200.00	2146.34	417.21	22°00′	5200.00	4514.97	2202.10	52°00′
2300.00	2238.72	455.47	23°00′	5300.00	4575.84	2281.43	53°00′
2400.00	2330.43	495.35	24°00′	5400.00	4635.33	2361.82	54°00′
2500.00	2421.42	536.82	25°00′	5500.00	4693.40	2443.23	55°00′
2600.00	2511.68	579.87	26°00′	5600.00	4750.04	2525.64	56°00′
2700.00	2601.17	624.49	27°00′	5700.00	4805.23	2609.03	57°00′
2800.00	2689.87	670.66	28°00′	5800.00	4858.96	2693.36	58°00′
2900.00	2777.75	718.38	29°00′	5900.00	4911.21	2778.63	59°00′
3000.00	2864.79	767.62	30°00′	6000.00	4961.96	2864.79	60°00′

UNIFORM 2°00′ INCREASE IN DRIFT PER 100 FT OF HOLE DRILLED

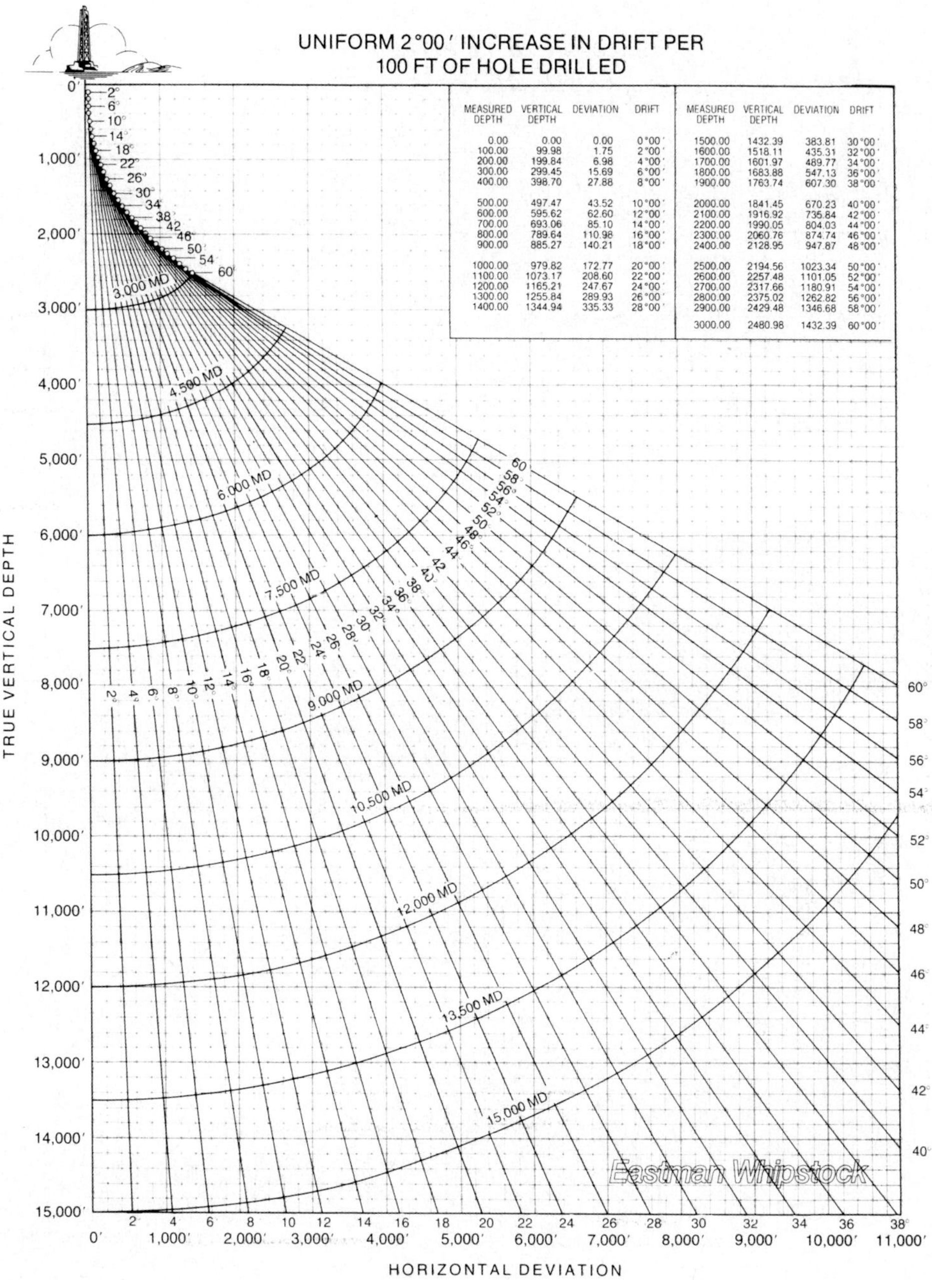

MEASURED DEPTH	VERTICAL DEPTH	DEVIATION	DRIFT
0.00	0.00	0.00	0°00′
100.00	99.98	1.75	2°00′
200.00	199.84	6.98	4°00′
300.00	299.45	15.69	6°00′
400.00	398.70	27.88	8°00′
500.00	497.47	43.52	10°00′
600.00	595.62	62.60	12°00′
700.00	693.06	85.10	14°00′
800.00	789.64	110.98	16°00′
900.00	885.27	140.21	18°00′
1000.00	979.82	172.77	20°00′
1100.00	1073.17	208.60	22°00′
1200.00	1165.21	247.67	24°00′
1300.00	1255.84	289.93	26°00′
1400.00	1344.94	335.33	28°00′

MEASURED DEPTH	VERTICAL DEPTH	DEVIATION	DRIFT
1500.00	1432.39	383.81	30°00′
1600.00	1518.11	435.31	32°00′
1700.00	1601.97	489.77	34°00′
1800.00	1683.88	547.13	36°00′
1900.00	1763.74	607.30	38°00′
2000.00	1841.45	670.23	40°00′
2100.00	1916.92	735.84	42°00′
2200.00	1990.05	804.03	44°00′
2300.00	2060.76	874.74	46°00′
2400.00	2128.95	947.87	48°00′
2500.00	2194.56	1023.34	50°00′
2600.00	2257.48	1101.05	52°00′
2700.00	2317.66	1180.91	54°00′
2800.00	2375.02	1262.82	56°00′
2900.00	2429.48	1346.68	58°00′
3000.00	2480.98	1432.39	60°00′

UNIFORM 3°00′ INCREASE IN DRIFT PER 100 FT OF HOLE DRILLED

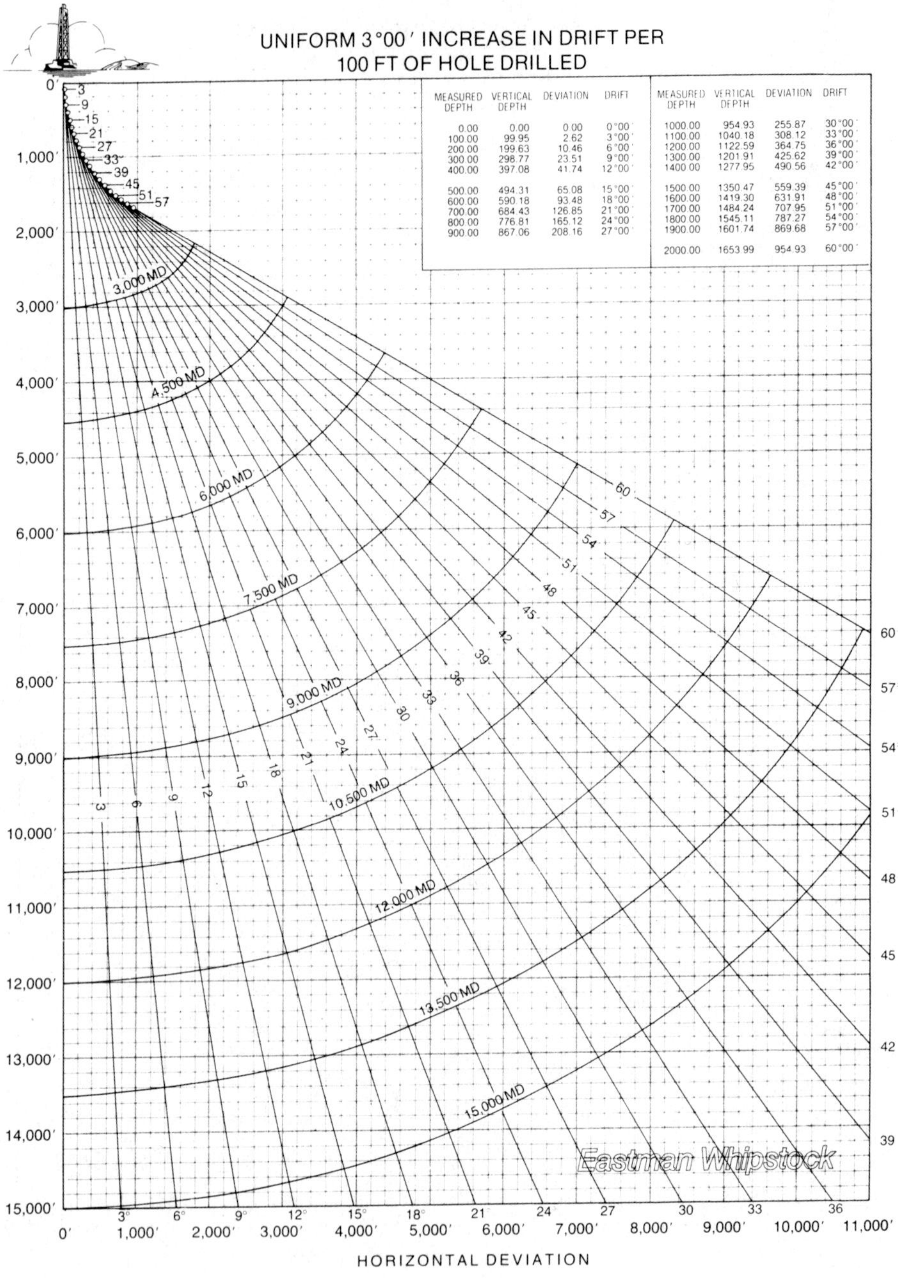

MEASURED DEPTH	VERTICAL DEPTH	DEVIATION	DRIFT
0.00	0.00	0.00	0°00′
100.00	99.95	2.62	3°00′
200.00	199.63	10.46	6°00′
300.00	298.77	23.51	9°00′
400.00	397.08	41.74	12°00′
500.00	494.31	65.08	15°00′
600.00	590.18	93.48	18°00′
700.00	684.43	126.85	21°00′
800.00	776.81	165.12	24°00′
900.00	867.06	208.16	27°00′

MEASURED DEPTH	VERTICAL DEPTH	DEVIATION	DRIFT
1000.00	954.93	255.87	30°00′
1100.00	1040.18	308.12	33°00′
1200.00	1122.59	364.75	36°00′
1300.00	1201.91	425.62	39°00′
1400.00	1277.95	490.56	42°00′
1500.00	1350.47	559.39	45°00′
1600.00	1419.30	631.91	48°00′
1700.00	1484.24	707.95	51°00′
1800.00	1545.11	787.27	54°00′
1900.00	1601.74	869.68	57°00′
2000.00	1653.99	954.93	60°00′

UNIFORM 5°00′ INCREASE IN DRIFT PER
100 FT OF HOLE DRILLED

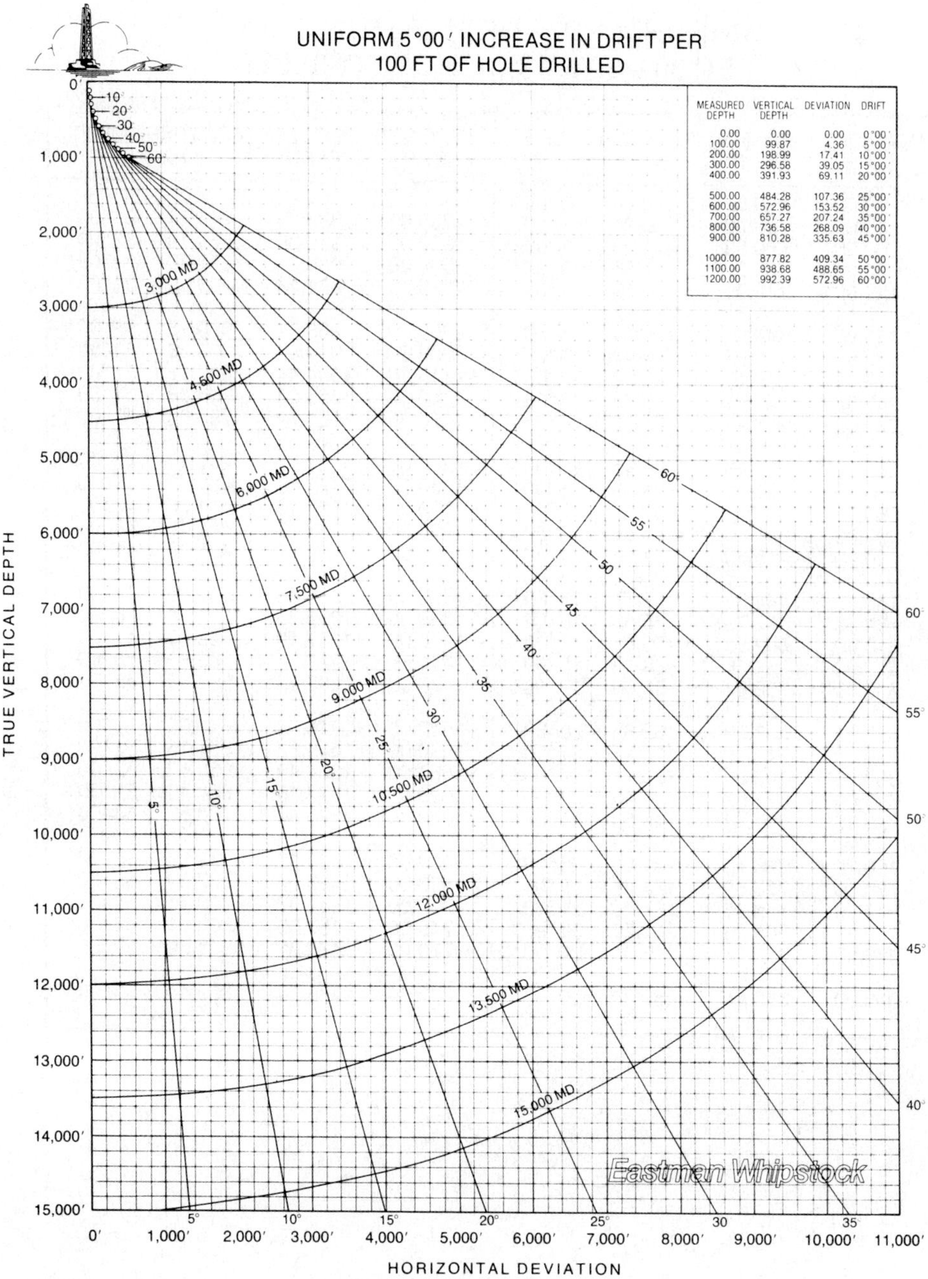

MEASURED DEPTH	VERTICAL DEPTH	DEVIATION	DRIFT
0.00	0.00	0.00	0°00′
100.00	99.87	4.36	5°00′
200.00	198.99	17.41	10°00′
300.00	296.58	39.05	15°00′
400.00	391.93	69.11	20°00′
500.00	484.28	107.36	25°00′
600.00	572.96	153.52	30°00′
700.00	657.27	207.24	35°00′
800.00	736.58	268.09	40°00′
900.00	810.28	335.63	45°00′
1000.00	877.82	409.34	50°00′
1100.00	938.68	488.65	55°00′
1200.00	992.39	572.96	60°00′

UNIFORM 1°00′ INCREASE IN DRIFT PER 30 METERS OF HOLE DRILLED

MEASURED DEPTH	VERTICAL DEPTH	DEVIATION	DRIFT	MEASURED DEPTH	VERTICAL DEPTH	DEVIATION	DRIFT
0.00	0.00	0.00	0°00′				
30.00	30.00	0.26	1°00′	930.00	885.29	245.51	31°00′
60.00	59.99	1.05	2°00′	960.00	910.86	261.19	32°00′
90.00	89.96	2.36	3°00′	990.00	936.17	277.30	33°00′
120.00	119.90	4.19	4°00′	1020.00	961.18	293.86	34°00′
150.00	149.81	6.54	5°00′	1050.00	985.91	310.85	35°00′
180.00	179.67	9.42	6°00′	1080.00	1010.33	328.28	36°00′
210.00	209.48	12.81	7°00′	1110.00	1034.44	346.12	37°00′
240.00	239.22	16.73	8°00′	1140.00	1058.24	364.38	38°00′
270.00	268.89	21.16	9°00′	1170.00	1081.72	383.06	39°00′
300.00	298.48	26.11	10°00′	1200.00	1104.87	402.14	40°00′
330.00	327.98	31.58	11°00′	1230.00	1127.68	421.62	41°00′
360.00	357.37	37.56	12°00′	1260.00	1150.15	441.50	42°00′
390.00	386.66	44.05	13°00′	1290.00	1172.27	461.77	43°00′
420.00	415.83	51.06	14°00′	1320.00	1194.03	482.42	44°00′
450.00	444.88	58.57	15°00′	1350.00	1215.43	503.45	45°00′
480.00	473.79	66.59	16°00′	1380.00	1236.45	524.84	46°00′
510.00	502.55	75.11	17°00′	1410.00	1257.10	546.60	47°00′
540.00	531.16	84.13	18°00′	1440.00	1277.37	568.72	48°00′
570.00	559.61	93.65	19°00′	1470.00	1297.25	591.19	49°00′
600.00	587.89	103.66	20°00′	1500.00	1316.73	614.00	50°00′
630.00	615.99	114.17	21°00′	1530.00	1335.82	637.15	51°00′
660.00	643.90	125.16	22°00′	1560.00	1354.49	660.63	52°00′
690.00	671.62	136.64	23°00′	1590.00	1372.75	684.43	53°00′
720.00	699.13	148.60	24°00′	1620.00	1390.60	708.55	54°00′
750.00	726.43	161.05	25°00′	1650.00	1408.02	732.97	55°00′
780.00	753.50	173.96	26°00′	1680.00	1425.01	757.69	56°00′
810.00	780.35	187.35	27°00′	1710.00	1441.57	782.71	57°00′
840.00	806.96	201.20	28°00′	1740.00	1457.69	808.01	58°00′
870.00	833.33	215.51	29°00′	1770.00	1473.36	833.59	59°00′
900.00	859.44	230.29	30°00′	1800.00	1488.59	859.44	60°00′

UNIFORM 1°30′ INCREASE IN DRIFT PER 30 METERS OF HOLE DRILLED

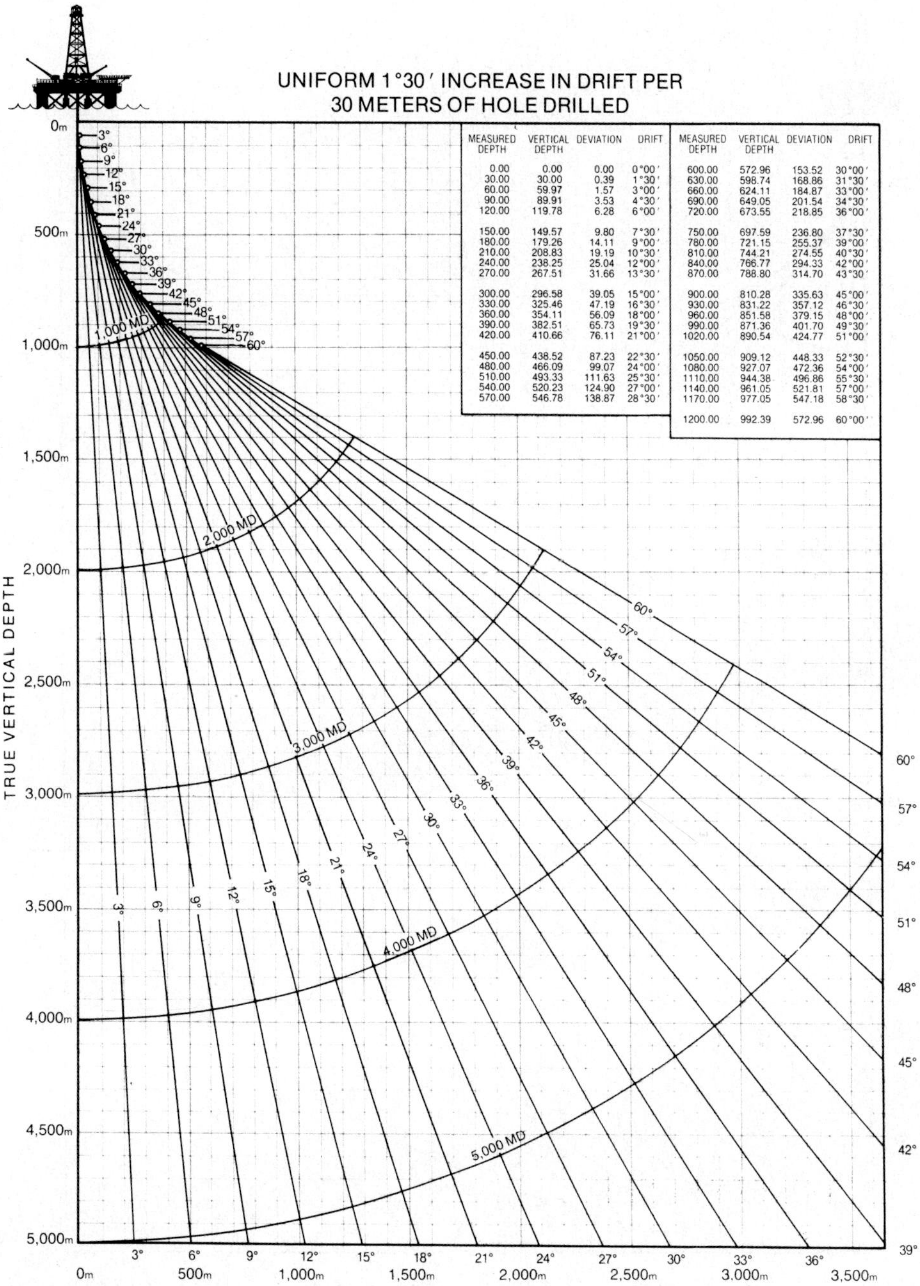

MEASURED DEPTH	VERTICAL DEPTH	DEVIATION	DRIFT
0.00	0.00	0.00	0°00′
30.00	30.00	0.39	1°30′
60.00	59.97	1.57	3°00′
90.00	89.91	3.53	4°30′
120.00	119.78	6.28	6°00′
150.00	149.57	9.80	7°30′
180.00	179.26	14.11	9°00′
210.00	208.83	19.19	10°30′
240.00	238.25	25.04	12°00′
270.00	267.51	31.66	13°30′
300.00	296.58	39.05	15°00′
330.00	325.46	47.19	16°30′
360.00	354.11	56.09	18°00′
390.00	382.51	65.73	19°30′
420.00	410.66	76.11	21°00′
450.00	438.52	87.23	22°30′
480.00	466.09	99.07	24°00′
510.00	493.33	111.63	25°30′
540.00	520.23	124.90	27°00′
570.00	546.78	138.87	28°30′

MEASURED DEPTH	VERTICAL DEPTH	DEVIATION	DRIFT
600.00	572.96	153.52	30°00′
630.00	598.74	168.86	31°30′
660.00	624.11	184.87	33°00′
690.00	649.05	201.54	34°30′
720.00	673.55	218.85	36°00′
750.00	697.59	236.80	37°30′
780.00	721.15	255.37	39°00′
810.00	744.21	274.55	40°30′
840.00	766.77	294.33	42°00′
870.00	788.80	314.70	43°30′
900.00	810.28	335.63	45°00′
930.00	831.22	357.12	46°30′
960.00	851.58	379.15	48°00′
990.00	871.36	401.70	49°30′
1020.00	890.54	424.77	51°00′
1050.00	909.12	448.33	52°30′
1080.00	927.07	472.36	54°00′
1110.00	944.38	496.86	55°30′
1140.00	961.05	521.81	57°00′
1170.00	977.05	547.18	58°30′
1200.00	992.39	572.96	60°00′

UNIFORM 2°30′ INCREASE IN DRIFT PER 30 METERS OF HOLE DRILLED

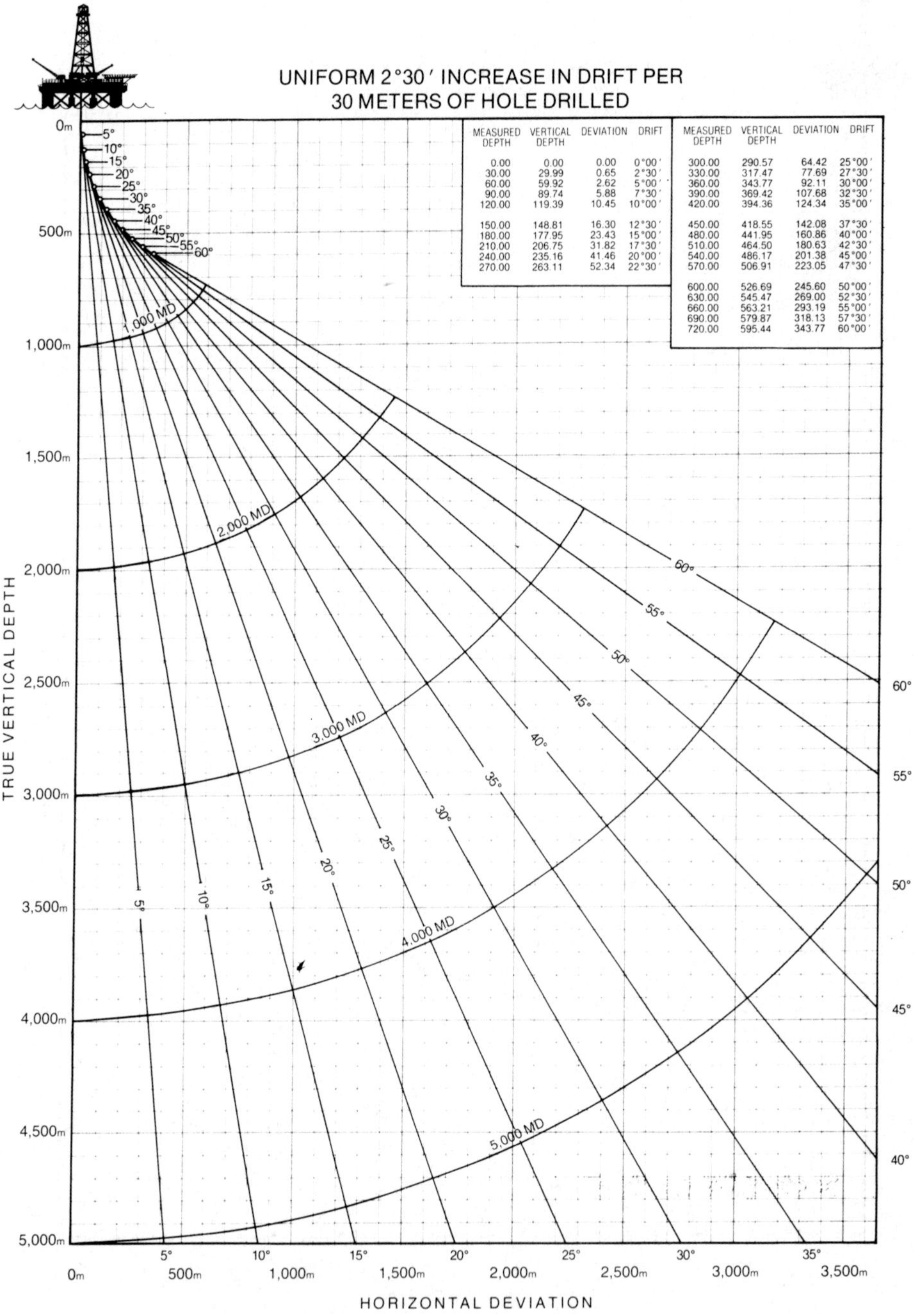

MEASURED DEPTH	VERTICAL DEPTH	DEVIATION	DRIFT
0.00	0.00	0.00	0°00′
30.00	29.99	0.65	2°30′
60.00	59.92	2.62	5°00′
90.00	89.74	5.88	7°30′
120.00	119.39	10.45	10°00′
150.00	148.81	16.30	12°30′
180.00	177.95	23.43	15°00′
210.00	206.75	31.82	17°30′
240.00	235.16	41.46	20°00′
270.00	263.11	52.34	22°30′

MEASURED DEPTH	VERTICAL DEPTH	DEVIATION	DRIFT
300.00	290.57	64.42	25°00′
330.00	317.47	77.69	27°30′
360.00	343.77	92.11	30°00′
390.00	369.42	107.68	32°30′
420.00	394.36	124.34	35°00′
450.00	418.55	142.08	37°30′
480.00	441.95	160.86	40°00′
510.00	464.50	180.63	42°30′
540.00	486.17	201.38	45°00′
570.00	506.91	223.05	47°30′
600.00	526.69	245.60	50°00′
630.00	545.47	269.00	52°30′
660.00	563.21	293.19	55°00′
690.00	579.87	318.13	57°30′
720.00	595.44	343.77	60°00′

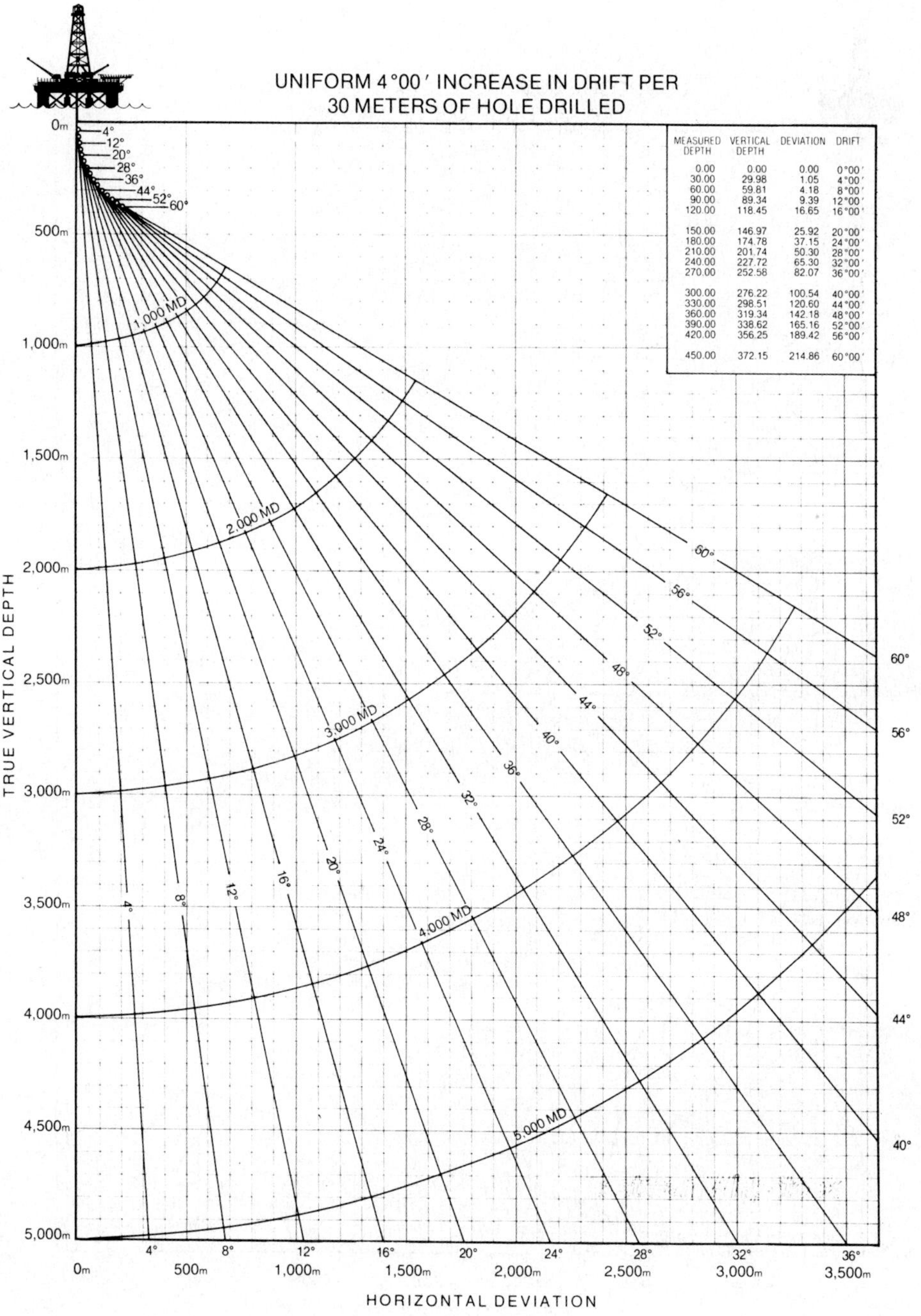

MEASURED DEPTH	VERTICAL DEPTH	DEVIATION	DRIFT
0.00	0.00	0.00	0°00′
30.00	29.98	1.05	4°00′
60.00	59.81	4.18	8°00′
90.00	89.34	9.39	12°00′
120.00	118.45	16.65	16°00′
150.00	146.97	25.92	20°00′
180.00	174.78	37.15	24°00′
210.00	201.74	50.30	28°00′
240.00	227.72	65.30	32°00′
270.00	252.58	82.07	36°00′
300.00	276.22	100.54	40°00′
330.00	298.51	120.60	44°00′
360.00	319.34	142.18	48°00′
390.00	338.62	165.16	52°00′
420.00	356.25	189.42	56°00′
450.00	372.15	214.86	60°00′

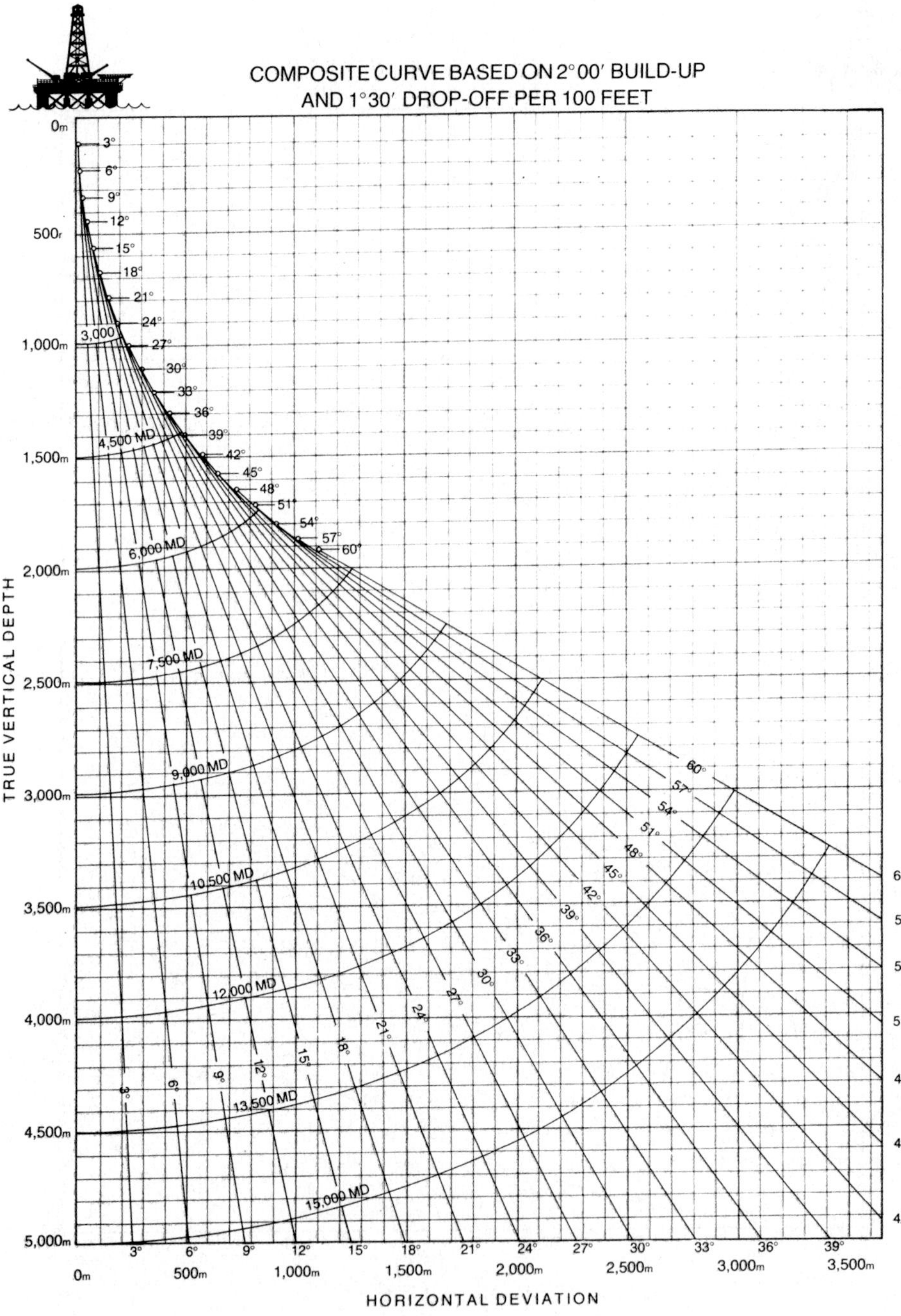
COMPOSITE CURVE BASED ON 2°00′ BUILD-UP
AND 1°30′ DROP-OFF PER 100 FEET
TRUE VERTICAL DEPTH
HORIZONTAL DEVIATION
0m
500m
1,000m
1,500m
2,000m
2,500m
3,000m
3,500m
4,000m
4,500m
5,000m
3,000
4,500 MD
6,000 MD
7,500 MD
9,000 MD
10,500 MD
12,000 MD
13,500 MD
15,000 MD
3°
6°
9°
12°
15°
18°
21°
24°
27°
30°
33°
36°
39°
42°
45°
48°
51°
54°
57°
60°
500m
1,000m
1,500m
2,000m
2,500m
3,000m
3,500m

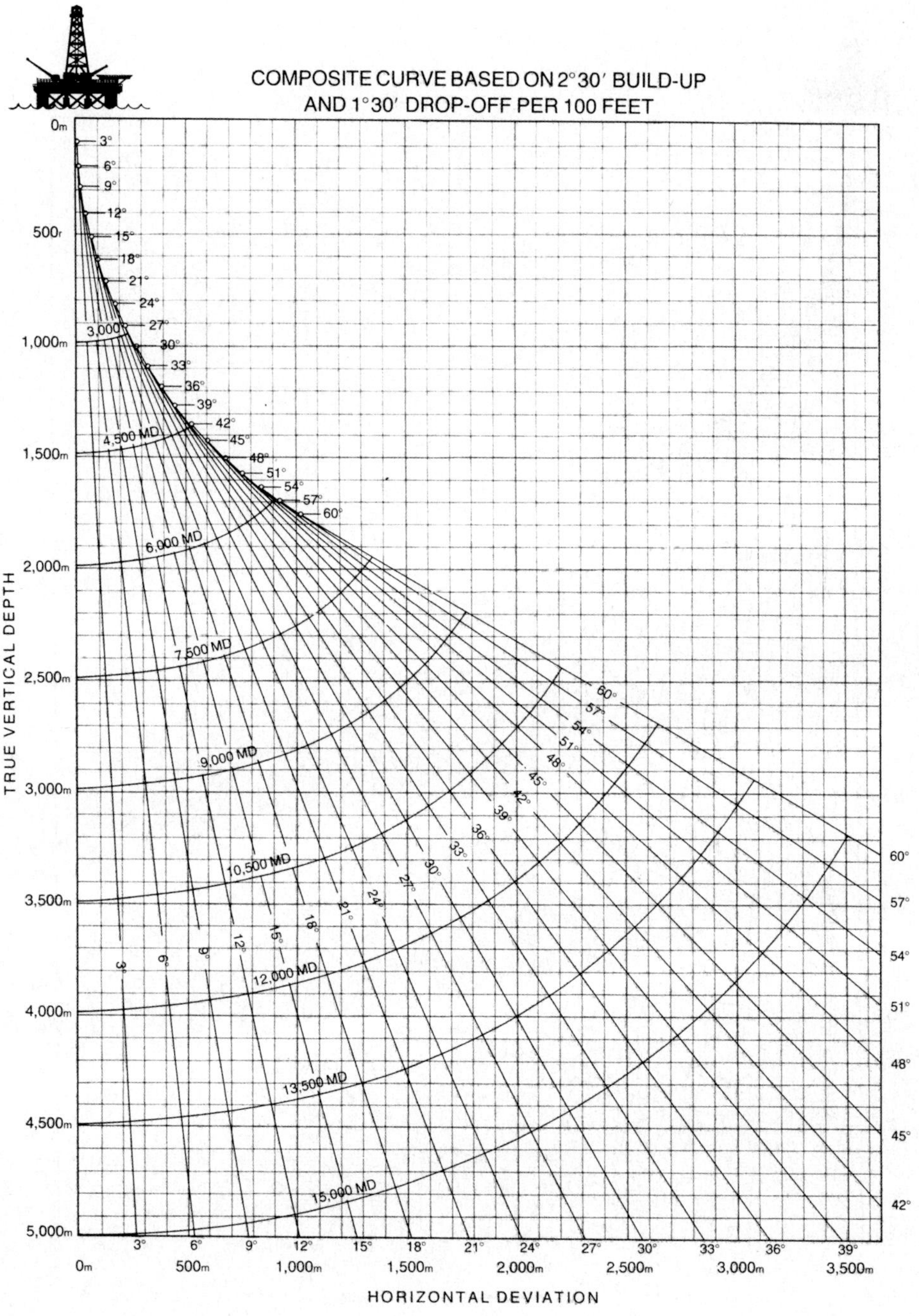
COMPOSITE CURVE BASED ON 2°30′ BUILD-UP
AND 1°30′ DROP-OFF PER 100 FEET
TRUE VERTICAL DEPTH
HORIZONTAL DEVIATION
0m
500m
1,000m
1,500m
2,000m
2,500m
3,000m
3,500m
4,000m
4,500m
5,000m
3,000
4,500 MD
6,000 MD
7,500 MD
9,000 MD
10,500 MD
12,000 MD
13,500 MD
15,000 MD
0m
500m
1,000m
1,500m
2,000m
2,500m
3,000m
3,500m
60°
57°
54°
51°
48°
45°
42°
39°
36°
33°
30°
27°
24°
21°
18°
15°
12°
9°
6°
3°

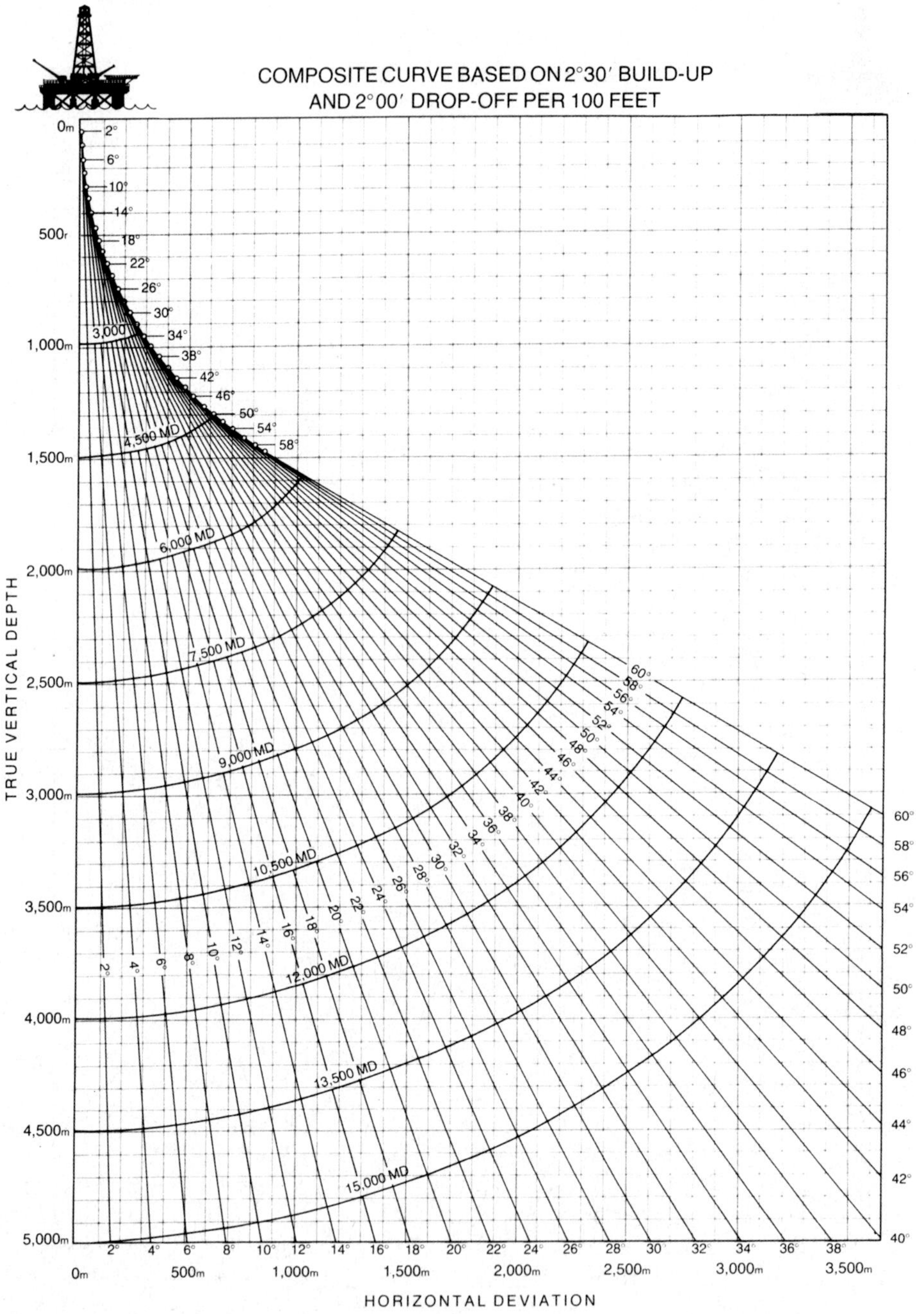
COMPOSITE CURVE BASED ON 2°30′ BUILD-UP
AND 2°00′ DROP-OFF PER 100 FEET
TRUE VERTICAL DEPTH
HORIZONTAL DEVIATION
0m
500m
1,000m
1,500m
2,000m
2,500m
3,000m
3,500m
4,000m
4,500m
5,000m
3,000
4,500 MD
6,000 MD
7,500 MD
9,000 MD
10,500 MD
12,000 MD
13,500 MD
15,000 MD
0m
500m
1,000m
1,500m
2,000m
2,500m
3,000m
3,500m

Appendix A

ENGINEERING DATA AND CALCULATIONS *

MAGCOBAR, DRESSER INDUSTRIES, INC.

INTRODUCTION

Data essential for mud engineering calculations presented in this Appendix closely follow the order as it appears on the Drilling Mud Report form shown in Fig. A-1. The following information must be available: (1) volume of mud in the system, (2) pump output, (3) mud cycle, (4) annular velocity of the mud, and (5) hydrostatic head of the mud.

VOLUME OF THE FLUID SYSTEM

The mud engineer must calculate the volume of mud in the system, which consists of the mud in the pits and the mud in the hole. (Volume of mud in the system = Volume of mud in the pits + Volume of mud in the hole.) In some cases mud is stored in a reserve pit or tank and although it is useful to know this reserve volume, it is not considered in the regular treating program.

Volume of fluid in pits

In straight-sided rectangular steel pits, the volume of mud in barrels may be calculated as follows:

$$V_m = (L \times W \times D)/5.61 \qquad \text{(A-1)}$$

where L = length in ft, W = width in ft, D = depth in ft, and 5.61 = conversion factor, cu ft/bbl.

In the case of earthen pits, the slopes at the ends and sides of the pits must be taken into consideration to get the average dimensions.

The method of measuring an earthen pit is presented in Fig. A-2. The effective length A includes measurement of only one sloped end. For all practical purposes the end not measured is balanced by the remaining tapered section and, therefore, is not considered a part of the over-all length. If the sides also slope appreciably, the same type of measurement should be used to determine the width.

*Edited by George V. Chilingarian

DRILLING MUD REPORT

Magcobar
DRILLING FLUID SERVICES
OILFIELD PRODUCTS DIVISION
Dresser Industries, Inc.

P. O. BOX 6504
HOUSTON, TEXAS 77005

DRILLING MUD REPORT — REPORT NO. 169

DATE 12/23 1971 — DEPTH 18649

API WELL NO.	STATE	COUNTY	WELL	S/T

OPERATOR: MAJOR OIL COMPANY	CONTRACTOR: LARGE BROS DRLG Co.	RIG NO. 18
ADDRESS: RIG	ADDRESS: RIG	SPUD DATE: 5/15/70
REPORT FOR MR. BILL EATON	REPORT FOR MR. JOE LESEBERSON	SECTION, TOWNSHIP, RANGE: 18/11N/18W

WELL NAME AND NO.	FIELD OR BLOCK NO.	COUNTY, PARISH OR OFFSHORE AREA	STATE OR PROVINCE
JAMES SMITH NO. 1	ELK CITY	BECKHAM	OKLA.

OPERATION		CASING	MUD VOLUME (BBL)	CIRCULATION DATA	
Present Activity: DRLG.		Surface 13⅜ in. at 9285 Ft.	Hole 1245 — Pits 750	Pump Size 5½ X 18 in.	Annular Vel. (Ft./Min.) Opposite DP 220′
Bit Size (in.): 8 3/8″	No. 103	Intermediate 9⅝ in. at 17,285 Ft.	Total Circulating Volume 1995	Pump Make EMSCO; Model D.1350	Opposite Collar 320′
Drill Pipe Size: 5″	Type X.H.	Production or Liner in. at Ft.	In Storage 400	Bbl/Stroke 0.142; Stroke/Min. 60	Circulating Pressure psi 2800; Bottoms Up (Min.) 138
Drill Collar Size: 6½″	Length 780	Mud Type: SPERSENE-XP-20		Bbl/Min. 8.5	System Total (Min.) 238

Sample from ☒ Flowline ☐ Pit
Flowline Temperature 155 °F

MUD PROPERTIES			
Time Sample Taken	3:00 AM		
Depth (ft.)	18,649		
Weight ☒ (ppg) ☐ (lb./cu. ft.)	13.8		
Mud Gradient (psi/ft.)	.7176		
Funnel Viscosity (sec./qt.) API at ___ °F	77 SUCTION		
Plastic Viscosity cps at 115 °F	51		
Yield Point (lb./100 sq. ft.)	4		
Gel Strength (lb./100 sq. ft.) 10 sec./10 min.	5/11	/	/
pH ☒ Strip ☐ Meter	9.3		
Filtrate API (ml./30 min.)	6.0		
API HP-HT Filtrate (ml/30 min.) 300 °F	17.8		
Cake Thickness 32nd in. API ☐ HP-HT ☒	3/32		
Alkalinity, Mud (Pm)	0.8		
Alkalinity, Filtrate (Pf/Mf)	0.3/1.2	/	/
Salt ☐ppm ☐gpg Chloride ☒ppm ☐gpg	1600		
Calcium ☒ ppm ☐ Gyp (ppb)	80		
Sand Content (% by Vol.)	3/4 / 1%		
Solids Content (% by Vol.)	22.8%		
Oil Content (% by Vol.)	—		
Water Content (% by Vol.)	77.2%		
Methylene Blue Capacity ☐(ml/ml mud) ☒(equiv. #/bbl bent.)	18		
E.C.D.	14#/gal		

EQUIPMENT

	SIZE	HRS./TOUR		SIZE	HRS./TOUR
Centrifuge	✓		Desilter	✓	
Degasser	✓		Shaker	✓	80 ASTM MESH
Desander	✓		Other		

DAILY COST	CUMULATIVE COST

MUD PROPERTIES SPECIFICATIONS

WEIGHT	VISCOSITY	FILTRATE

BY AUTHORITY: ☐ OPERATOR'S WRITTEN ☐ OPERATOR'S REPRESENTATIVE ☐ DRILLING CONTRACTOR ☐ OTHER

MUD CONTAINS		TOUR TREATMENT
☑ Magcobar	sx	
☑ Magcogel	sx	
☐ Salt Gel	sx	
☑ Spersene	sx	
☑ XP-20	sx	
☐ TannAthin	sx	
☑ Caustic Soda	sx	
☐ My-Lo-Jel	sx	
☐		
☐		
☐		
☐		
☐		
☐		

REMARKS—Give operation, depth and nature of any problems encountered

☑ Drilling @ 18,649 — MAINTAIN WGT AT 13.8-14.0#/gal at suction.
☐ Salt Water flow @ — " VISCOSITY AT 77 SEC/QT WITH
☐ Gas kick @ — MAGCOGEL MIXED SLOWLY THROUGH
☐ Stuck pipe @ — THE HOPPER. RUN THE CENTRIFUGE WHILE
☐ Tight hole @ — DRILLING.
☐ Sloughing Shale @
☐ Lost Returns @

DRESSER MAGCOBAR ENGINEER: Peter Good	ADDRESS: Jack Rohlit, OKLA.	PHONE	562-3063
REPRESENTATIVE	HOME ADDRESS: "	PHONE	"
MOBILE UNIT	WAREHOUSE LOCATION: ELK CITY, OKLA.	PHONE	562-3989

PRINTED IN U.S.A. — THIS REPORT IS SUBJECT TO THE TERMS AND CONDITIONS AS SET FORTH ON THE REVERSE SIDE

Fig. A-1. Drilling mud report.

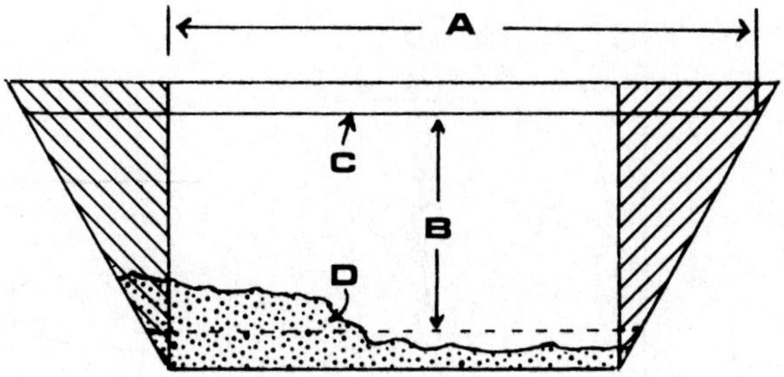

Fig. A-2. Measurements of an irregular pit: A = length or width, B = depth, C = mud level, and D = cutting fill.

Working mud pits and tanks are seldom clean, because cuttings and sand settle out and make the bottom of the pit irregular. It is, therefore, necessary to use a stick, shovel, plumb line, or some other device to determine the depth at different points. The average depth of several soundings should be considered as the depth of the pits (B).

Four other types of pits are in more or less common use in various parts of the world. In Kansas and West Texas, U.S.A., there are pits with sloping sides and different widths at the top and bottom (Fig. A-3). In pits of this type the volume may be calculated by taking the average width, by adding the widths at top and bottom together and dividing by two. Volume of mud in bbl, V_m, is equal to (see eq. A-1 and Fig. A-2):

$$V_m = \frac{A \times W \times B}{5.61} \qquad \text{(A-2)}$$

In some areas of the Gulf Coast, U.S.A., pits have vertical sides for part of the depth and sloping sides for the remainder of the depth (Fig. A-4). This results in certain advantages of cleaning and moving mud from one pit to

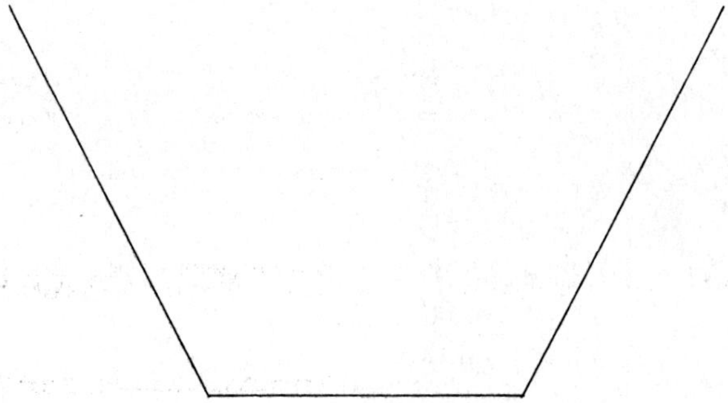

Fig. A-3. Mud pit with sloping sides.

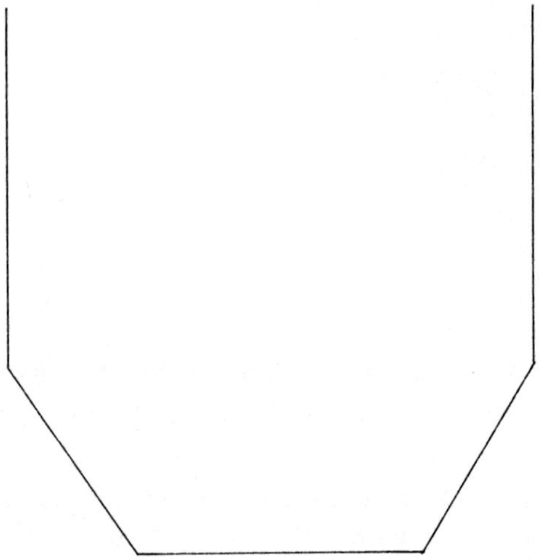

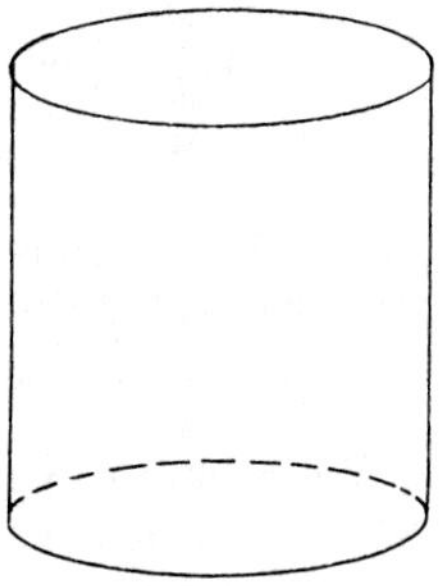

Fig. A-4. Mud pit with vertical and sloping sides.

Fig. A-5. Cylindrical mud pit.

another. The volume of such pits must be calculated in two parts:

(1) The volume of mud in the vertical section is equal to:

$$V_m = \frac{\text{length} \times \text{width} \times \text{depth of vertical sec.}}{5.61} \tag{A-3}$$

(2) The sloping section is measured as a triangle and the volume of this portion of the pit is equal to the length × $\frac{1}{2}$ width × depth. The total volume is equal to the sum of these two volumes [(1) + (2)].

Circular pits in the form of tanks (Fig. A-5) are sometimes used for storing mud and they are becoming increasingly important for storage of oil-base and water-in-oil emulsion drilling fluids. The volume of such circular tanks in

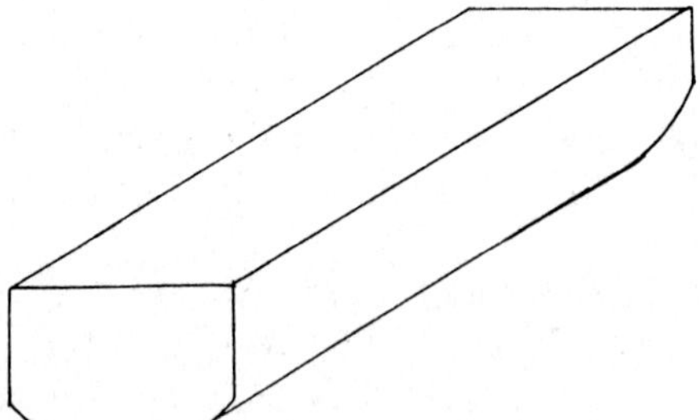

Fig. A-6. Half-cylindrical mud pit.

bbl may be calculated as follows:

$$V_m = \frac{d^2 \times 3.1416 \times D}{4 \times 5.61} \quad \text{(A-4)}$$

where: d = diameter in ft, and D = depth in ft.

If diameter is not known, it is necessary to measure the circumference, C, with a tape and determine diameter, d, as follows:

$$d = \frac{C}{3.1416} \quad \text{(A-5)}$$

In a few cases, large circular drums are cut in half and used for open pits as shown in Fig. A-6. The volume of such pits should be found as the volume of a cylinder (in bbl) divided by two:

$$V_m = \frac{d^2 \times 3.1416 \times L}{5.61 \times 8} \quad \text{(A-6)}$$

Capacity of hole

The most frequently used rule-of-thumb formula for determining the volume of mud in the hole (a cylinder) is:

Volume in bbl/1000 ft of hole = (diameter in inches)^2 (A-7)

For example, a $9\frac{7}{8}$-inch bit will cut approximately a 10-inch hole. Using eq. A-7, 10×10 gives 100 bbl/1000 ft of depth. If the hole is made up of $10\frac{3}{4}$-inch surface pipe and $9\frac{7}{8}$-inch open hole, the square of the approximate diameter of 10 inches (10×10) again gives a close approximation of the actual volume of 100 bbl/1000 ft of hole. If the diameter includes a fraction of an inch, the interpolation between the squares of the nearest whole numbers is readily made. In such approximate calculations, the volume occupied by the steel of the drillpipe is generally disregarded.

The diameter of the open drilled hole is rarely equal to that of the bit. Salt, shales, and other soluble or soft formations generally give rise to overgauge holes, whereas boreholes in sandstones and carbonates may be close to bit size in diameter. This fact is generally disregarded in mud volume calculations in which bit size is used to determine the approximate volume of open hole. Some engineers add from 25 to 50% of the volume of the open hole to allow for overgauge portions of the borehole; however, this practice is generally not acceptable.

Because of the uncertainty as to the exact volume of the open hole, volume calculations are rarely carried out exactly. Experience proved that practical and reasonable approximations are proper, because they are certainly as (or even more) accurate as measurements of the volume using exact gauge. Occasionally, it is quite important to know the exact time of the mud cycle.

TABLE A-I

Capacities of open boreholes

Diameter (inches)	Capacity (bbl/100 ft)	Diameter (inches)	Capacity (bbl/100 ft)
$3^{7}/_{8}$	1.46	$8^{1}/_{2}$	7.02
$4^{1}/_{4}$	1.75	$8^{5}/_{8}$	7.23
$4^{3}/_{4}$	2.19	$8^{3}/_{4}$	7.45
$5^{5}/_{8}$	3.08	$9^{5}/_{8}$	9.00
6	3.50	$9^{7}/_{8}$	9.49
$6^{1}/_{8}$	3.65	11	11.75
$6^{1}/_{4}$	3.80	$12^{1}/_{4}$	14.58
$6^{3}/_{4}$	4.43	15	21.86
$7^{3}/_{8}$	5.28	$17^{1}/_{2}$	29.75
$7^{5}/_{8}$	5.65	18	31.47
$7^{7}/_{8}$	6.02	20	38.86
$8^{3}/_{8}$	6.81	24	55.95

TABLE A-II

Capacities of API standard casings

Size, nominal OD (inches)	Weight (lb/ft)	Capacity (bbl/100 ft)	Size, nominal OD (inches)	Weight (lb/ft)	Capacity (bbl/100 ft)
5	11.50	2.02	$7^{5}/_{8}$	39.00	4.26
5	13.00	1.96	$8^{5}/_{8}$	24.00	6.37
5	15.00	1.89	$8^{5}/_{8}$	28.00	6.24
5	18.00	1.78	$8^{5}/_{8}$	32.00	6.10
$5^{1}/_{2}$	14.00	2.44	$8^{5}/_{8}$	36.00	5.95
$5^{1}/_{2}$	15.50	2.38	$8^{5}/_{8}$	40.00	5.79
$5^{1}/_{2}$	17.00	2.32	$8^{5}/_{8}$	44.00	5.65
$5^{1}/_{2}$	20.00	2.22	$9^{5}/_{8}$	23.30	7.87
$5^{1}/_{2}$	23.00	2.11	$9^{5}/_{8}$	36.00	7.73
6	18.00	2.85	$9^{5}/_{8}$	40.00	7.58
6	20.00	2.78	$9^{5}/_{8}$	43.50	7.45
6	23.00	2.66	$9^{5}/_{8}$	47.00	7.32
$6^{5}/_{8}$	20.00	3.55	$9^{5}/_{8}$	53.50	7.08
$6^{5}/_{8}$	24.00	3.41	$10^{3}/_{4}$	32.75	10.09
$6^{5}/_{8}$	28.00	3.26	$10^{3}/_{4}$	40.50	9.81
7	17.00	4.15	$10^{3}/_{4}$	45.50	9.62
7	20.00	4.05	$10^{3}/_{4}$	51.00	9.43
7	23.00	3.93	$10^{3}/_{4}$	55.50	9.25
7	26.00	3.83	$13^{3}/_{4}$	48.00	15.71
7	29.00	3.71	$13^{3}/_{8}$	54.50	15.46
7	32.00	3.60	$13^{3}/_{8}$	61.00	15.22
7	35.00	3.50	$13^{3}/_{8}$	68.00	14.97
7	38.00	3.40	$13^{3}/_{8}$	72.00	14.81
$7^{5}/_{8}$	24.00	4.79	16	65.00	22.59
$7^{5}/_{8}$	26.40	4.72	16	75.00	22.22
$7^{5}/_{8}$	29.70	4.59	16	84.00	21.89
$7^{5}/_{8}$	33.70	4.45	20	94.00	35.53

TABLE A-III

Capacities and displacement volumes of API standard drillpipes

Size, nominal (inches)	ID (inches)	Weight (lb/ft)	Capacity (bbl/100 ft)	Displ. (bbl/100 ft)
$2^3/_8$	1.815	6.65	0.320	0.287
$2^7/_8$	2.151	10.40	0.450	0.406
$3^1/_2$	2.764	13.30	0.742	0.513
$3^1/_2$	2.602	15.50	0.659	0.585
4	3.340	14.00	1.113	0.566
$4^1/_2$	3.820	16.60	1.422	0.643
$4^1/_2$	3.640	20.00	1.288	0.771
5	4.276	19.50	1.780	0.750
$5^1/_2$	4.778	21.90	2.220	0.871
$5^1/_2$	4.670	24.70	2.110	0.964
$6^5/_8$	5.965	25.20	3.456	0.984

TABLE A-IV

Capacities and displacement volumes of drillcollars

OD (inches)	ID (inches)	Weight (lb/ft)	Capacity (bbl/100 ft)	Displ. (bbl/100 ft)
3.5	1.5	26.64	0.219	0.971
4.125	2	34.68	0.389	1.361
4.75	2.25	46.70	0.489	1.701
5.75	2.25	74.70	0.489	2.721
6	2.25	82.50	0.489	3.011
6.25	2.25	90.60	0.489	3.301
6.5	2.8125	91.56	0.766	3.334
6.75	2.25	108.00	0.489	3.941
6.75	2.8125	100.08	0.766	3.664
7	2.8125	109.68	0.766	3.994
7.25	2.8125	119.28	0.766	4.344
7.75	2.8125	138.48	0.766	5.064
8	2.8125	150.48	0.766	5.454

TABLE A-V

Capacities of API standard tubing

Size, nominal (inches)	Size, OD (inches)	ID (inches)	Weight (lb/ft)	Capacity (bbl/100 ft)
$1^1/_2$	$1^{15}/_{16}$	1.610	2.75	0.252
2	$2^3/_8$	1.995	4.60	0.387
$2^1/_2$	$2^7/_8$	2.441	6.40	0.579
3	$3^1/_2$	2.992	10.20	0.829
$3^1/_2$	4	3.476	11.00	1.174
4	$4^1/_2$	3.958	12.60	1.522

TABLE A-VI

Physical properties of API standard casing grades *

	Casing grade			
	F-25	H-40	J-55	N-80
Yield strength (minimum) in tension (lb/sq inch)	25,000	40,000	55,000	80,000
Tensile strength (minimum) (lb/sq inch)	40,000	60,000	75,000	100,000
Elongation in 2 inch (minimum) (%)	40	27	20	18

* Four different grades of steel and iron for manufacture of casings have been established by the API. These are designated F-25, H-40, J-55 and N-80. Casings are further classified according to: (1) type of tool joint used, (2) material used in manufacture, and (3) method of manufacture.

TABLE A-VII

Comparison chart of 3-cutter rock bits

Security sealed bearing	Type	Manufacturer			
		Security	Hughes	Reed	Smith
S33S		S3S	OSC-3A	YT3A	DS
		S3	OSC-3	YT3	DT
S33	soft	S3T		YT3T	
	formation	S4	OSC-16	YT1A	DG
S44		S4T	CIC	TY1T	
		S6	OSC	YTL	K2
M44N	medium	M4N	OWV	YSI	V2
M44		M4L	OWC	YM	T2
		H7	W7	YH	
H77		H7T	W7C		L4
		H7U	W7R2	YHW2	W4
H77S	hard	H7SG	WD7		
H77C		HC	WR	YBVG	WC
*S88			J44	SCM	3
**M88	insert	M8	X-55R	YC5G	4
H88	bits	H8	RG7	YC4G	5-6-7
H100		H10	RG2B	YC2G	9

*S88 for S44—M44 drilling.

**M88 for M44N—H77 drilling.

Realizing that both the calculations of hole volume and pump volume may be in error, it is expedient to insert easily-seen solids such as cotton seed, corn, or oats in the drillpipe and check the exact time required for this material moving with the mud stream to complete the cycle and reappear again on the shaker screen. When close figures are desired, the following tables on the capacities of casing, open hole, and drillpipe are included here (Tables A-I, A-II, and A-III). These tables are also necessary for accurate calculation of annular velocity of the mud in the hole. Capacities of tubing and rock-bit sizes may also be found useful (Tables A-IV, A-V, A-VI, and A-VII.)

Time required for the return of cuttings from bottom (bottoms up) can be calculated as follows:

$$\text{Bottoms up (min)} = \frac{HV - (CDP + DDP)}{PO} \qquad \text{(A-8)}$$

where: HV = hole volume (bbl), CDP = capacity of drillpipe (bbl), DDP = displacement of drillpipe (bbl), and PO = pump output (bbl/min).

ANNULAR VELOCITY

Design of hydraulic programs should be based on the distribution of hydraulic horsepower rather than the return velocity in the annulus. Inasmuch as a minimum annular velocity is required, however, a lower limit of the circulation rate must be established.

Table A-VIII gives suggested minimum annular velocities by hole size. These values, however, should be used with caution, because local drilling conditions, such as high penetration rate and unusual hole conditions, may dictate higher velocities.

$$\text{Annular velocity in 100 ft/min} = \frac{\text{Pump output (bbl/min)}}{\text{Annular volume (bbl/100 ft)}} \qquad \text{(A-9)}$$

TABLE A-VIII

Minimum required annular velocities

Hole size (inches)	Annular velocity (ft/min)
15	80
$12^1/_4$	90
$10^5/_8$	110
$8^3/_4$	120
$7^7/_8$	130
6	140

where: Annular volume/100 ft, bbl = (Capacity of open hole in bbl) − (Capacity of drillpipe, bbl/100 ft + Displacement of drillpipe, bbl/100 ft).

For example, the volume of $9\frac{7}{8}$-inch open hole in bbl/100 ft is 9.49. The volume of $4\frac{1}{2}$-inch drillpipe (1.422 bbl/100 ft) plus the displacement of the drillpipe (0.643 bbl/ft) is equal to 2.065. The volume of the annulus/100 ft is, therefore, equal to 7.425 bbl (= 9.49—2.065).

A $6\frac{1}{2}'' \times 16''$ pump pumping 0.2025 bbl/stroke at 45 strokes/min will move 9.11 bbl/min (= 0.2025 × 45). Thus annular velocity is equal to: (9.11 : 7.43) × 100 = 123 ft/min.

HYDROSTATIC HEAD OF THE DRILLING FLUID

One of the primary functions of drilling fluid is the control of all subsurface pressures, both normal and abnormal. Hydrostatic head is the pressure exerted by a fluid at any point of depth, equally in all directions. The hydrostatic pressure, p, of a mud column is determined as follows:

$$p = M \times D \times K \qquad \text{(A-10)}$$

where M = specific weight of mud, D = depth, and K = conversion constant.
K = 0.052 when D is in ft, M is in lb/gal, and p is in psi.
K = 0.00695 when D is in ft, M is expressed in lb/cu ft, and p is in psi.
K = 0.098 when D is in m, M is expressed as specific gravity, and p is in atm.

MISCELLANEOUS CALCULATIONS

There are some methods of making approximate rapid calculations for field use that should be mentioned. In mixing clays, for example, it is easy to remember that one sack of high-yield clay gives two bbl of mud. One sack of bentonite or of attapulgite (salt gel) gives five bbl of mud.

In adding barite to increase mud weight, for specific weights below 12 ppg (lb/gal), 60 lb of barite will increase the weight of one bbl of mud by 1 ppg. The addition of 60 sacks of barite will increase the weight of 100 bbl of mud by 1 ppg.

The rate of weight increase imparted to mud is not constant: it decreases as the specific weight (or specific gravity) of the mud increases and also decreases because of the increase in the volume of the mud. This is also true in the case of addition of clay. For weights above 12 ppg, the desired weight must be divided by 0.2 to find the number of sacks of barite necessary to raise the weight of 100 bbl of mud by 1 ppg. For example, in order to change the specific weight of 100 bbl of mud from 18 ppg to 19 ppg, it is necessary to add 95 sacks of barite: 19.0/0.2 = 95 sacks.

Table A-IX facilitates calculation of the amount of barite necessary to raise any mud weight to any other desired specific weight.

TABLE A-IX

Weighting and increase in volume table for barite

MAGCOBAR WEIGHTING TABLE

DESIRED WEIGHT

PRESENT WEIGHT	8.0	8.5	9.0	9.5	10.0	10.5	11.0	11.5	12.0	12.5	13.0	13.5	14.0	14.5	15.0	15.5	16.0	16.5	17.0	17.5	18.0	PRESENT WEIGHT
7.35	35.5	64.0	93.5	124.0	156.0	189.0	223.5	260.0	297.5	337.0	377.5	421.0	466.0	514.0	562.0	614.5	670.0	728.0	787.0	852.0	920.0	
8.0		28.0	56.5	86.5	117.5	150.0	184.0	218.5	255.5	294.5	334.0	376.0	420.0	465.0	515.0	565.0	619.0	674.0	735.0	798.0	865.0	
8.5			28.0	57.5	88.0	119.0	153.0	188.0	224.0	261.5	301.0	342.0	385.5	430.0	477.0	527.0	580.0	636.0	694.0	756.0	821.0	
8.7	3.0		17.0	45.5	76.5	108.0	141.0	175.0	211.0	248.0	287.5	328.5	371.0	416.0	464.0	512.0	565.0	620.0	678.0	739.5	803.0	17.5
9.0	6.0	3.0		29.0	59.0	90.0	122.5	156.5	192.0	229.0	267.5	308.0	350.0	394.5	441.0	490.0	542.0	596.0	653.0	714.0	779.0	17.0
9.5	9.0	5.5	3.0		29.5	60.0	92.0	125.0	160.0	196.0	234.0	274.0	315.0	358.5	404.0	452.0	502.0	556.0	613.5	672.0	735.0	16.5
10.0	12.0	8.5	5.5	2.5		30.0	61.0	93.5	128.0	163.5	200.0	239.5	280.5	323.0	368.0	415.0	464.0	517.0	572.0	630.0	692.0	16.0
10.5	14.5	11.5	8.5	5.5	2.5		30.5	62.5	95.5	131.0	167.0	205.5	245.0	287.0	330.5	376.5	426.0	477.0	531.0	588.0	649.0	15.5
11.0	17.5	14.5	11.0	8.0	5.5	2.5		31.0	64.0	98.0	133.5	171.0	210.0	251.0	294.0	339.5	387.5	437.0	490.0	546.0	606.0	15.0
11.5	20.5	17.0	14.0	11.0	8.0	5.0	2.5		32.0	65.5	100.0	137.0	175.0	215.0	257.0	302.0	348.0	398.0	449.0	504.0	562.0	14.5
12.0	23.5	20.0	16.0	13.5	10.5	7.5	5.0	2.5		32.5	67.0	102.5	140.0	179.5	220.5	264.0	309.5	357.5	408.0	462.0	519.0	14.0
12.5	26.5	23.0	19.5	16.0	13.0	10.5	7.5	5.0	2.5		33.5	68.5	105.0	145.0	187.5	228.0	271.0	318.5	367.5	420.0	476.0	13.5
13.0	29.5	25.5	22.0	19.0	16.0	13.0	10.0	7.5	5.0	2.5		34.0	70.0	107.5	147.0	188.5	232.0	278.0	327.0	378.0	433.0	13.0
13.5	32.5	28.5	25.0	21.5	18.5	15.5	12.5	10.0	7.0	4.5	2.5		35.0	71.5	110.5	151.0	194.0	238.5	285.5	336.0	389.0	12.5
14.0	35.5	31.5	27.5	24.5	21.0	18.0	15.0	12.0	9.5	7.0	4.5	2.0		36.0	73.5	113.0	155.0	199.0	238.5	294.0	346.0	12.0
14.5	38.0	34.0	30.5	27.0	23.5	20.5	17.5	14.5	12.0	9.5	7.0	4.5	2.0		36.5	75.5	116.0	159.0	204.0	252.0	303.0	11.5
15.0	41.0	37.0	33.5	29.5	26.5	23.0	20.0	17.0	14.5	11.5	9.0	6.5	4.5	2.0		37.5	77.5	120.0	163.5	210.0	260.0	11.0
15.5	44.0	40.0	36.0	32.5	29.0	25.5	22.5	19.5	16.5	14.0	11.5	9.0	6.5	4.0	2.0		38.5	79.5	122.5	168.0	216.0	10.5
16.0	47.0	43.0	39.0	35.0	31.5	28.0	25.0	22.0	19.0	16.5	13.5	11.0	8.5	6.5	4.0	2.0		40.0	81.5	126.0	173.0	10.0
16.5	50.0	45.5	41.5	38.0	34.0	31.0	27.5	24.5	21.5	18.5	16.0	13.5	11.0	8.5	6.0	4.0	2.0		41.0	84.0	130.0	9.5
17.0	53.0	48.5	44.5	40.5	37.0	33.5	30.0	27.0	24.0	21.0	18.0	15.5	13.0	10.5	8.5	6.0	4.0	2.0		42.0	86.5	9.0
17.5	54.5	50.0	46.0	42.0	38.5	35.5	31.5	28.5	25.0	22.5	19.5	17.0	14.5	12.0	9.5	7.5	5.0	3.0	1.0		43.0	8.7
	56.0	52.0	47.0	43.0	39.5	36.0	32.5	29.0	26.0	23.0	20.5	18.0	15.0	13.0	10.5	8.0	6.0	4.0	2.0			8.5
	59.0	54.5	50.0	46.0	42.0	38.5	35.0	31.5	28.5	25.5	22.5	20.0	17.5	15.0	12.5	10.0	8.0	6.0	4.0	2.0		8.0
	62.5	58.0	53.5	49.5	45.5	41.5	38.0	35.0	31.5	28.5	25.5	23.0	20.0	17.5	15.0	13.0	10.5	8.5	6.5	4.5	2.5	7.35
	18.0	17.5	17.0	16.5	16.0	15.5	15.0	14.5	14.0	13.5	13.0	12.5	12.0	11.5	11.0	10.5	10.0	9.5	9.0	8.5	8.0	

DESIRED WEIGHT

VOLUME INCREASE TABLE

EXPLANATION

I. Upper right hand section gives the number of 100 pound bags of Barite to be added to 100 barrels of present weight mud when raising the weight to any desired new weight.

EXAMPLE:

To raise 10 pounds per gallon mud to 14 pounds per gallon mud: Find 10.0 in left hand margin and read across horizontally to the vertical column under 14. The number 280.5 is the number of 100 pound bags needed to raise the weight of 100 barrels of 10 pounds per gallon mud to 14 pounds per gallon.

II. Lower left hand section gives resulting volume increase in barrels upon adding 100 pound bags of Barite to reach a desired weight.

EXAMPLE:

To find volume increase due to adding 280.5 bags of Barite to 100 barrels of mud find 10 in right hand margin and read left horizontally to column above 14 found at bottom of table. The figure 19.0 is the volume increase in barrels resulting from the addition of 280.5 bags of weighting material.

Some important formulas

$$(1)\ \text{Annular velocity (cm/sec)} = \frac{35{,}353\,Q}{D_h^2 - D_p^2} \qquad \text{(A-11)}$$

where: D = diameter in mm, and Q = flow rate in m^3/hr.

$$(2)\ \text{Shear rate } (\text{sec}^{-1}) = \frac{120\,V}{D_h - D_p} \qquad \text{(A-12)}$$

where: D = diameter in mm, and V = velocity in cm/sec.

$$(3)\ \text{Weight increase using barite: } X = \frac{4250(W_2 - W_1)}{4250 - W_2} \qquad \text{(A-13)}$$

where: X = amount of barite (kg/m^3); W_1 = present specific weight (kg/m^3); and W_2 = desired specific weight (kg/m^3). To convert barite requirements, X, from kg/m^3 to 100-lb sacks/m^3, one can use the following formula:

100-lb sacks/m^3 = kg/m^3/45.36

(4) Volume increase due to barite additions:

Volume increase (m^3) = kg of barite : 4250 kg/m^3, or = 100-lb sacks of barite : 93.7. (A-14)

$$(5)\ \text{Weight reduction with water: } X = \frac{V(W_1 - W_2)}{W_2 - 1000} \qquad \text{(A-15)}$$

where: X = m^3 of water required, V = original mud volume in m^3, W_1 = initial mud weight in kg/m^3, and W_2 = desired mud weight in kg/m^3.

For weight reduction with diesel oil, 840 should be substituted for 1000 in the denominator of this equation.

$$(6)\ \text{Hydrostatic pressure } (\text{kg/cm}^2) = \frac{\gamma(\text{kg/m}^3) \times D(\text{m})}{10{,}000} \qquad \text{(A-16)}$$

where γ_f = specific weight of fluid, and D = depth.

PUMPS

Although there are many types of pumps, the reciprocating type pump shown in Fig. A-7 is of primary importance in the drilling industry. They are used almost exclusively for mud circulation. The double-acting duplex pump is the most common one (Fig. A-7). One stroke * consists of a complete for-

* Stroke is considered as one movement in each direction, forward and backward, of each piston; and count is based on one piston only. Pumps are usually 90% (or less) efficient depending upon age, maintenance, and other factors.

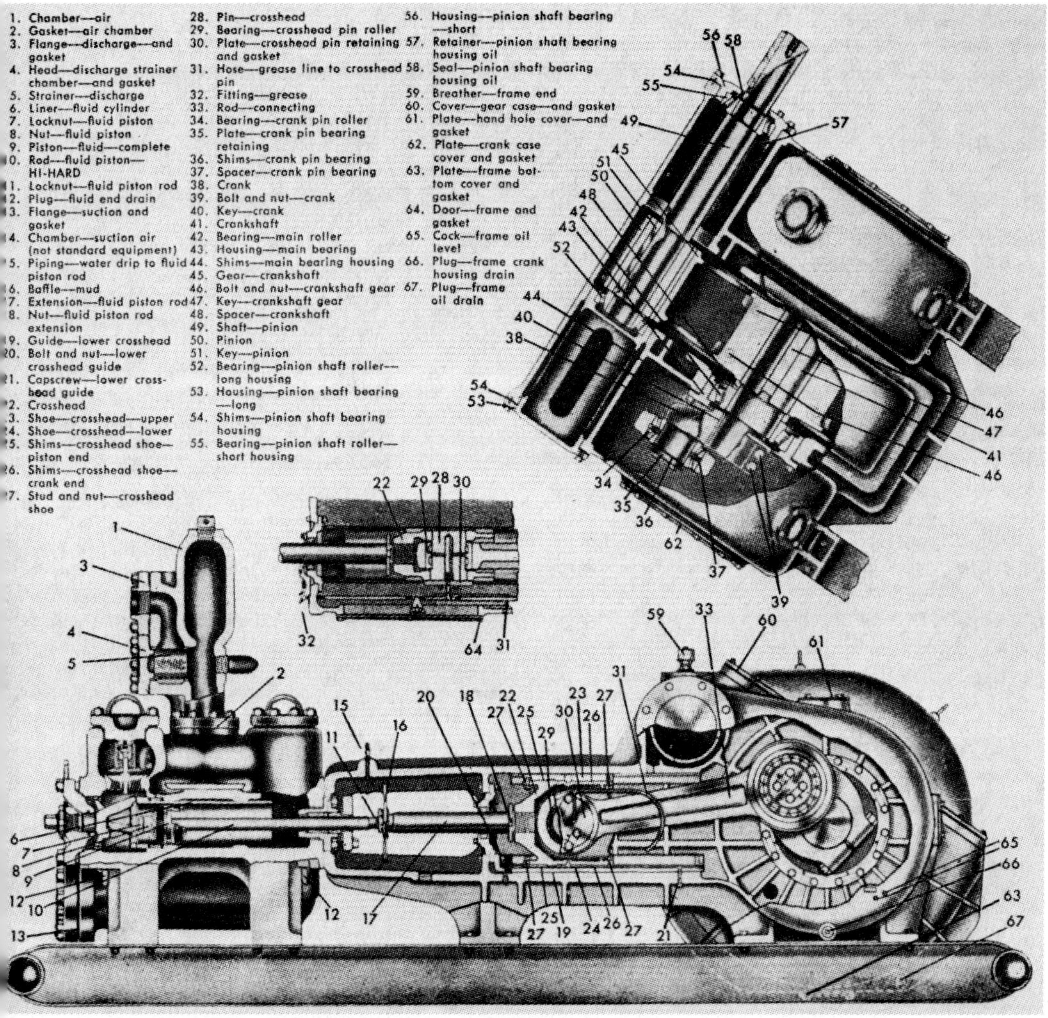

Fig. A-7. Reciprocating double-acting duplex pump.

TABLE A-X

Single acting triplex slush pump specifications (output-bbl/stroke at 100% efficiency)

		2-1/2	2-5/8	2-3/4	3	3-1/4	3-1/2	3-3/4	4	4-1/4	4-1/2	4-3/4	5	5-1/4	5-1/2	5-3/4	6	6-1/4	6-1/2	6-3/4	7
STROKE	7	.0105	.0117	.0129	.0152	.0179	.0207	.0238	.0271												
	8								.0310	.0350	.0393	.0438	.0486	.0536	.0588	.0643	.0698				
	8.5								.0330	.0373	.0418	.0466	.0516	.0569	.0624	.0682	.0743				
	10										.0493	.0548	.0607	.0669	.0736	.0802	.0874	.0948			
	12														.0881	.0964	.1050	.1138	.1229	.1326	.1426
PO-7							.0417		.0543		.0688		.0850		.1029		.1224				
PC-9							.0533		.0698		.0883		.1093		.1321		.1571				

TABLE A-XI

Double-acting duplex pump displacement tables; pump efficiency = 100%

Stroke (inches)	Linear size (inches)	strokes/bbl	bbl/stroke
10	$3^{1}/_{2}$	27.77	0.0360
10	4	20.76	0.0481
10	$4^{1}/_{2}$	16.15	0.0619
10	5	13.43	0.0745
12	4	17.3	0.0578
12	5	11.2	0.0894
12	$5^{1}/_{2}$	9.13	0.1095
12	6	7.58	0.1320
12	$6^{1}/_{4}$	6.96	0.1437
12	$6^{1}/_{2}$	6.39	0.1565
12	$6^{3}/_{4}$	5.92	0.1690
12	7	5.48	0.1825
14	5	9.6	0.1040
14	$5^{1}/_{2}$	7.81	0.1280
14	6	6.50	0.1540
14	$6^{1}/_{4}$	5.96	0.1675
14	$6^{1}/_{2}$	5.48	0.1825
14	$6^{3}/_{4}$	5.08	0.1970
14	7	4.70	0.2130
14	$7^{3}/_{4}$	4.46	0.2240
14	$7^{1}/_{2}$	4.15	0.2410
14	$7^{3}/_{4}$	3.88	0.2575
16	5	8.81	0.1135
16	$5^{1}/_{2}$	7.12	0.1405
16	6	5.86	0.1705
16	$6^{1}/_{4}$	5.40	0.1855
16	$6^{1}/_{2}$	4.93	0.2025
16	$6^{3}/_{4}$	4.65	0.2200
16	7	4.21	0.2375
16	$7^{1}/_{4}$	3.91	0.2560
16	$7^{1}/_{2}$	3.64	0.2750
16	$7^{3}/_{4}$	3.39	0.2950
18	5	7.84	0.1275
18	$5^{1}/_{2}$	6.33	0.1580
18	6	5.21	0.1920
18	$6^{1}/_{4}$	4.78	0.2090
18	$6^{1}/_{2}$	4.38	0.2280
18	$6^{3}/_{4}$	4.05	0.2465
18	7	3.74	0.2670
18	$7^{1}/_{4}$	3.47	0.2880
18	$7^{1}/_{2}$	3.22	0.3100
18	$7^{3}/_{4}$	3.02	0.3310

ward and backward movement, or one complete revolution of the crank shaft. As the moving member is forced to move to the left, the suction valve opens resulting in the filling of the cylinder with liquid. Then, as this member reverses and moves to the right, the suction valve is closed, the discharge valve is forced open by the built up pressure, and one unit of liquid volume is delivered.

Occasionally, triplex pumps are also used. In such cases, the output volume is determined by counting the strokes (forward and backward) as in the duplex pump and multiplying by $1\frac{1}{2}$ (see Table A-X).

For convenience in determining the pump output, one can use Table A-XI. These tables are based on 100% efficiency. Pumps, even when in good condition, however, have an efficiency of only about 90%. Frequently, badly worn pumps may be only 60—75% efficient.

Pump output and the drilling fluid cycle

In mud conditioning, it is important to know the time required for the mud to make a complete cycle (i.e., from pump suction to pump suction). This complete mud cycle is noted on the mud report form). All chemicals and/or weight material are generally added at such a rate that the mud will make at least one complete cycle during the treatment. Two factors are involved in cycling time: (1) the volume of mud under consideration, and (2) the volume output of the pump.

Cycle time

A complete mud cycle (in minutes) can be calculated as follows:

$$\text{Complete mud cycle} = TM/PO \tag{A-17}$$

where: TM = total mud in system (bbl) and PO = pump output (bbl/min).

The hole cycle in minutes (down the hole and back) may be calculated by using the following formula:

$$\text{Hole cycle} = \frac{HV - DDP}{PO} \tag{A-18}$$

where: HV = hole volume (bbl), and DDP = displacement of drillpipe (bbl).

SOLIDS DETERMINATION

Inasmuch as the solids content determines in large measure the properties and the performance of drilling fluids, it may be necessary to determine not only the percent of solids in a mud but also the character of these solids. The percentages of high-gravity and of low-gravity solids must be known.

An approximate, but sometimes accurate, determination of the solids in a field mud is possible using the Fann mud still. This information, together with other mud properties, may give an indication of the increase in the low-gravity solids and help to improve mud control techniques.

The total solids content can be determined in any drilling fluid, with or without oil. The same procedure as outlined for determining the oil content should be followed. After the distillation is complete, the volumes of oil and water collected are recorded. The volume percent of solids is then calculated by using the following formula:

$$V_s = 100 - (V_0 + V_w) \qquad \text{(A-19)}$$

where: V_s = vol% solids, V_0 = vol% oil (is equal to zero in non-oil mud), and V_w = vol% water.

For accuracy, it is necessary to measure the specific weight of mud with precision. Consequently, a recently calibrated mud balance should be used. Assuming an oil specific gravity of 0.8, the average gravity of the solids can be calculated as follows:

$$S_s = \frac{(100 \times S_m) - (V_w + 0.8V_0)}{V_s} \qquad \text{(A-20)}$$

where: S_s = average specific gravity of solids, and

$$S_m = \text{specific gravity of mud} = \frac{\text{mud weight, lb/gal}}{8.33\ \text{lb/gal}}$$

An approximate measure of the relative amounts of high-gravity and low-gravity solids can be obtained from Fig. A-8 (see Sample Problem No. 3). The high-gravity portion is assumed to have a specific gravity of 4.25 and the specific gravity of low-gravity solids is assumed to be 2.6. The low-gravity fraction, however, contains both clay and salts, causing the weight per cent to be high. Inasmuch as in highly saline conditions this error may be substantial, corrections should be made by subtracting the salt content in the test mud sample from the total amount of solids collected. Thus, for muds containing high chloride content, the specific gravity of the solids is determined by using the following formula:

$$S_s V_s = S_m - V_w + 0.8V_0 + \frac{V_w \times \%\ \text{NaCl}}{100 - \%\ \text{NaCl}} \qquad \text{(A-21)}$$

where: % NaCl = ppm NaCl/10,000, and V_s is determined from eq. A-19. Then S_s can be calculated and the relative amounts of high-gravity and low-gravity solids can be determined from Fig. A-8. At high salt concentrations (100,000 ppm or above), corrections should be applied to the NaCl content for density changes. Relationship between the percent by weight of salt and salt content in ppm is presented in Fig. A-9. Sometimes it is helpful to plot ppm vs lb/bbl of salt as shown in Fig. A-10.

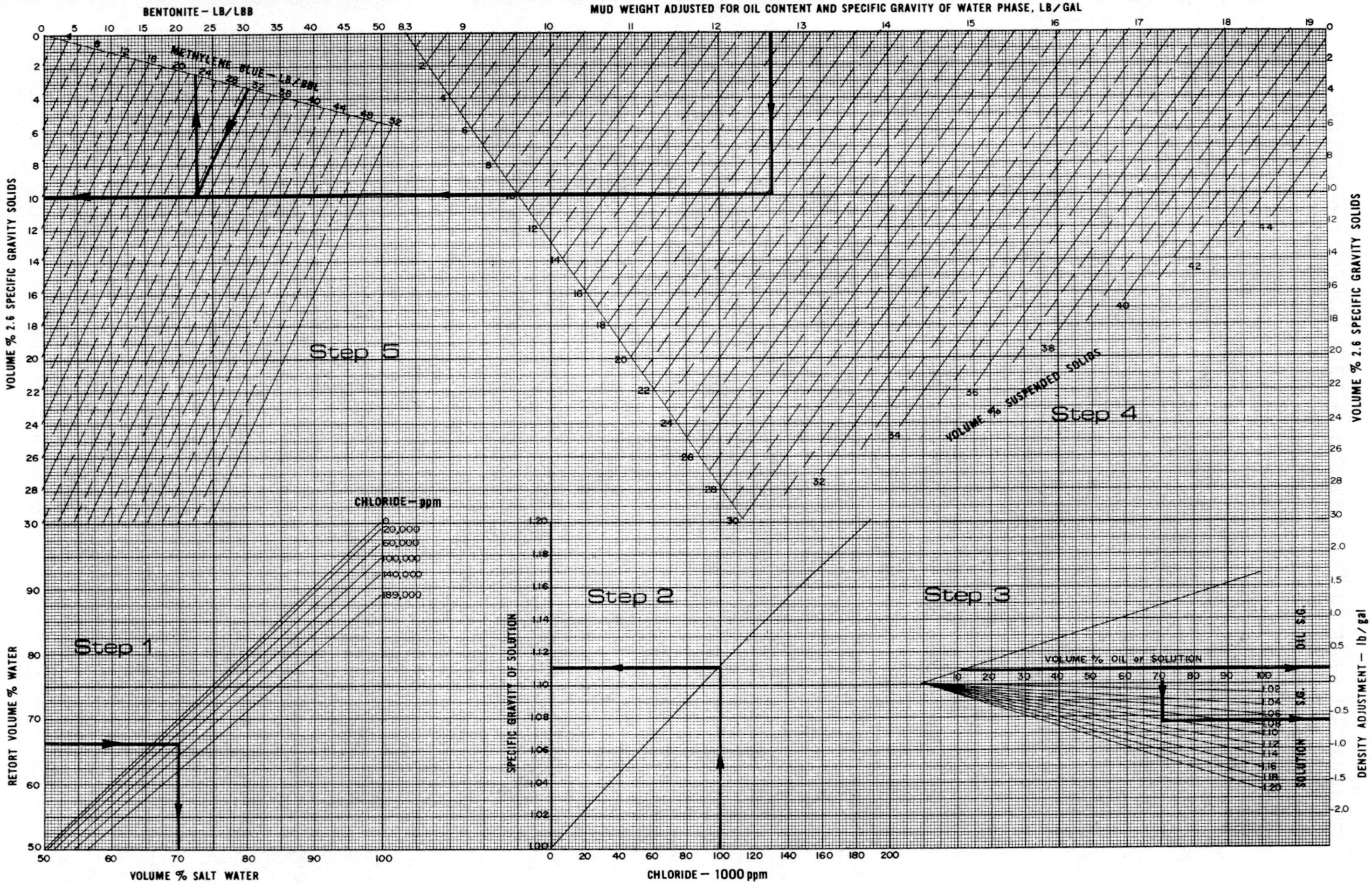

Fig. A-8. Nomogram for determining the amounts of high-gravity and low-gravity solids in drilling fluids.

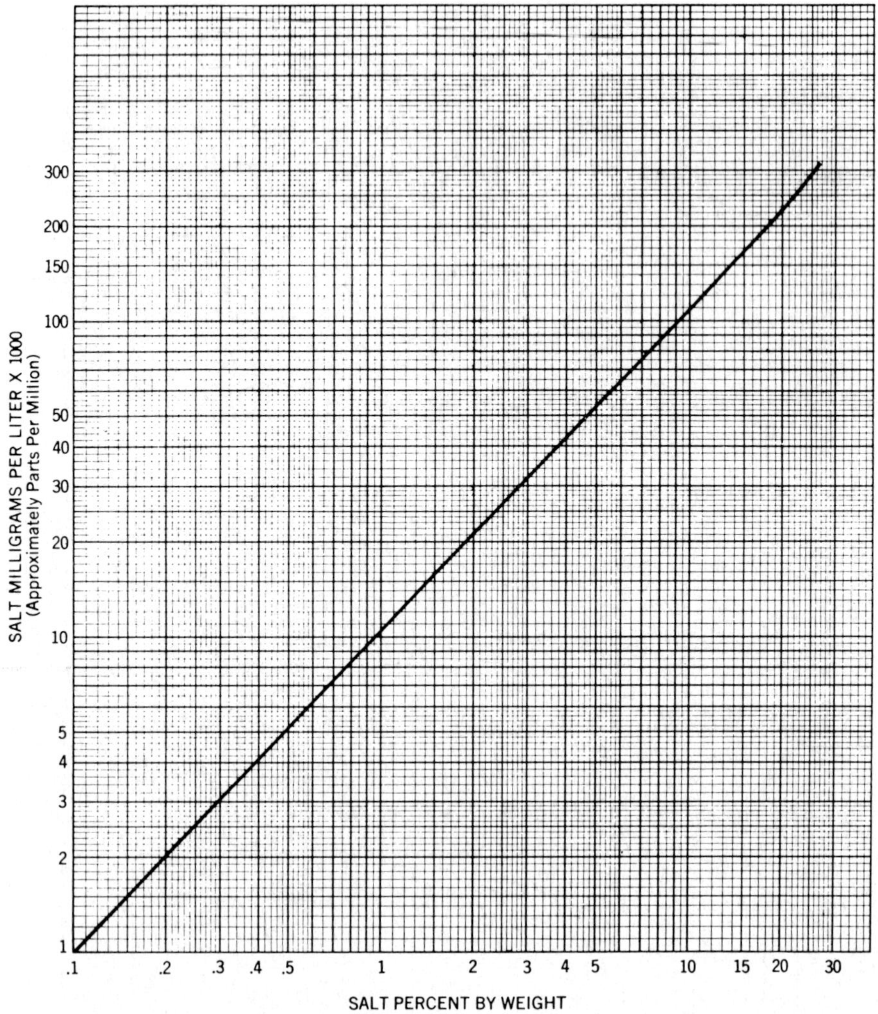

Fig. A-9. Relationship between the percent by weight of salt and salt content in ppm.

Sample problem No. 1

Given (see Fig. A-11)

Surface casing	— 1850 ft of $13\frac{3}{8}$-inch, 48 lb/ft.
Intermediate casing	— 8643 ft of $9\frac{5}{8}$-inch, 32.30 lb/ft.
Liner	— from 8300 ft to 14,500 ft, 7-inch, 20 lb/ft.
Bit size	— $6\frac{1}{8}$ inches
Total depth	— 17,800 ft.
Tapered string	— 5-inch drillpipe, 19.50 lb/ft, to 8000 ft.

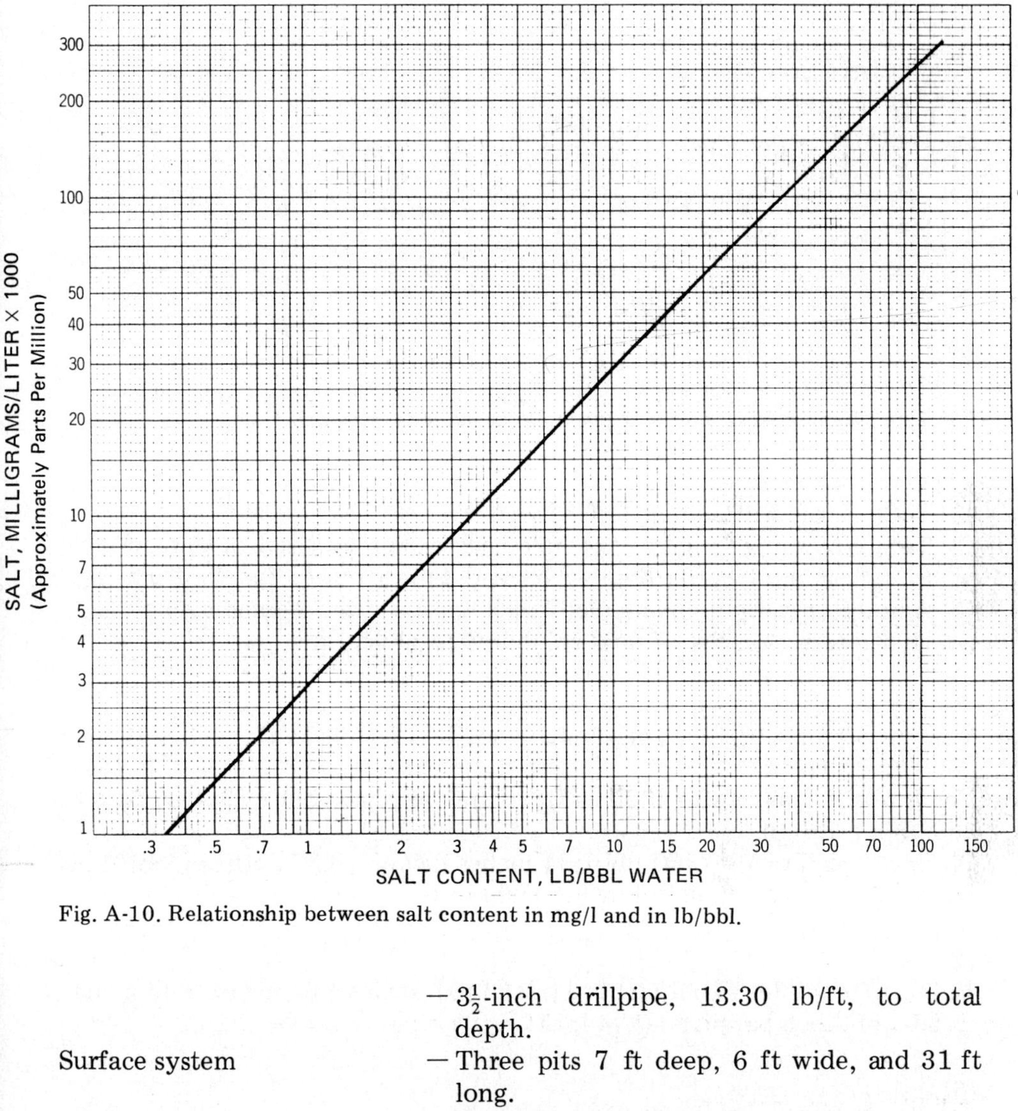

Fig. A-10. Relationship between salt content in mg/l and in lb/bbl.

	— $3\frac{1}{2}$-inch drillpipe, 13.30 lb/ft, to total depth.
Surface system	— Three pits 7 ft deep, 6 ft wide, and 31 ft long. Two pits have 64 inches of mud and the other has 46 inches of mud in it with the drillpipe in the hole.
Mud specific weight	— 16.3 lb/gal.
Pump (double-acting duplex)	— 6 × 16 making 34 stks/min, 90% efficiency.

Calculate

(1) Total capacity of surface system in bbl, bbl/inch, and bbl/ft.

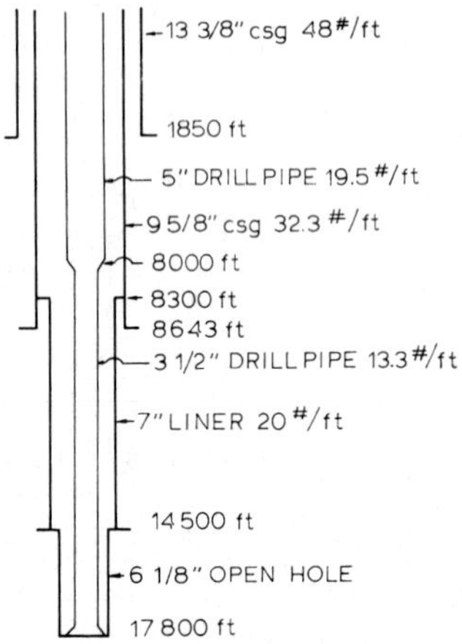

Fig. A-11. Borehole diagram for engineering calculations of Sample Problem No. 1.

(2) Total amount of mud in surface system.
(3) Total amount of mud in the hole with drillpipe in and out of the hole.
(4) Complete mud cycle, hole cycle, and time required for "bottoms up".
(5) Annular velocity in open hole, 7-inch liner, and the $9\frac{5}{8}$-inch casing.
(6) Hydrostatic pressure of mud column at total depth.

(1)
Volume per pit (cu ft) = length (ft) × width (ft) × depth (ft) = 6 × 31 × 7 = 1300 cu ft per pit.
For 3 pits = 1300 × 3 = 3900 cu ft; volume of 3 pits, bbl = 3900 cu ft: 5.61 cu ft/bbl = 700 bbl.
Bbl/inch per 3 pits = (700 bbl)/(84 inches deep) = 8.33 bbl/inch; bbl/ft per 3 pits = 12 inches × 8.33 bbl/inch = 100 bbl/ft.

(2)
Volume (bbl) = (8.33 bbl/inch) (3 pits) × (64 inches + 64 inches + 46 inches) = (2.77 bbl/inch per pit) × (174 inch) = 484 bbl.

(3)
Volume in hole without drillpipe, DP, in bbl = bbl in casing + bbl in open hole.
For $9\frac{5}{8}$-inch casing = 8300 ft × 7.87 bbl/100 ft = 654 bbl; for 7-inch liner = 6200 ft × 4.05 bbl/100 ft = 251 bbl; and for $6\frac{1}{8}$-inch hole = 3300 ft × 3.65 bbl/100 ft = 120 bbl.
Volume without drillpipe in hole = 654 + 251 + 120 = 1025 bbl.
Volume with DP in hole (bbl) = Vol. without pipe (bbl) — Displacement of DP (bbl).

DP displacement (bbl) = (8000 ft × 0.75 bbl/100 ft) + 9800 ft × 0.513 bbl/100 ft) = 60 bbl + 50 bbl = 110 bbl.
Volume in hole (DP in hole) = 1025 — 110 = 915 bbl.

(4)
Total mud cycle (min)

$$= \frac{\text{Hole volume (DP in hole)(bbl) + Surface volume (bbl)}}{\text{Pump output (bbl/min)}} .$$

Pump output (bbl/min) = (34 stk/min × 0.1705 bbl/stk) × (0.90) = 5.21 bbl/min.
Total mud cycle = (915 bbl + 484 bbl)/(5.21 bbl/min) = 1399 bbl/5.21 bbl/min = 269 min.

$$\text{Hole cycle (min)} = \frac{HV - DDP}{PO} = \frac{1025 - 110}{5.21} = 177 \text{ min} .$$

Bottoms up (min)

$$= \frac{\text{Hole volume (bbl)} - [\text{Displacement DP (bbl)} + \text{Capacity DP (bbl)}]}{\text{Pump output (bbl/min)}} .$$

Displacement of DP in bbl from (3) is equal to 110 bbl.
Capacity DP (bbl) = (8000 ft × 1.78 bbl/100 ft) + (9800 ft × 0.742 bbl/100 ft) = 142 bbl + 73 bbl = 215 bbl.

$$\text{Bottoms up (min)} = \frac{1025 \text{ bbl} - (110 \text{ bbl} + 215 \text{ bbl})}{5.21 \text{ bbl/min}}$$

$$= 700 \text{ bbl} : 5.21 \text{ bbl/min} = 134 \text{ min} .$$

(5)
Annular velocity, (100) × ft/min = Pump output (bbl/min) : Annular volume (bbl/100 ft).
Annular volume (bbl/100 ft) = Volume hole (bbl/100 ft) — [Displacement DP (bbl/100 ft) + Capacity DP (bbl/100 ft)].
Annular volume of open hole = 3.65 bbl/100 ft — (0.513 bbl/100 ft + 0.742 bbl/100 ft) = 2.395 bbl/100 ft.
Annular velocity in open hole (ft/min) = (5.21 bbl/min) : (2.395 bbl/100 ft) = 218 ft/min.
Annular volume of 7-inch casing (bbl) = 4.05 bbl/100 ft — (0.513 bbl/100 ft + 0.742 bbl/100 ft) = 2.79 bbl/100 ft.
Annular velocity in 7-inch casing (ft/min) = (5.21 bbl/min) : (2.79 bbl/100 ft) = 187 ft/min.
Annular volume of $9\frac{5}{8}$-inch casing (bbl) = 7.87 bbl/100 ft — (0.75 bbl/100 ft + 178 bbl/100 ft) = 5.34 bbl/100 ft.
Annular velocity in $9\frac{5}{8}$-inch casing (ft/min) = (5.21 bbl/min) : (5.34 bbl/100

ft) = 98 ft/min. On considering 300 ft of $3\frac{1}{2}$-inch DP, effective annular velocity will be equal to 97 ft/min.

(6)
Depth (ft) × 0.052 × wt (lb/gal) = Hydrostatic pressure (psi).
17,800 ft × 0.052 × 16.3 lb/gal = 15,200 psi.

Sample problem No. 2

Given

Surface casing	— $13\frac{3}{8}$ inches (339.7 mm), 2000 ft (609.6 m), 48 lb/ft (71.4 kg/m).
Total depth	— 10,000 ft (3048 m).
Bit size	— $9\frac{7}{8}$ inches (250.8 mm).
Drillpipe (DP)	— $4\frac{1}{2}$ inches (114.3 mm), 16.60 lb/ft (24.7 kg/m).
Surface system	— three pits 7 ft × 6 ft × 31 ft with 60 inches of mud in all three pits. (2.13 m) × (1.83 m) × (9.45 m) with 1.52 m of mud
Mud weight	— 12.5 lb/gal (1497 kg/m^3 or 1.497 SG)
Pump	— 6 inches (152 mm) × 16 inches (406 mm) at 50 spm; 90% efficiency.

Calculate
(1) Total capacity of surface system in m^3, m^3/m, and m^3/cm; (2) total amount of mud in surface system (m^3); (3) total amount of mud with and without DP in hole (m^3); (4) complete mud cycle, bottoms up (min); (5) annular velocity in open hole and in casing (cm/sec); and (6) hydrostatic pressure at total depth (kg/cm^2).

(1)
Volume per pit, m^3 = $L \times W \times D$ = (2.13 m)(1.83 m)(9.45 m) = 36.8 m^3 per pit.
Volume in 3 pits = 3 × 36.8 m^3 (for 3 pits) = 110.4 m^3 (for 3 pits).

$$\text{Thus capacity is equal to } \frac{110.4 \text{ m}^3}{1.83 \text{ m}} = 60.3 \text{ m}^3/\text{m for 3 pits},$$

or

$$\frac{110.4 \text{ m}^3}{183 \text{ cm}} = 0.603 \text{ m}^3/\text{cm for 3 pits}.$$

(2)
Surface volume, m^3 = (60.3 m^3/m)(1.52 m) = 91.6 m^3.

(3)
Volume in hole without DP (m^3) = Volume in csg (m^3) + Volume in open hole (m^3).
For $13\frac{3}{8}$-inch (339.7-mm) csg = 0.1570 bbl/ft × 2000 ft = 314 bbl, or 314 bbl (0.1590) = 49.9 m^3.
For $9\frac{7}{8}$-inch (250.8-mm) hole = (0.0947 bbl/ft) (8000 ft) = 757.6 bbl, or 757.6 bbl (0.1590) = 120.4 m^3.
Total volume without DP in hole = 120.4 + 49.9 = 170.3 m^3.
Volume with DP in hole = Volume without DP -- Volume of displacement of DP.
DP displacement = (0.00643 bbl/ft)(10,000) = 64.3 bbl, or 64.3 bbl (0.1590) = 10.2 m^3.
Volume with DP in hole = 170.3 m^3 — 10.2 m^3 = 160.1 m^3.

(4)

Total mud cycle

$$= \frac{\text{Hole volume (DP in hole) + Surface volume}}{\text{Pump output}}$$

Pump output = (50 spm)(0.1705 bbl/stk)(0.90) = 7.67 bbl/min, or = (7.67 bbl/min)(0.159) = 1219.9 dm^3/min or 1.22 m^3/min.
Total mud cycle = (160.1 m^3 + 91.6 m^3) : 1.22 m^3/min = 206 min.

Bottoms up

$$= \frac{\text{Hole volume} - \text{(Displacement of DP + Capacity of DP)}}{\text{Pump output}}$$

Capacity of DP (0.01422 bbl/ft)(10,000 ft) = 142.2 bbl, or = 142.2 bbl (1.1590) = 22.6 m^3.

$$\text{Bottoms up} = \frac{170.3\ \text{m}^3 - (10.2\ \text{m}^3 + 22.6\ \text{m}^3)}{1.22\ \text{m}^3/\text{min}} = 112.7\ \text{min}\ .$$

(5)

$$\text{Annular velocity}\ \frac{1030(\text{bbl/min})}{D_h^2 - D_p^2}$$

$$\text{Annular velocity in open hole} = \frac{1030(7.67\ \text{bbl/min})}{(9\frac{7}{8})^2 - (4\frac{1}{2})^2} = 102\ \text{ft/min}\ ,$$

or = (102 ft/min)(0.508) = 51.82 cm/sec .

$$\text{Annular velocity in csg} = \frac{(1030)(7.67\ \text{bbl/min})}{(13\frac{3}{8})^2 - (4\frac{1}{2})^2} = 49.8\ \text{ft/min}\ ,$$

or (49.8 ft/min)(0.508) = 25.3 cm/sec.

(6)
Hydrostatic pressure = wt (lb/gal) × depth (ft) × (0.052) = (12.5 lb/gal) × (10,000 ft)(0.052) = 6500 psi, or = (6500 psi)(0.0703) = 456 kg/cm².
Alternate method:

$$\text{Hydrostatic pressure} = \frac{(\text{kg/m}^3)(\text{m})}{10{,}000} = \frac{(1497)(3048)}{10{,}000} = 456\ \text{kg/cm}^2\,.$$

Sample problem No. 3

Given

Specific wt of mud	= 13.1 lb/gal.
Chloride content	= 100,000 ppm.
Mud retort test (% vol water)	= 66%.
Mud retort test (% vol oil)	= 9%.
Mud retort test (% vol solids)	= 25%.
Methylene blue test result	= 30 lb/bbl.

Procedure (use Fig. A-8)
Step (1) Vol% salt water:
a. Retort vol. % water = 66% (ordinate).
b. Chloride content = 100,000 ppm (inclined lines).
c. Volume % salt water = 70% as read on the abscissa.
Step (2) Specific gravity of solution:
a. Chloride content = 100,000 ppm.
b. Specific gravity of solution = 1.11 as read on the ordinate.
Step (3) Density adjustment (lb/gal):
a. From 9% oil by volume (retort) move up to 0.8 specific gravity line.
b. Move horizontally to the right and read 0.15 lb/gal density adjustment for oil.
c. Proceed down from 70% salt water solution to SG of solution (1.11) line, read —0.65 lb/gal adjustment for salt.
Total density adjustment = —0.65 + 0.15 = —0.50 lb/gal.
Step (4) Volume % of low-specific-gravity solids (SG = 2.6):
a. Mud weight adjusted for oil and SG of water phase = 13.1 — 0.50 = 12.6 lb/gal.
b. Determine vol. % of suspended solids; vol. % of salt water = 70%.
Water + 9% oil = 79% fluids.
100 — 79% = 21 % suspended solids.
c. Proceed down from 12.6 lb/gal to intersect volume % of suspended solids of 21%.
d. Proceed horizontally from intersection to volume % of 2.6-SG solids (= 10%).

Step (5) Bentonite content (lb/bbl):
a. Methylene blue test result = 30 lb/bbl.
b. Proceed along inclined line to intersect 10% volume line of 2.6-SG solids.
c. Proceed vertically to read 22.5 lb/bbl bentonite.
d. Assuming 9 lb/bbl of low-gravity solids = 1 vol. % solids, then:
Vol. % bentonite = (22.5 : 9) lb/bbl bentonite = 2.5%.
Vol. % drilled solids = 10% — 2.5% = 7.5%.
Wt drilled solids = 7.5 × 9 = 67.5 lb.

MUD REPORT

Some sort of permanent record should be made every time the mud engineer runs a mud check. The crews, the tool pushers, and management look to the mud engineer for direction and control of the drilling fluid. Every service company that is concerned with the well from spudding-in to completion must have the help of the mud engineer and the mud report.

Mud reports should be made as complete and accurate as possible. They constitute not only a permanent record of engineer's service and responsibility to the company, but also serve as a principal contact with the management. Inasmuch as they are commonly prepared in quintuplicate, the use of a well-sharpened medium-hard pencil is recommended. A sample mud report is presented in Fig. A-1.

The first copy of the mud report is sent to the operating company, whereas the second copy is kept at the rig. The third copy is sent to the district manager and is filed later in the division office. The fourth copy is a personal copy of the engineer and the fifth copy is an extra copy, unless the operator wants an extra copy. In the latter case, fourth copy is given to him. The fifth copy should be retained in the engineer's active file until the well is completed and then transferred to his inactive file and filed under areas or companies. Thus, the longer an engineer remains in a territory, the more valuable he can become to his own company and to the operating companies in his area.

Appendix B

SAMPLE EXPERIMENTS

GEORGE V. CHILINGARIAN and PAUL VORABUTR

EXPERIMENT No. 1 — PROPERTIES OF FRESH-WATER MUDS

Procedure

(1) Prepare a mud with Na-montmorillonite clay (e.g., aquagel), having a viscosity of about 45 sec Marsh funnel. Use tap water and mix for at least 15 minutes.
(2) Prepare a mud with a native clay or Ca-montmorillonite *, having a Marsh funnel viscosity of about 45 sec.
(3) Add 1% by weight of sand to one 500-cm^3 sample.
(4) Test the following properties of these muds:
(1) viscosity (both plastic and apparent); (2) yield point (lb/100 sq ft); (3) initial and 10-min gel strengths; (4) corrected 30-min water loss (cm^3); (5) corrected 30-min cake thickness in $\frac{1}{32}$ inch; (6) pH; (7) conductivity; (8) sand content of one sample; (9) initial and 10-min Shearometer gel strength (lb/100 sq ft); (10) specific weight in lb/cu ft and lb/gal.

Questions

(1) Discuss operation of various mud testing equipment.
(2) How does one calibrate rotational viscometer (e.g., V.G. meter)?
(3) Explain in detail differences between apparent and plastic viscosities. What is the significance of each?
(4) Assuming that the volumes of water and of clay are additive and that the specific gravity of clay is equal to 2.6, compute the wt % and vol % of clay in the two muds tested. Determine the yield of these two muds in bbl of mud/ton of clay, assuming a specific gravity of 2.6 for clays.
(5) What is the relationship between the viscosity as measured by the rotational viscometer and by the Marsh funnel?
(6) Leave samples of these muds overnight and determine their properties again. Explain all observed differences.

---

* Note that Ca-montmorillonite swells considerably less than Na-montmorillonite.

EXPERIMENT No. 2 — EFFECT OF WEIGHTING MATERIAL (BARITE, $BaSO_4$) ON PROPERTIES OF FRESH-WATER MUDS

Procedure

(1) Prepare a mud with bentonite, having a viscosity of about 45 sec Marsh funnel. Measure the specific weight of this mud.
(2) Add enough $BaSO_4$ to raise the specific weight of mud to about 75 lb/cu ft. Determine the weight of $BaSO_4$ added.
(3) Determine *all* the properties of the initial mud and final (weighted) mud.
Half of the class can determine properties of the initial mud, whereas the other half can measure properties of the weighted mud. The results should be recorded on the blackboard and discussed by the instructor.

Questions

(1) Discuss and explain differences between the properties of the initial and weighted muds.
(2) Calculate the weight of $BaSO_4$ necessary to raise the specific weight of unit volume of the initial bentonite mud to 75 lb/cu ft. Compare this calculated figure with the actual weight of the $BaSO_4$ added.
(3) What effect would the size of ground $BaSO_4$ have on (a) the apparent and plastic viscosities of the mud measured with rotational viscometer and (b) viscosity measured with Marsh funnel?

EXPERIMENT No. 3 — EFFECT OF THINNING AGENTS

Procedure

(1) Prepare Na-montmorillonite fresh-water mud having a viscosity of about 55 Marsh funnel seconds (original mud).
(2) Add sodium-acid-pyrophosphate (SAPP) or sodium-hexametaphosphate in the following concentrations to three different samples of the original mud: 0.1 g/100 cm^3 mud, 1 g/100 cm^3 mud, and 2 g/100 cm^3 mud.
(3) Add quebracho (tannin) in the following concentrations: 1 lb/bbl of original mud, 2 lb/bbl mud, and 3 lb/bbl mud. Adjust pH to 9.6 with NaOH (commonly 1 : 2 caustic/quebracho ratio). Caustic soda and quebracho should be added simultaneously.
(4) Add lignite in the following concentrations: 1 lb/bbl mud, 2 lb/bbl mud, and 3 lb/bbl mud. Adjust pH to 9.5 with NaOH.
(5) Add ferrochrome-lignosulfonate in the following concentrations: 1 lb/bbl, 2 lb/bbl, and 5 lb/bbl of initial thick mud. Adjust pH to 9 with NaOH.

(6) Determine viscosity (both plastic and apparent), yield point, initial and 10-min gel strengths, water loss, cake thickness, and pH for each sample. Groups of 2 to 3 students should be able to test two samples. Then the results should be exchanged by recording them on the blackboard.
(7) Plot viscosity, yield point, and gel strength versus the thinner concentration.

Questions

(1) Explain theory behind reduction of viscosity by various thinners.
(2) Why some thinners do not work at low pH?
(3) What is the advantage of using a 50 : 50 mixture of quebracho and lignite?
(4) What is the theory behind the reduction of water loss on addition of lignite to clay—water muds?

EXPERIMENT No. 4 — EFFECT OF NaCl AND CEMENT [$Ca(OH)_2$] CONTAMINATION ON PROPERTIES OF FRESH-WATER MUDS

Procedure

(1) Prepare a fresh-water mud, having a viscosity of about 45 Marsh funnel seconds, with bentonite clay.
(2) Contaminate four samples of original mud with 0.1, 1.0, 5.0, and 10.0 g of NaCl per 100 cm^3 of mud.
(3) Contaminate four more samples of original mud with 0.1, 1.0, 2.0, and 3.0 g of $Ca(OH)_2$ per 100 cm^3 of mud.
(4) Determine the following properties of the original mud and of the contaminated mud samples: viscosity (both plastic and apparent), yield point (lb/100 sq ft), gel strength (initial and 10-min), corrected 30-min water loss, corrected 30-min cake thickness, and pH.
(5) Plot various properties of muds versus content of contaminant in percent of original mud.

Each group of 2 to 3 students should be able to determine properties of 2 to 3 samples. Mix samples for at least 15 minutes. All results should be then reported on the blackboard and discussed by the instructor.

Questions

(1) How many g per 100 ml of mud is equivalent to 1 lb/bbl? Show all computations.
(2) Discuss the effects of NaCl and $Ca(OH)_2$ on various properties of fresh-

water muds. Explain the reasons (theoretical).

(3) Prescribe treatments for these contaminated muds, giving reasons for your recommendations.

Test your recommendations during a separate laboratory period (Experiment No. 5). Keep contaminated samples for the next laboratory period.

EXPERIMENT No. 6 — INTERRELATIONSHIPS AMONG VISCOSITY (BOTH PLASTIC AND APPARENT), GEL STRENGTH, YIELD POINT, WATER LOSS, YIELD (BBL/TON), SPECIFIC WEIGHT, AND PERCENTAGE OF SOLIDS

Procedure

(1) Determine viscosity (both plastic and apparent), gel strength, yield point, water loss, yield (bbl/ton), and specific weight of the muds containing the following percentages of (1) Na-montmorillonite — 2% wt, 5% wt, 8% wt, and 12% wt; and (2) some native clay or Ca-montmorillonite — 10% wt, 15% wt, 30% wt, and 40% wt.

(2) Plot viscosity (both plastic and apparent) versus percentage of solids by weight on both semilog and cartesian papers. Add the scales for the following properties on the graph: corrected 30-min water loss, cm^3 (ordinate), percentage of solids by volume (abcissa), specific weight in ppg and pcf (abcissa), and yield in bbl of mud per ton of clay (abcissa).

(3) Each group of students (2 or 3 per group) can test at least two samples containing different concentrations of clay and then exchange results with other groups. It is better, however, to report results on the blackboard and discuss them. Assume that sp. gr. of clay is equal to 2.6.

Questions

(1) Which one of the clays would give the desired mud viscosity with a lower specific weight? What would be the advantage of such a mud in a depleted field?

(2) In drilling a well to a depth of 5000 ft, it is desired to have an approximate plastic viscosity of 25 cP. Which one of the clays tested will fulfill the necessary requirements best? Determine the lb of clay per bbl of mud that will be required.

(3) Calculate the volume of fresh water in bbl that must be added per bbl of mud having a density of 95 lb/cu ft, to just offset a formation pressure of 3570 psi at 6250 ft.

(4) Compute the percentage of solids by weight from the specific weight measurements of muds with mud balance, assuming additive volumes and

weights. Then compare these values with the actual weight percentages of clays added. Discuss the differences!
(5) Discuss interrelationships between the various properties of muds tested.

EXPERIMENT No. 7 — SATURATED SALT-WATER MUDS

Procedure

(1) Prepare saturated salt water (≈315,000 ppm NaCl).
(2) Add 4% by weight of zeolitic clay (attapulgite). Prepare 6 samples, 350-cm^3 each.
(3) Add pregelatinized starch in the following concentrations: 1 lb/bbl, 3 lb/bbl, 5 lb/bbl, and 10 lb/bbl of mud. Use bactericide such as paraformaldehyde — 1/3 to 1/2 lb/bbl of mud.
(4) Add 7 lb/bbl of some Iranian gum, such as Tragacanth and Shiraz, or Indian gum Ghatti. If Guar gum is used, its concentration should be around 2—3 lb/bbl of mud.
(5) Add 3 lb/bbl of CMC (carboxy-methyl-cellulose).
(6) Determine viscosity (both plastic and apparent), yield strength, initial and ten-min gel strengths, water loss, cake thickness, and pH of all samples.
(7) Plot concentration of pregelatinized starch versus viscosity, water loss, yield point, and gel strengths.

Questions

(1) What is the theory behind the reduction of water loss of drilling muds through the use of starch and gums?
(2) Compare the function of bentonite in fresh-water muds versus that of attapulgite (zeogel) in salt-water muds.
(3) What is fermentation (use formulas)? How does one prevent fermentation (list all methods)?
(4) In drilling in salt dome areas, what other muds can be used successfully? Why?

EXPERIMENT No. 8 — RED LIME INHIBITIVE DRILLING MUD

Procedure

(1) Prepare a red-lime mud by adding the following constituents to 1000 cm^3 of tap water in the order indicated:
(a) 2.5 lb/bbl of $Ca(OH)_2$.

(b) 6% by weight of Na-montmorillonite.
(c) 3 lb/bbl quebracho plus 3 lb/bbl NaOH.
(d) 10 lb/bbl of pregelatinized starch.

(2) Determine *all* the properties of this mud.

(3) Contaminate one sample of this mud with cement (2 lb/bbl of $Ca(OH)_2$) and the other with 3 lb/bbl of NaCl. Determine *all* the properties again. Exchange data and compare results.

Questions

(1) Explain why the red-lime mud constituents are added in the order given above.
(2) What is the purpose of adding $Ca(OH)_2$?
(3) Where are calcium inhibitive muds used?
(4) What effect does lime concentration have on the temperature stability of these muds?

Note

Caustic soda (lb/bbl)	Resulting pH
$\frac{1}{4}$	8 – 8.5
$\frac{1}{3}$	8.5
$\frac{1}{2}$	9.5–10.0
$\frac{3}{4}$	10.0–10.5
1	11 –11.5
2	12 +
3	12.5 +
4	12.5 +

EXPERIMENT No. 9 — OIL-BASE AND EMULSION MUDS

Procedure

(1) Prepare the following muds and test *all* their properties:
(a) High-pH emulsion mud — first prepare 4% by weight of Na-montmorillonite mud and then add 5 lb/bbl of NaOH and 15 lb/bbl of Carbonox *. This is followed by adding 30% by volume of diesel oil and, finally, 2 lb/bbl of lime.
(b) Low-pH emulsion mud — prepare the mud as above except for using 10% by volume of oil and 1.5 lb/bbl of caustic. Lime should not be added.

* Carbonox (organic acid) is a partially refined lignite which is used to emulsify oil in fresh-water muds.

(c) Add diesel oil (182 lb/bbl) to water (50 lb/bbl), followed by $CaCl_2$ (26 lb/bbl), emulsifier (soap, polyamide — 15 lb/bbl), filtrate reducer (amine lignite — 10 lb/bbl), gellant (amine clay — 2 lb/bbl), and finally, barite (325 lb/bbl) to obtain a mud having specific weight of 14 ppg.
(d) Add 60—80% by volume of aqueous phase saturated with either NaCl or $CaCl_2$ to 20—40% by volume of oil, together with 2—10 lb/bbl of anhydrosorbitol monooleate or of polyoxyethylene sorbitol tetraoleate (water-in-oil emulsifier). Then add 20 lb/bbl of limestone having a particle size of minus 325 mesh screen.

Group of 2 to 3 students should prepare and test one particular drilling fluid. Results should then be recorded on the blackboard and discussed by the instructor. Suggest addition of other chemicals if properties of a drilling fluid are unsatisfactory.

Questions

(1) Distinguish between the chemical and mechanical emulsification.
(2) What are the advantages and disadvantages of oil-in-water and water-in-oil (invert) emulsions.
(3) How does one increase the viscosity of emulsion muds?

EXPERIMENT No. 10 — SETTLING OF SAND IN DRILLING FLUID

Object

In drilling an oil well a great deal of care is taken to prevent a complete shut-down while the drillstring is in the hole. The reason for this is that the mud is carrying cuttings that may be dropped when the velocity is reduced or becomes zero. Settling of these particles around the tools may result in a costly fishing job. The object of this experiment is to determine the amount of particles that can be held in suspension and their rate of settling as related to the size of particles and the shear strength of the mud. It is not inferred, however, that the shear strength and the particle size are the only factors affecting the rate of settling.

In this experiment, sand particles are used because of the ease of regulating their size and convenience of handling.

Procedure

First week

(1) Bentonite (or aquagel) is mixed with tap water to a viscosity of about 45 Marsh funnel sec API, and the mud is hydrated thoroughly for a period of 24 hours.

(2) Sand is screened to provide sufficient amounts in 14—28 and 28—48 mesh sizes.

Second week

(1) By the addition of chemicals, two samples of drilling mud are adjusted to a viscosity of about 45 Marsh funnel sec API and a shear strength of 0 at 0 time, but about 4 lb/100 sq ft at 10 min for the first and about 8 lb/100 sq ft for the second sample.
(2) Sand content of this drilling fluid is then determined.
(3) A volume of 450 cm^3 of this mud, with a shear strength of about 4 lb/100 sq ft, and 50 cm^3 of 14—28 mesh sand are placed in the mixing jar and thoroughly mixed. This gives rise to about 10% sand plus the original sand content of the mud.
(4) The thoroughly mixed mud and sand is poured into three 100-ml graduated cylinders. These samples are allowed to stand quiescent and the time is recorded.
(5) After 5 min, separate the top one-third of the sample and introduce it into a small beaker. Then separate the middle one-third and introduce into another beaker. The bottom one-third of the sample may remain in the graduated cylinder.
(6) Determine the sand content of the top one-third, and place the wet sand in the appropriately labeled jar. Repeat this test on the other portions of the mud.
(7) After 10 min, make a similar test on the mud in one of the other graduated cylinders.
(8) After 20 min (from the start), make a similar test on the mud in the remaining graduated cylinder.
(9) Repeat steps 3 to 8 using 28—48 mesh sand.
(10) Repeat steps 3 to 8 using mud with a 10-min shear strength of about 8 lb/100 sq ft.

Note:

Steps 7 and 8 probably will not be necessary in the case of the 14—28 mesh sand.

Questions

(1) Compare degree of sand settling in the top, middle, and bottom one-third portions of mud samples. Describe the method you used.
(2) How do the properties of mud and particles affect the rate of settling (include those mentioned in this experiment)?
(3) Based on the obtained results, what would occur when circulation is stopped and this type of mud contains (a) sand of 28-mesh size and larger, (b) 28- to 48-mesh sand, and (c) sand smaller than 48 mesh in size?

(4) May one assume these results to be true for any drilling fluid? State briefly the reasons for your answer.
(5) If the mud is allowed to stand in the slush pit at least 30 min before being circulated again and it is desirable to have a minimum of sand content in the mud being pumped into the hole, under what conditions would screening be most necessary?
(6) After what period of time equilibrium conditions are reached in the shut-down wellbore?

Problems

(1) Consider the motion of spherical particles settling through a fluid. The Reynolds number, N_R, is equal to $dv\rho/\mu$, the symbols being defined below. For a Reynolds number from 2 to 1000, the gravitational settling velocity for spherical particles may be expressed as follows:

$$v = 0.153\, g^{0.714}\, d^{1.142}\, (\rho_s - \rho_f)^{0.714} / \rho^{0.286}\, \mu^{0.428}$$

where v = gravitational settling velocity of particles relative to fluid (cm/sec); g = local acceleration due to gravity, in (cm/sec)/sec; d = diameter of particle (cm); ρ_s = true (not bulk) density of spherical particle (g/cm^3); ρ_f = density of dispersion medium or fluid (g/cm^3); and μ = viscosity of dispersion medium or fluid, in poises or g/(cm)(sec).
Compute the minimum size of water droplets which will settle completely from . . . . ft of . . . .° API oil having a viscosity of . . . . cP, during a . . . . hr period.
(2) Spherical particles with a specific gravity of . . . . and diameters close to . . . . mm are settled in water having a specific gravity of . . . . and a viscosity of . . . . cP. Calculate the settling velocity and the Reynolds number.
(3) The spherical particles described in Problem 2 are settled in air having a specific weight of . . . . lb/cu ft and a viscosity of . . . . cP. Compute the settling velocity and the Reynolds number.
(4) For the systems in Problems 2 and 3, calculate the settling velocities and the Reynolds numbers from the following expression of Stokes' law (see derivation below)

$$v = g\, d^2(\rho_s - \rho_f)/18\mu$$

where the symbols used are the same as in the Problem 1.

Derivation of Stokes' law equation *

For Reynolds number, N_R, below about 0.4, the drag coefficient, C_D, for a sphere is equal to $24/N_R$. Thus, for a laminar or viscous flow the drag force

* By George V. Chilingar (see "Surface Operations in Petroleum Production" by G.V. Chilingar and C.M. Beeson. Am. Elsevier Publ. Co., Inc., 1969).

in lb is equal to:

$$D = C_D \rho \frac{v^2}{2} A = \frac{24}{\rho v d/\mu} \rho \frac{v^2}{2} \frac{\pi d^2}{4} = 3\mu v \pi d \tag{1}$$

where ρ is mass per unit volume in slugs/cu ft, v is velocity in ft/sec, d is diameter of sphere in ft, A is the largest projected area in sq ft, and μ is viscosity in lb-sec/sq ft (or slug/ft-sec). *

Inasmuch as the buoyant force, B, and drag force, D, are acting in upward direction, whereas the weight, W, in lb is acting down:

$$B + D = W \tag{2}$$

or

$$\left(\frac{\pi d^3}{6}\right) \gamma_{\text{fluid}} + 3\mu v \pi d = \left(\frac{\pi d^3}{6}\right) \gamma_{\text{solid}} \tag{3}$$

where γ is specific weight in lb/cu ft ($\gamma = \rho g$); g (gravitational acceleration) = 32.174 ft/sec sec.

Thus:

$$\left(\frac{\pi d^3}{6}\right) (\gamma_s - \gamma_f) = 3\mu v \pi d \tag{4}$$

and

$$v = \frac{d^2(\gamma_s - \gamma_f)}{18\mu} \tag{5}$$

References

McCabe, W.L. and Smith, J.C., 1956. *Unit Operations of Chemical Engineering.* McGraw-Hill, New York, N.Y., 361 pp.

Lapple, C.E., 1954. *Fluid and Particle Mechanics.* Edwards Brothers, Inc., Ann Arbor, Mich., 284 pp.

EXPERIMENT No. 11 — WELL SURVEYING

Object

This experiment is intended to familiarize the student with the operation of a well-survey instrument and with calculations involving the data obtained.

---

* 1 poise = $2.089 \cdot 10^{-3}$ lb-sec/sq ft.

Discussion

The Eastman single shot is a directional survey instrument which records simultaneously the magnetic direction of the course of the borehole and the deviation of the well bore from the vertical. This is accomplished by photographing the location of a plumb bob with reference to a compass card within an angle-indicating unit.

The single shot is an important aid in controlled directional drilling. A survey made with the instrument is dependable because it is composed of a large number of separate and individually taken readings. Such a survey avoids many types of cumulative errors. The survey is conveniently obtained and is easily made without the help of specially trained personnel.

The single-unit instrument comprises a battery section at the top with flashlight batteries, a timing control section, a camera section, and a lower section containing the angle unit. The range of the instrument used can vary from 0°—18° drift to 0°—90° from vertical.

It is suggested that the instructions accompanying the instrument be read carefully and be understood completely before the experiment is begun, because the instruments are expensive.

The directional base, used as a holder for the instrument in carrying out the experiment, can be built to revolve through 360° on the base, and to enable swinging of the barrel 90° from vertical in two directions. It is important that the barrel is not swung through an angle greater than the maximum range (from vertical) of the instrument.

The direction of magnetic north indicated by the compass in the directional base may differ radically from the earth's magnetic north, because of the magnetic attractions of materials which surround the laboratory. In setting the drift angle, the bottom of the barrel should point in the direction of the drift. The barrel should not be moved without first loosening the lock-nut. At no time after the instrument has been placed in the stand, should any part of the stand or instrument be moved until after the picture has been taken.

Procedure

After the base is properly oriented, take pictures using any five different drift angles and drift directions.

Problems

(1) Using the values supplied by the instructor, compute the true (total) vertical depth and the total course deviation (horizontal distance from wellbore origin) at each point. Plot these points in a plan view and in a section view (side elevation). Label the points with the corresponding

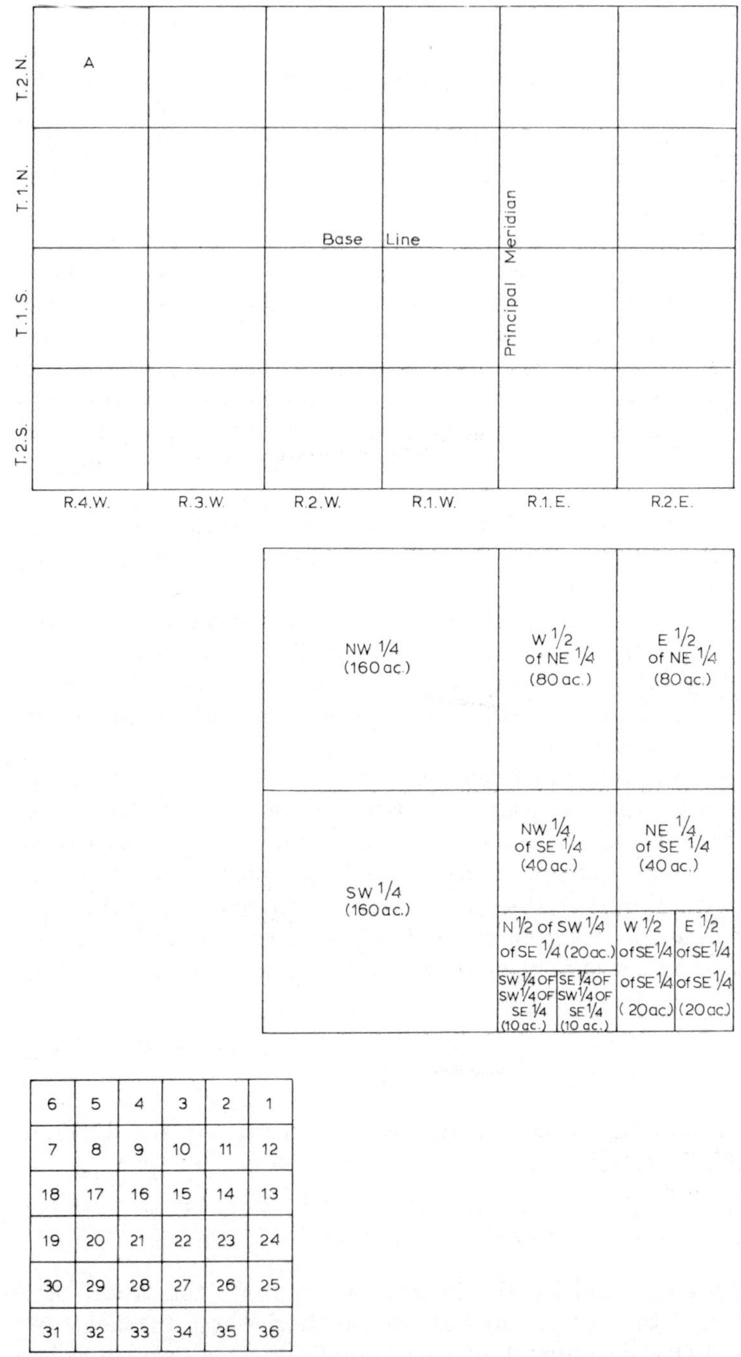

Fig. B-1. Methods of designating townships and ranges, numbering sections within a township, and designating parts of a section. (See F.H. Lahee, 1941. *Field Geology*, McGraw-Hill, New York and London.)
Township marked *A* = T.2N., R.4W. (township 2 north, range 4 west).

measured depths, and connect the points to show the well course in both views.

Tabulate the data supplied and your computations in the form given below:

Station No.	Measured depth	Course length	Drift angle	Drift direction
1	. . . . . .	. . . . . .	. . . . . .	. . . . . .
2	. . . . . .	. . . . . .	. . . . . .	. . . . . .
3	. . . . . .	. . . . . .	. . . . . .	. . . . . .
4	. . . . . .	. . . . . .	. . . . . .	. . . . . .
5	. . . . . .	. . . . . .	. . . . . .	. . . . . .
6	. . . . . .	. . . . . .	. . . . . .	. . . . . .

(2) A well is located . . . . ft west and . . . . ft north from the center of NW$\frac{1}{4}$ of section . . . .. A vertical hole was drilled to a depth of . . . . ft, at which depth a whipstock was set to deflect the hole S . . . .°E. Thereafter, the hole picked up inclination gradually (assume that the curved part is an arc of a circle) at the rate of . . . .° per . . . . ft of drilled depth for a drilled distance of . . . . ft from the "kick-off" point. The inclination remained constant after becoming . . . .°. If the measured depth to the top of a particular formation is . . . . ft, the measured depth to the bottom of the formation is . . . . ft, and the dip and strike of the formation are . . . .° and N . . . .°W, respectively, calculate the formation thickness.

(3) The following data apply to an exploratory well:
Location: . . . . ft east and . . . . ft south from the center of NW$\frac{1}{4}$ of SE$\frac{1}{4}$ of section . . . ..
Elevation: . . . . ft above sea level.
Deviation: hole is vertical.
It is planned to redrill the well, and the target is located at a depth (vertical) of . . . . ft below sea level (vertical subsea) and . . . . ft N . . . .°S from the present site. Hole angle will be increased gradually (assume that the curved part is an arc of a circle) at the rate of . . . .° per . . . . ft of drilled depth to a maximum of . . . .°. Thereafter, the hole will be maintained at a deviation of . . . .° until the target is reached.
(a) Compute the changes in horizontal distance and vertical depth during buildup of angle and while the angle is maintained.
(b) At what measured depth should the redrill start, that is, at what measured depth should the increase of angle begin?
(c) What will be the measured depth at the target in the redrilled hole?
(d) What are the horizontal coordinates of the target location, with reference to the southwest corner of section . . . .?

References

Instruction Bulletin of a company whose apparatus is being used.
Billings, M.P., 1954. *Structural Geology*. Prentice-Hall, Englewood Cliffs, N.J., 853 pp.

Appendix C

SAMPLE QUESTIONS AND PROBLEMS

GEORGE V. CHILINGARIAN and PAUL VORABUTR

Questions

(1) What is the relative effect of barite ($BaSO_4$) on viscosity of a water-base mud as measured by Marsh funnel and rotational viscosimeter? Explain why.
(2) Why is $NaHCO_3$ commonly used instead of Na_2CO_3 as a treating agent for cement contamination?
(3) Is the ten-minute gel strength measured by the Shearometer higher or lower than the true ten-minute gel strength? Explain why.
(4) Explain why quebracho will not work in a mud of low pH. Illustrate with a diagram.
(5) List three types of clay (with formulas) and explain their relative degree of swelling.
(6) What is a surfactant mud and where could it be used best? How does it compare to high-pH red lime-base mud?
(7) Give amounts of various components in water-in-oil and oil-in-water drilling fluids (choose examples you are most familiar with). Give reasons for adding each component. In what order are they added and why?
(8) Explain why the muds have thixotropic properties, i.e., develop gels on standing. Why do surfactant muds have low gel strengths?
(9) What are the significance of n and K values in Power law model? Discuss in detail.
(10) How would you prepare a *conductive* oil mud?
(11) What are the methods of predicting an abnormal pressure buildup and what causes it?
(12) Some researchers proved that polyphosphates such as sodium-acid-pyrophosphate reduce viscosity of muds as a result of plating them with negative charges. How did they prove it (explain in detail)?
(13) Will hydrophilic solids stabilize (a) an oil-in-water emulsion? (b) water-in-oil emulsion? Explain the reasons.
(14) How do electric, induction, acoustic, and density logs are used in detecting abnormally high-pressure zones? Explain using schematic diagrams.
(15) How do you calculate KWM?
(16) An aquagel (bentonite) mud was contaminated with 7 g/100 cm^3 of

NaCl. As a result, apparent viscosity increased from the initial value of 15 cP to 55 cP and water loss from 16 cm^3 to 58 cm^3 (corrected, 30-min). How would you treat this mud?
On adding 2 g/100 cm^3 of quebracho plus 1 g/100 cm^3 of NaOH, the viscosity increased further to 93 cP and water loss to 62 cm^3. Explain this behavior and prescribe future course of action. pH = 11.4.

(17) An initial aquagel (bentonite) mud having viscosity of 47 sec Marsh funnel and corrected 30-min water loss of 16 cm^3 was contaminated with 2 g/100 cm^3 of $Ca(OH)_2$. As a result water loss increased to 75 ml and viscosity to 150 sec M.F. On treating this mud with 2 g/100 cm^3 of $NaHCO_3$, the water loss dropped to 46 ml, whereas viscosity increased further to 170 sec M.F. Explain why.

(18) Explain lost circulation due to an overbalanced mud system.

(19) List five facts for improper solids control.

(20) Explain how to use modified *d*-exponent.

(21) Explain why water in shales is much fresher than that in associated sandstones.

(22) What effect(s) would SAPP have on water loss? Why?

(23) List typical components of an oil-base mud. Discuss each component in detail (functions, etc.)

(24) Why does pH of the mud decrease as a result of salt (NaCl) contamination?

(25) Describe methods of increasing viscosity of emulsion muds.

(26) How many g/100 ml of mud are equivalent to 1 lb/bbl? Show calculations.

(27) Describe theories behind the reduction of water loss of drilling muds through the use of starch, gums, and polymers.

(28) List all types of inhibitive (inhibited) muds.

(29) What are the typical components of (a) invert emulsion muds and (b) oil muds.

(30) Draw schematic yield value-versus-plastic viscosity curves for (1) clay + water system without chemicals, (2) stabilized (with chemicals) system, and (3) flocculated system. Which way would a point on curve (2) move on addition of clay? Which way would a point on curve (1) move on addition of water, clay, or dispersant? Which way would a point on curve (3) move on addition of water and clay?

(31) Using Smoluchowski's equation estimate the increase in viscosity due to charges on clay particles.

(32) Is water loss proportional to pressure if the filter cake is compressible or not? Why?

(33) Prescribe pretreatment for cement contamination (at least three). Which one is better and why?

(34) Why is the turbulent flow better than laminar flow for carrying the cuttings to the surface?

(35) What are the relative salinities of interstitial fluids in undercompacted and well-compacted shales and in their associated sandstones? Explain.

(36) Should mud filtrate into the formation have a higher or lower Na/Ca ratio than the formation water? Explain why.

(37) What is the effect of NaOH on various properties of clay-water muds? Draw curves showing relationship between NaOH concentration and various drilling fluid properties.

(38) Describe the following clay-water systems: (1) non-flocculated but aggregated, (2) aggregated and flocculated, (3) dispersed but flocculated, (4) nonaggregated and completely dispersed (draw schematic diagrams). Assign order of magnitude for the following properties of the above-named systems with explanation: $\overline{PV}$, $\overline{YP}$, initial and ten-minute gel strength, and fluid loss.

(39) What are the functions of the following in drilling fluids: (a) linear hydrocarbon chain polymers or copolymers; (b) latex (rubber); (c) sodium-hexametaphosphate; and (d) sulphates of cellulose.

(40) What are the functions of the following in oil base muds: CaO, asphalt, soap, $CaCO_3$, and $BaSO_4$?

(41) Explain why on approaching overpressured formations produced water often becomes fresher?

(42) Explain in detail differences between apparent and plastic viscosities. What is the significance of each?

(43) What effect would the size of ground $BaSO_4$ have on (a) the apparent and plastic viscosity of the mud measured with rotational viscometer and (b) viscosity measured with Marsh funnel?

(44) In a laboratory experiment it was observed that pH of the mud decreased as a result of salt (NaCl) contamination. Why?

Problems

(1) Calculate the yield of a mud in barrels of mud per ton of clay if the specific weight of the mud is 79.5 lb/ft^3 and the sp. gr. of the clay is 2.6.

(2) Given R = 30 ft/hr, N = 100 rev/min, W = 30,000 lb, and D = $9\frac{7}{8}$ inches, what is the value of d-exponent?

(3) Compute the cost of weighting material per bbl of original mud and per bbl of final mud, in order to make the hydrostatic pressure equal to formation pressure. The following information is given: depth of formation = 4000 ft; pressure of formation = 2500 psig; specific weight of original mud = 66 lb/cu ft; specific gravity of weighting material = 4.0; and cost of weighting material = 0.50 \$/1b.

(4) The following information is available: diameter of DP = 5 inches; hole size = $12\frac{1}{4}$ inches; $\overline{YP}$ = 16 lb/100 ft^2; original mud weight = 89.30 lb/

ft^3; system pressure loss = 600 psi; shut-in drillpipe pressure = 800 psi; shut-in casing pressure = 950 psi; and depth = 16,000 ft. In order to control the abnormal pressure buildup find: (1) annular pressure loss, (2) new mud weight required, (3) initial circulating pressure, and (4) final circulating pressure.

(5) Calculate the uncorrected and corrected 30-min fluid loss if: corrected fluid loss at 15 min = 8 ml and uncorrected fluid loss at 2 min = 8 ml.

(6) A jet of drilling fluid (sp. gr. = 1.3) 3 inches in diameter is deflected through an angle of 70° by a vane (in a series). The jet velocity is 30 ft/sec and the vane velocity is one third of jet velocity in the horizontal direction. Determine the total force acting on the vane, assuming that there is no friction force along the vane surface.

(7) A weighting material having specific gravity of 4 was added to a mud with a specific weight of 67 lb/cu ft in order to offset the formation pressure of 3780 psig. The cost of adding this weighting material is $ 70.00/bbl and one pound of weighting material costs $ 0.80/lb. What is the formation depth?

(8) Crude oil is to be emulsified in a water-base mud. Compute the number of tons of weighting material that must be added to maintain the original density of the mud. Given: volume of original mud = 110 bbl, specific weight of original mud = 69.4 lb/cu ft, volume of crude oil = 20 bbl, specific gravity of crude oil = 0.8, and specific gravity of weighting material = 4.1.

(9) Determine the "Swelling Index" (namely, ratio of volume of hydrated clay/volume of original dry clay), given the following data:

Vol. H_2O (cm^3)	Weight of clay (g)	Measured viscosity (cp)	Specific gravity of clay
90.0	9.1775	4.2	2.5

What the value of viscosity should have been if Einstein's equation was assumed to be valid:

$$\mu_s = \mu_0 (1 + 2.5\phi)$$

(10) Drilling fluid is deflected through an angle of 80° by a single blade in a turbodrill. Determine total force on the blade in lb, if specific weight of drilling fluid = 65 lb/cu ft, velocity = 20 ft/sec, and diameter of jet = 1.5 inches.

(11) Calculate the yield in bbl of mud per ton of clay given the following data: specific weight of mud = 86 lb/cu ft; specific gravity of clay = 2.8; and specific gravity of water = 1. Also calculate how many bbl of fresh water must be added per bbl of mud to just offset a formation pressure of 3500 psi at a depth of 6000 ft.

(12) Fluid loss may be estimated from the relation which student should derive. In the original test of a mud, the 68-min fluid loss was 27.3 cm^3.

(a) Estimate the 120-min fluid loss if the pressure difference were changed from 245 psi to 125 psi, other experimental conditions remaining the same as in the original test.
(b) Estimate the 55-min fluid loss in the case the filter press is changed so that the diameter of the cake is 2.28 inches instead of 3.17 inches, other experimental conditions remaining the same as in the original test.
(c) Estimate the 145-min fluid loss if the viscosity of the filtrate were changed from 1.38 cP to 1.24 cP, other experimental conditions remaining the same as in the original test.

(13) A cylinder 5 cm in diameter and 12 cm long rotates inside of (and concentric with) a fixed cylinder having a 10-cm diameter. Drilling fluid fills the space between these two cylinders, and a torque of 0.5 ft-lb is applied to the inner cylinder. Calculate the velocity gradients at the cylinder walls and the resulting rpm, after attainment of constant velocity, ignoring end effects. Apparent viscosity is equal to . . . . cP, whereas plastic is equal to . . . . cP.

(14) Given the following information, determine the cost of drilling per foot.

$8\frac{3}{4}$-inch bit OSC-1G	Footage = 200 ft
Rig cost = $ 500 per hour	Rotating time = 14 hr
Bit cost = $ 2,100	Weight = 40,000 lb
Trip time = 0.8 hr/1000 ft	Rotary speed = 110 rpm
Depth = 10,000 ft	Dull condition = $\frac{3}{8}$
Formation type = firm shale	Bearing = $\frac{3}{4}$

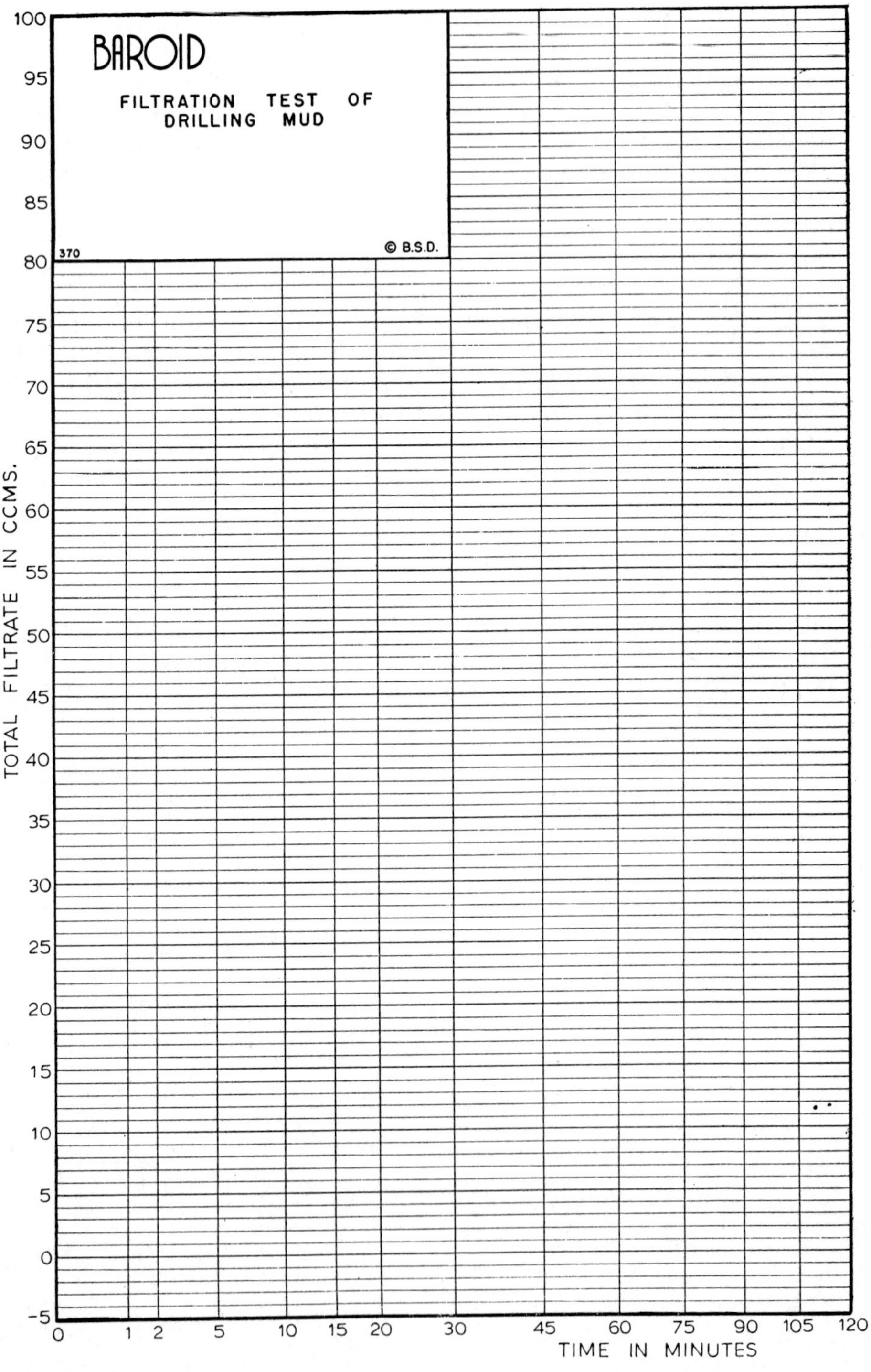
BAROID
FILTRATION TEST OF DRILLING MUD
370
© B.S.D.
TOTAL FILTRATE IN CCMS.
100
95
90
85
80
75
70
65
60
55
50
45
40
35
30
25
20
15
10
5
0
-5
0 1 2 5 10 15 20 30 45 60 75 90 105 120
TIME IN MINUTES

Fig. C-1.

Appendix D

GLOSSARY OF DRILLING-FLUID AND ASSOCIATED TERMS

(API Bull, D11; reprinted with courtesy of American Petroleum Institute, Washington, D.C.; the help extended by C.O. Liles and Bruce Whalen is greatly appreciated.)

—A—

ABSORPTION—The penetration or apparent disappearance of molecules or ions of one or more substances into the interior of a solid or liquid. For example, in hydrated bentonite, the planar water that is held between the mica-like layers is the result of absorption.

ACID—Any chemical compound containing hydrogen capable of being replaced by positive elements or radicals to form salts. In terms of the dissociation theory, it is a compound which, on dissociation in solution, yields excess hydrogen ions. Acids lower the pH. Examples of acids or acidic substances are: hydrochloric acid, tannic acid, sodium acid pyrophosphate.

ACIDITY—The relative acid strength of liquids as measured by pH. A pH value below 7. See pH.

ADHESION—The molecular attraction or force exerted across the surfaces of contact between unlike liquids and solids which resist their separation.

ADSORPTION—A surface phenomenon exhibited by a solid (adsorbent) to hold or concentrate gases, liquids, or dissolved substances (adsorptive) upon its surface, a property due to adhesion. For example, that water held to the outside surface of hydrated bentonite is absorbed water. See Adsorption in Section 2.

AERATION—The technique of injecting air or gas in varying amounts into a drilling fluid for the purpose of reducing hydrostatic head. See Air Cutting. See Aeration in Section 2.

AGGLOMERATE—A mass of particles or substances closely associated and clustered together.

AGGLOMERATION—The grouping of individual particles.

AGGREGATE—A group of two or more individual particles held together by strong forces. Aggregates are stable to normal stirring, shaking, or handling as powder or a suspension. They may be broken by drastic treatment such as ball milling a powder or by shearing a suspension.

AGGREGATION—Formation of aggregates. In drilling fluids, aggregation results in the stacking of the clay platelets face to face. The viscosity and gel strength decrease in consequence.

AIR CUTTING—The inadvertent mechanical incorporation and dispersion of air into a drilling-fluid system. See Aeration in Sections 1 and 2.

ALKALI—Any compound having marked basic properties. See Base.

ALKALINITY—The combining power of a base measured by the maximum number of equivalents of an acid with which it can react to form a salt. In water analysis, it represents the carbonates, bicarbonates, hydroxides, and occasionally the borates, silicates, and phosphates in the water. It is determined by titration with standard acid to certain datum points. See *API RP 13B** for specific directions for determination of phenolphthalein (P_f) and methyl orange (M_f) alkalinities of the filtrate in drilling fluids and the alkalinity of the mud itself (P_m). Also, see P_f, M_f, and P_m.

**API RP 13B: Recommended Practice for Standard Procedure for Testing Drilling Fluids* is available from API Production Department, 211 North Ervay, Ste. 1700, Dallas, Texas.

ALUMINUM STEARATE—An aluminum salt of stearic acid used as a defoamer. See Stearate.

ANALYSIS, DRILLING-FLUID OR MUD — Examination and testing of the drilling fluid to determine its physical and chemical properties and condition.

ANHYDRITE—See Calcium Sulfate. Anhydrite is often encountered while drilling. It may occur as thin stringers or massive formations.

ANHYDROUS—Without water.

ANILINE POINT—The lowest temperature at which equal volumes of freshly distilled aniline and an oil which is being tested are completely miscible. This test gives an indication of the character (paraffinic, naphthenic, asphaltic, aromatic, mid-continent, etc.) of the oil. The aniline point of diesels or crudes used in drilling fluid is also an indication of the deteriorating effect these materials may have on natural or synthetic rubber. The lower the aniline point of an oil the more severe it usually is in damaging rubber parts.

ANION—A negatively charged atom or radical, such as Cl^-, OH^-, $SO_4^=$, etc., in solution of an electrolyte. Anions move toward the anode (positive electrode) under the influence of an electrical potential.

ANNULAR VELOCITY—The velocity of a fluid moving in the annulus.

ANNULUS OR ANNULAR SPACE—The space between the drill string and the wall of the hole or casing.

ANTIFOAM—A substance used to prevent foam by greatly increasing the surface tension. See Defoamer.

API GRAVITY—The gravity (weight per unit volume) of crude oil or other related fluids as measured by a system recommended by the American Petroleum Institute. It is related to specific gravity by the following formula:

$$Deg\ API = \frac{141.5}{sp\ gr\ 60\ F/60\ F} - 131.5$$

APPARENT VISCOSITY—The viscosity a fluid appears to have on a given instrument at a stated rate of shear. It is a function of the fluid's plastic viscosity and the yield point. The apparent viscosity in centipoises, as determined by the direct-indicating viscometer (see Viscometer Direct-indicating), is equal to ½ the 600-rpm reading. In a Newtonian fluid, the apparent viscosity is numerically equal to the plastic viscosity. See also Viscosity, Plastic Viscosity, and Yield Point.

ASBESTOS — Term applied to many fibrous silicate minerals, some forms of which are used in certain drilling fluids. See Asbestos in Section 2.

ASPHALT—A natural or mechanical mixture of solid or viscous bitumens found in natural beds or obtained as a residue from petroleum. Asphalt, blends containing asphalt, and altered asphaltic materials (e.g., air-blown, chemically modified, etc.) have been added to certain drilling fluids for such widely different purposes as a component in oil-base drilling fluids, lost-circulation material, emulsifier, fluid-loss-control agent, wall-plastering agent, etc.

ATOM—According to atomic theory, the smallest quantity of an element which is capable of entering into chemical combination or that can exist alone.

ATOMIC NUMBER—A number, characteristic of an element, which designates the place of the element in the periodic table. This number represents the net positive charge on the nucleus of an atom and is also equal to the number of protons within the nucleus. In a neutral atom, the atomic number necessarily equals the number of electrons outside the nucleus.

ATTAPULGITE CLAY—A colloidal, viscosity-building clay used principally in salt-water drilling fluids. Attapulgite, a special fullers earth, is a hydrous magnesium aluminum silicate.

—B—

BALANCE, MUD—A beam-type balance used in determining drilling fluid density. It consists primarily of a base, graduated beam with constant-volume cup, lid, rider, knife edge, and counterweight.

BARITE, BARYTES, OR HEAVY SPAR—Natural barium sulfate used for increasing the density of drilling fluids. If required, it is usually upgraded to a specific gravity of 4.20. The barite mineral occurs in white, grayish, greenish, and reddish ores or crystalline masses.

BARIUM SULFATE—$BaSO_4$. See Barite.

BARREL—A volumetric unit of measure used in the petroleum industry consisting of 42 gallons.

BARREL EQUIVALENT—A laboratory unit used for evaluating or testing drilling fluids. One gram of material, when added to 350 ml of fluid, is equivalent to 1 lb of material when added to one 42-gal barrel of fluid.

BASE—A compound of a metal, or a metal-like group, with hydrogen and oxygen in the proportion to form an OH radical, which ionizes in aqueous solution to yield excess hydroxyl ions. Bases are formed when metallic oxides react with water. Bases increase the pH. Examples are caustic soda and lime.

BASE EXCHANGE—The replacement of cations associated with the clay surface by those of another species, e.g., the conversion of sodium clay to calcium clay.

BASICITY—pH value above 7. Ability to neutralize or accept protons from acids.

BENTONITE—A plastic, colloidal clay largely made up of the mineral sodium montmorillonite; a hydrated aluminum silicate. For use in drilling fluids, bentonite has a yield in excess of 85 bbl/ton. The generic term "bentonite" is neither an exact mineralogical name nor is the clay of definite mineralogical composition.

BICARB—See Sodium Bicarbonate.

BLOOIE LINE—Flow line for air or gas drilling.

BLOWOUT—An uncontrolled flow of well fluids and/or formation fluids from the wellbore caused by the formation pressure being greater than the hydrostatic head of the fluid in the hole. See Underground Blowout.

BOILERHOUSE—To make up a report on a condition as fact without knowledge of its accuracy. Sometimes referred to as "doghouse."

BRACKISH WATER—Water containing low concentrations of any soluble salts.

BREAK CIRCULATION—To start movement of the drilling fluid after it has been quiescent in the hole.

BREAKOUT, OIL—Oil that has risen to the surface of the drilling fluid which previously had been combined in the fluid as emulsion.

BRIDGE—An obstruction in a well formed by intrusion of subsurface formations.

BRINE—Water saturated with or containing a high concentration of common salt (sodium chloride); hence, any strong saline solution containing such other salts as calcium chloride, zinc chloride, calcium nitrate, etc.

BROMINE VALUE—The number of centigrams of bromine which are absorbed by 1 g of oil under certain conditions. This is a test for the degree of unsaturatedness of a given oil.

BROWNIAN MOVEMENT—Continuous, irregular motion exhibited by particles suspended in a liquid or gaseous medium, usually as a colloidal dispersion.

BS OR BS & W—Common abbreviation used for base sediment, or base sediment and water.

BUFFER—Any substance or combination of substances which, when dissolved in water, produces a solution which resists a change in its hydrogen ion concentration upon the addition of acid or base.

—C—

CABLE-TOOL DRILLING—A method of drilling a well by allowing a weighted bit at the bottom of a cable to fall against the formation being penetrated. See Rotary Drilling.

CAKE CONSISTENCY—According to *API RP 13B*,* such notations as "hard," "soft," "tough," "rubbery," "firm," etc., may be used to convey some idea of cake consistency.

CAKE THICKNESS—The measurement of the thickness of the filter cake deposited by a drilling fluid against a porous medium, most often following the standard API filtration test. Cake thickness is usually reported in 32nd of an inch. See Filter Cake and Wall Cake.

CALCIUM—One of the alkaline earth elements with a valence of 2 and an atomic weight of about 40. Calcium compounds are a common cause of the hardness of water. It is also a component of lime, gypsum, limestone, etc.

CALCIUM CARBONATE—$CaCO_3$. An insoluble calcium salt sometimes used as a weighting material (limestone, oyster shell, etc.), in specialized drilling fluids. It is also used as a unit and/or standard to report hardness.

*Available from API Production Department, 211 North Ervay, Suite 1700, Dallas, Texas 75201.

CALCIUM CHLORIDE—$CaCl_2$. A very soluble calcium salt sometimes added to drilling fluids to impart special properties, but primarily to increase the density of the fluid phase.

CALCIUM CONTAMINATION—Dissolved calcium ions in sufficient concentration to impart undesirable properties in a drilling fluid, such as flocculation, reduction in yield of bentonite, increase in fluid loss, etc. See also Calcium Sulfate, Gyp, Anhydrite, Lime, Calcium Carbonate.

CALCIUM HYDROXIDE—$Ca(OH)_2$. The active ingredient of slaked lime. It is also the main constituent in cement (when wet). This material is referred to as "lime" in field terminology.

CALCIUM-TREATED MUDS—Drilling fluids to which quantities of soluble calcium compounds have been added or allowed to remain from the formation drilled in order to impart special properties.

CALCIUM SULFATE — (Anhydrite: $CaSO_4$; plaster of paris: $CaSO_4 \cdot 1/2\ H_2O$; and gypsum: $CaSO_4 \cdot 2H_2O$). Calcium sulfate occurs in drilling fluids as a contaminant or may be added to certain drilling fluids to impart special properties.

CATION—The positively charged particle in the solution of an electrolyte which, under the influence of an electrical potential, moves toward the cathode (negative electrode). Examples are: Na^+, H^+, NH_4^+, Ca^{++}, Mg^{++}, Al^{+++}.

CAUSTIC OR CAUSTIC SODA—See Sodium Hydroxide.

CAVE-IN—See Sloughing. Cave-in is a severe form of sloughing.

CAVERNOUS FORMATION — A formation having voluminous voids, usually the result of dissolving by formation waters which may or may not be still present.

CC OR CUBIC CENTIMETER—A metric system unit for the measure of volume. It is essentially equal to the milliliter and commonly used interchangeably. One cubic centimeter of water at room temperature weighs approximately 1 g.

CEMENT—A mixture of calcium aluminates and silicates made by combining lime and clay while heating. Slaked cement contains about 62.5 percent calcium hydroxide, which is the major source of trouble when cement contaminates drilling fluids.

CENTIPOISE (CP)—A unit of viscosity equal to 0.01 poise. A poise equals 1 g per meter-second, and a centipoise is 1 g per centimeter-second. The viscosity of water at 20 C is 1.005 cp (1 cp = 0.000672 lb/ft-sec).

CENTRIFUGE—A device for the mechanical separation of high specific gravity solids from a drilling fluid. Usually used on weighted fluids to recover weight material and discard drill solids. The centrifuge uses high-speed mechanical rotation to achieve this separation, as distinguished from the cyclone-type separator in which the fluid energy alone provides the separating force. See Cyclone and Desander.

CHEMICALS—In drilling-fluid terminology, a chemical is any material that produces changes in the viscosity, yield point, gel strength, and fluid loss, as well as surface tension.

CHEMICAL BARREL — A container in which various chemicals are mixed prior to addition to the drilling fluid.

CHROMATE—A compound in which chromium has a valence of 6, e.g., sodium bichromate. Chromate may be added to drilling fluids either directly or as a constituent of chrome lignites or chrome lignosulfonates. In certain areas, chromate is widely used as an anodic corrosion inhibitor, often in conjunction with lime.

CHROME LIGNITE—Mined lignite, usually leonardite, to which chromate has been added and/or reacted. The lignite can also be causticized with either sodium or potassium hydroxide.

CIRCULATION—The movement of drilling fluid from the suction pit through pump, drill pipe, bit, annular space in the hole, and back again to the suction pit. The time involved is usually referred to as circulation time.

CIRCULATION, LOSS OF (OR LOST) — The result of drilling fluid escaping into the formation by way of crevices or porous media.

CIRCULATION RATE—The volume flow rate of the circulating drilling fluid usually expressed in gallons or barrels per minute.

CLABBERED—A slang term commonly used to describe moderate to severe flocculation of drilling fluid due to various contaminants; also called "gelled-up."

CLAY—A plastic, soft, variously-colored earth, commonly a hydrous silicate of alumina, formed by the decomposition of feldspar and other aluminum silicates. See also Attapulgite, Bentonite, High Yield, Low Yield, and Natural Clays. Clay minerals are essentially insoluble in water but disperse under hydration, shearing forces such as grinding, velocity effects, etc., into the extremely small particles varying from submicron to 100-micron sizes.

CLAY EXTENDER—Any of several substances, usually high molecular weight organic compounds that, when added in low concentrations to a bentonite or to certain other clay slurries, will increase the viscosity of the system, e.g., polyvinyl acetate-maleic anhydride copolymer. See Low-solids Muds.

CMC—See Sodium Carboxymethylcellulose.

COAGULATION—In drilling-fluid terminology, a synonym for flocculation. See Coagulation in Section 2.

COALESCENCE—The change from a liquid to a thickened curdlike state by chemical reaction. Also, the combination of globules in an emulsion caused by molecular attraction of the surfaces.

COHESION—The attractive force between the same kind of molecules, i.e., the force which holds the molecules of a substance together.

COLLOID—A state of subdivision of matter which consists either of single large molecules or of aggregations of smaller molecules dispersed to such a degree that the surface forces become an important factor in determining its properties. The size and electrical charge of the particles determine the different phenomena observed with colloids, e.g., Brownian movement. The sizes of colloids range from 1×10^{-7} cm to 5×10^{-5} cm (0.001 to 0.5 microns) in diameter, although the particle size of certain emulsoids can be in the micron range.

COLLOIDAL COMPOSITION—A colloidal suspension containing one or more colloidal constituents.

COLLOIDAL SUSPENSION—Finely divided particles of ultramicroscopic size swimming in a liquid.

CONDUCTIVITY—A measure of the quantity of electricity transferred across unit area per unit potential gradient per unit time. It is the reciprocal of resistivity. Electrolytes may be added to the drilling fluid to alter its conductivity for logging purposes.

CONNATE WATER — Water that probably was laid down and entrapped with sedimentary deposits, as distinguished from migratory waters that have flowed into deposits after they were laid down.

CONSISTENCY—The viscosity of a non-reversible fluid, in poises, for a certain time interval at a given pressure and temperature.

CONSISTOMETER—A thickening-time tester having a stirring apparatus to measure the relative thickening time for drilling fluid or cement slurries under predetermined temperatures and pressures. See *API RP 10B.**

CONTAMINATION—The presence in a drilling fluid of any foreign material that may tend to produce detrimental properties of the drilling fluid.

CONTINUOUS PHASE — The fluid phase which completely surrounds the dispersed phase that may be colloids, oil, etc.

CONTROLLED AGGREGATION—A condition in which the clay platelets are maintained stacked by a polyvalent cation, such as calcium, and are deflocculated by use of a thinner.

CONVENTIONAL MUD—A drilling fluid containing essentially clay and water.

COPOLYMER—A substance formed when two or more substances polymerize at the same time to yield a product which is not a mixture of separate polymers but a complex having properties different from either polymer alone. See Polymer. Examples are polyvinyl acetate-maleic anhydride copolymer (clay extender and selective flocculant), acrylamide-carboxylic acid copolymer (total flocculant), etc.

CORROSION—The adverse chemical alteration on a metal or the eating away of the metal by air, moisture, or chemicals; usually an oxide is formed.

CRATER — The formation of a large funnel-shaped cavity at the top of a hole resulting from a blowout or occasionally from caving.

CREAMING OF EMULSIONS—The settling or rising of the particles of the dispersed phase of an emulsion as observed by a difference in color shading of the layers formed. This can be either upward or downward creaming, depending upon the relative densities of the continuous and dispersed phases.

CREATED FRACTURES—Induced fractures by means of hydraulic or mechanical pressure exerted on the formation.

**API RP 10B: Recommended Practice for Testing Oil Well Cements and Cement Additives* is available from API Production Department, 211 North Ervay, Ste. 1700, Dallas, Texas 75201.

CUTTINGS — Small pieces of formation that are the result of the chipping and/or crushing action of the bit. See Samples.

CYCLE TIME, DRILLING-FLUID—The time of a cycle, or down the hole and back, is the time required for the pump to move the drilling fluid in the hole. The cycle in minutes equals the barrels of drilling fluid in the hole divided by barrels per minute.

CYCLONE—A device for the separation of various particles from a drilling fluid, most commonly used as a desander. The fluid is pumped tangentially into a cone, and the fluid rotation provides enough centrifugal force to separate particles by mass weight. See Centrifuge.

—D—

DARCY—A unit of permeability. A porous medium has a permeability of 1 Darcy when a pressure of 1 atm on a sample 1 cm long and 1 sq cm in cross section will force a liquid of 1-cp viscosity through the sample at the rate of 1 cc per sec.

DEFLOCCULATION—Breakup of flocs of gel structures by use of a thinner.

DEFOAMER OR DEFOAMING AGENT — Any substance used to reduce or eliminate foam by reducing the surface tension. See Antifoam.

DEHYDRATION—Removal of free or combined water from a compound. See Dehydration in Section 2.

DELIQUESCENCE—The liquefaction of a solid substance due to the solution of the solid by adsorption of moisture from the air, e.g., calcium chloride.

DENSITY—Matter measured as mass per unit volume expressed in pounds per gallon (ppg), pounds per square inch per 1,000 ft of depth (psi/1,000 ft), and pounds per cubic ft (lb/cu ft). Density is commonly referred to as "weight."

DESANDER—See Cyclone.

DIATOMACEOUS EARTH — An infusorial earth composed of siliceous skeletons of diatoms and being very porous. Sometimes used for combatting lost circulation and as an additive to cement; also has been added to special drilling fluids for a particular purpose.

DIESEL-OIL PLUG—See Gunk Plug.

DIFFERENTIAL PRESSURE—The difference in pressure between the hydrostatic head of the drilling-fluid column and the formation pressure at any given depth in the hole. It can be positive, zero, or negative with respect to the hydrostatic head.

DIFFERENTIAL-PRESSURE (WALL) STICKING—Sticking which occurs because part of the drill string (usually the drill collars) becomes embedded in the filter cake resulting in a non-uniform distribution of pressure around the circumference of the pipe. The conditions essential for sticking require a permeable formation and a pressure differential across a nearly impermeable filter cake and drill string.

DIFFUSION—The spreading, scattering, or mixing of a material (gas, liquid, or solid).

DILATANT FLUID—A dilatant or inverted plastic fluid is usually made up of a high concentration of well-dispersed solids which exhibits a non-linear consistency curve passing through the origin. The apparent viscosity increases instantaneously with increasing rate of shear. The yield point, as determined by conventional calculations from the direct-indicating viscometer readings, is negative; however, the true yield point is zero.

DILUENT—Liquid added to dilute or thin a solution.

DIRECT-INDICATING VISCOMETER—See Viscometer, Direct-indicating.

DISPERSANT—Any chemical which promotes dispersion of the dispersed phase. See Dispersant in Section 2.

DISPERSED PHASE—The scattered phase (solid, liquid, or gas) of a dispersion. The particles are finely divided and completely surrounded by the continuous phase.

DISPERSION (OF AGGREGATES)—Subdivision of aggregates. Dispersion increases the specific surface of the particle; hence results in an increase in viscosity and gel strength.

DISPERSOID—A colloid or finely divided substance.

DISSOCIATION—The splitting up of a compound or element into two or more simple molecules, atoms, or ions. Applied usually to the effect of the action of heat or solvents upon dissolved substances. The reaction is reversible and not as permanent as decomposition; i.e., when the solvent is removed, the ions recombine.

DISTILLATION—Process of first vaporizing a liquid and then condensing the vapor into a liquid (the distillate), leaving behind non-volatile substances, the total solids of a drilling fluid. The distillate is the water and/or oil content of a fluid.

DOG-LEG—The "elbow" caused by a sharp change of direction in the well bore.

DRILLING FLUID OR MUD—A circulating fluid used in rotary drilling to perform any or all of various functions required in the drilling operation.

DRILLING IN—The operation during the drilling procedure at the point of drilling into the pay formation.

DRILLING OUT — The operation during the drilling procedure when the cement is drilled out of the casing before further hole is made or completion attempted.

DRILL-STEM TEST (DST) — A test to determine whether oil and/or gas in commercial quantities has been encountered in the well bore.

DYNAMIC—The state of being active or in motion; opposed to static.

—E—

ELECTRIC LOGGING—Electric logs are run on a wire-line to obtain information concerning the porosity, permeability, fluid content of the formations drilled, and other information. The drilling-fluid characteristics may need to be altered to obtain good logs.

ELECTROLYTE — A substance which dissociates into charged positive and negative ions when in solution or a fused state and which will then conduct an electric current. Acids, bases, and salts are common electrolytes.

EMULSIFIER OR EMULSIFYING AGENT — A substance used to produce an emulsion of two liquids which do not mix. Emulsifiers may be divided, according to their behavior, into ionic and non-ionic agents. The ionic types may be further divided into anionic, cationic, and amphoteric, depending upon the nature of the ion-active groups.

EMULSION—A substantially permanent heterogeneous liquid mixture of two or more liquids which do not normally dissolve in each other but which are held in suspension or dispersion, one in the other, by mechanical agitation or, more frequently, by adding small amounts of substances known as emulsifiers. Emulsions may be mechanical, chemical, or a combination of the two. They may be oil-in-water or water-in-oil types.

EMULSOID—Colloidal particles which take up water.

END POINT—Indicates the end of some operation or when a definite change is observed. In titration, this change is frequently a change in color of an indicator which has been added to the solution or the disappearance of a colored reactant.

ENGINEER, MUD OR DRILLING-FLUID—One versed in drilling fluids whose duties are to manage, carry through, and maintain the various types of oil-well drilling fluid programs.

EP ADDITIVE—See Extreme-pressure Lubricant.

EPM OR EQUIVALENTS PER MILLION—Unit chemical weight of solute per million unit weights of solution. The epm of a solute in solution is equal to the ppm (parts per million) divided by the equivalent weight. Refer also to PPM.

EQUIVALENT CIRCULATING DENSITY—For a circulating fluid, the equivalent circulating density in lb/gal equals the hydrostatic head (psi) plus the total annular pressure drop (psi) divided by the depth (ft) and by 0.052.

EQUIVALENT WEIGHT OR COMBINING WEIGHT — The atomic or formula weight of an element, compound, or ion divided by its valence. Elements entering into combination always do so in quantities proportional to their equivalent weights.

EXTREME-PRESSURE LUBRICANT — Additives which, when added to the drilling fluid, impart lubrication to the bearing surfaces when subjected to extreme pressure conditions.

—F—

FAULT—Geological term denoting a formation break, upward or downward, in the subsurface strata. Faults can significantly affect the area drilling fluid and casing programs.

FERMENTATION—Decomposition process of certain organic substances, e. g., starch in which a chemical change is brought about by enzymes, bacteria, or other micro-organisms. Often referred to as "souring."

FIBER OR FIBROUS MATERIALS—Any tough stringy material used to prevent loss of circulation or to restore circulation. In field use, fiber generally refers to the larger fibers of plant origin.

Filling the Hole—Pumping drilling fluid continuously or intermittently into the well bore to maintain the fluid level in the hole near the surface. The purpose is to avoid danger of blowout, water intrusion, and/or caving of the well bore, e.g., as the pipe is withdrawn.

Fillup Line—The line through which fluid is added to the hole.

Filter Cake — The suspended solids that are deposited on a porous medium during the process of filtration. See also Cake Thickness.

Filter-cake Texture—The physical properties of a cake as measured by toughness, slickness, and brittleness. See also Cake Consistency.

Filter-cake Thickness — A measurement of the solids deposited on filter paper in 32nd of an inch during the standard 30-min API filter test. See Cake Thickness. In certain areas the filter-cake thickness is a measurement of the solids deposited on filter paper for a 7½-min duration.

Filter Loss—See Filtrate Volume.

Filter Paper—Porous unsized paper for filtering liquids. API filtration test specifies one thickness of 9-cm filter paper Whatman No. 50, S&S No. 576, or equivalent.

Filter Press—A device for determining fluid loss of a drilling fluid having specifications in accordance with *API RP 13B*.*

Filtrate—The liquid that is forced through a porous medium during the filtration process. See Filtrate Volume for recommended test procedure.

Filtrate Volume—Measure of the volume of fluid lost through filter media (usually filter paper) when drilling fluid is subjected to a differential pressure in accordance with the filtration procedure contained in *API RP 13B*.*

Filtration — The process of separating suspended solids from their liquid by forcing the latter through a porous medium. Two types of fluid filtration occur in a well: dynamic filtration while circulating, and static filtration when at rest.

Filtration Qualities—The filtration characteristics of a drilling fluid. Generally these qualities are inverse to the thickness of the filter cake deposited on the face of a porous medium and the amount of filtrate allowed to escape from the drilling fluid into or through the medium.

Filtration Rate—See Filtrate Volume.

Fishing—Operations on the rig for the purpose of retrieving from the well bore sections of pipe, collars, junk, or other obstructive items which are in the hole.

Flat Gel — A condition wherein the 10-min gel strength is substantially equal to the initial gel strength.

Flipped—When the opposite occurs of what is intended in a drilling fluid. In an invert water-in-oil emulsion, the emulsion is said to be flipped when the continuous and dispersed phases reverse.

*Available from API Production Department, 211 North Ervay, Suite 1700, Dallas, Texas 75201.

Flocculates—Groups of aggregates or particles in suspension subject to being broken up by normal shaking and stirring and reforming on standing.

Flocculating Agent — Substances, such as most electrolytes, some polysaccharides, certain natural or synthetic polymers, that bring about the thickening of the consistency of a drilling fluid. In Bingham plastic fluids, the yield point and gel strength increase.

Flocculation — Loose association of particles in lightly bonded groups, non-parallel association of clay platelets. In concentrated suspensions, such as drilling fluids, flocculation results in gelation. In some drilling fluids, flocculation may be followed by irreversible precipitation of colloids and certain other substances from the fluid, e.g., red beds.

Flocs—See Flocculates.

Fluid—A fluid is a substance readily assuming the shape of the container in which it is placed. The term includes both liquids and gases. It is a substance in which the application of every system of stresses (other than hydrostatic pressure) will produce a continuously increasing deformation without any relation between time rate of deformation at any instant and the magnitude of stresses at that instant. Drilling fluids are usually Newtonian and plastic, seldom pseudoplastic, and rarely dilatant fluids.

Fluid Flow—The state of fluid dynamics of a fluid in motion is determined by the type of fluid (e.g., Newtonian, plastic, pseudoplastic, dilatant), the properties of the fluid such as viscosity and density, the geometry of the system, and the velocity. Thus, under a given set of conditions and fluid properties, the fluid flow can be described as plug flow, laminar (called also Newtonian, streamline, parallel, or viscous) flow, or turbulent flow. See above terms and Reynolds number.

Fluidity—The reciprocal of viscosity. The measure of rate with which a fluid is continuously deformed by a shearing stress. Ease of flowing.

Fluid Loss—See Filtrate Volume.

Fluorescence—Instantaneous re-emission of light of a greater wave length than that light originally absorbed.

Foam—A foam is a two-phase system, similar to an emulsion, where the dispersed phase is a gas or air.

Foaming Agent—A substance that produces fairly stable bubbles at the air-liquid interface due to agitation, aeration, or ebullition. In air or gas drilling, foaming agents are added to turn water influx into aerated foam. This is commonly called "mist drilling."

Formation Damage—Damage to the productivity of a well resulting from invasion into the formation by drilling fluid particles or drilling fluid filtrates. Asphalt from crude oil will also damage some formations. See Mudding Off.

Formation Sensitivity — The tendency of certain producing formations to adversely react with invading mud filtrates.

FUNCTIONS OF DRILLING FLUIDS—The most important function of drilling fluids in rotary drilling is to bring cuttings from the bottom of the hole to the surface. Some other important functions are: control subsurface pressures, cool and lubricate the bit and drill string, deposition of an impermeable wall cake, etc.

FUNNEL VISCOSITY—See Marsh Funnel Viscosity.

—G—

GALENA—Lead sulfide (PbS). Technical grades (specific gravity about 7) are used for increasing the density of drilling fluids to points impractical or impossible with barite.

GAS CUT—Gas entrained by a drilling fluid. See Air Cutting.

GEL — A state of a colloidal suspension in which shearing stresses below a certain finite value fail to produce permanent deformation. The minimum shearing stress that will produce permanent deformation is known as the shear or gel strength of the gel. Gels commonly occur when the dispersed colloidal particles have a great affinity for the dispersing medium, i.e., are lyophilic. Thus gels commonly occur with bentonite in water. For their measurement, see Gel Strength, Initial and 10-min.

GEL—A term used to designate highly colloidal, high-yielding, viscosity-building commercial clays, such as bentonite and attapulgite clays.

GELATION—Association of particles to form a continuous structure.

GEL CEMENT—Cement having a small to moderate percentage of bentonite added as a filler and/or to reduce the slurry weight. See Gunk Plug.

GELLED UP—Oil-field jargon usually referring to any fluid with high gel strength and/or highly viscous properties. Often a state of severe flocculation.

GEL STRENGTH—The ability or the measure of the ability of a colloid to form gels. Gel strength is a pressure unit usually reported in lb/100 sq ft. It is a measure of the same interparticle forces of a fluid as determined by the yield point except that gel strength is measured under static conditions, yield point under dynamic conditions. The common gel-strength measurements are initial and the 10-min gels (see Gel Strength, Initial and 10-min). See also Shear and Thixotropy.

GEL STRENGTH, INITIAL—The measured initial gel strength of a fluid is the maximum reading (deflection) taken from a direct-reading viscometer after the fluid has been quiescent for 10 sec. It is reported in lb/100 sq ft. See *API RP 13B** for details of test procedure.

GEL STRENGTH, 10-MIN — The measured 10-min gel strength of a fluid is the maximum reading (deflection) taken from a direct-reading viscometer after the fluid has been quiescent for 10 min. The reading is reported in lb/100 sq ft. See *API RP 13B** for details of test procedure.

GPG OR GRAINS PER GALLON—PPM (see PPM) equals GPG × 17.1.

GRAVITY, SPECIFIC—The weight of a particular volume of any substance compared to the weight of an equal volume of water at a reference temperature. For gases, air is usually taken as the reference substance, although hydrogen is sometimes used.

GREASING OUT—Certain organic substances, usually fatty-acid derivatives, which are added to drilling fluids as emulsifiers, extreme pressure lubricants, etc., may react with such ions as calcium and magnesium that are in or will subsequently come into the system. An essentially water-insoluble greasy material separates out.

GUAR GUM—A naturally occurring hydrophilic polysaccharide derived from the seed of the guar plant. The gum is chemically classified as a galactomannan. Guar gum slurries made up in clear fresh or brine water possess pseudoplastic flow properties.

GUM—Any hydrophilic plant polysacchrides or their derivatives which, when dispersed in water, swell to produce a viscous dispersion or solution. Unlike resins, they are soluble in water and insoluble in alcohol.

GUMBO — Any relatively sticky formation, such as clay, encountered in drilling.

GUNK PLUG—A slurry in crude or diesel oil containing any of the following materials or combinations: bentonite, cement, attapulgite, and guar gum (never with cement). Used primarily in combatting lost circulation. The plug may or may not be squeezed.

GUNNING THE PITS — Mechanical agitation of the drilling fluid in a pit by means of a mud gun, electric mixer, or agitator.

GYP OR GYPSUM—See Calcium Sulfate. Gypsum is often encountered while drilling. It may occur as thin stringers or massive formations.

—H—

HARDNESS (OF WATER)—The hardness of water is due principally to the calcium and magnesium ions present in the water and is independent of the accompanying acid ions. The total hardness is measured in terms of parts per million of calcium carbonate or calcium and sometimes equivalents per million of calcium. For hardness tests, see *API RP 13B.**

HEAVING—The partial or complete collapse of the walls of a hole resulting from internal pressures due primarily to swelling from hydration or formation gas pressures. See Sloughing.

HETEROGENEOUS—A substance that consists of more than one phase and is not uniform, such as colloids, emulsions, etc. It has different properties in different parts.

HIGH-PH DRILLING FLUID—A drilling fluid with a pH range above 10.5. A high-alkalinity drilling fluid.

HIGH-YIELD DRILLING CLAY—A classification given to a group of commercial drilling-clay preparations having a yield of 35 to 50 bbl/ton and intermediate between bentonite and low-yield clays. High-yield drilling clays are usually prepared by peptizing low-yield calcium montmorillonite clays or, in a few cases, by blending some bentonite with the peptized low-yield clay.

HOMOGENEOUS — Of uniform or similar nature throughout; or a substance or fluid that has at all points the same property or composition.

*Available from API Production Department, 211 North Ervay, Suite 1700, Dallas, Texas 75201.

Hopper, Jet—See Mud-mixing Devices. A device to hold or feed drilling fluid additives.

Humic Acid—Organic acids of indefinite composition in naturally occurring leonardite lignite. The humic acids are the most valuable constituent. See Lignite.

Hydrate—A substance containing water combined in the molecular form (such as $CaSO_4 \cdot 2H_2O$). A crystalline substance containing water of crystallization.

Hydration—The act of a substance to take up water by means of absorption and/or adsorption.

Hydrogen Ion Concentration—A measure of the acidity or alkalinity of a solution, normally expressed as pH. See pH.

Hydrolysis — Hydrolysis is the reaction of a salt with water to form an acid and base. For example, soda ash (Na_2CO_3) hydrolyzes basically, and hydrolysis is responsible for the increase in the pH of water when soda ash is added.

Hydrometer—A floating instrument for determining the specific gravity or density of liquids, solutions, and slurries. A common example is the Mudwate hydrometer used to determine the density of drilling fluid.

Hydrophile—A substance usually in the colloidal state or an emulsion, which is wetted by water; i.e., it attracts water or water adheres to it.

Hydrophilic—A property of a substance having an affinity for water or one that is wetted by water.

Hydrophilic-Lipophilic Balance (HLB)—The hydrophilic-lipophilic balance (HLB) is one of the most important properties of emulsifiers. It is an expression of the relative attraction of an emulsifier for water and oil, determined largely by the chemical composition and ionization characteristics of a given emulsifier. The HLB of an emulsifier is not directly related to solubility, but it determines the type of an emulsion that tends to be formed. It is an indication of the behavior characteristics and not an indication of emulsifier efficiency.

Hydrophobe—A substance, usually in the colloidal state, not wetted by water.

Hydrophobic—Descriptive of a substance which repels water.

Hydrostatic Head—The pressure exerted by a column of fluid, usually expressed in pounds per square inch. To determine the hydrostatic head at a given depth in psi, multiply the depth in feet by the density in pounds per gallon by 0.052.

Hydroxide — A designation that is given for basic compounds containing the OH^- radical. When these substances are dissolved in water, they increase the pH of the solution. See Base.

Hygroscopic—The property of a substance enabling it to absorb water from the air.

—I—

Indicator—Substances in acid-base titrations which, in solution, change color or become colorless as the hydrogen ion concentration reaches a definite value, these values varying with the indicator. In other titrations such as chloride, hardness, and other determinations, these substances change color at the end of the reaction. Common indicators are phenolphthalein, potassium chromate, etc.

Inhibited Drilling Fluid—A drilling fluid having an aqueous phase with a chemical composition that tends to retard and even prevent (inhibit) appreciable hydration (swelling) or dispersion of formation clays and shales through chemical and/or physical means. See Inhibitor (Drilling Fluid).

Inhibitor (Corrosion) — Any agent which, when added to a system, slows down or prevents a chemical reaction or corrosion. Corrosion inhibitors are used widely in drilling and producing operations to prevent corrosion of metal equipment exposed to hydrogen sulfide, carbon dioxide, oxygen, salt water, etc. Common inhibitors added to drilling fluids are filming amines, chromates, and lime.

Inhibitor (Drilling Fluid)—Substances generally regarded as drilling fluid contaminants, such as salt and calcium sulfate are called inhibitors when purposely added to drilling fluid so that the filtrate from the drilling fluid will prevent or retard the hydration of formation clays and shales.

Initial Gel—See Gel Strength, Initial.

Interfacial Tension—The force required to break the surface between two immiscible liquids. The lower the interfacial tension between the two phases of an emulsion, the greater the ease of emulsification. When the values approach zero, emulsion formation is spontaneous. See Surface Tension.

Interstitial Water—Water contained in the interstices or voids of formations.

Invert Oil-emulsion Drilling Fluid — An invert emulsion is a water-in-oil emulsion where fresh or salt water is the dispersed phase and diesel, crude, or some other oil is the continuous phase. Water increases the viscosity and oil reduces the viscosity.

Iodine Number—The number indicating the amount of iodine absorbed by oils, fats, and waxes, giving a measure of the unsaturated linkages present. Generally, the higher the iodine number, the more severe the action of the oil on rubber.

Ion—Acids, bases, and salts (electrolytes) when dissolved in certain solvents, especially water, are more or less dissociated into electrically charged ions or parts of the molecules, due to loss or gain of one or more electrons. Loss of electrons results in positive charges producing a cation. A gain of electrons results in the formation of an anion with negative charges. The valence of an ion is equal to the number of charges borne by it.

—J—

Jetting—The process of periodically removing a portion of, or all of, the water, drilling fluid, and/or solids from the pits, usually by means of pumping through a jet nozzle arrangement.

Jones Effect—The net surface tension of salt solutions first decreases with an increase of concentration, passes through a minimum, and then increases as the concentration is raised.

—K—

Kelly or Kelly Joint — A heavy square pipe or other configuration that works through a like hole in the rotary table and rotates the drill stem.

KEY SEAT—That section of a hole, usually of abnormal deviation and relatively soft formation, which has been eroded or worn by drill pipe to a size smaller than the tool joints or collars. This keyhole type configuration will not allow these members to pass when pulling out of the hole.

KILLING A WELL—Bringing a well under control that is blowing out. Also, the procedure of circulating water and drilling fluids into a completed well before starting well servicing operations.

KILL LINE—A line connected to the annulus below the blowout preventers for the purpose of pumping into the annulus while the preventers are closed.

KINEMATIC VISCOSITY—The kinematic viscosity of a fluid is the ratio of the viscosity (e.g., cp in g/cm-sec) to the density (e.g., g/cc) using consistent units. In several common commercial viscometers the kinematic viscosity is measured in terms of the time of efflux (in seconds) of a fixed volume of liquid through a standard capillary tube or orifice. See Marsh Funnel Viscosity.

—L—

LAMINAR FLOW—Fluid elements flowing along fixed streamlines which are parallel to the walls of the channel of flow. In laminar flow, the fluid moves in plates or sections with a differential velocity across the front which varies from zero at the wall to a maximum toward the center of flow. Laminar flow is the first stage of flow in a Newtonian fluid; it is the second stage in a Bingham plastic fluid. This type of motion is also called parallel, streamline, or viscous flow. See Plug and Turbulent Flow.

LEONARDITE—A naturally occurring oxidized lignite. See Lignite.

LIGNITE—A collective term designating a broad group of naturally occurring amorphous, acidic materials derived from decayed vegetation, e.g., leonardite. These materials are obtained by strip mining from selected lignite deposits. The active component is humic acid. Lignites, which may or may not be combined with chemical compounds such as sodium or potassium hydroxide and/or chromates, are used as thinners, emulsifiers, and for fluid loss control.

LIGNOSULFONATES — Organic drilling fluid additives derived from by-products of sulfite paper manufacturing process from coniferous woods. Some of the common salts, such as the ferrochrome, chrome, calcium, and sodium, are used as universal deflocculants while others are used selectively for calcium-treated systems. In large quantities, the ferro-chrome and chrome salts are sometimes used for fluid-loss control and shale inhibition.

LIME—Commercial form of calcium hydroxide.

LIME-TREATED DRILLING FLUIDS—Commonly referred to as "lime-base" muds. These high-pH systems contain most of the conventional fresh-water additives to which slaked lime has been added to impart special properties. The alkalinities and lime contents vary from low to high.

LIMESTONE—See Calcium Carbonate.

LIPOPHILE—A substance usually colloidal and easily wetted by oil.

LIPOPHILIC—Having an affinity for oil.

LIVE OIL—Crude oil that contains gas and has not been stabilized or weathered. This oil can cause gas cutting when added to drilling fluid and is a potential fire hazard.

LOGGING—See Mud Logging and Electric Logging.

LOSS OF CIRCULATION—See Circulation, Loss of.

LOSS OF HEAD OR FRICTION LOSS—See Pressure-drop Loss.

LOST-CIRCULATION ADDITIVES — Materials added to drilling fluid to control or prevent lost circulation. These materials are added in varying amounts and are classified as fiber, flake, or granular.

LOST RETURNS—See Circulation, Loss of.

LOW-SOLIDS DRILLING FLUIDS—A designation given to any type of drilling fluid where high performing additives, e.g., CMC, have been partially or wholly substituted for commercial or natural clays. For comparable viscosity and densities (weighted with barite), a low-solids drilling fluid will have a lower volume-percent solids content.

LOW-YIELD CLAYS—Commercial clays chiefly of the calcium montmorillonite type having a yield of approximately 15 to 30 bbl/ton.

LYOPHILIC—Having an affinity for the suspending medium, such as bentonite in water.

LYOPHILIC COLLOID—A colloid that is not easily precipitated from a solution and is readily dispersible after precipitation by an addition of the solvent.

LYOPHOBIC COLLOID—A colloid that is readily precipitated from a solution and cannot be redispersed by an addition of the solution.

—M—

MARSH FUNNEL—An instrument used in determining the Marsh funnel viscosity. The Marsh funnel is a container with a fixed orifice at the bottom so that when filled with 1,500 cc fresh water, 1 qt (946 ml) will flow out in 26 ± 0.5 sec. For 1,000 cc out, the efflux time for water is 27.5 ± 0.5 sec. See *API RP 13B** for specifications.

MARSH FUNNEL VISCOSITY — Commonly called the funnel viscosity. The Marsh funnel viscosity is reported as the number of seconds required for a given fluid to flow 1 qt through the Marsh funnel. In some areas, the efflux quantity is 1,000 cc. See *API RP 13B** for instructions. See also Kinematic Viscosity.

MENISCUS—The curved upper surface of a liquid column, concave when the containing walls are wetted by the liquid and convex when not.

MESH—A measure of fineness of a woven material, screen, or sieve; e.g., a 200-mesh sieve has 200 openings per linear inch. A 200-mesh screen with a wire diameter of 0.0021 in. (0.0533 mm) has an opening of 0.074 mm, or will pass a particle of 74 microns. See Micron.

M_f—The methyl orange alkalinity of the filtrate, reported as the number of milliliters of 0.02 Normal (N/50) acid required per milliliter of filtrate to reach the methyl orange end point (pH 4.3).

*Available from API Production Department, 211 North Ervay, Suite 1700, Dallas, Texas 75201.

MICA—A naturally occurring flake material of varying size used in combatting lost circulation. Chemically, an alkali aluminum silicate.

MICELLES—Organic and inorganic molecular aggregates occurring in colloidal solutions. Long chains of individual structural units chemically joined to one another and laid side by side to form bundles. When bentonite hydrates, certain sodium or other metallic ions go into solution, the clay particle plus its atmosphere of ions is technically known as a micelle.

MICRON μ = MU—A unit of length equal to one millionth part of a meter, or one thousandth part of a millimeter.

MILK EMULSION—See Oil-emulsion Water.

MILLIDARCY—1/1000 Darcy. See Darcy.

MIST DRILLING—A method of rotary drilling whereby water and/or oil is dispersed in air and/or gas as the drilling fluid.

ML OR MILLILITER—A metric system unit for the measure of volume. Literally 1/1000th of a liter. In drilling fluid analysis work, this term is used interchangeably with cubic centimeter (cc). One quart is about equal to 946 ml.

MOLECULAR WEIGHT—The sum of the atomic weights of all the constituent atoms in the molecule of an element or compound.

MOLECULE—When atoms combine they form a molecule. In the case of an element or a compound, a molecule is the smallest unit which chemically still retains the properties of the substance in mass.

MONTMORILLONITE—A clay mineral commonly used as an additive to drilling fluids. Sodium montmorillonite is the main constituent in bentonite. The structure of montmorillonite is characterized by a form which consists of a thin platey-type sheet with the width and breadth indefinite, and thickness that of the molecule. The unit thickness of the molecule consists of three layers. Attached to the surface are ions that are replaceable. Calcium montmorillonite is the main constituent in low-yield clays.

MUD—A water- or oil-base drilling fluid whose properties have been altered by solids, commercial and/or native, dissolved and/or suspended. Used for circulating out cuttings and many other functions while drilling a well. Mud is the term most commonly given to drilling fluids. See Drilling Fluid.

MUD ADDITIVE — Any material added to a drilling fluid to achieve a particular purpose.

MUDDING OFF — Commonly thought of as reduced productivity caused by the penetrating, sealing, or plastering effect of a drilling fluid.

MUDDING UP—Process of mixing drilling fluid additives to achieve some desired purpose not possible with the former fluid, which usually has been water, air, or gas.

MUD HOUSE—A structure at the rig to store and shelter sacked materials used in drilling fluids.

MUD LOGGING—A method of determining the presence or absence of oil or gas in the various formations penetrated by the drill bit. The drilling fluid and the cuttings are continuously tested on their return to the surface, and the results of these tests are correlated with the depth or origin.

MUD-MIXING DEVICES—The most common device for adding solids to the drilling fluid is by means of the jet hopper. Some other devices for mixing are: eductors, paddle mixers, electric stirrers, mud guns, chemical barrels, etc.

MUD PIT—Earthen or steel storage facilities for the surface drilling fluid system. Mud pits which vary in volume and number are of two types: circulating and reserve. Drilling fluid testing and conditioning is normally done in the circulating pit system.

MUD PROGRAM—A proposed or followed plan or procedure for the type(s) and properties of drilling fluid(s) used in drilling a well with respect to depth. Some factors that influence the mud program are the casing program and such formation characteristics as type, competence, solubility, temperature, pressure, etc.

MUD PUMPS — Pumps at the rig used to circulate drilling fluids.

MUD SCALES—See Balance, Mud.

MUD STILL—An instrument used to distill oil, water, and other volatile material in a drilling fluid to determine oil, water, and total solids contents in volume-percent.

—N—

NATURAL CLAYS—Natural clays, as opposed to commercial clays, are clays that are encountered when drilling various formations. The yield of these clays varies greatly, and they may or may not be purposely incorporated into the drilling fluid system.

NEAT CEMENT—A slurry composed of portland cement and water.

NEUTRALIZATION—A reaction in which the hydrogen ion of an acid and the hydroxyl ion of a base unite to form water, the other ionic product being a salt.

NEWTONIAN FLOW—See Newtonian Fluid.

NEWTONIAN FLUID — The basic and simplest fluids from the standpoint of viscosity consideration in which the shear force is directly proportional to the shear rate. These fluids will immediately begin to move when a pressure or force in excess of zero is applied. Examples of Newtonian fluids are water, diesel oil, and glycerine. The yield point as determined by direct-indicating viscometer is zero.

NON-CONDUCTIVE DRILLING FLUID — Any drilling fluid, usually oil-base or invert-emulsion types, whose continuous phase does not conduct electricity, e.g., oil. The spontaneous potential (SP) and normal resistivity cannot be logged, although such other logs as the induction, acoustic velocity, etc., can be run.

NORMAL SOLUTION—A solution of such a concentration that it contains 1 gram-equivalent of a substance per liter of solution.

—O—

OIL-BASE DRILLING FLUID—The term "oil-base drilling fluid" is applied to a special type drilling fluid where oil is the continuous phase and water the dispersed phase. Such fluids contain blown asphalt and usually 1 to 5 percent water emulsified into the system with caustic soda or quick lime and an organic acid. Silicate, salt, and phosphate may also be present. Oil-base drilling fluids are differentiated from invert-emulsion drilling fluids (both water-in-oil emulsions) by the amounts of water used, method of controlling viscosity and thixotropic properties, wall-building materials, and fluid loss.

OIL CONTENT—The oil content of any drilling fluid is the amount of oil in volume-percent.

OIL-EMULSION WATER (MILK EMULSION)—A drilling fluid in which the oil content is usually kept between 3 to 7 percent and seldom over 10 percent (it can be considerably higher). The oil is emulsified into fresh or salt water with a chemical emulsifier. Sometimes CMC, starch, or gum may be added to the fresh- and salt-water systems.

OIL-IN-WATER EMULSION DRILLING FLUID—Commonly called "emulsion mud." Any conventional or special water-base drilling fluid to which oil has been added. The oil becomes the dispersed phase and may be emulsified into the drilling fluid either mechanically or chemically.

—P—

PACKER FLUID—Any fluid placed in the annulus between the tubing and casing above a packer. Along with other functions, the hydrostatic pressure of the packer fluid is utilized to reduce the pressure differentials between the formation and the inside of the casing and across the packer itself.

PARTICLE—A minute unit of matter, usually a single crystal, or of regular shape with a specific gravity approximating that of a single crystal.

PARTS PER MILLION—See PPM.

PARALLEL FLOW—See Laminar Flow.

PAY ZONE OR PAY FORMATION—The formation drilled into that contains oil and/or gas in commercial quantities.

PENETRATION, RATE OF—The rate in feet per hour at which the drill proceeds to deepen the well bore.

PEPTIZATION—An increased dispersion due to the addition of electrolytes or other chemical substances. See Deflocculation and Dispersion.

PEPTIZED CLAY—A clay to which an agent has been added to increase its initial yield. For example, soda ash is frequently added to calcium montmorillonite clay.

PERCENT—For weight-percent, see PPM. Volume-percent is the number of volumetric parts of any liquid or solid constituent per 100 like volumetric parts of the whole. Volume-percent is the most common method of reporting solids, oil, and water contents of drilling fluids.

PERMEABILITY—Normal permeability is a measure of ability of a rock to transmit a one-phase fluid under conditions of laminar flow. Unit of permeability is the Darcy. See Permeability in Section 2.

P_f—The phenolphthalein alkalinity of the filtrate, reported as the number of milliliters of 0.02 Normal (N/50) acid required per milliliter of filtrate to reach the phenolphthalein end point.

pH—An abbreviation for potential hydrogen ion. The pH numbers range from 0 to 14, 7 being neutral, and are indices of the acidity (below 7) or alkalinity (above 7) of the fluid. The numbers are a function of the hydrogen ion concentration in gram ionic weights per liter which, in turn, is a function of the dissociation of water as given by the following expression:

$$\frac{(H)\,(OH)}{(H_2O)} = K \qquad H_2O = 1 \times 10^{-14}$$

The pH may be expressed as the logarithm (base 10) of the reciprocal (or the negative logarithm) of the hydrogen ion concentration. The pH of a solution offers valuable information as to the immediate acidity or alkalinity, as contrasted to the total acidity or alkalinity (which may be titrated).

PHOSPHATE — Certain complex phosphates, usually sodium tetraphosphate ($Na_6P_4O_{13}$) and sodium acid pyrophosphate (SAPP, $Na_2H_2P_2O_7$), are used either as drilling fluid thinners or for treatment of various forms of calcium and magnesium contamination.

PILOT TESTING—A method of predicting behavior of drilling fluid systems by mixing small quantities of drilling fluid and additives, then testing the results.

PLASTIC FLOW—See Plastic Fluid.

PLASTIC FLUID—A complex, non-Newtonian fluid in which the shear force is not proportional to the shear rate. A definite pressure is required to start and maintain movement of the fluid. Plug flow is the initial type of flow and only occurs in plastic fluids. Most drilling fluids are plastic fluids. The yield point as determined by direct-indicating viscometer is in excess of zero.

PLASTICITY—The property possessed by some solids, particularly clays and clay slurries, of changing shape or flowing under applied stress without developing shear planes or fractures. Such bodies have yield points, and stress must be applied before movement begins. Beyond the yield point, the rate of movement is proportional to the stress applied, but ceases when the stress is removed. See Fluid.

PLASTIC VISCOSITY — The plastic viscosity is a measure of the internal resistance to fluid flow attributable to the amount, type, and size of solids present in a given fluid. It is expressed as the number of dynes per sq cm of tangential shearing force in excess of the Bingham yield value that will induce a unit rate of shear. This value, expressed in centipoises, is proportional to the slope of the consistency curve determined in the region of laminar flow for materials obeying Bingham's Law of Plastic Flow. When using the direct-indicating viscometer, the plastic viscosity is found by subtracting the 300-rpm reading from the 600-rpm reading.

PLUG FLOW—The movement of a material as a unit without shearing within the mass. Plug flow is the first type of flow exhibited by a plastic fluid after overcoming the initial force required to produce flow.

P_m—The phenolphthalein alkalinity of the drilling fluid reported as the number of milliliters of 0.02 Normal (N/50) acid required per milliliter of drilling fluid.

POLYMER—A substance formed by the union of two or more molecules of the same kind linked end to end into another compound having the same elements in the same proportion but a higher molecular weight and different physical properties, e.g., paraformaldehyde. See Copolymer.

Porosity—The amount of void space in a formation rock, usually expressed as percent voids per bulk volume. Absolute porosity refers to the total amount of pore space in a rock, regardless of whether or not that space is accessible to fluid penetration. Effective porosity refers to the amount of connected pore spaces, i.e., the space available to fluid penetration. See Permeability.

Potassium—One of the alkali metal elements with a valence of 1 and an atomic weight of about 39. Potassium compounds, most commonly potassium hydroxide (KOH), are sometimes added to drilling fluids to impart special properties, usually inhibition.

Pound Equivalent—A laboratory unit used in pilot testing. One gram or pound equivalent, when added to 350 ml of fluid, is equivalent to 1 lb/bbl.

PPM or Parts per Million—Unit weight of solute per million unit weights of solution (solute plus solvent), corresponding to weight-percent except that the basis is a million instead of a hundred. The results of standard API titrations of chloride, hardness, etc. are correctly expressed in milligrams (mg) of unknown per liter but not in ppm. At low concentrations, mg/l is about numerically equal to ppm. A correction for the solution specific gravity or density in g/ml must be made as follows:

$$ppm = \frac{mg/l}{soln\ den,\ g/ml}$$

$$\%\ by\ wt. = \frac{mg/l}{(10{,}000)\ (soln\ den,\ g/ml)} = \frac{ppm}{10{,}000}$$

Thus, 316,000 mg/l salt is commonly called 316,000 ppm or 31.6 percent, which correctly should be 264,000 ppm and 26.4 percent, respectively.

Precipitate—Material that separates out of solution or slurry as a solid. Precipitation of solids in a drilling fluid may follow flocculation or coagulation, such as the dispersed red-bed clays upon addition of a flocculation agent to the fluid.

Preservative—Usually paraformaldehyde. Any material used to prevent starch or any other substance from fermenting through bacterial action.

Pressure-drop Loss—The pressure lost in a pipeline or annulus due to the velocity of the liquid in the pipeline, the properties of the fluid, the condition of the pipe wall, and the alignment of the pipe. In certain drilling fluid mixing systems, the loss of head can be substantial.

Pressure Surge—A sudden, usually short-duration increase in pressure. When pipe or casing is run into a hole too rapidly, an increase in the hydrostatic pressure results, which may be great enough to create lost circulation.

Pseudoplastic Fluid — A complex non-Newtonian fluid that does not possess thixotropy. A pressure or force in excess of zero will start fluid flow. The apparent viscosity or consistency decreases instantaneously with increasing rate of shear until at a given point the viscosity becomes constant. The yield point as determined by direct-indicating viscometer is positive, the same as in Bingham plastic fluids; however, the true yield point is zero. An example of a pseudoplastic fluid is guar gum in fresh or salt water.

—Q—

Quebracho — A drilling fluid additive used for thinning or dispersing to control viscosity and thixotropy. It is a crystalline extract of the quebracho tree consisting essentially of tannic acid.

Quicklime—Calcium oxide, CaO. Used in certain oil-base drilling fluids to neutralize the organic acid.

Quiescence — The state of being quiet or at rest (being still). Static.

—R—

Radical—Two or more atoms behaving as a single chemical unit, i.e., as an atom; e.g., sulfate, phosphate, nitrate.

Rate of Shear—The rate at which an action, resulting from applied forces, causes or tends to cause two adjacent parts of a body to slide relatively to each other in a direction parallel to their plane of contact. Commonly given in rpm.

Red Mud—A clay, water-base drilling fluid containing sufficient amounts of caustic soda and tannates to give a pronounced red appearance. Normally a high-pH drilling fluid.

Red-lime Mud—A red mud which has been converted to a lime-treated drilling fluid. The pH is usually 12.0 to 13.0.

Resin—Semisolid or solid complex, amorphous mixture of organic compounds having no definite melting point nor tendency to crystalize. Resins may be a component of compounded materials that can be added to drilling fluids to impart special properties to the system, wall cake, etc.

Resistivity—The electrical resistance offered to the passage of a current, expressed in ohm-meters; the reciprocal of conductivity. Fresh-water drilling fluids are usually characterized by high resistivity; salt-water drilling fluids by a low resistivity.

Resistivity Meter — An instrument for measuring the resistivity of drilling fluids and their cakes.

Reverse Circulate—The method by which the normal flow of a drilling fluid is reversed by circulating down the annulus and up and out the drill string.

Reynolds Number — A dimensionless number, *Re*, that occurs in the theory of fluid dynamics. The diameter, velocity, density, and viscosity (consistent units) for a fluid flowing through a cylindrical conductor are related as follows:

$$Re = (diameter)(velocity)(density)/(viscosity)$$

or

$$= DV\rho/\mu.$$

The number is important in fluid hydraulics calculations for determining the type of fluid flow, i.e., whether laminar or turbulent. The transitional range occurs approximately from 2,000 to 3,000; below 2,000 the flow is laminar, above 3,000 the flow is turbulent.

Rheology—The science that deals with deformation and flow of matter.

Rotary Drilling—The method of drilling wells that depends on the rotation of a column of drill pipe to the bottom of which is attached a bit. A fluid is circulated to remove the cuttings.

—S—

SALT—In drilling fluid terminology, the term salt is applied to sodium chloride, NaCl. Chemically, the term salt is also applied to any one of a class of similar compounds formed when the acid hydrogen of an acid is partly or wholly replaced by a metal or a metallic radical. Salts are formed by the action of acids on metals, or oxides and hydroxides, directly with ammonia, and in other ways.

SALT-WATER CLAY—See Attapulgite Clay.

SALT-WATER DRILLING FLUIDS—A drilling fluid containing dissolved salt (brackish to saturated). These fluids may also include native solids, oil, and/or such commercial additives as clays, starch, etc.

SAMPLE MUD—A drilling fluid possessing properties to bring up suitable samples.

SAMPLES—Cuttings obtained for geological information from the drilling fluid as it emerges from the hole. They are washed, dried, and labeled as to the depth.

SAND—A loose granular material resulting from the disintegration of rocks, most often silica.

SAND CONTENT—The sand content of a drilling fluid is the insoluble abrasive solids content rejected by a 200-mesh screen. It is usually expressed as the percentage bulk volume of sand in a drilling fluid. This test is an elementary type in that the retained solids are not necessarily silica nor may not be altogether abrasive. For additional information concerning the kinds of solids retained on the 200-mesh screen, more specific tests would be required. See Mesh.

SATURATED SOLUTION—A solution is saturated if it contains at a given temperature as much of a solute as it can retain. At 68 F it takes 126.5 lb/bbl salt to saturate 1 bbl of fresh water. See Supersaturated.

SCREEN ANALYSIS — Determination of the relative percentages of substances, e.g., the suspended solids of a drilling fluid, passing through or retained on a sequence of screens of decreasing mesh size. Analysis may be by wet or dry methods. Referred to also as "sieve analysis." See Mesh.

SEALING AGENTS—Any of many materials added to drilling fluids to restore circulation.

SEA-WATER DRILLING FLUIDS—A special class of salt-water drilling fluids where sea water is used as the fluid phase.

SECONDS API—A unit of viscosity as measured with a Marsh funnel according to API procedure. See *API RP 13B** and Marsh Funnel Viscosity.

SEQUESTRATION—The formation of a stable, soluble complex by combining a metallic ion, such as calcium, magnasium, or iron, with a suitable agent thereby modifying the action of the ion. Representative requestering agents are ethelenediamine tetraacetic acid (EDTA) or its sodium salts, pyrophosphates, tripolyphosphates, and citrates.

SET CASING—The installation of pipe or casing in a well bore. Usually requires mudding up, reconditioning, or at least checking the drilling fluid properties.

SHALE—Fine-grained clay rock with slate-like cleavage, sometimes containing an organic oil-yielding substance.

SHALE SHAKER — Any of several mechanical devices for removing cuttings and other large solids from the drilling fluid. Common examples are vibrating screen, rotating cylindrical screen, etc.

SHEAR (SHEARING STRESS) — An action, resulting from applied forces, which causes or tends to cause two contiguous parts of a body to slide relatively to each other in a direction parallel to their plane of contact.

SHEAROMETER — A device used as an alternative method for measuring gel strengths. See *API RP 13B** for specifications and procedure.

SHEAR STRENGTH—A measure of the shear value of the fluid. The minimum shearing stress that will produce permanent deformation. See Gel Strength.

SIDETRACKING—See Whipstock.

SIEVE ANALYSIS—See Screen Analysis.

SILICA GEL—A porous substance consisting of SiO_2. Used on occasion as a dehydrating agent in air or gas drilling where small amount of water is encountered.

SILT — Materials that exhibit little or no swelling whose particle size generally falls between 2 microns and 74 microns (200-mesh). A certain portion of dispersed clays and barite for the most part also fall into this same particle-size range.

SKID — Moving a rig from one location to another, usually on tracks where little dismantling is required.

SLIP VELOCITY—The difference between the annular velocity of the fluid and the rate at which a cutting is removed from the hole.

SLOUGHING—The partial or complete collapse of the walls of a hole resulting from incompetent, unconsolidated formations, high angle or repose, and wetting along internal bedding planes. See Heaving and Cave-in.

SLUG THE PIPE—A procedure before pulling the drill pipe whereby a small quantity of heavy drilling fluid is pumped into the top section to cause an unbalanced column. As the pipe is pulled, the heavier column in the drill pipe will fall, thus keeping the inside of the drill pipe dry at the surface when the connection is unscrewed.

SOAP—The sodium or potassium salt of a high-molecular-weight fatty acid. When containing some metal other than sodium or potassium, they are called "metallic" soaps. Soaps are commonly used in drilling fluids to improve lubrication, emulsification, sample size, defoaming, etc.

SODA ASH—See Sodium Carbonate.

SODIUM—One of the alkali metal elements with a valence of 1 and an atomic weight of about 23. Numerous sodium compounds (all of which see) are used as additives to drilling fluids.

---

*Available from API Production Department, 211 North Ervay, Suite 1700, Dallas, Texas 75201.

SODIUM BICARBONATE—$NaHCO_3$. A material used extensively for treating cement contamination and occasionally other calcium contamination in drilling fluids. It is the half-neutralized sodium salt of carbonic acid.

SODIUM BICHROMATE—$Na_2Cr_2O_7$. Also correctly called "sodium dichromate." See Chromate.

SODIUM CARBONATE—Na_2CO_3. A material used extensively for treating out various types of calcium contamination. It is commonly called "soda ash." When sodium carbonate is added to a fluid, it increases the pH of the fluid by hydrolysis. Sodium carbonate can be added to salt (NaCl) water to increase the density of the fluid phase.

SODIUM CARBOXYMETHYLCELLULOSE—Commonly called CMC. Available in various viscosity grades and purity. An organic material used to control filtration, suspend weighting material, and build viscosity in drilling fluids. Used in conjunction with bentonite where low-solids drilling fluids are desired.

SODIUM CHLORIDE—NaCl. Commonly known as salt. Salt may be present in the drilling fluid as a contaminant or may be added for any of several reasons. See Salt.

SODIUM CHROMATE—Na_2CrO_4. See Chromate.

SODIUM HYDROXIDE—NaOH. Commonly referred to as "caustic" or "caustic soda." A chemical used primarily to impart a higher pH.

SODIUM POLYACRYLATE—A synthetic high-molecular-weight polymer of acrylonitrile used primarily as a fluid-loss-control agent.

SODIUM SILICATE MUDS—Special class of inhibited chemical drilling fluids using as their bases sodium silicate, salt, water, and clay.

SOLIDS CONCENTRATION OR CONTENT — The total amount of solids in a drilling fluid as determined by distillation includes both the dissolved and the suspended or undissolved solids. The suspended-solids content may be a combination of high and low specific gravity solids and native or commercial solids. Examples of dissolved solids are the soluble salts of sodium, calcium, and magnesium. Suspended solids make up the wall cake; dissolved solids remain in the filtrate. The total suspended and dissolved solids contents are commonly expressed as percent by volume, and less commonly as percent by weight.

SOL—A general term for colloidal dispersions, as distinguished from true solutions.

SOLUBILITY—The degree to which a substance will dissolve in a particular solvent.

SOLUTE—A substance which is dissolved in another (the solvent).

SOLUTION—A mixture of two or more components that form a homogeneous single phase. Example solutions are solids dissolved in liquid, liquid in liquid, gas in liquid.

SOLVENT—Liquid used to dissolve a substance (the solute).

SOURING—A term commonly used to mean fermentation. See Fermentation.

SPECIFIC GRAVITY—See Gravity, Specific.

SPECIFIC HEAT—The number of calories required to raise 1 g of a substance 1 deg Centigrade. The specific heat of a drilling fluid gives an indication of the fluid's ability to keep the bit cool for a given circulation rate.

SPUDDING IN—The starting of the drilling operations of a new hole.

SPUD MUD—The fluid used when drilling starts at the surface, often a thick bentonite-lime slurry.

SPURT LOSS—See Surge Loss.

SQUEEZE—A procedure whereby slurries of cement, drilling fluid, gunk plug, etc. are forced into the formation by pumping into the hole while maintaining a back pressure, usually by closing the rams.

STABILITY METER — An instrument to measure the breakdown voltage of invert emulsions.

STACKING A RIG—Storing a drilling rig upon completion of a job when the rig is to be withdrawn from operation for a period of time.

STARCH — A group of carbohydrates occurring in many plant cells. Starch is specially processed (pregelatinized) for use in drilling fluids to reduce filtration rate and occasionally to increase the viscosity. Without proper protection, starch can ferment.

STATIC—Opposite of dynamic. See Quiescence.

STEARATE—Salt of stearic acid, which is a saturated, 18-carbon fatty acid. Certain compounds, such as aluminum stearate, calcium stearate, zinc stearate, have been used in drilling fluids for one or more of the following purposes: defoamer, lubrication, air drilling in which a small amount of water is encountered, etc.

STREAMING POTENTIAL—The electrokinetic portion of the SP (spontaneous potential) electric-log curve which can be significantly influenced by the characteristics of the filtrate and mud cake of the drilling fluid that was used to drill the well.

STREAMLINE FLOW—See Laminar Flow.

STUCK—A condition whereby the drill pipe, casing, or other devices inadvertently become lodged in the hole. May occur while drilling is in progress, while casing is being run in the hole, or while the drill pipe is being hoisted. Frequently, a fishing job results.

SUPERSATURATION—If a solution contains a higher concentration of a solute in a solvent than would normally correspond to its solubility at a given temperature, this constitutes supersaturation. This is an unstable condition, as the excess solute separates when the solution is seeded by introducing a crystal of the solute. The term "supersaturation" is frequently used erroneously for hot salt drilling fluids.

SURFACE-ACTIVE MATERIALS—See Surfactant.

SURFACTANT—A material which tends to concentrate at an interface. Used in drilling fluids to control the degree of emulsification, aggregation, dispersion, interfacial tension, foaming, defoaming, wetting, etc. See Surfactant in Section 2.

SURFACTANT MUD—A drilling fluid which contains a surfactant. Usually refers to a drilling fluid containing surfactant material to effect control over degree of aggregation and dispersion or emulsification.

SURFACE TENSION—Generally, the force acting within the interface between a liquid and its own vapor which tends to maintain the area of the surface at a minimum and is expressed in dynes per centimeter. Since the surface tension of a liquid is approximately equal to the interfacial tension between the liquid and air, it is common practice to refer to values measured against air as surface tension, and to use the term "interfacial tension" for measurements at an interface between two liquids, or a liquid and a solid.

SURGE LOSS—The flux of fluids and solids which occurs in the initial stages of any filtration before pore openings are bridged and a filter cake is formed. Also called "spurt loss."

SUSPENSOID—A mixture consisting of finely divided colloidal particles floating in a liquid. The particles are so small that they do not settle but are kept in motion by the moving molecules of the liquid (Brownian movement).

SWABBING—When pipe is withdrawn from the hole in a viscous drilling fluid or if the bit is balled, a suction is created.

SWELLING—See Hydration.

SYNERGISM, SYNERGISTIC PROPERTIES—Term describing an effect obtained when two or more products are used simultaneously to obtain a certain result. Rather than the results of each product being additive to the other, the result is a multiple of the effects. See Synergism in Section 2.

—T—

TANNIC ACID—The active ingredient of quebracho and other quebracho substitutes such as mangrove bark, chestnut extract, hemlock, etc.

TEMPERATURE SURVEY — An operation to determine temperatures at various depths in the hole. This survey is used to find the location of inflows of water into the hole, where doubt exists as to proper cementing of the casing and for other reasons.

TEN-MINUTE GEL—See Gel Strength, 10-min.

THERMAL DECOMPOSITION—The chemical breakdown of a compound or substance by temperature into simple substances or into its constituent elements. Starch thermally decomposes in drilling fluids as the temperature approaches 300 F.

THINNER—Any of various organic agents (tannins, lignites, lignosulfonates, etc.) and inorganic agents (pyrophosphates, tetraphosphates, etc.) that are added to a drilling fluid to reduce the viscosity and/or thixotropic properties.

THIXOTROPY — The ability of fluid to develop gel strength with time. That property of a fluid which causes it to build up a rigid or semi-rigid gel structure if allowed to stand at rest, yet can be returned to a fluid state by mechanical agitation. This change is reversible.

TIGHTEN UP EMULSION OR DRILLING FLUID—Drilling fluid jargon to describe a condition in some systems to which oil has been added and the oil breaks out and rises to the surface. Any chemical or mechanical means which will emulsify the free oil is known as "tightening up."

TITRATION — A method, or the process of using a standard solution, for the determination of the amount of some substance in another solution. The known solution is usually added in a definite quantity to the unknown until a reaction is complete.

TOOL JOINT—A drill pipe coupler consisting of a pin and box of various designs and sizes. The internal design of tool joints has an important effect on drilling fluid hydrology.

TORQUE—A measure of the force or effort applied to a shaft causing it to rotate. On a rotary rig this applies especially to the rotation of the drill stem in its action against the bore of the hole. Torque reduction can usually be accomplished by the addition of various drilling fluid additives.

TOTAL DEPTH (OR TD)—The greatest depth reached by the drill bit.

TOTAL HARDNESS—See Hardness of Water.

TOUR—A person's turn in an orderly schedule. The word, which designates the shift of a drilling crew, is pronounced as if it were spelled t-o-w-e-r.

TURBULENT FLOW—Fluid flow in which the velocity at a given point changes constantly in magnitude and the direction of flow; pursues erratic and continually varying courses. Turbulent flow is the second and final stage of flow in a Newtonian fluid; it is the third and final stage in a Bingham plastic fluid. See Critical Velocity and Reynolds Number.

TWIST-OFF—The severing in two of a joint of drill pipe or other drill stem elements by excessive force applied by the rotary table.

—U—

ULTRAVIOLET LIGHT—Light waves shorter than the visible blue-violet waves of the spectrum. Crude oil, colored distillates, residuum, a few drilling fluid additives, and certain minerals and chemicals fluoresce in the presence of ultraviolet light. These substances, when present in drilling fluid, may cause the fluid to fluoresce.

UNDERGROUND BLOWOUT—An uncontrolled flow of well fluids and/or formation fluids into lower pressured subsurface zones. See Blowout.

UNIVALENT—Monovalent. See Valence.

—V—

VALENCE OR VALENCY—A number representing the combining power of an atom, i.e., the number of electrons lost, gained, or shared by an atom in a compound. It is also a measure of the number of hydrogen atoms with which an atom will combine or replace, e.g., an oxygen atom combines with two hydrogens, hence has a valence of 2. Thus, there are mono-, tri-, etc. valent ions.

VALENCE EFFECT—In general, the higher the valence of an ion the greater the loss of stability to emulsions, colloidal suspensions, etc., these polyvalent ions will impart.

*Available from API Production Department, 211 North Ervay, Suite 1700, Dallas, Texas 75201.

VELOCITY—Time rate of motion in a given direction and sense. It is a measure of the fluid flow and may be expressed in terms of linear velocity, mass velocity, volumetric velocity, etc. Velocity is one of the factors which contribute to the carrying capacity of a drilling fluid.

VELOCITY, CRITICAL—That velocity at the transitional point between laminar and turbulent types of fluid flow. This point occurs in the transitional range of Reynolds numbers of approximately 2,000 to 3,000.

V-G METER OR VISCOSITY-GRAVITY VISCOMETER—The name commonly used for the direct-indicating viscometer. See Viscometer, Direct-indicating.

VIBRATING SCREEN—See Shale Shaker.

VISCOMETER (VISCOSIMETER)—An apparatus to determine the viscosity of a fluid or suspension. Viscometers vary considerably in design and methods of testing.

VISCOMETER, DIRECT-INDICATING—Commonly called a "V-G meter." The instrument is a rotational-type device powered by means of an electric motor or handcrank, and is used to determine the apparent viscosity, plastic viscosity, yield point, and gel strengths (all of which see) of drilling fluids. The usual speeds are 600 and 300 rpm. See *API RP 13B** for operational procedures.

VISCOMETER, STORMER—A rotational shear viscometer used for measuring the viscosity and gel strength of drilling fluids. This instrument has been largely superseded by the direct-indicating viscometer. See Viscometer, Direct-indicating.

VISCOSIMETER—See Viscometer.

VISCOSITY—The internal resistance offered by a fluid to flow. This phenomenon is attributable to the attractions between molecules of a liquid, and is a measure of the combined effects of adhesion and cohesion to the effects of suspended particles, and to the liquid environment. The greater this resistance, the greater the viscosity. See Apparent Viscosity and Plastic Viscosity.

VISCOSITY, FUNNEL—See Funnel Viscosity.

VISCOUS FLOW—See Laminar Flow.

VOLATILE MATTER—Normally gaseous products, except moisture, given off by a substance, such as gas breaking out of live crude oil that has been added to a drilling fluid. In distillation of drilling fluids, the volatile matter is the water, oil, gas, etc., that are vaporized, leaving behind the total solids which can consist of both dissolved and suspended solids.

—W—

WALL CAKE—The solid material deposited along the wall of the hole resulting from filtration of the fluid part of the drilling fluid into the formation.

WALL STICKING—See Differential-pressure Sticking.

WATER-BASE DRILLING FLUID — Common conventional drilling fluids. Water is the suspending medium for solids and is the continuous phase, whether or not oil is present.

WATER BLOCK—Reduction of the permeability of a formation caused by the invasion of water into the pores (capillaries). The decrease in permeability can be caused by swelling of clays, thereby shutting off the pores, or in some cases by a capillary block of the pores due to surface tension phenomena.

WATER-IN-OIL EMULSION — See Invert Oil-emulsion Mud.

WATER LOSS—See Fluid Loss.

WEIGHT—In drilling fluid terminology, this refers to the density of a drilling fluid. This is normally expressed in either lb/gal, lb/cu ft, or psi hydrostatic pressure per 1,000 ft of depth.

WEIGHT MATERIAL—Any of the high specific gravity materials used to increase the density of drilling fluids. This material is most commonly barite but can be galena, etc. In special applications limestone is also called a weight material.

WELL LOGGING—See Electric Logging and Mud Logging.

WETTING—The adhesion of a liquid to the surface of a solid.

WETTING AGENT—A substance or composition which, when added to a liquid, increases the spreading of the liquid on a surface or the penetration of the liquid into a material. See Wetting Agent in Section 2.

WHIPSTOCK—A device inserted in a well bore used for deflecting or for directional drilling.

WILDCAT—A well in unproved territory.

WORKOVER FLUID—Any type of fluid used in the workover operation of a well.

—Y—

YIELD—A term used to define the quality of a clay by describing the number of barrels of a given centipoise slurry that can be made from a ton of the clay. Based on the yield, clays are classified as bentonite, high-yield, low-yield, etc., types of clays. Not related to yield value below. See *API RP 13B** for procedures.

YIELD POINT — In drilling-fluid terminology, yield point means yield value (see Yield Value). Of the two terms, yield point is by far the most commonly used expression.

YIELD VALUE — The yield value (commonly called "yield point") is the resistance to initial flow, or represents the stress required to start fluid movement. This resistance is due to electrical charges located on or near the surfaces of the particles. The values of the yield point and thixotropy, respectively, are measurements of the same fluid properties under dynamic and static states. The Bingham yield value, reported in lb/100 sq ft, is determined using the direct-indicating viscometer by subtracting the plastic viscosity from the 300-rpm reading.

—Z—

ZERO-ZERO GEL — A condition wherein the drilling fluid fails to form measurable gels during a quiescent time interval (usually 10 min).

ZETA POTENTIAL—Electrokinetic potential of a particle as determined by its electrophoretic mobility. This electric potential causes colloidal particles to repel each other and stay in suspension.

ZINC CHLORIDE—$ZnCl_2$. A very soluble salt used to increase the density of water to points more than double that of water. Normally added to a system first saturated with calcium chloride.

*Available from API Production Department, 211 North Ervay, Suite 1700, Dallas, Texas 75201.

SECTION 2
ENVIRONMENTAL TERMS

—A—

ABATEMENT—The act or process of reducing the intensity of pollution; also, the use of some method of abating pollution.

ACCLIMATION—The process of adjusting or adapting to changes in climate or environment.

ACUTE TOXICITY—Any poisonous effect produced within a short period of time, usually up to 24-96 hrs., resulting in severe biological harm and often death.

ADSORPTION—See Adsorption in Section 1.

AERATION—The process of being supplied or permeated with air. See Aeration in Section 1.

AEROBIC—The condition associated with the presence of free oxygen in an environment; living, active, or occurring only in the presence of oxygen.

AEROSOL—Suspension of liquid or solid particles in air or gas.

AIR CURTAIN—A method for mechanical containment of oil spills. Air is bubbled through a perforated pipe causing an upward water flow that retards the spreading of oil. Air curtains may also be used as barriers to prevent fish from entering a polluted body of water.

AIR POLLUTION—The presence of contaminants in the air in concentrations that interfere directly or indirectly with human health, safety, comfort, or with the full use and enjoyment of property.

AIR-QUALITY CRITERIA — The levels of pollution and lengths of exposure at which adverse effects on health and welfare occur.

AIR-QUALITY STANDARDS — The prescribed level of pollutants in the outside air that cannot be exceeded legally during a specified time in a specified geographic area.

ANAEROBIC—Refers to life or processes that occur in the absence of oxygen.

ANTIDEGRADATION CLAUSE—A provision in air-quality and water-quality laws that prohibits deterioration of air or water quality in areas where the pollution levels are presently below those allowed.

AQUATIC PLANTS—Plants that grow in water either growing up from the body of water, growing under the surface, or floating on the surface of the water.

AQUIFER—An underground bed or stratum of earth, gravel, or porous stone that contains water.

ASBESTOS—Mineral fiber (as amphibole) with numerous industrial uses; a hazardous air pollutant when inhaled. See Asbestos in Section 1.

ASSIMILATION—The transformation and incorporation of nutrient by an organism or ecosystem.

ATMOSPHERE—The layer of air surrounding the earth.

AUDIOMETER—An instrument for measuring hearing sensitivity and calibrated in decibels.

—B—

BACKGROUND LEVEL — With respect to pollution, amounts of pollutants present due to natural sources.

BACTERIA— Single-celled microorganisms that lack chlorophyll. Some bacteria are capable of causing human, animal, or plant diseases; others are essential in pollution control processes because they break down organic matter in water and air.

BIOASSAY—An assessment or test made using living organisms as the sensors; e.g., a fish toxicity test.

BIOCHEMICAL OXYGEN DEMAND (BOD)—A standardized measure of the amount of oxygen consumed in the biological processes that break down organic matter in water. It is measured as the quantity of dissolved oxygen (mg/l) required during stabilization of the decomposable organic matter by aerobic biochemical action.

BIODEGRADABLE—Decomposable as a result of the action of microorganisms.

BIOLOGICAL OXIDATION—The process by which bacterial and other microorganisms feed on complex organic materials and decompose them. The process is also called biochemical oxidation.

BIOMONITORING—The use of living organisms to test the suitability of effluent for discharge into receiving waters and to test the quality of such waters downstream from a discharge.

BOD_5—The amount of dissolved oxygen consumed in five days by biological processes breaking down organic matter in an effluent. See Biochemical Oxygen Demand in Section 2.

BOOM—A floating fence-like device that is used to contain oil on a body of water.

BUFFER CAPACITY—The ability of a solution to maintain a definite pH when subjected to the action of certain chemicals.

—C—

CARBON DIOXIDE (CO_2)—A colorless, odorless, nonpoisonous gas that is a normal part of the ambient air. Dissolved in the aqueous phase of drilling fluids, CO_2 may contribute to corrosivity of these fluids or to unstable rheological properties.

CARCINOGEN—A substance or agent producing or inciting cancerous growths in living tissues.

CENTRATE—Clarified liquid discharged from a centrifuge.

CHEMICAL OXYGEN DEMAND (COD)—A measure of the amount of oxygen required to oxidize organic and oxidizable inorganic compounds in water. The COD test, like the BOD test, is used to determine the degree of pollution in an effluent.

CHEMOTAXIS—The orientation or movement of a living organism in response to chemical agents.

CHRONIC BIOASSAY—A test involving a substantial portion of the life span of a fish or other organisms.

CLARIFICATION—In waste-water treatment, the removal of turbidity and suspended solids by settling, often aided by centrifugal action and chemically induced coagulation.

CLARIFIER—In waste-water treatment, a settling tank and/or centrifugal acceleration device which mechanically removes settleable solids from wastes.

COAGULATION—The clumping of particles in order to settle out impurities; often induced by chemicals such as lime or alum. See Coagulation in Section 1.

COASTAL ZONE—Coastal waters and the adjacent lands that exert a measurable influence on the uses of the sea and its ecology.

COLOR BODIES—Those complex molecules which impart color (usually undesirable) to a solution.

COLOR UNIT—Measures of the intensity of coloration of water using a platinum-cobalt standard.

CONSERVATIVE POLLUTANT—A pollutant that is relatively persistent and quite resistant to degradation, such as parachlorobiphenyls.

CONTAMINANT—A harmful or undesirable constituent; any substance that might constitute a health hazard or adversely affect desirable properties of drilling fluids.

COVER MATERIAL—Soil that is used to cover compacted solid waste in a sanitary landfill.

—D—

DECIBEL (db)—A unit of sound intensity, loudness, or pressure level in which 0.0002 microbar of pressure equals zero db sound-pressure level; the threshold of hearing.

DEHYDRATION—Removal of water and water vapor by any means from a gas, liquid, or solid. Dehydration is an essential step in proper purification of compressed air or gas for respiratory use. See Dehydration in Section 1.

DERMAL TOXICITY—The ability of a chemical to poison an animal or human by skin absorption.

DESALINIZATION—Salt removal from sea or brackish water.

DETERGENT—A substance that has cleansing action due to a combination of properties including lowering of surface tension, wetting action, emulsifying and dispersing action, and foam formation in some cases. The present connotation indicates a synthetic derivative in distinction to soap, which is derived from natural fats and oils.

DILUTION RATIO—The ratio of the volume of water of a stream to the volume of incoming waste. The capacity of a stream to assimilate waste is partially dependent upon the dilution ratio.

DISPERSANT—A chemical agent used to break up or disperse concentrations of various liquid or solid materials. In cleaning oil spills, dispersants are used to disperse oil on the water surface. See Dispersant in Section 1.

DISPOSAL—The orderly placement or discarding of unwanted material.

DISSOLVED OXYGEN (DO) — The oxygen dissolved in water or sewage. Adequately dissolved oxygen is necessary for the life of fish and other aquatic organisms and for the prevention of offensive odors. Low dissolved oxygen concentrations generally are due to discharge of excessive organic solids having high BOD, the result of inadequate waste treatment.

DISSOLVED SOLIDS—The total amount of dissolved material, organic and inorganic, contained in water or wastes. Excessive dissolved solids make water unpalatable for drinking and unsuitable for industrial use. Measurements are expressed as ppm or mg/liter (mg/l).

—E—

ECOLOGICAL IMPACT—The total effect of an environmental change, either natural or man-made, on the ecology of the area.

ECOLOGY—The interrelationships of living things to one another and to their environment, or the study of such interrelationships.

ECOSYSTEM—The interacting system of a biological community and its non-living environment.

EFFLUENT—A discharge of liquids and/or solids into the environment, partially or completely treated or in their natural state. Generally used in regard to discharges into waters.

EFFLUENT SEEPAGE — Diffuse discharge onto the ground of liquids that have percolated through solid waste or another medium; they contain dissolved or suspended materials.

EMISSION FACTOR—The average amount of a pollutant emitted from each type of polluting source in relation to a specific amount of material processed.

EMISSION STANDARD—The maximum amount of a pollutant legally permitted to be discharged from a single source either mobile or stationary.

ENVIRONMENT—The sum of all external conditions and influences affecting the life, development, an ultimately the survival of an organism.

ENVIRONMENTAL IMPACT STATEMENT—A document prepared by a Federal agency on the environmental impact of its proposals for legislation and other major actions significantly affecting the quality of the human environment. Environmental impact statements are used as tools for decision-making and are required by the National Environmental Policy Act.

EPA—The Environmental Protection Agency, a Federal agency.

ERDA—Energy Research and Development Administration, a Federal agency.

ESTUARIES—Areas where the fresh water meets salt water. For example, bays, mouths of rivers, salt marshes, and lagoons. Estuaries are delicate ecosystems; they serve as nurseries or spawning and feeding grounds for a large group of marine life and provide shelter and food for birds and wildlife.

EVAPORATION PONDS—Shallow, artificial ponds into which liquids or aqueous suspensions are pumped, permitted to dry, and either removed or buried by more added materials.

—F—

FOSSIL FUELS—Coal, oil, and natural gas; so-called because they are derived from the remains of ancient plant and animal life.

—G—

GERMICIDE (BACTERICIDE)—A chemical or agent that kills microorganisms such as bacteria. Such compounds must be registered as pesticides with EPA.

GROUNDWATER—Water present in the saturated zone of an aquifier.

—H—

HABITAT—The sum total of environmental conditions of a specific place that is occupied by an organism, a population, or a community.

HALF-LIFE—The period of time required for a radioactive substance to lose half of its active characteristics; used especially in radiological work.

HAZARDOUS AIR POLLUTANT—According to law, a pollutant to which no ambient air quality standard is applicable and that may cause or contribute to an increase in mortality or in serious illness. For example, asbestos, beryllium, and mercury have been declared hazardous air pollutants.

HEAVY METALS—Metallic elements with high atomic weights, which may be toxic to plant and animal life depending on their oxidation and chemical state. Such metals may be residual in the environment and exhibit biological accumulation. Examples include arsenic, cadmium, chromium, mercury, and lead.

HYDROCARBONS—A vast family of compounds containing carbon and hydrogen in various combinations, found especially in fossil fuels. Some hydrocarbons are major air pollutants, some may be carcinogenic, and others may contribute to photochemical smog.

HYDROGEN SULFIDE (H_2S)—A malodorous gas made up of hydrogen and sulfur with the characteristic odor of rotten eggs. Hydrogen sulfide may be present in some produced crude oils and formation waters. The gas, in the presence of an aqueous phase, is corrosive to tubular and wellhead goods. The gas is toxic and, in very low concentrations, can cause death.

HYDROLOGY—The science dealing with the properties, distribution, and circulation of water and snow.

—I—

IMPLEMENTATION PLAN—A document of the steps to be taken to ensure attainment of environmental quality standards within a specific time period. Implementation plans are required by various laws.

IMPOUNDMENT—A body of water, such as a pond, confined by a dam, dike, floodgate, or barrier.

INCINERATION—The controlled process by which solid, liquid, or gaseous combustible wastes are burned and changed into gases; the residue produced contains little or no combustible material.

INCINERATOR—An engineered apparatus used to burn waste substances and in which all the combustion factors, temperature, retention time, turbulence, and combustion air can be controlled.

INERT GAS—A gas that does not react with other substances under ordinary conditions.

INFILTRATION—The flow of a fluid into a substance through pores or small openings. Commonly used in hydrology to denote the flow of water into soil material.

INTERSTATE WATERS—According to law, waters defined as: (1) rivers, lakes, and other waters that flow across or form a part of state or international boundaries; (2) waters of the Great Lakes; (3) coastal waters whose scope has been defined to include ocean waters seaward to the territorial limits and waters along the coastline (including inland streams) influenced by the tide.

ION EXCHANGE—A reversible exchange of ions (electrically charged atoms or groups of atoms) between a solid and a liquid. In water and wastewater treatment, the solids (ion-exchange resins) exchange the polluting ions in the water for ions originally on the resin.

—J—

JTU (JACKSON TURBIDITY UNITS)—A measure of the turbidity of water, proportional to the ppm silica, where 100 ppm silica equals 21.5 JTU.

—L—

LC_{50}—See Median Lethal Concentration in Section 2.

LD_{50}—See Median Lethal Dose in Section 2.

LEACHATE—Liquid that has percolated through solid waste or other medium and has extracted dissolved or suspended materials from the medium.

LEACHING—The process by which soluble materials in the soil, such as nutrients, pesticide chemicals, or contaminants are washed into a lower layer of soil or are dissolved and carried away by water.

LIFE CYCLE—The phases, changes, or stages an organism passes through during its lifetime.

—M—

MEDIAN LETHAL CONCENTRATION (LC_{50})—A standard measure of toxicity. LC_{50} indicates the concentration of a substance or test material that causes death to fifty percent (50%) of a population within a given time period.

MEDIAN LETHAL DOSE (LD_{50})—The dose of a test material ingested or injected that kills fifty percent (50%) of a group of test organisms.

METHANE—A colorless, nonpoisonous, and flammable gaseous hydrocarbon. Methane (CH_4) is emitted by marshes and by dumps undergoing anaerobic decomposition.

MONITORING—Periodic or continuous determination of the amount of pollutants or contamination present in the environment or in drilling fluids.

—N—

NATURAL GAS—A fuel gas occurring naturally in certain geologic formations. Natural gas is usually a combustible mixture of methane and other hydrocarbons.

NIOSH—The National Institute for Occupational Safety and Health.

NUTRIENTS—Mineral elements including nitrogen, phosphorous, potassium, calcium, sulfur, magnesium, and iron which encourage plant and animal growth in soil or natural bodies of water.

—O—

OCEAN DISPOSAL—The deposition of waste into an ocean or estuarine body of water.

OIL SPILL—The accidental discharge of oil into oceans, bays, or inland waterways. Methods of oil-spill control include chemical dispersion, combustion, mechanical containment, and absorption.

OSHA—The Occupational Safety and Health Administration, a Federal agency.

OXIDATION—A chemical change or reaction in which oxygen unites or combines with other elements or substances. Organic matter may be oxidized or broken down by the action of aerobic bacteria as in certain waste-water treatment procedures.

—P—

PARTICULATES—Finely divided solid or liquid particles in the air or in an emission. Particulates include dust, smoke, fumes, mist, spray, and fog.

PERCOLATION — Downward flow or infiltration of water through the pores or spaces of a rock or soil.

PERMEABILITY—The capacity of a porous medium to conduct or transmit fluids. Normal permeability is a measure of ability of a rock to transmit a one-phase fluid under conditions of laminar flow. Unit of permeability is the Darcy. See Permeability in Section 1.

PHENOLS—A group of aromatic organic compounds containing one or more hydroxyl functions similar in structure to phenol (C_6H_5OH). Such compounds may produce a taste and odor problem in water at very low concentrations. In higher concentrations, they are toxic to aquatic life.

POLLUTANT—Any introduced gas, liquid, or solid that makes a resource unfit for a specific purpose.

POLLUTION—The presence of matter or energy whose nature, location, or quantity produces undesired environmental effects.

POTABLE WATER—Water suitable for drinking or cooking purposes from both health and aesthetic considerations.

PRIMARY TREATMENT—The first stage in waste-water treatment in which substantially all floating or settleable solids are mechanically removed by screening and sedimentation.

—R—

RECEIVING WATERS—Rivers, lakes, oceans, or other bodies that receive treated or untreated waste waters.

RECYCLING—The process by which materials pass through a cycle and undergo change or treatment, thus permitting their reclamation in some form to further use.

REFRACTORY—Resisting ordinary treatment and difficult to degrade.

RESERVOIR—A pond, lake, tank, or basin, natural or man-made, used for the storage, regulation, and control of water or other fluids.

RIPARIAN RIGHTS—Rights of a land owner to the water on or bordering his property, including the right to prevent diversion or misuse of upstream water.

RUNOFF—The portion of rainfall, melted snow, or irrigation water that flows across ground surface and eventually is returned to streams. Runoff can pick up pollutants from the air or the land and carry them to the receiving waters.

—S—

SALINITY—The degree of salt in water, expressed as ppm or as mg/l.

SALT-WATER INTRUSION—The invasion of salt water into a body of fresh water, occurring in either surface or ground-water bodies. When this invasion is caused by oceanic waters, it is called sea-water intrusion.

SCREENING—The removal of relatively coarse floating and suspended solids by straining through racks or screens.

SECONDARY TREATMENT—Waste-water treatment, beyond the primary state, in which bacteria consume the organic parts of the wastes. This biochemical action is accomplished by use of trickling filters or the activated-sludge process. Effective secondary treatment removes virtually all floating and settleable solids and approximately 90% of both BOD_5 and suspended solids. Customarily, disinfection by chlorination is the final stage of the secondary-treatment process.

SEDIMENTATION — In waste-water treatment, the settling out of solids by gravity.

SEEPAGE—Water that flows through the soil.

SETTLEABLE SOLIDS—Bits of debris and fine matter heavy enough to settle out of waste water.

SINKING—A method of controlling oil spills that employs an agent to entrap oil droplets and sink them to the bottom of the body of water. The oil and sinking agent are eventually biologically degraded.

SKIMMING—The mechanical removal of oil or scum from the surface of water.

SORPTION—A term including both adsorption and absorption. Sorption is basic to many processes used to remove gaseous and particulate pollutants from an emission and to clean up oil spills.

SPOIL—Dirt or rock that has been removed from its original location.

STRATIFICATION—The phenomenon occurring when a body of water becomes divided into distinguishable layers.

SURFACTANT—A surface-active agent that tends to concentrate at an interface. Such agents lower the surface tension of water and aid in emulsifying oil and dispersing materials in water. Used in drilling fluids to control the degree of emulsification, aggregation, dispersion, interfacial tension, foaming, defoaming, etc. See Surfactant in Section 1.

SURVEILLANCE SYSTEM—A monitoring system to determine environmental quality. Surveillance systems are established to monitor all aspects of progress toward attainment of environmental standards and to identify potential episodes of high pollutant concentrations in time to take preventive action.

SUSPENDED SOLIDS (SS)—Small particles of solid pollutants that contribute to turbidity and that resist separation by conventional means. The examination of suspended solids and the BOD test constitute the two main determinations for water quality performed at waste-water facilities.

SYNERGISM—The cooperative action of separate substances so that the total effect is greater than the sum of the effects of the substances acting independently. See Synergism in Section 1.

—T—

TERTIARY TREATMENT—Waste-water treatment, beyond the secondary or biological stage, that includes removal of nutrient such as phosphorous, nitrogen, and a high percentage of suspended solids. Tertiary treatment, also known as advanced waste treatment, produces a high-quality effluent.

THRESHOLD DOSE—The minimum dose of a given substance necessary to produce a measurable physiological or psychological effect.

TOLERANCE—The relative capability of an organism to endure an unfavorable environmental factor.

TOTAL SOLIDS—The total amount of solids in a waste water, both in solution and suspension.

TOXICANT—A substance that kills or injures an organism through its chemical or physical action, or by altering its environment; for example, cyanides, phenols, pesticides, or heavy metals.

TOXICITY—The quality or degree of being poisonous or harmful to plant or animal life.

—V—

VARIANCE—Sanction granted by a government body for delay or exception in the application of a given law, ordinance, or regulation.

VOLATILE—Evaporating readily at a relatively low temperature.

—W—

WASTE WATER — Water carrying wastes from homes, businesses, and industries that is a mixture of water and dissolved or suspended solids.

WATER POLLUTION—The addition of sewage, industrial wastes, or other harmful or objectionable material to water in concentrations or in sufficient quantities to result in measurable degradation of water quality.

WATER QUALITY CRITERIA—The levels of pollutants that affect the suitability of water for a given use; generally, water use classification includes: public water supply, recreation, propagation of fish and other aquatic life, agricultural use, and industrial use.

WATER QUALITY STANDARD—A plan for water quality management containing four major elements: the use (recreation, drinking water, fish and wildlife propagation, industrial, or agricultural) to be made of the water; criteria to protect those uses; implementation plans (for needed industrial-municipal waste treatment improvements); and enforcement plans and an antidegradation statement to protect existing high quality waste waters.

WATER TABLE—The upper level of ground water.

WETTING AGENT—A chemical or composition which, when added to a liquid, reduces the surface tension and increases the spreading of the liquid on a surface or the penetration of the liquid into a material. See Wetting Agent in Section 1.

REFERENCES

1. Studdard, G. J., *Common Environmental Terms*, U. S. Environmental Protection Agency, Washington, D. C., 1973.
2. Avants, J. B., "What Those Ecology Terms Mean," *Oil, Gas, & Petrochem Equipment*, Sept.-Dec., 1974.
3. Lund, H. F. (ed), *Industrial Pollution Control Handbook*, McGraw-Hill, 1971.
4. Liptak, B. G. (ed), *Environmental Engineer's Handbook, Vol. I, Water Pollution*, Chilton Book Co., 1974.
5. *Webster's New Collegiate Dictionary*, Second Edition, G & C Merriam & Co., 1959.

5M, 5/79, Egan
4½M, 8/81, Egan

Appendix E

CONVERSION AND INTERNATIONAL ATOMIC TABLES

The engineer has many occasions for converting one unit to another. This may involve length, volume, and weights within a system of weights and measures, and may also include conversions from one system to another. To facilitate such conversions, special conversion tables are included here (Table E-I and E-II).

Although few mud engineers are chemists, they are constantly dealing with complex chemical systems existing in drilling fluids. For convenience, therefore, the chart of international atomic weights is included in Table E-III.

TABLE E-I

Specific weight and pressure gradient of fluids

Specific weight			Pressure gradient	Specific weight			Pressure gradient
(lb/gal)	(lb/cu ft)	(g/cm^3) sp. gr.	(psi/ft)	(lb/gal)	(lb/cu ft)	(g/cm^3) sp. gr.	(psi/ft)
6.25	46.75	0.75	0.325	13.00	97.24	1.56	0.675
6.50	48.62	0.78	0.338	13.25	99.11	1.59	0.688
6.68	50.00	0.80	0.347	13.37	100.00	1.60	0.694
6.75	50.49	0.81	0.351	13.50	100.98	1.62	0.701
7.00	52.36	0.84	0.364	13.75	102.85	1.65	0.714
7.25	54.23	0.87	0.377	14.00	104.72	1.68	0.727
7.50	56.10	0.90	0.390	14.25	106.59	1.71	0.740
7.75	57.97	0.93	0.403	14.50	108.46	1.74	0.753
8.00	59.84	0.96	0.416	14.71	110.00	1.76	0.764
8.02	60.00	0.96	0.417	14.75	110.33	1.77	0.766
8.25	61.71	0.99	0.429	15.00	112.20	1.80	0.779
8.35	62.46	1.00	0.434	15.25	114.07	1.83	0.792
8.50	63.58	1.02	0.442	15.50	115.94	1.86	0.805
8.56	64.00	1.03	0.445	15.75	117.81	1.89	0.818
8.75	65.45	1.05	0.455	16.00	119.68	1.92	0.831
9.00	67.32	1.08	0.467	16.04	120.00	1.92	0.833
9.25	69.19	1.11	0.480	16.25	121.55	1.95	0.844
9.36	70.00	1.12	0.486	16.50	123.42	1.98	0.857
9.50	71.06	1.14	0.493	16.75	125.29	2.01	0.870
9.625	72.00	1.15	0.500	17.00	127.16	2.04	0.883
9.75	72.93	1.17	0.506	17.25	129.03	2.07	0.876
10.00	74.80	1.20	0.519	17.38	130.00	2.08	0.903
10.25	76.67	1.23	0.532	17.50	130.90	2.10	0.909
10.50	78.54	1.26	0.545	17.75	132.77	2.13	0.922
10.70	80.00	1.28	0.556	18.00	134.64	2.16	0.935
10.75	80.41	1.29	0.558	18.25	136.51	2.19	0.948
11.00	82.28	1.32	0.571	18.50	138.38	2.22	0.961
11.25	84.15	1.35	0.584	18.72	140.00	2.24	0.972
11.50	86.02	1.38	0.597	18.75	140.25	2.25	0.974
11.75	87.89	1.41	0.610	19.00	142.12	2.28	0.987
12.00	89.76	1.44	0.623	19.25	144.00	2.31	1.000
12.03	90.00	1.44	0.625	19.50	145.86	2.34	1.013
12.25	91.63	1.47	0.636	19.75	148.73	2.37	1.026
12.50	93.50	1.50	0.649	20.00	149.12	2.38	1.039
12.75	95.37	1.52	0.662	20.05	150.00	2.48	1.041

TABLE E-II

Conversion table (Courtesy of Magcobar)

(Standard Abbreviations)

Acres x 43,560 = sq ft
Acres x 40471 = sq m
Acres x 4840 = sq yd
Acre-ft x 43,560 = cu ft
Acre-ft x 325,851 = gal
Acre-ft x 1233.49 = cu m
Atm x 76 = cm of Hg
Atm x 29.92 = in. of Hg
Atm x 33.90 = ft of water
Atm x 14.7 = lb/sq in.
Atm x 1.058 = tons/sq ft
Bbl oil x 42 = gal oil
Bbl cement x 376 = lb cement
Bags or sacks of cement X 94 = lb cement
Bars x 10^6 = baryes
Bars x 75.006 = cm Hg (OC)
Bars x 14.504 = psi
Bars x 0.987 = atm
Baryes x 1 - dynes/sq cm
Board ft x 144 sq in. x lin. = cu in.
B x 0.2520 = kg cal
B x 778 = ft-lb
B x 3.927 x 10^{-4} = hp-hr
B x 2.930 x 10^{-4} = kwhr
B/min x 12.96 = ft-lb/sec
B/min x 0.02356 = hp
B/min x 0.01757 = kw
Bu (US) x 8 = gal (US) dry
Bu (US) x 9.3092 = gal (US) liq.
Cal (g) x 0.00397 = B
Cal (g) x 4.187 x 10^7 = ergs
Cal (g) x 0.001 = kg-cal
C x 1.8 + 32 = F
Cm of Hg x 0.01316 = atm
Cm of Hg x 0.4461 = ft water
Cm of Hg x 136.0 = kg/sq m
Cm of Hg x 27.85 = lb/sq ft
Cm of Hg x 0.1934 = lb/sq in.
Cm/sec x 1.969 = ft/min
Cm/sec x 0.036 = km/hr
Cm/sec x 0.02237 = miles/hr
Cm/sec/sec x 0.03281 = ft/sec/sec
Chains (Gunter's) x 66 = ft
Chains (Ramden's) x 100 = ft
Cu cm x 3.531 x 10^{-5} = cu ft
Cu cm x 6.102 x 10^{-2} = cu in.
Cu cm x 10^{-6} = cu m
Cu cm x 1.308 x 10^{-6} = cu yd
Cu cm x 2.642 x 10^{-4} = gal
Cu cm x 2.11 x 10^{-3} = pt (liq.)
Cu cm x 1.057 x 10^{-3} = qt (liq.)
Cu ft/min x 472.0 = cu cm/sec
Cu ft/min x 0.1247 = gal/sec
Cu ft/min x 0.4720 = 1/sec •
Cu ft/min x 62.43 = lb water/min
Cu ft/sec x 0.6463 = million gal/day
Cu ft/sec x 448.83 = gal/min
Cu in. x 16.39 = cu cm
Cu in. x 5.787 x 10^{-4} = cu ft
Cu in. x 1.639 x 10^{-5} = cu m
Cu in. x 2.143 x 10^{-5} = cu yd
Cu in. x 4.329 x 10^{-3} = gal
Cu in. x 1.639 x 10^{-2} = 1
Cu in. x 0.03463 = pt (liq.)
Cu in. x 0.01732 = qt (liq.)
Cu m x 10^6 = cu cm
Cu m x 35.31 = cu ft
Cu m x 61.023 = cu in.
Cu m x 1.308 = cu yd
Cu m x 264.2 = gal
Cu m x 10^3 = 1
Cu m x 2113 = pt (liq.)
Cu m x 1057 = qt (liq.)
Cu yd x 7.646 x 10^5 = cu cm
Cu yd x 27 = cu ft
Cu yd x 46,656 = cu in.
Cu yd x 0.7646 = cu m
Cu yd x 202.0 = gal
Cu yd x 764.6 = 1
Cu yd x 1616 = pt (liq.)
Cu yd x 807.9 = qt (liq.)
Cu yd/min x 0.45 = cu ft/sec
Cu yd/min x 3.367 = gal/sec
Cu yd/min x 12.74 = 1/sec
Deg (angle) x 60 = min
Deg (angle) x 0.01745 = radians
Deg (angle) x 3.600 = sec
Deg/sec x 0.01745 = radians/sec
Deg/sec x 0.1667 = rpm
Deg/sec x 0.002778 = rps
Dr x 27.3475 - grains
Dr x 0.0625 = oz avdp
Dr x 1.771845 = g
(F 32) 0.555 - C
Fathoms x 6 = ft
Ft x 30.48 = cm
Ft x 0.3048 - m
Ft x 0.33333 = yd
Ft water x 0.0295 = atm
Ft water x 0.8826 = in. Hg
Ft water x 304.8 = kg/sq m
Ft water x 62.43 = lb/sq ft
Ft water x 0.4335 = lb/sq in.
Ft/min x 0.508 = cm/sec
Ft/min x 0.01667 = ft/sec
Ft/min x 0.01829 = km/hr
Ft/min x 0.3048 = m/min
Ft/min x 0.01136 = miles/hr
Ft/sec x 30.48 = cm/sec
Ft/sec x 1.097 = km/hr
Ft/sec x 0.5921 = knots
Ft/sec x 18.29 = m/min
Ft/sec x 0.6818 = miles/hr
Ft/sec x 0.01136 = miles/min
Ft/sec/sec x 30.48 = cm/sec/sec
Ft/sec/sec x 0.3048 = m/sec/sec
Ft-lb x 1.28 x 10^{-3} = B
Ft-lb x 5.050 x 10^{-7} = hp-hr
Ft-lb x 3.24 x 10^{-4} = kg-cal
Ft-lb x 0.1383 = kg-m
Ft-lb x 3.766 x 10^{-7} = kwhr
Ft-lb/min x 1.286 x 10^{-3} = B/min
Ft-lb/min x 0.01667 = ft-lb/sec
Ft-lb/min x 3.030 x 10^{-5} = hp
Ft-lb/min x 3.241 x 10^{-4} = kg-cal/min
Ft-lb/min x 2.260 x 10^{-5} = kw
Ft-lb/sec x 7.717 x 10^{-2} = B/min
Ft-lb/sec x 1.818 x 10^{-3} = hp
Ft-lb/sec x .945 x 10^{-2} = kg-cal/min
Furlongs x 660 = ft
Gal x 3785 = cu cm
Gal x 0.1337 = cu ft
Gal x 231 = cu in.
Gal x 3.785 x 10^{-3} = cu m
Gal x 4.951 x 10^{-3} = cu yd
Gal x 3.785 = 1
Gal x 8 = pt (liq.)
Gal x 4 = qt (liq)
Gal Imperial x 1.20095 = gal (US)
Gal (US) x 0.83267 = gal Imperial
Gal water (US) x 8.3453 = lb water
Gal/min x 2.228 x 10^{-3} = cu ft/sec
Gal/min x 0.06308 = 1/sec
Gal/min x 8.0208 = cu ft/hr
Gal water/min x 6.0086 = tons/day
Grains troy x 1.0 = grains avdp
Grains troy x 0.06480 = g
Grains troy x 0.04167 = dwt troy
Grains troy x 2.0833 x 10^{-3} = oz troy
Grains/US gal x 17.118 = ppm
G x 980.7 = dynes
G x 10^{-3} = kg
G x 10^3 = mg
G x 0.03527 = oz avdp
G x 0.03215 = oz troy
G/cm x 5.6 x 10^{-3} = lb/in.
G/cu cm x 62.43 = lb/cu ft
G/cu cm x 0.03613 = lb/cu in.
G/1 x 58.417 = grains/gal (US)
G/1 x 0.062427 = lb/cu ft
Hp x 42.44 = B/min
Hp x 33,000 = ft-lb/min
Hp x 550 = ft-lb/sec
Hp x 1.014 = hp metric
Hp x 10.7 = kg-cal/min
Hp x 0.7457 = kw
Hp x 745.7 = W
Hp (boiler) x 33,479 = B/hr
Hp-hr x 2547 = B
Hp-hr x 1.98 x 10^6 = ft-lb
Hp-hr x 641.7 = kg-cal
Hp-hr x 2.737 x 10^5 = kg-m
Hp-hr x 0.7457 = kw-hr
In. x 2.54 = cm
In. Hg x 0.03342 = atm
In. of Hg x 1.133 = ft water
In. of Hg x 345.3 = kg/sq m
In. of Hg x 70.73 = lb/sq ft
In. of Hg x 0.4912 = lb/sq in.
In. water x 0.002458 = atm
In. water x 0.07355 = in. Hg
In. water x 25.4 = kg/sq m
In. water x 0.5781 = oz/sq in.
In. water x 5.202 = lb/sq ft
In. water x 0.03613 = lb/sq in.
Kg x 980,665 = dynes
Kg x 2.205 = lb avdp
Kg x 1.102 x 10^{-3} = tons (short)
Kg-cal x 3.968 = B
Kg-cal x 3086 = ft-lb
Kg-cal x 1.558 x 10^{-3} = hp-hr
Kg-cal x 1.162 x 10^{-3} = kw-hr
Kg-cal/min x 51.43 = ft-lb/sec
Kg-cal/min x 0.09351 = hp
Kg-cal/min x 0.06972 = kw
Kg/m x 0.6720 = lb/ft
Kg/sq m x 9.678 x 10^{-5} = atm
Kg/sq m x 3.281 x 10^{-3} = ft water
Kg/sq m x 2.896 x 10^{-3} = in. Hg
Kg/sq m x 0.2048 = lb/sq ft
Kg/sq m x 0.2048 = lb/sq ft
Kg/sq m x 1.422 x 10^{-3} = psi
Kg/sq m x 10^6 = kg/sq m
Km x 10^5 = cm
Km x 3281 = ft
Km x 0.6214 = miles
Km x 1094 = yd
Km/hr x 27.78 = cm/sec
Km/hr x 54.68 = ft/min
Km/hr x 0.9113 = ft/sec
Km/hr x 0.5396 = knots
Km/hr x 16.67 = m/min
Km/hr x 0.6214 = miles/hr
Km/hr/sec x 27.78 = cm/sec/sec
Km/hr/sec x 0.9113 = ft/sec/sec
Km/hr/sec x 0.2778 = m/sec/sec
Kw x 56.92 = B/min
Kw x 4.425 x 10^4 = ft-lb/min
Kw x 737.6 = ft-lb/sec
Kw x 1.341 = hp
Kw x 14.34 = kg-cal/min
Kw x 10^3 = w
Kw-hr x 3415 = B
Kw-hr x 2.655 x 10^6 = ft-lb
Kw-hr x 1.341 = hp-hr
Kw-hr x 860.5 = kg-cal
Kw-hr x 3.671 x 10^5 = kg-m
Lambert x 3183.1 = candela/m^2
Links (Gunter's) x 0.66 = ft
Links (Ramden's) x 1 = ft
L x 10^3 = cu cm
L x 0.03531 = cu ft
L x 61.02 = cu in.
L x 10^{-3} = cu m
L x 1.308 x 10^{-3} = cu yd
L x 0.2642 = gal
L x 2.113 = pt (liq.)
L x 1.057 = qt (liq.)
L/min x 5.866 x 10^{-4} = cu ft/sec
L/min x 4.403 x 10^{-3} = gal/sec
Lumber width (ft) x thickness (in.) x length (ft) = board ft
Lux x 1.00 = lumen/sq m
M x 100 = cm
M x 3.281 = ft
M x 29.37 = in.
M x 10^{-3} = km
M x 10^3 = mm
M x 1.094 = yd
M/min x 1.667 = cm/sec
M/min x 3.281 = ft/min
M/min x 0.05468 = ft/sec
M/min x 0.06 = km/hr
M/min x 0.03728 = miles/hr
M/sec x 196.8 = ft/min
M/sec x 3.281 = ft/sec
M/sec x 3.6 = km/hr
M/sec x 0.06 = km/min
M/sec x 2.237 = miles/hr
M/sec x 0.03728 = miles/min
Mu x 10^{-6} = m
Miles x 1.609 x 10^5 = cm
Miles x 5280 = ft
Miles x 1.609 = km
Miles x 1760 = yd
Miles (naut.) x 6,080.2 = ft.
Miles/hr x 44.70 = cm/sec
Miles/hr x 88 = ft/min
Miles/hr x 1.467 = ft/sec
Miles/hr x 1.609 = km/hr
Miles/hr x 0.8684 = knots
Miles/hr x 26.82 = m/min
Miles/min x 2682 = cm/sec
Miles/min x 88 = ft/sec
Miles/min x 1.609 = km/min
Miles/min x 60 = miles/hr
Mu x 10^{-6} = m
Millibars x 0.001 = bars
Millibars x 1,000 = bayres
Millibars x 1,000 = dynes/sq cm
Millibars x 2.0886 - psf
Mg x 10^{-3} = g
Ml x 10^{-3} = l
Ml x 0.1 = cm
Ml x 0.03937 = in.
Mg/1 x 1 = ppm
Million gal/day x 1.54723 = cu ft/sec
Min (angle) x 2.909 x 10^{-4} = radians
Oz avdp x 16 = drams avdp
Oz avdp x 437.5 = grains
Oz avdp x 0.0626 = lb avdp
Oz avdp x 28.349527 = g
Oz avdp x 0.9115 = oz troy
Oz avdp x 2.79 x 10^{-5} - tons (long)
Oz avdp x 2.835 x 10^{-5} = tons (metric)
Oz troy x 480 = grains
Oz troy x 20 = dwt (troy)
Oz troy x 0.08333 = lb troy
Oz troy x 31.103481 = g
Oz troy x 1.09714 = oz avdp
Oz fluid x 1.805 = cu in.
Oz fluid x 0.02957 = 1
Oz/sq in. x 0.0625 = lb/sq in.
Parts/million x 0.0584 = grains/US gal
Parts/million x 0.07016 = grains/imperial gal
Parts/million x 8.345 = lb/million gal
Dwt troy x 24 = grains
Dwt troy x 1.55517 = g
Dwt troy x 0.05 = oz troy
Dwt troy x 4.1667 x 10^{-3} = lb troy
Pica (printers) x 12 - points
Pica (printers) x 0.16666 = in.
Lb avdp x 16 = oz avdp
Lb avdp x 256 = drams
Lb avdp x 7,000 - grains
Lb avdp x 0.0005 = tons (short)
Lb avdp x 453.5924 = g
Lb avdp x 1.21528 = lb troy
Lb avdp x 14.5833 = oz troy
Lb troy x 5760 = grains
Lb troy x 240 = dwt troy
Lb troy x 12 = oz troy
Lb troy x 373.24177 = g
Lb troy x 0.822857 = lb avdp
Lb troy x 13.1657 = oz avdp
Lb troy x 3.6735 x 10^{-4} - tons (long)
Lb troy x 4.1143 x 10^{-4} - tons (short)
Lb troy x 3.7324 x 10^{-4} = tons (metric)
Lb water x 0.01602 = cu ft
Lb water x 27.68 = cu in.
Lb water x 0.1198 = gal US
Lb water/min x 2.67 x 10^{-4} = cu ft/sec
Lb/cu ft x 0.01602 = g/cu cm
Lb/cu ft x 16.02 = kg/cu m
Lb/cu ft x 5.787 x 10^{-4} = lb/cu in.
Lb/cu in. x 27.68 = g/cu cm
Lb/cu in. x 2.768 x 10^{-4} = kg/cu m
Lb/cu in. x 1728 = lb/cu ft
Lb/ft x 1.488 = kg/m
Lb/in. x 178.6 = g/cm
Psf x 0.01602 = ft water
Psf x 4.883 = kg/sq m
Psf x 6.945 x 10^{-3} = psi
Psi x 0.06804 = atm
Psi x 2.307 = ft water
Psi x 2.036 = in. Hg
Psi x 703.1 = kg/sq m
Quadrants (angle) x 90 = deg
Quadrants (angle) x 5400 = min
Quadrants (angle) x 1.571 = radians
Qt (dry) x 67.20 = cu in.
Qt (liq.) x 57.75 = cu in.
Radians x 57.30 = deg
Radians 3438 = min
Radians x 0.637 = quadrants
Radians/sec x 57.30 = deg/sec
Radians/sec x 0.1592 = rps
Radians/sec x 9.549 = rpm
Radians/sec/sec x 573 = rpm/min
Radians/sec/sec x 0.159 = rps/sec
Reams x 500 = sheets
Revolutions x 360 = deg
Revolutions x 4 = quadrants
Revolutions x 6.283 = radians
Rpm x 6 = deg/sec
Rpm x 0.1047 = radians/sec
Rpm x 0.01667 = rps
Rpm/min/min x 1.745 x 10^{-3} = radians/sec/sec
Rpm/min x 2.778 x 10^{-4} = revolutions/sec/sec
Rps/sec x 360 = deg/sec
Rps/sec x 6.283 = radians/sec
Rps/sec x 5.283 = radians/sec/sec
Rps/sec x 3600 = rpm/min
Seconds (angle) x 4.848 x 10^{-6} = radians
Sq cm x 1.076 x 10^{-3} = sq ft
Sq cm x 0.1550 = sq in.
Sq cm x 10^{-4} = sq m
Sq ft x 2.296 x 10^{-5} = acres
Sq ft x 929.0 = sq cm
Sq ft x 144 = sq in.
Sq ft x 0.0929 = sq m
Sq ft x 3.587 x 10^{-8} = sq miles
Sq in. x 6.452 = sq cm
Sq in. x 6.944 x 10^{-3} = sq ft
Sq in. x 645.2 = sq mm
Sq km x 247.1 = acres
Sq km x 10.76 x 10^6 = sq ft
Sq km x 10^6 = sq m
Sq km x 0.3861 = sq miles
Sq km x 1.196 x 10^6 = sq yd
Sq m x 2.471 x 10^{-4} = acres
Sq m x 10.76 = sq ft
Sq m x 3.861 x 10^{-7} = sq miles
Sq m x 1.196 = sq yd
Sq miles x 640 = acres
Sq miles x 27.88 x 10^6 - sq ft
Sq miles x 2.59 - sq km
Sq miles x 3.098 x 10^6 = sq yd
Sq mm x 0.01 = sq cm
Sq mm x 1.55 x 10^{-3} = sq in.
Sq yd x 2.066 x 10^{-4} = acres
Sq yd x 9 = sq ft
Sq yd x 0.8361 = sq m
Sq yd x 3.228 x 10^{-7} = sq miles
Slugs x 32.174 = lb avdp
Statamp x 3.3359 x 10^{-10} = amp
Tablespoon x 15 = cc
Teaspoon x 0.3333 = tablespoons
Ton (assay) x 29.166 = g
Tons (long) x 1016 = kg
Tons (long) x 2240 = lb avdp
Tons (long) x 1.12 - tons(short)
Tons (metric) x 10^3 = kg
Tons (metric) x 2205 = lb avdp
Tons (short) x 2000 = lb avdp
Tons (short) x 32,000 = oz avdp
Tons (short) x 907.18486 = kg
Tons (short) x 2430.56 = lb troy
Tons (short) x 0.89287 = tons (long)
Tons (short) x 29166.66 = oz troy
Tons (short) x 0.90718 = tons (metric)
Tons water/day x 83.333 = lb water/hr
Tons water/day x 0.16643 = gal/min
Tons water/day x 1.3349 = cu ft/hr
W x 0.05692 = B/min
W x 44.26 = ft-lb/min
W x 0.7376 = ft-lb/sec
W x 1.341 x 10^{-3} = hp
W x 0.01434 = kg-cal/min
Whr x 3.415 = B
Whr x 2655 = ft-lb
Whr x 1.341 x 10^{-3} = hp-hr
Whr x 0.8605 = kg-cal
Whr x 367.1 = kg-m
Whr x 10^{-3} - kwhr
Yd x 91.44 = cm
Yd x 36000 - mils (US)
Yr (mean solar) x 365.24 = days (ms)

TABLE E-III

International atomic table

Element	Symbol	Atomic number	Atomic weight	Valence
Actinium	Ac	89	227.0	—
Aluminum	Al	13	26.97	3
Antimony	Sb	51	121.76	3, 5
Argon	A	18	39.944	0
Arsenic	As	33	74.91	3, 5
Barium	Ba	56	137.36	2
Beryllium	Be	4	9.02	2
Bismuth	Bi	83	209.00	3, 5
Boron	B	5	10.82	3
Bromine	Br	35	79.916	1, 3, 5, 7
Cadmium	Cd	48	112.41	2
Calcium	Ca	20	40.08	2
Carbon	C	6	12.01	2, 4
Cerium	Ce	58	140.13	3, 4
Cesium	Cs	55	132.91	1
Chlorine	Cl	17	35.457	1, 3, 5, 7
Chromium	Cr	24	52.01	2, 3, 6
Cobalt	Co	27	58.94	2, 3
Columbium	Cb	41	92.91	3, 5
Copper	Cu	29	63.57	1, 2
Dysprosium	Dy	66	162.46	3
Erbium	Er	68	167.2	3
Europium	Eu	63	152.0	2, 3
Fluorine	F	9	19.000	1
Gadolinium	Gd	64	156.9	3
Gallium	Ga	31	69.72	2, 3
Germanium	Ge	32	72.60	4
Gold	Au	79	197.2	1, 3
Hafnium	Hf	72	178.6	4
Helium	He	2	4.003	0
Holmium	Ho	67	164.94	3
Hydrogen	H	1	1.0080	1
Indium	In	49	104.76	3
Iodine	I	53	126.92	1, 3, 5, 7
Iridium	Ir	77	193.1	3, 4
Iron	Fe	26	55.85	2, 3
Krypton	Kr	36	83.7	0
Lanthanum	La	57	138.92	3
Lead	Pb	82	207.21	2, 4
Lithium	Li	3	6.940	1
Lutecium	Lu	71	174.99	3
Magnesium	Mg	12	24.32	2
Manganese	Mn	25	54.93	2, 3, 4, 6, 7
Masurium	Ma	43	—	—
Mercury	Hg	80	200.61	1, 2
Molybdenum	Mo	42	95.95	3, 4, 6

TABLE E-III (continued)

Element	Symbol	Atomic number	Atomic weight	Valence
Neodymium	Nd	60	144.27	3
Neon	Ne	10	20.183	0
Nickel	Ni	28	58.69	2, 3
Nitrogen	N	7	14.008	3, 5
Osmium	Os	76	190.2	2, 3, 4, 8
Oxygen	O	8	16.000	2
Palladium	Pd	46	106.7	2, 4
Phosphorus	P	15	30.98	3, 5
Platinum	Pt	78	195.23	2, 4
Polonium	Po	84	210.0	—
Potassium	K	19	39.096	1
Praseodymium	Pr	59	140.92	3
Protoactinium	Pa	91	231.0	—
Radium	Ra	88	226.05	2
Radon	Rn	86	222.0	0
Rhenium	Re	75	186.31	—
Rhodium	Rh	45	102.91	3
Rubidium	Rb	37	85.48	1
Ruthenium	Ru	44	101.7	3, 4, 6, 8
Samarium	Sm, Sa	62	150.43	3
Scandium	Sc	21	45.10	3
Selenium	Se	34	78.96	2, 4, 6
Silicon	Si	14	28.06	4
Silver	Ag	47	107.880	1
Sodium	Na	11	22.997	1
Strontium	Sr	38	87.63	2
Sulphur	S	16	32.06	2, 4, 6
Tantalum	Ta	73	180.88	5
Tellurium	Te	52	127.61	2, 4, 6
Terbium	Tb	65	159.2	3
Thallium	Tl	81	204.39	1, 3
Thorium	Th	90	232.12	4
Thulium	Tm	69	169.4	3
Tin	Sn	50	118.70	2, 4
Titanium	Ti	22	47.90	3, 4
Tungsten	W	74	183.92	6
Uranium	U	92	238.07	4, 6
Vanadium	V	23	50.95	3, 5
Virginium	Vi	87	224.0	1
Xenon	Xe	54	131.3	0
Ytterbium	Yb	70	173.04	3
Yttrium	Yt	39	88.92	3
Zinc	Zn	30	65.38	2
Zirconium	Zr	40	91.22	4

ACKNOWLEDGEMENT

The help extended by James L. Bryan and Farhad Ghassemi is greatly appreciated by the authors. The contribution of James L. Bryan was indeed invaluable.

Appendix F

NEW DEVELOPMENTS IN DRILLING-FLUIDS TECHNOLOGY

RYEN CAENN and GEORGE V. CHILINGARIAN

INTRODUCTION

Since the first printing of the *Drilling and Drilling Fluids* book, there have been some important developments in the technology of drilling fluids and some drilling techniques that affect the drilling fluids. All technology is in a constant state of flux. The usual case in engineering technology is that new developments do not appear overnight. There is a gradual building upon prior knowledge, techniques, and formulations. The following descriptions of new systems and techniques are the result of logical steps that, in hindsight, can be seen developing in the pages of the rest of this book. These new developments owe their existence to the many investigators who have worked and are working in the drilling industry. The authors wish to express a debt of gratitude to all the research workers and practicing engineers who publish books and manuals on the subject bringing all available material together. Preparation of periodic reviews in the form of books and manuals is just as great a contribution to science and technology as the original research work.

Considerable time and effort of drilling fluid technologists over the last few years have been devoted to eliminating, or at least minimizing, some fairly serious drawbacks to the use of oil-base drilling fluids. These developments in the field of oil-base drilling fluids have been significant to the petroleum industry. They have revived a technology that appeared to be heading towards extinction. This was mainly due to the high cost and adverse environmental impact of oil-base drilling fluids.

The products, that have been developed to any extent during the last few years in the field of water-base drilling fluids, are a few "new" weighting materials. They are not really new in that iron oxides, for example, have been used in drilling fluids very nearly from the inception of drilling fluid technology. The products, however, have been newly introduced and differ from previous products of this type. They reportedly have some significant advantages over using barite as weighting agent.

Other major developments that affect drilling fluid technologists lie in the field of drilling techniques. The use of a new polycrystalline diamond bit (see Appendix G) and the development of reliable monitoring while drilling (MWD) for directional surveys, for example, have had a significant impact on drilling fluid technology. All drilling fluid engineers must be aware of these developing technologies.

The last item of discussion in this appendix is the environmental impact of drilling fluids. Again, it behooves all drilling fluid engineers to be aware of and be responsible for any actions that may affect the quality of the environment. There have been many irresponsible and inaccurate statements concerning environmental impact both from within and without the industry. The only hope to remedy this is for knowledgeable individuals to speak up and, more importantly, to develop better definitions of the parameters involved in protecting the environment.

RHEOLOGY

Chapter 7 of the *Drilling and Drilling Fluids* book gave a complete description of the state of the art of drilling fluid rheology. Since the first printing of the book, however, a new development has been published that allows the use of the effective viscosity plot, along with a technique called the critical viscosity plot to determine the pump speed needed to maintain turbulent flow in the hole. The critical viscosity plot is in actuality a method of plotting the Reynolds numbers at two different pumping rates. The intersection of this plot with the effective viscosity plot, shows the point which is the transition between laminar and turbulent flow. The value of keeping a drilling fluid in laminar flow is well known in the industry, primarily to avoid excessive washouts of the borehole.

Use of the critical velocity graph (Fig. F-1) provides a field procedure for determining the maximum annular velocity or pump rate that will allow laminar flow conditions to be maintained in the annulus. This procedure utilizes a viscosity plot, such as the Kelco Rotary Viscoso-Graph, a standard two-speed viscometer, and a simple calculator. This method is based on the drilling fluid's viscosity curve and the basic Reynolds number equation to determine the annular velocity at which the Reynolds number equals 2000. Utilizing a Reynolds number of 2000 to indicate the transition from laminar to turbulent flow provides a conservative critical velocity which may allow for variations in temperature, hole diameter, off-center drillpipe, and deviated hole. The actual transition from laminar to turbulent flow may occur at Reynolds numbers varying from 2000 to 3500, depending upon the specific fluid properties and the hole conditions. A highly shear-thinning drilling fluid may have a critical velocity 75% higher than is calculated by this procedure. When hole stability is critical, however, it is normally advisable to use the lower calculated critical velocity to insure laminar flow.

In some cases, hole instability can be attributed to mechanical erosion of the wellbore due to turbulent flow of the drilling fluid in the annulus. Under these conditions, maintaining laminar flow around the largest O.D. or downhole assembly has proven to be a significant step in improving borehole stability. This can be accomplished by both changing the flow rate and adjusting rheological properties of the mud. The rheological property that can

10/82
KT-4-5

CRITICAL VELOCITY VISCOSO-GRAPH®

KelcoRotary

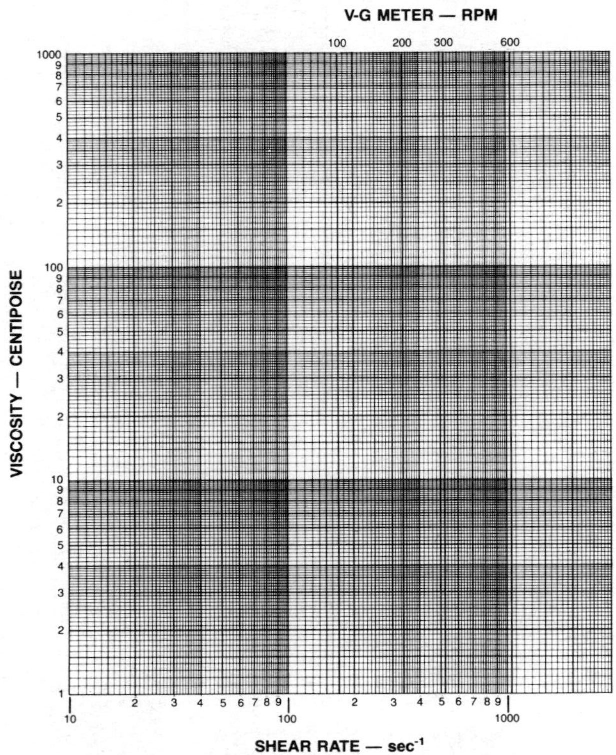

WELL DATA

WELL ______________ DATE ______________

LOCATION ______________ DEPTH __________ ft.

FLUID TYPE ______________

CASING, O.D. ________" I.D. ________" SET AT __________ ft.

DRILL PIPE, O.D. ________" COLLARS, O.D. ________"

HOLE I.D. OR BIT SIZE ________"

PUMP ______________ LINER SIZE __________"

RATE __________ GPM @ __________ SPM = __________ gal/stk

MUD PROPERTIES

Shear Rate		Dial Reading	Viscosity Multiplier	Viscosity (cps.)
RPM	sec^{-1}			
600	1020		0.5	
300	510		1	

P.V. __________ Y.P. __________

GELS __________ TEMP. __________

WT. __________ SOLIDS __________

VISCOSITY POINTS FOR CRITICAL INTERCEPT LINE

A. 100 RPM = $\frac{\text{Mud Wt. (Hole I.D. - Pipe O.D.)}^2}{1.833}$ = __________ cP

B. 200 RPM = 2 X 100 RPM Point = __________ cP

Critical Shear Rate = __________ sec^{-1}

C. Critical Velocity = $\frac{\text{Critical Shear Rate } (D_h - D_p)}{2.4}$ = __________ ft./min.

D. Critical Pump Rate = $\frac{\text{Critical Velocity } (D_h^2 - D_p^2)}{24.5}$ = __________ GPM

$\frac{\text{GPM}}{\text{gal/stk}}$ = __________ SPM

D_h = Hole Diameter, ins.
D_p = Drill Pipe or Collar O.D., ins.

Fig. F-1A. Critical velocity *viscoso-graph*®. (Courtesy of Kelko Rotary.)

CRITICAL VELOCITY VISCOSO-GRAPH®

KelcoRotary

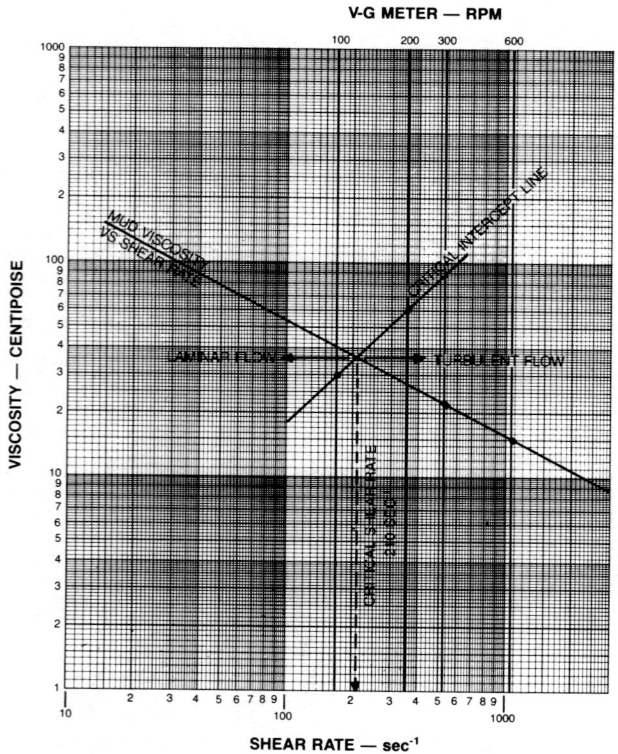

WELL DATA

WELL ________ DATE ________

LOCATION ________ DEPTH 9500 ft.

FLUID TYPE BEN-EX

CASING, O.D. 10¾" I.D. 9.95" SET AT 4175 ft.

DRILL PIPE, O.D. 4½" COLLARS, O.D. 6¼"

HOLE I.D. OR BIT SIZE 8¾"

PUMP NATIONAL K-700 LINER SIZE 6X18"

RATE 345 GPM @ 50 SPM = 6.9 gal/stk

MUD PROPERTIES

Shear Rate		Dial Reading	Viscosity Multiplier	Viscosity (cps.)
RPM	sec^{-1}			
600	1020	30	0.5	15
300	510	22	1	22

P.V. 8 Y.P. 14

GELS 4/12 TEMP. 92°F

WT. 9.0 SOLIDS 5%

VISCOSITY POINTS FOR CRITICAL INTERCEPT LINE

A. **100 RPM** = $\frac{9.0(8.75-6.25)^2}{1.833}$ = 31 cP

B. **200 RPM** = 2 X 100 RPM Point = 62 cP

Critical Shear Rate = 210 sec^{-1}

C. **Critical Velocity** (around collars) = $\frac{210(8.75-6.25)}{2.4}$ = 219 ft./min.

D. **Critical Pump Rate** = $\frac{219(76.6-39.0)}{24.5}$ = 336 GPM

$\frac{336}{6.9}$ = 49 SPM

D_h = Hole Diameter, ins.
D_p = Drill Pipe or Collar O.D., ins.

Fig. F-1B. Sample calculations using critical velocity *viscoso-graph*®. (Courtesy of Kelko Rotary.)

be adjusted is the effective viscosity of the fluid. Increasing the fluid viscosity will reduce the Reynolds number and, therefore, allow higher annular velocities before exceeding a Reynolds number of 2000. For example, in the case of a 30-cP, 9.0-lb/gal mud in a 10¾-inch hole (8-inch drillcollars), the critical velocity is 156 ft/min, whereas a 50 cP viscosity provides a critical velocity of 261 ft/min (based on a Reynolds number of 2000).

The procedure for using the critical velocity technique involves obtaining well information, such as hole size, drillpipe size, and the 600- and 300-rpm dial readings from a standard viscometer. The dial readings are converted to viscosities and plotted on a log—log plot (Fig. F-1A). The 100 rpm intercept point, I_{100}, is calculated from the following equation:

$$I_{100} = \frac{\gamma_f [d_h - d_p]^2}{1.833} \qquad \text{(F-1)}$$

where γ_f = specific weight of drilling fluid in lb/gal, d_h = hole diameter in inches, and d_p = outside diameter of drillpipe in inches. The 200-rpm intercept point, I_{200}, is calculated from the following equation:

$$I_{200} = 2 \times I_{100} \qquad \text{(F-2)}$$

These intercept points are plotted on the graph as if they were viscosities at 100 and 200 rpm. A straight line is then drawn between these two points and the critical shear rate, $\dot{\gamma}_c$ (in sec^{-1}), is read from the graph at the point of intersection of the fluid viscosity curve and the previously drawn critical velocity line. The critical shear rate for a specific fluid is then converted to the critical velocity, v_c (in ft/min), by the following formula:

$$v_c = \frac{\dot{\gamma}_c (d_h - d_p)}{2.4} \qquad \text{(F-3)}$$

The critical velocity can be, in turn, converted to a critical volumetric pump rate, q_{pc} (in gal/min), by the use of the following formula:

$$q_{pc} = \frac{v_c (d_h^2 - d_p^2)}{24.5} \qquad \text{(F-4)}$$

Figure F-1B shows an example of a completed critical viscosity graph. In this example, the 600- and 300-rpm dial readings have been converted to viscosities (15 cP and 22 cP) and plotted on the graph. The line designated as drilling fluid viscosity versus shear rate is the plot from the viscometer. The 100- and 200-intercept points are then calculated and plotted on the graph, using diameter of the drillhole and outside diameter of the drillcollars. A line is then drawn between these two points and is designated as the critical intercept line on the graph. The critical shear rate, in this case

210 sec^{-1}, is read from the graph at the intersection of these two lines. This intersection represents the shear rate at a Reynolds number of 2000 between the drillcollars and the open hole. The graph then shows that for all shear rates and corresponding pump rates to the left of this point the flow is laminar, whereas for all points to the right of the intercept the flow is turbulent around the drillcollars. In this example, maintaining the flow rate at or below 336 gal/min (or 49 strokes/min) will assure that the flow is laminar.

NEW DEVELOPMENTS IN THE FIELD OF OIL-BASE DRILLING FLUIDS

Oil-base drilling fluids have been recognized for many years as being superior to water-base drilling fluids in handling the problems of high temperatures, hole problems, and corrosion (especially, H_2S corrosion). These attributes of oil-base drilling fluids are discussed in detail in Chapter 10. There are some problems, however, that are specifically associated with the use of oil-base drilling fluids that severely limit their application. The most obvious limitation is the impact of using materials such as diesel oil and asphaltenes that could have a severe impact on the environment if used or disposed of improperly. In some areas, such as in the national forests, they are strictly forbidden. Other limitations include their comparatively high cost and relatively slow drilling rates. The cost of oil-base drilling fluids will probably always remain higher than that of water-base systems, but mitigation of the other drawbacks will make oil-base drilling fluids viable alternatives for drilling fluids for years to come.

Fontenot and Simpson (1974) studied the effect of oil-base drilling fluids on the relative drilling rates. The relative drilling rates were determined from a micro-bit drilling machine using sandstones and carbonate rocks. Typical oil-base drilling fluids in use at that time contained from 60 to 90% (volume) oil and 40 to 10% of water, containing $CaCl_2$ salt (20% to saturation). Included therein were also various emulsifiers, oil dispersible bentonite, and significant quantities of either oil dispersible lignite or asphaltene blends. The latter ingredients were added in sufficient amounts to lower the fluid loss to relatively low levels. The standard API ambient temperature fluid loss was usually zero for these drilling fluids. Fontenot and Simpson (1974) showed that the colloidal content of the oil-base drilling fluids, mainly the lignite, was a significant factor in the slower relative drilling rates of these drilling fluids.

Simpson (1979) related the surface measurements of filtrate and viscosity of oil-base drilling fluids to the actual downhole conditions. Simpson (1979) also reported some field data supporting the fact that these drilling fluids low in colloidal content drill significantly faster than the previously formulated oil muds. Simpson recommended the following: (1) use of low-viscosity oil for the continuous liquid phase, usually diesel oil; (2) low content of emulsified water, usually less than 15% (volume) of the liquid phase; (3) high salinity of the emulsified water phase, usually 300,000—

TABLE F-I

Performance of low-colloid oil-base drilling fluid in record well in Wyoming [16,000 ft (5060 m) to total depth] (after Simpson, 1979; courtesy of the SPE of AIME)

Well	Interval		Mud type	Days	Drilling rate	
	(ft)	(m)			(ft/D)	(m/D)
A	2000	610	low-colloid oil-base mud	29	69	21
B	2050	620	water-base mud	40	51	16
C	1800	550	water-base mud	37	49	15
D	1800	550	water-base mud	63	29	9

TABLE F-II

Performance of low-colloid drilling fluid in a well in Louisiana (16,400 ft (5000 m) to total depth) (after Simpson, 1979; courtesy of the SPE of AIME)

Well	Interval		Mud type	Bits	Days	Drilling rate	
	(ft)	(m)				(ft/D)	(m/D)
A	4173	1270	low-colloid oil-base mud	10	45	93	28
B	3817	1160	water-base mud	11	52	74	22
C	4020	1230	water-base mud	12	61	66	20

400,000 ppm of $CaCl_2$; (4) oil dispersible bentonite for the lifting of cuttings and the suspension of barite; and (5) temperature-stable amide for oil wetting the solids and for emulsification. He also pointed out that efficient solids control techniques must be used with oil-base drilling fluids to insure optimum drilling rates.

The relaxed-filtrate oil-base drilling fluids are used more than other oil-base drilling fluids. From their initial uses in West Texas (O'Brien et al., 1977, 1978), the relaxed-filtrate oil muds are now being used worldwide. One problem encountered with their use was that in drilling highly-permeable formations with high differential pressures there was a tendency toward differential sticking of the pipe. In such cases, operators have found it necessary to add more fluid-loss-control additives to the drilling fluid. Overall, however, the relaxed-filtrate oil-base drilling fluids have been very successful and have allowed their use even where once considered uneconomical.

The other major development in oil-base drilling fluids technology in the past few years has been in the area of the so-called non-polluting or environmentally clean fluids. It is generally accepted that the liquid constituent of oil-base drilling fluids (diesel oil, crude oil, along with asphaltenes in some cases) can be damaging to the environment if the problem is not corrected.

These materials have long life and essentially are non-biodegradable. When they show up as oil slicks or sheens and are released in quantities great enough to coat plants and animals, they are obvious pollutants. A more subtle pollution may take place when these materials become part of the food chain and endanger the surrounding plant and animal life. Recently, an environmentally acceptable substitute to formulate the liquid phase of oil-base drilling fluids was developed.

The initial discovery and use of mineral oil as a substitute for diesel oil was made by Chromalloy Drilling Fluids Company. These mineral-oil-based drilling fluids are being used extensively in the North Sea and in the Gulf Coast of the U.S.A.

The main reason for using the mineral oil as a substitute for diesel oil is the fact that the mineral oil is less harmful in bio-assay tests and does not develop a significant oil sheen (Clements and Hinds, 1982). Mineral oil was originally selected for testing in this application due to its long history of use for medicinal purposes. It had been used in foodstuffs and has been taken internally with little to no adverse effects. Various mineral oils can have different viscosities, which have a profound effect on the drilling rate. Thicker oils tend to reduce the drilling rate if other variables remain the same. Selection of the oil having proper viscosity is important in formulating oil-base drilling fluids.

Due to some very real environmental concerns over the use of oil-base drilling fluids, their use during the middle 1970's dropped to nearly zero. They were used only as a last resort and, in some cases, such as offshore and in the national forests, were being totally banned. Now, however, in spite of the higher cost of the mineral oils, the use of oil-base drilling fluids is back up to about 10—15% of the total mud systems being used.

SOME NEW DEVELOPMENTS IN DRILLING TECHNOLOGY

Two notable developments in the last 2—3 years that directly relate to drilling fluids are the polycrystalline diamond (PCD) bits (see Appendix G) and monitoring while drilling (MWD) techniques using mud pressure pulses. These developments are not actually new, but have evolved from existing technology, usually after the development of a particular product or upon the improvement of existing equipment.

The PCD bits were developed after a new technique was discovered that enabled manufacturers to bond diamonds to tungsten carbide making an extremely strong cutting surface for a drillbit. Of interest to mud technologists is that the use of the PCD bits in conjunction with oil-base drilling fluids can result in a considerable cost savings to the operator, especially with the relaxed-filtrate oil-base drilling fluids. It has been stated that the performance of a tricone bit falls off when used in an oil-base drilling fluid (Madigan and Caldwell, 1981). Inasmuch as the PCD bits fail the rock in shear, however, their performance is relatively unaffected by oil-base drilling fluids. A complete description of the PCD bit and its use in drilling operations was published by Madigan and Caldwell in 1981. This paper lists the various basins of the world in which the PCD bits have been used, gives a detailed economic and technical study of the bits, and covers the new designs and testing that are currently being conducted to expand the applications of these bits.

Another fairly recent technological development that impinges upon drilling fluid technology is the development of the techniques for downhole directional surveys without disrupting the drilling operation, i.e., measurement while drilling or MWD. These systems are also called mud pulse telemetry, because there is no wireline involved in transmitting the data to the surface.

One system that has been successfully used during the last few years transmits the information from the bottom of the hole by the use of a downhole valve which induces coded pressure pulses into the drilling fluid. These pressure pulses form a binary code that conducts the downhole directional information (direction, inclination, and tool facing) to a sensor on the surface, where it is decoded and displayed on a digital monitor. Extensive field trials of these systems have shown the advantages of MWD over the conventional wireline techniques in (1) time saved in retrieving downhole information, (2) evaluating bottomhole assemblies, and (3) trips (Roberts et al., 1982).

TABLE F-III

Survey time summary (after Roberts et al., 1982)

	Survey time (hr/well)		Direct survey time savings (hr/well)
	MWD wells	Conventional wells	
With steering tools		73.9	68.2
Without steering tools		41.7	36.0
Average	5.7	57.2	51.5

The significance of this new technique from a drilling fluid engineering standpoint is in abating some problems that may result from not optimizing directional drilling. The use of this tool (see Fig. F-2) also tends to lower the severity of doglegs. Inasmuch as the pumps are not shut off and the pipe is stopped only for a short period of time, the risk of encountering differential sticking is minimized (Roberts et al., 1982).

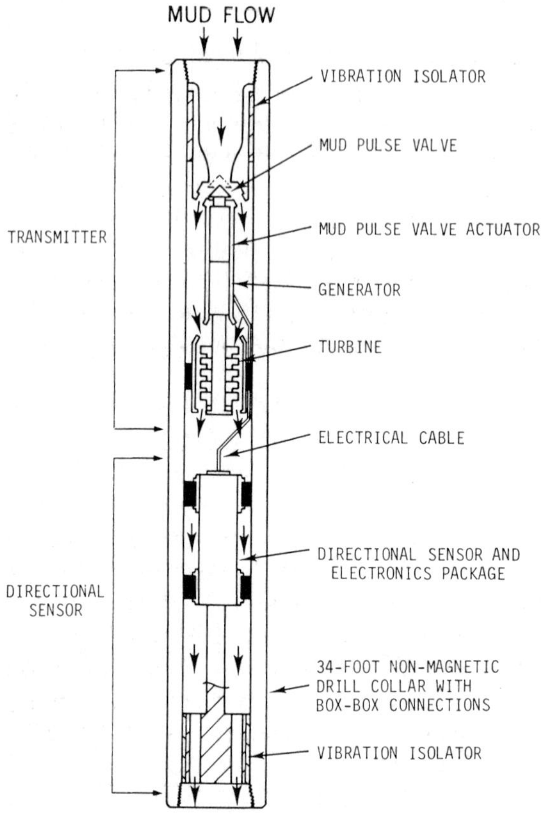

Fig. F-2. Internal components of the downhole MWD tool. (After Roberts et al., 1982; courtesy of the SPE of AIME.)

DRILLING FLUID DENSITY CHANGES

Sorelle et al. (1982) introduced an accurate and reliable method of predicting the influence of downhole conditions on the density of a static drilling fluid column. Their mathematical model requires a minimum of input data. The greatest increase in density occurs in the case of low thermal gradient in a high-weight, low-solids drilling fluid column, whereas the greatest reduction in density occurs in conditions of high thermal gradient in a low-weight, low-solids drilling fluid column. Water-base drilling fluids consistently showed lower densities at depth than oil-base drilling fluids of the same surface density, providing the thermal gradient was the same in both cases (Sorelle et al., 1982, p. 6).

ALTERNATE WEIGHTING MATERIALS

Some concern has been expressed recently over the continued availability of API-quality barite. Because of this and other economic reasons, two high-density mineral products are being marketed as alternates to barite: (1) a synthetic iron oxide (mostly hematite, Fe_2O_3) and (2) ilmenite ($FeO \cdot TiO_3$). The advantage of both materials is that they have higher densities than barite. This could result in considerable cost savings, especially in highly-weighted drilling fluids. They have also been found to be more durable than barite, resulting in lower buildup of submicron particles in the drilling fluid over a long period of time. Being more durable, however, means that these minerals tend to be more abrasive than barite (Blomberg et al., 1982, and Walker, 1982).

ENVIRONMENTAL CONCERNS

Considerable concern has been expressed in recent years about the effect of drilling fluids and drilling fluid constituents upon the environment. These concerns have resulted in some basic studies that were funded by both the Environmental Protective Agency and the American Petroleum Institute. The studies have been designed to determine the effects of drilling fluids on plants and animals. The problem of drilling fluid and drilled cuttings disposal is also being investigated. In recent years, there have been two joint API/EPA conferences on the effects of drilling fluids and cuttings on the environment (American Petroleum Institute, 1980). The major thrust of these meetings was to develop terminology and to set up the basic framework in which to work. In addition, most of the studies mainly addressed themselves to the offshore environment. Studies were performed on various marine plants and animals, and some field work was conducted to examine the fate of excess drilling fluids discarded from drilling rigs.

The API has also funded work on the effects of drilling fluids in a land-based environment. This included some laboratory and field studies at Utah State University (Miller, 1977) and Purdue University (Nelson et al., 1980) on the effects of drilling fluid additives on plant growth. In addition, the API is sponsoring work groups to develop (1) standard test procedures for determining trace-metal contents and (2) standard bio-assay tests for drilling fluids.

In general, the work to date showed that although there are some individual toxic components in drilling fluids, any evaluation of the effects of a drilling fluid on the environment must be done using the entire drilling fluid system. Taking one particular component of a drilling fluid and running a bio-assay or making a plant growth study does not necessarily indicate the true effect of that material when it is in combination with other ingredients of the drilling fluid. This component may be so tied up with other components in the drilling fluid that it is not available to affect plants or animals. The workers in this field, however, point out that there are potential environmental hazards associated with the disposal of drilling fluids. It must be remembered that the industry must act responsibly in order to keep the control over determination of allowable drilling fluids. Otherwise it may be taken out of their hands (Ayers et al., 1982, and Reeder et al., 1978).

Some specific components that are potentially hazardous to the environment include the following: (1) oil (especially from oil-base drilling fluids); (2) salts, such as sodium and calcium chlorides; and (3) soluble trace elements, such as zinc, lead, copper, cadmium, nickel, mercury, arsenic, barium, and chromium. If any of the above materials are present in a drilling fluid, then those in charge of the operation should be prudent in the disposal of the waste fluids (see Nesbitt and Sanders, 1981).

ACKNOWLEDGEMENT

The help extended by Ron Bolin and J. "Mac" Seheult is greatly appreciated by the authors.

REFERENCES

American Petroleum Institute, 1980. *Research on Environmental Fate and Effects of Drilling Fluids and Cuttings, Proc. Symp*., Lake Buena Vista, Fla., Jan. 21—24.

Ayers, R.C., Jr., Meek, R.P., Sauer, T.C., Jr. and Stuebner, D.O., 1982. An environmental study to assess the effect of drilling fluids on water quality parameters during high rate volume discharges to the ocean. *J. Pet. Technol.*, 34(1): 165—173.

Blomberg, N. et al., 1982. *Evaluation of illmenite as weight material in drilling fluids*. SPE—AIME Annu. Tech. Conf., New Orleans, La., Sept. 27—29; SPE, Pap. 11085: 6 pp.

Clements, W.R. and Hinds, A.A., 1982. *New Oil Mud Passes Environmental Tests*. SPE—AIME Annu. Tech. Conf., New Orleans, La., Sept. 27—29, SPEE, Pap. 11113.

Fontenot, J.E. and Simpson, J.P., 1974. A microbit investigation of the potential for improving the drilling rate of oil-based muds in low-permeability rocks. *J. Pet. Technol.*, (May): 507—514.

Madigan, J.A. and Caldwell, R.H., 1981. Applications for polycrystalline diamond compact bits from analysis of carbide insert and steel tooth bit performance. *J. Pet. Technol.*, 33 (July): 1171—1179.

Miller, R.W., 1977. *The Effects of Complete Drilling Fluid Mixtures in Different Soils*. Utah State Univ., Logan, Utah (March 1, 1977).

Nelson, D.W., Liu, S. and Sommers, L.E., 1980. *Plant Uptake and Accumulation of Metals Derived from Drilling Fluids*. Purdue Univ., West Lafayette, Ind. (Dec. 31).

Nesbitt, L.E. and Sanders, J.A., 1981. Drilling fluid disposal. *J. Pet. Technol.*, 33(12): 2377—2381.

O'Brien, T.B., Stinson, J.P. and Brownson, F., 1977, 1978. Relaxed fluid loss controls on invert muds increase ROP. *World Oil* (Aug. 1977, Mar. 1978): 75.

Reeder, L.B. and Associates, 1978. *Impact of Oil and Gas Production on Salinity in Major Fresh-Water Aquifers*. Paper prepared for Interstate Oil Compact Commission, Oklahoma City, Okla., July.

Roberts, A., Newton, R. and Stone, F., 1982. *MWD Field Use and Results in the Gulf of Mexico*. 57th Annu. Fall Tech. Conf. and Exhibition of SPE of AIME, New Orleans, Sept. 26—29, SPE, Pap. 11226: 7 pp.

Simpson, J.P., 1979. A new approach to oil-based muds for lower cost drilling. J. Pet. Technol., 31(5): 643—650.

Sorelle, R.R. (Falcon Engineering); Jardiolin, R.A. and Buckley, P. (Magcobar Data Operations, Dressor Industries, Inc.); and Barrios, J.R. (GHK Company), 1982. *Mathematical Field Model Predicts Downhole Density Changes in Static Drilling Fluids*. 57th Annu. Fall Tech. Conf. and Exhibition of SPE of AIME, New Orleans, La., Sept. 26—29, SPE, Pap. 11118: 7 pp.

Walker, C.O., 1982. *Alternate Weighting Material*. SPE—AIME Annu. Tech. Conf., New Orleans, La., Sept. 27—29, SPE, Pap. 11116: 5 pp.

Appendix G

NEW DEVELOPMENTS IN DRILLING TECHNOLOGY

CLAUDE L. THIBAULT and GEORGE V. CHILINGARIAN

INTRODUCTION

The objective of drilling is to reach the payzone at a lowest cost, maximum degree of safety, and minimum damage to the formation. To achieve this, two general requirements must be satisfied:

(1) Proper design of well programs, including the exploration of different formation layers, formation evaluation (various logs), coring, and testing operations.

(2) Proper choice of a drilling rig, which will reach the payzone most rapidly and cheaply, while observing the best safety conditions.

Well programs constitute about 40% of the well cost. In general, the distribution of costs for a land-based rig is as follows:

(1) Location (site preparation, etc.) 8–12%

(2) Fixed costs (moving, casing, service companies, cementing, formation evaluation (logs), etc.) 23–27%

(3) Completion (perforating, cleaning, etc.) 4–6%

(4) Cost proportional to time (drilling, fishing, trip time, etc.) 55–65%

Items 1, 2, and 3 are direct functions of the well program, and are practically independent of the rig cost.

The basic factors that affect the rate of penetration are: (1) bit selection, (2) bit weight, (3) bit rotary speed, (4) bit hydraulics, (5) drilling fluid properties, and (6) formation properties. These parameters are discussed in Chapter 3. Various formulas used in calculations are presented in Table G-I.

EXTENDED NOZZLE ROCK BITS

The study and development of new concepts and designs of rock bits to achieve the lowest cost per foot drilled is a major concern of rock bit manufacturers. The two-cone and three-cone extended nozzle designs are part of this overall effort.

A two-cone tungsten carbide insert bit (TCI) has been specifically designed for soft formations. Two extended nozzle tubes in conjunction with a special center jet are used to resist bit "balling" (Fig. G-1).

Three-cone milled tooth and TCI bits have been designed with three

TABLE G-I

Various equations used in drilling industry

A. BIT SELECTION EQUATIONS

1. Cost per foot

$$C_1 = \frac{B_1 + R_1 (T_1 + t)}{F_1}$$

2. Breakeven time, at constant rate of penetration

$$T_2 = \frac{B_2 + R_1 (t)}{C_1 \left(\frac{F_1}{T_1}\right) - R_1}$$

B. BIT WEIGHT-ROTATIONAL SPEED EQUATIONS

3. Drilling rate (soft formation)

$$ROP = k_{f_1} WR$$

4. Drilling rate (hard formation)

$$ROP = k_{f_1} W^{1.2} R^{0.5}$$

5. Bit size versus penetration rate
   a. Up to 17½"

$$ROP_2 = ROP_1 \left(\frac{D_1}{D_2}\right)$$

   b. 17½" to 36"

$$ROP_2 = ROP_1 \left(\frac{D_1}{D_2}\right) \times 1.25$$

6. Bearing wear constant

$$C_B = \frac{\text{Hours}}{\left(\dfrac{100}{\dfrac{W}{1000} \times \dfrac{R}{100}}\right) - \dfrac{W}{1000}}$$

7. Tooth wear constant

$$C_T = \frac{\text{Hours} \times e^{(.01 R + .0032 W)}}{189.2}$$

8. Mechanical horsepower at bit

$$HP = k_b W_b^{1.5} D_b^{2.5} R$$

9. Bit weight-rpm relationship to bit pressure drop

From Fullerton, for $(W_bR) < 250$

$$\triangle P_b = 0.678 \varrho D_h (W_bR)^{0.5}$$

for $250 < (W_bR) < 350$

$$\triangle P_b = 0.044 \varrho D_h (W_bR)$$

for $(W_bR) > 350$

$$\triangle P_b = 0.80 \varrho D_h (W_bR)^{0.5}$$

C. HYDRAULIC CALCULATION EQUATIONS

10. Drill stem bore pressure losses (turbulent flow)
    a. From Fanning

$$\triangle P = \frac{f \varrho V^2 L}{25.8 d}$$

    b. From Security

$$\triangle P = \frac{0.000061 \varrho L Q^{1.86}}{d^{4.86}}$$

    c. From Smith Tool

$$\triangle P = \frac{.0000765 \overline{PV}^{0.18} \varrho^{0.82} Q^{1.82} L}{d^{4.82}}$$

11. Bit hydraulic horsepower

$$BHHP = \frac{\triangle P_b \times Q}{1714}$$

12. Jet nozzle pressure loss

$$\triangle P_b = \frac{\varrho Q^2}{10858 (A_n)^2}$$

13. Total nozzle area

$$A_n = \frac{Q}{104.2} \left(\frac{\varrho}{\triangle P_b}\right)^{0.5}$$

14. Jet velocity

$$V_n = \frac{0.32 Q}{A_n}$$

15. Jet impact force

$$I_f = 0.000516 \varrho Q V_n$$

$$I_f = 0.0173 Q \sqrt{\triangle P_B \times \varrho}$$

16. Bottom hole pressure

$$BHP = 0.052 \varrho L$$

17. Bottom hole circulating pressure

$$BHCP = BHP + \triangle P_a$$

18. Annular pressure losses
    a. From Hagan-Poiseuille for Newtonian laminar flow

$$\triangle P_a = \frac{\mu LV}{1500 (D_h - D_p)^2}$$

    b. From Beck, Nuss and Dunn for plastic laminar flow

$$\triangle P_a = \frac{L\overline{YP}}{225(D_h - D_p)} + \frac{VL\overline{PV}}{1500(D_h - D_p)^2}$$

    c. From Fanning, turbulent flow

$$\triangle P_a = \frac{f \varrho L V^2}{25.8(D_h - D_p)}$$

---

*Courtesy of Sii Smith Tool, Division of Smith International, Inc.

C. HYDRAULIC CALCULATION EQUATIONS *(continued)*

d. From Security, turbulent flow

$$\triangle P_a = \frac{(1.4327 \times 10^{-7})\rho L \overline{AV}^2}{D_h - D_p}$$

e. From Smith Tool, turbulent flow

$$\triangle P_a = \frac{0.0000765 \overline{PV}^{0.18} \rho^{0.82} Q^{1.82} L}{(D_h - D_p)^3 (D_h + D_p)^{1.82}}$$

19. Equivalent circulating density

$$ECD = \frac{BHCP}{.052L}$$

20. Reynolds' number

a. Newtonian fluids

$$R_e = \frac{928\rho Vd}{\mu}$$

b. Plastic fluids (to determine "f")

$$R_e = \frac{2970\rho Vd}{\overline{PV}}$$

21. Average annulus flow velocity

$$\overline{AV} = \frac{24.5Q}{(D_h^2 - D_p^2)}$$

22. Annulus critical velocity

$$V_c = \frac{1.08\overline{PV} + 1.08\sqrt{\overline{PV}^2 + 9.3(D_h - D_p)^2\overline{YP}\rho}}{\rho(D_h - D_p)}$$

23. Optimum annular velocity
From Fullerton

$$\overline{AV}_o = \frac{11{,}800}{\rho D_h}$$

24. Optimum flow rate

a. From Fullerton

$$Q_o = \frac{[385 - 10(12 - D_h)]D_h}{\rho} = \frac{265D_h + 10D_h^2}{\rho}$$

b. From Fullerton

$$Q_o = \frac{482(D_h^2 - D_p^2)}{D_h X \rho}$$

25. Rock chip slip velocity

a. From Stokes, laminar flow, spherical chips

$$V_s = \frac{8310 d_c^2(\rho_c - \rho)}{\mu}$$

b. From Pigott, laminar flow, spherical chips

$$V_s = \frac{3226 d_c^2(\rho_c - \rho)}{\mu}$$

c. From Rittinger, turbulent flow, spherical chips

$$V_s = 155.9\sqrt{\frac{d_c(\rho_c - \rho)}{\rho}}$$

d. From Pigott, turbulent flow, flat chips

$$V_s = 60.6\sqrt{\frac{d_c(\rho_c - \rho)}{\rho}}$$

C. HYDRAULIC CALCULATION EQUATIONS *(continued)*

26. Effective viscosity

a. Viscosity definition

$$\mu = \frac{S_s}{S_r}$$

b. Bingham Plastic

$$\mu = \overline{PV} + \frac{399\,\overline{YP}\,(D_h - D_p)}{\overline{AV}}$$

c. Shear stress, Power Law fluids

$$S_s = kS_r^n$$

d. Effective viscosity, Power Law

$$\mu_e = kS_r^{n-1}$$

e. Annular shear rate

$$S_r = \frac{2.4\,\overline{AV}}{D_h - D_p}$$

f. Consistency index

$$k = \frac{511\,(\overline{YP} + \overline{PV})}{511^n}$$

g. Power Law index

$$n = 3.32 \log_{10}\left(\frac{\overline{YP} + 2\overline{PV}}{\overline{YP} + \overline{PV}}\right)$$

27. Total system losses or pump discharge pressure

$$\triangle P_t = \triangle P_s + \triangle P_c + \triangle P_p + \triangle P_b + \triangle P_{ca} + \triangle P_{pa}$$

D. DRILLING FLUID PROPERTY EQUATIONS

28. Effects of plastic viscosity

$$ROP_2 = ROP_1 \times 10^{.003(\overline{PV}_1 - \overline{PV}_2)}$$

29. Bentonitic clay effects

$$ROP_2 = ROP_1 \times e^{.051(Vol\%_1 - Vol\%_2)}$$

30. Total solids effects

$$ROP_2 = ROP_1 \times 10^{.0066(Vol\%_1 - Vol\%_2)}$$

31. Effects of water loss

$$ROP_2 = ROP_1 \times \frac{WL_2 + 35}{WL_1 + 35}$$

32. Oil content effects (Vol% Oil<30%)

$$ROP_2 = ROP_1\left[\frac{\sin(10.6\ Vol\%_2 - 4.83) + 10.33}{\sin(10.6\ Vol\%_1 - 4.83) + 10.33}\right]$$

33. Total drilling fluid effects (density, viscosity, solids, pressure loss)
For depths from 8000 feet to 12,000 feet

$$ROP_2 = ROP_1 e^{0.382(\rho_1 - \rho_2)}$$

34. From Fullerton, density effects

$$\log_{10} k_{f_2} = .000208(BHP_1 - BHP_2) + \log_{10} k_{f_1}$$

NOMENCLATURE

A_n = Total nozzle area, inches²

$\overline{AV}$, $\overline{AV}_o$ = Average, optimum annulus velocity, fpm

BHP = Bottom hole pressure, psi

BHCP = Bottom hole circulating pressure, psi

BHHP = Bit hydraulic horsepower, hp

B_1, B_2 = Cost of control, proposed bit, dollars

C_1 = Cost per foot of control bit, dollars/ft

C_B = Bearing wear constant

C_n = Nozzle coefficient, 95%

C_T = Tooth wear constant

d = Inside pipe diameter, inches

D_b = Bit diameter, inches

d_c = Chip diameter, inches

D_1, D_2 = Smaller, larger bit diameter, inches

D_h = Hole diameter, inches

D_p = Outside pipe diameter, inches

e = 2.718 Naperian base

ECD = Equivalent circulating density, ppg

f = Fanning friction factor

F_1 = Footage drilled, feet

k = Consistency index

k_b = Formation factor for horsepower calculation ranging from 4×10^{-5} for very hard to 14×10^{-5} for very soft formations

k_{f_1}, k_{f_2} = Apparent, corrected formation drillability factor

L = Pipe length or hole depth, feet

n = Power Law index

$\overline{PV}$ = Plastic viscosity, cps

$\overline{PV}_1$, $\overline{PV}_2$ = Initial, final plastic viscosity, cps

ϱ = Mud density, ppg

ϱ_c = Density of cuttings, ppg

ϱ_1, ϱ_2 = Initial, final mud desity, ppg

$\triangle P$ = Pressure loss, psi

$\triangle P_a$ = Annulus pressure loss, psi

$\triangle P_c$, $\triangle P_{ca}$ = Drill collar bore, annulus pressure loss, psi

$\triangle P_p$, $\triangle P_{pa}$ = Drill pipe bore, pipe annulus pressure loss, psi

$\triangle P_s$ = Surface connection pressure loss, psi

$\triangle P_t$ = Total system losses, psi

$\triangle P_b$ = Bit pressure loss, psi

Q = Pump volume, gpm

Q_o = Optimum flow rate, gpm

R = Bit rotational speed, rpm

R_1 = Rig cost or operating rate, dollars/hour

R_e = Reynolds number, dimensionless

ROP = Rate of penetration, feet/hour

ROP_1, ROP_2 = Initial, final rate of penetration, feet/hour

S_s = Shear stress, dynes/cm²

S_r = Shear rate, sec⁻¹

t = Round trip time, hours

T_1, T_2 = Rotating time for control, proposed bit, hours

μ, μ_e = Apparent, effective viscosity, cps

V = Fluid velocity, fps

V_c = Critical velocity in annulus, fps

V_n = Nozzle velocity, fps

V_s = Chip velocity, fpm

W = Weight per inch of bit diameter, lb/inch

W_b = Bit weight, 1000 lb/inch of bit diameter

WL_1, WL_2 = Initial, final water loss, cc/30 min

$\overline{YP}$ = Yield point, lb/100 ft²

Fig. G-1. Two-cone tungsten carbide insert bit (TCI), with extended nozzle tubes. (Courtesy of Sii Smith Tool, Division of Smith International, Inc.)

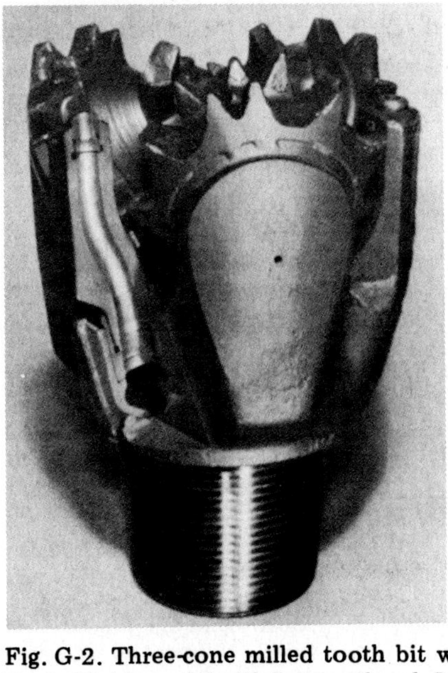

Fig. G-2. Three-cone milled tooth bit with extended nozzle tubes. (Courtesy of Sii Smith Tool, Division of Smith International, Inc.)

extended nozzle tubes and a center jet similar to the one in two-cone bits (see Fig. G-2).

Two-cone bits offer availability of space not provided by conventional three-cone bits.

Field test results of more than 90 extended nozzle — center jet three-cone bits indicate a 15—40% increase in the drilling rate (Pratt, 1977). The center jet used to reduce the balling tendency of the cutters has a special diffuser nozzle to reduce the exit velocity of the drilling fluid.

Jet nozzle studies

The word "jet" usually implies a nozzle which converts pressure to kinetic energy. High fluid pressure inside the bit gives rise to high velocity of fluid through the nozzle. Conversely, converting velocity into pressure is referred to as "diffusion". The center jet of the extended-nozzle bit utilizes the latter process.

Cross-section of a diffusion nozzle is presented in Fig. G-3. Flow lines in this figure do not imply laminar flow and are symbolic of averaged pathlines of fluid flow. For all drilling applications, the flow is turbulent through the center jet. At point *A* (see Fig. G-3), the pressure is maximum within the bit and is greater than the pressures at points *B*, *C*, *D*, and *E*. As the fluid passes point *B*, through the relatively sharp-edged passageway to the nozzle throat at point *C*, there is a large increase in fluid velocity. The maximum velocity is at point *C*. The actual nozzle entrance is shown in Fig. G-4.

Most of the total pressure drop of the center jet occurs at points *B* and *C*

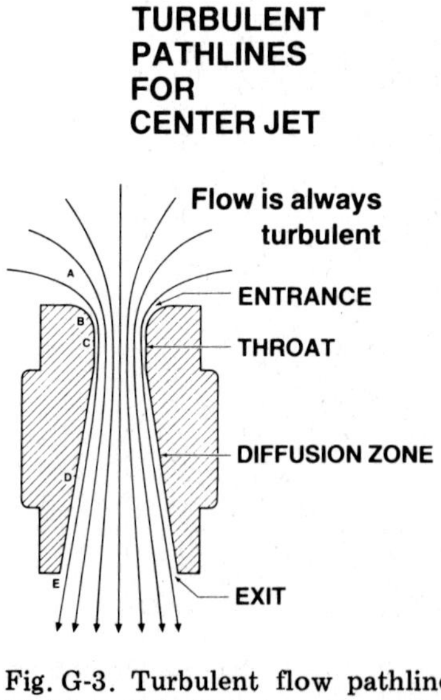

Fig. G-3. Turbulent flow pathlines through a center jet. (Courtesy of Sii Smith Tool, Division of Smith International, Inc.)

due to the sharp-edged and reduced flow area. From point C, fluid passes through the diffusion section (point D) to the nozzle exit at point E.

As the area increases in the diffusion section, the velocity decreases and "pressure recovery" occurs. As the fluid is ejected at point E, there is further pressure drop. The fluid velocity has both axial and radial components, but shortly thereafter, the radial component dissipates as the flow stream combines with surrounding fluid in the dome area of the bit (Fig. G-5).

If the nozzle did not have a diffusion section and the velocity impinged directly on top of the cones, excessive cone erosion would have occurred.

The following formula (Morris and Worrell, 1981) enables determination of the pressure recovery through the diffusion section:

$$\Delta p = k' \gamma_f (v_t - v_{ex})^2/1239.2 \qquad \text{(G-1)}$$

Fig. G-4. Nozzle entrance. (Courtesy of Sii Smith Tool, Division of Smith International, Inc.)

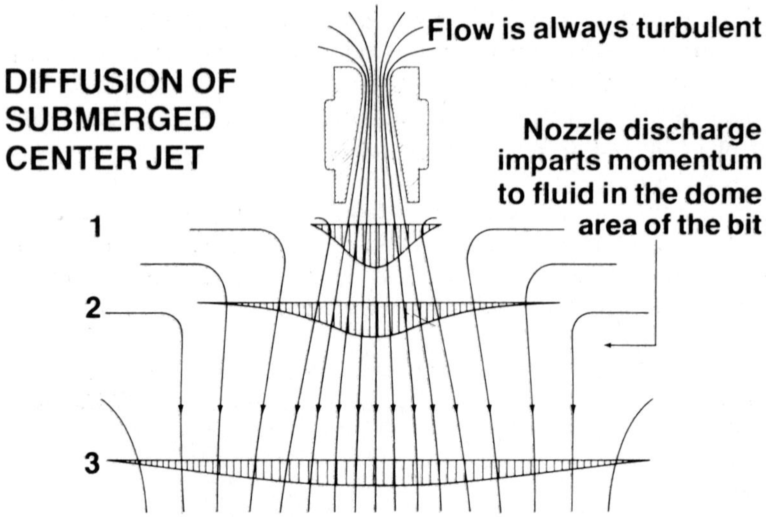

Fig. G-5. Diffusion area of submerged center jet. (Courtesy of Sii Smith Tool, Division of Smith International, Inc.)

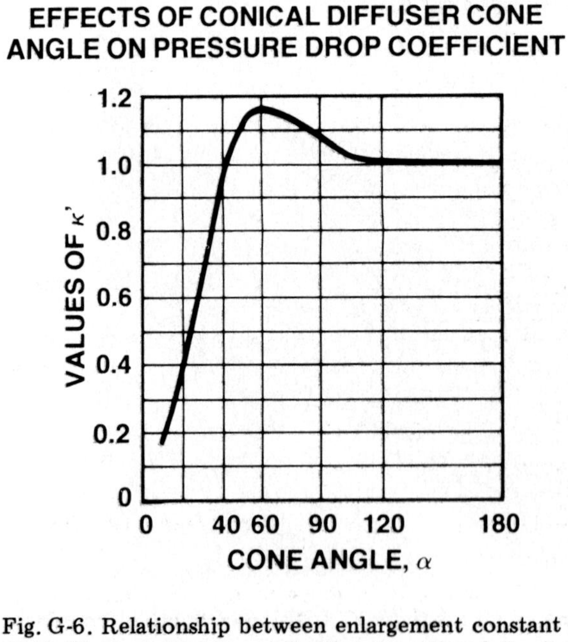

Fig. G-6. Relationship between enlargement constant k' and cone angle, α. (Courtesy of Sii Smith Tool, Division of Smith International, Inc.)

where Δp = pressure recovery (psi); k' = enlargement constant as obtained from Fig. G-6, dimensionless; γ_f = fluid specific weight (lb/gal); v_t = throat velocity (ft/sec); and v_{ex} = exit velocity (ft/sec). The constant k' is found from Fig. G-6 for a cone angle of 18°.

When the pressure drops and recoveries are combined across the center jet, the net pressure drop is less than the pressure drop through an equalized standard nozzle.

Inasmuch as the center jet has less pressure drop for a given nozzle flow rate, more fluid will pass through the center jet than through a standard jet nozzle of the same size. An excess of flow rate and hydraulic horsepower is available for cone cleaning, but the extended nozzles (two or three) will suffer.

Benefits of the extended nozzles are better realized when a large percentage of total flow volume is delivered through the extended nozzles for bottomhole cleaning (see Baker, 1979, 1981).

Laboratory and field tests

Both laboratory and field tests have been conducted to determine the actual nozzle characteristics. At the drilling research laboratory in Salt Lake City, Utah, drilling fluid was circulated through the center jet in a wellbore simulator. Up to 180 gal/min of 8.8-lb/gal drilling fluid were pumped

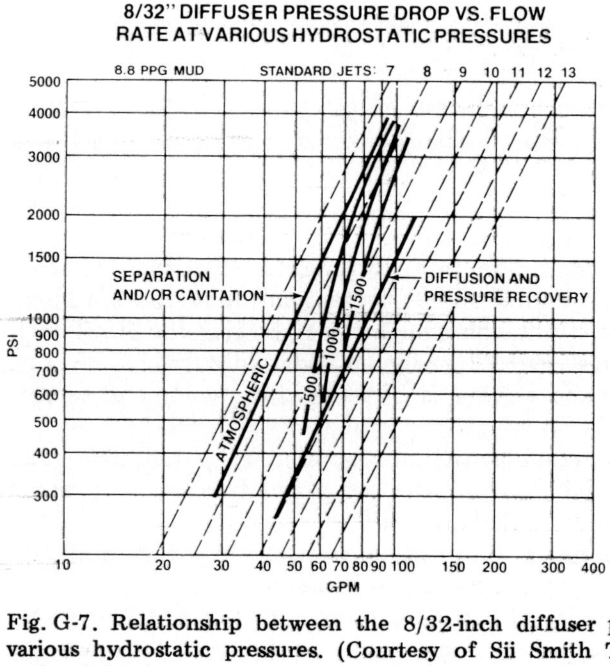

Fig. G-7. Relationship between the 8/32-inch diffuser pressure drop and flow rate at various hydrostatic pressures. (Courtesy of Sii Smith Tool, Division of Smith International, Inc.)

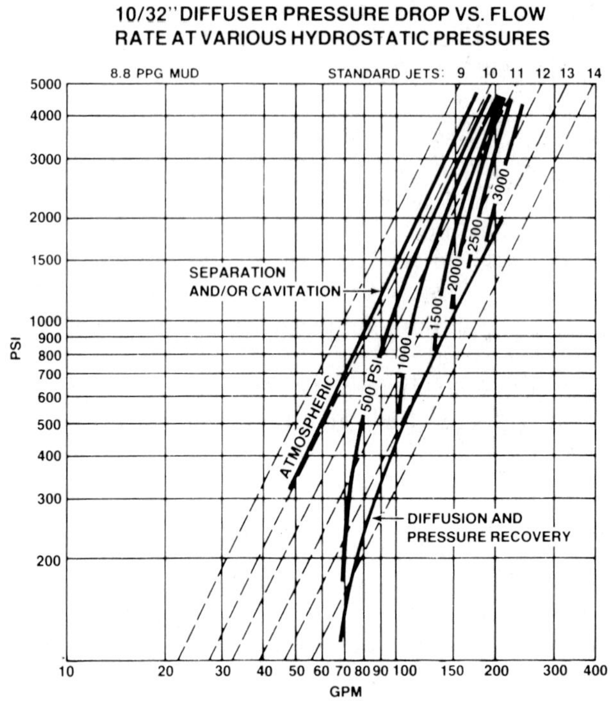

Fig. G-8. Relationship between the 10/32-inch diffuser pressure drop and flow rate at various hydrostatic pressures. (Courtesy of Sii Smith Tool, Division of Smith International, Inc.)

on two different nozzle sizes, with center jet throat sizes of 8/32 inch and 10/32 inch.

Hydrostatic pressure was varied to simulate several drilling depths. Results of these tests are plotted in Figs. G-7 and G-8. The diffusion line on the right depicts the pressure versus flow rate relationship when the center jet is in a steady state diffusion condition. At these conditions, the 8/32-inch center jet is equivalent to a 10/32-inch standard jet, whereas the 10/32-inch exhibited the flow characteristics of 13/32-inch standard nozzle.

Figures G-7 and G-8 also show the area of separation regime and separation-induced cavitation.

In the case of extended nozzle bits, the most economical performance can be obtained if the center diffusion nozzles are properly sized to provide ample fluid flow through the extended nozzles.

Cavitation phenomena exist in the center jet at shallow and moderate depths. The characteristics of the nozzle vary with depth.

At depth, where cavitation is not occurring to any great extent, the center jet gives rise to a pressure drop lower than that of a standard jet nozzle of the same nominal throat dimension.

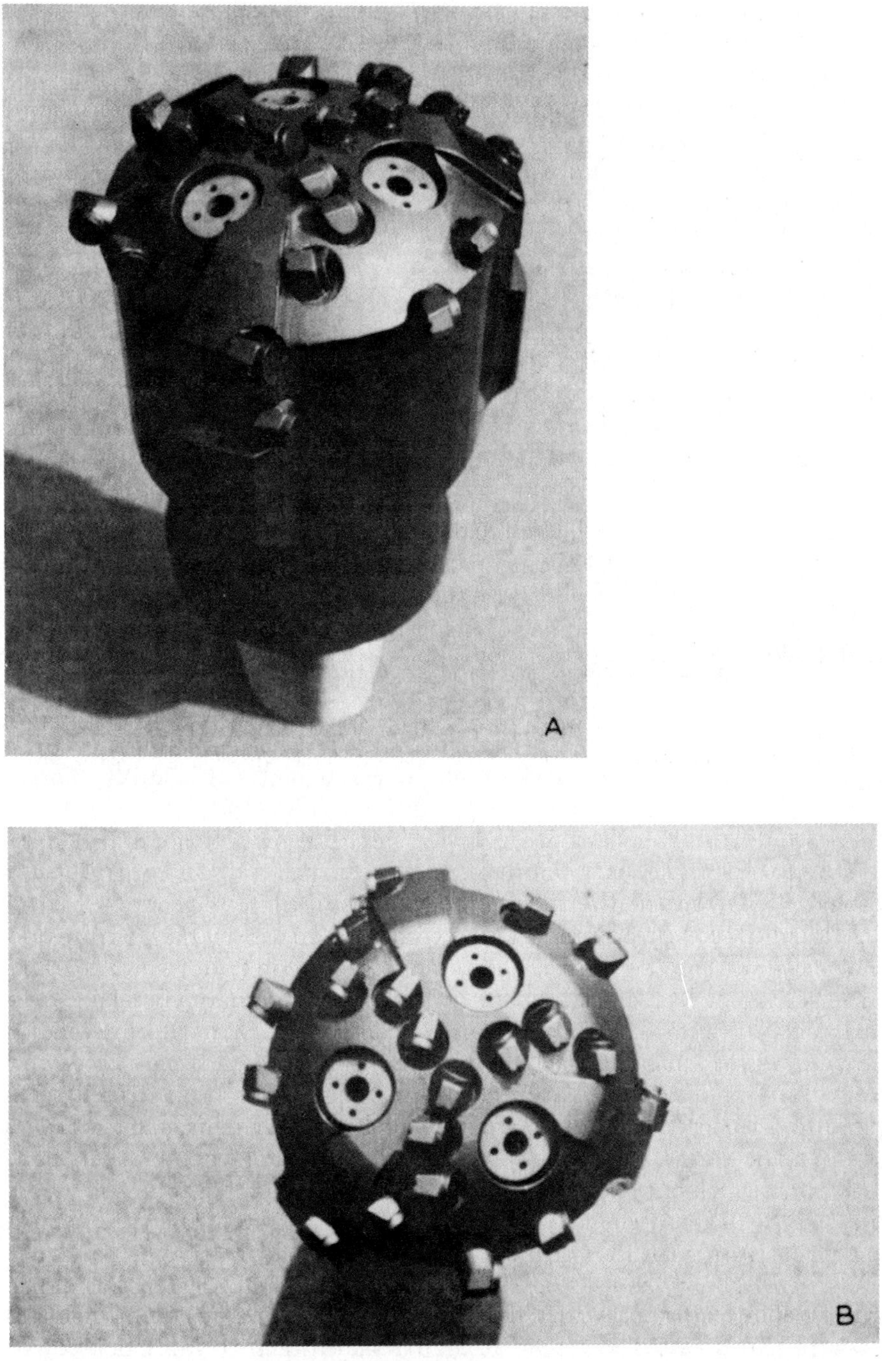

Fig. G-9. 6¾-inch SMITHPAX polycrystalline diamond drill bit. (Courtesy of Sii Smith Tool, Division of Smith International, Inc.)

POLYCRYSTALLINE DIAMOND CUTTER BITS

According to Kuhn and Forrest (1982), the design and development of a drag bit that utilizes polycrystalline diamond drill blanks require careful attention to hydraulics. For optimum performance and service life, the polycrystalline diamond drill blanks must be kept cool and clean. The critical hydraulic requirements were satisfied by optimizing flow velocities across the cutters, as well as by proper distribution of fluid volume. This was accomplished with strategically located flow restrictions and the incorporation of numerous interchangeable carbide jet nozzles (Kuhn and Forrest, 1982, p. 1).

In 1973, the specialty materials department of General Electric Company commercially introduced the first polycrystalline diamond tool blanks. Since that time, the drilling industry has been developing, first on an experimental basis and then as a proven drilling tool, the use of a polycrystalline diamond cutter (PDC) on a wide variety of drill bit designs. Bit body designs and component manufacture processes vary with each manufacturer. Primary development practices have focused upon the steel body with pressed-in stud-type drill blanks (Kuhn and Forrest, 1982, p. 1).

A 0.020-inch thick layer of polycrystalline synthetic diamond, supported by an integral blank of tungsten carbide, has an upper thermal threshold limit of 700°C (Offenbacher, 1979). Above this temperature, bond strength between the diamond crystals degenerates, increasing diamond wear rate.

The importance of maximizing the hydraulic design for these bits played a significant role in the systematic development of Smith Tool's SMITHPAX(TM) polycrystalline diamond drill bit (Fig. G-9).

Many experimental designs were tested for different projects (e.g., see Figs. G-9A and G-9B). During the development period, a careful analysis was undertaken to determine the ideal hydraulic conditions required for PDC bits. Of the hydraulic designs tested, the enhanced crossflow bit was selected for further development.

Crossflow restrictions

The current configuration utilizes baffles to separate the bit head into three regions, with ample room remaining for flow across the center. The height of baffles is a function of the anticipated penetration rate to maintain minimum clearance above the hole rock bottom.

Drilling hydraulics

When applied to drill bits, the term "hydraulics" describes the conditions of flow rate and pressure available at the bit that influence the efficiency of

---

(TM) **Trademark of Sii Smith Tool.**

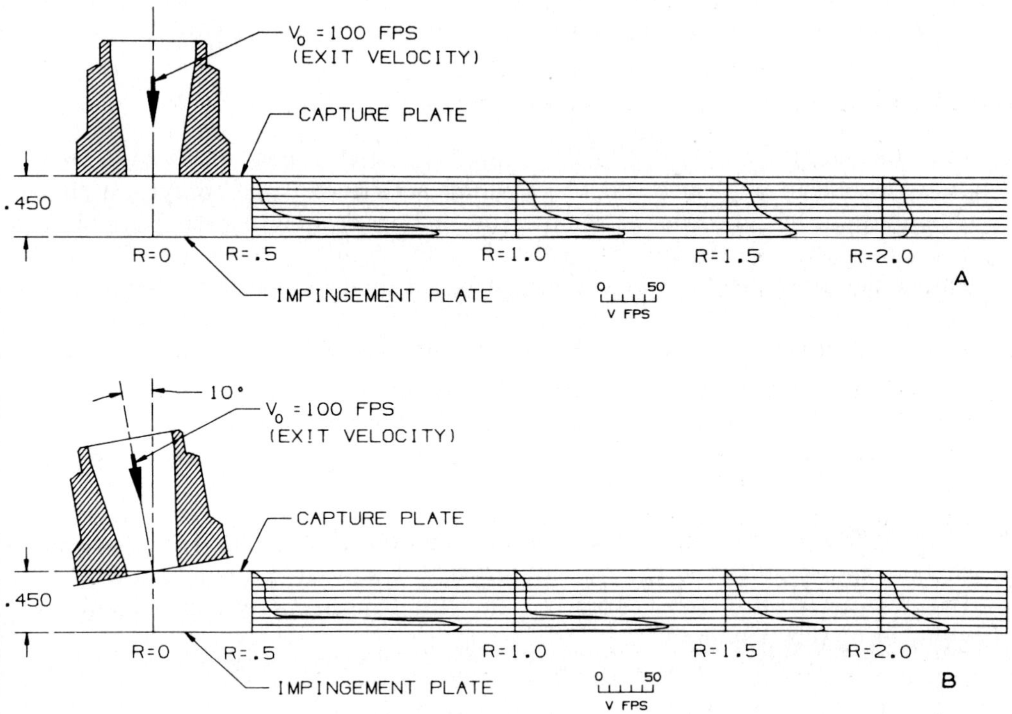

Fig. G-10. Velocity profile emanating from the jet centerline. (After Kuhn and Forrest, 1982, fig. 9; courtesy of Sii Smith Tool, Division of Smith International, Inc.) A. Crossflow velocity between plates resulting from a jet impinging normal to the surfaces. B. Crossflow velocity between plates resulting from a jet inclined at 80° to the surface.

bottomhole cleaning for a particular type of bit. The PDC bits differ significantly from roller cone bits in this respect. Optimized cutting removal from beneath a roller cone bit requires a directional, high-velocity jet to penetrate the turbulent fluid between the cones at the bottom of the borehole. This jet imposes some control on disturbance and indirectly scavenges cuttings off the bottom (Kuhn and Forrest, 1982, p. 3). Inasmuch as it was eventually realized that these submerged jets become entrained by the surrounding fluid, special tubes were added to concentrate fluid horsepower near the rock (extended nozzles) (Feenstra, 1968).

Two difficulties, which arise with jet placement, can be overcome through design innovations. The first one is the potential for erosion posed by the jet. The second drawback of high-velocity jets in PDC bits is that their placement normal to the rock face settles for a rather inefficient velocity profile emanating from the jet centerline (Fig. G-10A). Pressure drop is wasted in turbulence unless velocity is given some favorable direction in which to work. This condition can be corrected by inclining the jet with respect to the rock face (bottom of hole) at some low angle (e.g., 10°), which would result in a velocity profile similar to the one shown in Fig. G-10B.

Conclusion

Kuhn and Forrest (1982, p. 4) pointed out that a great deal of information has been published on the drilling action of single polycrystalline diamond cutters. In order to remove the cuttings from under an entire bit in a timely fashion, however, one must consider simultaneously the effects of chip generation and fluid cleaning, which is a much more difficult problem to overcome. Continued study and testing by rock bit manufacturers is rapidly increasing, and as the flow characteristics of PDC bits improve, the drilling rate will increase and thus lower the overall drilling cost per foot.

ACKNOWLEDGEMENTS

The authors wish to express their appreciation to Sii Smith Tool, Division of Smith International, Inc. for its support. The help extended by William Baker, Karl Kuhn, Scott Forrest, and Jim Kingsolver was invaluable in completing this Appendix.

REFERENCES

Baker, W.A., 1979. *Extended Nozzle Bit Require Precise Nozzle Sizing*. Sii Smith Tool Publ. Also: *Oil and Gas J.*, Mar. 19, 1979.

Baker, W., III, 1981. *Variable Wear Pad for Cross-Flow Drag Bit.* U.S. Patent No. 4,253,533, Mar. 3.

Feenstra, R., 1968. *Bit with Extended Jet Nozzles*. U.S. Patent No. 3,363,706, Jan. 16.

Kuhn, K.O. and Forrest, S., 1982. *Flow Enhancement Characterizes Design of Polycrystalline Diamond Cutter Bits*. SPE of AIME, Annu. Fall Meet., New Orleans, La., Sept. 26—29: 5 pp.

Morris, J.F. and Worrell, R.V., 1981. *Eccentric Counterbore for Diamond Insert Studies*. U.S. Patent No. 4,265,324, May 5.

Offenbacher, L.A., 1979. *Recent Developments in Stratapax Blank Bits*. ASME Pet. Mech. Engr. Conf., Tulsa, Okla., Oct.

Pratt, C.A., 1977. *Increased Penetration Rates Achieved with New Extended Nozzle Bits*. SPE of AIME Meet., Denver, Colo., Oct.

Sii Smith Tool, Division of Smith International, Inc., 1979. *How to Convert "Wags" to "Swags"*, 4 pp.

REFERENCES INDEX *

Numbers in italics refer to references lists

*Prepared by Shahrouzeh Saneie, Paul Vorabutr and George V. Chilingarian.

SUBJECT INDEX